ENCYCLOPEDIA OF MATHEMATICS AND ITS APPLICATIONS

G.-C. ROTA (Founding Editor)

Volume 46

Oriented Mat

(Second Editio

ENCYCLOPEDIA OF MATHEMATICS AND ITS APPLICATIONS

ENCYCLOPEDIA OF MATHEMATICS AND ITS APPLICATIONS

Oriented Matroids
(Second Edition)

Anders Björner
Royal Institute of Technology, Stockholm

Michel Las Vergnas
CNRS and Université Paris 6

Bernd Sturmfels
University of California at Berkeley

Neil White
University of Florida, Gainesville

Günter M. Ziegler
Technische Universität Berlin

CAMBRIDGE
UNIVERSITY PRESS

PUBLISHED BY THE PRESS SYNDICATE OF THE UNIVERSITY OF CAMBRIDGE
The Pitt Building, Trumpington Street, Cambridge, United Kingdom

CAMBRIDGE UNIVERSITY PRESS
The Edinburgh Building, Cambridge CB2 2RU, UK   http://www.cup.cam.ac.uk
40 West 20th Street, New York, NY 10011–4211, USA   http://www.cup.org
10 Stamford Road, Oakleigh, Melbourne 3166, Australia
Ruiz de Alarcón 13, 28014 Madrid, Spain

First published 1993
Second edition 1999

Printed in the United Kingdom at the University Press, Cambridge

A catalogue record for this book is available from the British Library

ISBN 0 521 77750 X paperback

Contents

Preface

Oriented matroids are a very natural mathematical concept which presents itself to us in many different guises, and which has connections and applications to many different areas. These areas include discrete and computational geometry, combinatorics, convexity, topology, algebraic geometry, operations research, computer science, and theoretical chemistry.

This book is intended for a diverse audience: graduate students who wish to learn the subject from scratch, researchers in the various fields of application who want to concentrate on certain aspects of the theory, specialists who need a thorough reference work, and all others between these extremes. There presently exists no comprehensive or accessible exposition of the field, and remedying this was our primary motivation for writing this book. It contains several new results that were developed in the course of its preparation, and we are confident that its appearance will stimulate many additional discoveries.

A list of problems and exercises is included after each of the ten chapters. These collections contain much information in addition to the main text. Unsolved problems are marked with a star.

We wish to thank L. Billera, R. Bland, K. Brown, W. Fenton, K. Fukuda, J.E. Goodman, G. Gordon, T. Havel, V. Klee, J. Lawrence, C. Lee, R. Pollack, S. Sandvik, P. Shor, T. Terlaky, Th. Wanner, T. Zaslavsky, and numerous others who helped us at various stages of this project. We (or some subset of us) worked together on the manuscript at some time at each of our home institutions, and at IMA, RISC–Linz, Augsburg, DIMACS, Oberwolfach and the Mittag–Leffler Institute. We are grateful to all of these institutions for their support.

Preface to the Second Edition

In preparing the second edition, we have made only minor updates and corrections to the main text. However, an appendix "Some Current Frontiers of Research" has been added by the first and last authors, the bibliography has been greatly expanded based on material from Ziegler (1996), and page pointers from it into the main text have been added. We take the addition of hundreds of new references to be a sign of a great and increasing vitality of the subject, and hope that this second, paperback, edition can again help to support and stimulate further research.

For help with the revision we thank in particular C. Athanasiadis, J. Bokowski, K. Brown, J. De Loera, R. Pollack, J. Rambau, J. Richter-Gebert, F. Santos, M. Ring and E. Pose. Thanks to Laura Anderson for help with Section A.2. Thanks for special support by Torsten Heldmann.

Notation

Sets:

$E_n = [n] = \{1, 2, \ldots, n\}$, with the natural ordering assumed

$\binom{E_n}{k} = \{1 \leq i_1 < \ldots < i_k \leq n\}$, the set of ordered k-subsets

$E \setminus A$, set difference

$E \setminus a$, $E \cup a$, etc. denote $E \setminus \{a\}$, $E \cup \{a\}$, etc.   102

$A \uplus B$, disjoint union

Signed sets/Sign vectors:

$X, X^1, X^2, X', X'', Y, Z, W, \ldots,$

     written as pairs $X = (X^+, X^-)$

     or as sign vectors $X \in \{+, -, 0\}^E$ 101

$\underline{X} = X^+ \cup X^- = E \setminus X^0$, support   101

$X^0 = E \setminus \underline{X} = z(X)$, zero set

$X_e = X(e)$, components   102

$X \circ Y = (X^+ \cup (Y^+ \setminus X^-), X^- \cup (Y^- \setminus X^+))$, composition   102

$X \perp Y$, orthogonality   115

Oriented Matroids:

$\mathcal{M} = \mathcal{M}(E) = (E, \mathcal{C})$, circuits   103

           $= (E, \mathcal{C}^*)$, cocircuits   115

           $= (E, \mathcal{V})$, vectors   143

           $= (E, \mathcal{L})$, covectors   143

           $= (E, \mathcal{T})$, topes (maximal covectors)   148

           $= (E, \mathcal{W})$, maximal vectors   148

           $= (E, \chi)$, chirotope   126

Bases:

Constructions:

Underlying Matroid:

Lattices:

Arrangements:

1

A First Orientation Session

Oriented matroids can be thought of as a combinatorial abstraction of point configurations over the reals, of real hyperplane arrangements, of convex polytopes, and of directed graphs. The creators of the theory of oriented matroids have, in fact, drawn their motivation from these diverse mathematical theories (see the historical sketch in Section 3.9), but they have nevertheless arrived at equivalent axiom systems – which manifests the fact that oriented matroids are "the right concept".

We will start out by illustrating these different aspects of oriented matroids. Doing this, we will present a number of examples while at the same time introducing the main concepts and terminology of oriented matroids. This should assist the reader who wishes to access the later chapters in a non-linear order, or who first wants a quick idea of what is going on. It should also provide intuition and motivation both for the axiomatics and the further development of the theory.

Hence, our first two chapters will avoid an extensive discussion of the axiom systems for oriented matroids, which are treated in Chapter 3. We will also minimize dependence on background from ordinary matroids. Furthermore, extensive attributions will not be given in these introductory chapters; we refer to later chapters and the bibliography.

1.1 Oriented matroids from directed graphs

Let us consider the simple cycles of a directed graph $D = (V, E)$ with arc set E, together with an orientation of each such cycle. Then every arc of a cycle is either a forward (positive) arc or a backward (negative) arc in the cycle. This allows us to consider the cycle as a *signed subset* of E, which consists of a positive and a negative part. Such a signed subset coming from a cycle is called a *signed circuit* of D.

For example, let D_0 be an acyclic orientation of the graph K_4, on a vertex set $V = \{v_1, v_2, v_3, v_4\}$, and with arcs labeled $E = \{1, 2, \ldots, 6\}$ as in Figure 1.1.1. Then the cycle $\{1, 2, 5, 6\}$ contains, if the corresponding cycle in the drawing of Figure 1.1.1 is oriented clockwise, the arcs 2 and 6 as positive elements and the arcs 1 and 5 as negative elements. We would write this as a signed circuit $X = (\{2, 6\}, \{1, 5\})$.

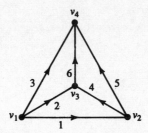

Figure 1.1.1: D_0, an acyclic orientation of K_4.

In general, we will write $X = (X^+, X^-)$ for such signed sets, where $X^+ \cap X^- = \emptyset$ is understood. We will call X^+ the set of *positive elements* of X, and X^- the set of *negative elements*. Note that if we reverse the orientation of the cycle considered, then we obtain the "opposite" signed set $-X := (X^-, X^+)$.

The signed circuits obtained from a digraph D in this way form the collection

$$\mathcal{C} = \{X = (X^+, X^-) : X \text{ is a signed circuit of } D \}.$$

The *oriented matroid* $\mathcal{M} = \mathcal{M}_D$ of D is the pair $(E, \mathcal{C})$, which is also denoted by

$$\mathcal{M} = \mathcal{M}(E).$$

Note that it contains a pair of opposite signed circuits $\pm X$ for every circuit of the *underlying matroid* $\underline{\mathcal{M}}(E) = (E, \underline{\mathcal{C}})$, where

$$\underline{\mathcal{C}} = \{\underline{X} : X \in \mathcal{C}\}$$

is simply the collection of (undirected) simple cycles of the graph. Here and in the following the notation $\underline{X} := X^+ \cup X^-$ will be used for the set $\underline{X}$ *underlying* a signed set X.

By a minimal cut of D, we mean a partition $V = V^1 \cup V^2$ of the vertices such that removing the arcs between V^1 and V^2 increases by one the number of components of the underlying undirected graph. Letting Y^+ be the "forward" arcs of this cut, that is, the arcs in D from V^1 to V^2, and Y^- the "backward arcs" from V^2 to V^1, we again get signed sets $Y = (Y^+, Y^-)$. We will call them the *signed cocircuits* of D, and write

$$\mathcal{C}^* = \{Y = (Y^+, Y^-) : Y \text{ is a signed cocircuit of } D \}.$$

Later we will prove that in fact $\mathcal{C}^*$ determines $\mathcal{C}$ (and conversely), which allows us to write

$$\mathcal{M} = (E, \mathcal{C}^*)$$

for the oriented matroid $\mathcal{M}$, given by its cocircuits.

For example, in the digraph D_0 shown in Figure 1.1.1, the partition $V^1 = \{v_1, v_3\}$, $V^2 = \{v_2, v_4\}$ determines the signed cocircuit $Y = (\{1, 3, 6\}, \{4\})$ of its oriented matroid $\mathcal{M}_0$.

It is interesting to study in more detail how properties of the digraph D are reflected in the collections $\mathcal{C}$ and $\mathcal{C}^*$ of signed circuits and cocircuits. For example, the digraph D_0 of our example was acyclic: it contains no oriented cycle. This is reflected in the fact that $\mathcal{C}$ contains no *positive circuit*, that is, no circuit X with $X^- = \emptyset$. This is used to define an *acyclic oriented matroid*. An elementary graph-theoretic argument shows that equivalently, D is acyclic if every arc is contained in a positive cocircuit, which translates into the corresponding property for oriented matroids: every $e \in E$ is contained in Y^+ for some positive cocircuit $Y = (Y^+, \emptyset)$. Also, observe how the reversal of arcs in D leads to change of signs of the corresponding elements in every circuit or cocircuit of $\mathcal{M}$, leaving the underlying matroid unchanged. This operation will be called a *reorientation* of $\mathcal{M}$.

Before we continue our discussion of oriented matroids, we will now explain two good alternatives to our quite impractical notation for signed sets.

First we can list a signed set as a string of elements, putting a bar over the negative elements. Thus $X = (\{2, 6\}, \{1, 5\})$ becomes $X = \overline{12}5\overline{6}$, and $Y = (\{1, 3, 6\}, \{4\})$ becomes $Y = 13\overline{4}6$.

This compact notation allows us to give the lists of *all* circuits and cocircuits for reasonably small oriented matroids. So our example $\mathcal{M}_0$ (the oriented matroid of the digraph in Figure 1.1.1) is the oriented matroid whose circuits are $1\overline{2}4$, $1\overline{3}5$, $2\overline{3}6$, $4\overline{5}6$, $1\overline{2}5\overline{6}$, $13\overline{4}6$, $2\overline{3}45$ and their negatives, and whose cocircuits are 123, $1\overline{4}5$, $24\overline{6}$, 356, $12\overline{5}6$, $13\overline{4}6$, 2345 and their negatives. For listings like this every circuit or cocircuit can be regarded as essentially interchangeable with its negative, like $\overline{12}4$ with $1\overline{2}4$, hence the negatives need not be listed.

The second useful notation records every signed set by its "signed incidence vector" in $\{+1, -1, 0\}^E$, or, shorter, in $\{+, -, 0\}^E$. Thus $X = (\{2, 6\}, \{1, 5\})$ is now represented by $X = (-+00-+)$, whereas $Y = (\{1, 3, 6\}, \{4\})$ now reads $Y = (+0+-0+)$.

With this notation a list of all circuits can be given by a matrix, such as

$$\mathcal{C}: \begin{pmatrix} +-0+0\,0 \\ +0-0+0 \\ 0+-0\,0+ \\ 0\,0\,0+-+ \\ +-0\,0+- \\ +0-+0+ \\ 0+--+0 \end{pmatrix},$$

where every circuit of $\mathcal{M}_0$ is either a row or the negative of a row of this matrix. Similarly the cocircuits can be given by a matrix – and the matrices for circuits and cocircuits are closely related, as we will see. In vector notation, the coordinate X_e is either $+$ or $-$ or 0, depending on whether e is contained in X positively or negatively or not at all.

Note that from now on we will often delete the adjective "signed" when referring to the circuits (or cocircuits, see below) of an oriented matroid, since the circuits of an oriented matroid are always signed sets (or sign vectors, depending on the notation).

We have now seen how to get the collections of signed circuits and signed cocircuits from a digraph D. It is then easy to derive the analogues of the usual matroid axioms for signed circuits and cocircuits. Namely, the *circuit axioms for oriented matroids* will be

(C0) $\emptyset$ is not a signed circuit.

(C1) If X is a signed circuit, then so is $-X$.

(C2) No proper subset of a circuit is a circuit.

(C3) If X_0 and X_1 are circuits with $X_1 \neq -X_0$ and $e \in X_0^+ \cap X_1^-$, then there is a
      third circuit $X \in \mathcal{C}$ with $X^+ \subseteq (X_0^+ \cup X_1^+)\backslash\{e\}$ and $X^- \subseteq (X_0^- \cup X_1^-)\backslash\{e\}$.

Here (C2) in fact refers to the underlying circuits $\underline{X}$: it requires that if $\underline{X}_1 \subseteq \underline{X}_2$ for two circuits, then $\underline{X}_1 = \underline{X}_2$, and, in fact, $X_1 = \pm X_2$.

The reader should check that these axioms are satisfied in digraphs. In particular, for (C3), one could even require a "strong elimination axiom": demand $f \in X^+$ for some $f \in X_0^+ \backslash \underline{X}_1$.

The next thing to check is that the collection of signed cocircuits $\mathcal{C}^*$ of a digraph satisfies exactly the same set of axioms. This naturally leads to the concept of the *dual oriented matroid*, whose collection of circuits is given by $\mathcal{C}^*$. For this, note that if D is a planar digraph, then there is a canonical way to orient the dual graph D^*, such that the circuits $\mathcal{C}(D^*)$ of the dual graph are exactly the cocircuits $\mathcal{C}^*(D)$, and conversely: for this draw both in the plane, and direct every arc of D^* so that it arises from a counter-clockwise turn of the corresponding arc in D. For our example D_0, this is done in Figure 1.1.2.

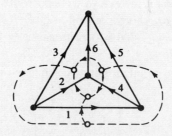

Figure 1.1.2: The dual digraph for the acyclic orientation D_0 of K_4.

Finally, we observe the important orthogonality property: if a directed cycle traverses a cut, then it also has to traverse it in the opposite direction. This translates into the following property for oriented matroids:

($\perp$) If $X \in \mathcal{C}$ is a circuit and $Y \in \mathcal{C}^*$ is a cocircuit of an oriented matroid with
      $(X^+ \cap Y^+) \cup (X^- \cap Y^-) \neq \emptyset$, then also

$$(X^+ \cap Y^-) \cup (X^- \cap Y^+) \neq \emptyset,$$

and conversely.

Perhaps this is a little less confusing with better notation: let

$$S(X,Y) := (X^+ \cap Y^-) \cup (X^- \cap Y^+)$$

denote the *separation set* for two sign vectors X and Y. Then the orthogonality condition can be written as

($\perp$) If $X \in \mathcal{C}$ is a circuit and $Y \in \mathcal{C}^*$ is a cocircuit, then $S(X, -Y) \neq \emptyset$ if and only if $S(X,Y) \neq \emptyset$.

Translating into $+/-/0$-vectors, this says that if X and Y agree in a non-zero coordinate, then they also disagree in a coordinate that is non-zero in both vectors, and conversely.

For two arbitrary sign vectors $X, Y \in \{+, -, 0\}^E$, we will write $X \perp Y$ and say that X and Y are *orthogonal* if the sets $S(X,Y)$ and $S(X, -Y)$ are either both empty or both non-empty. Note that this is a symmetric relation: $X \perp Y$ if and only if $Y \perp X$. It finally allows us to reformulate the *orthogonality* property in its simplest possible form:

($\perp$) $X \perp Y$ for all $X \in \mathcal{C}$ and all $Y \in \mathcal{C}^*$.

1.2 Point configurations and hyperplane arrangements

There are several closely related ways in which oriented matroids arise from geometric situations in real vector spaces. These situations provide alternative ways to view oriented matroids. In this section, we consider the two most important ones: point configurations and hyperplane arrangements. They provide models for oriented matroids that are far less restrictive than digraphs. In fact they are "nearly universal" in the sense that it is possible to think of oriented matroids as *generalized point configurations* or as *generalized hyperplane arrangements*.

(a) Vector configurations

Given a finite set of vectors that spans a vector space of dimension r over an arbitrary field, the minimal linear dependences yield the circuits of a matroid of rank r.

To get an *oriented* matroid, consider a finite set $E = \{\mathbf{v}_1, \ldots, \mathbf{v}_n\}$ of vectors that spans a vector space of dimension r over an *ordered* field, $\{\mathbf{v}_1, \ldots, \mathbf{v}_n\} \subseteq \mathbb{R}^r$, say. In this case a minimal linear dependence looks like

$$\sum_{i=1}^{n} \lambda_i \mathbf{v}_i = 0$$

with $\lambda_i \in \mathbb{R}$. Here the sets $\underline{X} := \{i : \lambda_i \neq 0\}$ corresponding to the minimal linear dependences are the circuits of the (unoriented) matroid. For the associated *oriented matroid*, we consider the signed sets

$$X = (X^+, X^-)$$

given by

$$X^+ := \{i : \lambda_i > 0\}, \qquad X^- := \{i : \lambda_i < 0\}$$

for all the minimal dependences among the $\mathbf{v}_i$. This yields the *oriented matroid* $\mathcal{M} = (E, \mathcal{C})$ *of a vector configuration* E in terms of its collection $\mathcal{C}$ of signed circuits. Note that the coefficients of a minimal linear dependence are unique up to a common scalar. This means that for a given circuit $\underline{X}$ of the matroid, there are exactly two signed circuits in $\mathcal{M}$, namely $X = (X^+, X^-)$ and $-X = (X^-, X^+)$. We will say that this oriented matroid has *rank* r, because the underlying unoriented matroid has rank r.

It is easy to check the axioms stated in the last section: in particular $X \in \mathcal{C}$ implies $-X \in \mathcal{C}$ because the negative of any linear dependence is again a linear dependence; (C2) is clear by definition; and the elimination axiom (C3) follows by taking a suitable linear combination of the two dependences in question, although finding a minimal resulting dependence requires some work.

Interesting vector configurations to be studied from this point of view are given, for example, by the vertices of polytopes and by the root systems of semisimple Lie algebras.

For a more specific example, let $\mathbf{v}_1, \ldots, \mathbf{v}_6$ be the vectors in $\mathbb{R}^3$ given by the columns of the matrix

$$A = \begin{pmatrix} 1 & 1 & 1 & 0 & 0 & 0 \\ 0 & 1 & 1 & 1 & 1 & 0 \\ 0 & 0 & 1 & 0 & 1 & 1 \end{pmatrix}.$$

From A we get an oriented matroid of rank 3 on $E = \{1, 2, \ldots, 6\}$, for which the linear dependence $\mathbf{v}_1 - \mathbf{v}_2 + \mathbf{v}_4 = 0$ translates into the circuit $1\bar{2}4$, whereas $\mathbf{v}_1 - \mathbf{v}_2 + \mathbf{v}_5 - \mathbf{v}_6 = 0$ yields the circuit $1\bar{2}5\bar{6}$, and so on.

In fact, we get the *same* list of signed circuits and hence the same oriented matroid as in our example in Section 1.1. Thus we will consider the matrix A as a *(vector) realization* of the oriented matroid $\mathcal{M}_0$.

The bases of the matroid corresponding to a vector configuration are the subsets that are vector space bases. Now it is natural to consider the determinants: after all, $\det(\mathbf{v}_{i_1}, \ldots, \mathbf{v}_{i_r}) \neq 0$ if and only if $\{\mathbf{v}_{i_1}, \ldots, \mathbf{v}_{i_r}\}$ is a basis.

For a vector configuration over $\mathbb{R}$, we consider the signs of the determinants of ordered r-subsets of $\{\mathbf{v}_1, \ldots, \mathbf{v}_n\}$. This leads us to define the *basis orientation* or *chirotope* of a vector configuration, by

$$\chi(i_1, \ldots, i_r) := \text{sign } \det(\mathbf{v}_{i_1}, \ldots, \mathbf{v}_{i_r}) \in \{+, -, 0\}.$$

Note that χ is an antisymmetric function. Thus we get $\chi(1, 2, 3) = +$, $\chi(1, 2, 4) = 0$ and $\chi(1, 3, 4) = -$ for our example, but $\chi(1, 3, 2) = -$ and so on.

In addition to antisymmetry, the determinants of a configuration of vectors also satisfy further relations, known as the Grassmann-Plücker relations. If these relations are reduced to the sign patterns they imply, then one obtains the *basis axioms* or *chirotope axioms* for oriented matroids. For example, in rank 3 one gets for every set of five vectors (denoting determinants by square brackets, and labeling the points 1 to 5) the relation

$$[123][145] - [124][135] + [125][134] = 0.$$

The corresponding oriented matroid axiom thus requires that the six signs of the brackets on the left side are such that the equality is at least possible for this sign pattern, when actual scalars are not given: for example, the six signs could be $+, +, +, +, +, +$, but not $+, +, -, +, +, +$.

The important observation is that the signed circuits of an oriented matroid do determine an associated basis orientation (unique up to a reversal of all the signs). Conversely, the signed circuits can be reconstructed from the chirotope. However, the correspondence is not quite straightforward, and the equivalence of the basis axioms with the circuit axioms for oriented matroids is much harder to see and prove than in the unoriented case.

We can also interpret the cocircuits of a vector configuration: for this consider any linear hyperplane H spanned by some subset of $\{v_1, \ldots, v_n\}$, and let Y_H^+ be the set of vectors on one side of H, and let Y_H^- be the set of vectors on the other side of H. Then the cocircuits of $\mathcal{M}$ are exactly the signed sets that arise as $Y := (Y_H^+, Y_H^-)$ in this way.

However, this is not easy to visualize even for small examples of rank 3 – which motivates the passage to affine space, in order to cut the dimension of our problem down by one. In affine space, vector configurations are best viewed as point configurations, which are treated next.

(b) Point configurations

Every vector configuration in $\mathbb{R}^r \backslash \{0\}$ corresponds to an affine point configuration in an $(r-1)$-dimensional affine space: for this choose a linear form ℓ_0 such that $\ell_0(v_i) \neq 0$ for all i, define

$$\mathbb{A}^{r-1} := \{x \in \mathbb{R}^r : \ell_0(x) = 1\}$$

as our affine space of dimension $r-1$, and associate with every vector v_i the point $\frac{1}{\ell_0(v_i)} v_i \in \mathbb{A}^{r-1}$. Here vectors v_i with $\ell_0(v_i) < 0$ determine "reoriented affine points", or "points with negative weight".

These "negative points" are somewhat annoying to have to deal with, although sometimes indispensable. However, if the vector configuration does not contain any positive linear dependences ($\sum_i \lambda_i v_i = 0$ with $\lambda_i \geq 0$), then we can choose ℓ_0 such that $\ell_0(v_i) > 0$ for all i, which results in an honest affine point configuration. This corresponds to the situation where the oriented matroid is *acyclic* in the sense mentioned in Section 1.1: it does not contain any positive circuits. Of course, we could easily achieve the acyclic situation by replacing some of the vectors v_i by their negatives, which would result in a *reorientation* of the oriented matroid.

Now every point configuration in a real affine space gives rise to an acyclic oriented matroid, whose circuits are given by the signs of the coefficients in the minimal *affine* dependences ($\sum_i \lambda_i v_i = 0$ with $\sum_i \lambda_i = 0$). This leads us to an important class of examples: the vertices of a convex polytope describe an acyclic oriented matroid, and this is how oriented matroids enter polytope theory.

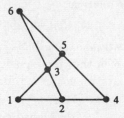

Figure 1.2.1: The affine point configuration corresponding to $\mathcal{M}_0$.

Starting from the configuration of six vectors in $\mathbb{R}^3$ discussed in Part (a) above, with $\ell_0(\mathbf{x}) = x_1 + x_2 + x_3$, one gets the point configuration of Figure 1.2.1 in the affine plane.

Affine coordinates for it are given as the column vectors of

$$\overline{A} = \begin{pmatrix} 0 & \frac{1}{2} & \frac{1}{3} & 1 & \frac{1}{2} & 0 \\ 0 & 0 & \frac{1}{3} & 0 & \frac{1}{2} & 1 \end{pmatrix},$$

which are obtained by deleting the first coordinates from the vectors $\frac{1}{\ell(\mathbf{v}_i)}\mathbf{v}_i$.

Note here the simple geometric interpretation of signed circuits in the affine picture: every circuit is a minimal signed set of points such that the convex hull of the positive elements intersects the convex hull of the negative elements. Thus the signed circuits of an affine point configuration are exactly the *minimal Radon partitions* of the point set.

For example, from $1\overline{2}4$ we see that point 2 must lie on the open line segment 14, and from $2\overline{3}\overline{4}5$ that the open line segments 25 and 34 must intersect (in the ambient real affine space, not necessarily in the oriented matroid itself).

The sign patterns of arbitrary (not necessarily minimal) affine dependences can be derived from the circuits via "composition". For this one defines the composition $X \circ Y$ of two signed sets as

$$X \circ Y = (X^+ \cup (Y^+ \backslash X^-), X^- \cup (Y^- \backslash X^+)),$$

which in vector notation reads as

$$(X \circ Y)_e = \begin{cases} X_e & \text{if } X_e \neq 0, \\ Y_e & \text{otherwise.} \end{cases}$$

The signed sets obtained by repeated compositions of circuits in this way are called the *vectors* of the oriented matroid. They correspond to the *Radon partitions* of the point configuration. A dependent set in the underlying matroid may correspond to several distinct vectors of the oriented matroid.

Now we observe that also signed cocircuits have a geometric interpretation in the oriented matroid of an affine point configuration, which is a direct translation of the analogous description in the linear case. An unsigned matroid cocircuit is the complement of a hyperplane of the matroid, and in the oriented matroid we record additionally on which side of the hyperplane each element of the cocircuit lies.

Vectors of $\mathcal{M}_0$

Circuits				Maximal vectors	
1$\overline{2}$4	1$\overline{2}$5$\overline{6}$	1$\overline{2}$345	1$\overline{2}$45$\overline{6}$	12$\overline{3}$456	123$\overline{4}$5$\overline{6}$
1$\overline{3}$5	1$\overline{3}$46	1$\overline{2}$345	1$\overline{2}$45$\overline{6}$	12$\overline{3}$456	123$\overline{4}$5$\overline{6}$
2$\overline{3}$6	2$\overline{3}$45	1$\overline{2}$345	1$\overline{2}$45$\overline{6}$	12$\overline{3}$456	123$\overline{4}$5$\overline{6}$
4$\overline{5}$6		1$\overline{2}$346	1$\overline{3}$456	12$\overline{3}$456	123$\overline{4}$5$\overline{6}$
		1$\overline{2}$346	1$\overline{3}$456	123$\overline{4}$56	1234$\overline{5}$6
		1$\overline{2}$346	1$\overline{3}$456	123$\overline{4}$56	1234$\overline{5}$6
		12$\overline{3}$56	2$\overline{3}$456		
		12$\overline{3}$56	2$\overline{3}$456		
		12$\overline{3}$56	2$\overline{3}$456		

Covectors of $\mathcal{M}_0$

Cocircuits				Maximal covectors	
123	1$\overline{2}$5$\overline{6}$	12345	1$\overline{2}\overline{4}$56	123456	123$\overline{4}\overline{5}$6
1$\overline{4}$5	1$\overline{3}$46	12345	1245$\overline{6}$	1$\overline{2}$3456	123456
24$\overline{6}$	2345	1$\overline{2}$345	12456	1$\overline{2}$3456	123456
356		1$\overline{2}$346	13456	1$\overline{2}$3456	123$\overline{4}$56
		1$\overline{2}$346	13456	123456	123$\overline{4}$56
		1$\overline{2}$346	13$\overline{4}$56	123456	123456
		12356	23456		
		123$\overline{5}$6	234$\overline{5}$6		
		12$\overline{3}$56	2$\overline{3}$456		

Figure 1.2.2: The vectors and covectors of $\mathcal{M}_0$.

Cutting the (affine) point configuration by an *arbitrary* hyperplane (not necessarily spanned by points of the configuration) we can still record which points are on the positive or on the negative side. The resulting sign patterns are called the *covectors* of the oriented matroid: they are given as the repeated compositions of the cocircuits. In particular, for every finite point configuration we can find a hyperplane in the ambient space for which all points are on the same side. Therefore $T = (E, \emptyset) = (++\ldots+)$ is a maximal covector, and in fact every point is contained in a positive cocircuit: for this consider the hyperplanes spanned by the facets of conv(E).

For our example $\mathcal{M}_0$, the covector 123$\overline{45}$ indicates that there is a hyperplane which contains the point 6, which has the points $1, 2$ and 3 on one side, and which has the points 4 and 5 on the other side. In Figure 1.2.2 we give a full listing of the vectors and covectors of $\mathcal{M}_0$, except that the negative of each listed vector or covector has not been listed.

There is additional information on the relative orientation of the points that can be gleaned from the above affine representation. In this rank 3 example, each ordered basis is an ordered triple of non-collinear points. We record whether the triple is placed in counter-clockwise or clockwise order in the plane by a + or − sign, respectively. Similar conventions apply to oriented matroids of higher rank. For example, in rank 4, a + indicates a right-hand ordered quadruple of points. In

Figure 1.2.3 we list a sign for one ordering of each basis of the underlying matroid, with the understanding that the sign for any other ordering of a particular basis is inferred according to the sign of the appropriate permutation.

$$
\begin{array}{cccccccc}
123+ & 125+ & 126+ & 134- & 136+ & 145+ & 146+ & 156+ \\
234- & 235- & 245+ & 246+ & 256+ & 345+ & 346+ & 356+
\end{array}
$$

Figure 1.2.3: Basis orientations for $\mathcal{M}_0$.

This sign so assigned is again given as the sign of a determinant: given affine coordinates, and writing them into column vectors $\overline{\mathbf{v}}_1, \ldots, \overline{\mathbf{v}}_n$, we have

$$
\chi(i_1, i_2, \ldots, i_r) = \text{sign det} \begin{pmatrix} 1 & 1 & \cdots & 1 \\ \overline{\mathbf{v}}_{i_1} & \overline{\mathbf{v}}_{i_2} & \cdots & \overline{\mathbf{v}}_{i_r} \end{pmatrix}.
$$

Thus, for example, we get $\chi(2,3,4) = -$ for our six point configuration from

$$
\begin{vmatrix} 1 & 1 & 1 \\ \frac{1}{2} & \frac{1}{3} & 1 \\ 0 & \frac{1}{3} & 0 \end{vmatrix} < 0.
$$

Perhaps we should caution the reader at this point that our example $\mathcal{M}_0$ is a very small and quite special oriented matroid (being derived from an acyclic digraph). Nevertheless the facts discussed about $\mathcal{M}_0$ so far all generalize to arbitrary oriented matroids. In particular it is true that any of the collections of signed circuits $\mathcal{C}$, cocircuits $\mathcal{C}^*$, vectors $\mathcal{V}$, covectors $\mathcal{L}$, basis orientations χ, maximal vectors $\mathcal{W}$, or of maximal covectors $\mathcal{T}$ determines the oriented matroid (and hence any other of the listings mentioned).

However, as the reader will have guessed by now, the concept of an oriented matroid is more general than that of a realized oriented matroid, that is, of a vector or point configuration. We will see later in this chapter examples of oriented matroids which cannot be realized. In Chapter 3, we give the equivalent (cryptomorphic) definitions of oriented matroids in terms of axioms for each of the above concepts. The relations between these concepts agree with the corresponding relations between the unsigned concepts in the underlying matroid where appropriate.

(c) Hyperplane arrangements

Let $\mathcal{A} = \{H_1, \ldots, H_n\}$ be a real hyperplane arrangement, that is, a finite set of hyperplanes through the origin in $\mathbb{R}^r$. Where is the matroid? Every hyperplane is given by a linear function $\ell_i(\mathbf{x}) = \sum_{j=1}^r a_{ij} x_j$, so we can write $H_i = \{\mathbf{x} \in \mathbb{R}^r : \ell_i(\mathbf{x}) = 0\}$. Now we can interpret the ℓ_i as vectors in the dual space $(\mathbb{R}^r)^*$. So the ℓ_i form a vector configuration in $(\mathbb{R}^r)^*$, which determines an oriented matroid (as before): this is the oriented matroid $\mathcal{M}_\mathcal{A}$ we consider in this subsection.

In fact, with the choice of the ℓ_i we have distinguished a positive side of every hyperplane H_i, given by $H_i^+ = \{\mathbf{x} \in \mathbb{R}^r : \ell_i(\mathbf{x}) \geq 0\}$, and the oriented matroid corresponds to the arrangement of halfspaces $\{H_i^+ : 1 \leq i \leq n\}$ in $\mathbb{R}^r$. Equivalently, via $S_i^+ := H_i^+ \cap S^{r-1}$ we get an arrangement of hemispheres

$\mathcal{S} = \{S_i^+ : 1 \leq i \leq n\}$ on the sphere $S^{r-1} = \{\mathbf{x} \in \mathbb{R}^r : ||\mathbf{x}|| = 1\}$ that represents the oriented matroid.

It is clear that an alternative to the choice of linear functions ℓ_i for each hyperplane H_i is simply the choice of a positive side H_i^+, and that at least this choice is necessary to fully specify an oriented matroid from a given hyperplane arrangement. One customary way to do this is to specify one *region* (full-dimensional cell in the subdivision of $\mathbb{R}^r$ by $\mathcal{A}$) to be on the positive side of every hyperplane in $\mathcal{A}$. It can be shown that this results in an acyclic oriented matroid.

The hyperplane arrangement $\mathcal{A}$ or the arrangement $\{S_i : 1 \leq i \leq n\}$ of $(r-2)$-spheres $S_i := H_i \cap S^{r-1}$ on the $(r-1)$-sphere alone determine an oriented matroid only up to reorientation: that is, they determine a *reorientation class* of oriented matroids.

Now arrangements of halfspaces give rise to vector configurations (up to positive scalar multiples), and conversely every configuration of vectors $\{\mathbf{v}_1, \ldots, \mathbf{v}_n\}$ gives the corresponding arrangement of hyperplanes (halfspaces) via $H_i^+ := \{\ell \in (\mathbb{R}^r)^* : \ell(\mathbf{v}_i) \geq 0\}$. So why do we study hyperplane arrangements in connection with oriented matroids, if they are only a trivial reformulation of vector configurations?

The answer is that there are at least two remarkable aspects of this way of viewing (realizable) oriented matroids. The first is that not only does every real arrangement of hyperplanes give rise to an oriented matroid, but the converse is also "nearly true": this is the content of the Topological Representation Theorem for oriented matroids, which we will discuss in Section 1.4 below.

The second reason is that the oriented matroid encodes a lot of geometric information about the hyperplane arrangement in a very simple and explicit way. For this, observe that every hyperplane arrangement decomposes its ambient space in a collection of pieces (that are in fact relatively open topological cells of various dimensions). Each of these *cells* is exactly determined by the information whether for the hyperplane H_i, the cell is on its positive side, its negative side, or on the hyperplane itself. This leads us to associate a sign vector with every cell.

These sign vectors are exactly the covectors of the oriented matroid. In fact, assuming $\bigcap \mathcal{A} = \{0\}$, the origin gets the zero string as its covector, while the one-dimensional rays correspond to the cocircuits of the oriented matroid. Every other cell gets the sign vector that is the composition of the cocircuit vectors in its boundary.

For our example $\mathcal{M}_0$, this looks as follows. We get an arrangement of six halfspaces in $\mathbb{R}^3$. To draw this, we consider the arrangement of hemispheres (each bounded by a great circle) this induces in S^2. Note that using the antipodal symmetry of this arrangement, it suffices to draw the upper hemisphere of S^2, which is bounded by $S_6 = H_6 \cap S^2$, see Figure 1.2.4. Here the positive hemisphere with respect to every line is indicated by a small arrow. Note that Figure 1.2.4 could also be thought of as picturing a projective line arrangement in the real projective plane, with S_6 representing the "line at infinity".

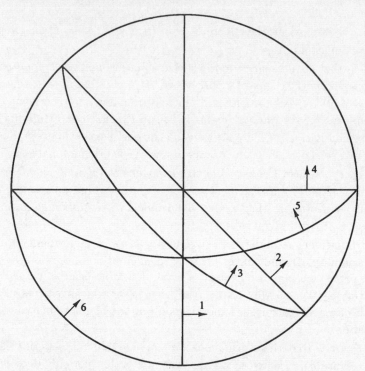

Figure 1.2.4: An arrangement of lines corresponding to $\mathcal{M}_0$.

The cocircuits of $\mathcal{M}_0$ now correspond to the vertices of this arrangement, vertex Y getting a set of labels that includes label i if it lies strictly on the positive side of H_i, and label $\bar{i}$ if it lies strictly on the negative side of H_i. If in a similar way we associate labels to every cell of this line arrangement in S^2, then we get a complete list of covectors (Figure 1.2.2) as shown for $\mathcal{M}_0$ in Figure 1.2.5. Note that the negatives of covectors on the upper hemisphere lie on the lower hemisphere and are therefore not shown in Figure 1.2.5.

It is also easy to derive the signed circuits from the line arrangement picture: $1\bar{2}4$ is a signed circuit because $\{S_1^+, S_2^-, S_4^+\}$ is a minimal set of hemispheres whose union is S^2 (see Section 1.4). Of course, one could also consider the corresponding set of halfspaces in $\mathbb{R}^3$.

To read off the basis orientations is equally simple (when $r = 3$). The rule is that $[i, j, k] = +$ if i, j and k bound the triangle $S_i^+ \cap S_j^+ \cap S_k^+$ in counterclockwise order, $[i, j, k] = -$ if they occur in clockwise order, and $[i, j, k] = 0$ if they do not determine a triangle, but go through a common point. So for $\mathcal{M}_0$ from Figure 1.2.4 we can read off $[1, 2, 6] = +$, $[2, 3, 5] = -$ and $[2, 3, 6] = 0$, in accordance with Figure 1.2.3.

(d) Vector subspaces

Every oriented matroid has a uniquely determined dual, as illustrated for directed graphs in Section 1.1. This is not the projective polarity (often called "projective duality") of point and hyperplane configurations. A different repre-

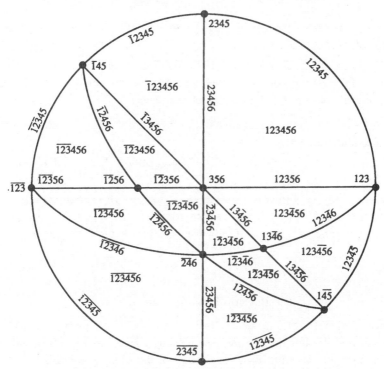

Figure 1.2.5: Labeling the cells of $\mathcal{M}_0$ by covectors.

sentation of hyperplane arrangements, however, makes the construction of the arrangement that corresponds to the dual oriented matroid more obvious.

For this let ξ be a vector subspace of dimension r of $\mathbb{R}^n$ (think of n as being large). Then the standard coordinate planes $\hat{H}_i := \{\mathbf{y} \in \mathbb{R}^n : y_i = 0\}$ cut out a hyperplane arrangement $\mathcal{A} = \{H_1, \ldots, H_n\}$ in the space ξ, with $H_i = \hat{H}_i \cap \xi = \{\mathbf{y} \in \xi : y_i = 0\}$. The oriented matroid associated with this hyperplane arrangement in ξ we denote by $\mathcal{M}(\xi)$. If $\xi \subseteq H_i$ for some i, then i is a *loop* of the oriented matroid, i.e., it defines a one-element circuit.

Conversely, if $\mathcal{A} = \{H_1, \ldots, H_n\}$ is given as a hyperplane arrangement in $\mathbb{R}^r$ by $\ell_i(\mathbf{x}) = \Sigma_j a_{ij} x_j$ as above, then we can identify $\mathbb{R}^r$ with the subspace

$$\xi := \{\mathbf{y} \in \mathbb{R}^n : y_i = \ell_i(\mathbf{x}) \text{ for all } i, \text{ for some } \mathbf{x} \in \mathbb{R}^r\}$$

of $\mathbb{R}^n$. Then $\mathcal{M}(\xi)$ is the same oriented matroid as $\mathcal{M}_\mathcal{A}$, the one directly given by the vectors ℓ_i.

The key observation now is that in the same way $\mathcal{M}(\xi)$ is associated to ξ, the dual matroid $\mathcal{M}(\xi)^*$ is associated to the orthogonal subspace $\xi^\perp$, that is,

$$(\mathcal{M}(\xi))^* = \mathcal{M}(\xi^\perp).$$

This is hard to picture, since it happens in a high-dimensional space. We can make it plausible, however, by the orthogonality property ($\perp$) discussed at the end of Section 1.1. If $\mathbf{x} \in \xi$ and $\mathbf{y} \in \xi^\perp$, then $\mathbf{x}^T\mathbf{y} = \sum_{i=1}^n x_i y_i = 0$. This means that if $\mathbf{x}$ and $\mathbf{y}$ have the same sign on a non-zero coordinate, then they also have opposite signs in some non-zero coordinate, and conversely. Now if X is the sign vector of **x** (that is, a covector of $\mathcal{M}(\xi)$) and Y is the sign vector of **y** (that is, a covector of $\mathcal{M}(\xi^\perp)$, or a vector of $\mathcal{M}(\xi)$), then since $\mathbf{x}^T\mathbf{y} = 0$ we get $X \perp Y$.

In the oriented matroid framework not only circuits and cocircuits are orthogonal, but in fact vectors and covectors are always orthogonal, and this characterizes either of these families in terms of the other:

($\perp$) $X \in \{+, -, 0\}^E$ is a vector of $\mathcal{M}$ if and only if $X \perp Y$ for all covectors Y of $\mathcal{M}$, and similarly

($\perp$) $Y \in \{+, -, 0\}^E$ is a covector of $\mathcal{M}$ if and only if $X \perp Y$ for all vectors X of $\mathcal{M}$.

To see the "only if" direction, observe that for any three sign vectors V^1, V^2 and Y with $V^1 \perp Y$ and $V^2 \perp Y$ we also have $(V^1 \circ V^2) \perp Y$, directly from the definitions. The "if" direction, however, requires some work, see Proposition 3.7.12.

1.3 Pseudoline arrangements

Oriented matroids are closely related to a number of questions in discrete geometry in real Euclidean and projective spaces. Here we give an initial examination of their relationship to arrangements of lines and pseudolines.

A finite collection of straight lines in the real projective plane, such as in Figure 1.3.1, is called an *arrangement of lines*. As we have already seen in Section 1.2(c), such an arrangement is a realization of an oriented matroid of rank 3. For non-degeneracy, we will also require that the intersection of all the lines is empty. Several questions have long been studied in the literature about combinatorial properties of such arrangements, particularly concerning the induced cell decomposition of the plane: numbers of vertices, edges, and faces of different kinds, and their incidences and isomorphism types.

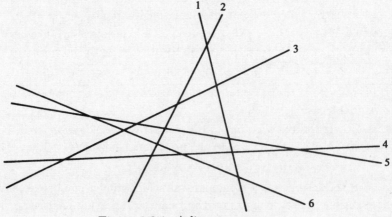

Figure 1.3.1: A line arrangement.

Many of the arguments used in investigations of arrangements of lines do not

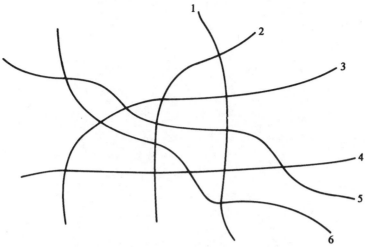

Figure 1.3.2: A pseudoline arrangement.

in any way use the fact that the lines are straight (rectilinear). This suggests that perhaps one should also study arrangements of "curved" lines, to develop a more general theory and to better understand the exact role that straightness plays; see Grünbaum's (1970) charming article "The importance of being straight".

A *pseudoline* is any simple closed curve in $\mathbb{P}^2$ (the real projective plane) whose complement is connected. By an *arrangement of pseudolines* we shall mean any finite collection of pseudolines in $\mathbb{P}^2$ satisfying the following two properties: (1) any two pseudolines intersect in exactly one point (this can be shown to imply that they "cross" each other there), (2) the intersection of all of the pseudolines is empty. Figure 1.3.2 shows an arrangement of pseudolines.

A basic question one faces when dealing with an arrangement of pseudolines is: Can it be *straightened,* or *stretched*? That is, does there exist some self-homeomorphism of $\mathbb{P}^2$ that moves each pseudoline to a straight line? Equivalently (see Section 6.3), is the induced cell decomposition of $\mathbb{P}^2$ combinatorially isomorphic to the cell decomposition induced by some arrangement of straight lines? Figure 1.3.1 is such a straightening of Figure 1.3.2. Figure 1.3.3 shows a non-stretchable arrangement. If all nine lines in Figure 1.3.3 could be made straight, then by Pappus's Theorem the intersection point p would have to lie on the line l, showing the impossibility of an isomorphic straight-line representation.

This example is minimal in the sense that every arrangement of at most eight pseudolines *is* stretchable (see Theorem 8.2.4(1)).

How can the stretchability question be tackled? Grünbaum (1970) wrote:

"Concerning arrangements, no significant tools seem to be known that discriminate between lines and pseudolines. But there are a large number of problems which – I believe – would become solvable if such tools were found."

Much progress has been made since 1970 via the theory of oriented matroids. The connection hinges on the following two facts:

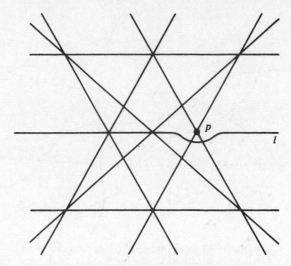

Figure 1.3.3: A non-stretchable pseudoline arrangement
(the "non-Pappus" oriented matroid).

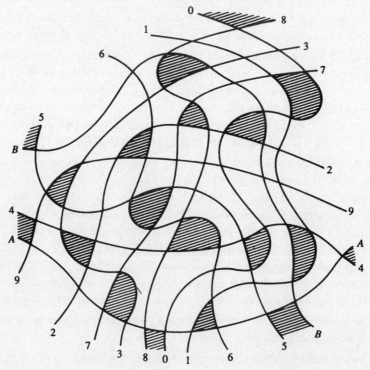

Figure 1.3.4: A pseudoline arrangement without adjacent triangles.

(1) There is a one-to-one correspondence between (equivalence classes of) arrangements of pseudolines and (reorientation classes of) simple rank 3 oriented matroids.

This is the rank 3 case of the Topological Representation Theorem of Folkman and Lawrence (1978), presented in Sections 1.4 and 5.2.

(2) Under this correspondence, stretchability of the arrangement is equivalent to realizability over $\mathbb{R}$ of the oriented matroid of rank 3.

Hence the stretchability question for pseudoline arrangements is a cryptomorphic version of the realizability question for oriented matroids. Realizability, in turn, can be dealt with by algebraic algorithmic methods (see Section 8.4).

To illustrate the progress that the oriented matroid point of view has led to in the study of arrangements, let us mention the conjecture of Grünbaum (1970) that in every arrangement of lines some two triangles have a common vertex. This was disproved by Ljubić, Roudneff and Sturmfels (1989), who found the arrangement of pseudolines depicted in Figure 1.3.4. To settle the conjecture, this arrangement must be accompanied by a stretchability proof. Such a proof is provided by the authors by way of an oriented matroid argument, which gives a complete description of all realizations over $\mathbb{R}$. From this information it can be gleaned that no easily readable straight-line representation of the arrangement in Figure 1.3.4 is possible, so this particular counter-example to Grünbaum's conjecture would hardly have been found by anyone working with a pen and ruler.

Arrangements of pseudolines are discussed in Chapter 6, and the realizability question for oriented matroids is discussed in Chapter 8.

1.4 Topological Representation Theorem

One of the cornerstones of the theory of oriented matroids is the Topological Representation Theorem of Folkman and Lawrence (1978), which states that every oriented matroid has a pseudosphere representation. This in a precise way describes how general oriented matroids differ from realized ones by a certain measure of topological deformation. It is the generalization to arbitrary dimension of the relationship between pseudoline arrangements and straight-line arrangements, as discussed in Section 1.3. It is also the generalization of the hyperplane arrangement model, mentioned in Section 1.2, to arbitrary oriented matroids. Chapters 4 and 5 are devoted to a detailed treatment of the Topological Representation Theorem. Here we will give only a brief preview of their content. We first review the realized case.

Let E be a finite, parallel-free, spanning set of non-zero vectors in $\mathbb{R}^{d+1}$, and let $\mathcal{C} \subseteq \{+, -, 0\}^E$ be the set of signed circuits of the associated realized oriented matroid, as in Section 1.2(a). For each $e \in E$, let $S_e = \{x \in \mathbb{R}^{d+1} : \langle x, e \rangle = 0, \|x\| = 1\}, S_e^+ = \{x \in \mathbb{R}^{d+1} : \langle x, e \rangle \geq 0, \|x\| = 1\}$, and $S_e^- = -S_e^+$. These sets are subsets of the unit d-sphere $S^d = \{x \in \mathbb{R}^{d+1} : \|x\| = 1\}$, S_e is a linear (or flat) $(d-1)$-sphere, and S_e^+ and S_e^- are its two closed hemispheres (see Figure 1.4.1). Of course, S_e is just the intersection of S^d with the hyperplane orthogonal

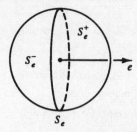

Figure 1.4.1: A signed hemisphere.

to the vector e, so the *arrangement of spheres* $\mathcal{A} = (S_e)_{e \in E}$ is equivalent to the arrangement of hyperplanes discussed in Section 1.2(c).

From a matroid point of view, the arrangement of spheres $\mathcal{A}$, together with the information about how the signs $+$ and $-$ are distributed to the hemispheres, carries the same information as E itself, because the vectors in E, up to length, can be reconstructed from it. Thus, realized rank $d + 1$ oriented matroids can be identified with such signed arrangements of $(d - 1)$-spheres in S^d. A little thought will show that the signed circuits $\mathcal{C}$ in this case are precisely those vectors $X \in \{+, -, 0\}^E$ such that

(C$_1$) $\bigcup_{e \in \underline{X}} S_e^{X_e} = S^d$, where $S_e^{X_e}$ denotes S_e^+ or S_e^-, as appropriate, and

(C$_2$) the support $\underline{X} = \{e \in E : X_e \neq 0\}$ is minimal with property (C$_1$).

Indeed, (C$_1$) is equivalent to the existence of a linear dependence $\sum \lambda_e \cdot e = 0$ with $sign(\lambda_e) = X_e$. Geometrically, no open halfspace determined by a hyperplane through $\mathbf{0}$ contains $X_e \cdot e$ for all $e \in \underline{X}$, which is in turn equivalent to $\mathbf{0}$ being in the convex hull of $\{X_e \cdot e : e \in \underline{X}\}$.

Let us review what all of this looks like in the case $d = 2$. Then E is a set of vectors in $\mathbb{R}^3$, and $\mathcal{A}$ is a collection of great circles on a 2-sphere. In the usual double-covering of the real projective plane by the 2-sphere, the image of each great circle is a projective line, giving the corresponding line arrangement.

A subset S of S^d will be called a *pseudosphere* if $S = h(S^{d-1})$ for some homeomorphism $h : S^d \to S^d$, where $S^{d-1} = \{x \in S^d : x_{d+1} = 0\}$. A pseudosphere S, as depicted in Figure 1.4.2, has two sides (hemispheres) S^+ and S^-, and is, of course, topologically indistinguishable from a linear $(d - 1)$-subsphere.

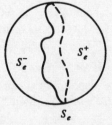

Figure 1.4.2: A pseudosphere.

We define an *arrangement of pseudospheres* $\mathcal{A} = (S_e)_{e \in E}$ to be a finite set of pseudospheres S_e in S^d such that

(A1) Every non-empty intersection $S_A = \bigcap_{e \in A} S_e$ is (homeomorphic to) a sphere of some dimension, for $A \subseteq E$.

(A2) For every non-empty intersection S_A and every $e \in E$ such that $S_A \not\subseteq S_e$, the intersection $S_A \cap S_e$ is a pseudosphere in S_A with sides $S_A \cap S_e^+$ and $S_A \cap S_e^-$.

In particular, if S_A consists of two points, then S_e separates those two points. This definition is due to Folkman and Lawrence (1978). They actually required more, but the additional assumptions were shown to be redundant by Edmonds and Mandel (1982). For simplicity we will also require here that $S_e \neq S_f$ if $e \neq f$ and that $\bigcap_{e \in E} S_e = \emptyset$. This will not be assumed in later sections (cf. Definition 5.1.3).

If $\mathcal{A}$ is a *signed arrangement of pseudospheres*, that is, a choice of a positive and negative side has been made for each pseudosphere, then let $\mathcal{C}(\mathcal{A})$ be the family of all those sign vectors $X \in \{+, -, 0\}^E$ that satisfy the conditions (C$_1$) and (C$_2$) above.

1.4.1 Topological Representation Theorem   (Folkman and Lawrence 1978).

(1) *If $\mathcal{A} = (S_e)_{e \in E}$ is a signed arrangement of pseudospheres in S^d, then $\mathcal{C}(\mathcal{A})$ is the family of circuits of a rank $d + 1$ simple oriented matroid on E.*

(2) *If $(E, \mathcal{C})$ is a rank $d + 1$ simple oriented matroid, then there exists a signed arrangement of pseudospheres $\mathcal{A}$ in S^d such that $\mathcal{C} = \mathcal{C}(\mathcal{A})$.*

(3) *$\mathcal{C}(\mathcal{A}) = \mathcal{C}(\mathcal{A}')$ for two signed arrangements $\mathcal{A}$ and $\mathcal{A}'$ in S^d if and only if $\mathcal{A}' = h(\mathcal{A})$ for some self-homeomorphism h of S^d.*

For a more detailed statement and proof of this theorem, see Section 5.2. One consequence of the theorem is that

> *there is a one-to-one correspondence between (equivalence classes of) arrangements of pseudospheres in S^d and (reorientation classes of) simple rank $d+1$ oriented matroids.*

Another consequence is that

> *under this correspondence stretchability of the arrangement (i.e., existence of a self-homeomorphism of S^d bringing each pseudosphere into linear position) is equivalent to realizability of the oriented matroid.*

The $d = 2$ case was already exemplified in Section 1.3 in the projective version, where, of course, the pseudospheres are pseudolines.

Because of the Topological Representation Theorem, the theory of oriented matroids has a distinctly more geometrical flavor than the theory of ordinary matroids. A non-coordinatizable matroid of abstract origin may be thought of as a geometric object only in a purely formal way, whereas an oriented matroid may always be thought of as a geometric-topological configuration on the d-sphere (or in projective space). The fact that rectilinear representation of the configuration is equivalent to realizability of the oriented matroid has the important consequence that questions of "straightening" in geometry (e.g., to distinguish convex polytope boundaries among combinatorial spheres, see Section 9.5) can be dealt with algorithmically.

The use of pseudosphere arrangements for the representation of oriented matroids suggests seeking a similar representation of oriented matroids as "pseudoconfigurations of points". For example, in the rank 3 case we could consider a set of vertices P in a pseudoline arrangement, such that any two vertices of P lie on a common pseudoline. This describes a rank 3 oriented matroid on P (Exercise 1.11).

However, the straightforward generalization to higher ranks breaks down. We can easily define *pseudoconfigurations of points* of rank r as certain vertex sets in pseudosphere arrangements. Then we find that every pseudoconfiguration of points determines an oriented matroid, but not every oriented matroid arises this way. The analysis of those oriented matroids that arise as pseudoconfigurations of points leads directly into the structure theory of oriented matroids – see Sections 5.3 and 7.5.

1.5 Realizability

Oriented matroids have the feature of "local realizability" in common with several other concepts in geometry. For instance, topological manifolds, differential manifolds and general varieties (or schemes) are all defined in terms of some property that guarantees that *locally* (i.e., in some neighborhood of each point) they look like some canonical type of model space that is being generalized (e.g., Euclidean space). The question of whether the geometric object as a whole can be embedded (or *realized*) in a model space is then an important problem, e.g., if a general variety is projective. The local structure of such geometric objects can be studied irrespective of whether these objects are globally realizable or not.

In what sense do oriented matroids fit this pattern? By Corollary 3.6.3 an acyclic rank r oriented matroid can be viewed as an abstract $(r-1)$-dimensional affine configuration of points such that every subconfiguration of $r+2$ of them can be coordinatized. In other words, there is a choice of local coordinate system for each such subconfiguration. As is made precise in Section 3.6, this characterizes oriented matroids.

Here we will discuss a certain acyclic rank 4 oriented matroid RS(8) on the set $E = \{1, 2, 3, 4, 5, 6, 7, 8\}$, which is globally non-realizable. (This oriented matroid was first constructed by contraction from a rank 5 oriented matroid AB(9) that we will encounter in Proposition 9.5.4.) We will think of RS(8) as an abstract configuration of eight points in general position in 3-dimensional affine space. We do not give a diagram for this oriented matroid yet because we wish to emphasize the interplay between abstract geometric concepts and pure sign pattern reasoning. We define RS(8) by prescribing the chirotope, that is, a sign for each ordered 4-tuple. In this case, every 4-tuple receives a non-zero value, which means every four-element subset of E is a basis of the underlying matroid. In the theory of ordinary matroids, this matroid is denoted $U_{4,8}$, and called the uniform rank 4 matroid on E. Hence we call RS(8) a *uniform* oriented matroid. The chirotope $\chi : \binom{E}{4} \to \{-,+\}$ of RS(8) is then defined by the following list.

1234 +	1235 +	1236 +	1237 +	1238 +	1245 +	1246 +
1247 −	1248 +	1256 +	1257 −	1258 −	1267 −	1268 −
1278 +	1345 −	1346 +	1347 −	1348 −	1356 +	1357 +
1358 +	1367 −	1368 −	1378 +	1456 +	1457 −	1458 −
1467 +	1468 −	1478 +	1567 +	1568 +	1578 −	1678 +
2345 −	2346 −	2347 −	2348 −	2356 +	2357 +	2358 +
2367 −	2368 +	2378 +	2456 +	2457 −	2458 +	2467 −
2468 +	2478 +	2567 +	2568 +	2578 −	2678 −	3456 +
3457 −	3458 −	3467 −	3468 −	3478 +	3567 +	3568 +
3578 −	3678 −	4567 −	4568 +	4578 −	4678 −	5678 +

We parenthetically note that such an oriented matroid may be very economically encoded by just the list of signs, $+++++++-++----\ldots$. Using the chirotope axioms given in Chapter 3 we can verify that this list does define an oriented matroid. The elements of E are called *points*, the two-element subsets of E are called *line segments*, the three-element subsets of E are called *triangles* and the four-element subsets of E are called *tetrahedra* of RS(8).

Many order and convexity properties can be read off from the chirotope χ. As an example we consider the subconfiguration of the first five points, which is given by the oriented tetrahedra $1234 +$, $1235 +$, $1245 +$, $1345 -$, and $2345 -$. In general, the signed circuit supported by five points $i_1 i_2 i_3 i_4 i_5$ is given in terms of the chirotope by $X_{i_1 i_2 i_3 i_4 i_5} = (X^+, X^-)$, where $X^+ = \{i_k : \chi(i_1 \ldots \hat{i}_k \ldots i_5) = (-1)^{k+1}\}$ and $X^- = \{i_k : \chi(i_1 \ldots \hat{i}_k \ldots i_5) = (-1)^k\}$. We see that

$$X_{12345} = (\{1,4\}, \{2,3,5\}).$$

In the language of convex geometry, $X_{12345} = (\{1,4\}, \{2,3,5\})$ is the unique minimal *Radon partition* on the point subset $\{1,2,3,4,5\}$, that is, the line segment $\{1,4\}$ intersects the triangle $\{2,3,5\}$ in the presumed ambient affine space (i.e., in local coordinates). In oriented matroid notation, $X_{12345} = \overline{1}23\overline{4}5$. The calculation of signed circuits from the chirotope and vice versa are discussed much more thoroughly in Section 3.5. From the above chirotope list, we compute the set $\mathcal{C}$ as the following list of signed circuits of RS(8), together with their negatives:

$1\overline{2}3\overline{4}5$	$1\overline{2}3\overline{4}6$	$1\overline{2}3\overline{4}7$	$1\overline{2}3\overline{4}8$	$1\overline{2}3\overline{5}6$	$1\overline{2}3\overline{5}7$	$1\overline{2}3\overline{5}8$
$1\overline{2}3\overline{6}7$	$1\overline{2}3\overline{6}8$	$1\overline{2}3\overline{7}8$	$1\overline{2}4\overline{5}6$	$1\overline{2}4\overline{5}7$	$1\overline{2}4\overline{5}8$	$1\overline{2}4\overline{6}7$
$1\overline{2}4\overline{6}8$	$1\overline{2}4\overline{7}8$	$1\overline{2}5\overline{6}7$	$1\overline{2}5\overline{6}8$	$1\overline{2}5\overline{7}8$	$1\overline{2}6\overline{7}8$	$1\overline{3}4\overline{5}6$
$1\overline{3}4\overline{5}7$	$1\overline{3}4\overline{5}8$	$1\overline{3}4\overline{6}7$	$1\overline{3}4\overline{6}8$	$1\overline{3}4\overline{7}8$	$1\overline{3}5\overline{6}7$	$1\overline{3}5\overline{6}8$
$1\overline{3}5\overline{7}8$	$1\overline{3}6\overline{7}8$	$1\overline{4}5\overline{6}7$	$1\overline{4}5\overline{6}8$	$1\overline{4}5\overline{7}8$	$1\overline{4}6\overline{7}8$	$1\overline{5}6\overline{7}8$
$2\overline{3}4\overline{5}6$	$2\overline{3}4\overline{5}7$	$2\overline{3}4\overline{5}8$	$2\overline{3}4\overline{6}7$	$2\overline{3}4\overline{6}8$	$2\overline{3}4\overline{7}8$	$2\overline{3}5\overline{6}7$
$2\overline{3}5\overline{6}8$	$2\overline{3}5\overline{7}8$	$2\overline{3}6\overline{7}8$	$2\overline{4}5\overline{6}7$	$2\overline{4}5\overline{6}8$	$2\overline{4}5\overline{7}8$	$2\overline{4}6\overline{7}8$
$2\overline{5}6\overline{7}8$	$3\overline{4}5\overline{6}7$	$3\overline{4}5\overline{6}8$	$3\overline{4}5\overline{7}8$	$3\overline{4}6\overline{7}8$	$3\overline{5}6\overline{7}8$	$4\overline{5}6\overline{7}8$

Let $\mathcal{F} \subset \binom{E}{3}$ denote the set of those triangles ijk such that $X^+ \not\subset \{i,j,k\}$ for all circuits X (cf. Corollary 9.1.3). This condition means geometrically that we are taking those triangles whose plane is not intersected by the interior of any

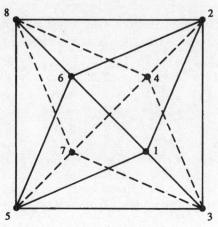

Figure 1.5.1: The facets of RS(8).

line segment. We find that

$$\mathcal{F} \;=\; \{\,123,\;126,\;135,\;156,\;234,\;248,\;268,$$
$$347,\;357,\;478,\;568,\;578\,\}.$$

These 12 triangles are the *facets* of our abstract point configuration. Note that a subset $\{i,j,k\}$ is a facet if and only if it is a facet of the convex hull (in local coordinates) of the point set $\{i,j,k,l,m\}$ for all choices of l and m.

Since all points $1,2,\ldots,8$ are extreme vertices, we say that RS(8) is a *matroid polytope*. Since we now know the facets, we can give a "diagram" of RS(8). This diagram cannot be a true depiction of this oriented matroid, since the latter is non-realizable (Proposition 1.5.1).

Each triangle of RS(8) spans a (hyper)plane which divides the point set E into two *open halfspaces*. For instance, for the facet triangle 123 we get the open halfspaces $Y^+ = \{4,5,6,7,8\}$ and $Y^- = \emptyset$. The hyperplane 237 separates the open halfspaces 468 and 15. In oriented matroid language, we get the signed cocircuit $Y_{237} = (\{4,6,8\},\{1,5\})$, denoted $\overline{1}4\overline{5}68$. In general, the signed cocircuit of a hyperplane ijk is computed as $Y_{ijk} = (Y^+,Y^-)$ where $Y^+ = \{e \in E : \chi(i,j,k,e) = +\}$ and $Y^- = \{e \in E : \chi(i,j,k,e) = -\}$. Again, more details may be found in Section 3.5. Recall that if i,j,k,e are out of order, we adjust the sign from the above chirotope listing according to the sign of the appropriate permutation. We list the resulting set $\mathcal{C}^*$ of all signed cocircuits of RS(8):

$\overline{1}234\overline{5}$	12346	$\overline{1}234\overline{7}$	$\overline{1}234\overline{8}$	$\overline{1}235\overline{6}$	$1\overline{2}357$	$\overline{1}235\overline{8}$
$1\overline{2}3\overline{6}7$	12368	$\overline{1}23\overline{7}8$	$\overline{1}24\overline{5}6$	$1\overline{2}457$	$12\overline{4}5\overline{8}$	$\overline{1}2467$
$\overline{1}2468$	$\overline{1}2478$	12567	12568	$\overline{1}257\overline{8}$	12678	$\overline{1}3456$
13457	$1\overline{3}4\overline{5}8$	$134\overline{6}7$	13468	$\overline{1}3478$	$\overline{1}3567$	$1\overline{3}568$
$\overline{1}3578$	$1\overline{3}678$	14567	$14\overline{5}68$	$\overline{1}4578$	$\overline{1}4678$	$\overline{1}5678$
$234\overline{5}6$	$2\overline{3}457$	23458	$234\overline{6}7$	$234\overline{6}8$	23478	$2\overline{3}567$
$2\overline{3}568$	23578	$2\overline{3}6\overline{7}8$	24567	24568	$24\overline{5}7\overline{8}$	24678
$2\overline{5}678$	$3\overline{4}567$	34568	$345\overline{7}8$	$3467\overline{8}$	35678	45678

Note that it is easy to determine the facet triangles from this list: they are exactly the complements of positive (or negative) cocircuits.

A realization of RS(8) is a labeled subset $\{x_1, \ldots, x_8\}$ of $\mathbb{R}^4$ such that (x_i, x_j, x_k, x_l) is a basis of $\mathbb{R}^4$ whose orientation (i.e., sign of determinant) equals $\chi(i, j, k, l)$ for all $1 \leq i < j < k < l \leq 8$. We interpret the x_i as homogeneous coordinate vectors of points in affine 3-space, and we list the x_i as the column vectors of a real 4×8-matrix X.

We will now show that the above sign patterns describe an abstract configuration which cannot be globally realized.

1.5.1 Proposition. *The oriented matroid* RS(8) *is not realizable.*

Proof. Our proof is elementary algebraic and is attained by contradiction.

Suppose that X is a 4×8-matrix whose maximal minors have the signs prescribed by the chirotope χ. After left multiplication with a suitable 4×4-matrix with positive determinant, we may assume that

$$X = \begin{pmatrix} 1 & 0 & 0 & 0 & a & e & i & m \\ 0 & 1 & 0 & 0 & b & f & j & n \\ 0 & 0 & 1 & 0 & c & g & k & o \\ 0 & 0 & 0 & 1 & d & h & l & p \end{pmatrix}$$

where the 16 indeterminates $a, b, c, d, \ldots, p$ satisfy the determinantal inequalities given by χ. More precisely, writing $[ijkl]$ for the 4×4-subdeterminant of X with column indices i, j, k, l, the chirotope χ induces the following system of 69 inequalities:

$$[1235] = d > 0, \quad [1236] = h > 0, \quad [1237] = l > 0, \quad [1238] = p > 0,$$

$$[1245] = -c > 0, \quad [1246] = -g > 0, \quad [1247] = -k < 0,$$

$$[1248] = -o > 0, \quad [1256] = \begin{vmatrix} c & d \\ g & h \end{vmatrix} > 0,$$

$$[1257] = \begin{vmatrix} c & d \\ k & l \end{vmatrix} < 0, \quad \ldots \quad \ldots \quad \ldots$$

$$\ldots \quad \ldots \quad [4578] = - \begin{vmatrix} a & b & c \\ i & j & k \\ m & n & o \end{vmatrix} < 0,$$

$$[4678] = - \begin{vmatrix} e & f & g \\ i & j & k \\ m & n & o \end{vmatrix} < 0, \quad [5678] = \begin{vmatrix} a & b & c & d \\ e & f & g & h \\ i & j & k & l \\ m & n & o & p \end{vmatrix} > 0$$

We will show that these 69 inequalities have no common solution in $\mathbb{R}^{16}$. To this end, consider the subset of inequalities arising from the oriented tetrahedra

1235, 1236, 1237, 1245, 1246, 1248, 1256, 1345, 1347,

1348, 1357, 1458, 2346, 2347, 2348, 2367, 2468, 3478.

We obtain

$$b < 0,\ c < 0,\ d > 0,\ e > 0,\ g < 0,\ h > 0,$$
$$i > 0,\ j < 0,\ l > 0,\ m > 0,\ n < 0,\ o < 0,$$

$$\begin{vmatrix} c & d \\ g & h \end{vmatrix} > 0, \quad \begin{vmatrix} b & d \\ j & l \end{vmatrix} < 0, \quad \begin{vmatrix} b & c \\ n & o \end{vmatrix} < 0$$

$$\begin{vmatrix} e & h \\ i & l \end{vmatrix} < 0, \quad \begin{vmatrix} e & g \\ m & o \end{vmatrix} < 0, \quad \begin{vmatrix} i & j \\ m & n \end{vmatrix} > 0.$$

These 18 inequalities imply the following relations among the absolute values:

$$|c||h| < |d||g|,\ |d||j| < |b||l|,\ |b||o| < |c||n|,$$
$$|e||l| < |h||i|,\ |g||m| < |e||o|,\ |i||n| < |j||m|.$$

Since the product of the right hand sides equals the product of the left hand sides, we see that this system is inconsistent. Therefore RS(8) is not realizable.

$$\square$$

Systematic procedures to obtain such non-realizability proofs will be discussed in Chapter 8. We will see that there exists a unified way of encoding non-realizability proofs for oriented matroids into so-called *final polynomials*. Such a final polynomial for RS(8) will be constructed in Example 8.5.4.

A more geometric non-realizability proof for this oriented matroid follows from the results of Sections 2.1 and 7.3. We can represent RS(8) as an arrangement of eight pseudo planes (labeled by E) in projective 3-space, using the Topological Representation Theorem. Among the maximal cells in the resulting cell decomposition of projective 3-space we find exactly seven tetrahedra, namely

$$1256 \quad 1357 \quad 1458 \quad 1467 \quad 2367 \quad 2468 \quad 3478.$$

On the other hand, by Theorem 2.1.5, the cell decomposition of 3-space given by any arrangement of eight "straight" planes has at least eight tetrahedra. Consequently RS(8) is not realizable. This proof appeared in Roudneff and Sturmfels (1988).

1.6 Combinatorial convexity

Oriented matroids are a powerful tool for studying higher-dimensional polytopes (the word "polytope" always means convex polytope). We wish to support this claim with an easy matroidal proof for a theorem about 4-dimensional polytopes. It is intended as a motivation for Chapter 9 where convexity in oriented matroids and its applications are studied in some detail.

A facet F of a polytope P is said to be *arbitrarily prescribable* if, for all polytopes F' combinatorially equivalent to F, there is a polytope P' combinatorially equivalent to P such that F' is the image of F under the isomorphism between P and P'. It is a known consequence of the proof of Steinitz's theorem that facets of 3-polytopes are arbitrarily prescribable. As the same result holds for d-polytopes with up to $d+3$ vertices, it becomes an interesting question whether the facets of all polytopes can have arbitrary shapes.

This question was answered in the negative by Kleinschmidt (1976a) (see also Barnette (1987)) who constructed a 4-polytope with eight vertices which has a facet of non-arbitrary shape. Kleinschmidt's polytope has 14 facets, and one can ask whether this number is minimal for polytopes with non-arbitrarily shaped facets. By checking all possible combinatorial types, it can be seen that no 4-polytope with six or fewer facets has this property.

Here we describe an example due to Sturmfels (1988a) of a 4-polytope P with seven facets, one of which is not arbitrarily prescribable. To this end we first construct a 4-polytope Q with seven vertices such that one vertex figure cannot be prescribed arbitrarily. The reader who is familiar with polytope theory will have no difficulty rephrasing our argument in terms of Gale transforms (cf. Proposition 9.1.4).

Let $\mathcal{M}$ be the rank 2 oriented matroid on $E = \{1, 2, 3, 4, 5, 6, 7\}$ defined by the vector configuration in Figure 1.6.1.

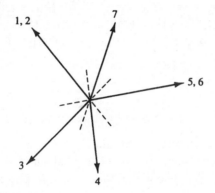

Figure 1.6.1: The dual $\mathcal{M}$ of a matroid polytope $\mathcal{M}^*$ whose vertex figure $\mathcal{M}^*/7$ cannot be prescribed arbitrarily.

The dual $\mathcal{M}^*$ of $\mathcal{M}$ is a rank 5 oriented matroid whose facets by definition are the complements of the positive circuits of $\mathcal{M}$. From Figure 1.6.1 we find that $\mathcal{M}^*$ has the following 13 facets:

$$1245 \quad 1246 \quad 1256 \quad 1356 \quad 1357 \quad 1367 \quad 1457$$
$$1467 \quad 2356 \quad 2357 \quad 2367 \quad 2457 \quad 2467.$$

For instance, 1245 is a facet because the vectors $3, 6$ and 7 are positively dependent. $\mathcal{M}^*$ is actually a matroid polytope in the sense of Section 9.1. Let Q be any 4-polytope whose vertices form a realization of $\mathcal{M}^*$. Then Q is a simplicial polytope whose facets are indexed by the facets of $\mathcal{M}^*$. Hence the vertex figure of Q at the vertex 7 is an octahedron, as can be seen from the list of facets above. This octahedron is denoted G_7 and labeled as in Figure 1.6.2.

Suppose now that Q' is a polytope combinatorially equivalent to Q and that the corresponding vertex figure G_7' of Q' is a *regular* octahedron. In that case the points indexed by $1, 2, 5$ and 6 lie in a two-dimensional affine subspace. On the other hand, these four points must span an affine three-dimensional

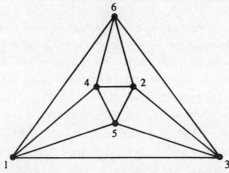

Figure 1.6.2: The vertex figure G_7 of the simplicial 4-polytope Q.

hyperplane because 1256 is a facet of Q. This contradiction proves that the vertex figure G_7 of the 4-polytope Q cannot be prescribed arbitrarily.

Now let P be a polar polytope of Q. Then P is a simple 4-polytope with 13 vertices and seven facets. The facet F_7 of P is a cube because it is polar to the octahedron G_7. By the above argument, there exists no polytope P' combinatorially equivalent to P such that the corresponding facet F_7' is a *regular* cube.

1.7 Linear programming

It was Rockafellar's (1969) basic insight that the key properties of the simplex method for linear programming mostly depend on sign patterns, and should therefore be part of a purely combinatorial theory. This led to the formulation and proof of linear programming duality in oriented matroids by Bland (1974) and Lawrence (1975), and finally to Bland's (1977a, 1977b) development of the simplex method for oriented matroids.

Here we sketch – for a simple, 2-dimensional example – how to translate a linear program and its basic properties into oriented matroid terms. Consider the linear program

$$\text{max } x + y - 3$$
$$-x + y \le 1$$
$$(LP_0) \qquad\qquad x - 2y \le 2$$
$$x + 2y \le 4$$
$$x, y \ge 0.$$

It asks for the maximization of an affine function ($f = x+y-3$) over a polyhedron in 2-dimensional affine space. The geometric situation is depicted in Figure 1.7.1. (The feasible polyhedron is shaded. The line of points with objective function value 0 is dashed.)

Now a standard procedure is to lift this to 3-dimensional linear space, whose coordinates we call x, y and z, and into which we embed the affine plane as the subset $\{z = 1\}$. Here the linear program is encoded by a set of linear halfspaces,

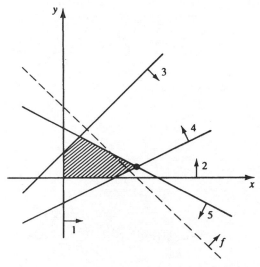

Figure 1.7.1: The linear program (LP_0).

which we label

$$
\begin{array}{rl}
e_1: & -x + y - z \leq 0 \\
e_2: & x - 2y - 2z \leq 0 \\
e_3: & x + 2y - 4z \leq 0 \\
e_4: & x \qquad\qquad \geq 0 \\
e_5: & \quad y \qquad\;\; \geq 0,
\end{array}
$$

with the objective function given by the halfspace

$$
f: \qquad x + y - 3z \geq 0,
$$

and the embedding of the affine plane being indicated by

$$
g: \qquad\qquad z \geq 0.
$$

Thus the whole linear program is given by an arrangement of seven halfspaces in $\mathbb{R}^3$.

The corresponding oriented matroid $\mathcal{M}$ has the ground set $E = \{e_1, \ldots, e_5, f, g\}$, where f and g are distinguished. A triple $(\mathcal{M}, g, f)$ of this type is called an *oriented matroid program*, see Definition 10.1.3. Since oriented matroids are more general than arrangements of halfspaces, we get *non-realizable* oriented matroid programs and a theory that is more general than linear programming.

We want to emphasize that linear programming duality has its natural abstract setting in this framework, with surprising simplicity and scope.

A geometric picture for a rank 3 oriented matroid program is obtained as follows. Represent the affine plane by an open disc containing the cells/covectors X with $X_g = +$ (corresponding to $z > 0$), whose boundary is the "line at infinity" ($X_g = 0$, $z = 0$). Halfspaces now correspond to pseudolines, with one *positive* side distinguished and marked by an arrow. Thus we get a picture such as Figure 1.7.2 for the oriented matroid program associated to LP_0.

Oriented matroid programming will be described in the following paragraphs.

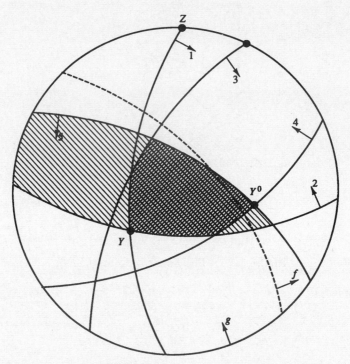

Figure 1.7.2: An oriented matroid program of rank 3.

The *feasible polyhedron* is the set of covectors Y in affine space $\{Y_g = +\}$ that satisfy $Y_{e_i} \geq 0$ for all i. Its *vertices* are the cocircuits that satisfy the same conditions. For our example, $Y = (00{+}{+}{+}|{-}{+})$ is a feasible vertex, see Figure 1.7.2.

Since the objective function is given only by the halfplane $f \geq 0$, we cannot compare arbitrary pairs of vertices with respect to f. But for pairs of points on the same pseudoline of the arrangement, we can in most cases deduce a direction of increasing objective function. The pseudoline in question has two intersection points with the "line at infinity". The corresponding cocircuits $Z, -Z$ with $Z_g = 0$ are called *directions* (cf. Definition 10.1.7). A direction Z is *increasing* if $Z_f = +$. Note that it is possible that neither direction is increasing. This (partially) directs the edges of the graph G_f of the program, defined as follows: the vertices of G_f are the cocircuits Y with $Y_g = +$, and the edges are the 1-cells connecting them.

The goal of oriented matroid programming is to find an *optimal* cocircuit (vertex), that is, one that is feasible and for which there is no feasible increasing direction. In our example, $Y^0 = (+{+}{+}{+}00|{+}{+})$ is the unique optimal cocircuit. In contrast, $Y = (00{+}{+}{+}|{-}{+})$ is feasible but not optimal: the increasing direction $Z = (0{+}{-}{+}{-}|{+}0)$ is a *feasible direction* for Y, which means that $Y \circ Z = (0{+}{+}{+}{+}|{-}{+})$ is again feasible.

Observe that feasible cocircuits such as Y describe lower bounds for linear programs. Dually, *bounded cones*, which by definition contain all feasible vectors but no increasing direction, yield upper bounds. (For example, the bounded cone $P(\{2,5\}) = \{Y \in C^* : Y_g = +, Y_2 \geq 0, Y_5 \geq 0\}$ is lightly shaded in Figure 1.7.2. Thus the oriented matroid program of that Figure is both feasible and bounded.) Both feasible cocircuits and bounded cones can be described by *bases* of the oriented matroid, and this leads to the duality theory for oriented matroid programs.

The Main Theorem of Oriented Matroid Programming (Theorem 10.1.13) states that if a program has both a feasible cocircuit and a bounded cone, then it has an optimal solution. In fact, it then has an *optimal basis*: a basis that describes both a bounded cone and a feasible cocircuit. In our example, we get the basis $N = \{4, 5, g\}$ of the oriented matroid that both describes the feasible cocircuit Y^0 and the bounding cone $P(\{4,5\}) = \{Y \in C^* : Y_g = +, Y_4 \geq 0, Y_5 \geq 0\}$.

The *algorithmic* approach of the *simplex method* is to construct a sequence of vertices, each "better" than the preceding one, until the process stops with optimality. This will trace a path which proceeds from vertex to vertex in the graph G_f, along increasing edges, from any feasible starting vertex to an optimal vertex Y^0.

1.8 Computational geometry

Computational geometry is a field of research on the borderline between computer science and mathematics. Its goal is to deal with and provide algorithms for "real life" geometric problems. Many of these can be modeled by or reduced to problems about point configurations and hyperplane arrangements, where a lot of attention is centered on the planar case – see Edelsbrunner (1987) and de Berg, van Kreveld, Overmars and Schwarzkopf (1997) for expositions.

Already the examples discussed so far suggest that oriented matroids may help in this study. They provide a combinatorial model and good ways to organize the combinatorial data, and perhaps they can offer a better understanding of what certain algorithms do, why they work, and how to improve them. See also Section 8.7.

In this section, we will study a class of oriented matroids of rank 4, defined on point configurations in the Euclidean plane. From the point of view of ordinary matroids, this will treat some minors of the "Möbius geometries" of Cheung and Crapo (1976), with their canonical orientations introduced by Bland and Las Vergnas (1978). These oriented matroids construct the *Delaunay triangulations* of point configurations. Such triangulations are important because they are polar to the *Voronoi diagrams*, which are used for many algorithms that solve nearest neighbor and point location problems in computational geometry.

In this context, we mention the algorithm due to Edelsbrunner and Seidel (1986) for constructing d-dimensional Voronoi diagrams (and various generalizations) from arrangements of (pseudo)hyperplanes in $\mathbb{R}^{d+1}$. See also the recent

work of Knuth (1992), which emphasizes the use of oriented matroid axiomatics
for questions in computational geometry.

Let $P = \{p_1, p_2, \ldots, p_n\}$ be a set of points in the Euclidean plane, $p_i = (x_i, y_i)$,
not all on one line or circle. To simplify our discussion, let us assume that P
is in general position, that is, no three points lie on the same line. Then every
three points p_i, p_j and p_k determine a unique circle, which we denote by C_{ijk}.

We now define an oriented matroid $\mathcal{M}_P$ of rank 4 on the set $E = \{1, \ldots, n\}$
by specifying its chirotope. For this we set $\chi(i, j, k, l) = 0$ if the four points
p_i, p_j, p_k, p_l lie on a common circle. We set $\chi(i, j, k, l) = +$ if p_i, p_j and p_k lie
in clockwise order on C_{ijk} with p_l inside C_{ijk}, or if p_i, p_j and p_k lie in counter-
clockwise order on C_{ijk} and p_l lies outside. Otherwise we put $\chi(i, j, k, l) = -$.
For example, for the point configuration in Figure 1.8.1 we set $\chi(1, 2, 3, 4) = +$,
but $\chi(1, 4, 5, 6) = -$.

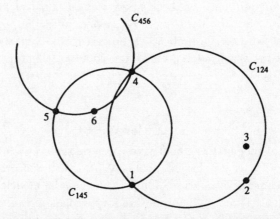

Figure 1.8.1: A circle geometry P in the plane.

Now the cocircuits of $\mathcal{M}_P$ are obtained as follows: they are given by the
complements (in P) of the circles determined by triples of points in P. Here two
elements of a cocircuit get the same sign if the corresponding points lie on the
same side of the circle or line. In our example (Figure 1.8.1), 123, $23\overline{6}$ and $3\overline{5}6$
are some cocircuits.

We can also describe the circuits of $\mathcal{M}_P$. They are the sets of four points on
a circle, with alternating signs, and the sets of five points of which no four lie
on the same circle, with signs attached so that if $\underline{X} = ijklm$ is a circuit, then i
and j get the same sign if they lie on opposite sides of the circle C_{klm}. In Figure
1.8.1, this means that, for example, $12\overline{3}4\overline{5}$ and $1\overline{3}45\overline{6}$ are signed circuits.

We now check that $\mathcal{M}_P$ is an oriented matroid by producing a realization. The
circle C_{ijk} through p_i, p_j, and p_k is given by $C_{ijk} = \{(x, y) \in \mathbb{R}^2 : c_{ijk}(x, y) = 0\}$,
where

$$c_{ijk}(x, y) = \begin{vmatrix} 1 & 1 & 1 & 1 \\ x_i & x_j & x_k & x \\ y_i & y_j & y_k & y \\ x_i^2 + y_i^2 & x_j^2 + y_j^2 & x_k^2 + y_k^2 & x^2 + y^2 \end{vmatrix}.$$

In fact, the equation $c_{ijk}(x, y) = 0$ is certainly satisfied by the points p_i, p_j and p_k. These three points are not collinear, hence the cofactor of $x^2 + y^2$ in the determinant does not vanish, and thus $c_{ijk}(x, y) = 0$ is the equation of the circle through the three points.

Now note that $c_{ijk}(x, y) > 0$ describes the interior of C_{ijk} if the three points p_i, p_j, p_k lie in clockwise order on C_{ijk}, and the outside otherwise. Thus $\chi(i, j, k, l)$ is the sign of the determinant $c_{ijk}(x_l, y_l)$.

An alternative geometric interpretation of the oriented matroid $\mathcal{M}_P$ is as follows. It is the rank 4 oriented matroid of the affine point configuration $\hat{P} = \{\hat{p}_1, \ldots, \hat{p}_n\}$ in $\mathbb{A}^3$, where $\hat{p}_i = (x_i, y_i, x_i^2 + y_i^2)$. Geometrically, $\mathcal{M}_P$ is thus obtained by interpreting P as a point set in the affine plane $\mathbb{A}^2 = \{(x, y, z) \in \mathbb{A}^3 : z = 0\}$ and then lifting every point p_i to the point $\hat{p}_i$ on the paraboloid $S = \{(x, y, z) \in \mathbb{A}^3 : z = x^2 + y^2\}$ directly above it.

This interpretation of $\mathcal{M}_P$ implies a lot of important properties.

1.8.1 Proposition. *Let P be a finite set of points in the plane, and let $\mathcal{M}_P$ be the associated rank 4 oriented matroid as above. Then $\mathcal{M}_P$ is a realizable matroid polytope.*

If no four points of P lie on the same circle, then $\mathcal{M}_P$ is uniform and thus the matroid polytope is simplicial.

In order to understand the facets of $\text{conv}(\hat{P})$, we observe that circles $C \subseteq \mathbb{A}^2$ in the plane lift exactly to the intersections of the paraboloid S with planes in $\mathbb{A}^3$: in fact, the circle C_{ijk} is lifted to an ellipse

$$S \cap H_{ijk} = \{(x, y, z) \in S : (x, y) \in C_{ijk}\},$$

where H_{ijk} is the affine plane through $\hat{p}_i, \hat{p}_j$ and $\hat{p}_k$ in $\mathbb{A}^3$. The equation for H_{ijk} is

$$h_{ijk}(x, y, z) = \begin{vmatrix} 1 & 1 & 1 & 1 \\ x_i & x_j & x_k & x \\ y_i & y_j & y_k & y \\ x_i^2 + y_i^2 & x_j^2 + y_j^2 & x_k^2 + y_k^2 & z \end{vmatrix} = 0.$$

Now by definition the facets of a matroid polytope are in bijection with the positive cocircuits. Thus if the points of P are in general position, then $\text{conv}(\hat{P})$ is a simplicial 3-polytope. Projecting this polytope down to $\mathbb{A}^2$, we get two triangulations of $\text{conv}(P)$ with very special properties, which are easily derived from $\mathcal{M}_P$. These two are lifting triangulations in the sense of Section 9.6.

1.8.2 Proposition. *Let P be a finite configuration of points in general position in the Euclidean plane, such that no four points of P lie on the same circle.*

(1) *The set of triples p_i, p_j, p_k such that no point of P lies inside C_{ijk} defines a triangulation of $\text{conv}(P)$, whose vertex set is P. It is called the Delaunay triangulation $D(P)$ of P.*

(2) *The set of triples p_i, p_j, p_k such that all other points of P lie inside C_{ijk} defines a triangulation of $\text{conv}(P)$, whose vertex set is the set of vertices*

of conv(P). *It is called the furthest site Delaunay triangulation $D_f(P)$ of* conv(P).

Proof. By our description of the cocircuits of $\mathcal{M}_P$ above we know that the circles C_{ijk} for which the other points of P lie either all inside or all outside C_{ijk} correspond to the positive cocircuits of $\mathcal{M}_P$ and hence to the facets of conv($\hat{P}$). Our intuition is that $D(P)$ is the image of the "lower hemisphere" of conv($\hat{P}$), whereas $D_f(P)$ comes from the facets on the "upper hemisphere" of conv($\hat{P}$). We have to show that this identification is correct.

To see this, let conv($\hat{p}_i, \hat{p}_j, \hat{p}_k$) be a facet of conv($\hat{P}$), and relabel so that p_i, p_j and p_k are in counter-clockwise order, which means that the coefficient of z in $h_{ijk}(x, y, z)$ is positive. Then conv($\hat{p}_i, \hat{p}_j, \hat{p}_k$) is a "lower facet" if and only if $h_{ijk}(\hat{p}) \geq 0$ for all $\hat{p} \in$ conv($\hat{P}$), or, equivalently, for all $\hat{p} = \hat{p}_l \in \hat{P}$. This is equivalent to $c_{ijk}(p_l) \geq 0$ for all $p_l \in P$, which means that all p_l in $P\backslash\{p_i, p_j, p_k\}$ lie outside C_{ijk}. Therefore conv($\hat{p}_i, \hat{p}_j, \hat{p}_k$) is a "lower facet" if and only if $[p_i, p_j, p_k]$ is a Delaunay triangle. Analogously, we can identify the triangles in the "upper hemisphere" of the polytope conv($\hat{P}$) with the triangles of $D_f(P)$.

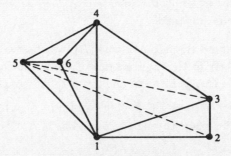

Figure 1.8.2: The Delaunay triangulation constructed from
the point configuration in Figure 1.8.1.

Finally, if p_i is an interior point of conv(P), then by convexity of S it is clear that $\hat{p}_i$ must lie on the "lower hemisphere" of conv($\hat{P}$). This implies that the vertices of the furthest site Delaunay triangulation are precisely the vertices of conv(P). This proves the claims on the vertex sets of $D(P)$ and $D_f(P)$. $\square$

Figure 1.8.2 depicts the Delaunay triangulation and the furthest site Delaunay triangulation for the point configuration in Figure 1.8.1. Here the edges of $D(P)$ are drawn as lines, whereas the interior edges of $D_f(P)$ are drawn dashed. By our discussion above, this drawing is the projection of a 3-polytope whose vertices fit on a paraboloid.

Note that if we allow for P to contain four points on a circle, then we get polygonal subdivisions of conv(P) with analogous properties, which can easily be completed to triangulations. We can even permit more than two points on a line (defining $\mathcal{M}_P$ by $\chi(i, j, k, l) :=$ sign $C_{ijk}(p_l)$) if we do not mind that possibly "vertical" faces of $\mathrm{conv}(\hat{P})$ get collapsed by projecting down into the plane, so that they do not appear as faces of either of the two decompositions.

The two types of Delaunay triangulations have various applications in computational geometry. For example, if we search for a point in $\mathrm{conv}(P)$ that has maximal distance from the point set P, then it is clear that it is the center of the circle of one of the Delaunay triangles. In particular, this proves that there are only finitely many points that achieve the maximum. Similarly, if we search for a point that minimizes the maximal distance from P, then it has to be the center of the circle for a furthest site Delaunay triangle.

These properties make the Delaunay triangulation of a point set combinatorially polar to the Voronoi diagram of the configuration, which is obtained by decomposing the plane into the regions R_i of points for which p_i is the closest point in P. Similarly, the furthest site Delaunay triangulation is polar to the furthest site Voronoi diagrams.

Our discussion can be generalized to higher dimension. To do so, one defines an oriented matroid of rank $d + 2$ and a Delaunay polytope of dimension $d + 1$ for every affine point configuration in $\mathbb{A}^d$.

It would be interesting to know which oriented matroids arise this way, and which polytopes are combinatorially equivalent to Delaunay polytopes. Although a complete answer is probably out of reach even for the case $d = 2$, some non-trivial observations can be made.

First, one can observe that not every polytope is a Delaunay polytope. One can show that every 3-polytope with at most 9 vertices is a Delaunay polytope, whereas there is a counter-example with 10 vertices (Exercise 1.13). But how restricted is the combinatorial structure of Delaunay 3-polytopes? It had been conjectured by Crapo and Laumond (1989) that every such 3-polytope is hamiltonian, but that has turned out to be false – see Dillencourt (1989). A positive result was obtained by Seidel (1987): every cyclic polytope (of arbitrary dimension) is a Delaunay polytope. This result was applied by him to give sharp upper bounds for the number of faces in Voronoi diagrams that in turn imply lower bounds for the complexity of algorithms that construct them.

1.9 Chirality in molecular chemistry

In good approximation of "reality", we may think of a chemical molecule as a configuration of points in real 3-dimensional affine space. The associated acyclic rank 4 oriented matroid or chirotope may be regarded as a mathematical encoding of the chemical concept of *chirality*. This connection, which accounts etymologically for the word *chirotope*, was first observed by Dreiding and Wirth (1980), and by Dreiding, Dress and Haegi (1982). It was later independently rediscovered by Tratch and Zefirov (1987), see Klin, Tratch and Zefirov (1990). See Crippen and Havel (1988) for a discussion of oriented matroids from this point of view, and for much of the material of this section.

Actually, there is already a problem with the oriented matroid model, since not all molecules are *rigid*. So, one part of the molecule may be able to rotate about some bond with respect to another part of the molecule and change some of the orientation (a rigid rotation of the whole molecule is not a problem, since that does not change the oriented matroid). However, if we consider the molecule at a fixed instant in time, then we are looking at a particular oriented matroid. Alternatively, we could consider a *partial chirotope,* by paying attention to only tetrahedra whose orientations are unaffected by rotations. Incidentally, Crippen and Havel give an algorithm to list all chirotopes which extend a given partial chirotope. Note that we are here considering the orientation, or *chirality,* of any ordered quadruple of points, rather than only those quadruples which are adjacent to a fifth tetravalent atom, as is customary in chemistry.

A classical example in which the chirality is important is that of tartaric acid, which has three distinct molecular forms, known as dextro, levo, and meso. It is this compound which Pasteur in 1848 first separated into the dextro and levo forms by inspecting crystals with a magnifying glass and sorting them into two mirror-image groups; see any standard organic chemistry text such as Morrison and Boyd (1987). We show in Figure 1.9.1 the three forms, where in the first diagram the dark bonds are intended to show the H and OH groups rising above the page, and the light bonds show the COOH groups below the page. Note that each of the two central carbon atoms is adjacent to four other atoms, which in the actual molecule will be located at the four vertices of a tetrahedron around the carbon atom. Let a denote one of the central carbon atoms, and b, c, d, and e its four neighbors. The crucial issue is whether the chirotope sign of $bcde$ is $+$ or $-$, that is, whether the four neighbors in a certain order have right-handed or left-handed orientation. Dextro-tartaric acid has both of these central bases right-handed, levo has both left-handed, and meso has one of each (these different molecules are referred to as *stereoisomers* of each other). It follows that dextro and levo are mirror images of each other, while meso is mirror-symmetric. A fourth version, racemic tartaric acid, is simply an equal mixture of the dextro and levo forms. The four varieties of tartaric acid have distinct physical and biological properties. Typically, when an organic compound has stereoisomerism possible, only one of the stereoisomers occurs in living organisms.

Chemists have much better methods (such as NMR spectroscopy) for determining distances between atoms in a molecule than for determining chirality. As mentioned above, internal rotations may cause some of these distances and chiralities not to be fixed. In the case of a rigid molecule, however, it is possible to compute the full chirotope (up to mirror image) from the inter-atomic distance data. This is done by means of the Cayley-Menger bideterminant. Let D_{ij} denote the square of the distance from point (or atom) i to point j, let $V(a, b, c, d)$ be the oriented volume of the tetrahedron (a, b, c, d), and let a, b, c, d, w, x, y, z

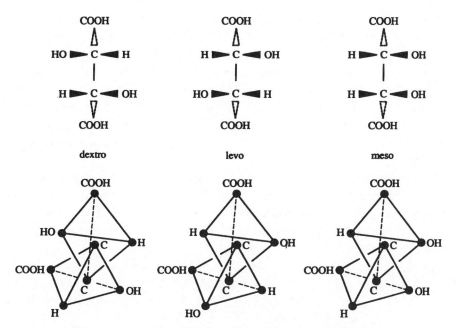

Figure 1.9.1: The three stereoisomers of tartaric acid.

be eight points in $\mathbb{R}^3$. Then it is known that

$$36V(a,b,c,d)V(w,x,y,z) = \det \begin{vmatrix} 0 & 1 & 1 & 1 & 1 \\ 1 & D_{aw} & D_{ax} & D_{ay} & D_{az} \\ 1 & D_{bw} & D_{bx} & D_{by} & D_{bz} \\ 1 & D_{cw} & D_{cx} & D_{cy} & D_{cz} \\ 1 & D_{dw} & D_{dx} & D_{dy} & D_{dz} \end{vmatrix},$$

and the Cayley-Menger bideterminant, $D(a,b,c,d;w,x,y,z)$, is defined to be this determinant (Blumenthal 1953). Thus

$$\text{sign}(D(a,b,c,d;w,x,y,z)) = \chi(a,b,c,d)\chi(w,x,y,z),$$

and to compute the full chirotope we need only fix the value (arbitrarily) of one quadruple.

Although described here for 3 dimensions, this calculation can be generalized to any dimension. Furthermore, the Cayley-Menger bideterminants can be used to determine the minimum dimension Euclidean space in which a given set of distance data can be realized (assuming that it can be so realized). That dimension is the largest number n for which some of the Cayley-Menger determinants $D(a_1, a_2, \ldots, a_{n+1}) := D(a_1, a_2, \ldots, a_{n+1}; a_1, a_2, \ldots, a_{n+1})$ are non-zero. In fact, these data can be realized in Euclidean space of dimension n for the n just determined if and only if $D(b_1, b_2, \ldots, b_{n+1})$ is non-negative for every $n+1$-tuple $b_1, b_2, \ldots, b_{n+1}$.

1.10 Allowable sequences

Point and line configurations in the plane suggest many interesting combinatorial and computational problems. It is natural to look at some of these from a matroid or oriented matroid point of view. In this section we will take a first look at a useful combinatorial tool for the study of point configurations, the allowable sequences of Goodman and Pollack (1980a, 1982b). Allowable sequences are closely related to, and indeed a refinement of, rank 3 oriented matroids, somewhat similar to the refinement that oriented matroids provide to ordinary matroids. From another point of view, to be explained in Section 6.4, allowable sequences are equivalent to a particular class of (larger) oriented matroids.

In the next section, we will describe a particularly elegant application of allowable sequences: Ungar's (1982) proof that n points in the plane determine at least $n-1$ different slopes (if they are not all on a line). Other uses of the same technique can be found in Alon and Győry (1986), Edelsbrunner and Welzl (1985) and Welzl (1986).

Let C be a configuration of n distinct points in the real plane $\mathbb{R}^2$ labeled by $E_n = \{1, 2, \ldots, n\}$. Let L be a directed line in the plane, and project the points of C orthogonally onto L. If L is not orthogonal to any line containing two or more points of C, then the distinct images of the points determine a permutation of E_n. We assume for simplicity (relabeling if necessary) that this is the identity permutation on $\{1, 2, \ldots, n\}$. If we now rotate L counter-clockwise 180 degrees, we get a change in the permutation every time L moves through a direction orthogonal to a line containing two or more points of C. In this way we get a sequence of permutations, called the *allowable sequence of permutations associated to* C.

These sequences have been thoroughly studied by Goodman and Pollack (1980b, 1984b, and other papers, see the survey 1993). We give an example in Figure 1.10.1, showing the sequence of permutations as well as the *moves* separating them.

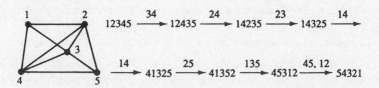

Figure 1.10.1: An allowable sequence of permutations.

Let Σ be the sequence of permutations associated to C. Then we can easily verify that Σ has the following properties:

(1) Σ begins with the identity permutation $12\ldots n$ and ends with the reverse permutation $n(n-1)\ldots 1$,

(2) the move between consecutive terms of Σ consists of reversing one or more non-overlapping increasing substrings.

If Σ is an arbitrary sequence of permutations of E_n satisfying properties (1) and

(2), then we say that Σ is an *allowable sequence of permutations*. There are allowable sequences which are not realizable in terms of a configuration of points in the plane.

However, given an allowable sequence Σ, the following construction yields an oriented matroid of rank 3 on the set E_n. Using the terminology of Goodman and Pollack (1984b), we define a *semispace* of the allowable sequence Σ to be any initial or terminal segment of a permutation in Σ. Consider any *line L*, that is, a segment which is reversed in some move of Σ. There are two disjoint semispaces S_1 and S_2, which precede and follow L in either of the two permutations connected by the move in question. Then we define $S_1 \cup S_2$ to be a cocircuit with positive part S_1 and negative part S_2, and symmetrically we get the opposite cocircuit with the roles of S_1 and S_2 exchanged. For example, in Figure 1.10.1, if we pick the line 12, we see that $S_1 = \{3, 4, 5\}$ and S_2 is empty. The collection of all cocircuits obtained in this manner defines an oriented matroid on E_n. There are several other ways to obtain the same oriented matroid, by defining the covectors, maximal covectors, bases, or the pseudoline representation (Exercise 1.10) in terms of Σ. Let us call this oriented matroid the *little* oriented matroid determined by Σ.

Next, we observe that different allowable sequences can determine the same oriented matroid on E_n. In other words, the allowable sequence carries more information than the little oriented matroid. An example of this is given in Figure 1.10.2, where a realized allowable sequence is shown which has the same oriented matroid as the sequence in Figure 1.10.1. In this example, the oriented matroid does not tell us whether the lines 14 and 25 intersect above or below the configuration, or whether they are parallel, whereas the allowable sequences do contain this information. We can, however, record all of this information in the following larger oriented matroid.

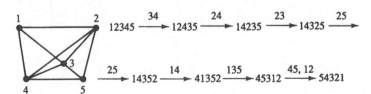

Figure 1.10.2: Different allowable sequence, same oriented matroid.

Given an allowable sequence Σ, let $E = E_n \cup K$, where K is the set of moves of Σ. We will define three types of cocircuits.

Type 1: Let L be a line, belonging to a move $k \in K$. Let

$$C_1 = S_1 \cup \{j : j \in K, j \text{ follows } k\},$$
$$C_2 = S_2 \cup \{j : j \in K, j \text{ precedes } k\}.$$

Then we define $C_1 \cup C_2$ to be a cocircuit with positive part C_1 and negative part C_2.

Type 2: Let p be a point in E_n and k a move in K, with p not in any line switched during move k. Now examine σ, either one of the two permutations adjacent to the move k. Let S_1 be the points of E_n preceding p in σ, and S_2 the points following p in σ. Then define C_1, C_2, and the resulting cocircuit exactly as above.

Type 3: Define E_n itself to be a positive cocircuit.

The oriented matroid $\mathcal{M}$ on E is defined to have its cocircuits consisting of all three of the above types and their negatives. This is the *big* oriented matroid determined by Σ. In order to visualize this oriented matroid, it is useful to think of the elements of K as points at infinity adjoined to the lines of Σ, with lines of the same move having a common such point at infinity. The cocircuits of a rank 3 oriented matroid, ignoring signs, are the complements of lines of the underlying matroid. Cocircuits of $\mathcal{M}$ of the first type are complements of extended lines of Σ, those of the second type are complements of two-point lines, one point in E_n and one in K, and the last cocircuit is the complement of the line K at infinity. See Figure 1.10.3 for an affine diagram of the big oriented matroid derived from the allowable sequence of Figure 1.10.1. In this example, the line 23 determines the type 1 cocircuit with positive part $C_1 = \{1, 4, d, e, f, g\}$ and negative part $C_2 = \{5, a, b\}$, and the line $3d$ determines the type 2 cocircuit with $C_1 = \{1, 4, e, f, g\}$, $C_2 = \{2, 5, a, b, c\}$. The proof that $\mathcal{M}(E)$ is an oriented matroid is left as Exercise 1.14.

The big oriented matroid determined by an allowable sequence has special structure, in that it has a *modular* line K (i.e., K meets every other line), and furthermore, every point on K is on a line spanned by (at least two) points of $E - K$. In Section 6.4 we will show that every oriented matroid of rank three with such special structure is the big oriented matroid of an allowable sequence. Thus it turns out that allowable sequences in fact are a disguised form of a particular class of rank 3 oriented matroids.

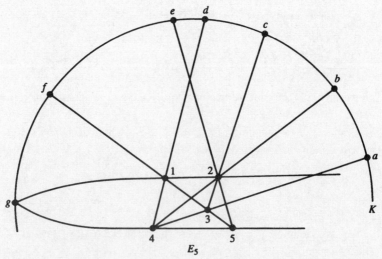

Figure 1.10.3: The big oriented matroid derived from Figure 1.10.1.

1.11 Slope problems

This section will discuss some combinatorial "slope problems" that arise from point configurations in the plane. These and other problems about planar point configurations can be solved using matroid techniques, and for each such problem we have to decide whether it is best modeled (with increasing specialization and refinement) by

- matroids,
- oriented matroids,
- allowable sequences.

To illustrate these distinctions, let us discuss the solutions to the following three famous problems. Given a configuration C of n points in the real affine plane, not all on one line,

(A)   *show that at least n lines are spanned by (points of) C,*

(B)   *show that some line spanned by C contains exactly two points of C,*

(C)   *show that at least $2\lfloor n/2 \rfloor$ slopes are determined by the lines spanned by C.*

The first problem, posed by Erdős and Motzkin in the 1930s, was solved by de Bruijn, Erdős, and Szekeres in the 1940s. Subsequent research by several authors has shown that this result extends to any matroid of arbitrary rank, leading to the so-called "Hyperplane Theorem" for matroids (for notes and references, see White, 1987, p. 158): *Every simple matroid on n elements has at least n hyperplanes.*

The second problem was stated by Sylvester in 1893 and solved by Gallai and others (1930s and 1940s). Later Kelly and Moser (1958) obtained the sharper answer that there must be at least $3n/7$ two-point lines. Even sharper lower bounds have been conjectured, and this is still an active area of research, see Grünbaum (1972) and Erdős and Purdy (1993). None of these results generalize to matroids, since many rank 3 matroids such as the finite projective planes have *no* two-point lines. However, they *do* generalize to *orientable* matroids, as shown by Kelly and Rottenberg (1972), who extended the $3n/7$ lower bound. See Theorem 6.5.3.

The third problem was posed by Scott (1970) and solved by Ungar (1982). Ungar's solution is in terms of allowable sequences, and will be explained now. We prove:

Any n points in the plane, not all on a line, determine at least n slopes if n is even.

The assumption that n is even is for ease of exposition; in the odd case, the correct bound is $n - 1$. If n is even, an example realizing the bound is the regular n-gon. In the odd case, an extremal configuration is the $(n - 1)$-gon, plus its central point.

It is obvious from the discussion in Section 1.10 that each parallelism class in the given point configuration C (i.e., each slope determined by C) corresponds to exactly one move in any allowable sequence associated to C. Hence, the result follows from the stronger purely combinatorial statement:

For even n every nontrivial allowable sequence of permutations (whether realizable or not) has at least n moves.

The *trivial allowable sequence* is the one that in one move proceeds from $12\ldots n$ to its reverse; its geometric realization is given by n consecutively labeled points on a line.

1.11.1 Theorem   (Ungar 1982). *Every nontrivial allowable sequence of permutations of E_n, n even, has at least n moves.*

Proof. Each move consists of reversing a collection of disjoint increasing substrings in the current permutation. Call a move *crossing* if one of these substrings contains the elements in positions $n/2$ and $n/2+1$. The idea is to focus on what happens at the center "barrier" separating the initial and the final substrings of length $n/2$. A move is crossing precisely if it forces some elements to cross this barrier. Let $m_1, m_2, \ldots, m_t$ be the crossing moves, and let d_i be the distance from the center barrier to the nearest end of the substring straddling it which is reversed in move m_i. See Figure 1.11.1, where brackets show substrings reversed in move m_i.

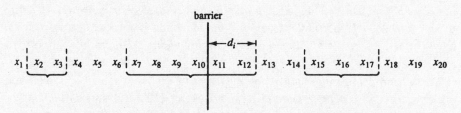

Figure 1.11.1: A crossing move. (Brackets show the three
substrings simultaneously reversed in move m_i).

Now, in each crossing move m_i at least $2d_i$ elements cross the barrier. Since at the end all n elements have crossed, we conclude that

(1) $\sum_{i=1}^{t} 2d_i \geq n$.

Furthermore,

(2)   there are at least $d_i + d_{i+1} - 1$ non-crossing moves between m_i and m_{i+1}, for $1 \leq i < t$.

In order to prove (2), we first note that between the crossing moves m_i and m_{i+1} there is a unique non-crossing move m which changes the center barrier from being inside a decreasing substring to being inside an increasing substring. There are at least $d_i - 1$ non-crossing moves between m_i and m, because one single move can shorten a decreasing string by at most one element at each end. After the move m the nearer end of the newly created maximal increasing substring straddling the barrier has distance one from the barrier, because m must involve reversing a substring which is adjacent to but not crossing the barrier. Next we see that there are at least $d_{i+1} - 1$ non-crossing moves between m and m_{i+1}, because one single move can lengthen an increasing string by at most one element at each end. Altogether, therefore, there are at least $(d_i - 1) + 1 + (d_{i+1} - 1)$ non-crossing moves between m_i and m_{i+1}.

We will need one more fact:

(3)   there are at least $d_t + d_1 - 1$ non-crossing moves after m_t and before m_1.

This fact is a bit harder to see at the level of permutations, but is easy to verify in the realizable case. If the reference line L is rotated beyond 180 degrees, then the allowable sequence of permutations associated to a point configuration C starts to repeat itself, but with each original permutation replaced by its reverse. The combinatorics of crossing moves and their associated numbers d_i is, however, unchanged, and one sees that statement (3) is implied by statement (2) via rotational continuation. This argument can be formalized and does not depend on the suggested visualization.

Adding up the number of crossing and non-crossing moves, inequalities (1)-(3) imply that the total number of moves is at least

$$t + d_t + d_1 - 1 + \sum_{i=1}^{t-1}(d_i + d_{i+1} - 1) \quad = \quad \sum_{i=1}^{t} 2d_i \geq n,$$

completing the proof of Theorem 1.11.1.   □

Much additional work has been done which is related to Scott's slope problem in $\mathbb{R}^2$. Higher-dimensional analogues have also been considered. See Jamison (1985) for a survey. The analogous slope problem in affine planes over finite fields has also been studied, by L. Rédei and others. Rédei proved for all primes p that p points in the affine plane over $GF(p)$, not all on a line, determine at least $\frac{p+3}{2}$ slopes (parallelism classes). A characterization of the configurations which achieve this lower bound as a specific algebraic curve (in a suitable coordinate system) was given by Lovász and Schrijver (1981). The situation for non-prime fields is more complicated. See Lovász and Schrijver (1981) for further information and references.

We conclude this section with another slope problem, posed by Ringel (1956).

(D)   "Ringel's Conjecture": *The slopes in every simple arrangement of lines can be arbitrarily prescribed.*

More precisely, this may be stated:

Suppose we are given an arrangement $\mathcal{A} = \{L_1, L_2, \ldots, L_n\}$ of straight lines in $\mathbb{R}^2$ in general position (no three of them meet at any point). Suppose furthermore that line L_i has slope a_i, and that $a_1 < a_2 < \ldots < a_n$. Then for every sequence $b_1 < b_2 < \ldots < b_n$ of real numbers there exists an arrangement $\mathcal{A}' = \{L'_1, L'_2, \ldots, L'_n\}$ of straight lines, combinatorially isomorphic to $\mathcal{A}$, such that line L'_i has slope b_i, for $1 \leq i \leq n$.

The exact meaning of combinatorial isomorphism of line and pseudoline arrangements will be discussed in Section 6.3. Las Vergnas (1986a) gave a counterexample to Ringel's conjecture, based on oriented matroid theoretic considerations. His construction yields a counter-example with 32 lines, which he could reduce to 13 lines.

A fruitful setting for this problem is that of realization spaces of oriented ma-troids – see Chapter 8. The sequence of slopes can be represented by a point configuration on the line at infinity. We want to know whether all such config-urations "at infinity" do in fact come from a line arrangement of the prescribed combinatorial type. Equivalently, this asks whether the map of realization spaces corresponding to a contraction of a rank 3 oriented matroid is always surjective. This is not true, and the smallest example of such a failure was constructed by Richter and Sturmfels (1991). Their construction, also discussed in Exam-ple 8.1.3, leads to the following minimal counter-example to Ringel's conjecture. (See also Felsner and Ziegler (1999) for a new proof.)

Example 1.11.2. *Let $\mathcal{A}$ be the arrangement of six lines shown in Figure 1.11.2.*

In this arrangement we have slopes $a_1 < a_2 < \ldots < a_6$. Let us select slopes $b_1 = -1, b_2 = 0, b_3 = 1, b_5 = 9, b_6 = 10$. We are done if we can show that b_4 cannot be assigned arbitrarily in the open interval $(1,9)$. Without loss of generality, we may assume that point P is the origin, and by a similarity transformation, that point Q has coordinates $(1,10)$. Let point R have co-ordinates $(c,0)$. Then we can compute that point S must have coordinates $((9+9c)/8, (81+9c)/8)$. However, from $-1 = b_1 > \text{slope of } QR$, we see that $c < 11$. Hence $b_4 > \text{slope of } PS = (9+c)/(1+c) > 5/3$, as desired.

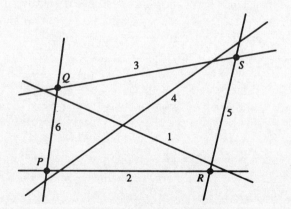

Figure 1.11.2: A line arrangement whose slopes are not
arbitrarily prescribable.

Exercises

1.1   Direct each edge of the complete bipartite graph $K_{3,3}$ arbitrarily. Then list the circuits, cocircuits, and chirotope of the corresponding oriented matroid.

1.2   Let D be a digraph and $\mathcal{M}_D = (E, \mathcal{C})$ its oriented matroid.

(a) Show that the vectors of $\mathcal{M}_D$ are exactly the disjoint unions of strongly connected orientations of subgraphs of D.

(b) Show that every strongly connected digraph is a union of directed circuits – so every vector of $\mathcal{M}_D$ is the conformal composition of all the (signed) circuits it contains.

(c) Identify the covectors of $\mathcal{M}_D$ in graph-theoretic terms.

1.3 Let $\mathcal{M} = \mathcal{M}(E)$ be the oriented matroid of a vector configuration in $\mathbb{R}^r$. Show that we get all the covectors of $\mathcal{M}$ from the linear hyperplanes in $\mathbb{R}^r$, by considering which vectors of E lie "on the positive side" respectively "on the negative side" of the hyperplane.

1.4 Consider the vertex set E of a regular 3-polytope (the icosahedron, say). What is the relationship between the oriented matroids that are obtained if E is considered as

(a) an affine point configuration (of rank 4),

(b) a vector configuration (of rank 3)?

1.5 Find an example of a non-orientable matroid, that is, a matroid which is not the underlying matroid of any oriented matroid.

1.6 Let $x_1, x_2, \ldots, x_{10}$ be the midpoints of the edges of a 4-dimensional (rank 5) simplex in real affine space. Let $\mathcal{M}$ be the oriented matroid determined by this point configuration.

(a) Find a vector realization of $\mathcal{M}$.

(b) Compute some of the signed circuits, cocircuits, maximal vectors, maximal covectors, and basis signatures of $\mathcal{M}$.

(c) Show that $\mathcal{M}$ is the unique orientation of $\underline{\mathcal{M}}$, up to reorientation.

1.7 Pick your favorite point configuration of between seven and ten points in the real affine plane, and construct the corresponding pseudoline arrangement. A challenging example, which happens to be non-realizable, is shown in Figure 8.3.2; an additional challenge is to preserve the 5-fold symmetry in the pseudoline arrangement. Hint: list the cocircuits first.

1.8 How can non-simple oriented matroids (i.e., having parallel elements and loops) be represented by pseudosphere arrangements? (Edmonds and Mandel 1982)

1.9 Given a pseudosphere arrangement $\mathcal{A}$, how can the cocircuits $\mathcal{C}^*(\mathcal{A})$ and the covectors $\mathcal{L}(\mathcal{A})$ be read off from the geometry?

1.10 Given an allowable sequence, show how to directly construct the pseudoline arrangement corresponding to the little oriented matroid.

1.11 Let $\mathcal{A}$ be an arrangement of pseudolines, and P a set of vertices in $\mathcal{A}$, such that any two vertices in P are connected by a pseudoline in $\mathcal{A}$.

(a) How does this determine an oriented matroid on P?

(b) How can the cocircuits, circuits and basis orientation of this oriented matroid be obtained from the geometry?

1.12 Can you find an example of a non-stretchable arrangement of 11 pseudolines such that every proper subarrangement is stretchable?

1.13 (a) Construct a 3-polytope that is not (combinatorially equivalent to) a Delaunay polytope.

(b) Show that every 3-polytope with at most nine vertices is Delaunay. Find a non-Delaunay 3-polytope with ten vertices.

1.14 Prove that the little and the big oriented matroids of an allowable sequence *are* oriented matroids.

1.15 Prove that condition (C_1) in Section 1.4 is equivalent to the statement that the intersection of the interiors of the hemispheres $S_e^{X_e}$, $e \in \underline{X}$, is empty.

1.16 Let $t_1 < t_2 < \ldots < t_n$ and let $\mathcal{M}$ be the oriented matroid of affine dependences on $\{(t_i^1, t_i^2, \ldots, t_i^{r-1}) \in \mathbb{A}^{r-1} : i = 1, \ldots, n\}$. Determine the chirotope, all circuits and all cocircuits of $\mathcal{M}$.

1.17 Using oriented matroid duality, list all combinatorial types of simplicial 4-polytopes with seven vertices. In each case, determine the number of edges, triangles and facets.

1.18 Let $0 \le r \le n$ be integers, let A be a real $r \times n$-matrix of rank r, let B be a real $(n-r) \times n$-matrix of rank $n-r$, and suppose that the row spaces of A and B are orthogonal complements in $\mathbb{R}^n$.

(a) Find a formula which relates the $r \times r$-minors of A to the $(n-r) \times (n-r)$-minors of B.

(b) Find a formula which relates the rank r chirotope of A to the rank $n-r$ chirotope of B.

1.19 Two vector subspaces ξ and ξ' of $\mathbb{R}^n$ determine the same matroid if and only if $\dim(\xi \cap L) = \dim(\xi' \cap L)$ for each coordinate subspace L of $\mathbb{R}^n$. Can you find a similar condition for ξ and ξ' to determine the same oriented matroid?

1.20 Let ξ be an r-dimensional linear subspace of $\mathbb{R}^n$, and let $\mathcal{M} = \mathcal{M}(\xi)$ be its oriented matroid, as defined in Section 1.2(d). A non-zero vector $v \in \xi$ is called *elementary* if its support $supp(v) := \{i : v_i \ne 0\}$ is minimal with respect to inclusion.

(a) Show that every vector of ξ is a conformal linear combination of elementary vectors.

(b) Show that the circuits of $\mathcal{M}$ are precisely the signed supports $sign(v)$, where v ranges over all elementary vectors of ξ.

1.21 In this book we encounter the following correspondences between a geometric object and an associated object:

(a) matroid $\longleftrightarrow$ dual matroid

(b) oriented matroid $\longleftrightarrow$ dual oriented matroid

(c) linear program $\longleftrightarrow$ dual linear program   (Chapter 10)

(d) convex polytope $\longleftrightarrow$ Gale transform   (Chapter 9)

(e) convex polytope $\longleftrightarrow$ polar convex polytope

(f) projective point configuration $\longleftrightarrow$ polar projective hyperplane arrangement

(g) Delaunay triangulation $\longleftrightarrow$ Voronoi diagram.

We have used the term "duality" for the cases (a)–(d) and "polarity" for (e)–(g). Formulate the similarities and distinctions that motivate this use of terminology. How does the adjoint of an oriented matroid (Section 5.3) fit into this pattern?

2

A Second Orientation Session

This chapter continues the discussion of general topics related to oriented matroids. More precisely, the geometric, algebraic and topological topics treated here are related to *realizable* oriented matroids, i.e., real matrices, point configurations and hyperplane arrangements. Nevertheless, the general point of view of oriented matroids seems to be relevant in many cases, and some understanding of these topics is important for a balanced view of oriented matroids within mathematics.

2.1 Real hyperplane arrangements

Arrangements of hyperplanes in $\mathbb{R}^d$ arise as fundamental objects in various mathematical theories: from inequality systems in linear programming, from facets of convex polytopes, from reflection groups in Lie theory, from geometric search in computational geometry, from questions in singularity theory, to name a few. Real hyperplane arrangements have also been studied for a long time by discrete geometers, particularly with respect to their combinatorial structure, that is, how they partition space.

In Section 1.2 it was explained how a hyperplane arrangement $\mathcal{A}$ gives rise to an oriented matroid $\mathcal{M}(\mathcal{A})$, and it follows from the discussion there (see also Section 1.4) that hyperplane arrangements correspond bijectively to *realizable* oriented matroids(up to reorientations). Here we will take a second look at hyperplane arrangements. Some basic definitions will be reviewed, the translation from geometric language to oriented matroid terminology will be explained in a few cases, and a theorem about the number of simplicial regions will be shown, which illustrates that the combinatorial behavior of oriented matroids can differ in the realizable and unrealizable cases.

Four variants of *linear* arrangements are presented in the following definition.

In Chapter 5 we will encounter a generalization to topologically deformed "pseu-dohyperplanes" and "pseudospheres", and arrangements ôf such.

A *hyperplane* in $\mathbb{R}^d$ is a $(d-1)$-dimensional linear subspace, and an *affine hy-perplane* is the translation of a hyperplane by some vector. Thus the hyperplanes are precisely those affine hyperplanes that are *central* (or *linear*), that is, contain the origin. Real projective $(d-1)$-space will be denoted by $\mathbb{P}^{d-1}$. A *hyperplane* in $\mathbb{P}^{d-1}$ is a projective subspace of codimension one.

2.1.1 Definition.

(1) A *central arrangement of hyperplanes* in $\mathbb{R}^d$ is a finite family $\mathcal{A} = (H_e)_{e \in E}$ of hyperplanes in $\mathbb{R}^d$. Its *rank* is $r = r(\mathcal{A}) := \operatorname{codim}(\cap \mathcal{A})$. It is *essential* if $\cap \mathcal{A} = \{0\}$, that is, if $r = d$.

(2) A *projective arrangement* in $\mathbb{P}^{d-1}$ is a finite family $\mathcal{A} = (H_e)_{e \in E}$ of hyper-planes in $\mathbb{P}^{d-1}$. Its *rank* is $r = \operatorname{codim}(\cap \mathcal{A})$. $\mathcal{A}$ is *essential* if $\cap \mathcal{A} = \emptyset$, that is, if $r = d$.

(3) A *spherical arrangement* on the $(d-1)$-sphere S^{d-1} is a finite set $\mathcal{A} = (S_e)_{e \in E}$ of linear codimension-one subspheres (that is, intersections of S^{d-1} with central hyperplanes). Its *rank* is $r = \operatorname{codim}(\cap \mathcal{A})$. It is *essential* if $\cap \mathcal{A} = \emptyset$, that is, if $r = d$.

(4) An *affine arrangement* in $\mathbb{R}^{d-1}$ is a finite set $\mathcal{A} = (H_e)_{e \in E}$ of affine hyper-planes in $\mathbb{R}^{d-1}$. It is *essential* if it contains a subset of hyperplanes whose intersection is one point.

The central arrangements (1) are the basic model we use. So, in the following, if we talk about an "arrangement" without further qualifications, we mean a central arrangement of hyperplanes in $\mathbb{R}^d$.

It is immediately clear that central, projective and spherical arrangements are equivalent, and also how to pass between arrangements of types (1), (2) and (3). The equivalence between central and spherical arrangements is discussed in some detail in Sections 1.4 and 5.1.

Affine arrangements in $\mathbb{R}^{d-1}$ are equivalent to "pointed" central arrangements in $\mathbb{R}^d$, that is, central arrangements with a choice of one of its hyperplanes. This correspondence is explained and exemplified at the beginning of Section 4.5, and from a different angle in Section 10.1.

An arrangement of any of the above types determines a decomposition of the ambient space into open topological cells. We will now describe this for central and affine arrangements. The cell decompositions induced by spherical arrangements (and also by projective arrangements) are thoroughly treated in Chapter 5.

Let $\mathcal{A} = (H_e)_{e \in E}$ be a central or affine arrangement in $\mathbb{R}^d$. Then we can choose a "positive side" for every hyperplane in the arrangement. This leads to a *sign vector* $X \in \{+, -, 0\}^E$ for every point $x \in \mathbb{R}^d$, where X_e denotes whether x is on the positive side of H_e, on its negative side, or lies on H_e. The set of all points $x \in \mathbb{R}^d$ having the same sign vector X forms a *cell* in the decomposition of $\mathbb{R}^d$ induced by $\mathcal{A}$. Every cell is *relatively open* and *convex*. Here "relatively open" means that the cell is an open subset of the linear subspace (respectively affine subspace) spanned by the cell.

2.1.2 Definition. The *face poset* $\mathcal{F}(\mathcal{A})$ is the set of all cells induced by $\mathcal{A}$, ordered by inclusion of their (topological) closures. The maximal cells (all d-dimensional) are called the *regions* (or *chambers*) of $\mathcal{A}$.

Two arrangements $\mathcal{A}$ and $\mathcal{A}'$ in $\mathbb{R}^d$ are said to be *combinatorially equivalent* if $\mathcal{F}(\mathcal{A}) \cong \mathcal{F}(\mathcal{A}')$.

The order relation of $\mathcal{F}(\mathcal{A})$ has an easy combinatorial description in terms of the sign vectors X of the cells. For this see Sections 4.1 (central case) and 4.5 (affine case). In terms of these sign vectors, the face poset $\mathcal{F}(\mathcal{A})$ of a central arrangement $\mathcal{A}$ is just the big face lattice of the oriented matroid $\mathcal{M}(\mathcal{A})$. Consequently, combinatorial equivalence of hyperplane arrangements is a concept identical to having isomorphic oriented matroids. A systematic study of face lattices $\mathcal{F}(\mathcal{A})$ will be undertaken in Chapter 4. The relation of having the same matroid can be similarly described.

2.1.3 Definition. The *intersection poset* $L(\mathcal{A})$ of a central or affine arrangement $\mathcal{A}$ is the set of intersections of subfamilies of $\mathcal{A}$, ordered by reverse inclusion.

If $\mathcal{A}$ is central, then $L(\mathcal{A})$ is a geometric lattice of rank $r(\mathcal{A})$. In fact, $L(\mathcal{A})$ is the geometric lattice of the matroid determined by $\mathcal{A}$, which is the underlying matroid of the oriented matroid $\mathcal{M}(\mathcal{A})$. In terms of sign vectors, $L(\mathcal{A})$ is the set of supports of sign vectors in $F(\mathcal{A})$, ordered by inclusion.

If $\mathcal{A}$ is affine, then $L(\mathcal{A})$ is a *geometric semilattice*: a geometric lattice from which an upper interval $[x, \hat{1}]$ has been removed. For details about geometric semilattices, see Zaslavsky (1981) and Wachs and Walker (1986). In both the central and the affine cases, the least element of $L(\mathcal{A})$ is $\mathbb{R}^d$ and $\mathcal{A}$ is the set of atoms.

From now on, we will only consider essential arrangements. From a combinatorial point of view, the specialization to essential arrangements causes no loss of generality, because for each central arrangement $\mathcal{A} = (H_e)_{e \in E}$ of rank r in $\mathbb{R}^d$ there exists an *associated essential arrangement* $\mathcal{A}'$ in $\mathbb{R}^r$ such that $\mathcal{F}(\mathcal{A}) \cong \mathcal{F}(\mathcal{A}')$. For this, just take the $(d - r)$-dimensional subspace $V = \cap \mathcal{A}$ and let $\mathcal{A}'$ be the arrangement of hyperplanes H_e/V in $\mathbb{R}^d/V$. There exist associated essential arrangements also for affine and other types of arrangements, see Exercise 2.1 and also Exercise 5.7. An affine arrangement is essential if and only if all minimal cells in the induced decomposition of $\mathbb{R}^{d-1}$ are points, see Exercise 2.2.

Let $\mathcal{A}$ be a central and essential arrangement in $\mathbb{R}^d$. Then the regions induced by $\mathcal{A}$ are precisely the connected components of $\mathbb{R}^d \setminus \cup \mathcal{A}$. The closure of a region is a d-dimensional pointed convex polyhedral cone with its vertex at the origin. The *facets* of this cone are the $(d - 1)$-dimensional intersections with supporting hyperplanes. Any intersection of facets is called a *face*. The collection of all faces of all regions, ordered by inclusion, gives precisely the face lattice $\mathcal{F}(\mathcal{A})$, in terms of closed cells rather than of open cells. Similar remarks apply to affine essential arrangements, except that here the regions may be bounded (convex polytopes) or unbounded polyhedra.

Readers seeking information about real arrangements in this book should be alerted to the fact that here regions are usually referred to as "topes". This name, coined by Edmonds and Mandel (1982), has become established in the oriented matroid literature. Actually, we try to separate the usage of the terms "regions" and "topes", so that the former refers to the geometric objects (as in Definition 2.1.2), while the latter refers to the sign vectors that encode them (as in Definition 4.1.2).

2.1.4 Definition. The *tope graph* (or *graph of regions*) $T(\mathcal{A})$ of a central or affine arrangement $\mathcal{A}$ has as vertices the set of regions of $\mathcal{A}$ and as edges the pairs of regions which share a common facet. Choosing a base region B and directing the edges of the tope graph away from B, we get the *tope poset* (or *poset of regions*) $T(\mathcal{A}, B)$.

Tope graphs and tope posets are systematically studied in Sections 4.2 and 4.4. For the affine case, see Section 4.5. For instance, it is shown that the graph-theoretic distance in $T(\mathcal{A})$ between two regions equals the number of hyperplanes that separate them (Proposition 4.2.3) and that $T(\mathcal{A}, B)$ is a graded poset (Proposition 4.2.10).

A region R of a central and essential arrangement in $\mathbb{R}^d$ is called *simplicial* if its facial structure is that of a simplex (equivalently, if R has precisely d facets). The existence of simplicial regions is guaranteed by a theorem of Camion (1968, Section 4, Theorem 3, p. 347) and the following stronger version of Shannon (1979). For convenience of the proof we will work with the projective version.

2.1.5 Theorem (Shannon 1979). *Let $\mathcal{A}$ be an essential arrangement of n hyperplanes in d-dimensional real projective space $\mathbb{P}^d$. Then*

(i)   *$\mathcal{A}$ has at least n simplicial regions,*

(ii) *for every $H \in \mathcal{A}$ there exist at least $d+1$ simplicial regions having a facet contained in H.*

2.1.6 Lemma. *Let $\mathcal{A}$ be as before and $H \in \mathcal{A}$. Let H_∞ denote a hyperplane in $\mathbb{P}^d$ which contains no vertex of $\mathcal{A}$. Then, in the affine space $\mathbb{P}^d \backslash H_\infty$ there exists at least one bounded simplicial region of $\mathcal{A}$ with a facet contained in H.*

Proof. The result is trivial for $d = 1$. Continuing by induction, we assume that the lemma is true for dimensions less than d. Choose a vertex x of $\mathcal{A}$, $x \notin H$, such that the distance from x to H is minimal. Let $H_1, H_2, \ldots, H_k$ denote the hyperplanes of $\mathcal{A}$ that contain x. Then $\mathcal{A}' = \{H \cap H_i : 1 \leq i \leq k\}$ is an arrangement of k hyperplanes in the $(d-1)$-dimensional projective space H, and $H \cap H_\infty$ contains no vertex of $\mathcal{A}'$. By the induction hypothesis there exists a bounded $(d-1)$-simplex in $\mathcal{A}'$ with vertices $x_1, x_2, \ldots, x_d$ (say), see Figure 2.1.1.

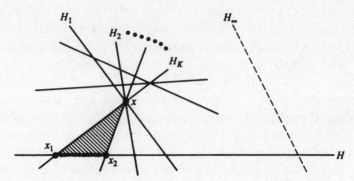

Figure 2.1.1: Sketch for the proof of Lemma 2.1.6.

By the choice of x as having minimal distance to H, every segment $[x, x_i]$ is an edge of $\mathcal{A}$. Hence $x, x_1, \ldots, x_d$ are the vertices of a bounded d-simplex of $\mathcal{A}$ having the facet $\{x_1, \ldots, x_d\}$ in H. □

Proof of Theorem 2.1.5. Call a simplicial region Δ of $\mathcal{A}$ *related* to $H \in \mathcal{A}$ if Δ has a facet in H. Each Δ is related to exactly $d + 1$ hyperplanes (it has $d + 1$ facets), and by (ii) each hyperplane is related to at least $d+1$ simplices Δ. Hence (ii) implies (i).

To prove (ii), let $\Delta_1, \Delta_2, \ldots, \Delta_k$ denote the d-simplices of $\mathcal{A}$ having a facet in H, and suppose that $k \leq d$. Then there exists a hyperplane H_∞ in $\mathbb{P}^d$ which contains no vertex of $\mathcal{A}$ and which intersects the interior of each Δ_i, $1 \leq i \leq k$. Then in the affine space $\mathbb{P}^d \backslash H_\infty$ there is no bounded d-simplex having a facet in H, contradicting Lemma 2.1.6. Hence $k \geq d + 1$. □

2.1.7 Corollary. *Let $\mathcal{A}$ be a central and essential arrangement of n hyperplanes in $\mathbb{R}^d$. Then $\mathcal{A}$ has at least $2n$ simplicial regions.*

Is Shannon's Theorem 2.1.5 true also for non-realizable oriented matroids? This is the case in rank 3 (Theorem 6.5.2), but it is false in rank 4: the oriented matroid RS(8) discussed in Section 1.5 corresponds to an arrangement of 8 pseudoplanes in projective 3-space which has only 7 simplicial regions, see also Roudneff and Sturmfels (1988). In fact, for oriented matroids of rank $r \geq 4$ it is not even known whether there is always at least one simplicial region (Problem 4.5).

2.2 Zonotopes

In this section we study a class of convex polytopes which provides a natural model for (realizable) oriented matroids. Our starting point is the observation that every convex polytope can be represented as the projection of a regular simplex. Let $e_1, e_2, \ldots, e_n$ denote the standard basis vectors in $\mathbb{R}^n$, and consider the regular $(n-1)$-simplex

$$\Delta_{n-1} := \text{conv}\{e_1, e_2, \ldots, e_n\}.$$

If P is any $(r-1)$-dimensional polytope with n vertices, then there exists a unique affine transformation, represented by an $r \times n$-matrix $\mathbf{X}$ of maximal rank, such that P is congruent to $\mathbf{X}(\Delta_{n-1})$, the projection of the regular simplex Δ_{n-1} under $\mathbf{X}$. Indeed, the i-th column of $\mathbf{X}$ is just the r-tuple of homogeneous coordinates of the i-th vertex of P.

Now consider another $r \times n$-matrix $\mathbf{Y}$ and its $(r-1)$-polytope $\mathbf{Y}(\Delta_{n-1})$. If the matrices $\mathbf{X}$ and $\mathbf{Y}$ define the same rank r oriented matroid, i.e., if corresponding maximal minors of $\mathbf{X}$ and $\mathbf{Y}$ have the same signs, then the polytopes $\mathbf{X}(\Delta_{n-1})$ and $\mathbf{Y}(\Delta_{n-1})$ are *combinatorially equivalent*, which means that their face lattices are isomorphic. Unfortunately, the converse does not hold: The polytopes $\mathbf{X}(\Delta_{n-1})$ and $\mathbf{Y}(\Delta_{n-1})$ may be combinatorially equivalent although the matrices $\mathbf{X}$ and $\mathbf{Y}$ have different oriented matroids. For example, this is the case when $\mathbf{X}$ and $\mathbf{Y}$ are the 4×6-matrices of homogeneous vertex coordinates for a regular octahedron and a non-regular octahedron respectively (cf. Section 9.1).

The situation is rectified if we replace the regular $(n-1)$-simplex Δ_{n-1} with the *regular n-cube*

$$C_n := \Big\{ \sum_{i=1}^{n} \lambda_i \mathbf{e}_i \ : \ -1 \le \lambda_i \le +1 \Big\} \quad \subset \quad \mathbb{R}^n.$$

It will be proved below that two $r \times n$-matrices $\mathbf{X}$ and $\mathbf{Y}$ of rank r have the same oriented matroid on their columns (up to reorientation) if and only if the projections $\mathbf{X}(C_n)$ and $\mathbf{Y}(C_n)$ of the n-cube C_n are combinatorially equivalent polytopes. The only drawback is that, while every polytope is a projection of a regular simplex, not every polytope is a projection of a regular cube. A polytope which happens to be a projection of a regular cube is called a *zonotope*. This discussion shows that the study of realizable oriented matroids is equivalent to the study of zonotopes (or projections of cubes).

Another reason why zonotopes are of importance for oriented matroid theory is that they furnish us with a polar perspective on the intuitive picture of an oriented matroid as a "slightly deformed" hyperplane arrangement. This picture will be made precise in Chapter 5, and in this setting the generalization of zonotopes to non-realizable oriented matroids is provided by the "opposite" *PL* spheres of Corollary 4.3.4.

Suppose $\mathcal{M}$ is the rank r oriented matroid defined by an $r \times n$-matrix $\mathbf{X}$. Then the linear hyperplanes normal to the column vectors $\mathbf{x}_1, \mathbf{x}_2, \ldots, \mathbf{x}_n$ of $\mathbf{X}$ constitute an r-dimensional central hyperplane arrangement $\mathcal{A}(\mathbf{X})$. Its face lattice $\mathcal{F}(\mathcal{A}(\mathbf{X}))$ is the *(big) face lattice* $\mathcal{F}_{big}(\mathcal{M})$ of the oriented matroid $\mathcal{M}$. We shall see that the hyperplane arrangement $\mathcal{A}(\mathbf{X})$ is geometrically polar to the zonotope $\mathcal{Z}(\mathbf{X}) := \mathbf{X}(C_n)$. This means that the regions of $\mathcal{A}(\mathbf{X})$ are precisely the cones of outer normals at the faces of $\mathcal{Z}(\mathbf{X})$. On the combinatorial side, it follows that the face lattice of a zonotope $\mathcal{Z}$ is anti-isomorphic to the (big) face lattice of its oriented matroid $\mathcal{M}$, that is, $\mathcal{F}(\mathcal{Z}) \simeq (\mathcal{F}_{big}(\mathcal{M}))^{op}$.

Our exposition of zonotopes from the oriented matroid point of view includes complete proofs for most of their basic properties. One objective is to furnish the reader with the background for studying the original literature on zonotopes. This includes the classical articles by Coxeter (1962), McMullen (1971a), Shephard (1974), see also Halsey (1971), as well as applications of zonotopes in the more recent literature; see e.g. Martini (1985), Bohne, Dress and Fischer (1989), Stanley (1991), Billera and Sturmfels (1992), Gritzmann and Sturmfels (1993), Richter-Gebert (1993d), and Crapo and Senechal (1995).

Let $\mathbf{X}$ be an $r \times n$-matrix of rank r with column vectors $\mathbf{x}_1, \mathbf{x}_2, \ldots, \mathbf{x}_n$. Throughout this section we will assume that all $\mathbf{x}_i$ are non-zero, which means that the rank r oriented matroid $\mathcal{M}$ of $\mathbf{X}$ has no loops. Recall from Section 1.2 that $\mathcal{M}$ can be interpreted in terms of the following geometric objects arising from $\mathbf{X}$: a configuration of n points in $(r-1)$-dimensional projective space, a signed configuration of n points in $(r-1)$-dimensional affine space, an arrangement of n hyperplanes in $(r-1)$-dimensional affine or projective space, or a central arrangement of n linear hyperplanes in $\mathbb{R}^r$. We shall relate these interpretations to the zonotope $\mathcal{Z}(\mathbf{X})$ of $\mathbf{X}$. By the definition given above, $\mathcal{Z}(\mathbf{X})$ is the projection of the regular n-cube C_n under $\mathbf{X}$, and we can write

$$\mathcal{Z}(\mathbf{X}) = \Big\{ \sum_{i=1}^{n} \lambda_i \mathbf{x}_i \ : \ -1 \leq \lambda_i \leq +1 \Big\} \quad \subset \quad \mathbb{R}^r$$

Equivalently, this zonotope can be expressed as the Minkowski sum of n line segments:

$$\mathcal{Z}(\mathbf{X}) \quad = \quad [-\mathbf{x}_1, \mathbf{x}_1] + [-\mathbf{x}_2, \mathbf{x}_2] + \ldots + [-\mathbf{x}_n, \mathbf{x}_n].$$

By linearity, every vertex of $\mathcal{Z}$ is the image under $\mathbf{X}$ of a vertex of the n-cube C_n and can therefore be written in the form $\sum_{i=1}^{n} \sigma_i \mathbf{x}_i$ for some vector $\sigma = (\sigma_1, \sigma_2, \ldots, \sigma_n) \in \{-, +\}^n$. (The signs "+" and "−" are as usual identified with the numbers "−1" and "+1".) While every vertex of $\mathcal{Z}$ can be encoded by a sign vector, in general, many sign vectors do not correspond to vertices but only to interior points of $\mathcal{Z}$. We will see in Proposition 2.2.2 that the sign vectors corresponding to vertices of $\mathcal{Z}$ are precisely the maximal covectors of the oriented matroid $\mathcal{M}$. Let us first look at an example.

2.2.1 Example (The rhombododecahedron and its oriented matroid).
Let $\mathcal{M} = (E_4, \mathcal{L})$ be the rank 3 oriented matroid defined by the matrix

$$\mathbf{X} \quad = \quad \begin{pmatrix} 1 & 0 & 0 & 1 \\ 0 & 1 & 0 & -1 \\ 0 & 0 & 1 & 1 \end{pmatrix}.$$

Then $\mathcal{M}$ is the unique rank 3 oriented matroid on $E_4 = \{1, 2, 3, 4\}$ having the signed circuit $1\bar{2}3\bar{4} \equiv (+ - + -)$. If we interpret the columns of $\mathbf{X}$ as homogeneous coordinate vectors, then we get the affine configuration of 4 points depicted in Figure 2.2.1. There are 14 distinct ways of separating the four points by an oriented affine line into positive and negative sides. These 14 "hyperplane sections" are indexed by the set $\mathcal{T}(\mathcal{M}) = \{+, -\}^4 \setminus \{(+ - + -), (- + - +)\}$ of maximal covectors of $\mathcal{M}$. For instance, the line having point 3 on its positive side and points $1, 2, 4$ on its negative side is encoded by the covector $(- - + -)$.

In the central arrangement of four (hyper)planes defined by $\mathbf{X}$, the maximal covectors correspond to the regions. Let us now consider the zonotope

$$\mathcal{Z} := \mathcal{Z}(\mathbf{X}) = \text{conv}\left\{ (\sigma_1 + \sigma_4, \sigma_2 - \sigma_4, \sigma_3 + \sigma_4) \in \mathbb{R}^3 \mid \sigma \in \{-,+\}^4 \right\}$$

which is the projection of the regular 4-cube under the linear map $\mathbf{X}$. The zonotope $\mathcal{Z}$, known to geometers as the *rhombododecahedron*, is depicted in Figure 2.2.1.

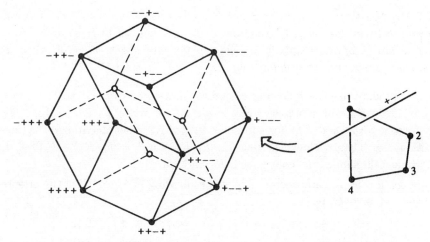

Figure 2.2.1:   The rhombododecahedron and an affine realization
of its oriented matroid.

Which choices of sign vectors σ correspond to vertices of $\mathcal{Z}$? It can easily be checked that a point $\mathbf{X} \cdot \sigma = (\sigma_1 + \sigma_4, \sigma_2 - \sigma_4, \sigma_3 + \sigma_4)$ is a vertex of $\mathcal{Z}$ if and only if σ is a maximal covector of $\mathcal{M}$. In other words, precisely the two vertices $\sigma = (+ - + -)$ and $\sigma = (- + - +)$ of the 4-cube are mapped into the interior of $\mathcal{Z}$ under the projection $\mathbf{X}$. The rhombododecahedron has 14 vertices, one for each maximal covector of $\mathcal{M}$, and it has 24 edges and 12 facets. Under the correspondence to be described below, these can be read off from the 36 non-maximal covectors of $\mathcal{M}$. As an example, consider the covectors $(+ - 0+)$ and $(0 - 0+)$. In the affine point diagram $(+ - 0+)$ represents the line through point 3 which has $1, 4$ on its positive side and 2 on its negative side, while the cocircuit $(0-0+)$ represents the line spanned by 1 and 3. The covector $(+-0+)$ encodes the edge joining the vertices $(2, -2, 0)$ and $(2, -2, 2)$ of $\mathcal{Z}$, which are encoded by $(+ - -+)$ and $(+ - ++)$, respectively. This edge is contained in the quadrangular facet conv $\{(0, -2, 0),\ (0, -2, 2),\ (2, -2, 0),\ (2, -2, 2)\}$ which is encoded by the cocircuit $(0 - 0+)$.

We now return to the case of a general $r \times n$-matrix $\mathbf{X}$ of maximal rank with column vectors $\mathbf{x}_1, \mathbf{x}_2, \ldots, \mathbf{x}_n$. Let $\mathcal{Z} = \mathcal{Z}(\mathbf{X})$ denote its zonotope and $\mathcal{M} = (E_n, \mathcal{L})$ its oriented matroid.

There is a natural labeling of the 3^n non-empty faces of the regular n-cube using the sign vectors $C \in \{-, 0, +\}^n$. Under the projection $\mathbf{X}$, the face of the cube labeled C is mapped onto the polytope

$$\text{face}(C) \quad := \quad \sum_{i \in C^+} \mathbf{x}_i - \sum_{j \in C^-} \mathbf{x}_j + \sum_{k \in C^0} [-\mathbf{x}_k, \mathbf{x}_k].$$

This polytope is a zonotope of some dimension $\leq r$ which is contained in the r-zonotope $\mathcal{Z}$. We will prove that $\text{face}(C)$ is a face of $\mathcal{Z}$ if and only if C is a covector of $\mathcal{M}$.

2.2.2 Proposition. *The assignment $C \mapsto \text{face}(C)$ defines an order-reversing bijection between the set $\mathcal{L}$ of covectors of $\mathcal{M}$ and the set of faces of $\mathcal{Z}$. Hence the face lattice of $\mathcal{Z}$ is anti-isomorphic to the face lattice $\widehat{\mathcal{L}} = \mathcal{F}(\mathcal{A})$ of the hyperplane arrangement $\mathcal{A}$ defined by $\mathbf{X}$.*

Proof. We shall use the following easy fact from convexity. Given two convex sets $K_1, K_2 \subset \mathbb{R}^r$ and a non-zero vector $\mathbf{a} \in \mathbb{R}^r$, we write $F_{\mathbf{a}}(K_i) = \{\mathbf{x} \in K_i : \mathbf{a} \cdot \mathbf{x} \geq \mathbf{a} \cdot \mathbf{y} \text{ for all } \mathbf{y} \in K_i\}$ for the closed face of K_i which maximizes the linear functional $\mathbf{a}$. Then we have the relation $F_{\mathbf{a}}(K_1 + K_2) = F_{\mathbf{a}}(K_1) + F_{\mathbf{a}}(K_2)$ for the Minkowski sum of the convex sets K_1 and K_2.

Now let $\mathbf{a} \in \mathbb{R}^r$ be any vector. The corresponding covector C of the oriented matroid $\mathcal{M}$ is given by

$$C^+ = \{i : \mathbf{a} \cdot \mathbf{x}_i > 0\}, \quad C^- = \{j : \mathbf{a} \cdot \mathbf{x}_j < 0\},$$

and $C^0 = \{k : \mathbf{a} \cdot \mathbf{x}_k = 0\}$. Using the above observation about Minkowski sums of convex sets, we find

$$\begin{aligned}
F_{\mathbf{a}}(\mathcal{Z}(\mathbf{X})) \quad &= \quad \sum_{i=1}^{n} F_{\mathbf{a}}([-\mathbf{x}_i, \mathbf{x}_i]) \\
&= \quad \sum_{i \in C^+} F_{\mathbf{a}}([-\mathbf{x}_i, \mathbf{x}_i]) + \sum_{j \in C^-} F_{\mathbf{a}}([-\mathbf{x}_j, \mathbf{x}_j]) \\
&\quad + \sum_{k \in C^0} F_{\mathbf{a}}([-\mathbf{x}_k, \mathbf{x}_k]) \\
&= \quad \text{face}(C).
\end{aligned}$$

Hence every face of the zonotope $\mathcal{Z}$ is equal to $\text{face}(C)$ for some covector C of $\mathcal{M}$. Conversely, $\text{face}(C)$ is a face of $\mathcal{Z}$ because each covector C of $\mathcal{M}$ can be represented by some vector $\mathbf{a} \in \mathbb{R}^r$ as above. Note that different covectors C_1 and C_2 give different faces, and that $\text{face}(C_2) \subseteq \text{face}(C_1)$ if and only if $C_1 \leq C_2$ in the big face lattice $\mathcal{F}_{big}(\mathcal{M})$. This proves that the map $\widehat{\mathcal{L}} \to \mathcal{F}(\mathcal{Z})$, $C \mapsto \text{face}(C)$ is an anti-isomorphism from the (big) face lattice of $\mathcal{M}$ to the face lattice $\mathcal{F}(\mathcal{Z})$ of the zonotope $\mathcal{Z}$. $\square$

Proposition 2.2.2 implies the remarkable fact, pointed out already at the beginning of this section, that there is a one-to-one correspondence between realizable oriented matroids and combinatorial types of zonotopes. At this point we need to interject a word of caution. In the statement of Corollary 2.2.3 and throughout Section 2.2 we will consider two oriented matroids the same if they are *reorientation equivalent*, which means in the realizable case that they define isomorphic hyperplane arrangements.

2.2.3 Corollary. *Two $r \times n$-matrices $\mathbf{X}, \mathbf{Y}$ have the same oriented matroid (up to reorientation) if and only if their zonotopes $\mathcal{Z}(\mathbf{X})$ and $\mathcal{Z}(\mathbf{Y})$ have the same face lattice.*

Most of the rest of this section illustrates the correspondence of Corollary 2.2.3 by setting up a dictionary which translates convexity properties and constructions of zonotopes into the language of oriented matroids and vice versa. As before, $\mathcal{M}$ is the realizable rank r oriented matroid of $\mathbf{X} = (\mathbf{x}_1, \ldots, \mathbf{x}_n)$ and $\mathcal{Z}$ its zonotope. We will think of $\mathbf{X}$ as a configuration of n points in affine $(r-1)$-space. In the terminology of Coxeter (1962) and Shephard (1974), $\mathcal{Z}$ is the *associated zonotope* of the point configuration $\mathbf{X}$.

The reader should be aware that any $(r-1)$-dimensional affine interpretation of the vectors $\mathbf{x}_1, \ldots, \mathbf{x}_n$ comes with the usual twists of projective geometry. If we draw an affine diagram, then this means we are choosing a specific embedding of affine $(r-1)$-space into projective $(r-1)$-space. In other words, we are making a specific choice of a hyperplane H_∞ at infinity. Here H_∞ can be any linear hyperplane which contains none of the vectors $\mathbf{x}_i$. All of these hyperplanes are created equal and there is no *a priori* reason for preferring a specific H_∞.

Once we have made our choice of H_∞, then we get the affine $(r-1)$-dimensional picture of $\mathbf{X}$ as follows. First replace each vector $\mathbf{x}_i$ on the negative side of H_∞ by its antipode $-\mathbf{x}_i$. By our tacit assumption that reorientation equivalent oriented matroids be considered the same, this operation does not change $\mathcal{M}$. Clearly, this operation also does not change the zonotope $\mathcal{Z}$. There exists an affine hyperplane $\mathbb{A}^{r-1}$ which strictly separates the origin from all $\mathbf{x}_i$. We mean the unique points $\lambda_i \mathbf{x}_i$, $0 < \lambda_i < 1$, contained in the $(r-1)$-dimensional affine space $\mathbb{A}^{r-1}$, whenever we are speaking of the affine configuration $\mathbf{X}$. Now, if we make some other choices along the way, then this means that our affine configuration $\mathbf{X}$ is subjected to an admissible projective transformation.

Let us now assume that $\mathbf{X}$ is given together with a fixed specific affine embedding as above. The hyperplane sections of $\mathbf{X}$ correspond to the covectors of $\mathcal{M}$ (up to sign-reversal), and, by Proposition 2.2.2, they correspond to the faces of $\mathcal{Z}$. Let H be any hyperplane which contains none of the $\mathbf{x}_i$. The sections of $\mathbf{X}$ by such hyperplanes correspond to the maximal covectors of $\mathcal{M}$, and, by Proposition 2.2.2, they correspond to the vertices of $\mathcal{Z}$.

With the hyperplane H we can associate a projective transformation which maps H_∞, the hyperplane at infinity, onto H. This defines a correspondence between the images of the point configuration $\mathbf{X}$ under admissible projective transformations with the maximal covectors of $\mathcal{M}$. As a result, we get a correspondence between the vertices of the zonotope $\mathcal{Z}$ and projective images of $\mathbf{X}$. We will come back to this construction in Proposition 2.2.7.

Let us now look at the other extreme case where the hyperplane H is spanned by points $\mathbf{x}_{i_1}, \mathbf{x}_{i_2}, \ldots, \mathbf{x}_{i_{r-1}}$ in our configuration. Such a hyperplane H is complementary to a cocircuit of $\mathcal{M}$, and by Proposition 2.2.2, it corresponds to a facet of the zonotope $\mathcal{Z}$. If H contains no other point $\mathbf{x}_j$, then its complementary cocircuit C is supported on $n-r+1$ elements. In this case the facet face(C) of $\mathcal{Z}$ is (an affine image of) an $(r-1)$-cube. In particular, if $\mathcal{M}$ is a uniform oriented matroid, then all facets of $\mathcal{Z}$ are cubes. Such a "generic" zonotope is called *cubical*. For instance, the rhombododecahedron is a cubical 3-zonotope.

2.2.4 Proposition.　*The combinatorial types of cubical zonotopes are in one-to-one correspondence with realizable uniform oriented matroids (up to reorientation).*

For a general zonotope $\mathcal{Z}$, the facet face(C) given by a cocircuit C is an $(r-1)$-dimensional zonotope. What is the realizable oriented matroid of rank $r-1$ corresponding to this zonotope ? It is clear from the definition that face(C) is a translate of $\sum_{k \in C^0} [-\mathbf{x}_k, \mathbf{x}_k]$. Thus face($C$) is the zonotope associated with the rank $r-1$ oriented matroid $\mathcal{M} \setminus C$, which is the restriction of $\mathcal{M}$ to the hyperplane complementary to C.

In the following example we will illustrate the correspondence between the covectors of an oriented matroid, the cells of its hyperplane arrangement and the faces of its zonotope. The zonotope to be investigated is the 4-permutohedron, an important example of a non-cubical 3-zonotope.

2.2.5 Example (The permutohedron).　Let $\mathcal{M}_0$ be the rank 3 oriented matroid on $\{1, 2, 3, 4, 5, 6\}$ defined by the point configuration in (I) or the line arrangement in (II) of Figure 2.2.2. This oriented matroid was studied in detail in Sections 1.1 and 1.2, where complete lists of its vectors and covectors were given in Figure 1.2.2. We have seen in Section 1.1 that $\mathcal{M}_0$ is the graphic oriented matroid arising from a specific acyclic orientation of the complete graph K_4 on four nodes (see Figure 1.1.1).

Up to sign-reversal, $\mathcal{M}_0$ has 12 maximal covectors, which are labeled a, b, $c, \ldots, k$, l in Figure 2.2.2. They correspond to the generic hyperplane sections or admissible projective transformations of the point configuration (I). They also correspond to the maximal cells of the line arrangement (II). The "front side" of its zonotope $\mathcal{Z}_0$ is depicted in (III).

The zonotope $\mathcal{Z}_0$ is known to combinatorialists as the 4-*permutohedron*. This choice of name is motivated by the following specific 4-dimensional embedding of $\mathcal{Z}_0$. Consider the six vectors $\mathbf{x}_1 = (1, -1, 0, 0)$, $\mathbf{x}_2 = (1, 0, -1, 0)$, $\mathbf{x}_3 = (1, 0, 0, -1)$, $\mathbf{x}_4 = (0, 1, -1, 0)$, $\mathbf{x}_5 = (0, 1, 0, -1)$, $\mathbf{x}_6 = (0, 0, 1, -1)$. These vectors lie in a linear hyperplane of $\mathbb{R}^4$, and, as is easily checked, they form a linear realization of the oriented matroid $\mathcal{M}_0$. Thus

$$\mathcal{Z}_0 = [-\mathbf{x}_1, \mathbf{x}_1] + [-\mathbf{x}_2, \mathbf{x}_2] + [-\mathbf{x}_3, \mathbf{x}_3] + [-\mathbf{x}_4, \mathbf{x}_4] + [-\mathbf{x}_5, \mathbf{x}_5] + [-\mathbf{x}_6, \mathbf{x}_6]$$

is the 3-dimensional zonotope depicted in Figure 2.2.2. We now translate this zonotope by the vector $(5, 5, 5, 5)$, and afterwards we scale it by a factor of $1/2$. The vertices of the resulting zonotope $\mathcal{Z}_0' := \frac{1}{2}(\mathcal{Z}_0 + (5, 5, 5, 5))$ are precisely the vectors of the form $(\pi_1, \pi_2, \pi_3, \pi_4)$ where π ranges over all 24 permutations of $\{1, 2, 3, 4\}$. In other words, the permutohedron $\mathcal{Z}_0'$ is the convex hull of all points in 4-space whose distinct coordinates are $1, 2, 3, 4$ in some order.

We now explain the structure of the permutohedron by describing the bijection between its vertices and the maximal covectors of $\mathcal{M}_0$. For instance, for the covector $a = (+ + - - - -)$ we compute the vertex

$$\text{face}(a) = \frac{1}{2}\left(\mathbf{x}_1 + \mathbf{x}_2 - \mathbf{x}_3 - \mathbf{x}_4 - \mathbf{x}_5 - \mathbf{x}_6 + (5, 5, 5, 5)\right) = (3, 1, 2, 4).$$

The adjacent vertex, labeled $b = (+ - - - - -)$, equals

$$\text{face}(b) = \frac{1}{2}\left(\mathbf{x}_1 - \mathbf{x}_2 - \mathbf{x}_3 - \mathbf{x}_4 - \mathbf{x}_5 - \mathbf{x}_6 + (5, 5, 5, 5)\right) = (2, 1, 3, 4).$$

The edge connecting vertex a and vertex b is parallel to $\mathbf{x}_2 = (1, 0, -1, 0)$, and in the symmetric group this corresponds to two permutations which differ by a transposition of adjacent values in the first and third positions. In a similar manner all 24 vertices of $\mathcal{Z}_0'$ are labeled with maximal covectors of $\mathcal{M}_0$ and can be identified with the permutations of $\{1, 2, 3, 4\}$, and all 36 edges can be identified with the pairs of permutations that differ by a transposition of i and $i+1$ for some i. The permutohedron $\mathcal{Z}_0'$ has 14 facets, corresponding to cocircuits of $\mathcal{M}_0$ (or cuts of the complete graph K_4 in Figure 1.1.1).

In analogy with this example, there exists an n-permutohedron for the symmetric group S_n on $\{1, 2, \ldots, n\}$. It is defined as the convex hull of all vectors $(\pi_1, \pi_2, \ldots, \pi_n)$ such that $\pi \in S_n$. This is an $(n-1)$-dimensional zonotope whose oriented matroid is the graphic matroid of the complete graph K_n, and whose associated arrangement is a special case of a reflection arrangement. As will be seen in Section 2.3, all reflection arrangements have the property that their maximal cells are simplices. By polarity, this implies that the n-permutohedron is a simple $(n-1)$-polytope. We can see in Figure 2.2.2 that the 4-permutohedron $\mathcal{Z}_0$ is indeed a simple 3-polytope.

Permutohedra were introduced and studied by Schoute (1911). See Exercises 2.9 and 2.10 for some additional details.

We return to our general discussion, and let $\mathbf{x}_1, \ldots, \mathbf{x}_n$, $\mathcal{Z}$ and $\mathcal{M}$ be as before. The next topic is a geometric description in terms of the r-zonotope $\mathcal{Z}$ for the oriented matroid operations of *contraction* and *deletion*. Consider the linear hyperplane H_i which is perpendicular to the vector $\mathbf{x}_i$, and let $\pi_i : \mathbb{R}^r \to H_i$ denote the orthogonal projection onto H_i. The projected vector configuration

$$\pi_i(\mathbf{x}_1), \ldots, \pi_i(\mathbf{x}_{i-1}), \pi_i(\mathbf{x}_{i+1}), \ldots, \pi_i(\mathbf{x}_n)$$

is a realization of the rank $r-1$ oriented matroid $\mathcal{M}/i$ by contraction of i. Its $(r-1)$-zonotope is the image $\pi_i(\mathcal{Z})$ of $\mathcal{Z}$ under the projection π_i. Note that the preimage in $\mathcal{Z}$ of the boundary of $\pi_i(\mathcal{Z})$ consists of all faces of $\mathcal{Z}$ which have $[-\mathbf{x}_i, \mathbf{x}_i]$ as a Minkowski summand. This subcomplex of the boundary of $\mathcal{Z}$ is called the i-th *zone*. We have proved part (a) of the following:

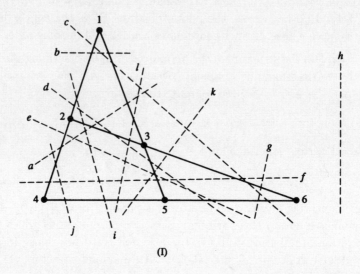

(I)

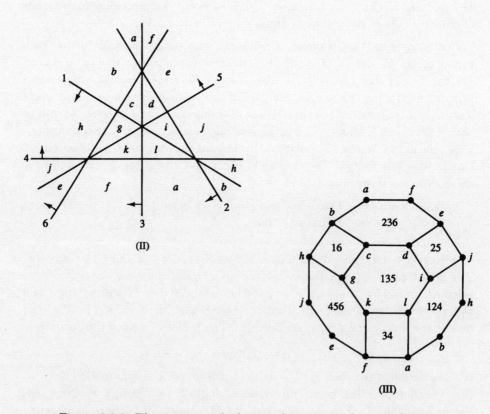

(II)

(III)

Figure 2.2.2: The 4-permutohedron and its oriented matroid $\mathcal{M}_0$.

2.2.6 Proposition.

(a)   *The oriented matroid operation of contracting an element i corresponds to the zonotope operation of projecting along the i-th zone.*

(b)   *The oriented matroid operation of deleting an element i corresponds to the zonotope operation of contracting the i-th zone.*

The operation of "contracting the i-th zone" in part (b) is understood as follows. If the point $\mathbf{x}_i$ is deleted from the set of generators, then the resulting zonotope $\mathcal{Z}_i$ has the property that $\mathcal{Z}_i + [-\mathbf{x}_i, \mathbf{x}_i] = \mathcal{Z}$. Geometrically speaking, the zonotope $\mathcal{Z}_i$ is obtained from $\mathcal{Z}$ by simultaneously contracting all edges which are parallel to $\mathbf{x}_i$.

The *vertex figures* of the r-zonotope $\mathcal{Z}$ have a nice interpretation in terms of projective geometry. As before, we fix an affine embedding of the point configuration associated with $\mathbf{X}$. (In oriented matroid language: "assume w.l.o.g. that $\mathcal{M}$ is acyclic".) A hyperplane having all points $\mathbf{x}_i$ on its positive side corresponds to the maximal covector $(+++\ldots+)$ of $\mathcal{M}$. Consider the vertex of $\mathcal{Z}$ which is labeled $(+++\ldots+)$. This vertex, which is thought of as the *base vertex* of the zonotope, plays the same role here as the *base tope* of an arrangement in Section 4.2. It follows from the above construction of affine space $\mathbb{A}^{r-1}$ that the vertex figure of $\mathcal{Z}$ at the base vertex is equal to the convex hull of the $(r-1)$-dimensional affine configuration associated with $\mathbf{X}$.

If we apply an admissible projective transformation to the affine point set $\mathbf{X}$, then the hyperplane at infinity H_∞ gets mapped to a hyperplane H not containing any point of $\mathbf{X}$. Let $C \in \{+, -\}^n$ be the maximal covector of $\mathcal{M}$ corresponding to the hyperplane H. The image of $\mathbf{X}$ under this projective transformation is a new affine point configuration $\mathbf{X}_C$ whose oriented matroid is obtained from $\mathcal{M}$ by reorientation of the set C^-. Consider the vertex of $\mathcal{Z}$ which is labeled C. Now, the vertex figure of $\mathcal{Z}$ at this vertex is equal to the convex hull of $\mathbf{X}_C$. We have derived the following result.

2.2.7 Proposition.   *The set of $(r-1)$-polytopes obtained from an affine point configuration $\mathbf{X}$ by applying all admissible projective transformations equals the set of vertex figures of its associated zonotope $\mathcal{Z}$.*

Proposition 2.2.7 can also be understood from the perspective of hyperplane arrangements. The choice of an affine embedding for $\mathbf{X}$ (i.e. the choice of H_∞) corresponds to the choice of a specific point p_0 in the interior of some region R_0 of the hyperplane arrangement polar to $\mathbf{X}$. The point p_0 has the property that the polytopes R_0 and $conv(\mathbf{X})$ are geometrically polar with respect to the center of polarity p_0. Now, consider any point p_1 in some other region R_1 of the arrangement. The polar polytope of R_1 with respect to the center of polarity p_1 is the vertex figure of the zonotope $\mathcal{Z}$ at the vertex labeled R_1. (Here R_1 is identified with a covector of $\mathcal{M}$.) This vertex figure is the $(r-1)$-polytope obtained from $\mathbf{X}$ by applying the projective transformation induced by the choice of p_1.

Many theorems about hyperplane arrangements have interesting reformulations for zonotopes. For instance, we can count the face numbers of zonotopes using the results for arrangements in Section 4.6. It is an interesting fact that the f-vector of a zonotope depends only on the *underlying matroid* of its associated oriented matroid (see Corollary 4.6.4 and also Proposition 9.3.11).

Here are two other examples of such reformulations. Recall that a vertex of an r-polytope P is called *simple* if it is contained in exactly r facets of P. Shannon's Theorem (via Corollary 2.1.7) implies the following:

2.2.8 Proposition.　*Every r-zonotope with n zones has at least $2n$ simple vertices.*

The tope graph of a hyperplane arrangement (Definition 2.1.4) equals the 1-skeleton of the associated zonotope. As a direct consequence of a result of Björner, Edelman and Ziegler (Theorem 4.2.14) we find that zonotopes are determined by their 1-skeletons.

2.2.9 Proposition.　*Two zonotopes are combinatorially equivalent if and only if their 1-skeletons are isomorphic graphs.*

The conclusion of Proposition 2.2.9 is also known to hold for all simple polytopes. This had been conjectured by M. Perles, and was proved by Blind and Mani (1987) and later by Kalai (1988a).

The following surprising connection between zonotopes and matroid theory was discovered by Shephard (1974) and McMullen (1975); see also White (1987, Prop. 3.3.4). Recall that an (oriented) matroid $\mathcal{M}$ is *unimodular* if there exists a realization matrix $\mathbf{X}$ of $\mathcal{M}$ such that all maximal minors of $\mathbf{X}$ are equal to -1, 0 or $+1$.

2.2.10 Proposition.　*An r-zonotope $\mathcal{Z}$ tiles $\mathbb{R}^r$ if and only if its oriented matroid $\mathcal{M}$ is unimodular.*

Here "$\mathcal{Z}$ tiles $\mathbb{R}^r$" means there exists a polyhedral subdivision of $\mathbb{R}^r$ all of whose maximal cells are translates of $\mathcal{Z}$. Note that Proposition 2.2.10 implies that 3-dimensional space can be tiled with the 4-permutohedron (Example 2.2.5). For, the oriented matroid $\mathcal{M}_0$ of the 4-permutohedron is graphic and hence unimodular. Similarly, $\mathbb{R}^r$ can be tiled with the $(r+1)$-permutohedron spanned by all permutations of $\{1, 2, \ldots, r+1\}$.

While in Proposition 2.2.10 the zonotope $\mathcal{Z}$ is used as a tile, we will now turn our attention to decompositions of $\mathcal{Z}$ itself. A polyhedral subdivision Δ of a zonotope $\mathcal{Z}$ is called a *zonotopal subdivision* if each cell of Δ is a zonotope whose edges are translates of edges of $\mathcal{Z}$. If all cells of Δ are (affine images of) cubes, then Δ is called a *cubical subdivision* of $\mathcal{Z}$. Two zonotopal subdivisions of a 2-zonotope with six zones are depicted in Figure 2.2.3. The decomposition of the 12-gon into parallelograms ($= $ 2-cubes) is a cubical subdivision.

We next describe a general method for producing zonotopal subdivisions of

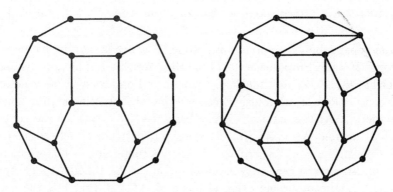

Figure 2.2.3: Two zonotopal subdivisions of a 2-zonotope.

an r-zonotope with n zones. Let $\mathcal{Z}'$ be any $(r+1)$-zonotope with $n+1$ zones such that $\mathcal{Z}$ is obtained from $\mathcal{Z}'$ by projection along the $(n+1)$-st zone. If $\mathcal{M}$ and $\mathcal{M}'$ denote the respective oriented matroids, then, by Proposition 2.2.6, $\mathcal{M} = \mathcal{M}'/(n+1)$ is the contraction of $\mathcal{M}'$ by $n+1$. Suppose that

$$\mathcal{Z} \;=\; [-\mathbf{x}_1, \mathbf{x}_1] + [-\mathbf{x}_2, \mathbf{x}_2] + \ldots + [-\mathbf{x}_n, \mathbf{x}_n].$$

Then we can write any $(r+1)$-zonotope $\mathcal{Z}'$ which projects along an edge onto $\mathcal{Z}$ as

$$\mathcal{Z}' = [-\mathbf{x}_1 - h_1 \mathbf{e}_{r+1}, \; \mathbf{x}_1 + h_1 \mathbf{e}_{r+1}] + \ldots$$
$$+ \, [-\mathbf{x}_n - h_n \mathbf{e}_{r+1}, \; \mathbf{x}_n + h_n \mathbf{e}_{r+1}] + [-\mathbf{e}_{r+1}, \mathbf{e}_{r+1}],$$

where $h_1, h_2, \ldots, h_n \in \mathbb{R}$. Here it is assumed that the $(n+1)$-st zone of $\mathcal{Z}'$ is parallel to the $(r+1)$-st standard basis vector in $\mathbb{R}^{r+1}$. Given any sign vector $\sigma = (\sigma_1, \ldots, \sigma_n) \in \{-, 0, +\}^n$, we write $(\sigma, +)$ for the extended sign vector $(\sigma_1, \ldots, \sigma_n, +)$. Using the abbreviation of Proposition 2.2.2, $\mathrm{face}(\sigma) \subset \mathbb{R}^r$ and $\mathrm{face}(\sigma, +) \subset \mathbb{R}^{r+1}$ are zonotopes inside of $\mathcal{Z}$ and $\mathcal{Z}'$ respectively.

2.2.11 Proposition. *The set of zonotopes*

$$\mathcal{T}(\mathcal{M}') \;=\; \{\, \mathrm{face}(\sigma) \mid (\sigma, +) \text{ is a covector of } \mathcal{M}' \,\}$$

is a zonotopal subdivision of $\mathcal{Z}$.

Proof. Every cell $\mathrm{face}(\sigma)$ is a zonotope whose edges are translates of $[-\mathbf{x}_i, \mathbf{x}_i]$. Hence we need only show that $\mathcal{T}(\mathcal{M}')$ is a polyhedral subdivision of $\mathcal{Z}$. For $\tau \in \{-, 0, +\}$ consider the following set of zonotopes in $\mathcal{Z}'$:

$$\mathcal{T}_\tau(\mathcal{M}') \;=\; \{\, \mathrm{face}(\sigma, \tau) \mid (\sigma, \tau) \text{ is a covector of } \mathcal{M}' \,\}.$$

By Proposition 2.2.2, $\mathcal{T}_-(\mathcal{M}') \cup \mathcal{T}_0(\mathcal{M}') \cup \mathcal{T}_+(\mathcal{M}')$ equals the boundary complex of $\mathcal{Z}'$. Its subcomplex $\mathcal{T}_0(\mathcal{M}')$ is the $(n+1)$-st zone of $\mathcal{Z}'$. By definition, this is the shadow boundary of the projection π which maps $\mathcal{Z}'$ onto $\mathcal{Z}$. The subcomplex $\mathcal{T}_+(\mathcal{M}')$ is a polyhedral r-ball consisting of all faces above the shadow boundary $\mathcal{T}_0(\mathcal{M}')$, while $\mathcal{T}_-(\mathcal{M}')$ consists of all faces below $\mathcal{T}_0(\mathcal{M}')$. Using standard and intuitively obvious results from polytope theory, this implies that the image of $\mathcal{T}_+(\mathcal{M}')$ under π is a polyhedral subdivision of $\mathcal{Z}$. The proof of Proposition 2.2.11 is completed by observing that $\pi(\text{face}(\sigma, +)) = \text{face}(\sigma)$. $\qquad\square$

A zonotopal subdivision of an r-zonotope $\mathcal{Z}$ which is induced by an $(r+1)$-zonotope $\mathcal{Z}'$ as in Proposition 2.2.11 is called *regular*. Let us now assume that the heights $h_1, \ldots, h_n$ are sufficiently generic real numbers. Using general facts about contractions of matroids, we can see that the resulting regular zonotopal subdivision $\mathcal{T}(\mathcal{M}')$ is cubical: Every basis $\{i_1, i_2, \ldots, i_r\}$ of the contracted (oriented) matroid $\mathcal{M} = \mathcal{M}'/(n+1)$ spans a hyperplane of $\mathcal{M}'$ not containing $n+1$. Conversely, every hyperplane of $\mathcal{M}'$ not containing $n+1$ is spanned by a basis of $\mathcal{M}$. By the genericity assumption on the h_i, the hyperplane spanned by $i_1, i_2, \ldots, i_r$ contains no other element of $\mathcal{M}'$. If $(\sigma, +)$ is the cocircuit complementary to this r-element hyperplane of $\mathcal{M}'$, then the facet $\text{face}(\sigma, +)$ of $\mathcal{Z}'$ is an r-cube. Its image $\text{face}(\sigma)$ under the projection π is also an r-cube, namely, it is a translate of $[-\mathbf{x}_{i_1}, \mathbf{x}_{i_1}] + [-\mathbf{x}_{i_2}, \mathbf{x}_{i_2}] + \ldots + [-\mathbf{x}_{i_r}, \mathbf{x}_{i_r}]$. Hence all maximal faces of the induced subdivision $\mathcal{T}(\mathcal{M}')$ are r-cubes, which means that $\mathcal{T}(\mathcal{M}')$ is a regular cubical subdivision of $\mathcal{Z}$. As a consequence of this construction, we obtain the following well-known volume formula for zonotopes. This formula appears as equation (57) in Shephard (1974) and is there attributed to McMullen.

2.2.12 Proposition (P. McMullen, see Shephard 1974). *The volume of the zonotope* $\mathcal{Z}(\mathbf{X})$ *equals*

$$2^r \cdot \sum_{1 \le i_1 < i_2 < \ldots < i_r \le n} |\det(\mathbf{x}_{i_1}, \mathbf{x}_{i_2}, \ldots, \mathbf{x}_{i_r})|.$$

An analogue of Proposition 2.2.12 for counting the number of lattice points in a lattice zonotope is found in Stanley (1980b), and is applied to the problem of counting degree sequences of graphs in Stanley (1991). See also Stanley (1981) for an interesting matroid-theoretic application of mixed volumes of zonotopes.

We have seen that the projections of a zonotope along its edges are in one-to-one correspondence with the one-element contractions of its associated oriented matroid. This implies that the map $\mathcal{M}' \to \mathcal{T}(\mathcal{M}')$ is a bijection between the single element liftings of $\mathcal{M}$ which are realizable for $\mathbf{X}$ and the regular zonotopal subdivisions of $\mathcal{Z}(\mathbf{X})$. Using realization spaces as in Chapter 8 (or strata as in Section 2.4), this statement can be rephrased as follows: The regular zonotopal subdivisions of $\mathcal{Z}$ are in one-to-one correspondence with the oriented matroids $\mathcal{M}'$ on E_{n+1} such that

(i) $\mathcal{M} = \mathcal{M}'/(n+1)$,

(ii) $\mathcal{M}'$ is realizable, and

(iii) $\mathbf{X}$ lies in the image of the contraction map $\gamma_{n+1} : \mathcal{R}(\mathcal{M}') \to \mathcal{R}(\mathcal{M})$ between the two realization spaces.

If we drop the hypothesis (iii) then we will get some zonotopal subdivisions $\mathcal{T}(\mathcal{M}')$ which are not regular. For instance, it can be shown that the subdivisions in Figure 2.2.3 are regular for certain 2-zonotopes with 6 zones, but *not* for all. More precisely, while (ii) is still satisfied for the respective rank 3 oriented matroids $\mathcal{M}'$ which induce the subdivisions (a) and (b), in both cases there are some realizations of $\mathcal{M}$ (= the uniform rank 2 oriented matroid on 6 elements) for which (iii) is violated. For instance, the zonotopal tiling (a) is regular if and only if its six edges, viewed as points on the projective line, are the projection of a quadrilateral set. This example is isomorphic to the one given in Figure 2.4.4.

This raises the question of whether there is a general combinatorial description of all zonotopal tilings. This question is answered in the affirmative by the following result of Dress [presented at the "Symposium on Combinatorics and Geometry", Stockholm, 1989]. Note that Theorem 2.2.13 has the flavor of a zonotopal analogue to the Topological Representation Theorem. The role played by non-realizable liftings shows the value of the oriented matroid point of view for the study of zonotopes.

2.2.13 Theorem (A. Dress, see Bohne 1992a and Section A.3) *Let* $\mathcal{Z}$ *be an r-zonotope with n zones, and let* $\mathcal{M}$ *be the corresponding oriented matroid on* E_n. *Then the map* $\mathcal{M}' \mapsto \mathcal{T}(\mathcal{M}')$ *is a bijection between the single element liftings* $\mathcal{M}'$ *of* $\mathcal{M}$ *and the zonotopal subdivisions of* $\mathcal{Z}$.

We will next take a look at how *oriented matroid duality* is expressed in the world of zonotopes. Given $\mathbf{X}$ as before, we let $\mathbf{X}^*$ be an $(n-r) \times n$-matrix whose row space is the orthogonal complement of the row space of $\mathbf{X}$. In addition, we assume that the rows of $\mathbf{X}$ and $\mathbf{X}^*$ are orthonormal bases for their respective row spaces. Then the *dual zonotope* to $\mathcal{Z} = \mathcal{Z}(\mathbf{X})$ is the $(n-r)$-zonotope $\mathcal{Z}^* = \mathcal{Z}(\mathbf{X}^*)$. Thus dual pairs of zonotopes are obtained by projecting the regular n-cube into perpendicular flats of complementary dimension. The duality of zonotopes was first studied and related to Gale transforms by McMullen (1971a). This technique is sometimes referred to as *zonal diagrams*.

By construction, the rank $n - r$ oriented matroid $\mathcal{M}^*$ of $\mathcal{Z}^*$ is dual to the rank r oriented matroid $\mathcal{M}$ of $\mathcal{Z}$. Since the covectors of $\mathcal{M}$ are the vectors of $\mathcal{M}^*$, this means that the faces of $\mathcal{Z}$ can be read off directly from the vectors of $\mathcal{M}^*$. (Recall that the vectors of $\mathcal{M}^*$ are the Radon partitions in its affine diagram.) As with Gale diagrams (cf. Chapter 9), the zonal diagram technique is particularly useful for studying r-dimensional zonotopes with at most $r + 3$ zones.

This combinatorial duality of zonotopes has a metric analogue which is easily understood from the exterior algebra point of view of Section 2.4. Up to signs, the $r \times r$-determinants occurring in the formula in Proposition 2.2.12 are the Plücker coordinates of both the r-flat corresponding to $\mathcal{Z}$ and the $(n - r)$-flat corresponding to $\mathcal{Z}^*$. This implies a result of McMullen (1984) which states that the r-dimensional volume of $\mathcal{Z}$ is equal to the $(n - r)$-dimensional volume of it dual zonotope $\mathcal{Z}^*$ (after dividing by the fudge factors of 2^r and 2^{n-r}, which stem from the fact that our unit n-cube has volume 2^n).

We close this section with a well-known result, mentioned in Coxeter (1962), whose proof is hard to find in the literature: zonotopes are characterized as the polytopes with centrally symmetric 2-faces.

2.2.14 Proposition.   *For a polytope P, the following conditions are equivalent:*

(1)   *P is a zonotope;*

(2)   *every edge of P is a Minkowski summand of P;*

(3)   *every 2-dimensional face of P is a zonotope;*

(4)   *every 2-dimensional face of P is centrally symmetric.*

Proof. By Proposition 2.2.2, every face of a zonotope $\mathcal{Z}$ is again a zonotope, and is a Minkowski summand of $\mathcal{Z}$. It follows directly from the definition that zonotopes are centrally symmetric. Hence (1) implies (2),(3), and (4).

The implication "(2) $\Rightarrow$ (1)" can be proved by induction on the number of faces of P. Suppose (2), and let $[\mathbf{u}_1, \mathbf{u}_2]$ be any edge of P. Then we can write $P = P' + [\mathbf{u}_1, \mathbf{u}_2]$ for some polytope P'. If $[\mathbf{v}_1, \mathbf{v}_2]$ is an edge of P', then, for either $i = 1$ or $i = 2$, $[\mathbf{u}_i + \mathbf{v}_1, \mathbf{u}_i + \mathbf{v}_2]$ is an edge of P. Since $[\mathbf{u}_i + \mathbf{v}_1, \mathbf{u}_i + \mathbf{v}_2]$ is a Minkowski summand of P, we find that $[\mathbf{v}_1, \mathbf{v}_2]$ is also a Minkowski summand of P'. Now, since P' is a proper Minkowski summand of P, the number of faces of P' is strictly less than the number of faces of P. By the induction hypothesis, we know that P' is a zonotope. Consequently P is a zonotope, completing the proof of "(2) $\Rightarrow$ (1)".

A polygon in the plane is centrally symmetric if and only if for every edge there is another parallel edge of the same length. But this means that every edge is a Minkowski summand, and therefore (4) implies (3).

It remains to prove the implication "(3) $\Rightarrow$ (2)". We use induction on the dimension. If P is a 2-polytope satisfying (3), then clearly (2) follows. Let P be a polytope of dimension $r \geq 3$, and suppose that "(3) $\Rightarrow$ (2)" holds for all polytopes of dimension $\leq r - 1$. Hence all facets (and all proper faces) of P are zonotopes. We need to show that under this assumption P must also be a zonotope.

Let e_0 be any edge of P, F_0 a facet of P containing e_0, P' the projection of P along e_0, and let F_0' be the projection of F_0 along e_0. We claim that every facet of P' is the projection of a facet of P which has e_0 as a Minkowski summand.

By construction, this claim is true for the facet F_0' of P'. Next consider some other facet F_1' of P' which intersects F_0'. Then $F_0' \cap F_1'$ contains a vertex v_1 of P'. Since $v_1 \in F_0'$, v_1 is the projection of an edge e_1 of P that is parallel to e_0. But this means that e_1 is an edge of the preimage F_1 of F_1', and hence e_0 is a summand of F_1. Now let F_2' be a facet of P' which intersects F_1'. We can apply the same argument as before to conclude that F_2' is the image of a facet F_2 of P which has e_1 and hence e_0 as a Minkowski summand. Since $r \geq 3$, the boundary of the $(r - 1)$-polytope P' is connected, whence the claim follows by induction.

The width of the polytope P in the direction of the edge e_0 over a given point in P' is a convex function on P'. We have shown that this function is

$\geq \text{length}(e_0)$ on the boundary of P', and hence it is $\geq \text{length}(e_0)$ on all of P'. This proves that e_0 is a Minkowski summand of P. $\qquad\square$

2.3 Reflection arrangements

The arrangements of reflecting hyperplanes of finite reflection groups have a lot of special structure. Due to their importance in algebra (in connection with root systems in the representation theory of semi-simple Lie algebras) these arrangements have long been studied, and an enormous amount of information is available. We will here review some special properties of reflection arrangements which are of interest from a geometric and combinatorial point of view, namely,

(a)   their classification,

(b)   the combinatorial structure of regions,

(c)   the face lattice (also known as the Coxeter complex),

(d)   the characteristic polynomial.

In what ways are reflection arrangements relevant for the study of oriented matroids? First of all, they provide interesting examples of highly structured hyperplane arrangements, and thus of oriented matroids. Second, part of their well-developed structure theory generalizes to oriented matroids and is an important source of motivation. The theory of reflection arrangements and finite reflection groups can be developed in the abstract setting of oriented matroids. It is shown in Björner and Ziegler (1992b) that such "reflection oriented matroids" are realizable.

As a general technical reference for reflection groups and reflection arrangements (root systems) we recommend Bourbaki (1968), which contains a wealth of information. More readable introductions to the subject are given in the first three chapters of Brown (1989), in Humphreys (1990), and also at a more leisurely pace in Benson and Grove (1985).

Throughout this section, we consider $\mathbb{R}^d$ equipped with a fixed positive definite inner product $\langle \cdot, \cdot \rangle$. Every hyperplane H determines a non-singular linear mapping σ_H by

$$(2.3.1) \qquad \sigma_H(\mathbf{x}) := \mathbf{x} - 2\frac{\langle \mathbf{x}, \mathbf{r} \rangle}{\langle \mathbf{r}, \mathbf{r} \rangle}\mathbf{r},$$

where $\mathbf{r}$ is any non-zero vector orthogonal to H. Then $\sigma_H(\mathbf{x}) = \mathbf{x}$ for all $\mathbf{x} \in H$ and $\sigma_H(\mathbf{r}) = -\mathbf{r}$, so we recognize that σ_H is the *orthogonal reflection* through H (see Figure 2.3.1).

Each reflection σ_H is an element of the orthogonal group $O(d)$. Thus, every arrangement of hyperplanes $\mathcal{A}$ in $\mathbb{R}^d$ determines a subgroup $G(\mathcal{A}) \subset O(d)$, namely the subgroup $\langle \sigma_H : H \in \mathcal{A} \rangle$ generated by the reflections determined by $\mathcal{A}$. Note that $G(\mathcal{A})$ may be infinite even if $\mathcal{A}$ is finite, and that $G(\mathcal{A})$ may contain other reflections in addition to those coming from $\mathcal{A}$.

2.3.1 Definition.   A finite group of the form $G(\mathcal{A})$ is called a (finite) *reflection group*. An arrangement of hyperplanes is called a *reflection arrangement* (or

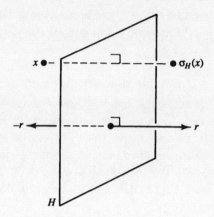

Figure 2.3.1: Orthogonal reflection through H.

Coxeter arrangement) if its associated reflections are all the reflections belonging to some reflection group.

The mapping $\mathcal{A} \mapsto G(\mathcal{A})$ clearly gives a one-to-one correspondence between reflection arrangements and reflection groups. Reflection groups have been classified (up to natural equivalence), and this is therefore true also for reflection arrangements. Before discussing this classification we will look at two examples.

2.3.2 Example. The $d = 2$ case is very instructive. Consider two lines ℓ_1 and ℓ_2 through the origin of $\mathbb{R}^2$, and assume that they form the angle α (see Figure 2.3.2).

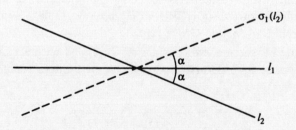

Figure 2.3.2: Sketch for Example 2.3.2.

Let σ_1 and σ_2 be the corresponding reflections. As a simple exercise in "anschauliche Geometrie" one computes that the product $\sigma_1\sigma_2$ is the counterclockwise rotation of the plane by an angle of 2α, and that $\sigma_1\sigma_2\sigma_1$ is the reflection through the line $\sigma_1(\ell_2)$. Hence, the group $G(\{\ell_1, \ell_2\})$ will contain rotations through all angles $2k\alpha$, and reflections through all lines at angles $k\alpha$ from ℓ_2, for all $k \in \mathbb{Z}$. It follows that $G(\{\ell_1, \ell_2\})$ is finite if and only if α is a rational multiple of 2π. We deduce the following description of reflection arrangements in $\mathbb{R}^2$:

*For each integer $m \geq 1$ there is a unique (up to rotation) reflection arrange-
ment called $\mathcal{I}_2(m)$ in $\mathbb{R}^2$ consisting of m lines through the origin forming successive angles of π/m.*

The corresponding reflection group $I_2(m) = G(\mathcal{I}_2(m))$ is the dihedral group
of order $2m$ (having m rotations and m reflections). Choose a region (such as
the shaded one in Figure 2.3.3), and let σ_1 and σ_2 be the reflections through its
bounding lines. Using the above geometric ideas, it can be shown that

$$(2.3.2) \qquad I_2(m) = <\sigma_1, \sigma_2 | \sigma_1^2 = \sigma_2^2 = (\sigma_1\sigma_2)^m = id>,$$

meaning that σ_1 and σ_2 generate $I_2(m)$ and that all relations are generated by
σ_1, σ_2 and $\sigma_1\sigma_2$ raised to their respective order. The general significance of this
will soon be seen.

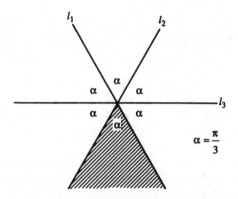

Figure 2.3.3: The 2-dimensional reflection arrangement $\mathcal{I}_2(3)$.

2.3.3 Example. For $1 \leq i < j \leq n$ we consider the hyperplane

$$H_{ij} := \{\mathbf{x} \in \mathbb{R}^n : x_i = x_j\}$$

in $\mathbb{R}^n$. The corresponding reflection σ_{ij} acts on a point $\mathbf{x} = (x_1, \ldots, x_n) \in \mathbb{R}^n$ by
switching its i-th and j-th coordinates. From this it follows that the arrangement
$\mathcal{A}_{n-1} = \{H_{ij} : 1 \leq i < j \leq n\}$ generates the symmetric group S_n, acting on
$\mathbb{R}^n$ by permuting coordinates. The group isomorphism $G(\mathcal{A}_{n-1}) \cong S_n$ is easy
to observe on the set $\{\mathbf{e}_1, \mathbf{e}_2, \ldots, \mathbf{e}_n\} \subset \mathbb{R}^n$ of standard basis vectors, which
is permuted by $G(\mathcal{A}_{n-1})$. No additional reflections belong to $G(\mathcal{A}_{n-1})$, since
every element of order two in the symmetric group is a product of transpositions
σ_{ij} corresponding to reflections in orthogonal hyperplanes. Hence $\mathcal{A}_{n-1}$ is a
reflection arrangement.

Note that the arrangement $\mathcal{A}_{n-1}$ is not essential since the line $x_1 = x_2 = \ldots =
x_n$ is contained in every H_{ij}. Let

$$H = \{\mathbf{x} \in \mathbb{R}^n : x_1 + x_2 + \ldots + x_n = 0\}$$

be the orthogonal complement to this line, and define $\mathcal{A}'_{n-1} := \{H'_{ij} = H \cap H_{ij} :
H_{ij} \in \mathcal{A}_{n-1}\}$. The $(n-1)$-dimensional arrangement $\mathcal{A}'_{n-1}$ is essential. (In fact,

the subscript "$n - 1$" stands to indicate that the true dimension (or *rank*) of $\mathcal{A}_{n-1}$ and $\mathcal{A}'_{n-1}$ is $n - 1$.)

The arrangement $\mathcal{A}'_{n-1}$ has $n!$ regions which are in a natural correspondence to the elements of its associated group S_n. More precisely, a permutation $\pi \in S_n$ corresponds to the region

$$R_\pi := \{\mathbf{x} \in H \,|\, x_{\pi(1)} < x_{\pi(2)} < \ldots < x_{\pi(n)}\}.$$

One also sees that two regions R_π and $R_{\pi'}$ are adjacent if and only if π and π' differ by a transposition of the values in two adjacent positions. The zonotope polar to the arrangement $\mathcal{A}'_{n-1}$ is the n-permutohedron described in Example 2.2.5.

Now choose a fixed region B, for instance

$$B := R_{id} = \{\mathbf{x} \in H \,|\, x_1 < x_2 < \ldots < x_n\}.$$

Then B is a simplicial cone whose $n - 1$ bounding hyperplanes are $H_{i,i+1}, 1 \leq i \leq n-1$. The corresponding reflections $\sigma_{i,i+1}$ are the so-called *adjacent transpositions*, which clearly generate S_n. The following presentation of the symmetric group in terms of these generators, which we abbreviate $\sigma_i := \sigma_{i,i+1}$, is easy to derive combinatorially. It also follows from the theory underlying Theorem 2.3.7 below.

$$(2.3.3) \qquad S_n = G(\mathcal{A}_{n-1}) = \langle \sigma_1, \ldots, \sigma_{n-1} : \sigma_i^2 = (\sigma_j \sigma_{j+1})^3 = (\sigma_k \sigma_l)^2 = id$$
$$\text{for all } i, j, k, l \text{ such that } |k - l| \geq 2 \rangle.$$

Reflection groups have a very useful abstract characterization, which we now explain. A *Coxeter diagram* is defined to be a finite graph for which some of the edges (or none) are labeled by integers ≥ 4 or by ∞. Figure 2.3.4 shows three Coxeter diagrams.

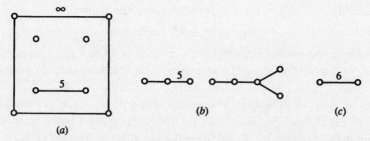

(a)

(b)   (c)

Figure 2.3.4. Examples of Coxeter diagrams.

Given a Coxeter diagram Γ with vertex set S, we define a function m_Γ : $S \times S \to \mathbb{Z}^+ \cup \{\infty\}$ by:

$$(2.3.4) \quad m_\Gamma(s, s') := \begin{cases} 1, & \text{if } s = s', \\ 2, & \text{if } s \neq s' \text{ and no edge joins } s \text{ and } s', \\ 3, & \text{if } s \neq s' \text{ and an unmarked edge joins } s \text{ and } s', \\ m, & \text{if } s \neq s' \text{ and an edge marked } m \text{ joins } s \text{ and } s'. \end{cases}$$

2.3.4 Definition. The *Coxeter group* W_Γ determined by a Coxeter diagram Γ as above has as generators the set S and as defining relations:

(1) $s^2 = id$ for all $s \in S$, and

(2) $(ss')^{m_\Gamma(s,s')} = id$ for all $s \neq s' \in S$.

In other words, it is defined by the presentation

$$(2.3.5) \qquad W_\Gamma \;=\; \langle S : (ss')^{m_\Gamma(s,s')} = id \text{ for all } s, s' \in S \rangle.$$

For instance, the diagram (c) in Figure 2.3.4 determines the dihedral group of order 12. It is clear from (2.3.2) that the dihedral groups are the Coxeter groups with diagram $\circ\!\overset{m}{-\!\!-}\!\circ$. Also, (2.3.3) shows that the symmetric group S_n is a Coxeter group with diagram $\circ\!\!-\!\!-\!\circ\!\!-\!\!-\!\circ\!-\ldots-\!\circ\!\!-\!\!-\!\circ$ ($n-1$ nodes).

Reflection groups and finite Coxeter groups are equivalent as classes of mathematical objects. Some details of the cryptomorphism between them will be sketched after the following two propositions.

2.3.5 Proposition. *Let $\mathcal{A}$ be a reflection arrangement with group $G = G(\mathcal{A})$. Then*

(i) $\gamma(\mathcal{A}) = \mathcal{A}$ *for every $\gamma \in G$;*

(ii) *for any pair R', R'' of regions of $\mathcal{A}$ there exists a unique element $\gamma \in G$ such that $\gamma(R') = R''$;*

(iii) *the number of regions equals the order of G.*

For a proof, see the cited references, e.g. Section 1.5 of Brown (1989). The property expressed by (i) is equivalent to being a reflection arrangement (Exercise 2.13).

2.3.6 Proposition. *Reflection arrangements are simplicial, i.e., every region is a simplicial cone.*

Proof. By Corollary 2.1.7 there exists *some* simplicial region B. Part (ii) of the preceding proposition shows that the group G provides isometries γ sending B onto every other region, hence all regions are isometric and in particular simplicial. $\qquad\square$

Let us now describe the cryptomorphisms. Start with a reflection group $G \subset O(d)$ with associated reflection arrangement $\mathcal{A}$, which we may assume to be essential in $\mathbb{R}^d$. Choose a base region B (often called a "fundamental chamber"). The choice is algebraically and geometrically irrelevant according to Proposition 2.3.5, and according to Proposition 2.3.6 the region B is supported by exactly d hyperplanes $H_1, H_2, \ldots, H_d$ such that $\dim(B \cap H_i) = d-1$. Let $\sigma_1, \sigma_2, \ldots, \sigma_d$ be the corresponding reflections. Then the set $S = \{\sigma_1, \sigma_2, \ldots, \sigma_d\}$ generates G as a Coxeter group. The Coxeter diagram Γ has vertex set S and an edge marked m_{ij} between σ_i and σ_j if and only if the dihedral angle between H_i and H_j equals π/m_{ij}. As is customary, $m_{ij} = 3$ is not written, and $m_{ij} = 2$ (i.e. H_i and H_j are orthogonal) is encoded by an omitted edge. It is clear from Example 2.3.2

that the angle between H_i and H_j must be of the form π/m with $m \in \{2, 3, \ldots\}$, since G is finite.

Conversely, start with a finite Coxeter group W_Γ of rank d. Identify the nodes $s_1, s_2, \ldots, s_d$ of the Coxeter diagram Γ arbitrarily with the standard basis vectors $\mathbf{e}_1, \mathbf{e}_2, \ldots, \mathbf{e}_d$ of $\mathbb{R}^d$ and define a symmetric bilinear form on $\mathbb{R}^d$ by

$$(2.3.6) \qquad\qquad b(\mathbf{e}_i, \mathbf{e}_j) := -\cos\left(\frac{\pi}{m_\Gamma(s_i, s_j)}\right)$$

for all $1 \leq i, j \leq d$. It turns out that this form $b(\cdot, \cdot)$ is positive definite, i.e., it is an inner product on $\mathbb{R}^d$. (In fact, this property of b is equivalent to W_Γ being finite.) Now, let $\mathcal{A}$ be the arrangement of hyperplanes orthogonal to $\mathbf{e}_1, \mathbf{e}_2, \ldots, \mathbf{e}_d$, and define reflections as in (2.3.1) but using the inner product $b(\cdot, \cdot)$. Let $G = G(\mathcal{A}) \subset O(d)$. Then $G \cong W_\Gamma$.

A more geometrical description of how to represent W_Γ as a reflection group is the following "kaleidoscope" method (Coxeter 1973). Take d "mirrors" in $\mathbb{R}^d$ and arrange them to form the boundary of a simplicial cone in such a way that the dihedral angle between the i-th and the j-th mirror is $\pi/m_\Gamma(s_i, s_j)$ for all $1 \leq i, j \leq d$. The reflections of this arrangement of mirrors will generate a group isomorphic to W_Γ, of which we can get an artistic impression by "looking" into the mirrored cone.

The previous discussion is summarized in the following important theorem.

2.3.7 Theorem.

(i) *Every reflection group G has a presentation as a Coxeter group W_Γ, such that the nodes of the Coxeter diagram Γ correspond to the reflections determined by the walls of a fixed chamber. Furthermore, Γ is uniquely determined by G up to isomorphism.*

(ii) *Every finite Coxeter group W_Γ has a faithful representation as a reflection group, such that the nodes of Γ correspond to the reflections determined by the walls of a fixed chamber. Furthermore, two such representations are conjugate as subgroups of $O(d)$.*

The identification expressed by Theorem 2.3.7 is a special case of a general correspondence between discrete reflection groups and (not necessarily finite) Coxeter groups. See Bourbaki (1968) or Humphreys (1990) for a general discussion, and Brown (1989) or Benson and Grove (1985) for the finite case.

After this general introduction we now come to the specific points mentioned at the beginning of the section.

(a) Classification

It was mentioned earlier that the finite irreducible reflection groups have been completely classified. This classification is due to Coxeter (1935). It directly translates into a classification of irreducible reflection arrangements. Here "irreducible" means that there is no non-trivial partition of the hyperplanes into two mutually orthogonal classes (equivalently, the Coxeter diagram is connected).

Observe from the discussion above that the Coxeter diagram of a reflection arrangement encodes all information about the shape of its regions, namely the angles of the facets of a simplicial cone isometric to all regions. The classification is given in Table 2.3.5. The subscript n indicates the number of nodes in the Coxeter diagram and hence the dimension of the corresponding essential arrangement.

All reflection arrangements in Table 2.3.5 arise "in nature" either from regular polytopes or from root systems, and sometimes from both.

A polytope is called *regular* if its symmetry group is transitive on the set of complete flags of faces (i.e., maximal chains in its face lattice). The classification of all regular polytopes has been known since antiquity for $d = 3$ (the Platonic solids) and since the 1800s in higher dimensions, see Coxeter (1973). It coincides with the list of *linear* Coxeter diagrams in Table 2.3.5, where the corresponding arrangement is the set of all reflectional symmetry planes of the regular polytope. Thus:

- $\mathcal{A}_n$ corresponds to the n-simplex (e.g., Figure 2.3.3 shows $\mathcal{A}_2$ which is the set of three lines of symmetry of an equilateral triangle and Figure 2.3.6 (a) shows $\mathcal{A}_3$, the six planes of symmetry of a regular tetrahedron);

- $\mathcal{BC}_n$ corresponds to the n-cube or its polar, the n-crosspolytope (Figure 2.3.6 (b) shows $\mathcal{BC}_3$, which is the set of nine planes of symmetry of the 3-cube);

- $\mathcal{F}_4$ corresponds to the 24-cell, a 4-dimensional self-polar polytope whose 24 facets are 3-crosspolytopes (= octahedra);

- $\mathcal{H}_3$ corresponds to the dodecahedron (having 12 pentagonal facets) and its polar, the icosahedron (having 20 triangular facets);

- $\mathcal{H}_4$ corresponds to the 120-cell and its polar, the 600-cell. These are 4-dimensional polytopes whose 120 (resp. 600) facets are dodecahedra (resp. tetrahedra);

- $\mathcal{I}_2(m)$, finally, corresponds to the regular convex m-gon in $\mathbb{R}^2$ (cf. Example 2.3.2). Note that $\mathcal{I}_2(3) = \mathcal{A}_2$ and $\mathcal{I}_2(4) = \mathcal{BC}_2$.

A *root system* is a finite spanning set Φ of non-zero vectors in $\mathbb{R}^d$ such that

(1)   if $\mathbf{v} \in \Phi$ then the only multiples of $\mathbf{v}$ in Φ are $\pm\mathbf{v}$;

(2)   if $\mathbf{v} \in \Phi$ then $\sigma_H(\Phi) = \Phi$, where $H = \mathbf{v}^\perp$;

(3)   if $\mathbf{v}, \mathbf{u} \in \Phi$ then $2\frac{\langle \mathbf{v}, \mathbf{u} \rangle}{\langle \mathbf{u}, \mathbf{u} \rangle}$ is an integer.

A root system is *irreducible* if there is no partition of the set Φ into two non-empty and mutually orthogonal classes. The irreducible root systems play a fundamental role in the classification and representation theory of semi-simple Lie algebras and Lie groups, see Bourbaki (1968) and Humphreys (1972).

The family of all hyperplanes orthogonal to the root vectors gives an arrangement $\mathcal{A}(\Phi)$ which contains all information about Φ except the length of the root vectors (which is sometimes important). Such an arrangement $\mathcal{A}(\Phi)$ is a reflection arrangement. The corresponding reflection group is called the *Weyl group* of Φ, the regions of $\mathcal{A}(\Phi)$ are called *Weyl chambers*, etc.

Name	Coxeter diagram	No. of hyperplanes	No. of regions
$\mathcal{A}_n$	o—o—o· · ·—o—o	$\binom{n+1}{2}$	$(n+1)!$
$\mathcal{BC}_n(n \geq 2)$	o—o—o· · ·—o—o $\overset{4}{—}$ o	n^2	$2^n \cdot n!$
$\mathcal{D}_n(n \geq 4)$	o—o—o· · ·—o—<	$n^2 - n$	$2^{n-1} \cdot n!$
$\mathcal{E}_6$	o—o—o—o—o (with branch)	36	$2^7 \cdot 3^4 \cdot 5$
$\mathcal{E}_7$	o—o—o—o—o—o (with branch)	63	$2^{10} \cdot 3^4 \cdot 5 \cdot 7$
$\mathcal{E}_8$	o—o—o—o—o—o—o (with branch)	120	$2^{14} \cdot 3^5 \cdot 5^2 \cdot 7$
$\mathcal{F}_4$	o—o $\overset{4}{—}$ o—o	24	1152
$\mathcal{H}_3$	o—o $\overset{5}{—}$ o	15	120
$\mathcal{H}_4$	o—o—o $\overset{5}{—}$ o	60	14400
$\mathcal{I}_2(m)(m \geq 5)$	o $\overset{m}{—}$ o	m	$2m$

Table 2.3.5: Classification of irreducible reflection arrangements

The classification of irreducible root systems coincides with the list of irreducible reflection arrangements in Table 2.3.5, except for $\mathcal{H}_3, \mathcal{H}_4$ and $\mathcal{I}_2(m)$ for $m \neq 3, 4, 6$ which do not arise from root systems. Actually there are two distinct dual root systems B_n and C_n that differ only in the length of roots and which both produce the reflection arrangement $\mathcal{BC}_n$. Very detailed information about all irreducible root systems, and hence about the corresponding reflection arrangements, can be found in the appendices to Bourbaki (1968), see also Humphreys (1972). We leave the subject by listing the four infinite families of

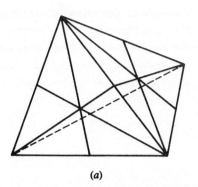
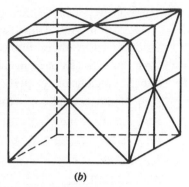

(a)   (b)

Figure 2.3.6: The regular polytopes of the reflection
arrangements $\mathcal{A}_3$ and $\mathcal{BC}_3$.

root systems in terms of the standard orthonormal basis $\{\mathbf{e}_1, \mathbf{e}_2, \dots, \mathbf{e}_n\}$ of $\mathbb{R}^n$:

$$
\begin{aligned}
A_{n-1} &= \{\mathbf{e}_i - \mathbf{e}_j : 1 \le i, j \le n, i \ne j\}; \\
D_n &= A_{n-1} \cup \{\pm(\mathbf{e}_i + \mathbf{e}_j) : 1 \le i, j \le n, i \ne j\}; \\
B_n &= D_n \cup \{\pm\mathbf{e}_i : 1 \le i \le n\}; \\
C_n &= D_n \cup \{\pm 2\mathbf{e}_i : 1 \le i \le n\}.
\end{aligned}
$$

(2.3.7)

(b) Regions

We have seen in Proposition 2.3.6 that all regions of a reflection arrangement $\mathcal{A}$
in $\mathbb{R}^d$ are simplicial cones. Hence, the tope graph $\mathcal{T}(\mathcal{A})$ of such an arrangement is
d-regular (cf. Definition 2.1.4). Observe that the tope graph $\mathcal{T}(\mathcal{A})$ of a reflection
arrangement $\mathcal{A}$ is the Cayley graph of the corresponding reflection group under
its standard representation as a Coxeter group, explained in connection with
Theorem 2.3.7.

Suppose now that $\mathcal{A}$ is a reflection arrangement in $\mathbb{R}^d$. Choosing a base region
B and directing the edges of the tope graph away from B, we get the tope poset
$\mathcal{T}(\mathcal{A}, B)$. By Proposition 4.2.10 $\mathcal{T}(\mathcal{A}, B)$ is a graded poset with rank function
$\rho_B(R)$ equal to the number of hyperplanes separating B and R, for all regions
R. Furthermore, it follows from Proposition 4.4.5 that the poset $\mathcal{T}(\mathcal{A}, B)$ is a
lattice. Note that the lattice $\mathcal{T}(\mathcal{A}, B)$ of a reflection arrangement is independent
(up to isomorphism) of the choice of the base region B, since the reflection group
acts vertex-transitively on its Cayley graph.

The rank-generating function of this graded lattice has a remarkable factorization property.

2.3.8 Theorem (Solomon 1966). *For every reflection arrangement $\mathcal{A}$ in $\mathbb{R}^d$
there exist positive integers $e_1 = 1, e_2, \dots, e_d$ such that*

$$
\sum q^{\rho_B(R)} = \prod_{i=1}^{d} (1 + q + q^2 + \dots + q^{e_i}),
$$

where the sum extends over all regions R of $\mathcal{A}$.

The integers $e_1, e_2, \ldots, e_d$ are called the *exponents* of the reflection arrangement. They are usually defined in terms of the invariant ring of the reflection group or of the eigenvalues of its Coxeter element (product of the Coxeter generators), see pp. 118–123 of Bourbaki (1968). It follows from Theorem 2.3.8 that the number of regions of $\mathcal{A}$ equals $\prod_{i=1}^{d}(1+e_i)$, while the number of hyperplanes is $\sum_{i=1}^{d} e_i$. Some examples of exponents are

$\mathcal{A}_n : 1, 2, 3, \ldots, n;$

$\mathcal{BC}_n : 1, 3, 5, \ldots, 2n - 1;$

$\mathcal{H}_3 : 1, 5, 9;$

$\mathcal{H}_4 : 1, 11, 19, 29,$

cf. Table 2.3.5.

As a partial ordering of the regions, the tope lattice $\mathcal{T}(\mathcal{A}, B)$ of a reflection arrangement $\mathcal{A}$ is often called *weak Bruhat order*. As the name reveals there is also a *strong Bruhat order* (usually called just *"Bruhat order"*), which we now define. In contrast to weak Bruhat order, which is a geometric phenomenon that without difficulty extends to all oriented matroids (Definition 4.2.9), Bruhat order does not seem to have any reasonable geometric meaning beyond reflection arrangements and Coxeter groups.

Let $\mathcal{A}$ be a reflection arrangement, fix a base region B, and as before let $\rho_B(R)$ denote the number of hyperplanes separating region R from B. Then the *Bruhat ordering* of the set of regions of $\mathcal{A}$ is defined as the transitive closure generated by the comparability relations $R < R'$ if $R = \sigma_H(R')$ for some $H \in \mathcal{A}$ and $\rho_B(R) < \rho_B(R')$.

Like its weak counterpart, Bruhat order of the regions is a graded poset with rank function ρ_B, least element B and greatest element $-B$. As a relation, weak Bruhat order is a subset of Bruhat order. Both versions of Bruhat order are usually defined as partial orderings of the elements of the corresponding reflection group, but the geometric translation is straightforward. Bruhat order arises in a number of important connections in algebraic geometry and representation theory. For more information and further references concerning these orderings see e.g. Hiller (1982) or Björner (1984b).

To illustrate these definitions we take a look at two examples. Figure 2.3.7 shows the dihedral arrangement $\mathcal{BC}_2 = \mathcal{I}_2(4)$, and the Hasse diagram of weak Bruhat order (solid lines) of the regions. Figure 2.3.8 shows weak Bruhat order (solid lines) of the arrangement $\mathcal{A}_3$ discussed in Example 2.3.3 and depicted in Figure 2.3.6(a).

In both figures the additional relations of Bruhat order are indicated by dashed lines. The regions R_π of $\mathcal{A}_3$ are marked by the corresponding permutation $\pi \in S_4$, and the description in Example 2.3.3 of adjacency between regions should be kept in mind.

The tope graph of an arrangement is isomorphic to the edge graph (1-skeleton) of the polar zonotope. This is illustrated by a comparison of Figure 2.3.8 with the 4-permutohedron in Figure 2.2.2. The edge graph of this zonotope equals

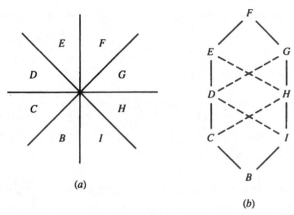

(a)

(b)

Figure 2.3.7: Weak Bruhat order and Bruhat order of $\mathcal{BC}_2$.

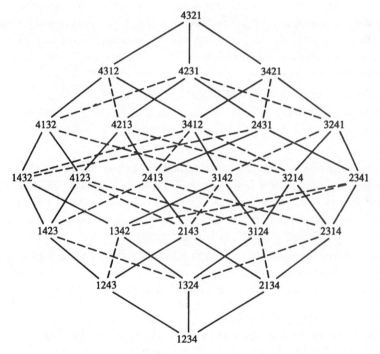

Figure 2.3.8: Weak Bruhat order and Bruhat order of $\mathcal{A}_3$.

the Hasse diagram (solid lines) in Figure 2.3.8. Observe also that the maximal chains in Figure 2.3.8 are precisely the simple allowable sequences in S_4. See (6.4.3) for additional information.

(c) Coxeter complex (face lattice)

The face lattice $\mathcal{F}(\mathcal{A})$ (Def. 2.1.2) of a reflection arrangement is, by Proposition 2.3.6, the face lattice of a simplicial complex, known as the *Coxeter complex* of $\mathcal{A}$. Taking the polar of the zonotope associated to $\mathcal{A}$ leads to the following observation.

2.3.9 Proposition.   *Every Coxeter complex is the boundary complex of a centrally symmetric simplicial polytope.*

For the reflection arrangements that come from regular polytopes, the Coxeter complex coincides with the barycentric subdivision of that polytope. For instance, Figure 2.3.6 shows the Coxeter complexes of type $\mathcal{A}_3$ and $\mathcal{BC}_3$ which are the barycentric subdivisions of the boundaries of the tetrahedron and the 3-cube.

Coxeter complexes have been studied in great detail by J. Tits. See Chapter 2 of Tits (1974), which contains a combinatorial characterization of them as a class of simplicial complexes as well as descriptions of several special properties. Many of these combinatorial properties (those which do not depend on the high degree of symmetry) generalize from Coxeter complexes to the big face lattice of arbitrary oriented matroids (Sections 4.1 and 4.2). Some of the cited work of Tits can be profitably read from an oriented matroid point of view, once the basic dictionary has been established:

> *chamber* $\longleftrightarrow$ *tope,*
>
> *gallery* $\longleftrightarrow$ *path in the tope graph,*
>
> *root* $\longleftrightarrow$ *halfspace,*
>
> $\mathrm{proj}_X(Y) \longleftrightarrow X \circ Y,$
>
> etc.

Coxeter complexes can also be defined in terms of systems of so-called "parabolic" subgroups of a reflection group and their cosets. Similar constructions for groups of Lie type lead to simplicial complexes called *buildings*, which contain many Coxeter complexes as subcomplexes. See Tits (1974) or the recent monographs by Brown (1989) and Ronan (1989).

Later work of Tits (1982) has emphasized the role of chambers (regions) in Coxeter complexes and buildings, and it bases the theory on certain tope graphs with colored edges, called *chamber systems*. See Ronan (1989) for an exposition.

(d) Characteristic polynomial

For an essential arrangement $\mathcal{A}$ of hyperplanes in $\mathbb{R}^d$ let $L(\mathcal{A})$ be its intersection lattice (cf. Definition 2.1.3). The *characteristic polynomial* of $\mathcal{A}$, and of $L = L(\mathcal{A})$, is defined by

$$(2.3.8) \qquad p(\mathcal{A}; \lambda) = p(L; \lambda) = \sum_{x \in L} \mu(\hat{0}, x) \lambda^{d - r(x)}.$$

Here $\mu(\cdot, \cdot)$ denotes the Möbius function of L. See Chapter 7 of White (1987) for basic facts concerning characteristic polynomials of geometric lattices.

The characteristic polynomial of $\mathcal{A}$ is interesting in connection with the complexified arrangement $\mathcal{A}^{\mathbb{C}}$, see Section 2.5 for the definition. It follows from (2.5.1) and Theorem 2.5.3 that $p(\mathcal{A}; \lambda)$ equals (up to a substitution) the Poincaré polynomial of the cohomology algebra of the complement of $\mathcal{A}^{\mathbb{C}}$ in $\mathbb{C}^d$.

Stanley (1972) showed that if L is supersolvable (has a maximal chain of modular elements), then $p(L; \lambda) = (\lambda - e_1)(\lambda - e_2) \dots (\lambda - e_d)$ for certain positive integers $e_1, e_2, \dots, e_d$. Since partition lattices are supersolvable, this implies that the characteristic polynomial of reflection arrangements of type $\mathcal{A}_n$ have such a complete factorization (cf. Exercise 2.14). However, the same factorization property turns out to hold for *all* reflection arrangements, although these in general have non-supersolvable intersection lattices (except for types $\mathcal{A}_n$, $\mathcal{BC}_n$ and $\mathcal{I}_2(m)$). The only known systematic proof for this factorization property uses H. Terao's theory of *T-free arrangements*, of which we will now give a sketch.

Let $\mathcal{A}$ be a central arrangement of hyperplanes in $\mathbb{R}^d$. Among all polynomial mappings $\mathbf{p} : \mathbb{R}^d \to \mathbb{R}^d, \mathbf{p} = (p_1, \dots, p_d), p_i \in \mathbb{R}[x_1, \dots, x_d]$, consider those that satisfy $\mathbf{p}(H) \subseteq H$ for all $H \in \mathcal{A}$. They form a graded module $T(\mathcal{A})$ over the polynomial ring $\mathbb{R}[x_1, \dots, x_d]$, the *Terao module*. The arrangement $\mathcal{A}$ is said to be *T-free* (free in the sense of Terao) if $T(\mathcal{A})$ is a free module. In that case $T(\mathcal{A})$ determines a sequence $e_1 = 1, e_2, \dots, e_d$ of positive integers, called the *exponents* of $\mathcal{A}$, as the degrees of the elements in a homogeneous free basis for $T(\mathcal{A})$. The theory of T-free arrangements has been developed in a series of papers by Terao, starting with Terao (1980). See also Stanley (1984b), Orlik (1989), Ziegler (1989a,b) and Yuzvinsky (1990). One early conjecture, which remains open, is that T-freeness of $\mathcal{A}$ can be decided from the intersection lattice $L(\mathcal{A})$ alone. One of the most important results of the theory is the following factorization theorem of Terao (1981), see also Solomon and Terao (1987).

2.3.10 Theorem (Terao 1981). *If the arrangement $\mathcal{A}$ is T-free with exponents $e_1, e_2, \dots, e_d$, then*
$$p(\mathcal{A}; \lambda) = (\lambda - e_1)(\lambda - e_2) \cdot \dots \cdot (\lambda - e_d).$$

Supersolvable arrangements are T-free (Jambu and Terao 1984, Stanley 1984b), so Theorem 2.3.10 subsumes Stanley's factorization result (in the $\mathbb{R}$-realizable case). Reflection arrangements are also T-free (Arnol'd 1979, Saito 1975), so we deduce:

2.3.11 Corollary. *If $\mathcal{A}$ is a reflection arrangement then*
$$p(\mathcal{A}; \lambda) = (\lambda - e_1)(\lambda - e_2) \cdot \dots \cdot (\lambda - e_d).$$

It is a remarkable fact that the exponents occurring here are the same as the exponents of Theorem 2.3.8. As was mentioned before, these same numbers arise algebraically in several other connections in the study of reflection groups. A satisfactory combinatorial explanation for their existence and important role is lacking.

2.4 Stratifications of the Grassmann variety

In recent years, there has been an exciting interaction between algebraic geometry and combinatorics, bringing concepts and results from each area to bear on problems of the other. This includes research on generalized hypergeometric functions and multivariate discriminants by I.M. Gel'fand and coworkers, algebraic geometry of point configurations and hyperplane arrangements, homology of toric varieties and enumeration of faces of convex polytopes, and connections of Gröbner bases to both classical and geometric invariant theory.

In this section we take a perspective on matroids and oriented matroids which is motivated by these developments. A common thread and prototypical object of study, connecting many facets of the recent advances in combinatorial algebraic geometry, is the Grassmann variety $G_r(V)$ of r-dimensional linear subspaces of an n-dimensional vector space V. It is the purpose of this section to familiarize the reader with the Grassmann variety and to show how it relates to both matroids and oriented matroids. We shall see that the (representable) rank r matroids on n elements correspond to certain constructible subsets of $G_r(V)$, and if the underlying field is ordered, then the (realizable) oriented matroids correspond to certain semi-algebraic subsets of $G_r(V)$. It is noteworthy that these "matroid strata" are the natural domains for the generalized hypergeometric functions in Gel'fand (1986). We will investigate the interplay of topology and combinatorics in the resulting matroid stratification (resp. oriented matroid stratification) of $G_r(V)$. In particular, we shall study the close relation between the matroid stratification and the decomposition of $G_r(V)$ into Schubert cells.

The recent work of MacPherson (1993) on *combinatorial differential manifolds*, generalizing the concept of a differential manifold, should also be mentioned in this connection. The idea is to let the local tangential data provided by the differential structure be replaced by a system of locally defined oriented matroids. Classically the Grassmannian plays the role of a classifying space for the vector bundles over differential manifolds. In this new set-up, we get an *OM-Grassmannian* (also called the "MacPhersonian") as the classifying space for *oriented matroid bundles*. See MacPherson (1993), Gel'fand and MacPherson (1992), Babson (1992) and Mnëv and Ziegler (1993) for further details, and also the comment at the end of Section 7.2

Let K denote a field of characteristic 0, such as the real numbers $\mathbb{R}$ or the complex numbers $\mathbb{C}$, and consider the vector space K^n equipped with its canonical basis $\{e_1, e_2, \ldots, e_n\}$. As a *set* we define the *Grassmann variety* $G_r(K^n)$ to be the collection of all r-dimensional linear subspaces in K^n. As a guide for the ensuing discussion let us mention that we aim to familiarize the reader with looking at the elements of $G_r(K^n)$ from the following inter-related points of view:

(a)   as the cosets of $GL(K^n)$ modulo a maximal parabolic subgroup P,

(b)   as the decomposable antisymmetric tensors in $\wedge_r K^n / K^*$,

(c)   as the points in $\left(\binom{n}{r} - 1\right)$-dimensional projective space whose coordinates satisfy the quadratic Grassmann-Plücker relations,

(d)   as the configurations of n vectors in K^r modulo the action of $GL(K^r)$.

There are several ways to see that $G_r(K^n)$ has the structure of a smooth projective variety of dimension $r(n-r)$. Let us start with the more general and less explicit point of view in (a). The algebraic group $\Gamma := GL(K^n)$ of invertible $n \times n$-matrices acts transitively on the set $G_r(K^n)$. The stabilizer P_ξ of a subspace $\xi \in G_r(K^n)$ is the subgroup of linear automorphisms of K^n which map ξ onto itself. By elementary group theory there is an identification $G_r(K^n) = \Gamma/P_\xi$, and this defines on $G_r(K^n)$ the structure of a homogeneous space for Γ. For $\xi = \operatorname{span}\{\mathbf{e}_1, \mathbf{e}_2, \ldots, \mathbf{e}_r\}$ we get the closed subgroup $P = P_\xi$ of invertible $n \times n$-matrices $M = (m_{ij})$ with $m_{ij} = 0$ for $i = r+1, \ldots, n$, $j = 1, \ldots, r$. The matrix group P is a parabolic subgroup of Γ because it contains the Borel subgroup B of upper triangular matrices. It follows from general results on algebraic groups (cf. Humphreys 1975) that $G_r(K^n) = \Gamma/P$ is a projective variety, embedded in the flag variety Γ/B. The dimension of the Grassmann variety equals

$$\dim G_r(K^n) = \dim \Gamma - \dim P = r(n - r).$$

In their paper *"On the general definition of a matroid and a greedoid"*, I.M. Gel'fand and V. Serganova (1987) take this set-up as the starting point for introducing "matroids" arising from strata in arbitrary compact homogeneous spaces Γ/P. As this general point of view constitutes a surprising and promising connection between matroid theory and algebraic groups, we suggest that it would be worthwhile to give "the general definition of an oriented matroid" along the same lines.

We will now take a closer and more concrete look at the Grassmann variety by describing an explicit embedding into projective space, called the *Plücker embedding* of $G_r(K^n)$. Let $\wedge_r K^n$ denote the r-fold exterior product of the vector space K^n. The elements of $\wedge_r K^n$ are called *antisymmetric tensors*. We can think of $\wedge_r K^n$ as the $\binom{n}{r}$-dimensional K-vector space which has the canonical basis

$$\{\, \mathbf{e}_{i_1} \wedge \mathbf{e}_{i_2} \wedge \ldots \wedge \mathbf{e}_{i_r} \mid 1 \le i_1 < i_2 < \ldots < i_r \le n \,\}.$$

For each basis vector we get a coordinate function, which is called a *bracket* and is denoted $[i_1 i_2 \ldots i_r]$. Technically, $[i_1 i_2 \ldots i_r]$ is an element of the dual vector space $(\wedge_r K^n)^*$, but, as our choice of basis identifies K^n with $(K^n)^*$, we also get an identification of $(\wedge_r K^n)^*$ and $\wedge_r K^n$. The ring of polynomial functions on $\wedge_r K^n$ is the *bracket ring*, and it is denoted

$$\begin{aligned} \mathcal{B}_{n,r} :&= \operatorname{Sym}(\wedge_r K^n) \\ &= K\big[\,\{\, [i_1 i_2 \ldots i_r] \;:\; 1 \le i_1 < i_2 < \ldots < i_r \le n \,\}\,\big]. \end{aligned}$$

Consider the canonical map

$$\phi \;:\; \underbrace{K^n \times K^n \times \ldots \times K^n}_{r \text{ times}} \;\longrightarrow\; \wedge_r K^n$$

$$(\mathbf{v}_1, \mathbf{v}_2, \ldots, \mathbf{v}_r) \;\longmapsto\;$$

$$\sum_{1 \le i_1 < \ldots < i_r \le n} \begin{vmatrix} v_{1 i_1} & v_{1 i_2} & \cdots & v_{1 i_r} \\ v_{2 i_1} & v_{2 i_2} & \cdots & v_{2 i_r} \\ \vdots & \vdots & \ddots & \vdots \\ v_{r i_1} & v_{r i_2} & \cdots & v_{r i_r} \end{vmatrix} \cdot \mathbf{e}_{i_1} \wedge \mathbf{e}_{i_2} \wedge \ldots \wedge \mathbf{e}_{i_r}.$$

This map assigns to each $n \times r$-matrix $\mathbf{X} = (v_{ji})$ the vector of length $\binom{n}{r}$ which consists of all $r \times r$-minors of $\mathbf{X}$. The vector space $\wedge_r K^n$ and the map ϕ are characterized by the following universal property.

2.4.1 Proposition (Universal property of the exterior product of vector spaces). *Let V be any K-vector space, and let $\psi : (K^n)^r \to V$ be any multilinear alternating map. Then there exists a unique linear map $\tilde{\psi} : \wedge_r K^n \to V$ such that $\psi = \tilde{\psi} \circ \phi$.*

An antisymmetric tensor $\xi \in \wedge_r K^n$ is said to be *decomposable* if it is non-zero and contained in the image of the map ϕ, i.e., if we can write it in the form

$$\xi = \mathbf{v}_1 \wedge \mathbf{v}_2 \wedge \ldots \wedge \mathbf{v}_r := \phi(\mathbf{v}_1, \mathbf{v}_2, \ldots, \mathbf{v}_r).$$

It turns out that the decomposable antisymmetric tensors are, up to scaling, in one-to-one correspondence with points on the Grassmann variety.

2.4.2 Proposition. *Consider two r-dimensional subspaces V and W in K^n, and let $\{\mathbf{v}_1, \mathbf{v}_2, \ldots, \mathbf{v}_r\}$ and $\{\mathbf{w}_1, \mathbf{w}_2, \ldots, \mathbf{w}_r\}$ be bases for V and W respectively. Then*

$$V = W \text{ if and only if } \mathbf{v}_1 \wedge \mathbf{v}_2 \wedge \ldots \wedge \mathbf{v}_r = c \cdot \mathbf{w}_1 \wedge \mathbf{w}_2 \wedge \ldots \wedge \mathbf{w}_r$$

for some non-zero scalar $c \in K^*$.

Proposition 2.4.2 states in other words that the map ϕ induces an embedding

$$\xi = \operatorname{span}(\mathbf{v}_1, \mathbf{v}_2, \ldots, \mathbf{v}_r) \mapsto \phi(\xi) = \mathbf{v}_1 \wedge \mathbf{v}_2 \wedge \ldots \wedge \mathbf{v}_r$$

of the Grassmann variety $G_r(K^n)$ into $(\binom{n}{r} - 1)$-dimensional projective space $\wedge_r K^n / K^*$. The values of the coordinate functions $\xi_{i_1 i_2 \ldots i_r} := [i_1 i_2 \ldots i_r](\phi(\xi))$ are said to be the *Plücker coordinates* of the subspace ξ. In the following we identify ξ with its Plücker coordinate vector, and we write with slight abuse of notation

$$G_r(K^n) = \{ \xi \in \wedge_r K^n / K^* : \xi = \mathbf{v}_1 \wedge \mathbf{v}_2 \wedge \ldots \wedge \mathbf{v}_r \text{ decomposable } \}.$$

Being defined as the image of a polynomial map, this projective subvariety is irreducible, and hence it is the zero set of a homogeneous prime ideal in the bracket ring $\mathcal{B}_{n,r}$. The following theorem gives an explicit description of a set of quadratic bracket polynomials which generate the ideal in question. See Hodge and Pedoe (1947) or Bokowski and Sturmfels (1989b) for a proof of Theorem 2.4.3, and see Sturmfels and White (1989) for an explicit Gröbner basis for the ideal of the Grassmann variety.

2.4.3 Theorem (Grassmann-Plücker relations). *Let K be a field of characteristic 0. The embedded Grassmann variety*

$$G_r(K^n) \hookrightarrow \wedge_r K^n / K^*$$

is the zero set of the bracket polynomials

$$[b_1 b_2 b_3 \ldots b_r][b_1' b_2' \ldots b_r'] - \sum_{i=1}^{r} [b_i' b_2 b_3 \ldots b_r][b_1' \ldots b_{i-1}' b_1 b_{i+1}' \ldots b_r'],$$

where

$$b_1, \ldots, b_r, b'_1, \ldots, b'_r \in \{1, 2, \ldots, n\}.$$

More strongly, these quadratic polynomials generate the prime ideal of $G_r(K^n)$.

In Theorem 2.4.3 the Grassmann-Plücker relations are stated in a form which immediately exhibits their close connection to the basis exchange axiom for matroids. This axiom, which is stated below, is the immediate combinatorial analogue to the above algebraic identities.

We recall that a *matroid* $M = (E, \mathcal{B})$ consists of a finite set E and a nonempty collection $\mathcal{B}$ of subsets of E, called *bases*, such that

(2.4.1) for any two bases B and B', and for all $b_1 \in B \setminus B'$, there exists a $b'_i \in B' \setminus B$ such that $(B \cup b'_i) \setminus b_1$ is a basis.

It is an easy consequence of (2.4.1) that all bases of a matroid M have the same cardinality, say r, which is the *rank* of M.

Theorem 2.4.3 implies that the index sets of non-zero Plücker coordinates of any r-flat $\xi \in G_r(K^n)$ are the bases of a rank r matroid M_ξ on $E_n = \{1, 2, \ldots, n\}$. Conversely, given any rank r matroid M on E_n, we can define its *stratum* (or *realization space*) as the set

$$\mathcal{R}(M) := \{ \xi \in G_r(K^n) \ : \ M_\xi = M \}.$$

The stratum $\mathcal{R}(M)$ is a constructible subset (see below) of the Grassmann variety, defined by the following system of polynomial equations and inequations

$$\begin{cases} [i_1 i_2 \ldots i_r] = 0 & \text{if } \{i_1, i_2, \ldots, i_r\} \text{ is not a basis of } M \\ [i_1 i_2 \ldots i_r] \neq 0 & \text{if } \{i_1, i_2, \ldots, i_r\} \text{ is a basis of } M. \end{cases}$$

The collection of strata $\{ \mathcal{R}(M) \ : \ M \text{ is a rank } r \text{ matroid on } E_n \}$ is called the *matroid stratification* of $G_r(K^n)$.

At this point we need to recall a few definitions from elementary algebraic geometry. The *Zariski topology* on K^n is the weakest topology such that all polynomial functions are continuous. In other words, the closed subsets of K^n in the Zariski topology are precisely the zero sets of systems of polynomials. Throughout this section all topological operations will be understood with respect to the Zariski topology, and we will use the terms *(algebraic) variety, isomorphism of varieties* and *birational isomorphism of varieties* as defined in the textbook of Mumford (1988). A subset of K^n is called *constructible* if it is a boolean combination of Zariski closed subsets, or, in other words, if it is defined by a finite set of polynomial equations and inequations. If $K = \mathbb{C}$, the complex numbers, then the Zariski topology is much weaker than the usual strong topology, so, in general, the Zariski closure of a subset of $\mathbb{C}^n$ is a proper superset of its strong closure. However, it is known that for any constructible subset $V \subset \mathbb{C}^n$ the Zariski closure is equal to the strong closure (Mumford 1988, Corollary I.10.1). For our discussion of the matroid stratification of the complex Grassmann variety, the reader may therefore think about closures in the usual strong sense.

We will now derive an alternative interpretation of the strata $\mathcal{R}(M)$ as linear equivalence classes of vector configurations in K^r, and we will use this picture later to address the following questions:

(1)   Is $\mathcal{R}(M) \neq \emptyset$ for every matroid ?

(2)   Is the matroid stratification *normal*, i.e., does $\overline{\mathcal{R}(M_1)} \cap \mathcal{R}(M_2) \neq \emptyset$ imply $\mathcal{R}(M_2) \subset \overline{\mathcal{R}(M_1)}$?

(3)   Is the "obvious" combinatorial characterization for adjacency of matroid strata correct ? (That is, is $\overline{\mathcal{R}(M_1)} \cap \mathcal{R}(M_2) \neq \emptyset$ equivalent to M_2 being a weak image of M_1 ?)

(4)   Which types of algebraic varieties occur as $\mathcal{R}(M)$ for some matroid M ?

The preimage of a point $\xi \in G_r(K^n)$ under the map ϕ consists of all $n \times r$-matrices $\mathbf{X}$ whose column space equals ξ. These matrices can be transformed into each other by right multiplication with an invertible $r \times r$-matrix.

2.4.4 Proposition.   *The points on the Grassmann variety $G_r(K^n)$ are in one-to-one correspondence with the configurations of n vectors in K^r modulo the action of the general linear group $GL(K^r)$.*

In the following we will interpret the rows of an $n \times r$-matrix $\mathbf{X}$ as a configuration of vectors $\mathbf{x}_1, \mathbf{x}_2, \ldots, \mathbf{x}_n$ in K^r. Then the matroid M_ξ is just the usual matroid of linear dependence on these vectors. The corresponding r-flat $\xi = \phi(\mathbf{X})$ has as its Plücker coordinate $\xi_{i_1 i_2 \ldots i_r}$ the signed volume of the parallelepiped spanned by the vectors $\mathbf{x}_{i_1}, \mathbf{x}_{i_2}, \ldots, \mathbf{x}_{i_r}$.

There is also an intrinsic, geometric way of describing the bijection of Proposition 2.4.4. Given any r-flat $\xi \in G_r(K^n)$, the point configuration corresponding to ξ is the image of the canonical basis of K^n under a projection $\pi_\xi : K^n \to K^n$ onto the subspace ξ.

In order to see that this geometric picture coincides with the above algebraic description, we pick any basis $(\mathbf{v}_1, \ldots, \mathbf{v}_r)$ of ξ. Let $\mathbf{X}$ denote the $n \times r$-matrix with column vectors $\mathbf{v}_i$, and define the *orthogonal projection* $\pi_\xi : K^n \to K^n$ with respect to the canonical basis $\{\mathbf{e}_1, \ldots, \mathbf{e}_n\}$ of K^n by the rank r matrix $\mathbf{X} \cdot (\mathbf{X} \cdot (\mathbf{X}^T \mathbf{X})^{-1})^T$. The columns of this matrix are the coordinates of the vectors $\pi_\xi(\mathbf{e}_i)$ in the basis $\{\mathbf{e}_1, \ldots, \mathbf{e}_n\}$ of K^n. (In the case of $K = \mathbb{R}$ with its usual inner product π_ξ is orthogonal projection, and if $(\mathbf{v}_1, \ldots, \mathbf{v}_r)$ is an orthonormal basis then of course $\mathbf{X}^T \mathbf{X} = \mathrm{Id}_r$.) One sees that the rows of the matrix $\mathbf{X} \cdot (\mathbf{X}^T \mathbf{X})^{-1}$ are the coordinates of the vectors $\pi_\xi(\mathbf{e}_i)$ in the basis $\{\mathbf{v}_1, \ldots, \mathbf{v}_r\}$ of the subspace ξ. As these n row vectors are linearly equivalent to the row vectors of $\mathbf{X}$ via the $r \times r$-matrix $\mathbf{X}^T \mathbf{X}$, we obtain the desired identification

$$\xi = \mathbf{v}_1 \wedge \ldots \wedge \mathbf{v}_r = \phi(\mathbf{X}) = \phi(\mathbf{X} \cdot (\mathbf{X}^T \mathbf{X})^{-1})$$
$$= \phi(\text{ "}\pi_\xi(\mathbf{e}_1), \ldots, \pi_\xi(\mathbf{e}_n) \text{ in } \xi\text{" }).$$

Every configuration ξ of n non-zero vectors in K^r gives rise to a configuration $\tilde{\xi}$ of n points in $(r-1)$-dimensional projective space $\mathbb{P}^{r-1}(K)$. If $\xi = \phi(\mathbf{X})$, then the rows of the matrix $\mathbf{X}$ are homogeneous coordinate vectors for these points in projective space. Thus $\tilde{\xi}$ is a projective realization of the matroid M_ξ of ξ.

This projective configuration has a nice interpretation in the algebraic group scenario mentioned at the beginning of this section. Let $H = (K^*)^n$ denote the group of $n \times n$-diagonal matrices, which is a maximal torus in $P \subset GL_n(K)$. The action of the torus H on the embedded Grassmann variety can be expressed as a linear action on the Plücker coordinates:

$$H \times G_r(K^n) \to G_r(K^n),$$

$$(\operatorname{diag}(t_1, t_2, \ldots, t_n), \xi_{i_1 i_2 \ldots i_r}) \mapsto t_{i_1} t_{i_2} \ldots t_{i_r} \cdot \xi_{i_1 i_2 \ldots i_r}.$$

This action of the n-dimensional torus corresponds to scaling up and down the n vectors in K^r which ξ represents. Therefore two vector configurations ξ and η in $G_r(K^n)$ are contained in the same H-orbit if and only if they are projectively equivalent. This means that we can identify each configuration $\tilde{\xi}$ of n points in projective $(r-1)$-space (modulo the projective group action) with an orbit $H\xi$ of the n-dimensional torus acting on the Grassmann variety $G_r(K^n)$. Since projectively equivalent point configurations give rise to the same matroid, we have the inclusion $H\xi \subseteq \mathcal{R}(M_\xi)$. As a consequence of this, we can now define the *projective realization space* $\mathcal{R}(M)/H$ of M as the algebraic quotient of the realization space $\mathcal{R}(M)$ modulo the action of the algebraic group H. The projective realization space $\mathcal{R}(M)/H$ is a singleton if and only if $H\xi = \mathcal{R}(M_\xi)$, in which case the matroid M is called *projectively unique*. At the end of this section we will briefly comment on connections to the theory of toric varieties.

The matroid stratification of the Grassmann variety $G_r(K^n)$ is a refinement of the classical decomposition of $G_r(K^n)$ into Schubert cells. In fact, every matroid stratum can be obtained as the intersection of Schubert cells induced by different flags. Before answering the questions (1) – (4) posed above, we want to describe this interesting aspect of matroid realization spaces, first pointed out by Gel'fand, Goresky, MacPherson and Serganova (1987). Elementary introductions to the Schubert calculus can be found in Kleiman and Laksov (1972) and Hiller (1982).

Fix a complete flag of subspaces $0 = F_0 \subset F_1 \subset \ldots \subset F_n = K^n$, with $\dim F_i = i$. For each r-dimensional subspace ξ look at the chain of subspaces $F_0 \cap \xi \subseteq F_1 \cap \xi \subseteq \ldots \subseteq F_n \cap \xi = \xi$, and let $\sigma(\xi)$ be the sequence $(i_1, i_2, \ldots, i_r)$ of indices where the dimension increases, i.e., where $\dim(F_{i_k-1} \cap \xi) < \dim(F_{i_k} \cap \xi) = k$, $1 \le k \le r$. This gives a surjective mapping

$$\sigma : G_r(K^n) \to \binom{E_n}{r} = \{1 \le i_1 < i_2 < \ldots < i_r \le n\}.$$

The inverse images

$$\Omega(i_1, i_2, \ldots, i_r) = \sigma^{-1}(i_1, i_2, \ldots, i_r)$$

are constructible subsets of $G_r(K^n)$, called *Schubert cells*. Their closures in $G_r(K^n)$ are called *Schubert varieties*. The Schubert varieties are projective subvarieties of $\wedge_r K^n / K^*$ under the Plücker embedding of the Grassmann variety.

The decomposition of $G_r(K^n)$ into Schubert cells is normal, i.e.,

$$\Omega(i_1, \ldots, i_r) \cap \overline{\Omega(j_1, \ldots, j_r)} \neq \emptyset \quad \text{implies} \quad \Omega(i_1, \ldots, i_r) \subseteq \overline{\Omega(j_1, \ldots, j_r)}.$$

Hence the boundary $\overline{\Omega(j_1, \ldots, j_r)} \backslash \Omega(j_1, \ldots, j_r)$ of a Schubert variety is a disjoint union of "smaller" Schubert cells. This partial ordering of Schubert cells, called *Bruhat order*, has a very simple combinatorial description:

$$(2.4.2) \qquad \Omega(i_1, \ldots, i_r) \subseteq \overline{\Omega(j_1, \ldots, j_r)} \quad \Longleftrightarrow \quad i_k \leq j_k, \, 1 \leq k \leq r.$$

The relationship of Bruhat order on Schubert cells to Bruhat order on the symmetric group as defined in Section 2.3 is outlined in Exercise 2.17. As a consequence of (2.4.2) we deduce the following description of the Schubert varieties:

$$\overline{\Omega(i_1, \ldots, i_r)} \quad = \quad \{\xi \in G_r(K^n) : \dim(F_{i_k} \cap \xi) \geq k, \, 1 \leq k \leq r\}.$$

For example, one sees that $\Omega(1, 2, \ldots, r)$ as well as its closure consists of a single point, namely F_r, whereas the "big cell" $\Omega(n - r + 1, n - r + 2, \ldots, n)$ is dense in $G_r(K^n)$, so its Schubert variety is all of $G_r(K^n)$.

It can be shown that $\Omega(i_1, \ldots, i_r)$ is isomorphic as an algebraic variety to affine $(\sum_{k=1}^r i_k - \binom{r+1}{2})$-dimensional K-space. In particular, in the case $K = \mathbb{C}$ with the strong topology, this Schubert cell is a topological $2(\sum_{k=1}^r i_k - \binom{r+1}{2})$-dimensional cell and the Schubert cells give a cell decomposition (in fact, a CW decomposition) of the Grassmann variety into even-dimensional cells.

Now, fix the particular reference flag $F_i = \text{span}\{e_n, e_{n-1}, \ldots, e_{n-i+1}\}, 0 \leq i \leq n$, where as before $\{e_1, e_2, \ldots, e_n\}$ is the canonical basis in K^n. Comparison of the definitions of Schubert cells and of matroid strata, and a few moments reflection, show that

$$\Omega(i_1, i_2, \ldots, i_r) = \bigcup \{\mathcal{R}(M) \, : \, M \text{ is a matroid with lexicographically smallest}$$
$$\text{basis } \{n + 1 - i_r, n + 1 - i_{r-1}, \ldots, n + 1 - i_1\}\}.$$

We define the *Schubert matroid* $M_{i_1 i_2 \ldots i_r}$ to be the freest of these matroids, i.e., an r-tuple $(n+1-j_r, n+1-j_{r-1}, \ldots, n+1-j_1)$ forms a basis of $M_{i_1 i_2 \ldots i_r}$ if and only if $i_k \geq j_k, 1 \leq k \leq r$. Then the Schubert variety $\overline{\Omega(i_1, i_2, \ldots, i_r)}$ equals the closure $\overline{\mathcal{R}(M_{i_1 i_2 \ldots i_r})}$ of the realization space of the corresponding Schubert matroid.

For instance the Schubert cell $\Omega(1, 4, 5)$ in $G_3(K^6)$ is decomposed into matroid strata $\mathcal{R}(M)$ for all matroids on $\{1, 2, \ldots, 6\}$ with lexicographically first basis $\{2, 3, 6\}$. There are 17 such matroids, the freest of which is the Schubert matroid M_{145} which has the point 6, the line 2345 and the loop 1. Figure 2.4.1 shows the Schubert cells in $G_3(K^6)$ in Bruhat order, where each cell is labeled by the corresponding Schubert matroid.

For each permutation σ of $\{1, 2, \ldots, n\}$, let $\mathbf{F}^\sigma$ denote the flag of coordinate flats $F_i^\sigma = \text{span}\{e_{\sigma(1)}, e_{\sigma(2)}, \ldots, e_{\sigma(i)}\}, \, 0 \leq i \leq n$, and let $\Omega^\sigma(i_1, i_2, \ldots, i_r)$ denote the Schubert cells in $G_r(K^n)$ induced by $\mathbf{F}^\sigma$. Clearly, $\Omega^\sigma(i_1, i_2, \ldots, i_r)$ again decomposes into matroid strata; the only difference is that the reading order of the ground set and hence the lexicographic order on bases has been changed by σ. Since every basis of a matroid M is lexicographically smallest with respect to some ordering of the ground set, we have derived the following result of Gel'fand *et al.* (1987).

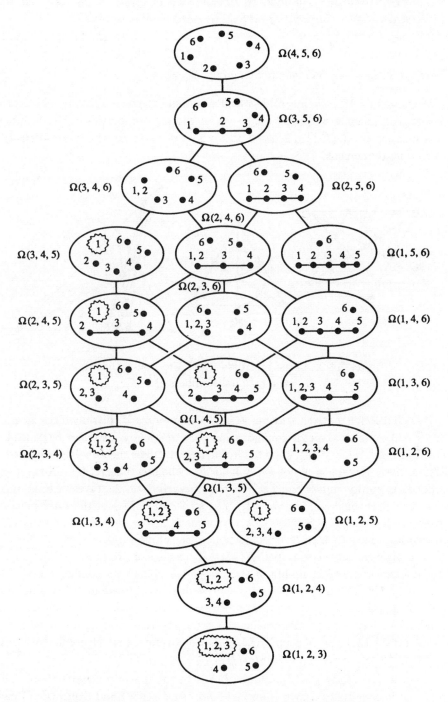

Figure 2.4.1: The Bruhat order on Schubert varieties
and Schubert matroids in $G_{3,6}$.

2.4.5 Proposition.   *The matroid stratification of $G_r(K^n)$ coincides with the coarsest common refinement of all Schubert cell decompositions*

$$\{\,\Omega^\sigma(i_1, i_2, \ldots, i_r) \;:\; (i_1, i_2, \ldots, i_r) \in \binom{E_n}{r}\,\},$$

where σ ranges over the symmetric group S_n.

We now return to the matroid stratification and the questions (1) – (4) stated earlier. Our results will show that each of the nice properties of the Schubert stratification of $G_r(K^n)$ discussed above does *not* hold for the matroid stratification, its refinement.

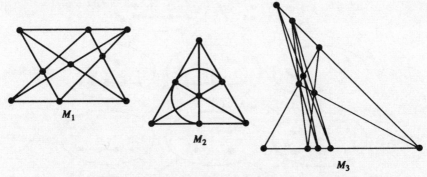

Figure 2.4.2: Three matroids which are non-realizable over
the rational numbers $\mathbb{Q}$.

By definition, a matroid stratum $\mathcal{R}(M)$ is non-empty if and only if the matroid M is realizable as a vector configuration over the field K. Consider the rank 3 matroids M_1, M_2, and M_3, depicted in Figure 2.4.2. For the *non-Pappus matroid* M_1 we have $\mathcal{R}(M_1) = \emptyset$ for every field K. The stratum $\mathcal{R}(M_2)$ of the *Fano matroid* is empty unless the field K has characteristic 2. (We have assumed char $K = 0$ throughout Section 2.4.) The stratum $\mathcal{R}(M_3)$ of the third matroid in Figure 2.4.2 is empty unless $\sqrt{2}$ is contained in the field K. The projective realization space $\mathcal{R}(M_3)/H$ is isomorphic to the affine variety $\mathcal{V}(x^2 - 2) \hookrightarrow K^1$. If K is algebraically closed, then $\mathcal{R}(M_3)/H$ consists of two points. When lifting this configuration of 11 points in the projective plane to a configuration of 11 vectors in K^3 we gain precisely $7 = 11 - 4$ degrees of freedom. This is because four of the points (or vectors) form a projective basis while the remaining seven can be scaled arbitrarily. Rephrasing this geometric reasoning in precise algebraic terms proves that the stratum $\mathcal{R}(M_3)$ is isomorphic as an algebraic variety to $(K^*)^7 \cup (K^*)^7 = (K^*)^7 \times \mathcal{R}(M_3)/H$.

We will now study the adjacency of cells in the matroid stratification. This material is essentially drawn from Gel'fand *et al.* (1987) and Sturmfels (1989). Let $r = 3, n = 5$ and consider the two matroids M_1 and M_2 which are depicted in Figure 2.4.3.

The matroid M_1 has as its bases all triples except 123, and M_2 has as its

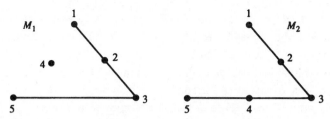

Figure 2.4.3: The matroid M_2 is a weak image of the matroid M_1.

bases all triples except 123 and 345. Hence M_2 is a *weak image* of M_1, which means that every basis of M_2 is also a basis of M_1 (cf. Section 7.7). The realization spaces (or strata) of both matroids are subsets of the six-dimensional Grassmann variety $G_3(K^5)$ of 3-flats in K^5. We recommend that the reader check the following facts about these matroid strata:

(i) The stratum $\mathcal{R}(M_1)$ is a smooth 5-dimensional manifold. It is dense in the hypersurface defined by the equation $[123] = 0$ in $G_3(K^5)$. *Hint:* Identify $\mathcal{R}(M_1)$ with the set of points $(x_1, x_2, x_3, x_4, x_5)$ in 5-space such that $x_i \neq 0$, for $1 \leq i \leq 5$, and $\begin{vmatrix} x_1 & x_2 \\ x_3 & x_4 \end{vmatrix} \neq 0$, via the coordinatization matrix

$$\begin{pmatrix} x_1 & x_2 & 0 \\ 1 & 0 & 0 \\ 0 & 1 & 0 \\ 0 & 0 & 1 \\ x_3 & x_4 & x_5 \end{pmatrix}.$$

(ii) The stratum $\mathcal{R}(M_2)$ is a smooth 4-dimensional manifold. It is dense in the projective subvariety of $G_3(K^5)$ which is defined by the equations $[123] = [345] = 0$. *Hint:* Similarly for (x_1, x_2, x_3, x_4), $x_i \neq 0$, and

$$\begin{pmatrix} x_1 & x_2 & 0 \\ 1 & 0 & 0 \\ 0 & 1 & 0 \\ 0 & 0 & 1 \\ 0 & x_3 & x_4 \end{pmatrix}.$$

(iii) The projective orbit spaces $\mathcal{R}(M_1)/H$ and $\mathcal{R}(M_2)/H$ are both one-dimensional. *Hint:* This follows from

$$\text{diag}(\frac{x_4}{x_2}, x_3, x_4, x_5, 1) \cdot \begin{pmatrix} x_1 & x_2 & 0 \\ 1 & 0 & 0 \\ 0 & 1 & 0 \\ 0 & 0 & 1 \\ x_3 & x_4 & x_5 \end{pmatrix} \cdot \text{diag}(\frac{1}{x_3}, \frac{1}{x_4}, \frac{1}{x_5})$$

$$= \begin{pmatrix} \frac{x_1 x_4}{x_2 x_3} & 1 & 0 \\ 1 & 0 & 0 \\ 0 & 1 & 0 \\ 0 & 0 & 1 \\ 1 & 1 & 1 \end{pmatrix}.$$

(iv) Every realization of M_2 can be approximated by realizations of M_1, that is, we have the inclusion $\mathcal{R}(M_2) \subset \overline{\mathcal{R}(M_1)}$. (This is clear from the coordinatization matrices shown.)

It will now be shown that the matroid stratification is not normal, that is, the answer to our question (2) is "no". Let $r = 3, n = 7$ and consider the two matroids M_1 and M_2 which are depicted in Figure 2.4.4.

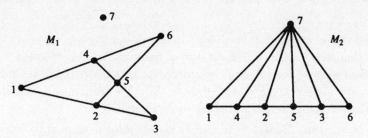

Figure 2.4.4: $\mathcal{R}(M_2) \cap \overline{\mathcal{R}(M_1)} \neq \emptyset$ but $\mathcal{R}(M_2)$ is not contained in $\overline{\mathcal{R}(M_1)}$.

The bases of M_1 are all triples except $123, 146, 256, 345$, and the bases of M_2 are all triples containing the point 7. Hence M_2 is a weak image of M_1. The stratum $\mathcal{R}(M_2)$ intersects the closure of the stratum $\mathcal{R}(M_1)$ because the quadrilateral set $\{1, 2, 3, 4, 5, 6\}$ can be deformed to lie arbitrary close to a line. We claim that $\mathcal{R}(M_2)$ is not contained in $\overline{\mathcal{R}(M_1)}$. Intuitively, the reason is that by "flattening" a quadrilateral set we can get only a special class of configurations of six points on a line. In fact, interpreting the "dummy point" 7 as the center of a projection, we get exactly those six-point configurations on a line which are projections of quadrilateral sets. Algebraically, it can be shown that the polynomial function $P := [137][257][467] - [237][567][147]$ is contained in the ideal generated by $[123], [146], [256]$ and $[345]$ in the coordinate ring of $G_3(K^7)$. This means that P vanishes on $\mathcal{R}(M_1)$ and therefore it vanishes on $\overline{\mathcal{R}(M_1)}$. On the other hand, it is easy to find a point ξ in $\mathcal{R}(M_2)$ where P is non-zero. As an example, we get $P(\xi) = 9$ if we take $\xi = \sum_{1 \leq i < j \leq 6}(j - i) \cdot \mathbf{e}_i \wedge \mathbf{e}_j \wedge \mathbf{e}_7$. This 3-flat corresponds to placing the points $1, 2, 3, 4, 5, 6$ in order and equidistantly on a line.

The following statement gives a partial answer to the question how the combinatorial notion of "being in more special position" relates to the adjacency relation in the matroid stratification.

2.4.6 Proposition. *Let M_1 and M_2 be rank r matroids on E_n such that $\mathcal{R}(M_2) \cap \overline{\mathcal{R}(M_1)} \neq \emptyset$. Then M_2 is a weak image of M_1.*

Proof. Let $\{i_1, i_2, \ldots, i_r\}$ be a non-basis of M_1. This means that $[i_1 i_2 \ldots i_r] = 0$ on $\mathcal{R}(M_1)$ and therefore also on its closure $\overline{\mathcal{R}(M_1)}$. Since the closure intersects the stratum of M_2, the coordinate function is zero on some points of $\mathcal{R}(M_2)$. But this means that it is zero on all of $\mathcal{R}(M_2)$, and $\{i_1, i_2, \ldots, i_r\}$ is a non-basis also of M_2. □

We answer question (3) by showing that the converse of Proposition 2.4.6 is not true.

2.4.7 Proposition.   *There exist rank 4 matroids M_1 and M_2 on E_8, both realizable over K, such that M_2 is a weak image of M_1 but $\mathcal{R}(M_2) \cap \overline{\mathcal{R}(M_1)} = \emptyset$.*

Proof. We describe two configurations M_1 and M_2 of eight points in projective 3-space which have the stated properties. For M_1 we take four lines $\overline{12}$, $\overline{34}$, $\overline{56}$, $\overline{78}$, which meet in a point in 3-space, but with no other dependencies among $1, 2, \ldots, 8$. For M_2 we collapse 7 and 8 into a single point, we move point 6 into the intersection point $\overline{34} \cap \overline{56}$, and we move $\overline{12}$ away from this intersection but into the plane 3456. More precisely, M_1 has the non-bases 1234, 1256, 1278, 3456, 3478, 5678, and M_2 has as its non-bases 3467, 3468, and all quadruples which contain 78 or are contained in 123456. Hence M_2 is a weak image of M_1.

Now consider the polynomial function

$$P := [1268][3458] - [1258][3468]$$

on the Grassmann variety $G_4(K^8)$. One can show that vanishing of this polynomial means that the planes 128, 348 and 568 meet in a common line in 3-space. The line $\overline{78}$ constitutes such a line for every realization of M_1, and consequently we have $P = 0$ on $\overline{\mathcal{R}(M_1)}$. On the other hand, on the stratum $\mathcal{R}(M_2)$ we have $P = [1268][3458] \neq 0$ because both quadruples are bases of M_2.   □

We refer to (White 1991) for algorithmic techniques for extracting the geometric content of bracket polynomials (such as the above P).

Concerning our last question (4), we have the following surprising universality result.

2.4.8 Universality Theorem.   *Given any affine algebraic variety V, defined over $\mathbb{Q}$, there exists a rank 3 matroid M whose projective stratum $\mathcal{R}(M)/H$ is birationally isomorphic to V.*

This assertion is a weak version of the *Universality Theorem for Oriented Matroids* of N.E. Mnëv (Theorem 8.6.6). Theorem 2.4.8 can be proved in a straightforward manner using the technique of projective addition and multiplication ("von-Staudt construction") which is described in White (1987, Chapter 1) or in Bokowski and Sturmfels (1989b, Section 2).

On the other hand, the substantially more powerful construction of Mnëv (1988) allows us to replace "birationally isomorphic" by "isomorphic" in the statement of Theorem 2.4.8. This implies in particular that arbitrary singularities of V can be encoded into the matroid stratum $\mathcal{R}(M)$, and it also implies that $\mathcal{R}(M)$ need not be a $K(\pi, 1)$-space. These two questions had been asked in Section 5.1 of Gel'fand *et al.* (1987).

We are now prepared to describe the refinement of the matroid stratification determined by *oriented matroids*. Here K must be an ordered field, and for simplicity we choose $K = \mathbb{R}$. Definition 3.5.3 describes an oriented matroid as a pair $\mathcal{M} = (E, \chi)$ consisting of a finite set E and a non-zero, alternating map $\chi : E^r \to \{-1, 0, +1\}$ which satisfies the following:

(2.4.3) For all $b_1, b_2, \ldots, b_r, b_1', b_2', \ldots, b_r' \in E$ such that

$$\chi(\, b_i', b_2, b_3, \ldots, b_r \,) \cdot \chi(b_1', \ldots, b_{i-1}', b_1, b_{i+1}', \ldots, b_r') \geq 0$$

for $i = 1, 2, \ldots, r$, we have

$$\chi(b_1, b_2, b_3, \ldots, b_r) \cdot \chi(b_1', b_2', \ldots, b_r') \geq 0.$$

While the basis exchange axiom (2.4.1) for matroids was interpreted earlier as a zero/non-zero combinatorial analogue to the Grassmann-Plücker relations, we now find that the signed basis axiom (2.4.3) for oriented matroids is the signed combinatorial analogue to the Grassmann-Plücker relations. It follows from Theorem 2.4.3 that the signs of the Plücker coordinates of any r-flat $\xi \in G_r(\mathbb{R}^n)$ define an oriented matroid $\mathcal{M}_\xi = (E_n, \chi)$ via $\chi(i_1, i_2, \ldots, i_r) := \mathrm{sign}(\xi_{i_1 i_2 \ldots i_r})$. Conversely, given any rank r oriented matroid $\mathcal{M} = (E_n, \chi)$, its *stratum* or *realization space* is the set $\mathcal{R}(\mathcal{M}) := \{\, \xi \in G_r(\mathbb{R}^n) \,:\, \mathcal{M}_\xi = \mathcal{M} \,\}$.

The stratum $\mathcal{R}(\mathcal{M})$ is a semi-algebraic subset of the Grassmann variety, defined by the system of polynomial equations and inequalities "$\mathrm{sign}\,[i_1 i_2 \ldots i_r] = \chi(i_1, i_2, \ldots, i_r)$". Since each oriented matroid stratum $\mathcal{R}(\mathcal{M})$ is contained in the stratum $\mathcal{R}(\underline{\mathcal{M}})$ of its underlying matroid $\underline{\mathcal{M}}$, the *oriented matroid stratification* of $G_r(\mathbb{R}^n)$ is a refinement of the matroid stratification of $G_r(\mathbb{R}^n)$. Note, for instance, that the matroid stratification has a unique open, top-dimensional, generic stratum, stemming from the uniform rank r matroid on E_n. This generic matroid stratum is decomposed in the oriented matroid stratification into many distinct generic strata, one for each realizable uniform oriented matroid.

Our use of similar symbols M_ξ and $\mathcal{M}_\xi$ (respectively $\mathcal{R}(M)$ and $\mathcal{R}(\mathcal{M})$) for matroids and oriented matroids is meant to stress the close analogy between both cases. Indeed, we can ask the same questions (1)–(4) also for oriented matroids, and we will obtain the same answers. Here an oriented matroid $\mathcal{M}_2$ is said to be a *weak image* of an oriented matroid $\mathcal{M}_1$ if every basis of $\mathcal{M}_2$ is also a basis with the same sign in $\mathcal{M}_1$, cf. Section 7.7. We will return to the discussion of realization spaces of oriented matroids in Chapter 8, where the following issues (and many more) will be addressed:

(a) combinatorial criteria and algebraic methods for deciding "$\mathcal{R}(\mathcal{M}) \neq \emptyset$?";

(b) disconnected strata $\mathcal{R}(\mathcal{M})$, corresponding to oriented matroids without the isotopy property;

(c) a "gap theorem": a tight bound for the size of coordinates of $\mathbb{Q}$-rational points in the open strata.

We close this section by outlining a connection between matroids and the theory of toric varieties (cf. Oda 1988) which arises naturally from what we have discussed so far. As before let $H = (\mathbb{C}^*)^n$ denote the n-dimensional complex torus, and write $H_+ := (\mathbb{R}_+)^n$ for the (positive) real torus contained in it. In the following we fix an arbitrary r-flat $\xi \in G_r(\mathbb{C}^n)$. The closure $\overline{H\xi}$ of the torus orbit

$$\Big\{ \sum_{1 \leq i_1 < \ldots < i_r \leq n} \xi_{i_1 i_2 \ldots i_r} \cdot t_{i_1} t_{i_2} \ldots t_{i_r} \cdot \mathbf{e}_{i_1} \wedge \mathbf{e}_{i_2} \wedge \ldots \wedge \mathbf{e}_{i_r}$$

$$\in G_r(\mathbb{C}^n) \mid t_1, t_2, \ldots, t_n \in \mathbb{C}^* \Big\}$$

is an irreducible subvariety of the Grassmann variety. The *projective toric variety* $\overline{H\xi}$ is the zero set of the following set of homogeneous bracket polynomials of degree 2:

$$\xi_{i_1 i_2 \ldots i_{d-2} j_1 k_1} \xi_{i_1 i_2 \ldots i_{d-2} j_2 k_2} \cdot [i_1 i_2 \ldots i_{d-2} j_1 j_2][i_1 i_2 \ldots i_{d-2} k_1 k_2]$$

$$- \; \xi_{i_1 i_2 \ldots i_{d-2} j_1 j_2} \xi_{i_1 i_2 \ldots i_{d-2} k_1 k_2} \cdot [i_1 i_2 \ldots i_{d-2} j_1 k_1][i_1 i_2 \ldots i_{d-2} j_2 k_2],$$

for all $i_1, \ldots, i_{d-2}, j_1, j_2, k_1, k_2$. This bracket polynomial expresses the "invariance of cross-ratio" along H-orbits. Recall from our earlier discussion that another r-flat η is contained in the H-orbit of ξ if and only if they define the same projective point configuration. This, however, is the case if and only if the cross-ratios $(j_1, j_2; k_1, k_2)$ around all colines $i_1 i_2 \ldots i_{d-2}$ are identical for ξ and η.

Both the coordinate ring of the toric variety $\overline{H\xi}$ and its associated lattice polytope are well-known objects in matroid theory. The coordinate ring of $\overline{H\xi}$ is isomorphic to the subring of $\mathbb{C}[t_1, t_2, \ldots, t_n]$ generated by all monomials of the form $t_{b_1} t_{b_2} \ldots t_{b_r}$ where $\{b_1, b_2, \ldots, b_r\}$ is a basis of the matroid $M = M_\xi$. This ring is the *basis monomial ring* of M which was introduced and studied by White (1977). If M is projectively unique, then the varieties $\overline{H\xi}$ and $\mathcal{R}(M)$ are equal, and the basis monomial ring is equal up to radical to the *bracket ring* of M (cf. White 1987). The lattice polytope associated with the toric variety $\overline{H\xi}$ is the *basis polytope* $\mathcal{P}_M$ of the matroid M which is defined as the convex hull of all incidence vectors of bases, i.e.,

$$\mathcal{P}_M = \text{conv}\big\{ \mathbf{e}_{b_1} + \mathbf{e}_{b_2} + \ldots + \mathbf{e}_{b_r} \mid \{b_1, b_2, \ldots, b_r\} \text{ is a basis of } M \big\}.$$

The set of closed H-orbits in $\overline{H\xi}$, ordered by inclusion, forms a lattice which is isomorphic to the face lattice of the basis polytope $\mathcal{P}_M$. This polytope was introduced and studied by Edmonds (1970).

In addition to this combinatorial correspondence there is also a highly nontrivial topological correspondence between the toric variety and the basis polytope. The closed orbit $\overline{H_+\xi}$ under the *real* torus H_+, called a *Grassmannian simplex* by Gel'fand and MacPherson (1987), is a topological ball of some dimension $\leq n - 1$. The results in Atiyah (1982) imply that the *moment map*

$$\mu : G_r(\mathbb{C}^n) \;\longrightarrow\; \mathbb{R}^n$$

$$\xi \;\longmapsto\; \frac{1}{|\xi|^2} \cdot \sum_{k=1}^{n} \sum_{\substack{i_1 < \ldots < i_r \\ k \in \{i_1, \ldots, i_r\}}} |\xi_{i_1 \ldots i_r}|^2 \cdot \mathbf{e}_k$$

defines a polynomial *homeomorphism* between the Grassmannian simplex $\overline{H_+\xi}$ and the basis polytope $\mathcal{P}_M$. Here $|\xi|^2$ denotes the sum of the squares of the Plücker coordinates of ξ. To round off the picture, we remark that, for $\xi \in$

$G_r(\mathbb{R}^n)$, the open Grassmannian simplex $H_+\xi$ is not only contained in the matroid stratum but even in the oriented matroid stratum of ξ.

2.5 Complexified arrangements

We have discussed in Sections 1.2 and 2.1 how every real hyperplane arrangement $\mathcal{A}$ determines an oriented matroid $\mathcal{M}(\mathcal{A})$, and how the oriented matroid encodes the combinatorial structure of the arrangement. We will now consider the complex arrangement naturally associated to a real one. It turns out that in this situation the oriented matroid of the real arrangement carries a lot of non-trivial information also about the complexification, and serves as a useful tool to describe and analyze it.

2.5.1 Definition.

(i)   A *complex hyperplane arrangement* is a finite collection $\mathcal{B} = (H_e)_{e \in E}$ of codimension-one linear subspaces in $\mathbb{C}^d$.

(ii) Let $\mathcal{A} = (H_e)_{e \in E}$ be a *central arrangement of hyperplanes* in $\mathbb{R}^d$. Then every hyperplane H_e is defined by a linear equation with real coefficients. The *complexification* of $\mathcal{A}$ is the complex hyperplane arrangement $\mathcal{A}^{\mathbb{C}} = (H_e^{\mathbb{C}})_{e \in E}$ in $\mathbb{C}^d$ defined by the same set of linear equations.

The relationship between a real arrangement $\mathcal{A}$ and its complexification $\mathcal{A}^{\mathbb{C}}$ is quite subtle, and the process of complexification has interesting combinatorial aspects. Here are some initial matroid-theoretic observations.

(1)   The matroid of an arrangement (real or complex) is given by its intersection lattice, or, equivalently, by the linear dependences between the equations that define the hyperplanes. These linear dependences are not changed by the complexification. Thus the arrangements $\mathcal{A}$ and $\mathcal{A}^{\mathbb{C}}$ have isomorphic intersection lattices $L(\mathcal{A}) \cong L(\mathcal{A}^{\mathbb{C}})$, and they have the same matroid $M(\mathcal{A}) = M(\mathcal{A}^{\mathbb{C}})$. It follows that the matroid $M(\mathcal{A}^{\mathbb{C}})$ of a complexified arrangement $\mathcal{A}^{\mathbb{C}}$ is the underlying matroid of the oriented matroid $\mathcal{M}(\mathcal{A})$.

(2)   Discussing linear subspaces of $\mathbb{C}^d$, one has to distinguish real and complex dimension. $\mathbb{C}^d$ is a real vector space of dimension $2d$. The elements of $\mathcal{B}$ are complex hyperplanes, hence subspaces of real codimension 2. Similarly, a flat $W \in L(\mathcal{B})$ of matroid rank k has complex codimension k, hence it is a vector space of real dimension $2d - 2k$.

(3)   A complex arrangement does not in general have an oriented matroid naturally associated with it, since $\mathbb{C}$ is not an ordered field. In fact, being (isomorphic to) a complexified arrangement is a serious restriction for the structure of a complex arrangement. It is known (see Example 6.6.2(3)) that there are complex matroids that have no real representations, so the complex arrangements that correspond to such matroids do not arise by complexification of real arrangements.

For any complex arrangement $\mathcal{B} = (H_e)_{e \in E}$ in $\mathbb{C}^d$ the union $V(\mathcal{B}) := \bigcup_{e \in E} H_e$

of its hyperplanes is an interesting singular complex variety in $\mathbb{C}^d$. Such varieties have been studied from several points of view, using the methods of analytic and of algebraic geometry. For instance, the Terao modules that were defined in Section 2.3(d) naturally arise in this setting.

On the other hand, the *complement* $Y(\mathcal{B}) := \mathbb{C}^d \setminus \bigcup_{e \in E} H_e$ of an arrangement $\mathcal{B}$ is an open submanifold of $\mathbb{C}^d$. Observe that the removal of the hyperplanes (which have real codimension 2) does not disconnect the space $\mathbb{C}^d$, so $Y(\mathcal{B})$ is a connected smooth orientable manifold. Techniques of differential and algebraic topology are suited for its study. See Orlik (1989) and Orlik and Terao (1992) for an extensive survey of work in this direction.

In the following we sketch a few results that show how much of the topological structure of the manifold $Y(\mathcal{B})$ is in fact determined by the matroid of the arrangement $\mathcal{B}$, and in the complexified case $\mathcal{B} = \mathcal{A}^{\mathbb{C}}$ by the oriented matroid of $\mathcal{A}$. This is of interest for the general program of understanding and classifying the manifolds of type $Y(\mathcal{B})$. Here we must of course expect different answers, depending on the classification scheme. For this, various topological equivalence relations for the complements $Y(\mathcal{B})$ suggest themselves, for example:

> diffeomorphic
>
> homeomorphic (denoted $\cong$)
>
> homotopy equivalent (denoted $\simeq$)
>
> isomorphic cohomology.

For general manifolds, each of the first three properties is stronger than the next one:

> *diffeomorphic* $\Longrightarrow$ *homeomorphic* $\Longrightarrow$ *homotopy equivalent*
> $\Longrightarrow$ *isomorphic cohomology*,

and all the reverse implications are false.

The first result showing the relevance of matroids for the topology of $Y(\mathcal{B})$ is due to Orlik and Solomon (1980). It implies that the cohomological structure is a matroid property. To be able to state their main result, we need to first define the Orlik-Solomon algebra $OS(M)$ of a matroid M. For more details about this algebra see Orlik and Solomon (1980), Orlik (1989) or Chapter 7 of White (1992).

2.5.2 Definition. Let $M = M(E)$ be a matroid of rank r. Consider the free abelian group $\mathbb{Z}^E$ generated by the elements of E, and let $\wedge \mathbb{Z}^E$ be the exterior $\mathbb{Z}$-algebra over this group. In $\wedge \mathbb{Z}^E$, let I_M be the $\mathbb{Z}$-ideal generated by the elements $\sum_{k=1}^{s}(-1)^k e_1 \wedge \dots \wedge \widehat{e_k} \wedge \dots \wedge e_s$, one for each circuit $\{e_1, \dots, e_s\}$ of M.

Then the *Orlik-Solomon algebra of* M is the quotient $OS(M) := \wedge \mathbb{Z}^E / I_M$.

The Orlik-Solomon algebra naturally inherits a grading from the exterior algebra. The Poincaré polynomial of this graded algebra turns out to be

$$(2.5.1) \qquad \sum_{i \geq 0} \operatorname{rank} OS(M)^i \lambda^i = (-\lambda)^r p(M; -\frac{1}{\lambda}),$$

where $p(M; \lambda)$ is the characteristic polynomial (2.3.8) of M.

2.5.3 Theorem (Orlik and Solomon 1980).   *For every complex arrange-ment* $\mathcal{B}$, *the singular cohomology algebra of* $Y(\mathcal{B})$ *with* $\mathbb{Z}$*-coefficients is isomor-phic (as a graded algebra) to the Orlik-Solomon algebra of its matroid:*

$$H^*(Y(\mathcal{B}), \mathbb{Z}) \cong OS(M(\mathcal{B})).$$

2.5.4 Corollary.   *If two complex arrangements have the same matroid, then their complements have isomorphic cohomology:*

$$M(\mathcal{B}_1) \cong M(\mathcal{B}_2) \Longrightarrow H^*(Y(\mathcal{B}_1), \mathbb{Z}) \cong H^*(Y(\mathcal{B}_2), \mathbb{Z}).$$

It is not known whether this implication can be strengthened to homotopy equivalence: $M(\mathcal{B}_1) \cong M(\mathcal{B}_2) \Longrightarrow Y(\mathcal{B}_1) \simeq Y(\mathcal{B}_2)$? However, it is known that $Y(\mathcal{B}_1)$ and $Y(\mathcal{B}_2)$ are diffeomorphic if $\mathcal{B}_1$ and $\mathcal{B}_2$ are connected by a smooth one-parameter family of arrangements having the same matroid throughout (Randell 1989). This connectivity condition does not always hold, because matroid strata in the complex Grassmannian can be disconnected, see Theorems 2.4.8 and 8.6.6.

In summary, for general complex arrangements $\mathcal{B}$ in $\mathbb{C}^d$, the cohomology of $Y(\mathcal{B})$ is completely determined by the matroid $M(\mathcal{B})$, but it is at present unclear whether all details about the homotopy type of $Y(\mathcal{B})$ can be gotten from its matroid alone.

If we specialize to complexified arrangements, the situation improves. Here it turns out that the *oriented matroid* $\mathcal{M}(\mathcal{A})$ contains enough information to determine $Y(\mathcal{A}^{\mathbb{C}})$ up to homeomorphism. A key step in this direction was the construction due to Salvetti (1987) of a regular cell complex $\Delta_{Sal}(\mathcal{A})$ having the homotopy type of $Y(\mathcal{A}^{\mathbb{C}})$. The definition of Salvetti's complex was later reformulated by Ziegler (1987) in the following way, here stated for arbitrary oriented matroids. We refer to Section 4.7 for details concerning regular cell complexes.

2.5.5 Definition. Let $\mathcal{L} \subseteq \{+, -, 0\}^E$ be the face poset (the set of covectors; corresponding to regions, see Section 4.1) of an oriented matroid $\mathcal{M}$, and let $\mathcal{T}$ be the subset of topes (maximal elements). Define a partial order on the set of all pairs (X, T) for which $X \in \mathcal{L}$, $T \in \mathcal{T}$, and $X \leq T$, by the following rule:

$$(X', T') \leq (X, T) \quad \text{if and only if} \quad X \leq X' \text{ and } X' \circ T = T'.$$

The *Salvetti complex* $\Delta_{Sal}(\mathcal{M})$ is the regular cell complex that has this poset as its face poset.

With the methods and results of Chapter 4 it can be seen that the poset of pairs (X, T) defined in this way *is* the face poset of a regular cell complex, so that $\Delta_{Sal}(\mathcal{M})$ is well-defined, see Exercise 2.21. A different description of the Salvetti complex is given in Björner and Ziegler (1992a), which generalizes its construction and the following result to arbitrary complex arrangements.

2.5.6 Proposition (Salvetti 1987).   *The complement of the complexifica-tion of a real arrangement is homotopy equivalent to the Salvetti complex of its oriented matroid:*

$$Y(\mathcal{A}^{\mathbb{C}}) \simeq \Delta_{Sal}(\mathcal{M}(\mathcal{A})).$$

It follows that if two real arrangements have the same *oriented* matroid, then the complements of their complexifications are homotopy equivalent. This conclusion can be sharpened as follows.

2.5.7 Proposition (Björner and Ziegler 1992a). *If two real arrangements have the same oriented matroid, then the complements of their complexifications are homeomorphic:*

$$\mathcal{M}(\mathcal{A}_1) \cong \mathcal{M}(\mathcal{A}_2) \Longrightarrow Y(\mathcal{A}_1^{\mathbb{C}}) \cong Y(\mathcal{A}_2^{\mathbb{C}}).$$

It is not clear whether $Y(\mathcal{A}_1^{\mathbb{C}})$ and $Y(\mathcal{A}_2^{\mathbb{C}})$ have to be diffeomorphic in this case. On the other hand, it is known that complexified arrangements with different matroids can have homotopy equivalent complements, see Falk (1990).

The construction of the Salvetti complex given above yields a regular cell complex $\Delta_{Sal}(\mathcal{M})$ of dimension r for every oriented matroid $\mathcal{M}$ of rank r. According to Gel'fand and Rybnikov (1990) this complex has the "right cohomology":

$$H^*(\Delta_{Sal}(\mathcal{M}), \mathbb{Z}) \cong OS(\underline{\mathcal{M}}).$$

Thus, the $\mathcal{A}^{\mathbb{C}}$-case of the Orlik-Solomon theorem 2.5.3 generalizes to (non-realizable) oriented matroids. One can expect similar "oriented matroid versions" of other results about the topology of complexified arrangements. For instance, the celebrated Deligne (1972) theorem (*"if $\mathcal{A}$ is simplicial, then $Y(\mathcal{A}^{\mathbb{C}})$ is a $K(\pi, 1)$-space"*) has been generalized to oriented matroids by Cordovil (1994a) and Salvetti (1993); see also Paris (1993). Oriented matroid tools are used for the study of fundamental groups of arrangements also in Cordovil and Fachada (1995) and in Cordovil, Guedes de Oliveira and Las Vergnas (1996).

Real sign vectors with entries from $\{0, +1, -1\}$ are used in oriented matroid theory to specify the combinatorial position of points with respect to some real hyperplane arrangement. Similarly one can use *complex sign vectors* with entries from $\{0, +1, -1, +i, -i\}$ as position vectors for points in $\mathbb{C}^d$ with respect to a complex hyperplane arrangement. An analysis of general complex arrangements based on such a complex sign vector encoding is given in Björner and Ziegler (1992a). It is shown there that complexification in the sense of Definition 2.5.1 can be described combinatorially as a process of converting a family of real sign vectors into a family of complex sign vectors. Therefore complexification can be defined for all oriented matroids. This formal concept of complexification in fact carries all the topological information inherent in the realizable case, and the previously mentioned results of Salvetti (1987) and Gel'fand and Rybnikov (1990) find a systematic explanation in this setting.

Exercises

2.1   Describe the construction of the *essential arrangement associated with $\mathcal{A}$*, for each of the four types of arrangements of Definition 2.1.1.

2.2   Let $\mathcal{A}$ be an affine arrangement.

(a) Show that the face poset $\mathcal{F}(\mathcal{A})$ is pure.

(b) Conclude that $\mathcal{A}$ is essential if and only if every minimal cell induced by $\mathcal{A}$ is a point.

(c) Define a reasonable concept of *rank* for affine arrangements.

2.3   Let $\mathcal{A}$ be an affine arrangement, and let $\hat{\mathcal{A}}$ be the associated central arrangement. Identify $\mathcal{F}(\mathcal{A})$ as a subposet of $\mathcal{F}(\hat{\mathcal{A}})$.

2.4   What does Shannon's Theorem 2.1.5 prove about affine hyperplane arrangements?

2.5   Let $\widehat{\mathcal{F}}(\mathcal{A}) = \mathcal{F}(\mathcal{A}) \cup \{\hat{1}\}$ be the face poset of a central hyperplane arrangement, augmented by a new top element $\hat{1}$. Show the following:

(a) $\widehat{\mathcal{F}}(\mathcal{A})$ is a lattice,

(b) $\widehat{\mathcal{F}}(\mathcal{A})$ is isomorphic to the face lattice of a centrally symmetric convex polytope.

2.6   Enumerate all combinatorial types of 3-dimensional zonotopes with at most 6 zones.

2.7   Prove Proposition 2.2.10: A zonotope tiles space if and only if its (oriented) matroid is unimodular.

2.8   Let $\Sigma(C_{10})$ denote the set of all cubical subdivisions of a regular 10-gon C_{10}.

(a) List all elements of $\Sigma(C_{10})$ up to congruence.

(b) Show that all cubical subdivisions of C_{10} are regular.

(c) We define a graph on the set $\Sigma(C_{10})$ by joining two subdivisions by an edge if they agree outside of a convex hexagon. Show that $\Sigma(C_{10})$ is the edge graph of a 3-dimensional zonotope with 10 zones (cf. Billera and Sturmfels 1992).

2.9   Let P_n be the *n-permutohedron*, which is defined as the convex hull of all points $\pi = (\pi_1, \pi_2, \ldots, \pi_n)$ in $\mathbb{R}^n$ where π ranges over all permutations of $E_n = \{1, 2, \ldots, n\}$ (see Example 2.2.5 for the case $n = 4$). For any chain $\emptyset \subset A_1 \subset A_2 \subset \ldots \subset A_k \subset E_n$, let $F(A_1, \ldots, A_k)$ be the convex hull of all vertices $\pi^{-1} \in P_n$ for permutations π which first list the set A_1 in some order, then $A_2 \setminus A_1$, then $A_3 \setminus A_2$, and so on. For instance, $(3, 2, 5, 1, 6, 7, 4) \in F(\{2, 4\}, \{1, 2, 3, 4, 7\}), n = 7$.

(a) Show that $F(A_1, \ldots, A_k)$ is an $(n - k - 1)$-dimensional face of P_n, and that every face of P_n is of this form.

(b) Conclude that the polar polytope to P_n is the barycentric subdivision of an $(n - 1)$-simplex.

(c) What is the number of facets of P_n? Edges?

2.10 Show that the n-permutohedron P_n is the solution set in $\mathbb{R}^n$ of the following system of linear inequalities:

(i) $\sum_{i \in E_n} x_i = \binom{n+1}{2}$

(ii) $\sum_{i \in J} x_i \leq n \cdot |J| - \binom{|J|}{2}$ for all $J \subseteq E_n$. (Rado 1952)

2.11 Use Proposition 6.5.4 to derive the following two results of McMullen (1971a):

(a) Every zonotope has at least one facet which is a prism.

(b) Every r-dimensional zonotope has at least one $\lfloor \frac{r+1}{2} \rfloor$-dimensional face which is a parallelotope.

2.12 A d-polytope is called *cubical* if all its facets are combinatorially equivalent to the $(d-1)$-cube.

(a) Show that for every facet in a cubical polytope there is some other facet from which it is disjoint. (Blind and Blind 1990)

(b) Deduce from (a) that all cubical d-polytopes have at least 3^d faces.

(c)* Determine the maximum number of facets of a cubical d-polytope with n vertices. (see Joswig and Ziegler 1999)

2.13 Show that an arrangement $\mathcal{A}$ of hyperplanes in $\mathbb{R}^d$ is a reflection arrangement if and only if $\sigma_H(\mathcal{A}) = \mathcal{A}$ for every $H \in \mathcal{A}$.

2.14 Show that the intersection lattice of the reflection arrangement $\mathcal{A}_{n-1}$ is isomorphic to the lattice of partitions of the set E_n.

2.15 Show that the tope graph of a reflection arrangement is isomorphic to the Cayley graph of the corresponding Coxeter group.

2.16* Is there any "Bruhat order" of the topes for oriented matroids more general than reflection arrangements?

2.17 For any permutation $\pi \in S_n$ let $inv(\pi)$ denote the number of *inversions*, that is, pairs $(i < j)$ with $\pi_i > \pi_j$.

(a) Show that the geometric description of Bruhat order given in Section 2.3 has the following meaning for the symmetric group S_n: $\pi < \pi'$ if and only if there exist transpositions $\tau_1, \tau_2, \ldots, \tau_k$ such that $\pi' = \pi\tau_1\tau_2\ldots\tau_k$ and $inv(\pi\tau_1\tau_2\ldots\tau_{i-1}) < inv(\pi\tau_1\tau_2\ldots\tau_i)$, for $1 \leq i \leq k$; (cf. Figure 2.3.8).

(b) Let $S_{n,r} = \{\pi \in S_n : \pi_i > \pi_{i+1}$ only if $i = r\}$. Show that Bruhat order restricted to the subset $S_{n,r}$ gives the same poset (distributive lattice) that is described by (2.4.2). For example, Bruhat order on $S_{6,3}$ is isomorphic to the poset in Figure 2.4.1.

2.18 Find two realizable (oriented) matroids $\mathcal{M}_1$ and $\mathcal{M}_2$ such that $\mathcal{M}_2$ is a weak image of $\mathcal{M}_1$ but $\dim(\mathcal{R}(\mathcal{M}_1)) < \dim(\mathcal{R}(\mathcal{M}_2))$.

2.19 Consider the Schubert cell $\Omega(2,4)$ in the Grassmannian $G_2(\mathbb{R}^5)$, and compute its decompositions into matroid strata and oriented matroid strata.

2.20 Choose a realization $\xi \in G_2(\mathbb{R}^4)$ of the uniform rank 2 matroid M on four elements, and draw it as a projective point configuration. Prove that the basis polytope $\mathcal{P}_M$ is an octahedron. Compute the Grassmannian simplex $\overline{H_+\xi}$, and give an explicit formula in coordinates for the homeomorphism (via the moment map) between $\overline{H_+\xi}$ and the octahedron $\mathcal{P}_M$.

2.21 Let P be the poset of pairs (X, T) of Definition 2.5.5.

(a) Show that $P_{\leq(X,T)} \cong \mathcal{L}_{\geq X}$.

(b) Conclude, using Corollary 4.3.4, that $P_{<(X,T)}$ is the face poset of a regular cell decomposition of a sphere.

(c) Conclude, using Proposition 4.7.23, that P is the face poset of a regular cell complex of dimension $r(\mathcal{M})$.

2.22 Let $\Delta = \Delta_{Sal}(\mathcal{M})$ be the Salvetti complex of a rank r oriented matroid $\mathcal{M}$. Suppose that the number of i-dimensional cells in Δ is f_i, for $0 \leq i \leq r$. Show the following:

(a) $f_0 = f_r = $ number of topes in $\mathcal{M}$.

(b) The f-vector $(f_0, f_1, \ldots, f_r)$ depends only on the underlying matroid $\underline{\mathcal{M}}$. (*Hint:* Use Corollary 4.6.3.)

(c) $f_0 - f_1 + f_2 - \ldots + (-1)^r f_r = 0$.

(d) $f_k < 2^k \binom{r-1}{k} f_0$, for $0 < k < r$. (*Hint:* Use Proposition 4.6.9.)

(e) $f_k \geq \binom{r}{k} f_0$, for $0 \leq k \leq r$.

(f) If $\mathcal{M}$ is simplicial, then $f_k = \binom{r}{k} f_0$, for $0 \leq k \leq r$.

(g) If $\mathcal{M}$ is uniform on n elements, then

$$f_k = 2^{k+1}\binom{n}{k} \sum_{i=0}^{r-k-1} \binom{n-k-1}{i}, \qquad \text{for } 0 \leq k < r.$$

(*Hint:* Use Exercise 4.32.)

(h) The boundary complex of every maximal cell in Δ is isomorphic to the sphere $\Delta^{op}(\mathcal{L})$ of Corollary 4.3.4.   (Björner and Ziegler 1992a)

2.23 Show that two arrangements of at most 8 hyperplanes in $\mathbb{C}^3$ have homotopy equivalent complements if they have isomorphic matroids.

(Falk and Sturmfels)

2.24 Let $\mathcal{M} = (E, \mathcal{C})$ be an oriented matroid of rank r, with set of topes $\mathcal{T}$. Define a commutative $\mathbb{Z}$-algebra

$$P(\mathcal{M}) := \mathbb{Z}[x_e : e \in E]/I_{\mathcal{M}},$$

the *Varchenko-Gel'fand ring* of $\mathcal{M}$, where $I_{\mathcal{M}}$ is the ideal generated by the relations

(i) $x_e^2 - x_e = 0$,   for all $e \in E$,

(ii) $\prod_{e \in X^+} x_e \cdot \prod_{e \in X^-} (1 - x_e) = 0$, for all $X \in \mathcal{C}$.

(a) Show that $P(\mathcal{M})$ is isomorphic to the ring of functions $\mathcal{T} \longrightarrow \mathbb{Z}$, for which multiplication is defined component-wise. Derive the rank of $P(\mathcal{M})$ as an Abelian group.

(b) If $\mathcal{A} = (S_e)_{e \in E}$ is a representation of $\mathcal{M}$ by an arrangement of pseudospheres in S^{r-1} (see Chapter 5), then $P(\mathcal{M})$ is isomorphic to the cohomology ring

$$H^*(S^{r-1}\backslash(\cup_{e \in E} S_e), \mathbb{Z}).$$

(c) Identify the ring homomorphisms $P(\mathcal{M}) \longrightarrow \mathbb{Z}$ with $\mathcal{T}$.

(d) For $0 \leq k \leq r$, let P^k be the subspace of elements in $P(\mathcal{M})$ that correspond to polynomials of degree at most k. Show that $P^r = P(\mathcal{M})$, and determine $\mathrm{rank}(P^k/P^{k-1})$.

(e) Show that the systems of all monomials that correspond to k-element subsets of E not containing any broken circuit is a basis for P^k/P^{k-1}. (For the definition of broken circuits, see White (1992, Chapter 7).)

(f) Construct a similar ring $Q(\mathcal{M})$, the *big Varchenko-Gel'fand ring* of $\mathcal{M}$, which encodes the functions $\mathcal{L}(\mathcal{M}) \longrightarrow \mathbb{Z}$. Study its structure.

(Varchenko and Gel'fand 1988; Gel'fand and Rybnikov 1990)

3

Axiomatics

3.1 Introductory remarks

The first two chapters were devoted to an informal introduction to oriented matroids and an overview of their place in pure and applied mathematics. In this chapter we begin the systematic development of oriented matroid theory based on an axiomatic foundation.

A characteristic feature of oriented matroids is the variety of different axiom systems. The lack of one leading axiomatization is not a deficiency but adds richness and widens the applicability of the theory.

There are four basic axiom systems for oriented matroids:

(a) circuit axioms (Definition 3.2.1),

(b) orthogonality axioms (Theorem 3.4.3),

(c) chirotopes, or basis orientations (Definition 3.5.3), and

(d) vector axioms (Theorem 3.7.5).

The axioms abstract natural properties of the examples which motivate the theory. The four basic axiom systems arise this way from

(a) directed graphs,

(b) orthogonal pairs of real vector subspaces,

(c) point configurations and convex polytopes, and

(d) real hyperplane arrangements.

The various axiom systems look quite different at first glance, and the proofs of equivalence are far from trivial. This is in contrast to the theory of ordinary matroids, which is even richer in axiom systems, but for which the proofs of equivalence are considerably easier.

All the equivalences will be established in this chapter, and they provide the basic links between different aspects of the theory, and between different subsequent chapters of this book. One additional axiomatization of oriented matroids

– equivalence classes of pseudosphere arrangements – will be deferred to Chapters 4 and 5, because of its different nature and because of the special tools its analysis and its equivalence proofs require.

Let us mention a few important aspects which follow quite naturally in the course of the equivalence proofs given in this chapter.

The idea of an oriented matroid as a locally realizable abstract configuration discussed in Section 1.5 is made precise by the technique of *reduction to corank 2*. For chirotopes this means that the three-term Grassmann-Plücker relations imply the general ones if the underlying structure is a matroid. Similarly, we get weaker versions of the circuit axioms and of the orthogonality axioms.

Implicit in the presentation in this chapter are algorithms to convert from one encoding of an oriented matroid to another. This yields important procedures for various geometric constructions, whose basic properties are frequently exploited. From a slightly different point of view, we obtain data structures for geometry together with methods for their conversion.

As was already mentioned, the proofs of equivalence in this chapter are technical and in some cases quite difficult. To help the reader gain an overview we list the main results:

– Theorem 3.2.5: Equivalence of strong and weak elimination for circuits.
– Theorem 3.3.7: Excluded minors for circuit signature.
– Theorem 3.4.3: Axioms for dual pairs of oriented matroids.
– Theorem 3.4.4: Painting axioms.
– Theorem 3.5.5: Equivalence of chirotopes and oriented matroids.
– Theorem 3.6.1: Modular circuit elimination (reduction to corank 2).
– Theorem 3.6.2: Three-term Grassmann-Plücker relations.
– Theorem 3.7.5: Vector axioms.

Some rather modest knowledge of the theory of ordinary matroids will be necessary at certain points, in this chapter and later. We will not explicitly review this material, instead we refer to Welsh (1976), Aigner (1979), White (1986, 1987, 1992), and Oxley (1992); see also Kingan (1999).

In this chapter we also establish notation. We have tried to standardize it as far as possible in a book with five authors. On the other hand, it is good to preserve a certain notational flexibility when presenting a subject so rich in different points of view.

Basic to the theory is the notion of a signed set or signed vector, and we now discuss the notation to be used. For a summary see also the index of notation. A *signed set* X is a set $\underline{X}$ together with a partition (X^+, X^-) of $\underline{X}$ into two distinguished subsets: X^+, the set of *positive elements* of X, and X^-, its set of *negative elements*. The set $\underline{X} = X^+ \cup X^-$ is the *support* of X. By $X = Y$ we mean $X^+ = Y^+$ and $X^- = Y^-$. A signed set X is a *restriction* of a signed set Y, or *conforms* to Y, if and only if $X^+ \subseteq Y^+$ and $X^- \subseteq Y^-$. If F is an unsigned set and X is a signed set, then $X \cap F$ denotes the signed set Y such that $Y^+ = X^+ \cap F$ and $Y^- = X^- \cap F$. We also say that Y is the *restriction of* X *to* F, denoted $X|_F$. A signed set X is called *positive* if $X^- = \emptyset$ and is called *negative* if $X^+ = \emptyset$. The *empty* signed set is $(\emptyset, \emptyset)$, and will again be denoted by $\emptyset$.

In contrast to equality, some other set-theoretic symbols, such as $\subseteq, \cup, \cap$, do not make clear sense when applied to signed sets. We will, however, use the symbols $\in, |.|$, and $\backslash$, to be understood as follows. Given signed sets X and Y, then $e \in X$ means $e \in X^+ \cup X^-$, $|X|$ means $|\underline{X}|$, the cardinality of $\underline{X}$, $X \backslash Y$ denotes the restriction of X to $\underline{X} \backslash \underline{Y}$, and similarly if Y is an unsigned set.

For both signed and unsigned sets, we will sometimes use such abbreviations as $X \backslash x$ for $X \backslash \{x\}$ and $X \cup x$ for $X \cup \{x\}$.

Let E be any set. A *signed subset* of E is a signed set whose support is contained in E. A signed subset of E can be identified with an element of $\{-1, 0, 1\}^E$, which we sometimes abbreviate by $\{-, 0, +\}^E$. We will write $X \in \{-1, 0, 1\}^E$ to abbreviate "X is a signed subset of E" and $\mathcal{S} \subseteq \{-1, 0, 1\}^E$ to abbreviate "$\mathcal{S}$ is a family of signed subsets of E." If $E = E_n = \{1, 2, \ldots, n\}$, and $X \in \{-, 0, +\}^E$, then in later chapters we frequently denote X as a vector of length n with entries $-, 0$, or $+$, called a *sign vector*. This was already exemplified in Section 1.1 together with another convention, namely that of denoting a signed set by the string of its elements with a bar over each negative element.

A signed set X can also be viewed as a set X together with a mapping $sg_X : X \to \{-1, 1\}$ such that $X^+ = \{e \in X : sg_X(e) = 1\}$ and $X^- = \{e \in X : sg_X(e) = -1\}$. The mapping sg_X is the *signature* of X. If no confusion results, we will simplify notation by considering X itself as a function, instead of sg_X. Thus $X(e) = 1$ if $e \in X^+$, $X(e) = -1$ if $e \in X^-$, and $X(e) = 0$ if $e \in E \backslash \underline{X}$. We also denote $X(e)$ by X_e when using the vector notation mentioned above.

For any $\mathcal{S} \subseteq \{-1, 0, 1\}^E$ we will denote by $\mathrm{Min}(\mathcal{S})$ the collection of non-empty signed sets in $\mathcal{S}$ with inclusion-minimal support and by $\mathrm{Max}(\mathcal{S})$ the collection of signed sets in $\mathcal{S}$ with inclusion-maximal supports.

The *composition* $X \circ Y$ of two signed sets X, Y is the signed set defined by $(X \circ Y)^+ = X^+ \cup (Y^+ \backslash X^-)$ and $(X \circ Y)^- = X^- \cup (Y^- \backslash X^+)$. This operation is associative, but not commutative in general. We have $X \circ Y = Y \circ X$ if and only if the restrictions of X and Y to their intersection are equal. The signed set X is a restriction of Y if and only if $X \circ Y = Y$. Sometimes this situation is also denoted as $X \leq Y$. For a discussion of composition and this order relation in terms of sign vectors, see Section 4.1.

The *opposite* of a signed set X, denoted by $-X$, is the signed set with $(-X)^+ = X^-$ and $(-X)^- = X^+$. More generally, given a signed set X and a set A, we denote by $_{-A}X$ the signed set with $(_{-A}X)^+ = (X^+ \backslash A) \cup (X^- \cap A)$ and $(_{-A}X)^- = (X^- \backslash A) \cup (X^+ \cap A)$. We say that the signed set $_{-A}X$ is obtained from X by *sign reversal* or *reorientation* on A.

3.2 Circuits

We begin with the definition of an oriented matroid in terms of its signed circuits.

3.2.1 Definition (Oriented matroid: circuit axioms). A collection $\mathcal{C}$ of signed subsets of a set E is the set of *signed circuits* of an *oriented matroid* on E if and only if it satisfies the following axioms:

(C0) $\emptyset \notin \mathcal{C}$,

(C1) $\mathcal{C} = -\mathcal{C}$,   (symmetry)

(C2) for all $X, Y \in \mathcal{C}$, if $\underline{X} \subseteq \underline{Y}$, then $X = Y$ or $X = -Y$,   (incomparability)

(C3) for all $X, Y \in \mathcal{C}$, $X \neq -Y$, and $e \in X^+ \cap Y^-$ there is a $Z \in \mathcal{C}$ such that
$$Z^+ \subseteq (X^+ \cup Y^+)\backslash\{e\} \text{ and}$$
$$Z^- \subseteq (X^- \cup Y^-)\backslash\{e\}.$$ (weak elimination)

3.2.2 Remarks.

(1) By forgetting signs, (C0), (C2), and (C3) reduce to the axioms of ordinary (non-oriented) matroid circuits. Thus, the circuit supports $\underline{\mathcal{C}} := \{\underline{X} : X \in \mathcal{C}\}$ in an oriented matroid $\mathcal{M}$ constitute the circuits of a matroid, denoted by $\underline{\mathcal{M}}$, called the *underlying matroid of* $\mathcal{M}$.

In this situation, $\mathcal{C}$ is called a *circuit orientation* of the matroid $M = \underline{\mathcal{M}}$.

(2) By a natural abuse of terminology, all objects in $\underline{\mathcal{M}}$ will be considered as objects in $\mathcal{M}$. For instance, a set F of elements of $\mathcal{M}$ will be called a *flat* of $\mathcal{M}$ if F is a flat of the underlying matroid $\underline{\mathcal{M}}$, etc. In particular, the *rank* of an oriented matroid is the rank of its underlying matroid.

(3) In the following, we will have to take care that matroid terminology and oriented matroid terminology are compatible.

For example, if $e \in E$ is such that $(\{e\}, \emptyset) \in \mathcal{C}$, we say that e is a *loop* of $\mathcal{M}$. If $e \notin X$ for every $X \in \mathcal{C}$, we say e is a *coloop* of $\mathcal{M}$.

This type of definition is consistent because $e \in E$ is a loop of $\mathcal{M}$ if and only if it is a loop of the underlying matroid $\underline{\mathcal{M}}$, and similarly for coloops.

(4) An ordinary matroid M is *orientable* if there is an oriented matroid $\mathcal{M}$ with underlying matroid M, that is, if it has a circuit orientation. Orientable matroids constitute a proper subclass of matroids. Questions such as deciding whether a given matroid is orientable or enumerating all orientations of an orientable matroid are difficult problems. Some results in this direction can be found in Section 7.9.

In the sequel, we will almost exclusively deal with *oriented* matroids. So, we will say 'circuit' instead of 'signed circuit' if no confusion results.

3.2.3 Remark. Let $\mathcal{M}$ be an oriented matroid on a set E with set of circuits $\mathcal{C}$. Let A be a subset of E. We set $_{-A}\mathcal{C} = \{_{-A}X : X \in \mathcal{C}\}$. It is immediate from the axioms that also $_{-A}\mathcal{C}$ is the set of circuits of an oriented matroid. We will denote this oriented matroid by $_{-A}\mathcal{M}$, and we say that $_{-A}\mathcal{M}$ is obtained from $\mathcal{M}$ by *sign reversal* or *reorientation on* A.

Many properties considered in the present chapter, such as the definition of oriented matroid circuits, are invariant under reorientation. To simplify notation, in proofs we will often consider only the case when certain signed sets are positive. The sentence "Using reorientation if necessary, we may suppose that (a certain signed set) is positive" will mean that the considered property is invariant under reorientation and that there is no loss in generality in restricting the proof to this special case.

Examples of oriented matroid circuits in various contexts have already been given in Chapter 1. Among the many geometric models for oriented matroids, (see Sections 1.2 and 1.3), we will follow two explicitly through this chapter.

The *first* is that of affine point configurations. It will be exemplified by giving a complete and explicit description of the (rank 4, 8-point) oriented matroid CUBE in the various axiom systems.

The *second* model considers a linear subspace of $\mathbb{R}^E$. This leads to high-dimensional situations (CUBE corresponds to the 4-dimensional subspace V_{CUBE} of $\mathbb{R}^8$), but it has the following advantages:

(i)   not only all the circuits, but also all the "vectors" are immediately read off,

(ii)   duality is apparent,

(iii)   loops, coloops, parallel elements etc. are allowed,

(iv)   the model is general: it covers *all* realizable oriented matroids (not only the acyclic ones).

So, let V be a linear subspace of $\mathbb{R}^E$. The *signed support* $s(u)$ of a vector $u \in V$ is the signed set defined by $s^+(u) = \{e \in E : u(e) > 0\}$ and $s^-(u) = \{e \in E : u(e) < 0\}$. Let $\mathcal{V} := \text{Supp}(V)$ be the set of signed supports of all vectors of V, and let $\mathcal{C} := \text{Min}(\mathcal{V})$, the subset of inclusion-minimal (non-empty) signed supports.

We will consider here the oriented matroid of affine dependences of the vertices of a 3-cube.

Example (The CUBE). Let $E = \{e_1, e_2, e_3, e_4, e_5, e_6, e_7, e_8\}$ be the set of vertices of a regular cube in $\mathbb{R}^3$ as shown in Figure 3.2.1. Let V_{CUBE} be the vector subspace of $\mathbb{R}^E$ consisting of all affine dependences of E over $\mathbb{R}$, that is,

$$V_{\text{CUBE}} \quad = \quad \{\lambda \in \mathbb{R}^E : \sum_{e \in E} \lambda(e)e = 0 \quad \text{and} \quad \sum_{e \in E} \lambda(e) = 0\}.$$

We will write CUBE for the oriented matroid of signed affine dependences of E.

Signed circuits can be interpreted geometrically as minimal Radon partitions. In the case of CUBE, a circuit is formed by either four coplanar points with no three collinear, or by five points with no four points coplanar, as shown in Figure 3.2.2. Each circuit has a unique partition into two subsets whose convex hulls

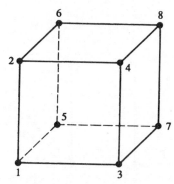

Figure 3.2.1: The rank 4 oriented matroid CUBE.

intersect (the Radon partition), and this determines the positive and negative parts of the circuit (in either order, thus determining a signed circuit and also its negative).

The oriented matroid CUBE has 40 signed circuits, namely (up to opposites),

$$
\begin{array}{cccc}
1\overline{2}\overline{3}4 & 1\overline{2}\overline{5}6 & 1\overline{2}\overline{7}8 & 1\overline{3}\overline{5}7 \\
1\overline{3}\overline{6}8 & 1\overline{4}\overline{5}8 & 2\overline{3}\overline{6}7 & 2\overline{4}\overline{5}7 \\
2\overline{4}\overline{6}8 & 3\overline{4}\overline{5}6 & 3\overline{4}\overline{7}8 & 5\overline{6}\overline{7}8 \\[6pt]
1\overline{2}3\overline{5}8 & 1\overline{2}467\overline{} & 1\overline{3}4\overline{6}7 & 1\overline{4}567\overline{} \\
1\overline{4}678\overline{} & 2\overline{3}45\overline{8} & 2\overline{3}56\overline{8} & 2\overline{3}57\overline{8}
\end{array}
$$

Convention: $1\overline{2}\overline{3}4$ denotes the signed circuit with positive part $\{e_1, e_4\}$ and negative part $\{e_2, e_3\}$. Only half of the circuits are displayed, the other circuits being obtained by taking opposites. Displayed circuits are such that the first element in numerical ordering is positive.

We now show that if $V \subseteq \mathbb{R}^E$ is a vector space, $\mathcal{V} = \mathrm{Supp}(V)$ and $\mathcal{C} = \mathrm{Min}(\mathcal{V})$, then $\mathcal{C}$ is the set of circuits of an oriented matroid.

Clearly $\mathcal{C}$ satisfies (C0) and (C1). To prove (C2), suppose there exist $X, Y \in \mathcal{C}$ such that $\underline{X} \subseteq \underline{Y}$ but $X \neq Y$ and $X \neq -Y$. By minimality X and Y have the same support. Let $u, v \in V$ such that $X = s(u)$ and $Y = s(v)$. Then u, v are two non-zero vectors with the same support which are not proportional. Let $e \in \underline{X}$. Then the vector $w = u - \frac{u(e)}{v(e)}v \in V$ is non-zero, and its support satisfies $s(w) \subseteq s(u)\backslash e$. This contradicts the minimality of $X = s(u)$.

Let us see that (C3) holds in $\mathcal{V}$. Let $X, Y \in \mathcal{V}$, $X \neq -Y$, and $e \in X^+ \cap Y^-$. By definition of $\mathcal{V}$ there are $u, v \in V$ such that $X = s(u)$ and $Y = s(v)$. We have $u(e) > 0$, $v(e) < 0$, and we set $\alpha := -\frac{u(e)}{v(e)} > 0$. Then the vector $w := u + \alpha v$ is a non-zero linear combination of u and v with positive coefficients such that $e \notin s(w)$. Information on signs of coordinates of w are summarized in the following tableau, where ? can be $+$, $-$ or 0.

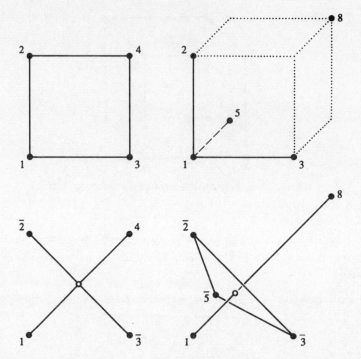

Figure 3.2.2: Two circuits of CUBE and their Radon partition.

	e			$\neq e$					
$X = s(u)$	$+$	$+$	$+$	$+$	$-$	$-$	$-$	0	0	0
$Y = s(v)$	$+$	$-$	$-$	0	$+$	$-$	0	$+$	$-$	0
$Z = s(w)$	$+$	0	$?$	$+$	$?$	$-$	$-$	$+$	$-$	0

Set $Z = s(w) \in \mathcal{V}$. From the tableau we read off that an element $x \neq e$ is positive (resp. negative) in Z only if x is positive (resp. negative) in at least one of X or Y. Thus the sign vector Z satisfies the property (C3). However, $Z = s(w)$ need not be a circuit, and so we have to replace $s(w)$ by a minimal dependence whose sign is compatible with $s(w)$. We will prove in Proposition 3.2.4(i) that this is always possible.

The information on Z in the above tableau can be expressed as follows:

(V3′) *For all $X, Y \in \mathcal{V}$ and $e \in X^+ \cap Y^-$, there is $Z \in \mathcal{V}$ such that*

$Z^+ \subseteq (X^+ \cup Y^+) \setminus \{e\}$,

$Z^- \subseteq (X^- \cup Y^-) \setminus \{e\}$ *and*

$(\underline{X \setminus Y}) \cup (\underline{Y \setminus X}) \cup (X^+ \cap Y^+) \cup (X^- \cap Y^-) \subseteq \underline{Z}$.   (vector elimination)

As will be seen in Corollary 3.7.9, this property is used as one of the axioms for oriented matroid "vectors". Various properties of $\mathcal{V}$ are obtained for other choices of α: see (V2) and (V3″) in Section 3.7.

In order to derive (weak) elimination (C3) for $\mathcal{C}$ from vector elimination (V3′) for $\mathcal{V}$, the following property of subspaces of $\mathbb{R}^E$ is needed:

For every $w \in V\backslash\{0\}$ there is a $w' \in V\backslash\{0\}$ with inclusion-minimal support such that $s^+(w') \subseteq s^+(w)$ and $s^-(w') \subseteq s^-(w)$.

We will prove this property in an abstract setting.

3.2.4 Proposition. *Let $\mathcal{S} \subseteq \{-1,0,1\}^E$ be such that $0 \notin \mathcal{S}$, $\mathcal{S} = -\mathcal{S}$, and $\mathcal{S}$ satisfies the signed elimination axiom (C3). Then*

(i)   *for any $X \in \mathcal{S}$ there is $Y \in \mathrm{Min}(\mathcal{S})$ with $Y^+ \subseteq X^+$ and $Y^- \subseteq X^-$;*

(ii) $\mathrm{Min}(\mathcal{S})$ *is the set of circuits of an oriented matroid.*

Proof.

(i) Using reorientation if necessary, we may suppose that X is positive. Let $Y \in \mathcal{S}$ such that Y is positive, $\underline{Y} \subseteq \underline{X}$ and $|Y|$ as small as possible with these properties. If $Y \in \mathrm{Min}(\mathcal{S})$ we are done. Otherwise consider $Z \in \mathrm{Min}(\mathcal{S})$ with $\underline{Z} \subseteq \underline{Y}$ such that $|Z^-|$ minimal. By definition of Z we have $Z^- \neq \emptyset$. Applying (C3) to Y, Z and $e \in Y^+ \cap Z^-$, there is $Z' \in \mathcal{S}$ with $\underline{Z'} \subseteq \underline{Y}\backslash e$ and $|Z'^-| < |Z^-|$, a contradiction.

(ii) We first prove (C2) for $\mathrm{Min}(\mathcal{S})$. Let $X, Y \in \mathrm{Min}(\mathcal{S})$, $X \neq Y$, $X \neq -Y$, such that $\underline{X} \subseteq \underline{Y}$. By minimality, $\underline{X} = \underline{Y}$, and hence there exists an $e \in (X^+ \cap Y^-) \cup (X^- \cap Y^+)$. By (C3) for $\mathcal{S}$ there is $Z \in \mathcal{S}$ such that $\underline{Z} \subseteq (\underline{X} \cup \underline{Y})\backslash\{e\}$, contradicting the minimality of both X and Y.

We now prove (C3) for $\mathrm{Min}(\mathcal{S})$. Let $X, Y \in \mathrm{Min}(\mathcal{S})$, $X \neq -Y$, and $e \in X^+ \cap Y^-$. Applying (C3) to X, Y and e, there is $Z' \in \mathcal{S}$ such that $Z'^+ \subseteq (X^+ \cup Y^+)\backslash\{e\}$ and $Z'^- \subseteq (X^- \cup Y^-)\backslash\{e\}$. By (i) there is $Z \in \mathrm{Min}(\mathcal{S})$ which is a restriction of Z'. We have $Z^+ \subseteq (X^+ \cup Y^+)\backslash\{e\}$ and $Z^- \subseteq (X^- \cup Y^-)\backslash\{e\}$ as desired.   □

We point out that a stronger property holds in vector spaces:

For every $w \in V$ and $e \in s(w)$ there is a $w' \in V$ with inclusion-minimal support such that $e \in s(w)$, $s^+(w') \subseteq s^+(w)$ and $s^-(w') \subseteq s^-(w)$.

This property also generalizes to oriented matroids, see Lemma 3.3.4 below.

The oriented matroid whose circuits are the minimal supports of V will be denoted by $\mathcal{M} = \mathrm{Supp}(V)$. An oriented matroid $\mathcal{M}$ on E is *realizable* if $\mathcal{M} = \mathrm{Supp}(V)$ for some vector subspace of $\mathbb{R}^E$. Note that for realizable oriented matroids the main axiom (C3) expresses very simple sign properties of linear combinations: the sum of two non-positive coordinates cannot be strictly positive.

3.2.5 Theorem (Strong elimination. Bland and Las Vergnas 1978, Folkman and Lawrence 1978). *Let $\mathcal{C}$ be a collection of signed subsets of a set E satisfying (C0), (C1) and (C2). Then (C3) is equivalent to*

(C3') *for all* $X, Y \in C$, $e \in X^+ \cap Y^-$ *and* $f \in (X^+ \backslash Y^-) \cup (X^- \backslash Y^+)$, *there is a* $Z \in C$ *such that*
$$Z^+ \subseteq (X^+ \cup Y^+) \backslash \{e\},$$
$$Z^- \subseteq (X^- \cup Y^-) \backslash \{e\}$$
and $f \in Z$.   (strong elimination)

Proof. The axiom (C3') clearly implies (C3). We shall prove that (C0), (C1), (C2) and (C3) together imply (C3'). We proceed by induction on $|\underline{X} \cup \underline{Y}|$, with the reduction accomplished by way of contraction. First we establish two auxiliary results regarding contraction.

(1)   *Let* M *be an oriented matroid on* E *with set of signed circuits* C, *and let* $p \in E$. *Then the collection of signed sets* $C_p = \{X \backslash p : X \in C, X \backslash p \neq \emptyset\}$ *satisfies* (C3).

Proof. Consider $X', Y' \in C_p$, $X' \neq -Y'$ and $e \in X'^+ \cap Y'^-$. Let $X, Y \in C$ such that $X' = X \backslash p$ and $Y' = Y \backslash p$. By (C3) applied to X, Y and e, there is $Z \in C$ such that $Z^+ \subseteq (X^+ \cup Y^+) \backslash \{e\}$ and $Z^- \subseteq (X^- \cup Y^-) \backslash \{e\}$. Set $Z' = Z \backslash p$. We have $Z' \neq \emptyset$, otherwise the support of Z equals $\{p\}$ and is thus strictly contained in the support of X, a contradiction with (C2). Hence $Z' \in C_p$ with the desired properties exists, proving (1).   ◁

The set C_p clearly satisfies (C1) too. By (ii) of Proposition 3.2.4,
$$C' = \text{Min}(\{X \backslash p : X \in C, X \backslash p \neq \emptyset\})$$
is the *set of circuits of an oriented matroid*. The oriented matroid with set of circuits C' is the *contraction* M/p *of* M *by* p. Oriented matroid minors obtained by contraction and deletion will be studied in detail in Section 3.3. Here we note that
$$C' \supseteq \{X \backslash p : X \in C, p \in X, X \backslash p \neq \emptyset\} \tag{$*$}$$
follows from (C2).

(2)   *For all* $X \in C$ *and* $e \in X \backslash p$ *there is* $X' \in C'$ *such that* $e \in X'$, $X'^+ \subseteq X^+$ *and* $X'^- \subseteq X^-$.

Proof. Using reorientation if necessary, it suffices to prove (2) when X is positive. If $p \in X$, then $X \backslash p \in C'$ by ($*$), and hence $X' = X \backslash p$ has the required properties. Suppose now that $p \notin X$ and $X = X \backslash p \notin C'$. We can assume that p is not a loop. There exists $Y' \in C'$ with support strictly contained in X. Let $Y \in C$ such that $Y' = Y \backslash p$. By (C2) we have $p \in Y$. Replacing Y by $-Y$ if necessary using (C1), we may suppose $p \in Y^+$.

Consider a signed set $Y_1 \in C$ with the properties $\underline{Y_1} \subseteq \underline{X} \cup p$, $p \in Y_1^+$ (for example, Y has this form) such that $|\underline{X} \cap Y_1^-|$ is minimal. Suppose $x \in \underline{X} \cap Y_1^-$. By (C3), there is $Z \in C$ such that $Z^+ \subseteq (X^+ \cup Y_1^+) \backslash x$ and $Z^- \subseteq (X^- \cup Y_1^-) \backslash x = Y_1^- \backslash x$, implying $|\underline{X} \cap Z^-| < |\underline{X} \cap Y_1^-|$. Now $p \in Z$, otherwise $\underline{Z}$ is strictly contained in $\underline{X}$, contradicting (C2). Thus $p \in Z^+$. Then Z contradicts the definition of Y_1. Hence $\underline{X} \cap Y_1^- = \emptyset$, so Y is positive.

Analogously, if we choose $Y_2 \in C$ with the properties $\underline{Y}_2 \subseteq \underline{X} \cup p$ and $p \in Y_2^-$ (for example, $-Y$ is of this form) and such that $|\underline{X} \cap \underline{Y}_2|$ is minimal, then by application of (C3) and (C2) we find $Y_2^- = \{p\}$.

Since p is not a loop, we have $Y_2 \neq -Y_1$. Applying (C3) to Y_1, Y_2 and p, there is $Z \in C$, with $\underline{Z}$ contained in $\underline{X}$. By (C2) we must have $Z = X$ or $Z = -X$. In particular $\underline{X} = (\underline{Y}_1 \backslash p) \cup (\underline{Y}_2 \backslash p)$. This proves (2) because both $Y_1 \backslash p$ and $Y_2 \backslash p$ are positive, and in C' by (*).   ◁

Now we will prove by induction on $|\underline{X} \cup \underline{Y}|$ that for any collection C of signed sets satisfying (C0), (C1), (C2), (C3), and any $X, Y \in C$, $e \in X^+ \cap Y^-$ and $f \in (X^+ \backslash Y^-) \cup (X^- \backslash Y^+)$, there is $Z \in C$ such that $Z^+ \subseteq (X^+ \cup Y^+) \backslash \{e\}$, $Z^- \subseteq (X^- \cup Y^-) \backslash \{e\}$ and $f \in Z$. By a slight abuse of terminology, we will say that the signed set Z satisfies (C3'). This property is void for $|\underline{X} \cup \underline{Y}| \leq 2$ and is immediate for $|\underline{X} \cup \underline{Y}| = 3$.

Suppose $|\underline{X} \cup \underline{Y}| \geq 4$ and that there is no $Z \in C$ satisfying (C3'). Using reorientation if necessary, we may suppose that X and $Y \backslash X$ are positive. Then we claim that

(3) $(X^+ \cap Y^-) = \{e\}$.

Proof. Otherwise, let $p \in (X^+ \cap Y^-)$ with $p \neq e$. Note that $p \neq f$. Both $X' = X \backslash p$ and $Y' = Y \backslash p$ are in $C' = \mathrm{Min}(\{Z \backslash p : Z \in C\})$. By (1), C' is the set of circuits of an oriented matroid. Since $|\underline{X'} \cup \underline{Y'}| < |\underline{X} \cup \underline{Y}|$, by the induction hypothesis applied to X', Y', e and f, there is $Z' \in C'$ such that $e \notin Z'$, $f \in Z'$, $Z'^+ \subseteq X'^+ \cup Y'^+$ and $Z'^- \subseteq X'^- \cup Y'^-$. Then $Z \in C$ such that $Z' = Z \backslash p$ satisfies (C3'), contradicting our assumption. This proves (3).   ◁

In particular (3) implies $Y^- = \{e\}$.

(4) $|\underline{Y} \backslash \underline{X}| \geq 2$.

Proof. Suppose $\underline{Y} \backslash \underline{X} = \{p\}$. By (3) and (C3) applied to X, Y and e, there is a positive Z in C such that $\underline{Z} \subseteq (\underline{X} \cup \underline{Y}) \backslash e$. Since Z does not satisfy (C3'), we have $f \notin \underline{Z}$. On the other hand $p \in Z$, otherwise $\underline{Z} \subseteq \underline{X}$, contradicting (C2). Apply (C3) to $Y, -Z$ and p. By (C2), the only possible resulting circuits are X and $-X$. Assuming we get X, then $X^+ \subseteq (Y^+ \cup Z^-) \backslash p$, but $e \in X^+$, $e \in Y^-$ and $e \notin Z$, a contradiction. If we get $-X$, then $f \in (-X)^-$, $f \in Y^+$ and $f \notin \underline{Z}$, again a contradiction.   ◁

(5)   *For all $p \in Y \backslash X$ there is a $Z \in C$ with $Z^- = \{p\}$ such that $f \in \underline{Z} \subseteq (\underline{X} \cup \underline{Y}) \backslash e$.*

Proof. By (2) there is X' positive in $C' = \mathrm{Min}(\{U \backslash p : U \in C\})$ such that $f \in \underline{X'} \subseteq \underline{X}$. If $e \notin \underline{X'}$, then we set $Z' = X'$. If $e \in \underline{X'}$, then for $Y \backslash p = Y'$ we have $X' \neq Y'$ by (4) and $|\underline{X'} \cup \underline{Y'}| < |\underline{X} \cup \underline{Y}|$, so by the induction hypothesis there is Z' positive in C' such that $f \in \underline{Z'} \subseteq (\underline{X'} \cup \underline{Y'}) \backslash e$. In both cases let $Z \in C$ be such that $Z' = Z \backslash p$. Since Z cannot satisfy (C3′), we have $Z^- = \{p\}$.   ◁

(6)   There is a positive $W \in C$ such that $\underline{W} \subseteq (\underline{X} \cup \underline{Y}) \backslash e$.

Proof. Let $p \in Y \backslash X$ and let Z be as in (5). By (C3) applied to Y, Z and p, there is $W \in C$ with $W^- \subseteq \{e\}$ such that $\underline{W} \subseteq (\underline{X} \cup \underline{Y}) \backslash p$. We have $|\underline{X} \cup \underline{W}| < |\underline{X} \cup \underline{Y}|$.

Hence $e \notin W$, otherwise by the induction hypothesis applied to X, W, e and f we get a circuit satisfying (C3′).   ◁

Let W be as in (6) and $q \in W \backslash X$. By (5), there is $Z \in C$ with $Z^- = \{q\}$ such that $f \in \underline{Z} \subseteq (\underline{X} \cup \underline{Y}) \backslash e$. We have $|\underline{W} \cup \underline{Z}| < |\underline{X} \cup \underline{Y}|$. Hence by the induction hypothesis applied to W, Z, q and f, there is a positive circuit in C satisfying (C3′), a final contradiction.   □

A still stronger "modular" elimination property (C3″) will be given in Theorem 3.6.1.

3.3 Minors

We recall the definitions of minors for ordinary matroids. Given a matroid M on a set E, then the *submatroid* of M induced by a subset $F \subseteq E$, denoted by $M(F)$, is the matroid on F whose circuits are the circuits of M contained in F. Alternatively, we write $M \backslash (E \backslash F) = M \backslash A$ for $M(F)$ when the emphasis is on the elements *deleted* from M. The *contraction* of M on F, denoted by $M/(E \backslash F)$, is the matroid on F whose circuits are all inclusion-minimal nonempty intersections of circuits of M with F. A *minor* of M is any matroid obtained from M by a sequence of deletions and contractions. For more details on these notions, see White (1986, Chapter 7). We will show in this section that they have natural counterparts for oriented matroids. The proof of the following statement is immediate.

3.3.1 Proposition (Deletion).   *Let M be an oriented matroid on a set E with set of (signed) circuits C, and let F be a subset of E. Then $\overset{\rightarrow}{C'} = \{X \in C : \underline{X} \subseteq F\}$, the set of circuits of M contained in F, is the set of circuits of an oriented matroid on F.*

This oriented matroid is called the submatroid of M induced on F, denoted by $M(F)$.

When the emphasis is on the set $A = E \backslash F$ of elements *deleted* from M, the alternate notation $M \backslash A$ is preferred to $M(E \backslash A)$. The set of circuits of $M \backslash A$ is denoted by $C \backslash A$. In order to avoid confusion note that in general $C \backslash A$ is different from $\{X \backslash A : X \in C\}$.

3.3.2 Proposition (Contraction).   *Let M be an oriented matroid on a set E with set of circuits C, and let F be a subset of E. Then $\mathrm{Min}(\{X|_F : X \in C\})$,* *the set of inclusion-minimal non-empty intersections of circuits of $\mathcal{M}$ with F, is the set of circuits of an oriented matroid on F.*

This matroid is called the contraction of $\mathcal{M}$ to F, denoted by $\mathcal{M}/A$, where $A = E \backslash F$.

Proof. Set $\mathcal{C}' = \{X|_F : X \in \mathcal{C}\}$ and $\mathcal{C}/A = \text{Min}(\mathcal{C}')$. Clearly $\mathcal{C}/A$ satisfies axioms (C0) and (C1) of oriented matroid circuits.

Suppose $\mathcal{C}/A$ did not satisfy (C2). Since supports of signed sets in $\mathcal{C}/A$ are pairwise incomparable with respect to inclusion, there are $X, Y \in \mathcal{C}$ such that $X|_F$ and $Y|_F$ are both in $\mathcal{C}/A$, are neither equal nor opposite and have the same support. Hence there exist $e \in ((X^+ \cap Y^-) \cup (X^- \cap Y^+)) \cap F$ and $f \in ((X^+ \cap Y^+) \cup (X^- \cap Y^-)) \cap F$. By strong elimination (Theorem 3.2.5), there is a $Z \in \mathcal{C}$ such that $f \in \underline{Z} \subseteq (\underline{X} \cup \underline{Y}) \backslash \{e\}$. The support of $Z|_F$ is non-empty and strictly contained in the common support of $X|_F$ and $Y|_F$, contradicting the minimality in the definition of $\mathcal{C}/A$.

In order to show that $\mathcal{C}/A$ satisfies (C3), it suffices, by Proposition 3.2.4, to show that $\mathcal{C}'$ satisfies (C3). Let $X', Y' \in \mathcal{C}'$, $X' \neq -Y'$ and $e \in X'^+ \cap Y'^-$. Since $X' \neq -Y'$, there is $f \in (X'^+ \cap Y'^+) \cup (X'^- \cap Y'^-) \cup (X' \backslash Y') \cup (Y' \backslash X')$. Let $X, Y \in \mathcal{C}$ such that $X' = X|_F$ and $Y' = Y|_F$. By strong elimination in $\mathcal{C}$ applied to X, Y, e and f, there is $Z \in \mathcal{C}$ such that $f \in Z$, $Z^+ \subseteq (X^+ \cup Y^+) \backslash \{e\}$ and $Z^- \subseteq (X^- \cup Y^-) \backslash \{e\}$. Since $f \in Z|_F$, the signed set $Z' = Z|_F \in \mathcal{C}'$ satisfies (C3). □

Interpretations of deletion and contraction for oriented matroids of directed graphs, of linear or affine dependences, or of supports of vectors are similar to the non-oriented case. If $\mathcal{M}$ is the rank $d+1$ oriented matroid of affine dependences of a configuration E in $\mathbb{R}^d$, the contraction $\mathcal{M}/e$ is again an oriented matroid of affine dependences only if e is an *extreme point* (see Chapter 9). Otherwise $\mathcal{M}$ is linear but not affine. If e is an extreme point, then $\mathcal{M}/e$ is the oriented matroid of affine dependences of the configuration obtained from E by central projection from e to a hyperplane in general position that meets all halflines joining e to the other points of E.

Example (The CUBE, continued). In the case of the rank 4 oriented matroid CUBE defined in the previous section, all points are extremal. The contraction CUBE/8 is the oriented matroid of affine dependences of NON-FANO, shown in Figure 3.3.1.

The rank 3 oriented matroid CUBE/8 has 34 circuits, given by the following list of 17 signed subsets of $\{1, 2, 3, 4, 5, 6, 7\}$ (and their opposites):

$$1\overline{2}7 \quad 1\overline{3}6 \quad 1\overline{4}5 \quad 2\overline{4}6 \quad 3\overline{4}7 \quad 5\overline{6}7$$

$$1\overline{2}34 \quad 1\overline{2}\overline{3}5 \quad 1\overline{2}\overline{5}6 \quad 1\overline{3}\overline{5}7 \quad 1\overline{4}67 \quad 2\overline{3}45$$

$$2\overline{3}56 \quad 2\overline{3}57 \quad 2\overline{3}67 \quad 2\overline{4}57 \quad 3\overline{4}56$$

3.3.3 Proposition. *Let $\mathcal{M}$ be an oriented matroid on a set E and A, B be*

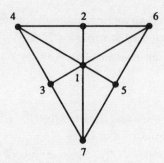

Figure 3.3.1. The rank 3 oriented matroid NON-FANO.

two disjoint subsets of E. We have

$$(\mathcal{M}\backslash A)\backslash B = \mathcal{M}\backslash(A \cup B)$$

$$(\mathcal{M}/A)/B = \mathcal{M}/(A \cup B)$$

$$(\mathcal{M}\backslash A)/B = (\mathcal{M}/B)\backslash A.$$

Proposition 3.3.3 follows easily from the definitions of contraction and deletion.

3.3.4 Lemma.   *Let $\mathcal{S} \subseteq \{+, -, 0\}^E$ with $\mathcal{S} = -\mathcal{S}$ satisfying the strong elimination property (C3'). Then for any $X \in \mathcal{S}$ and $e \in X$ there is a $Y \in \mathrm{Min}(\mathcal{S})$ such that $e \in Y$, $Y^+ \subseteq X^+$, and $Y^- \subseteq X^-$.*

Proof. Using reorientation if necessary, we may suppose X positive. By Proposition 3.2.4 there is a positive $Z \in \mathrm{Min}(\mathcal{S})$ such that $Z \subseteq X$. If $e \in Z$ we are done. Suppose $e \notin Z$.

By (C3') applied to $X, -Z, f \in Z$ and e, there is $Y \in \mathcal{S}$ such that $e \in \underline{Y} \subseteq \underline{X}$ and $Y^- \subseteq Z$. Consider Y with these properties such that $|Y^-|$ is minimal. Then $Y^- = \emptyset$. Otherwise, by (C3') applied to $Y, Z, f \in Y^-$ and e, there is $Y' \in \mathcal{S}$ such that $e \in \underline{Y'} \subset \underline{X}$ with $Y'^- \subset Y^-$, contradicting the choice of Y.

We have shown that if $X \in \mathcal{S}$ there is $Y \in \mathcal{S}$ positive such that $e \in \underline{Y} \subset \underline{X}$. Hence the desired property follows by induction.   □

3.3.5 Lemma.   *Let $\mathcal{M}$ be an oriented matroid on a set E with set of circuits $\mathcal{C}$. For any $a \in E$, $X \in \mathcal{C}$ and $e \in X\backslash a$, there is $X' \in \mathcal{C}/a$ with $e \in X'$, $X'^+ \subseteq X^+$ and $X'^- \subseteq X^-$.*

Proof. Set $\mathcal{S} = \{X\backslash a : X \in \mathcal{C}\}$. Since $\mathcal{C}/a = \mathrm{Min}(\mathcal{S})$, we only need to apply Lemma 3.3.4 to $e \in X\backslash a \in \mathcal{S}$.   □

Lemma 3.3.5 may be interpreted as saying that contraction is a of oriented matroids (see Section 7.7). The following theorem characterizes the excluded minors for circuit orientations. It is important, because it identifies the two base cases for proofs "by deletion and contraction" for matroid orientations.

3.3.6 Definition (circuit signatures). *A circuit signature of a matroid M assigns to each circuit C of M two opposite signed sets X and $-X$ supported by C.*

Note that by this definition, a circuit signature satisfies (C0), (C1) and (C2), but not necessarily (C3). Thus a matroid with a circuit signature is not yet an oriented matroid: not every circuit signature is a circuit orientation.

3.3.7 Theorem (Las Vergnas 1978b, 1984a). *Let M be a matroid on a set E, and let $\mathcal{S}$ be a circuit signature of M. Suppose that for all $e \in E$ the induced circuit signatures $\mathcal{S}\backslash e$ and $\mathcal{S}/e$ are circuit orientations of $M\backslash e$ and M/e respectively. Then either*

(i)   *$\mathcal{S}$ is a circuit orientation of $\mathcal{M}$, or*

(ii) *$|E| = 3$ and, up to reorientation and relabeling, $\mathcal{S} = \{12, 13, 23 \text{ and their opposites}\}$, or*

(iii) *$|E| = 4$ and, up to reorientation and relabeling, $\mathcal{S} = \{123, 1\overline{24}, 1\overline{3}4, 2\overline{3}4 \text{ and their opposites}\}$.*

Proof. Suppose $\mathcal{S}$ did not satisfy the elimination property (C3). Then there are $X, Y \in \mathcal{S}$ with $X \neq -Y$ and $e \in X^+ \cap Y^-$ such that no $Z \in \mathcal{S}$ satisfies $Z^+ \subseteq (X^+ \cup Y^+)\backslash\{e\}$ and $Z^- \subseteq (X^- \cup Y^-)\backslash\{e\}$. Using reorientation if necessary, we may suppose that $X^+ = X$ and $\underline{Y}\backslash\underline{X} \subseteq Y^+$.

(1) $\underline{X} \cup \underline{Y} = E$

Proof. Otherwise for $a \in E\backslash(\underline{X} \cup \underline{Y})$ we get a contradiction, since $\mathcal{S}\backslash a$ satisfies (C3). ◁

(2) $X^+ \cap Y^- = \{e\}$, *and for all* $a \in X^+ \cap Y^+$ *there is* $Z \in \mathcal{S}$ *such that* $\underline{Z} \subseteq (\underline{X} \cup \underline{Y})\backslash\{e\}$ *and* $Z^- = \{a\}$.

Proof. Let $a \in (\underline{X} \cap \underline{Y})\backslash\{e\}$ and set $X' = X\backslash a$, $Y' = Y\backslash a$. We have $X', Y' \in \mathcal{S}/a$, $X' \neq -Y'$ and $e \in X'^+ \cap Y'^-$. Since $\mathcal{S}/a$ satisfies the elimination property, there is $Z' \in \mathcal{S}/a$ such that $Z'^+ \subseteq (X'^+ \cup Y'^+)\backslash\{e\}$ and $Z'^- \subseteq (X'^- \cup Y'^-)\backslash\{e\}$. Choose $Z \in \mathcal{S}$ such that $Z' = Z\backslash a$.

If $a \in Y^-$, we have $Z^+ \subseteq (X^+ \cup Y^+)\backslash\{e\}$ and $Z^- \subseteq (X^- \cup Y^-)\backslash\{e\}$, contradicting the choice of X and Y. Hence $a \in Y^+$. This proves $X^+ \cap Y^- = Y^- = \{e\}$. For Z this yields $Z^- \subseteq \{a\}$, where $Z^- \neq \emptyset$ from the choice of X and Y. ◁

It follows from (2) that a signed set $Z \in \mathcal{S}$ satisfies $Z^+ \subseteq (X^+ \cup Y^+)\backslash\{e\}$ and $Z^- \subseteq (X^- \cup Y^-)\backslash\{e\}$ if and only if it is positive and $e \notin \underline{Z}$.

(3)   *If* $|E\backslash\underline{X}| \geq 2$, *then for all* $a \in E\backslash\underline{X}$ *there is* $X' \in \mathcal{S}/a$ *such that* $e \in \underline{X'} \subseteq \underline{X}$.

Proof. In the proof of Lemma 3.3.4 elimination is used only for signed sets $W \in \mathcal{S}$ with $\underline{W} \subseteq \underline{X}$, hence in Lemma 3.3.5 only for signed sets $W \in \mathcal{S}$ with $\underline{W} \subseteq \underline{X} \cup a$.

If $|E \backslash \underline{X}| \geq 2$, then elimination holds for $\mathcal{S}(\underline{X} \cup a)$ by induction, hence Property (3) follows by applying Lemma 3.3.5.   ◁

(4)   *If $|E \backslash \underline{X}| \geq 2$ then for all $a \in E \backslash \underline{X}$ there is $Z \in \mathcal{S}$ such that $e \notin \underline{Z}$ and $Z^- = \{a\}$.*

Proof. Let $a \in E \backslash \underline{X} = \underline{Y} \backslash \underline{X}$. Then $Y' = Y \backslash a \in \mathcal{S}/a$, and by (3) there is $X' \in \mathcal{S}/a$ such that $e \in \underline{X}' \subseteq \underline{X}$. We have $X' \neq -Y'$ since $\underline{Y}' \backslash \underline{X} \neq \emptyset$. Since $\mathcal{S}/a$ satisfies elimination, there is $Z' \in \mathcal{S}/a$, Z' positive with $e \notin \underline{Z}'$. Let $Z \in \mathcal{S}$ be such that $Z' = Z \backslash a$. Since Z cannot be positive by the choice of X and Y, we have $Z^- = \{a\}$.   ◁

Similarly, by symmetry:

(4')   *If $|E \backslash \underline{Y}| \geq 2$, for all $a \in E \backslash \underline{Y}$ there is $Z \in \mathcal{S}$ such that $e \notin \underline{Z}$ and $Z^- = \{a\}$.*

(5)   *If $|E \backslash \underline{X}| \geq 2$ and $|E \backslash \underline{Y}| \geq 2$, then in (4) we may suppose additionally that $\underline{Z} \backslash \underline{Y} \neq E \backslash \underline{Y}$.*

Proof. Let $a_1 \in \underline{Z} \backslash \underline{Y}$. By (4') there is $Z_1 \in \mathcal{S}$ such that $e \notin \underline{Z}_1$ and $Z_1^- = \{a_1\}$. If $Z_1 = -Z$, then we have $Z^- = \{a\}$, $Z^+ = \{a_1\}$, hence $\underline{Z} \backslash \underline{Y} \neq E \backslash \underline{Y}$. Suppose $Z_1 \neq -Z$. Since $\mathcal{S} \backslash e$ satisfies elimination, there is $Z_2 \in \mathcal{S} \backslash e$ such that $Z_2^+ \subseteq (Z^+ \cup Z_1^+) \backslash \{a_1\}$ and $Z_2^- \subseteq (Z^- \cup Z_1^-) \backslash \{a_1\} = \{a\}$. We have $e \notin \underline{Z}_2$, hence $Z_2^- = \{a\}$: otherwise Z_2 is positive. Since $\underline{Z}_2 \backslash \underline{Y} \neq E \backslash \underline{Y}$, we can take Z_2 for Z, to get (5).   ◁

(6) $|E \backslash \underline{X}| = 1$ *or* $|E \backslash \underline{Y}| = 1$

Proof. Suppose that $|E \backslash \underline{X}| \geq 2$ and $|E \backslash \underline{Y}| \geq 2$, and let $a \in E \backslash \underline{X}$. By (5) there is $Z \in \mathcal{S}$ such that $e \notin \underline{Z}$, $Z^- = \{a\}$ and $\underline{Z} \backslash \underline{Y} \neq E \backslash \underline{Y}$. By the latter condition we can choose $b \in E \backslash (\underline{Y} \cup \underline{Z})$. Since $\mathcal{S} \backslash b$ satisfies elimination, there is $X_1 \in \mathcal{S} \backslash b$ such that $\underline{X}_1 \subseteq (\underline{Y} \cup \underline{Z}) \backslash \{a\}$ and $X_1^- = \{e\}$. Since $\mathcal{S} \backslash a$ satisfies elimination, there is $Z_1 \in \mathcal{S} \backslash a$ such that $\underline{Z}_1 \subseteq (\underline{X}_1 \cup \underline{X}) \backslash \{e\}$ and Z_1 positive, a contradiction.

                                    ◁

(7) $|E \backslash \underline{X}| = 1$ *and* $|E \backslash \underline{Y}| = 1$

Proof. By (6) we may suppose that $|E \backslash \underline{X}| \geq 2$ and $|E \backslash \underline{Y}| = 1$, the case $|E \backslash \underline{X}| = 1$ and $|E \backslash \underline{Y}| \geq 2$ being equivalent after reorientation on $\underline{X} \cap \underline{Y}$. Let $a_1, a_2 \in E \backslash \underline{X}$, $a_1 \neq a_2$. By (4) there are $Z_1, Z_2 \in \mathcal{S}$ such that $\underline{Z}_i \subseteq (\underline{X} \cup \underline{Y}) \backslash \{e\}$ and $Z_i^- = \{a_i\}$ for $i = 1, 2$. Since necessarily $E \backslash \underline{Y} \subseteq Z_1^+ \cap Z_2^+$, otherwise $\underline{Z}_1$ or $\underline{Z}_2$ is contained in $\underline{Y} \backslash \{e\}$, we have also $Z_1 \neq -Z_2$. Hence the supports of Z_1, Z_2 are two different circuits of $\mathcal{M}$. Eliminating in $\underline{M}$ the element of $E \backslash \underline{Y}$ from $\underline{Z}_1$ and $\underline{Z}_2$, we get a new circuit $\underline{W}$ of $\underline{M}$ with $\underline{W} \subseteq \underline{X} \backslash \{e\}$: a contradiction.   ◁

 (It is here where we need the fact that $\underline{M}$ satisfies the circuit axioms for unoriented matroids.)

(8) $|\underline{X} \cap \underline{Y}| \leq 2$

Proof. Suppose $|\underline{X} \cap \underline{Y}| \geq 3$ and let $a_1, a_2 \in (\underline{X} \cap \underline{Y})\backslash\{e\}$, $a_1 \neq a_2$. By (2) there are $Z_1, Z_2 \in S$ such that $e \notin \underline{Z}_i$ and $Z_i^- = \{a_i\}$, $i = 1, 2$. We have $Z_1 \neq Z_2$, and since by (7) necessarily $(E\backslash\underline{X})\cup(E\backslash\underline{Y}) \subseteq Z_1^+ \cap Z_2^+$, also $Z_1 \neq -Z_2$. Hence again the supports of Z_1, Z_2 are two different circuits of $\mathcal{M}$ contained in $(\underline{X}\cup\underline{Y})\backslash\{e\}$, and elimination yields a contradiction, as in (7). ◁

By (7) and (8) we have $|E| \leq 4$. The rest of the proof is a straightforward case analysis. □

3.4 Duality

Let M be an (ordinary) matroid on a set E with set of circuits C. The set of non-empty subsets Y of E which are inclusion-minimal with respect to the property

$$\text{"}|X \cap Y| \neq 1 \quad \text{for all} \quad X \in C\text{"}$$

constitute the circuits of another matroid on E, called the *dual* or *orthogonal* of M and denoted by M^*; see White (1986, Chapter 5). The circuits of M^* are also called *cocircuits* of M. A subset $Y \subseteq E$ is a cocircuit of M if and only if $E\backslash Y$ is a hyperplane of M. A subset $B \subseteq E$ is a base of $\mathcal{M}^*$ if and only if $E\backslash B$ is a base of $\mathcal{M}$, and for $A \subseteq E$ we have

$$r_{M^*}(A) = |A| + r_M(E\backslash A) - r_M(E).$$

In particular, $r(M^*) + r(M) = |E|$.

In this section, we establish the existence and main properties of the *dual of an oriented matroid*. If $\mathcal{M}$ is an oriented matroid on E, B is a basis of $\underline{\mathcal{M}}$, and $e \in E\backslash B$, then there is a unique circuit $\underline{c}(e, B)$ of $\underline{\mathcal{M}}$ contained in $B \cup e$. Furthermore, $\underline{c}(e, B)$ supports a unique signed circuit of $\mathcal{M}$, up to sign reversal. Let $c(e, B)$ denote the *basic* (or *fundamental*) *circuit of e with respect to B*, the signed circuit supported by $\underline{c}(e, B)$ which has e in its positive part. Dually, if $e \in B$, then $\underline{c}^*(e, B)$ denotes the unique cocircuit of $\underline{\mathcal{M}}$ which is disjoint from $B\backslash e$, and $c^*(e, B)$ denotes the corresponding signed cocircuit of $\mathcal{M}$ which is positive on e, the *basic* (or *fundamental*) *cocircuit of e with respect to B*. The existence and uniqueness of $c^*(e, B)$ will follow from Proposition 3.4.1.

Two signed sets X, Y are said to be *orthogonal*, denoted by $X \perp Y$, if either $\underline{X} \cap \underline{Y} = \emptyset$, or the restrictions of X and Y to their intersection are neither equal nor opposite, i.e., there are $e, f \in \underline{X} \cap \underline{Y}$ such that $X(e)Y(e) = -X(f)Y(f)$.

3.4.1 Proposition (Dual. Bland and Las Vergnas 1978). *Let $\mathcal{M}$ be an oriented matroid on a set E with set of circuits C.*

(i) *There is a unique signature C^* of the cocircuits of $\underline{\mathcal{M}}$ such that*

$$(\perp) \qquad X \perp Y \text{ for all } X \in C \text{ and } Y \in C^*.$$

(ii) *The collection C^* is the set of (signed) circuits of an oriented matroid on E called the dual (or orthogonal) of $\mathcal{M}$ and denoted by $\mathcal{M}^*$.*

(iii) *We have $\mathcal{M}^{**} = \mathcal{M}$.*

The proof will be presented after a few comments and a lemma.

The definition of orthogonality of signed sets is motivated by sign properties of the scalar product of two vectors with real coordinates. Let $X = s(u)$ and $Y = s(v)$ be the signed supports of two orthogonal vectors $u, v \in \mathbb{R}^E$. We have $\langle u, v \rangle = \sum_{e \in E} u(e) \cdot v(e) = 0$. The orthogonality $X \perp Y$ of X and Y amounts to the fact that the non-zero terms of this sum (if any) cannot all have the same sign.

Let V be a vector subspace of $\mathbb{R}^E$, and let $V^\perp = \{v \in \mathbb{R}^E : \langle u, v \rangle = 0 \text{ for all } u \in V\}$ be the orthogonal complement of V. Let $\mathrm{Supp}(V)$ be the oriented matroid whose circuits are the minimal signed supports of V as defined in Section 3.2. As in the case of ordinary matroids (see White 1987, Proposition 1.3.1), we have the relation $(\mathrm{Supp}(V))^* = \mathrm{Supp}(V^\perp)$. In other words, dual pairs of realizable oriented matroids on E correspond to pairs of orthogonal subspaces in $\mathbb{R}^E$.

The fact that the combinatorial orthogonality condition "$X \perp Y$" is sufficient to determine uniquely the signed support of $V^\perp$ from the signed supports of V is a consequence of the following matroid-theoretic lemma.

3.4.2 Lemma. *Let M be a matroid. Given any circuit X of M and $e, f \in X$, $e \neq f$, there is a cocircuit Y of M such that $X \cap Y = \{e, f\}$.*

Proof. Let B be a base of M containing $X \backslash f$. We have $f \notin B$, otherwise B would contain the circuit X. Then the basic cocircuit $Y = \underline{c}^*(e, B)$ is such that $X \cap Y = \{e, f\}$. □

Signed cocircuits have the following geometric interpretation. Let E be a spanning set of vectors of $\mathbb{R}^d$, and let $\mathcal{M}$ be the rank d oriented matroid of linear dependences of E. Let H be a hyperplane of $\underline{\mathcal{M}}$, that is, a closed subset of E which spans a hyperplane in $\mathbb{R}^d$. Then $D = E \backslash H$ is a cocircuit of $\underline{\mathcal{M}}$. Up to scaling there exists a unique linear function h on $\mathbb{R}^d$ whose kernel is H. The signature of D in $\mathcal{M}^*$ is given by $D^+ = \{e \in D : h(e) > 0\}$ and $D^- = \{e \in D : h(e) < 0\}$.

In the affine case, let E be a configuration of points in affine $(d-1)$-space and let D be a cocircuit of the oriented matroid $\mathcal{M}$ of affine dependences of E. The signature of D in $\mathcal{M}^*$ is given by $D^+ = D \cap H^+$ and $D^- = D \cap H^-$ where H^+ and H^- are the two open halfspaces of $\mathbb{R}^{d-1}$ determined by any affine hyperplane H that contains $E \backslash D$.

Example (The CUBE, continued). The rank 4 oriented matroid CUBE (introduced in Section 3.2) has 40 signed cocircuits, namely (up to opposites),

$$
\begin{array}{cccc}
1234 & 1256 & 12\overline{78} & 1357 \\
13\overline{68} & 1\overline{458} & 23\overline{67} & 2457 \\
2468 & 3\overline{456} & 3478 & 5678 \\
\\
1235\overline{8} & 1246\overline{7} & 134\overline{67} & 1\overline{4567} \\
1\overline{4678} & 234\overline{58} & 2\overline{3568} & 23\overline{578}
\end{array}
$$

Actually, CUBE and CUBE* are equal up to reorientation:

$$_{-\{2358\}}\text{CUBE} = \text{CUBE*}.$$

The automorphism group of CUBE, which is isomorphic to the symmetry group of the geometric cube, is of order 48. The group of permutations σ such that $\sigma(\text{CUBE})$ is isomorphic to CUBE *up to reorientation* is of order 196. This group is the symmetry group of the arrangement in *projective* 3-space of the eight planes spanned by the facets of a regular octahedron.

Proof of Proposition 3.4.1.

(i) Fix a cocircuit $\underline{Y}$ of $\mathcal{M}$. For any pair $y_1, y_2 \in \underline{Y}$ we define a sign

$$\sigma(y_1, y_2) := -X(y_1)X(y_2) \in \{+, -\},$$

where $X \in \mathcal{C}$ is any circuit with $\underline{X} \cap \underline{Y} = \{y_1, y_2\}$. Such a circuit exists by Lemma 3.4.2 (dualized). If X' is another circuit of $\mathcal{M}$ such that $\underline{X}' \cap \underline{Y} = \{y_1, y_2\}$, then $X(y_1)X(y_2) = X'(y_1)X'(y_2)$; otherwise we could use (C3') to eliminate either y_1 or y_2 from X and X', thus obtaining a circuit X'' with $|\underline{X}'' \cap \underline{Y}| = 1$, which is impossible. Thus σ is well-defined.

Now we claim that

$$\sigma(y_1, y_2) \cdot \sigma(y_2, y_3) = \sigma(y_1, y_3),$$

for triples $y_1, y_2, y_3 \in \underline{Y}$. In fact, choose circuits $X_1, X_2 \in \mathcal{C}$ with $\underline{X}_1 \cap \underline{Y} = \{y_1, y_2\}$ and $\underline{X}_2 \cap \underline{Y} = \{y_2, y_3\}$. Strong elimination of y_2 from X_1 and ϵX_2 with $\epsilon := -X_1(y_2)X_2(y_2)$ yields a circuit $X \in \mathcal{C}$ with $y_1 \in \underline{X} \subseteq (\underline{X}_1 \cup \underline{X}_2) \backslash y_2$. Hence $y_1 \in \underline{X} \cap \underline{Y} \subseteq \{y_1, y_3\}$, which implies $\underline{X} \cap \underline{Y} = \{y_1, y_3\}$. But $X(y_1) = X_1(y_1)$ and $X(y_3) = \epsilon X_2(y_3) = -X_1(y_2)X_2(y_2)X_2(y_3)$, which implies

$$\sigma(y_1, y_3) = +X_1(y_1)X_1(y_2)X_2(y_2)X_2(y_3) = \sigma(y_1, y_2) \cdot \sigma(y_2, y_3).$$

Thus σ defines a bipartition $\underline{Y} = Y^+ \uplus Y^-$, if we choose some $e \in Y$ and put

$$Y^+ := \{y \in \underline{Y} \backslash e : \sigma(e, y) = +\} \cup \{e\} \text{ and}$$
$$Y^- := \{y \in \underline{Y} \backslash e : \sigma(e, y) = -\}.$$

This bipartition does not depend on the choice of e (except for possibly an exchange of Y^+ and Y^-), and it yields two opposite signed sets $\pm Y$ with $Y := (Y^+, Y^-)$, which satisfy $X \perp Y$ for all $X \in \mathcal{C}$ with $|\underline{X} \cap \underline{Y}| = 2$.

Consider now a circuit X of $\mathcal{M}$ such that X and Y are not orthogonal. Suppose $|\underline{X} \cap \underline{Y}|$ is minimal with these properties. By the above argument, we have $|\underline{X} \cap \underline{Y}| \geq 3$. Reorienting if necessary, we may suppose that $X(e) = Y(e) = 1$ for all $e \in \underline{X} \cap \underline{Y}$. Let e, y_1, y_2 be three elements of $\underline{X} \cap \underline{Y}$. By Lemma 3.4.2, there is a circuit X_1 of $\mathcal{M}$ such that $\underline{X}_1 \cap \underline{Y} = \{e, y_1\}$. We have $X_1(e)X_1(y_1) = -Y(e)Y(y_1)$, hence $X_1(e)X_1(y_1) = -1$. Replacing X_1 by $-X_1$ if necessary, we may suppose $X_1(y_1) = 1$, hence $X_1(e) = -1$. By strong elimination (C3') applied to X, X_1, e and y_2 there is a circuit X' of $\mathcal{M}$ such that $y_2 \in \underline{X}' \cap \underline{Y} \subseteq (\underline{X} \cap \underline{Y}) \backslash \{e\} \subseteq X'^+$. We have $|\underline{X}' \cap \underline{Y}| < |\underline{X} \cap \underline{Y}|$ and X', Y not orthogonal, a contradiction.

(ii) Let $\mathcal{C}$ and $\mathcal{C}^*$ be orthogonal circuit and cocircuit signatures of a matroid

$\mathcal{M}$ on a set E. We have to show that if $\mathcal{C}$ is a circuit orientation, that is, satisfies properties (C0), (C1), (C2) and (C3) of Section 3.1, then $\mathcal{C}^*$ is also a circuit orientation. The proof is by induction on $|E|$. The property being trivially true for $|E| = 1$, we may suppose $|E| \geq 2$.

For any $p \in E$, the circuit signatures $\mathcal{C}^*\backslash p$ and $\mathcal{C}^*/p$ of $\mathcal{M}^*\backslash p$ and $\mathcal{M}^*/p$ induced by $\mathcal{C}^*$ are orthogonal to $\mathcal{C}/p$ and $\mathcal{C}\backslash p$. Hence by the induction hypothesis, the circuit signatures $\mathcal{C}^*\backslash p$ and $\mathcal{C}^*/p$ are circuit orientations.

Suppose $\mathcal{C}^*$ is not a circuit orientation. Then by Theorem 3.3.7 either $|E| = 3$, say $E = \{1, 2, 3\}$, and up to reorientation $\mathcal{C}^* = \{12, 13, 23,$ and opposites$\}$, or $|E| = 4$, say $E = \{1, 2, 3, 4\}$, and up to reorientation $\mathcal{C}^* = \{123, 1\overline{2}4, 1\overline{3}4, 2\overline{3}4,$ and opposites$\}$. In the first case $\mathcal{C}$ is a signature of 123 orthogonal to 12, 13, and 23. Clearly, no such signature exists. In the second case, $\mathcal{C}$ is a signature of 123, 124, 134 and 234. The only signature of 123 which is orthogonal to $1\overline{2}4$, $1\overline{3}4$ and $2\overline{3}4$ is the all-positive signature 123. But $123 \in \mathcal{C}$ is not orthogonal to $123 \in \mathcal{C}^*$, a contradiction. Hence $\mathcal{C}^*$ is a circuit signature.

(iii) The existence and uniqueness of the cocircuit signature $\mathcal{C}^*$ implies that $\mathcal{M}^{**} = \mathcal{M}$.

□

3.4.3 Theorem (Axioms for dual pairs. Bland and Las Vergnas 1978).
Let M be a matroid, let $\mathcal{C}$ be a circuit signature of M and $\mathcal{C}^*$ a cocircuit signature of M. Then the following properties are equivalent:

($\perp$) *$\mathcal{C}$ and $\mathcal{C}^*$ are the circuit collections of a pair of dual oriented matroids,*
($\perp'$) *$X \perp Y$ for all $X \in \mathcal{C}$ and $Y \in \mathcal{C}^*$,*
($\perp''$) *$X \perp Y$ for all $X \in \mathcal{C}$ and $Y \in \mathcal{C}^*$ with $|\underline{X} \cap \underline{Y}| \leq 3$.*

(orthogonality axioms)

Proof. Clearly ($\perp$) implies ($\perp'$), and ($\perp'$) implies ($\perp''$). We will use induction on $|E|$ to show that ($\perp''$) implies ($\perp$). Suppose that $\mathcal{C}$ and $\mathcal{C}^*$ satisfy ($\perp''$). Then also the circuit signatures $\mathcal{C}\backslash e$ and $\mathcal{C}^*/e$ satisfy ($\perp''$). To see this, consider any $X' \in \mathcal{C}\backslash e$ and $Y' \in \mathcal{C}^*/e$ such that $|X' \cap Y'| \leq 3$. We have $X' \in \mathcal{C}$ and there is $Y \in \mathcal{C}^*$ such that $Y' = Y\backslash\{e\}$. Since $|X' \cap Y| = |X' \cap Y'| \leq 3$, by ($\perp''$) we have $X' \perp Y$. Therefore $X' \perp Y'$. Similarly, the circuit signatures $\mathcal{C}/e$ and $\mathcal{C}^*\backslash e$ satisfy ($\perp''$).

By the induction hypothesis, $\mathcal{C}\backslash e$ and $\mathcal{C}/e$ are circuit orientations for any $e \in E$. Hence by Theorem 3.3.7, either $\mathcal{C}$ is a circuit orientation, or $|E| = 3$, say $E = \{1, 2, 3\}$, and up to reorientation $\mathcal{C} = \{12, 13, 23,$ and opposites$\}$, or $|E| = 4$, say $E = \{1, 2, 3, 4\}$, and up to reorientation $\mathcal{C} = \{123, 1\overline{2}4, 1\overline{3}4, 2\overline{3}4,$ and opposites$\}$. Using an argument as in the proof of Theorem 3.4.2, it is easily seen that the two last cases are impossible. Hence $\mathcal{C}$ is a circuit orientation, and similarly $\mathcal{C}^*$ is a cocircuit orientation.   □

Cocircuits are circuits of the dual oriented matroid. Thus C^* may be defined by the same axiom systems as C. Furthermore, an oriented matroid M may be defined either by its signed circuits C or its signed cocircuits C^*, and either family may be constructed from the other by Theorem 3.4.3. Similarly, all other objects defined in oriented matroids can be dualized.

So, every reorientation of M also corresponds to a reorientation of M^*. A loop of M is a coloop of M^*, and conversely. A vector (Section 3.7) resp. maximal vector (Section 3.8) of M^* is also a *covector* resp. *maximal covector* of M. A base orientation (Section 3.5) of M^* is a *cobase orientation* of M. Since $M^{**} = M$, an oriented matroid M can be defined either directly or dually. In addition, we have the following self-dual axiomatics.

3.4.4 Theorem (Painting axioms. Bland and Las Vergnas 1978, 1979).
Let C and C^* be two collections of signed subsets of a set E. Then C is the set of circuits and C^* is the set of cocircuits of an oriented matroid on E if and only if they both satisfy (C0), (C1), (C2) and one (hence both) of the equivalent properties (3P) or (4P):

(3P) *for all 3-partitions $E = B \cup G \cup R$ and $e \in B$, either there is an $X \in C$ such that $e \in \underline{X} \subseteq B \cup G$ and $\underline{X} \cap B \subseteq X^+$, or there is a $Y \in C^*$ such that $e \in \underline{Y} \subseteq B \cup R$ and $\underline{Y} \cap B \subseteq Y^+$, but not both;*   (3-painting)

(4P) *for all 4-partitions $E = B \uplus W \uplus G \uplus R$ and $e \in B \uplus W$, either there is $X \in C$ such that $e \in \underline{X} \subseteq B \cup W \cup G$, $\underline{X} \cap B \subseteq X^+$ and $\underline{X} \cap W \subseteq X^-$, or there is $Y \in C^*$ such that $e \in \underline{Y} \subseteq B \cup W \cup R$, $\underline{Y} \cap B \subseteq Y^+$ and $\underline{Y} \cap W \subseteq Y^-$, but not both.*   (4-painting)

The partitions in Theorem 3.4.4 are traditionally called *colorings* or *paintings*: here B stands for black, W for white, G for green, R for red. Note that whereas (4P) is clearly invariant under reorientation, this invariance is not immediately apparent for (3P), but it results from Theorem 3.4.4.

Proof. (4P) is (3P) after reorientation on a subset of B which is then called W ("repainted white"). Thus for the "only if"-part it suffices to show that an oriented matroid satisfies (3P). Section 10.2 will offer an alternative, "constructive" proof for this part.

On the other hand, (3P) is the special case of (4P) where $W = \emptyset$. For the "if"-part it therefore suffices to show that if $C, C^* \subseteq \{+, -, 0\}^E$ satisfy (C0), (C1), (C2) and (3P) then they are the sets of circuits and cocircuits of an oriented matroid. More precisely, we will show that C, C^* are circuit signatures of two dual matroids and satisfy $X \perp Y$ for all $X \in C$ and $Y \in C^*$ with $|\underline{X} \cap \underline{Y}| \leq 3$, and then apply Theorem 3.4.3($\perp''$).

We start with the "only if" direction: every oriented matroid satisfies (3P). The two alternatives of (3P) are mutually exclusive by the orthogonality condition of Theorem 3.4.3($\perp'$). Using induction on $|B|$, we will prove that one of these alternatives must hold. Suppose on the contrary that for some 3-partition $E = B \cup G \cup R$ and $e \in B$ there is no circuit $X \in \mathcal{C}$ with $e \in \underline{X} \subseteq B \cup G$ such that $\underline{X} \cap B \subseteq X^+$ and no cocircuit $Y \in \mathcal{C}^*$ with $e \in \underline{Y} \subseteq B \cup R$ such that $\underline{Y} \cap B \subseteq Y^+$. Furthermore suppose that $|B|$ is minimal with these properties.

We first show $|B| \geq 2$. Suppose to the contrary that $B = \{e\}$, in which case the sign properties of (3P) are trivially satisfied. If there is no circuit X with $e \in \underline{X} \subseteq e \cup G$, we have $e \notin cl_M(G)$. There is a hyperplane H such that $G \subseteq H$ and $e \notin H$. Then $Y = E \backslash H$ is a cocircuit of $\mathcal{M}$ such that $e \in \underline{Y} \subseteq E \backslash G = B \cup R$. This is a contradiction and therefore $|B| \geq 2$.

Let $f \in B \backslash e$. Consider the 3-partition $B' = B \backslash f$, $G' = G \cup f$, $R' = R$ and distinguished element e. The hypothesis implies that there is no $Y \in \mathcal{C}^*$ such that $e \in \underline{Y} \subseteq B' \cup R'$ and $\underline{Y} \cap B' \subseteq Y^+$. Since (3P) is satisfied by the minimality of B, there is $X \in \mathcal{C}$ such that $e \in \underline{X} \subseteq B' \cup G' = B \cup G$ and $\underline{X} \cap (B \backslash f) \subseteq X^+$. Necessarily, we have $f \in X^-$, otherwise X satisfies (3P) applied to B, G, R, and e. Similarly, considering $B'' = B \backslash f$, $G'' = G$, $R'' = R \cup f$, and the distinguished element e, there is $Y \in \mathcal{C}^*$ such that $e \in \underline{Y} \subseteq B \cup R$, $\underline{Y} \cap (B \backslash f) \subseteq Y^+$ and $f \in Y^+$. Then X and Y contradict the orthogonality condition of Theorem 3.4.3($\perp'$). This completes the proof of the "only if"-direction.

For the "if" direction, let $\mathcal{C}, \mathcal{C}^* \subseteq \{+, -, 0\}^E$ satisfying (C0), (C1), (C2) and (3P). We will first show the following:

(1) $\mathcal{C}$ *is a circuit signature of a matroid* $\mathcal{M}$ *on* E, *and* $\mathcal{C}^*$ *is a cocircuit signature of* $\mathcal{M}$.

Proof. If we forget signs in (3P) then we get Minty's self-dual axiom for circuits and cocircuits of (non-oriented) matroids. For completeness, we include a proof that Minty's axioms imply the more familiar circuit axioms.

For any $X \in \mathcal{C}$ and $Y \in \mathcal{C}^*$ we have $|\underline{X} \cap \underline{Y}| \neq 1$. Otherwise, let $\underline{X} \cap \underline{Y} = \{e\}$. We get a contradiction by applying (3P) with $B = \{e\}$, $G = \underline{X} \backslash e$, $R = E \backslash \underline{X}$, since both alternatives hold.

Let $X_1, X_2 \in \mathcal{C}$, $e \in \underline{X}_1 \cap \underline{X}_2$ and $f \in \underline{X}_1 \backslash \underline{X}_2$. Applying (3P) with $B = \{f\}$, $G = (\underline{X}_1 \cup \underline{X}_2) \backslash \{e, f\}$, $R = (E \backslash (\underline{X}_1 \cup \underline{X}_2)) \cup \{e\}$, there is either $X \in \mathcal{C}$ such that $f \in \underline{X} \subseteq (\underline{X}_1 \cup \underline{X}_2) \backslash \{e\}$, or there is $Y \in \mathcal{C}^*$ such that $f \in \underline{Y} \subseteq (E \backslash (\underline{X}_1 \cup \underline{X}_2)) \cup \{e, f\}$. The second alternative cannot hold since $e \notin Y$ implies $\underline{X}_1 \cap \underline{Y} = \{f\}$ and since $e \in Y$ implies $\underline{X}_2 \cap \underline{Y} = \{e\}$. Therefore the supports of the signed sets in $\mathcal{C}$, which are non-empty and pairwise inclusion-incomparable by (C0) and (C2), are the circuits of a matroid M on E.

To establish that the supports of $\mathcal{C}^*$ are the cocircuits of $\mathcal{M}$, we now show that for any $X \in \mathcal{C}$ and $e, f \in X$, $e \neq f$, there is $Y \in \mathcal{C}^*$ such that $\underline{X} \cap \underline{Y} = \{e, f\}$. We apply (3P) with $B = \{e\}$, $G = \underline{X} \backslash \{e, f\}$, $R = (E \backslash \underline{X}) \cup \{f\}$. Since there is no circuit in $B \cup G = \underline{X} \backslash \{f\}$, there is a cocircuit $Y \in \mathcal{C}^*$ such that $e \in \underline{Y} \subseteq (E \backslash \underline{X}) \cup \{e, f\}$. We have $\{e\} \subseteq \underline{X} \cap \underline{Y} \subseteq \{e, f\}$, hence $\underline{X} \cap \underline{Y} = \{e, f\}$ by the previous observation that $|\underline{X} \cap \underline{Y}| = 1$ is impossible.   ◁

(2)   *We have* $X \perp Y$ *for all* $X \in \mathcal{C}$ *and* $Y \in \mathcal{C}^*$ *with* $|\underline{X} \cap \underline{Y}| = 2$.

Proof. Let $X \in \mathcal{C}$ and $Y \in \mathcal{C}^*$ with $\underline{X} \cap \underline{Y} = \{e, f\}$. Since $X \perp Y$ implies $X \perp (-Y)$, $(-X) \perp Y$, and $(-X) \perp (-Y)$, we may suppose without loss of generality that $e \in X^+ \cap Y^+$. Note that $(\underline{X} \backslash e) \cup (E \backslash \underline{Y})$ is spanning in $\mathcal{M}$, since this set contains $\{f\} \cup (E \backslash \underline{Y})$ where $E \backslash \underline{Y}$ is a hyperplane and $f \notin E \backslash \underline{Y}$. Let A be a base of $\mathcal{M}$ such that $\underline{X} \backslash e \subseteq A \subseteq (\underline{X} \backslash e) \cup (E \backslash \underline{Y})$. Apply (3P) with $B = \{e, f\}$, $G = A \backslash f$, $R = E \backslash (A \cup e)$. Then $\underline{X} = \underline{c}(e, A)$ is the only circuit contained in $B \cup G$. Also $\underline{Y} = \underline{c}^*(f, A)$ is the only cocircuit contained in $B \cup R = (E \backslash A) \cup f$. Since exactly one alternative holds in (3P), we have either $f \in X^+ \cap Y^-$ or $f \in X^- \cap Y^+$. Hence $X \perp Y$. $\triangleleft$

It remains to show the following.

(3) *We have $X \perp Y$ for all $X \in \mathcal{C}$ and $Y \in \mathcal{C}^*$ with $|\underline{X} \cap \underline{Y}| = 3$.*

Proof. Let $X \in \mathcal{C}$ and $Y \in \mathcal{C}^*$ with $\underline{X} \cap \underline{Y} = \{e, f, g\}$ and suppose that X and Y are not orthogonal. Apply (3P) with $B = \underline{X} \cap \underline{Y}$, $G = \underline{X} \backslash \underline{Y}$, $R = E \backslash \underline{X}$. Since $\underline{X} \subseteq B \cup G$ and $\underline{Y} \subseteq B \cup R$, by (3P) we cannot have both $\underline{X} \cap \underline{Y} \subseteq X^+$ and $\underline{X} \cap \underline{Y} \subseteq Y^+$. Considering similarly the non-orthogonal pairs $(X, -Y), (-X, Y)$ and $(-X, -Y)$ and permuting e, f and g, we see that without loss of generality we may suppose $X(e) = X(f) = Y(e) = Y(f) = 1$ and $X(g) = Y(g) = -1$.

Since $|\underline{X} \cap \underline{Y}| \geq 2$, there is a basis A_1 of $\mathcal{M}$ such that $\underline{X} \backslash g \subseteq A_1 \subseteq (\underline{X} \backslash g) \cup (E \backslash \underline{Y})$. We have $\{e, f\} \subseteq A_1$. Set $A = A_1 \backslash \{e, f\}$. Since $|A| = r(M) - 2 < r_M(E \backslash \underline{Y}) = r(M) - 1$, there is $h \in (E \backslash \underline{Y}) \backslash cl_{\mathcal{M}}(A)$. Note that $A \cup h$ is a base of the hyperplane $E \backslash \underline{Y}$. Finally, set $A' = (E \backslash A) \backslash \{e, f, g, h\}$.

Apply (3P) with $B = \{e, g\}$, $G = A \cup f$, $R = A' \cup h$, and distinguished element e. Since $\pm X$, with $X = c(g, A_1)$, are the only circuits contained in $B \cup G$, and satisfy $e \in X^+ \cap B$ and $g \in X^- \cap B$, there is $Y_1 \in \mathcal{C}^*$ such that $\underline{Y}_1 \subseteq A' \cup \{e, g, h\}$, $e \in Y_1^+$ and $g \notin Y_1^-$. We have $e \in \underline{X} \cap \underline{Y}_1 \subseteq \{e, g\}$, hence $g \in Y_1^+$. Similarly, exchanging e and f, there is $Y_2 \in \mathcal{C}^*$ such that $\underline{Y}_2 \subseteq A' \cup \{f, g, h\}$ and $\{f, g\} \subseteq Y_2^+$.

Apply (3P) with $B = \{f, g\}$, $G = A \cup h$, $R = A' \cup e$, and distinguished element f. Since $G = A \cup h$ is a base of the hyperplane $E \backslash \underline{Y}$, $\pm Y$ are the only cocircuits contained in $B \cup R$. Since Y has $f \in Y^+ \cap B$ and $g \in Y^- \cap B$, there is $X_1 \in \mathcal{C}$ such that $\underline{X}_1 \subseteq B \cup G = A \cup \{f, g, h\}$ and $f \in \underline{X}_1 \cap B \subseteq X_1^+$. Since $f \in \underline{X}_1 \cap \underline{Y} \subseteq \{f, g\}$, we have $\{f, g\} \subseteq X_1^+$. Similarly, exchanging f and e, there is $X_2 \in \mathcal{C}$ such that $\underline{X}_2 \subseteq A \cup \{e, g, h\}$ and $\{e, g\} \subseteq X_2^+$.

We have $g \in \underline{X}_1 \cap \underline{Y}_1 \subseteq \{g, h\}$ and $g \in \underline{X}_2 \cap \underline{Y}_2 \subseteq \{g, h\}$, hence $\underline{X}_1 \cap \underline{Y}_1 = \underline{X}_2 \cap \underline{Y}_2 = \{g, h\}$. By (2), we have $X_1 \perp Y_1$ and $X_2 \perp Y_2$, hence $X_1(h) = -Y_1(h) = \pm 1$ and $X_2(h) = -Y_2(h) = \pm 1$. On the other hand we have $\underline{X}_1 \cap \underline{Y}_2 = \{f, g, h\}$ and $\{f, g\} \subseteq X_1^+ \cap Y_2^+$. Applying (3P) with $B = \underline{X}_1 \cap \underline{Y}_2$, $G = \underline{X}_1 \backslash \underline{Y}_2$, and $R = E \backslash \underline{X}_1$, we see that $X_1^+ \cap Y_2^+ = \{f, g, h\}$ cannot hold. Hence we have either $X_1(h) = -Y_2(h)$ or $X_1(h) = Y_2(h) = -1$. Similarly, considering X_2 and Y_1, we have either $X_2(h) = -Y_1(h)$ or $X_2(h) = Y_1(h) = -1$. It follows that $X_1(h) = X_2(h) = -Y_1(h) = -Y_2(h) = \pm 1$.

Now applying strong elimination to the cocircuits $Y_1, -Y_2$ and elements $g \in \underline{Y}_1 \cap \underline{Y}_2$ and $f \in \underline{Y}_1 \backslash \underline{Y}_2$, there is $Y_3 \in \mathcal{C}^*$ such that $f \in \underline{Y}_3 \subseteq (\underline{Y}_1 \cup \underline{Y}_2) \backslash \{g\}$.

Replacing Y_3 by $-Y_3$ if necessary, we may assume that $f \in Y_3^+$. We have $f \in \underline{X} \cap \underline{Y}_3 \subseteq \{e, f\}$ and $X(f) = X(e) = Y_3(f) = 1$, hence (2) implies $Y_3(e) = -1$. We have $f \in \underline{X}_1 \cap \underline{Y}_3 \subseteq \{f, h\}$ and $X_1(f) = Y_3(f) = 1$, hence (2) implies $X_1(h) = -Y_3(h)$. We have $e \in \underline{X}_2 \cap \underline{Y}_3 \subseteq \{e, h\}$ and $X_2(e) = -Y_3(e) = 1$, hence (2) implies $X_2(h) = Y_3(h)$. Therefore $X_1(h) = -Y_3(h) = -X_2(h)$, contradicting $X_1(h) = X_2(h)$.   □

3.4.5 Remark. In the "if"-part of the above proof the property (3P) is only applied in the cases $|B| \leq 3$ and $|G| < 2r(M)$. Thus Theorem 3.4.4 can be strengthened accordingly.

Specializing (3P) to the case $G = R = \emptyset$, we get

3.4.6 Corollary.   *Any element of an oriented matroid belongs either to a positive circuit or to a positive cocircuit, but not to both.*

This corollary is a generalization of the Farkas Lemma to oriented matroids. Let $A \in \mathbb{R}^{d \times n}$ be a real matrix with d rows and n columns, and let $b \in \mathbb{R}^{d \times 1}$. The Farkas Lemma states that either

(i)   there is $x \in \mathbb{R}^{n \times 1}$ such that $Ax = b$ and $x \geq 0$ (where $x \geq 0$ means that all coordinates of x are ≥ 0), or

(ii)   there is $y \in \mathbb{R}^{1 \times d}$ such that $yA \geq 0$ and $yb < 0$, but not both.

To see that the Farkas Lemma follows from Corollary 3.4.6, let $e_1, e_2, \ldots, e_{n+1}$ denote the columns of the matrix $A' = (A, -b) \in \mathbb{R}^{d \times (n+1)}$. Let V be the vector space of linear dependences of the e_i's over $\mathbb{R}$, and let $\mathcal{M} = \text{Supp}(V)$ be the corresponding oriented matroid. We have (i) if and only if $(x, 1) \in \mathbb{R}^{(n+1) \times 1}$ is a linear dependence of the e_i's, hence if and only if there is a non-negative vector in V whose $(n + 1)$-st coordinate is strictly positive. Using Lemma 3.3.4 this is equivalent to the existence of a positive circuit of $\mathcal{M}$ containing e_{n+1}.

On the other hand we have (ii) if and only if the vector $yA' \in \mathbb{R}^{1 \times (n+1)}$ has non-negative coordinates, with the $(n + 1)$-st coordinate being strictly positive. Now $yA' \in V^\perp$, the row space of the matrix A'. Since $\text{Supp}(V^\perp) = \text{Supp}(V)^* = \mathcal{M}^*$, we have (ii) if and only if there is a positive cocircuit of $\mathcal{M}$ containing e_{n+1}.

See Schrijver (1986) for background on the Farkas Lemma, and Bachem, Dress and Wenzel (1992) for variations.

We now use Corollary 3.4.6 to isolate and characterize an important class of oriented matroids.

3.4.7 Definition (Acyclic oriented matroids).   An oriented matroid $\mathcal{M} = (E, \mathcal{C})$ is *acyclic* if it does not contain a positive circuit.

It is *totally cyclic* if every element is contained in a positive circuit.

The oriented matroids that arise from many geometric situations (like affine point configurations, polytopes, etc.) are acyclic. Also, if a (pseudo)hyperplane arrangement is oriented such that one specified region ("tope") is positive, then the corresponding oriented matroid is acyclic.

We collect the relevant characterizations in the following proposition, also for later reference. The reader may wish to skip the parts that involve "topes" and "extreme points", whose systematic development will be done in Sections 3.8, 4.2 and 9.1.

3.4.8 Proposition.   *Let M be an oriented matroid on E.*
(a)   *The following conditions are equivalent:*
  (i) M is acyclic (that is, it has no positive circuit),
  (ii) $T^+ := (+++\ldots+)$ is a tope of M,
  (iii) every $e \in E$ is contained in a positive cocircuit,
  (iv) M^* is totally cyclic.
(b)   *If M is acyclic, then so is every submatroid $M\backslash A$.*
  If M is acyclic and $e \in E$, then M/e is acyclic if and only if e is an extreme vertex of M.

Proof. (a) (i)$\Longleftrightarrow$(iii) is Corollary 3.4.6, and (iii)$\Longleftrightarrow$(iv) follows from the definitions of "totally cyclic" and of M^*.

(iii)$\Longrightarrow$(ii) follows by taking the conformal product of (all) positive cocircuits (because a tope is a maximal product of cocircuits).

(ii)$\Longrightarrow$(i) follows from orthogonality: all circuits have to be orthogonal to T^+, so they cannot be positive (see Section 3.8).

(b) The first part is clear. For the second, we refer to Proposition 9.1.2. $\square$

Finally, Theorem 3.4.3 allows us to identify the minors of M^* with the duals of the minors of M, as follows.

3.4.9 Proposition.   *Let M be an oriented matroid on a set E and A a subset of E. Then*

$$(M\backslash A)^* = M^*/A,$$
$$(M/A)^* = M^*\backslash A.$$

Proof. The set of signed circuits of $M\backslash A$ equals $\mathcal{C}\backslash A = \{X \in \mathcal{C} : X \cap A = \emptyset\}$. The set of signed circuits of M^*/A equals $\mathcal{C}^*/A = \text{Min}\{Y\backslash A : Y \in \mathcal{C}^*\}$. Consider any $X' \in \mathcal{C}\backslash A$ and $Y' \in \mathcal{C}^*/A$. Then $X' \in \mathcal{C}$ and there exists $Y \in \mathcal{C}^*$ such that $Y' = Y\backslash A$. We have $X' \perp Y$ and consequently $X' \perp Y'$. Using Theorem 3.4.3 this proves that the oriented matroids $M\backslash A$ and M^*/A are dual.

With the same argument we can see that the oriented matroids M/A and $M^*\backslash A$ are dual – in fact, the second statement is equivalent to the first, if applied to the dual matroid. $\square$

3.5 Basis orientations and chirotopes

In the previous sections we have discussed the axiom systems for oriented matroids arising from signing the circuits and cocircuits of a matroid. Now we prepare to give another cryptomorphic version of the definition of an oriented matroid via signing the bases of a matroid. First we start with the definition and construction of a basis orientation for every oriented matroid, which characterizes the oriented matroid in terms of signed bases (Las Vergnas 1975a, 1978a). After that we give a natural axiom system for the basis orientation of an oriented matroid, arising from Grassmann-Plücker relations (Lawrence 1982). The basis signature of an oriented matroid is also called a *chirotope*, a term which is derived from *chirality* in organic chemistry (see Section 1.9).

Let $\mathcal{M}$ be an oriented matroid with signed circuits $\mathcal{C}$. The *bases* of $\mathcal{M}$ are the maximal subsets of E which contain no circuit, that is, they are the bases of the underlying matroid $\underline{M}$.

3.5.1 Definition (Basis orientation). *A basis orientation of an oriented matroid $\mathcal{M}$ is a mapping χ of the set of ordered bases of $\mathcal{M}$ to $\{-1,1\}$ satisfying the following two properties:*

(B1) χ *is alternating,*

(PV) *for any two ordered bases of $\mathcal{M}$ of the form $(e, x_2, \ldots, x_r)$ and $(f, x_2, \ldots, x_r)$, $e \neq f$, we have*

$$\chi(f, x_2, \ldots, x_r) = -C(e)C(f)\chi(e, x_2, \ldots, x_r),$$

where C is one of the two opposite signed circuits of $\mathcal{M}$ in the set $\{e, f, x_2, \ldots, x_r\}$.   (basis orientation – pivoting property)

We will see in Proposition 3.5.2 that every oriented matroid $\mathcal{M}$ has exactly two basis orientations and these two basis orientations are opposite, χ and $-\chi$.

Note that if χ is a basis orientation of $\mathcal{M}$, then $\mathcal{M}$ is uniquely determined by $\underline{M}$ and χ. In fact, given any circuit $\underline{C}$ of $\underline{M}$, we can read off the signature of C in $\mathcal{M}$ from χ as follows:

Choose elements $x_1, \ldots, x_r, x_{r+1}$ of $\mathcal{M}$ such that

$$C \subset \{x_1, \ldots, x_r, x_{r+1}\}$$

and

$$\{x_1, \ldots, x_r\}$$

is a basis of $\underline{M}$. Then

$$C(x_i) = (-1)^i \chi(x_1, \ldots, x_{i-1}, x_{i+1}, \ldots, x_{r+1})$$

for all $x_i \in \underline{C}$. This alternating rule follows directly from (PV) and (B1).

Usually we extend χ to a mapping defined on E^r, where $r = r(\mathcal{M})$, with values in $\{-1, 0, 1\}$, by setting $\chi(x_1, x_2, \ldots, x_r) = 0$ if $\{x_1, x_2, \ldots, x_r\}$ is not a basis of $\mathcal{M}$. We will also need the dual version of the pivoting property, as follows.

(PV*) *for any two ordered bases of $\mathcal{M}$ of the form $(e, x_2, \ldots, x_r)$ and $(f, x_2, \ldots, x_r)$, $e \neq f$, we have*

$$\chi(f, x_2, \ldots, x_r) = D(e)D(f)\chi(e, x_2, \ldots, x_r),$$

where D is one of the two opposite signed cocircuits complementary to the hyperplane spanned by $\{x_2, \ldots, x_r\}$ in $\mathcal{M}$.   (dual pivoting property)

We note that (PV) and (PV*) are equivalent for a map χ, if $\mathcal{M}$ is an oriented matroid. To see this, consider D as in (PV*) and let C denote one of the two opposite signed circuits contained in $\{e, f, x_2, \ldots, x_r\}$. We have $\underline{C} \cap \underline{D} = \{e, f\}$. If $\mathcal{M}$ is an oriented matroid, then we have $C(e)C(f) = -D(e)D(f)$ by orthogonality of circuits and cocircuits. This implies in particular that (PV) could be replaced by (PV*) in the Definition 3.5.1 of basis orientations.

3.5.2 Proposition (Las Vergnas 1975a, 1978a). *Let $\mathcal{M}$ be an oriented matroid given as a matroid $M = \underline{\mathcal{M}}$ together with a signature $\mathcal{C}$ of its circuits that satisfies (C3), and let $(b_1, b_2, \ldots, b_r)$ be an ordered basis of M.*

There exists a unique alternating mapping χ from the set of ordered bases of M to $\{-1, 1\}$ satisfying the dual pivoting property (PV) and such that $\chi(b_1, b_2, \ldots, b_r) = 1$. Furthermore the mapping χ uniquely determines the oriented matroid $\mathcal{M}$.*

First proof. (Lawrence 1982) We proceed by induction on $|E|$, $|E| = r$ being trivial. Therefore suppose that $|E| > r$ and let

$$a \in E \backslash \{b_1, b_2, \ldots, b_r\}.$$

By induction, there is a unique alternating map χ from the set of ordered bases of M not containing a into $\{-1, 1\}$ satisfying (PV*) and $\chi(b_1, b_2, \ldots, b_r) = 1$.

Now we show that χ extends uniquely to all ordered bases containing a. By the alternating property, it suffices to consider an ordered basis of the form $B = (a, x_2, \ldots, x_r)$. Let $D = c^*(a, B)$ be the signed cocircuit complementary to the hyperplane spanned by $x_2, \ldots, x_r$ and such that $a \in D^+$. Since a is not a coloop of M, we have $|D| \geq 2$. By induction, χ satisfies the dual pivoting property (PV*) for all bases not containing a (see the remark before this proposition). Hence the expression $D(e)\chi(e, x_2, \ldots, x_r)$ does not depend on the choice of $e \in D \backslash \{a\}$, and we can define

$$\chi(a, x_2, \ldots, x_r) := D(a)D(e)\chi(e, x_2, \ldots, x_r)$$

where e is any element of $D \backslash \{a\}$.

Now, to see (PV), consider two ordered bases $(e, x_2, \ldots, x_r)$ and $(f, x_2, \ldots, x_r)$, $e \neq f$, of M. If a is not contained in $\{e, f, x_2, \ldots, x_r\}$ then (PV) holds by the induction hypothesis. If $a = e$ or $a = f$ then (PV) holds by the construction of χ in the previous paragraph. Hence $a \in \{x_2, \ldots, x_r\}$, and by the alternating property we may suppose that $a = x_2$.

To simplify notation, we set $[e, a] = \chi(e, a, x_3, \ldots, x_r)$, etc., where $[e, a] = -[a, e]$.

Let C be one of the two opposite signed circuits contained in $\{e, f, x_2, \ldots, x_r\}$. There are two subcases:

Case 1: $a \in \underline{C}$. In this case $\{e, f, x_3, \ldots, x_r\}$ is a basis. By the above construction of χ, we have $[a, e] = -C(a)C(f)[f, e]$ and $[a, f] = -C(a)C(e)[e, f]$. Therefore $[f, a] = -C(e)C(f)[e, a]$ as required.

Case 2: $a \notin \underline{C}$. In this case $\{e, f, x_3, \ldots, x_r\}$ is not a basis but spans a hyperplane in $\mathcal{M}$. Let D be one of the two opposite signed cocircuits complementary to that hyperplane. We have $a \in D$ and $|D| \geq 2$. Pick any $y \in D \backslash a$. Then both $\{e, y, x_3, \ldots, x_r\}$ and $\{f, y, x_3, \ldots, x_r\}$ are bases of $\mathcal{M}$, and we have $[f, y] = -C(e)C(f)[e, y]$ by the induction hypothesis. On the other hand we have $D(a)[a, e] = D(y)[y, e]$ and $D(a)[a, f] = D(y)[y, f]$ by the construction of χ. Therefore $[f, a] = -C(e)C(f)[e, a]$ as required. The fact that $\mathcal{M}$ is uniquely determined by χ follows from the observation after Definition 3.5.1. $\square$

We will return to this result later in this section with a sketch of a second proof.

3.5.3 Definition (Chirotopes).   Let $r \geq 1$ be an integer, and let E be a finite set.

A *chirotope* of rank r on E is a mapping $\chi : E^r \to \{-1, 0, 1\}$ which satisfies the following three properties:

(B0) χ *is not identically zero,*

(B1) χ *is alternating, that is,* $\chi(x_{\sigma_1}, x_{\sigma_2}, \ldots, x_{\sigma_r}) = \text{sign}(\sigma)\chi(x_1, x_2, \ldots, x_r)$
for all $x_1, x_2, \ldots, x_r \in E$ *and every permutation* σ,

(B2) *for all* $x_1, x_2, \ldots, x_r, y_1, y_2, \ldots, y_r \in E$ *such that*

$$\chi(y_i, x_2, x_3, \ldots, x_r) \cdot \chi(y_1, y_2, \ldots, y_{i-1}, x_1, y_{i+1}, y_{i+2}, \ldots, y_r) \geq 0$$

for $i = 1, 2, \ldots, r$, *we have*

$$\chi(x_1, x_2, x_3, \ldots, x_r) \cdot \chi(y_1, y_2, y_3, \ldots, y_r) \geq 0. \qquad \text{(chirotope axioms)}$$

In Theorem 3.5.5 we will see how this definition of a chirotope relates to our earlier axiomatizations of oriented matroids. If $\mathcal{M}$ is the rank r oriented matroid of linear dependences on a set of vectors $E \subset \mathbb{R}^r$, then the corresponding chirotope χ is given by

$$\chi(x_1, x_2, \ldots, x_r) \quad = \quad \text{sign}(\det(x_1, x_2, \ldots, x_r))$$

for all $x_1, x_2, \ldots, x_r \in E$. Here the axiom (B2) comes from abstracting sign properties in the Grassmann-Plücker relations for r-order determinants. These relations were already stated in Theorem 2.4.3, and we recall them here in slightly different form. For all

$$x_1, x_2, \ldots, x_r, y_1, y_2, \ldots, y_r \in \mathbb{R}^r$$

we have the identity

$$\det(x_1, x_2, x_3, \dots, x_r) \cdot \det(y_1, y_2, y_3, \dots, y_r)$$

$$= \sum_{i=1}^{r} \det(y_i, x_2, x_3, \dots, x_r) \cdot \det(y_1, \dots, y_{i-1}, x_1, y_{i+1}, \dots, y_r).$$

The proof of the Grassmann-Plücker relation is easy: observe that the difference of the two sides is an alternating multilinear form in the $r + 1$ arguments $x_1, y_1, y_2, \dots, y_r$, which are vectors in an r-dimensional vector space. Hence this form is zero.

Example (The CUBE, continued). The chirotope of the oriented matroid CUBE can be encoded by the following sequence of signs:

$$0+++++++0----0----+0+-0+-0+--+++--$$
$$----+++0+++0+-0+++--0----0++--++--0$$

This encoding is to be understood as follows. CUBE is a rank 4 chirotope on 8 elements. There are $\binom{8}{4} = 70$ four-element subsets of an 8-element set. The above sequence is a 70-letter word over the alphabet $\{+, -, 0\}$. The i-th letter gives the value of χ on the i-th four-subset in the lexicographic ordering of all 70 four-subsets of $\{1, 2, 3, 4, 5, 6, 7, 8\}$.

For instance, the first letter being 0 means that $\chi(e_1, e_2, e_3, e_4) = \chi(1234) = 0$; the second letter being $+$ means that $\chi(1235) = +1$; the third letter being $+$ means that $\chi(1236) = +1$, etc. The values of χ for non-increasing strings are obtained using the alternating property (B1). For example,

$$\chi(3241) = \chi(3421) = 0,$$

$$\chi(1253) = \chi(2513) = -1, \text{ or}$$

$$\chi(2163) = \chi(6321) = -\chi(6312) = +1.$$

Let us check one instance of (B2) for this chirotope χ. We consider

$$(x_1, x_2, x_3, x_4) = (1, 4, 6, 7) \text{ and } (y_1, y_2, y_3, y_4) = (2, 3, 4, 5).$$

The hypothesis of (B2) is satisfied:

$$\chi(2467)\chi(1345) = (-1)(-1) = 1$$
$$\chi(3467)\chi(2145) = -\chi(3467)\chi(1245) = -(-1)(1) = 1$$
$$\chi(4467)\chi(2315) = 0 \cdot \chi(2315) = 0$$
$$\chi(5467)\chi(2341) = \chi(4567)\chi(1234) = \chi(4567) \cdot 0 = 0$$

The conclusion of (B2) is that $\chi(1467)\chi(2345)$ is non-negative. Indeed we find

$$\chi(1467)\chi(2345) = (-1)(-1) = 1.$$

It is often convenient to work with the following equivalent form of (B2).

3.5.4 Lemma. *Let $\chi : E^r \to \{-1, 0, 1\}$ be a non-zero alternating map. Then (B2) is equivalent to*

> (B2') *for all* $x_1, x_2, \ldots, x_r, y_1, y_2, \ldots, y_r \in E$ *such that*
> $\chi(x_1, x_2, \ldots, x_r) \cdot \chi(y_1, y_2, \ldots, y_r) \neq 0,$
>   *there exists an* $i \in \{1, 2, \ldots, r\}$ *such that*
> $\chi(y_i, x_2, x_3, \ldots, x_r) \cdot \chi(y_1, y_2, \ldots, y_{i-1}, x_1, y_{i+1}, y_{i+2}, \ldots, y_r)$
> $= \chi(x_1, x_2, x_3, \ldots, x_r) \cdot \chi(y_1, y_2, y_3, \ldots, y_r).$

Proof. First note that (B2) and (B2') are both vacuous for $r = 1$.

For $r \geq 2$, let $x_1, x_2, \ldots, x_r, y_1, y_2, \ldots, y_r \in E$ and set
$$\epsilon_i(x_1, x_2, \ldots, x_r, y_1, y_2, \ldots, y_r) :=$$
$$\chi(y_i, x_2, x_3, \ldots, x_r) \cdot \chi(y_1, y_2, \ldots, y_{i-1}, x_1, y_{i+1}, y_{i+2}, \ldots, y_r)$$
for $i = 1, 2, \ldots, r$.

Then by interchanging y_1 and y_2, the sign vector $(\epsilon_1, \epsilon_2, \epsilon_3, \ldots, \epsilon_r)$ is changed into $(-\epsilon_2, -\epsilon_1, -\epsilon_3, \ldots, -\epsilon_r)$. Hence (B0), (B1) and (B2) together imply

(*) if $\chi(y_i, x_2, x_3, \ldots, x_r) \cdot \chi(y_1, y_2, \ldots, y_{i-1}, x_1, y_{i+1}, y_{i+2}, \ldots, y_r) \leq 0$
     for $i = 1, 2, \ldots, r$,
     then $\chi(x_1, x_2, \ldots, x_r) \cdot \chi(y_1, y_2, \ldots, y_r) \leq 0.$

On the other hand, condition (B2') is clearly equivalent to the conjunction of (B2) and (*). □

3.5.5 Theorem ("Chirotopes = oriented matroids". Lawrence 1982).
Let r be an integer ≥ 1 and E be a set. A mapping $\chi : E^r \longrightarrow \{-1, 0, 1\}$ is a basis orientation of an oriented matroid of rank r on E if and only if it is a chirotope.

We split the proof of the "if"-part of Theorem 3.5.5 into the following steps:

(i)   construction of a matroid M from χ (Lemma 3.5.6),
(ii)   construction of a circuit signature $\mathcal{C}$ of M from χ (Lemma 3.5.7),
(iii) construction of a cocircuit signature $\mathcal{D}$ of M from χ (Lemma 3.5.8),
(iv) $\mathcal{C}$ and $\mathcal{D}$ are orthogonal (Lemma 3.5.9),
(v) χ is a basis orientation of an oriented matroid $\mathcal{M}$ with signed circuits $\mathcal{C}$ (Lemma 3.5.10).

In Lemmas 3.5.6 to 3.5.10 we suppose that $\chi : E^r \longmapsto \{-1, 0, 1\}$ is a given chirotope. This means in particular that χ satisfies (B2') by Lemma 3.5.4.

3.5.6 Lemma.
Let $\mathcal{B}$ be the set of r-subsets B of E such that
$$\chi(x_1, x_2, \ldots, x_r) \neq 0$$
for some ordering $(x_1, x_2, \ldots, x_r)$ of B. Then $\mathcal{B}$ is the set of bases of a matroid on E.

Proof. First note that by (B1) if $B \in \mathcal{B}$ then $\chi(x_1, x_2, \ldots, x_r) \neq 0$ for *all* orderings $(x_1, x_2, \ldots, x_r)$ of B.

We will verify the basis exchange axiom for $\mathcal{B}$. Let $B, B' \in \mathcal{B}$ and $x \in B$. Let $x_1 = x$, and let $(x_1, x_2, \ldots, x_r)$ and $(y_1, y_2, \ldots, y_r)$ be orderings of B and B' respectively. Then

$$\chi(x_1, x_2, \ldots, x_r) \cdot \chi(y_1, y_2, \ldots, y_r) \neq 0$$

by the definition of $\mathcal{B}$. Now (B2′) implies that

$$\chi(y_i, x_2, \ldots, x_r) \cdot \chi(y_1, y_2, \ldots, y_{i-1}, x_1, y_{i+1}, \ldots, y_r) \neq 0$$

for some $i \in \{1, \ldots, r\}$. Hence for $y = y_i \in B'$ we have $(B \backslash x) \cup y \in \mathcal{B}$.   □

The proof of Lemma 3.5.6 shows that moreover $(B' \backslash y) \cup x \in \mathcal{B}$. In other words, (B2′) can be considered as an oriented version of the symmetric exchange property of matroid bases. See also the discussion following Theorem 2.4.3.

3.5.7 Lemma.   *Let C be a circuit of the matroid M constructed in* Lemma 3.5.6.

(i)   *Let $e, f \in C, e \neq f$. Set*

$$\sigma(e, f) := -\chi(e, x_2, \ldots, x_r) \cdot \chi(f, x_2, \ldots, x_r) \in \{-1, 1\}$$

where $X = (f, x_2, \ldots, x_r)$ is any ordered basis of M containing $C \backslash e$. Then $\sigma(e, f)$ does not depend on the choice of X.

(ii)   *Set $C^+ = \{e\} \cup \{f \in C \backslash e : \sigma(e, f) = 1\}$ and $C^- = \{f \in C \backslash e : \sigma(e, f) = -1\}$.*
       Then the pair of opposite signatures of C defined by (C^+, C^-) and the reverse does not depend on the choice of e.

Proof. (i) Let $Y = (f, y_2, \ldots, y_r)$ be another ordered basis of M containing $C \backslash e$, and let $y_1 = f$. We need to show that

$$\chi(e, x_2, \ldots, x_r) \cdot \chi(f, y_2, \ldots, y_r) = \chi(f, x_2, \ldots, x_r) \cdot \chi(e, y_2, \ldots, y_r).$$

Abbreviate $\epsilon := \chi(e, x_2, \ldots, x_r) \cdot \chi(f, y_2, \ldots, y_r)$. We have $\epsilon \neq 0$ because $\{e, x_2, \ldots, x_r\}$ and $\{f, y_2, \ldots, y_r\}$ are two bases of the matroid M defined in Lemma 3.5.6. Hence by (B2′) there is an index $i \in \{1, \ldots, r\}$ such that

$$\chi(y_i, x_2, \ldots, x_r) \cdot \chi(f, y_2, \ldots, y_{i-1}, e, y_{i+1}, \ldots, y_r) = \epsilon.$$

Suppose $y_i \notin C$. Then $C \subseteq \{f, y_2, \ldots, y_{i-1}, e, y_{i+1}, \ldots, y_r\}$, and hence we have $\chi(f, y_2, \ldots, y_{i-1}, e, y_{i+1}, \ldots, y_r) = 0$ by the definition of M. This is impossible because $\epsilon \neq 0$. If $y_i \in C \backslash f$ then $y_i \in \{x_2, \ldots, x_r\}$. Since χ is alternating, this implies $\chi(y_i, x_2, \ldots, x_r) = 0$. Also this is impossible because $\epsilon \neq 0$.

The only possibility left is $y_i = f$ and $i = 1$. Therefore $\chi(f, x_2, \ldots, x_r) \cdot \chi(e, y_2, \ldots, y_r) = \epsilon$, as desired.

(ii) It suffices to show that for any three pairwise different elements e, f, g of C, we have $\sigma(e, f) \cdot \sigma(f, g) = \sigma(e, g)$. Let $(f, g, x_3, \ldots, x_r)$ be an ordered basis of M containing $C \backslash e$. By (i), we have

$$\sigma(e, f) = -\chi(e, g, x_3, \ldots, x_r) \cdot \chi(f, g, x_3, \ldots, x_r)$$
$$\sigma(f, g) = -\chi(f, e, x_3, \ldots, x_r) \cdot \chi(g, e, x_3, \ldots, x_r)$$
$$\sigma(e, g) = -\chi(e, f, x_3, \ldots, x_r) \cdot \chi(g, f, x_3, \ldots, x_r)$$

Then $\sigma(e,f) \cdot \sigma(f,g) = \sigma(e,g)$ is an immediate consequence of (B1).   □

We say that the signature of the circuits C obtained in Lemma 3.5.7 is *induced by* the chirotope χ. By construction, χ satisfies the pivoting property (PV) for this signature. Now we will prove a dual version of Lemma 3.5.7, with a very similar proof.

3.5.8 Lemma.   *Let D be a cocircuit of the matroid M constructed in Lemma 3.5.6.*

(i)   *Let $e, f \in D$, $e \neq f$. Set*
$$\sigma(e,f) = \chi(e, x_2, \ldots, x_r) \cdot \chi(f, x_2, \ldots, x_r) \in \{-1, 1\}$$
where $X = (x_2, \ldots, x_r)$ is an ordered basis of the hyperplane $E \backslash D$ in M. Then $\sigma(e,f)$ does not depend on the choice of X.

(ii)   *Set $D^+ = \{e\} \cup \{f \in D \backslash e : \sigma(e,f) = 1\}$ and $D^- = \{f \in D \backslash e : \sigma(e,f) = -1\}$. Then the pair of opposite signatures of D defined by (D^+, D^-) and the reverse does not depend on the choice of e.*

Proof. (i) Let $Y = (y_2, \ldots, y_r)$ be another ordered basis of the hyperplane $E \backslash D$ in M, and let $y_1 := f$. We need to prove that
$$\chi(e, x_2, \ldots, x_r) \cdot \chi(f, y_2, \ldots, y_r) = \chi(f, x_2, \ldots, x_r) \cdot \chi(e, y_2, \ldots, y_r).$$
Abbreviate $\epsilon := \chi(e, x_2, \ldots, x_r) \cdot \chi(f, y_2, \ldots, y_r)$. We have $\epsilon \neq 0$ because $\{e, x_2, \ldots, x_r\}$ and $\{f, y_2, \ldots, y_r\}$ are two bases of M. Hence by (B2′) there is an index $i \in \{1, 2, \ldots, r\}$ such that
$$\chi(y_i, x_2, \ldots, x_r) \cdot \chi(f, y_2, \ldots, y_{i-1}, e, y_{i+1}, \ldots, y_r) = \epsilon.$$
Suppose $i > 1$. Then y_i lies in the hyperplane $E \backslash D$ which is spanned by $\{x_2, \ldots, x_r\}$. The definition of M implies $\chi(y_i, x_2, \ldots, x_r) = 0$. This is impossible because $\epsilon \neq 0$.

The only possibility left is $i = 1$ and $y_i = f$. Therefore $\chi(f, x_2, \ldots, x_r) \cdot \chi(e, y_2, \ldots, y_r) = \epsilon$, as desired.

(ii) It suffices to show that for any three pairwise different elements e, f, g of D, we have $\sigma(e,f) \cdot \sigma(f,g) = \sigma(e,g)$. Let $(x_2, \ldots, x_r)$ be any ordered basis of the hyperplane $E \backslash D$. By (i) we have
$$\sigma(e,f) = \chi(e, x_2, \ldots, x_r) \cdot \chi(f, x_2, \ldots, x_r)$$
$$\sigma(f,g) = \chi(f, x_2, \ldots, x_r) \cdot \chi(g, x_2, \ldots, x_r)$$
$$\sigma(e,g) = \chi(e, x_2, \ldots, x_r) \cdot \chi(g, x_2, \ldots, x_r)$$
Hence $\sigma(e,f) \cdot \sigma(f,g) = \sigma(e,g)$.   □

We say that the signature of the cocircuits D obtained in Lemma 3.5.8 is *induced by* the chirotope χ. By construction, χ satisfies the dual pivoting property (PV*).

Returning to the "if"-part of Theorem 3.5.5, we consider the matroid M defined in Lemma 3.5.6 and the signatures of its circuits and cocircuits defined in Lemmas 3.5.7 and 3.5.8 respectively.

3.5.9 Lemma. *The signatures of circuits and of cocircuits of M induced by χ are orthogonal.*

Proof. Let C be a circuit of M and D be a cocircuit of M such that $C \cap D \neq \emptyset$, with signatures induced by χ. Let $Y = \{y_2, \ldots, y_r\}$ be a basis of the hyperplane $E \backslash D$. Let $e \in C \cap D$ and $X = \{x_1, x_2, \ldots, x_r\}$ be a basis of M containing $C \backslash e$ and contained in $(C \backslash e) \cup Y$. Such a basis exists because $|C \cap D| \geq 2$ implies that $(C \backslash e) \cup Y$ is spanning.

Abbreviate $\epsilon = \chi(e, y_2, \ldots, y_r) \cdot \chi(x_1, x_2, \ldots, x_r)$. Since both $\{x_1, x_2, \ldots, x_r\}$ and $\{e, y_2, \ldots, y_r\}$ are bases of M, we have $\epsilon \neq 0$. Hence by (B2') there is an index $i \in \{1, \ldots, r\}$ such that

$$\chi(x_i, y_2, \ldots, y_r) \cdot \chi(x_1, x_2, \ldots, x_{i-1}, e, x_{i+1}, \ldots, x_r) = \epsilon.$$

Suppose x_i lies in the hyperplane $E \backslash D$ which is spanned then by Y. By the definition of M, we then have $\chi(x_i, y_2, \ldots, y_r) = 0$. But this is impossible because $\epsilon \neq 0$.

Therefore $x_i \in D$. It follows that $x_i \in C \cap D$, since $X \cap D \subseteq C \backslash e$ by the choice of X. Now the definitions of circuit and cocircuit signatures induced by χ imply

$$
\begin{aligned}
C(e) \cdot C(x_i) &= -\chi(x_1, x_2, \ldots, x_{i-1}, e, x_{i+1}, \ldots, x_r) \cdot \chi(x_1, x_2, \ldots, x_r) \\
&= -\chi(e, y_2, \ldots, y_r) \cdot \chi(x_i, y_2, \ldots, y_r) \\
&= -D(e) \cdot D(x_i).
\end{aligned}
$$

$\square$

3.5.10 Lemma. *The circuit signature $\mathcal{C}$ and the cocircuit signature $\mathcal{D}$ of M induced by χ are dual matroid orientations, and χ is a basis orientation of the oriented matroid $\mathcal{M}$ with set of signed circuits $\mathcal{C}$ and $\underline{\mathcal{M}} = M$.*

Proof. The first part of Lemma 3.5.10 follows from Lemma 3.5.9 and the self-dual axiomatics of oriented matroids (Theorem 3.4.3). The second part is an immediate consequence of the definition of the circuit signature induced by χ.

$\square$

This completes the "if"-part of the proof of Theorem 3.5.5. For the "only if"-part, we use Proposition 3.5.2. With Lemma 3.5.4, the following lemma supplies the last missing piece.

3.5.11 Lemma. *Let $\mathcal{M} = (E, \mathcal{C}^*)$ be an oriented matroid. Then the mapping χ constructed in Proposition 3.5.2 satisfies (B2').*

Proof. Let $x_1, x_2, \ldots, x_r, y_1, y_2, \ldots, y_r \in E$ be such that

$$\epsilon := \chi(x_1, x_2, \ldots, x_r) \cdot \chi(y_1, y_2, \ldots, y_r) \in \{-1, 1\}.$$

Abbreviate $B := \{x_1, x_2, \ldots, x_r\}$ and $B' := \{y_1, y_2, \ldots, y_r\}$.

If $x_1 \in B'$ then (B2') is trivially satisfied. Suppose $x_1 \notin B'$. Consider the basic circuit $C = c(x_1, B')$ and the basic cocircuit $D = c^*(x_1, B)$. We have $x_1 \in C \cap D$, hence for some $i, 1 \le i \le r$, we have $C(x_1)C(y_i) = -D(x_1)D(y_i)$ by orthogonality (cf. Section 3.4). Now since $\mathcal{M}$ is an oriented matroid, (PV*) (which χ satisfies by construction) implies (PV), as observed above. Thus we have

$$\chi(y_1, \ldots, y_{i-1}, x_1, y_{i+1}, \ldots, y_r) = -C(x_1)C(y_i)\chi(y_1, y_2, \ldots, y_r)$$

and

$$\chi(y_i, x_2, \ldots, x_r) = D(y_i)D(x_1)\chi(x_1, x_2, \ldots, x_r).$$

Hence $\chi(y_i, x_2, \ldots, x_r)\chi(y_1, \ldots, y_{i-1}, x_1, y_{i+1}, \ldots, y_r) = \epsilon$ as required. $\qquad\square$

We can restate Theorem 3.5.5 as follows.

3.5.12 Corollary. *The circuit axioms for oriented matroids are cryptomorphic to the chirotope axioms* (B0), (B1) *and* (B2) *via the translation given by the pivoting property* (PV).

The cryptomorphism from signed circuits to signed bases is in practice achieved by stepwise extension of the chirotope χ, using the basis graph of the underlying matroid. The *basis graph* BG_M of a matroid M has as vertices the bases of M and as edges the pairs of bases that differ in exactly one element. Such graphs were characterized by Maurer (1973). Note that BG_M is clearly connected.

Let $\mathcal{M} = (E, \mathcal{C})$ be an oriented matroid, presented in terms of its signed circuits $\mathcal{C}$. Assuming that the ground set E is ordered, we identify each basis $B = \{x_1, x_2, \ldots, x_r\}$ of $\underline{M}$ with its increasing permutation $B = (x_1, x_2, \ldots, x_r)$. For each edge $\{B_1, B_2\}$ of the basis graph $\mathrm{BG}_{\mathcal{M}}$ we define a sign $\eta(B_1, B_2)$ by the following rule: if $B_1 = (x_1, x_2, \ldots, x_r)$, $B_2 = (y_1, y_2, \ldots, y_r)$, $B_1 \triangle B_2 = \{x_i, y_j\}$, and if $C \in \mathcal{C}$ is one of the two opposite signed circuits contained in $B_1 \cup B_2 = B_1 \cup \{y_j\}$, then

$$(3.5.1) \qquad \eta(B_1, B_2) = (-1)^{1+i+j}C(x_i)C(y_j).$$

This way we get the *signed basis graph* $\mathrm{SBG}_{\mathcal{M}}$ of the oriented matroid $\mathcal{M}$. Note that $\mathrm{SBG}_{\mathcal{M}}$ depends on the linear ordering of the ground set E.

Second Proof of Proposition 3.5.2. (Las Vergnas 1975a, 1980a) To construct the chirotope $\chi : E^r \to \{-1, 0, +1\}$ from $\mathrm{SBG}_{\mathcal{M}}$ it is sufficient to determine the values $\chi(B)$ for increasingly ordered bases B. Put $\chi(B_0) = +1$ for some arbitrarily chosen basis B_0. Then, for any other basis B,

$$(3.5.2) \qquad \chi(B) = \prod_{i=1}^{k} \eta(B_{i-1}, B_i)$$

where $B_0, B_1, B_2, \ldots, B_k = B$ is an arbitrary path from B_0 to B in the signed basis graph $\mathrm{SBG}_{\mathcal{M}}$. We have to show that contradictions cannot arise, i.e., the value of (3.5.2) is independent of the choice of path from B_0 to B. Equivalently,

$$(3.5.3) \qquad \prod_{i=1}^{t} \eta(B_{i-1}, B_i) = 1$$

for every closed path $B_0, B_1, B_2, \ldots, B_t = B_0$ in $\text{SBG}_{\mathcal{M}}$. Now we use the homotopy theorem of Maurer (1973), which says that all such closed paths can be deformed into trivial paths (of length zero) via a sequence of "elementary homotopies" involving only the triangles and quadrilaterals of the basis graph. This way a direct verification of (3.5.3) is brought down to manageable size, since it needs only be done for $t = 2, 3, 4$, and the triangles and quadrilaterals in the basis graph are easily characterized. □

This second proof has the advantage of covering the case of modular cocircuit elimination as well – see Section 3.6.

Example. Let $\mathcal{C} = \{\pm 1\overline{3}4, \pm \overline{2}45, \pm \overline{1}235\}$. The signed basis graph of the oriented matroid $\mathcal{M} = (E_5, \mathcal{C})$ is shown in Figure 3.5.1.

For instance, $\eta(235, 345) = (-1)^{1+1+2}C(2)C(4) = -1$, which we get using $C = \overline{2}45$ in the definition (3.5.1). Now, if we put $\chi(124) = +1$, the chirotope χ of $\mathcal{M}$ can be quickly computed from Figure 3.5.1 using formula (3.5.2). For example, $\chi(235) = +1, \chi(345) = -1$, etc.

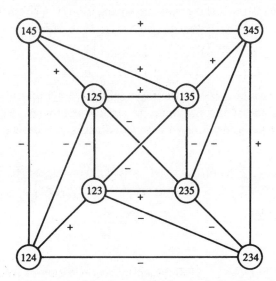

Figure 3.5.1. The signed basis graph of a rank 3 oriented matroid.

We will close this section by translating the oriented matroid operations of deletion, contraction, reorientation and duality into chirotope notation. In the following we will assume that $\mathcal{M}$ is a rank r oriented matroid on a set E with set of circuits $\mathcal{C}$ and with chirotope χ.

Deletion:

The oriented matroid $\mathcal{M} \backslash A$ obtained by deletion of a subset $A \subset E$ was defined by its set of circuits $\mathcal{C} \backslash A = \{C \in \mathcal{C} : C \cap A = \emptyset\}$. Suppose that $\mathcal{M} \backslash A$

has rank $s < r$, and pick any $a_1, \ldots, a_{r-s} \in A$ such that $E \backslash A \cup \{a_1, \ldots, a_{r-s}\}$ spans $\mathcal{M}$. We will show that, up to negation, the chirotope $\chi \backslash A$ of $\mathcal{M} \backslash A$ equals

$$\chi \backslash A : (E \backslash A)^s \quad \rightarrow \quad \{-1, 0, +1\}$$
$$(x_1, \ldots, x_s) \quad \mapsto \quad \chi(x_1, \ldots, x_s, a_1, \ldots, a_{r-s}).$$

We need to check that $\chi \backslash A$ is a basis orientation of $\mathcal{M} \backslash A$. Consider two bases of $\mathcal{M} \backslash A$ of the form $(e, x_2, \ldots, x_s)$ and $(f, x_2, \ldots, x_s)$, and let $C \in \mathcal{C} \backslash A$ be one of the two opposite circuits contained in $\{e, f, x_2, \ldots, x_s\}$. Since χ is a basis orientation of $\mathcal{M}$, we have

$$\chi(f, x_2, \ldots, x_s, a_1, \ldots, a_{r-s}) = -C(e)C(f)\chi(e, x_2, \ldots, x_r, a_1, \ldots, a_{r-s}),$$

and therefore $(\chi \backslash A)(f, x_2, \ldots, x_s) = -C(e) \cdot C(f) \cdot (\chi \backslash A)(e, x_2, \ldots, x_s)$, as desired. The uniqueness in Proposition 3.5.2 applied to $\mathcal{M} \backslash A$ now shows that our definition of $\chi \backslash A$ is independent of the choice of $a_1, \ldots, a_{r-s}$.

Contraction:

The oriented matroid $\mathcal{M}/A$ obtained by contraction of a subset $A \subset E$ was defined by its set of circuits $\mathcal{C}/A = \text{Min}\{C \backslash A : C \in \mathcal{C}\}$. Let $a_1, \ldots, a_{r-s}$ be a basis of A in $\mathcal{M}$. We will show that, up to negation, the chirotope χ/A of $\mathcal{M}/A$ equals

$$\chi/A : (E \backslash A)^s \quad \rightarrow \quad \{-1, 0, +1\}$$
$$(x_1, \ldots, x_s) \quad \mapsto \quad \chi(x_1, \ldots, x_s, a_1, \ldots, a_{r-s}).$$

Consider two bases of $\mathcal{M}/A$ of the form $(e, x_2, \ldots, x_s)$ and $(f, x_2, \ldots, x_s)$, and let $C \in \mathcal{C}/A$ be one of the two opposite circuits contained in $\{e, f, x_2, \ldots, x_s\}$. Then $C = C' \backslash A$ for some circuit C' of $\mathcal{M}$. Since χ is a basis orientation of $\mathcal{M}$, we have

$$\chi(f, x_2, \ldots, x_s, a_1, \ldots, a_{r-s}) = -C'(e)C'(f)\chi(e, x_2, \ldots, x_s, a_1, \ldots, a_{r-s}).$$

Clearly, $C(e) = C'(e)$ and $C(f) = C'(f)$. Therefore $(\chi \backslash A)(f, x_2, \ldots, x_s) = -C(e) \cdot C(f) \cdot (\chi \backslash A)(e, x_2, \ldots, x_s)$, as desired.

Reorientation:

The oriented matroid $_{-A}\mathcal{M}$ obtained by reorientation of a subset $A \subset E$ was defined by its set of circuits $_{-A}\mathcal{C} = \{_{-A}C : C \in \mathcal{C}\}$, where the signed set $_{-A}C$ is defined by $(_{-A}C)(x) = (-1)^{|A \cap \{x\}|} \cdot C(x)$. We will show that the chirotope $_{-A}\chi$ of $_{-A}\mathcal{M}$ equals

$$_{-A}\chi \quad : \quad E^r \quad \rightarrow \quad \{-1, 0, +1\}$$
$$(x_1, \ldots, x_r) \quad \mapsto \quad \chi(x_1, \ldots, x_r) \cdot (-1)^{|A \cap \{x_1, \ldots, x_r\}|}.$$

This function is easily seen to be alternating. By Proposition 3.5.2, we only have to verify (PV) to see that it is the chirotope of $_{-A}\mathcal{M}$. (It is also easy to directly verify that $_{-A}\chi$ is a chirotope.) Consider two bases of $\mathcal{M}$ of the form $(e, x_2, \ldots, x_r)$ and $(f, x_2, \ldots, x_r)$, and let $C \in \mathcal{C}$ be one of the two opposite circuits contained in $\{e, f, x_2, \ldots, x_s\}$. Then

$$
\begin{aligned}
_{-A}\chi(f, x_2, \ldots, x_r) &= (-1)^{|A \cap \{f, x_2, \ldots, x_r\}|} \chi(f, x_2, \ldots, x_r) \\
&= -(-1)^{|A \cap \{f, x_2, \ldots, x_r\}|} C(e) C(f) \chi(e, x_2, \ldots, x_r) \\
&= -(-1)^{|A \cap \{e, f\}|} C(e) C(f) (_{-A}\chi)(e, x_2, \ldots, x_r) \\
&= -{_{-A}}C(e) \, _{-A}C(f) \, (_{-A}\chi)(e, x_2, \ldots, x_r)
\end{aligned}
$$

Duality:

Suppose $E = \{1, 2, \ldots, n\}$. Given any $(n-r)$-tuple $(x_1, \ldots, x_{n-r})$ of elements in E, then we write $(x'_1, \ldots, x'_r)$ for some permutation of $E \backslash \{x_1, \ldots, x_{n-r}\}$. In particular, $(x_1, \ldots, x_{n-r}, x'_1, \ldots, x'_r)$ is a permutation of $(1, 2, \ldots, n)$, and we can compute $\text{sign}(x_1, \ldots, x_{n-r}, x'_1, \ldots, x'_r)$ as the parity of the number of inversions of this string. We will show – using the dual version of Proposition 3.5.2 – that the chirotope χ^* of the dual oriented matroid $\mathcal{M}^*$ equals

$$
\begin{aligned}
\chi^* \quad &: \quad E^{n-r} \quad \rightarrow \quad \{-1, 0, +1\} \\
(x_1, \ldots, x_{n-r}) \quad &\mapsto \quad \chi(x'_1, \ldots, x'_r) \cdot \text{sign}(x_1, \ldots, x_{n-r}, x'_1, \ldots, x'_r).
\end{aligned}
$$

Observe that this χ^* is alternating, and its definition does not depend on the ordering of the x'_i chosen above.

Now consider two bases of $\mathcal{M}^*$ of the form $(e, x_2, \ldots, x_{n-r})$ and $(f, x_2, \ldots, x_{n-r})$, and consider the complementary bases $(f, x'_2, \ldots, x'_r)$ and $(e, x'_2, \ldots, x'_r)$ of $\mathcal{M}$. Let $D \in \mathcal{C}^*$ be one of the two opposite cocircuits of $\mathcal{M}$ contained in $\{e, f, x_2, \ldots, x_{n-r}\}$. Then

$$
\begin{aligned}
\chi^*(f, x_2, \ldots, x_{n-r}) &= \chi(e, x'_2, \ldots, x'_r) \cdot \text{sign}(f, x_2, \ldots, x_{n-r}, e, x'_2 \ldots, x'_r) \\
&= D(e) D(f) \cdot \chi(f, x'_2, \ldots, x'_r) \cdot \text{sign}(f, x_2, \ldots, x_{n-r}, e, x'_2 \ldots, x'_r) \\
&= -D(e) D(f) \cdot \chi(f, x'_2, \ldots, x'_r) \cdot \text{sign}(e, x_2, \ldots, x_{n-r}, f, x'_2 \ldots, x'_r) \\
&= -D(e) D(f) \cdot \chi^*(e, x_2, \ldots, x_{n-r}).
\end{aligned}
$$

3.6 Modular elimination and local realizability

In this section we will see that oriented matroid axiomatics can be considerably sharpened ("reduced to a rank 2 situation") *if the underlying structure is known to be a matroid*: in this case, (co)circuit elimination for *modular pairs*, respectively the 3-term Grassmann-Plücker relations, are sufficient to define oriented matroids.

To explain the corresponding situation in the (co)circuit axiom systems properly, we have to review some simple geometry of matroids. For this let M be a matroid of rank r on the set E, and let M^* be the dual matroid.

○ The cocircuits of M (i.e., the circuits of M^*) are the complements of the hyperplanes (i.e., flats of rank $r - 1$) of M.

○ To interpret cocircuit exchange, we take complements and consider "hyperplane exchange":

(H3) If H_1, H_2 are two different hyperplanes of M and $e \notin H_1 \cap H_2$, then there is a hyperplane H of M that contains $(H_1 \cap H_2) \cup e$.

o   The intersection $H_1 \cap H_2$ can have rank $r - 2$ or smaller. In fact, H_1 and H_2 could even be "parallel", with $H_1 \cap H_2 = \emptyset$. However, the geometrically decisive case is that where $r(H_1 \cap H_2) = r - 2$, or, in other words, "H_1 and H_2 intersect in corank 2", or "$H_1 \cap H_2$ is a *coline*".

o   Observe that if H_1 and H_2 satisfy $r(H_1 \cap H_2) = r - 2$, and if $e \notin H_1 \cap H_2$, then H is *uniquely* determined as $H = cl_M((H_1 \cap H_2) \cup e)$. This uniqueness is restricted to this geometric situation – hyperplane and cocircuit elimination are not unique otherwise.

o   This is a special case of a more general situation. For this recall that for any subsets $A, B \subseteq E$, we have the *semimodular inequality*:
$$r(A \cap B) \le r(A) + r(B) - r(A \cup B).$$
We say that A and B form a *modular pair of M* (or are *comodular*) if this holds with equality. In particular, H_1 and H_2 are comodular exactly if $r(H_1 \cap H_2) = r - 2$.

o   Now let us reconsider cocircuit elimination. A basic observation is that $A, B \subseteq E$ are comodular in M if and only if $E \backslash A$ and $E \backslash B$ are comodular in M^*. This follows from a straightforward computation, using that the rank function r^* of M^* is given by $r^*(A) = (|A| - r) + r(E \backslash A)$ or $r^*(E \backslash A) = r^* - |A| + r(A)$.

o   In particular, the hyperplanes H_1, H_2 of M form a modular pair in M if and only if $E \backslash H_1$ and $E \backslash H_2$ are a modular pair of circuits of M^*, that is, if the cocircuits $Y_1 := E \backslash H_1$ and $Y_2 := E \backslash H_2$ of M are comodular in M^*.

(The usual convention is to simply talk about "a modular pair of hyperplanes resp. cocircuits resp. circuits", where it is understood that modular pairs of hyperplanes or circuits are considered in M, whereas modular pairs of cocircuits automatically refer to M^*.)

o   For oriented matroids, this leads us to a restricted version of the axiom system for cocircuits (and for circuits, after dualization), in which cocircuit elimination (C3) is postulated only between modular pairs. This situation is referred to as "modular cocircuit exchange", see below.

o   In the modular case, we only consider hyperplanes of M that contain the $(r - 2)$-flat $H_1 \cap H_2$ – equivalently, we could also pass

–   to modular pairs of cocircuits (see Theorem 3.6.1*), or

–   to the rank 2 contraction $M/(H_1 \cap H_2)$, or

–   to the restriction $M^* \backslash (H_1 \cap H_2)$ of rank $r^* = r(\mathcal{M}^*)$, with $r^* + 2$ elements (as in Corollary 3.6.3), or

–   to bases of the form $(e, f, x_3, \ldots, x_r)$, where $x_3, \ldots, x_r$ is a fixed basis of $H_1 \cap H_2$, (as considered by (B2″) in Theorem 3.6.2), or

–   to bases $\{e, f\}$ of the matroid $M/(H_1 \cap H_2)$.

o   If M has rank 2, then every pair of hyperplanes or cocircuits is automatically modular – so in rank 2 "modular cocircuit exchange" covers the general case.

After this, we are prepared to state and prove the modular elimination theorem. We will describe both the circuit and the cocircuit version – they are clearly equivalent, by the duality results of Section 3.4.

3.6.1 Theorem (Modular elimination. Las Vergnas 1978b, 1984a). *Let $\mathcal{C}$ be a circuit signature of a matroid M. Then* (C3) *is equivalent to*

(C3″) *for all $X, Y \in \mathcal{C}$, $\underline{X}$ and $\underline{Y}$ comodular in M, $X \neq -Y$, and $e \in X^+ \cap Y^-$,*
   there is a $Z \in \mathcal{C}$ such that
   $Z^+ \subseteq (X^+ \cup Y^+)\backslash e$ *and*
   $Z^- \subseteq (X^- \cup Y^-)\backslash e$.   (modular circuit elimination)

Proof. The proof is by induction on the cardinality of the underlying set E. Let $a \in E$. A key observation is that $\mathcal{C}\backslash a$ and $\mathcal{C}/a$ satisfy (C3″): this is clear for $\mathcal{C}\backslash a$. Now let $\bar{X}, \bar{Y} \in \mathcal{C}/a$, with $\bar{X}, \bar{Y}$ comodular in M/a, and $e \in \bar{X}^+ \cap \bar{Y}^-$. As is easily seen, the unique circuits X and Y such that $\bar{X} = X\backslash a$ and $\bar{Y} = Y\backslash a$ are comodular in M. Hence by (C3″) there is $\hat{Z} \in \mathcal{C}$ such that $Z^+ \subseteq (X^+ \cup Y^+)\backslash e$ and $Z^- \subseteq (X^- \cup Y^-)\backslash e$. By unique elimination of comodular circuits, we have necessarily $\bar{Z} = Z\backslash a \in \mathcal{C}/a$.

By induction hypothesis $\mathcal{C}\backslash a$ and $\mathcal{C}/a$ satisfy (C3). Now by Theorem 3.3.7 either $\mathcal{C}$ satisfies (C3) as required, or $|E| = 3$ or 4 and we are in case (ii) or (iii). As is readily checked, in these latter cases (C3″) is not satisfied.   □

The axiom (C3″) is obtained by restricting (C3) to modular pairs of circuits and requires significantly less verification in general. Note that in Theorem 3.6.1 we have to assume that the supports of signed sets in $\mathcal{C}$ constitute the circuits of a matroid – otherwise "comodular" is not defined.

Also, if we talk about "comodular cocircuits of $\mathcal{M}$", then this of course refers to a property of the underlying unsigned cocircuits.

3.6.1* Theorem (Las Vergnas 1978b, 1984a). *Let $\mathcal{C}^*$ be a cocircuit signature of a matroid M. Then* (C3) *is equivalent to*

(C3″) *for all $X, Y \in \mathcal{C}^*$, X and Y a modular pair of cocircuits, $X \neq -Y$, and*
   $e \in X^+ \cap Y^-$, there is a $Z \in \mathcal{C}^*$ such that
   $Z^+ \subseteq (X^+ \cup Y^+)\backslash e$ *and*
   $Z^- \subseteq (X^- \cup Y^-)\backslash e$.   (modular cocircuit elimination)

We now continue with the chirotope version of this result, proving that in Theorem 3.5.5 the axioms (B0)+(B1)+(B2) can be replaced by the seemingly weaker properties (B1′)+(B2″), which involve only three signs.

The practical importance of this stems from the fact that it substantially cuts down the number of conditions to be tested when deciding whether or not an alternating map χ is a chirotope.

3.6.2 Theorem (3-term Grassmann-Plücker relations).   *A mapping* $\chi : E^r \to \{-1,0,1\}$ *is a chirotope if and only if it satisfies the following two properties:*

(B1′) χ *is alternating, and the set of r-subsets $\{x_1, x_2, \ldots, x_r\}$ of E such that*

$$\chi(x_1, x_2, \ldots, x_r) \neq 0 \text{ is the set of bases of a matroid of rank } r \text{ on } E,$$

(B2″)   *for any* $x_1, x_2, \ldots, x_r, y_1, y_2 \in E$,

if $\chi(y_1, x_2, x_3, \ldots, x_r) \cdot \chi(x_1, y_2, x_3, \ldots, x_r) \geq 0$

and $\chi(y_2, x_2, x_3, \ldots, x_r) \cdot \chi(y_1, x_1, x_3, \ldots, x_r) \geq 0$

then $\chi(x_1, x_2, x_3, \ldots, x_r) \cdot \chi(y_1, y_2, x_3, \ldots, x_r) \geq 0$.

(3-term Grassmann-Plücker relations)

Property (B2″) is the special case of (B2) where $x_i = y_i$ for $i = 3, 4, \ldots, r$. Applying Lemma 3.5.4 to this special case, we see that for an alternating mapping $\chi : E^r \to \{-1,0,1\}$, the property (B2″) is equivalent to

---

(B2‴)   *for any* $x_1, x_2, \ldots, x_r, y_1, y_2 \in E$,

if $\epsilon := \chi(x_1, x_2, x_3, x_4, \ldots, x_r) \cdot \chi(y_1, y_2, x_3, x_4, \ldots, x_r) \in \{-1, 1\}$,

then either $\chi(y_1, x_2, x_3, x_4, \ldots, x_r) \cdot \chi(x_1, y_2, x_3, x_4, \ldots, x_r) = \epsilon$

or $\chi(y_2, x_2, x_3, x_4, \ldots, x_r) \cdot \chi(y_1, x_1, x_3, x_4, \ldots, x_r) = \epsilon$.

Proof of Theorem 3.6.2: The "only if"-part of Theorem 3.6.2 is implied by Theorem 3.5.5 because (B2″) is a special case of (B2).

The "if"-part of Theorem 3.6.2 can be obtained by slight modifications in the proof of Theorem 3.5.5. Note that the matroid assumption in (B1′) implies (B0). We observe that in the proofs of part (i) in Lemmas 3.5.7 and 3.5.8 the weaker property (B2″) suffices whenever $|X \Delta Y| = 2$. But – using basis exchange in M – it is sufficient to consider this case, and hence (B2″) suffices to prove Lemmas 3.5.7 and 3.5.8. In the proof of Lemma 3.5.9, the weaker property (B2″) implies that the signatures induced by χ of a circuit C and a cocircuit D of $\mathcal{M}$ are orthogonal whenever $|C \cap D| \leq 3$. But now we can apply part ($\perp''$) of Theorem 3.4.3 to complete the proof of Theorem 3.6.2. $\square$

Note that for the "if"-part of Theorem 3.6.2, the underlying matroid is not constructed as in Lemma 3.5.6, because it has been postulated in (B1′). The reason for this is simple: (B2″) alone does not imply the matroid property for $\mathcal{B}$ (as defined in Lemma 3.5.6). To see this, note that (B2″) is always trivially satisfied if $\mathcal{B}$ has the property $|B \cup B'| \geq r + 3$ for any two different $B, B' \in \mathcal{B}$. However, $\mathcal{B}$ with this property cannot be the collection of bases of a matroid if it contains at least two sets. The smallest non-matroidal $\mathcal{B}$ of this form is $\mathcal{B} = \{123, 456\}$.

We now discuss two other important interpretations of the 3-term Grassmann-Plücker condition. We will show first that (B2″) is very directly equivalent to the signed elimination property for comodular cocircuits. Later we will see that (B2″) amounts to an axiomatization of oriented matroids in terms of local realizability.

The key observation is that (B2″) reduces the checking to a rank 2 situation, because it only involves bases that contain a common $(r-2)$-set, corresponding to the bases of the contraction of the set. This allows us to derive the sufficiency of the the 3-term Grassmann-Plücker relations directly from Theorem 3.6.1* about modular cocircuit elimination.

Proof of Theorem 3.6.2 from Theorem 3.6.1*. ["(B2″) from (C3″)"].

Let M be a matroid with a cocircuit signature $\mathcal{C}^*$ that satisfies (C3″). We want to get a mapping $\chi : E^r \longrightarrow \{-1, 0, 1\}$ that satisfies (B1′) and (PV*).

Here we have two alternatives: either we observe that the existence of this χ is established by Theorem 3.6.1; or we argue that for the second (sketched) proof of Proposition 3.5.2, only modular cocircuit exchange as in (C3″) is needed, so we can derive the existence of χ from it.

We now show that this χ also satisfies (B2″): after this, we know that χ is a chirotope, and thus (by Lemma 3.5.10) $\mathcal{C}^*$ is an orientation of M, because it satisfies (PV*).

To prove (B2″), consider $x_1, x_2, \ldots, x_r, y_1, y_2 \in E$. All the bases whose orientations are considered by (B2″) contain $\overline{B} := \{x_3, \ldots, x_r\}$. Thus we set $\overline{E} := E \backslash \{x_3, \ldots, x_r\}$ and

$$\overline{\mathcal{C}^*} := \{X \in \mathcal{C}^* : \underline{X} \cap \overline{B} = \emptyset\}.$$

This is a cocircuit signature of the contracted matroid $M/\overline{B}$ of rank 2. We know that $\mathcal{C}^*$ satisfies (C3″), and therefore $M/\overline{B}$ satisfies (C3″) as well – but this is the same as (C3), because $M/\overline{B}$ has rank 2. So $\overline{\mathcal{C}^*}$ is a cocircuit orientation of $M/\overline{B}$.

Now consider

$$\overline{\chi} : \overline{E}^2 :\longrightarrow \{+1, -1, 0\},$$

$$(a, b) \longmapsto \chi(a, b, x_3, \ldots, x_r).$$

This $\overline{\chi}$ is an alternating map from the ordered bases of $M/\overline{B}$ to $\{+1, -1, 0\}$ that satisfies (PV*) with respect to the matroid $(M/\overline{B}, \overline{\mathcal{C}^*})$ – so $\overline{\chi}$ satisfies (B2′) by Lemma 3.5.11, which is equivalent to (B2‴) for χ and the $x_3, \ldots, x_r$ that we had fixed. □

We have claimed that modular cocircuit elimination (C3″) is in some sense equivalent to the 3-term Grassmann-Plücker relations (B2″). The last proof has substantiated half of this claim: it has shown how to derive (B2″) from (C3″). To establish the converse, we show how to derive modular cocircuit elimination (Theorem 3.6.1*) from the 3-term Grassmann-Plücker relations (Theorem 3.6.2).

Proof of Theorem 3.6.1* from Theorem 3.6.2. ["(C3″) from (B2″)"]. Let

$\chi : E^r \longrightarrow \{-1,0,1\}$ be an alternating map satisfying (B1$'$) and (B2$''$), and let M be the matroid defined from χ by (B1$'$).

Let $\mathcal{C}^*$ be the cocircuit signature of M induced by χ, via Lemma 3.5.8 – note that in the proof of this lemma, it suffices to consider *adjacent* bases $X = (x_2,\ldots,x_r)$ and $Y = (y_2,\ldots,y_r)$ of $E\backslash\underline{D}$, so that we may assume $x_3 = y_3, \ldots$ $,x_r = y_r$, and (B2$''$) is sufficient to prove Lemma 3.5.8.

Let $X,Y \in \mathcal{C}^*$ be a modular pair of circuits, $X \neq -Y$, and $e \in X^+ \cap Y^-$, which determine the coline $L := E\backslash(\underline{X}\cup\underline{Y})$. Now the contracted chirotope χ/L has rank 2, with underlying matroid M/L. It satisfies (B2$''$), which in rank 2 is equivalent to (B2).

But $\mathcal{C}^*/L = \{Y \in \mathcal{C}^* : \underline{Y} \cap L = \emptyset\}$ is the cocircuit signature of M/L that Lemma 3.5.8 derives from χ/L: and since χ/L is a chirotope (satisfying (B2)), we get from Lemma 3.5.10 that $\mathcal{C}^*/L$ satisfies (C3$''$): which produces the required cocircuit Z that eliminates e from X and Y, and does not meet L. □

Dualizing Theorem 3.6.1*, oriented matroids can be viewed as locally realizable vector configurations in the following sense.

3.6.3 Corollary. *A mapping $\chi : E^r \to \{-1,0,1\}$ is a chirotope if and only if χ satisfies* (B1$'$) *and*

> (LR) *the restriction of χ to any $(r+2)$-element subset of E is realizable.*
> (local realizability)

The local realizability axiom (LR) states in other words that for each subset $\{e_1,\ldots,e_{r+2}\}$ of E there exists $\{x_1,\ldots,x_{r+2}\} \subset \mathbb{R}^r$ such that

$$\chi(e_{i_1},e_{i_2},\ldots,e_{i_r}) = \text{sign}(\det(x_{i_1},x_{i_2},\ldots,x_{i_r}))$$

for all $1 \leq i_1 < i_2 < \ldots < i_r \leq r+2$.

Proof of Corollary 3.6.3. Since the 3-term Grassmann-Plücker condition is clearly satisfied for (realized) vector configurations, we have that (LR) implies (B2$''$). Using Theorem 3.6.2, this means (B1$'$) and (LR) together imply that χ is a chirotope.

Conversely suppose that χ is a chirotope. By Theorem 3.5.5, χ is the basis orientation of some rank r oriented matroid $\mathcal{M}$. The condition (B1$'$) being immediate, we only need to verify (LR). Let A be any $(r+2)$-element subset of E. We have to show that the restricted map $\chi|_A : A^r \to \{-1,0,+1\}$ is realizable. If A does not span $\mathcal{M}$, then $\chi|_A$ is identically zero and is hence realizable. Therefore we may assume that A spans $\mathcal{M}$, which means that $\chi|_A$ is the chirotope of the restricted oriented matroid $\mathcal{M}(A)$ (see below). Its dual $\mathcal{M}(A)^*$ is a rank 2 oriented matroid. It is easy to verify from any one of the axiom systems that all rank 2 oriented matroids are realizable (Corollary 8.2.3). Hence $\mathcal{M}(A)^*$ is realizable. Since realizability is preserved under duality, also $M(A)$ is realizable and hence $\chi|_A$ is realizable. □

This result can be simplified for uniform oriented matroids, in which case the condition (B1′) is automatically satisfied.

3.6.4 Corollary. *An alternating mapping* $\chi : E^r \to \{-1, +1\}$ *is a uniform chirotope if and only if* χ *satisfies* (LR).

Dualizing Corollary 3.6.3, we obtain that a cocircuit signature of a matroid defines an orientation if and only for every coline, the signature implies a consistent choice of a positive and a negative side for the hyperplanes that contain it. Equivalently, every contraction to rank 2 has to be a realizable oriented matroid. This describes *"oriented matroids as signed geometries real in corank 2"* (Las Vergnas 1984a).

3.7 Vectors and covectors

Vectors of oriented matroids were introduced in (Bland and Las Vergnas 1978) under the name of "signed spans". Edmonds and Mandel (1982) gave an intrinsic characterization of the set of vectors of an oriented matroid, and we will prove this result in Theorem 3.7.5. In the following chapters these vector axioms will be crucial for the Topological Representation Theorem. In this context it is customary to consider the vector axioms dually as axioms for the covectors of an oriented matroid, because of their interpretation as the sign vectors of the cells of the corresponding (pseudo)sphere arrangement.

Recall that the *composition* of the signed sets $X_1, \ldots, X_k$ is the signed set

$$Y = X_1 \circ X_2 \circ \ldots \circ X_k$$

which is defined as follows:

$$\underline{Y} = \underline{X}_1 \cup \underline{X}_2 \cup \ldots \cup \underline{X}_k,$$

$Y(e) = X_i(e)$, where i is the smallest index such that $e \in \underline{X}_i$.

The composition is called *conformal* provided $X_i(e)X_j(e) \geq 0$ for all indices i, j and all $e \in \underline{Y}$, and in this case the signed sets X_j are said to *conform* to Y.

3.7.1 Definition. A *vector* of an oriented matroid is any composition of circuits.

A *covector* is a vector of the dual oriented matroid, that is, any composition of cocircuits.

(Thus any result about vectors applies equally well to covectors.)

We first show that in Definition 3.7.1 it suffices to consider compositions which are conformal.

3.7.2 Proposition. *Any vector X of an oriented matroid is a composition of circuits conforming to X.*

Proof. Using reorientation if necessary, we may suppose that X is positive.

Choose X' as the conformal composition of all positive cocircuits that are contained in X, and assume $X' \neq X$. Using induction on $|X \backslash X'|$, we only have to treat the case that $X = X' \circ Y$, where Y is a circuit, and show (*):

(*) *There exists a positive circuit $Z \subseteq X$ such that $\underline{Z} \backslash \underline{X'} \neq \emptyset$.*

For this, pick any circuit Z such that $\underline{Z} \subseteq X$ and $\underline{Z} \backslash \underline{X'} \neq \emptyset$. After repeated elimination with the circuit Y if necessary, we may assume that $Z^- \subseteq \underline{X'}$. Furthermore assume that Z is a circuit with these properties such that $|Z^-|$ is minimal. Suppose $Z^- \neq \emptyset$. Then let $e \in Z^- \subseteq \underline{X'}$ and $f \in \underline{Z} \backslash \underline{X'}$. By the construction of X', there is a positive circuit X_1 such that $e \in \underline{X}_1 \subseteq \underline{X'}$. By (C3') applied to X_1, Z, e and f, there is a circuit Z' with $f \in Z'$, $\underline{Z}' \subseteq \underline{X}_1 \cup \underline{Z} \subseteq X$, and $Z'^- \subseteq Z^- \backslash \{e\}$. This is a contradiction to the choice of Z, and hence $Z^- = \emptyset$. This completes the proof of (*) and of Proposition 3.7.2. $\square$

Let V be a subspace of $\mathbb{R}^E$, and let $\mathcal{V}$ be the collection of all signed supports of vectors of V. We have seen in Section 3.2 that $\mathcal{C} = \text{Min}(\mathcal{V})$ is the set of circuits of an oriented matroid $\mathcal{M} = \text{Supp}(V)$. Let $\mathcal{V}_{\mathcal{M}}$ denote the set of vectors of $\mathcal{M}$. We now show that $\mathcal{V} = \mathcal{V}_{\mathcal{M}}$.

Let $X, Y \in \mathcal{V}$. There are $u, v \in V$ such that $X = s(u)$ and $Y = s(v)$, where s denotes signed support. Let $w = u + \alpha \cdot v$ where α is a sufficiently small positive real number. Then $w \in V$ and, as is easily seen, $X \circ Y = s(w) \in \mathcal{V}$. Hence $\mathcal{V}$ is closed under composition. This implies $\mathcal{V}_{\mathcal{M}} \subseteq \mathcal{V}$.

The reverse inclusion $\mathcal{V} \subseteq \mathcal{V}_{\mathcal{M}}$ is a consequence of the following property which strengthens Proposition 3.2.4:

For any $w \in V \backslash \{0\}$ and $e \in s(w)$ there is $w' \in V \backslash \{0\}$ with inclusion-minimal support such that $e \in s(w')$, $s^+(w') \subseteq s^+(w)$ and $s^-(w') \subseteq s^-(w)$.

From the discussion in Section 3.2, we know that $\mathcal{V}$ satisfies (C3') = (V3). Hence the desired property is a particular case of the Lemma 3.3.4.

3.7.3 Remark. Let $\mathcal{M}$ be the affine oriented matroid of a configuration E of points in Euclidean space. The vectors and covectors of $\mathcal{M}$ have the following geometric interpretation. A signed subset $X = (X^+, X^-)$ of E is a vector of $\mathcal{M}$ if and only if it is a *Radon partition*, that is,

$$\text{relint}(\text{conv}(X^-)) \cap \text{relint}(\text{conv}(X^+)) \quad \neq \quad \emptyset.$$

A signed subset $Y = (Y^+, Y^-)$ of E is a covector of $\mathcal{M}$ (i.e. a vector of the dual $\mathcal{M}^*$) if and only if it is a *hyperplane section*, that is, there exists an affine hyperplane H with corresponding open halfspaces H^- and H^+ such that $Y^- = E \cap H^-$ and $Y^+ = E \cap H^+$. We refer to Figure 1.2.2 for a complete list of all vectors and covectors of a configuration of six points in the Euclidean plane.

Furthermore, if $\mathcal{M}$ is represented by an essential hyperplane arrangement (as discussed in Section 1.2), then the covectors of $\mathcal{M}$ are in one-to-one correspondence with the cells of the arrangement. This suggests that we interpret the set $\mathcal{L}$ of covectors – with the right partial order – as the face lattice of the hyperplane arrangement.

3.7.4 Remark.   The set of vectors $\mathcal{V}$ of an oriented matroid $\mathcal{M}$ is most naturally viewed as a partially ordered set ("poset"), whose partial order is given by

$$Y \leq X \quad \text{if } Y \text{ is a restriction of } X.$$

The poset $(\mathcal{V}, \leq)$ of covectors is a pure poset of rank $r^* = r(\mathcal{M}^*)$, whose unique minimal element is 0, whose atoms (covering 0) are the circuits of $\mathcal{M}$, and whose maximal elements are the maximal vectors of $\mathcal{M}$.

These properties, in their dualized version for covectors, are basic in Chapter 4 in preparation for the topological representation of oriented matroids, which yields a much finer analysis of the (co)vector posets. We refer to Section 4.1 (see Theorem 4.1.14) for proofs and further discussion.

Of course, by duality we get from Remark 3.7.4 a similar statement about the covectors of $\mathcal{M}$, which form a pure poset of rank $r = r(\mathcal{M})$.

We now come to the first main result of this section, an axiomatic characterization of the collection of vectors of an oriented matroid.

3.7.5 Theorem (Bland and Las Vergnas 1978, Edmonds and Mandel 1982).   *A collection $\mathcal{V}$ of signed subsets of a set E is the set of vectors of an oriented matroid if and only if the following properties hold:*

(V0) $\emptyset \in \mathcal{V}$,

(V1) $\mathcal{V} = -\mathcal{V}$,   (symmetry)

(V2) *for all $X, Y \in \mathcal{V}$ we have $X \circ Y \in \mathcal{V}$,*   (composition)

(V3) *for all $X, Y \in \mathcal{V}$, $e \in X^+ \cap Y^-$ and $f \in (\underline{X \backslash Y}) \cup (\underline{Y \backslash X}) \cup (X^+ \cap Y^+) \cup (X^- \cap Y^-)$, there is $Z \in \mathcal{V}$ such that*

$$Z^+ \subseteq (X^+ \cup Y^+) \backslash e,$$
$$Z^- \subseteq (X^- \cup Y^-) \backslash e,$$

and $f \in \underline{Z}$.   (strong vector elimination)

In this case $\mathrm{Min}(\mathcal{V})$ *is the set of circuits of the oriented matroid with set of vectors* $\mathcal{V}$.

Proof. Let $\mathcal{V}$ be the set of vectors of an oriented matroid with set of circuits $\mathcal{C}$. Clearly $\mathcal{V}$ satisfies (V0), (V1) and (V2). We prove (V3). Let X, Y, e and f be as in the hypothesis of (V3).

Suppose $f \in \underline{X \backslash Y}$. By Proposition 3.7.2 there are circuits $X', Y' \in \mathcal{C}$ such that X' is a restriction of X, Y' is a restriction of Y, $f \in X'$, and $e \in Y'$. If $e \notin X'$, we are done, with $Z = X'$. Otherwise we are done by (C3') applied to the circuits X', Y', e, f. By symmetry, we are also done if $f \in \underline{Y \backslash X}$.

Hence we may assume $f \in (X^+ \cap Y^+) \cup (X^- \cap Y^-)$. By Proposition 3.7.2 there are circuits $X', Y' \in \mathcal{C}$ such that X' is a restriction of X, Y' is a restriction of Y, $f \in X'$, and $f \in Y'$. If $e \notin X'$ or $e \notin Y'$ we are done, with $Z = X'$

or $Z = Y'$ respectively. Suppose $e \in \underline{X}' \cap \underline{Y}'$. We have $e \in X'^+ \cap Y'^-$ and $f \in (\underline{X}' \backslash \underline{Y}') \cup (\underline{Y}' \backslash \underline{X}') \cup (X'^+ \cap Y'^+) \cup (X'^- \cap Y'^-)$. By (C3′) applied to X', Y', e, and f, there is $Z \in \mathcal{C}$ such that $Z^+ \subseteq (X'^+ \cup Y'^+) \backslash \{e\}$, $Z^- \subseteq (X'^- \cup Y'^-) \backslash \{e\}$ and $f \in Z$. We have $Z \in \mathcal{V}$, $Z^+ \subseteq (X^+ \cup Y^+) \backslash \{e\}$, $Z^- \subseteq (X^- \cup Y^-) \backslash \{e\}$ and $f \in Z$. Hence $\mathcal{V}$ satisfies (V3).

Conversely, consider $\mathcal{V} \subseteq \{-1, 0, +1\}^E$ satisfying (V0), (V1), (V2) and (V3). By Proposition 3.2.4, (V1) and (V3) imply that $\mathcal{C} = \mathrm{Min}(\mathcal{V})$ is the set of circuits of an oriented matroid $\mathcal{M}$. Let $\mathcal{V}_{\mathcal{M}}$ be the set of vectors of $\mathcal{M}$. It follows from (V2) that $\mathcal{V} \subseteq \mathcal{V}_{\mathcal{M}}$. It follows from Proposition 3.7.3 that $\mathcal{V}_{\mathcal{M}} \subseteq \mathcal{V}$. This completes the proof.   □

3.7.6 Corollary. *The circuit axioms for oriented matroids are cryptomorphic to the vector axioms* (V0), (V1), (V2) *and* (V3) *under the interpretations*

$\qquad \mathcal{C} = \mathrm{Min}(\mathcal{V}) \qquad$ *and*

$\qquad \mathcal{V} = \{V : V \text{ is a (conformal) composition of elements of } \mathcal{C}\}.$

Proposition 3.7.2 implies that these results can be strengthened as follows.

3.7.7 Corollary. *In* Theorem 3.7.5 *and in* Corollary 3.7.6, (V2) *can be replaced by the weaker axiom*

(V2′) *for all* $X, Y \in \mathcal{V}$ *with equal restriction on their intersection, we have*
$\qquad X \circ Y \in \mathcal{V}.$ (conformal composition)

3.7.8 Remark. If $\mathcal{M}$ is the oriented matroid associated with a hyperplane arrangement $\mathcal{H}$, then the conformal composition operation of Corollary 3.7.7 has the following geometric interpretation. The covectors X and Y correspond to faces of (the cell complex defined by) $\mathcal{H}$. They have equal restriction on their intersection if and only if the faces corresponding to X and Y are faces of a common proper cell of $\mathcal{H}$. In that case $X \circ Y$ is the covector corresponding to the smallest such cell.

Here are two more strengthenings of Theorem 3.7.5.

3.7.9 Corollary (Edmonds and Fukuda 1982, Edmonds and Mandel 1982). *In Theorem* 3.7.5 *the axiom* (V3) *can be replaced by*

(V3′) *for all* $X, Y \in \mathcal{V}$ *and* $e \in X^+ \cap Y^-$, *there is a* $Z \in \mathcal{V}$ *such that*
$\qquad Z^+ \subseteq (X^+ \cup Y^+) \backslash e,$
$\qquad Z^- \subseteq (X^- \cup Y^-) \backslash e$
and $(\underline{X} \backslash \underline{Y}) \cup (\underline{Y} \backslash \underline{X}) \cup (X^+ \cap Y^+) \cup (X^- \cap Y^-) \subseteq Z.$ (vector elimination)

Proof. Clearly (V3′) implies (V3). Thus vector elimination is in fact stronger than strong elimination. In spite of the possibility of confusion, the term "strong" is used for (V3) because it is closely related to strong elimination (C3$'$) for circuits.

Conversely assume (V3) and let $X, Y \in \mathcal{V}$ and $e \in X^+ \cap Y^-$. By (V3) for all $f \in (\underline{X} \backslash \underline{Y}) \cup (\underline{Y} \backslash \underline{X}) \cup (X^+ \cap Y^+) \cup (X^- \cap Y^-)$, there is $Z_f \in \mathcal{V}$ such that $Z_f^+ \subseteq (X^+ \cup Y^+) \backslash \{e\}$, $Z_f^- \subseteq (X^- \cup Y^-) \backslash \{e\}$ and $f \in \underline{Z}_f$. Let Z be any composition of the Z_f's for all $f \in (\underline{X} \backslash \underline{Y}) \cup (\underline{Y} \backslash \underline{X}) \cup (X^+ \cap Y^+) \cup (X^- \cap Y^-)$. The signed set Z is contained in $\mathcal{V}$ by (V2), and so it has all properties required in (V3$'$). $\square$

3.7.10 Proposition (Edmonds and Mandel 1982). *In Theorem 3.7.5 the axiom* (V3) *can be replaced by*

(V3$''$) *for all* $X, Y \in \mathcal{V}$ *with* $\underline{Y} \subseteq \underline{X}$ *and* $X^+ \cap Y^- \neq \emptyset$, *there is a proper restriction* Z *of* X *such that*

$$Z \in \mathcal{V} \text{ and } (X^+ \backslash Y^-) \cup (X^- \backslash Y^+) \subseteq \underline{Z}. \qquad \text{(Y-approximation of X)}$$

Before proving Proposition 3.7.10, let us verify the property (V3$''$) for realizable oriented matroids. Let V be a subspace of $\mathbb{R}^E$ and $u, v \in V$. Then $X = s(u)$ and $Y = s(v)$ are two vectors of the oriented matroid $\operatorname{Supp}(V)$.

Now (V2) is given by $X \circ Y = s(w)$ where $w = u + \alpha \cdot v$ with α any real number with $0 < \alpha < \operatorname{Min}\left(-\frac{u(e)}{v(e)} : e \in X^+ \cap Y^-\right)$. Under the additional hypothesis $\underline{Y} \subseteq \underline{X}$ and $X^+ \cap Y^- \neq \emptyset$, the approximation property (V3$''$) is given by $Z = s(w)$ where $w = u + \alpha \cdot v$ with $\alpha = \operatorname{Min}\left(-\frac{u(e)}{v(e)} : e \in X^+ \cap Y^-\right)$.

Proof of Proposition 3.7.10.

(1) (V2) and (V3) imply (V3$''$).

Let $X, Y \in \mathcal{V}$ with $\underline{Y} \subseteq \underline{X}$ and $X^+ \cap Y^- \neq \emptyset$. Using reorientation if necessary, we may suppose X positive. Let $Z \in \mathcal{V}$ such that $\underline{Z} \subseteq \underline{X}$, $Z \neq X$, $X^+ \backslash Y^- \subseteq Z^+$ and $|Z^-|$ is minimal. Such a Z exists because $\underline{Y \circ X} \subseteq \underline{X}$, $Y \circ X \neq X$, $X^+ \backslash Y^- = (Y \circ X)^+$, and $Y \circ X \in \mathcal{V}$ by (V2). Suppose $Z^- \neq \emptyset$ and pick $e \in Z^-$. By (V3$'$) applied to X, Z and e, there is $Z' \in \mathcal{V}$ such $X^+ \backslash Y^- \subseteq X^+ \backslash Z^- \subseteq Z'^+ \subset X^+ \backslash \{e\}$ and $Z'^- \subseteq Z^- \backslash \{e\}$. This is a contradiction to the choice of Z. Consequently Z is positive, hence a proper restriction of X, and contains $X^+ \backslash Y^-$, as desired.

(2) (V2) *and* (V3$''$) *imply* (V3$'$).

Let $X, Y \in \mathcal{V}$ and $e \in X^+ \cap Y^-$. There is $X' \in \mathcal{V}$ with $X'(e) = 1$ and $X'(f) = (X \circ Y)(f)$ for all $f \in (\underline{X} \backslash \underline{Y}) \cup (\underline{Y} \backslash \underline{X}) \cup (X^+ \cap Y^+) \cup (X^- \cap Y^-)$, namely $X' = X \circ Y$, as $X \circ Y \in \mathcal{V}$ by (V2). Choose X' with these properties such that $|(X'^+ \cap Y^-) \cup (X'^- \cap Y^+)|$ is minimal. We have $\underline{Y} \subseteq \underline{X'}$ and $e \in X'^+ \cap Y^-$. Let Z be given by (V3$''$) applied to X' and Y. We have $Z(f) = (X \circ Y)(f)$ for all $f \in (\underline{X} \backslash \underline{Y}) \cup (\underline{Y} \backslash \underline{X}) \cup (X^+ \cap Y^+) \cup (X^- \cap Y^-)$ and $|(Z^+ \cap Y^-) \cup (Z^- \cap Y^+)| < |(X'^+ \cap Y^-) \cup (X'^- \cap Y^+)|$. Now the choice of X' implies that $Z(e) = 0$. Therefore Z has the required properties for (V3$'$). $\square$

For other variations on axiomatics of vectors see Fukuda and Tamura (1990). We will close this section by describing the oriented matroid operations of deletion, contraction and duality in terms of vectors.

3.7.11 Proposition.   *Let M be an oriented matroid on E with set of circuits C and set of vectors V, and let $A \subseteq E$.*

(1)   *The set of vectors of the deletion $M\backslash A$ equals*

$$V\backslash A = \{X \in V \ : \ \underline{X} \cap A = \emptyset\}.$$

(2)   *The set of vectors of the contraction M/A equals*

$$V/A = \{X\backslash A \ : \ X \in V\}.$$

Proof. The expression for $V\backslash A$ follows immediately from Definition 3.7.1 and Proposition 3.3.1.

By Propositions 3.3.2 and 3.7.2, any $X\backslash A$ with $X \in V$ is a conformal composition of signed sets in $\mathrm{Min}(\{X\backslash A : X \in V\}) = C/A$, hence is in V/A. The converse is immediate. $\Box$

3.7.12 Proposition.   *Let M be an oriented matroid on E with set of circuits C and set of vectors V. Then the set of vectors of its dual M^* equals*

$$
\begin{aligned}
\mathcal{L} \ &= \ \{Y \in \{+,-,0\}^E \ : \ X \perp Y \text{ for all } X \in C\} \\
&= \ \{Y \in \{+,-,0\}^E \ : \ X \perp Y \text{ for all } X \in V\}.
\end{aligned}
$$

Proof. From Section 3.4 and Definition 3.7.1 it follows immediately that $X \perp Y$ for all $X \in V$ and $Y \in \mathcal{L}$.

Conversely, consider $Y \in \{+,-,0\}^E$ such that $X \perp Y$ for all $X \in C$. Apply the 4-Painting Theorem 3.4.4 with $B = Y^+$, $W = Y^-$, $G = E\backslash \underline{Y}$, $R = \emptyset$, and $e \in \underline{Y} = B \cup W$. If the first alternative holds, there is a circuit X such that $e \in X$ with $\underline{X} \cap Y^+ \subseteq X^+$ and $\underline{X} \cap Y^- \subseteq X^-$, contradicting $X \perp Y$. Hence the second alternative holds: there is a cocircuit Y_e such that $e \in \underline{Y_e} \subseteq \underline{Y}$, $\underline{Y_e} \cap Y^+ \subseteq Y_e{}^+$ and $\underline{Y_e} \cap Y^- \subseteq Y_e{}^-$. This means that Y_e is a restriction of Y. It follows that Y is a conformal composition of cocircuits Y_e, $e \in \underline{Y}$. Hence $Y \in \mathcal{L}$. $\Box$

3.8 Maximal vectors and topes

3.8.1 Definition (Maximal vectors and covectors, topes).   A *maximal vector* of an oriented matroid M is a vector whose support is maximal with respect to inclusion.

A *maximal covector* of an oriented matroid M is a covector whose support is maximal with respect to inclusion.

Maximal covectors are usually called *topes*, following Edmonds and Mandel

(1982) and Edmonds and Fukuda (1982). They correspond to the open cells (or 'regions, or 'chambers') in an arrangement of hyperplanes or pseudospheres. A cryptomorphic axiomatization of oriented matroids in terms of their maximal vectors (or, equivalently, their maximal covectors) is presented in Proposition 3.8.3.

The families of maximal vectors and covectors form further signed set systems that encode oriented matroids. They are not that well understood, but especially interesting. (See also the Exercises 3.27 to 3.32.) Unlike the other systems discussed before, they do *not* form a generalization of some unoriented matroid concept.

In fact, it is easy to see that all maximal vectors have the same support, namely $E\backslash I$, where I is the set of coloops of $\mathcal{M}$. Recall that $e \in E$ is a *coloop* of $\mathcal{M}$ if it appears in no circuit.

By duality, the maximal covectors all have support $E\backslash L$, where L is the set of loops of $\mathcal{M}$.

Example (The CUBE, continued). The 104 maximal vectors of the oriented matroid CUBE (defined in Section 3.2) are the following, together with their opposites:

$$
\begin{array}{cccccc}
1\overline{23}45678 & 12\overline{345}678 & 1\overline{2345}678 & 12\overline{345}678 & 1\overline{2345}678 & 12\overline{345}678 \\
\overline{123}45678 & 1\overline{2345}678 & 12\overline{345}678 & 12\overline{3}45678 & 12\overline{3}45678 & 12\overline{3}45678 \\
1\overline{23}45678 & 1\overline{2}3456\overline{7}8 & 1\overline{2}345\overline{6}78 & 12\overline{3}456\overline{7}8 & 12\overline{3}45\overline{6}78 & 12\overline{3}45\overline{6}78 \\
12\overline{34}5678 & 1\overline{234}5678 & 12\overline{34}5678 & 12\overline{34}5678 & 12\overline{34}5678 & 12\overline{34}5678 \\
12345\overline{6}78 & 12345\overline{6}78 & 12345\overline{6}78 & 12345\overline{6}78 & 12345\overline{6}78 & 12345\overline{6}78 \\
1\overline{2345}678 & 12\overline{345}678 & 12\overline{345}678 & 12345\overline{6}\overline{7}8 & 12345\overline{6}7\overline{8} & 12345\overline{6}7\overline{8} \\
1\overline{2}345\overline{6}\overline{7}8 & 12\overline{3}456\overline{7}8 & 12\overline{3}45\overline{6}78 & 12\overline{3}456\overline{7}8 & 12\overline{3}456\overline{7}8 & 12\overline{3}456\overline{7}8 \\
1\overline{2}345\overline{6}\overline{7}8 & 12\overline{345}6\overline{7}8 & 1\overline{2}345\overline{6}78 & 12\overline{345}678 & 1\overline{2}3456\overline{7}8 & 12\overline{3}456\overline{7}8 \\
1\overline{2}345\overline{6}78 & 1\overline{2345}678 & 12\overline{345}678 & 12345\overline{678} & & \\
\end{array}
$$

The topes (= maximal covectors) of CUBE are (up to opposites) given in the following list.

$$
\begin{array}{cccccc}
12345678 & 1\overline{2}345678 & 12\overline{3}45678 & 123\overline{4}5678 & 1\overline{23}45678 & 12\overline{34}5678 \\
1\overline{2}3\overline{4}5678 & 1234\overline{5}678 & 12345\overline{6}78 & 12\overline{3}45\overline{6}78 & 12\overline{34}5\overline{6}78 & 12345\overline{6}78 \\
1\overline{2}345\overline{6}78 & 1234\overline{56}78 & 12\overline{3}45\overline{6}78 & 12\overline{3}45\overline{6}78 & 12345\overline{6}78 & 12\overline{3}45\overline{6}78 \\
1234\overline{56}78 & 12345678 & 12\overline{3}45\overline{6}78 & 12\overline{3}45\overline{6}78 & 12345\overline{6}78 & 12\overline{3}45\overline{6}78 \\
123456\overline{78} & 12\overline{3}456\overline{78} & 12\overline{345}678 & 1\overline{2}345\overline{6}78 & 12\overline{3}45\overline{6}78 & 12345\overline{678} \\
1\overline{2}345\overline{6}78 & 1\overline{2}345\overline{78} & 12345678 & 12\overline{3}45678 & 12\overline{3}45678 & 12\overline{3}45678 \\
1\overline{23}45678 & 12345\overline{6}78 & 12\overline{345}678 & 12\overline{3}45678 & 12345\overline{6}78 & 12\overline{345}678 \\
1\overline{2}345\overline{6}78 & 12\overline{3}45678 & 12\overline{3}45678 & 12345678 & 1\overline{2}345678 & 12\overline{3}45678 \\
12\overline{3}45678 & 1\overline{2}345678 & 12\overline{3}45678 & 12\overline{3}45678 & & \\
\end{array}
$$

These topes correspond to

○ the ways to 2-partition the vertices of the cube by a hyperplane,

○ the maximal, 3-dimensional cells in the arrangement of eight facet planes of
a regular octahedron.

Here the first tope $12345678 = (++++++++)$ corresponds to the octahedron
itself. The next tope $\bar{1}2345678 = (+-++++++)$ corresponds to a 3-polytope
which meets the octahedron in a common facet in plane 2, etc. (see Exercise
3.26).

We now come to the basic fact that W determines V (and hence M), and
dually T determines L (and hence M).

3.8.2 Proposition (Mandel, see Cordovil 1985).

(a) *Every oriented matroid M is uniquely determined by the collection $W =$
$W(M)$ of its maximal vectors, via*

$$V = \{X \in \{+,-,0\}^E : X \circ W \in W \text{ for all } W \in W\}$$
$$= \{X \in \{+,-,0\}^E : X \circ W \subseteq W\}.$$

(b) *Any oriented matroid M is uniquely determined by the collection $T =$
$T(M)$ of topes, via*

$$L = \{X \in \{+,-,0\}^E : X \circ T \subseteq T\}.$$

Proof. The statements (a) and (b) are dual to each other – we will prove (b) as
Theorem 4.2.13.   □

In the following we will abbreviate the conditions in this proposition by $X \circ W \subseteq$
W and $Y \circ T \subseteq T$.

In order to state the known axiomatics of maximal vectors we introduce the
following notations. Let $W \subseteq \{-,0,+\}^E$ be any collection of signed sets, and
let $A \subseteq E$. Then we define

$$W/A := W(E \backslash A) := \{Z \backslash A : Z \in W\},$$

$$\text{cl}^*_W(A) := A \cup \{e \in E \backslash A : \text{there is } B \subseteq A \text{ such that } W(B) = \{+,-\}^B$$
$$\text{and } W(B \cup e) \neq \{+,-\}^{B \cup e}\},$$

$$W \backslash A := \{Z \in W/\text{cl}^*_W(A) : Z \circ W \subseteq W\}.$$

These definitions make sense, because when W is the set of maximal vectors of
an oriented matroid M, then $W \backslash A$ (resp. W/A) is the set of maximal vectors
of $M \backslash A$ (resp. M/A), the oriented matroid obtained from M by deleting (resp.
contracting) A. In that case the mapping $\text{cl}^*_W : 2^E \to 2^E$ is the closure operator
of the dual matroid $\underline{M}^*$.

The following result of da Silva strengthens an earlier theorem of Bienia and
Cordovil (1987). The uniform case is due to Lawrence (1983), see Exercise 3.28.

3.8.3 Proposition (da Silva 1988).   *Let $W \subseteq \{+,-,0\}^E$. Then W is the set*
of maximal vectors of an oriented matroid if and only if it satisfies the following
three axioms.

(W0) $W \neq \emptyset$, and for all $X, Y \in W$ we have $\underline{X} = \underline{Y}$,

(W1) $W = -W$,

(W2) for all $A, B \subseteq E$ with $A \cap B = \emptyset$ we have $(W \backslash A)/B = (W/B) \backslash A$.

In this case, $\mathcal{V} = \{X \subseteq \{+, -, 0\}^E : X \circ W \subseteq W\}$ is the set of vectors of the (unique) oriented matroid on E with set of maximal vectors $W = \mathrm{Max}(\mathcal{V})$.

Proof. The necessity of the axioms follows from Proposition 3.3.3, and by Proposition 3.8.2 we have $\mathcal{V} = \{X \subseteq \{+, -, 0\}^E : X \circ W \subseteq W\}$ if W comes from an oriented matroid.

Now we assume that W is a family of signed sets that satisfies (W0), (W1) and (W2) and prove sufficiency.

Clearly $\emptyset \in \mathcal{V}$, and from $W = -W$ it follows that $\mathcal{V} = -\mathcal{V}$, hence we have (V0) and (V1). Let $X, Y \in \mathcal{V}$. Since $(X \circ Y) \circ W = X \circ (Y \circ W)$, we have $X \circ Y \in \mathcal{V}$. Hence the set $\mathcal{V}$ also satisfies (V2).

We now show that $\mathcal{V}$ satisfies (V3). Let $X, Y \in \mathcal{V}$, $X \neq -Y$, and $e \in X^+ \cap Y^-$. Set $E_2 = (X^+ \cap Y^-) \cup (X^- \cap Y^+)$, $E_1 = (\underline{X} \cup \underline{Y}) \backslash E_2$ and $E_0 = E \backslash (\underline{X} \cup \underline{Y}) = E \backslash (E_1 \cup E_2)$.

Let $f \in E_1$. As is easily checked, the restrictions to the pair (e, f) of the four signed sets $X \circ Y, -(X \circ Y), Y \circ X, -(Y \circ X)$ of $\mathcal{V}$ are $(+, +), (-, -), (+, -), (-, +)$ (up to reordering). Hence for any $B \subseteq E_0$ such that $W(B) = \{+, -\}^B$ we have $W(B \cup \{e, f\}) = \{+, -\}^{B \cup \{e, f\}}$. It follows that $\mathrm{cl}^*_{W/(E_2 \backslash e)}(E_0 \cup e) = E_0 \cup e$.

We abbreviate $E_0' = E_0 \cup e$, $E_2' = E_2 \backslash e$, $W' = W/E_2'$ and $Z' = (X \circ Y) \backslash E_2 = (Y \circ X) \backslash E_2$. Since $X \circ Y \in \mathcal{V}$, we have

$$Z' = (X \circ Y) \backslash (E_2' \cup E_0') \quad \in \quad W'/E_0' \quad = \quad W'/\mathrm{cl}^*_{W'}(E_0').$$

Consider any $V' \in W'$. We have $V' = V \backslash E_2'$ for some $V \in W$. If $V'(e) = 1$, we have $Z' \circ V' = ((X \circ Y) \backslash E_2') \circ V' = (X \circ Y \circ V) \backslash E_2' \in W'$ since $X \circ Y \circ V \in W$. If $V'(e) = -1$, we have $Z' \circ V' = ((Y \circ X) \backslash E_2') \circ V' = (Y \circ X \circ V) \backslash E_2' \in W'$ since $Y \circ X \circ V \in W$. Therefore $Z' \in W' \backslash E_0' = (W/E_2') \backslash E_0'$.

By (W2), we have $Z' \in (W \backslash E_0')/E_2'$. Hence there is $Z \in W \backslash E_0'$ such that $Z' = Z \backslash E_2'$. By definition of $W \backslash E_0'$, we have $Z \circ W \subseteq W$, implying $Z \in \mathcal{V}$, and $\underline{Z} \subseteq (\underline{X} \cup \underline{Y}) \backslash \{e\}$. Since $Z' = Z \backslash E_2'$ we have $Z \neq \emptyset$, $Z^+ \subseteq X^+ \cup Y^+$ and $Z^- \subseteq X^- \cup Y^-$, proving (V3).

Let $X, Y \in W$. Since all signed sets in W have the same support by (W0), we have $X \circ Y = X$. This shows that $X \circ W \subseteq W$ and therefore $W \subseteq \mathcal{V}$. Moreover, W is the set of maximal elements of $\mathcal{V}$. This completes the proof of Proposition 3.8.3. $\square$

Hence each of the previously given axiom systems for oriented matroids is cryptomorphic to the axiom system (W0), (W1) and (W2) for maximal vectors. We omit the proof of the following strengthening of this result.

3.8.4 Proposition (Handa 1990).   *In* Proposition 3.8.3 *the axiom* (W2) *can be replaced by*

(W2′) *for all $A \subseteq E$ and all $X, Y \in \mathcal{W}/A$, $X \neq -Y$, there is an $e \in X^+ \cap Y^-$ such that*

$$-_B X \in \mathcal{W}/A, \text{ where } B = \mathrm{cl}^*_{\mathcal{W}/A}(e).$$

3.9 Historical sketch

This brief note attempts to give a review of the *origins* of oriented matroid theory, specifically concerning the early axiom systems. Obviously, many mathematicians in addition to those mentioned here have contributed greatly to the development of the field, as can be seen by reading the subsequent chapters and by perusing the bibliography.

It seems fair to say that the major credit for the origination of oriented matroid theory should be shared by Robert Bland, Jon Folkman, Michel Las Vergnas, and Jim Lawrence. Folkman began working on oriented matroids by 1967, in an attempt to prove the lower bound conjecture for polytopes by generalizing it. Tragically, he died before publishing his theory. His notes resided with Victor Klee and with Ray Fulkerson. Later, when Klee discovered that his doctoral student, Jim Lawrence, was already thinking along similar lines, and had made substantial progress, Klee gave him the notes. Lawrence completed the theory in his doctoral thesis (1975), and later published the results in a joint paper with Folkman (Folkman and Lawrence 1978). This paper included the Topological Representation Theorem, which was Lawrence's work. Meanwhile, Bland and Las Vergnas were each independently developing the notion of an oriented matroid. Las Vergnas's viewpoint was more graph theoretical and combinatorial whereas Bland's motivation was to abstract linear programming duality. Las Vergnas had preprints and was giving conference talks by 1974, and published an extensive announcement in Comptes Rendus de l'Académie des Sciences (Paris, 1975a). Bland had completed much of the basic theory before he learned from his thesis advisor, Fulkerson, of Folkman's work. However, his approach was quite disjoint from Folkman's. Bland's thesis appeared in 1974. Bland and Las Vergnas learned that they were working on the same subject, and they decided to write a joint paper on the intersection of their work. This caused some delay in the rewriting, and their joint paper appeared in the Journal of Combinatorial Theory, Series B, in 1978. The Folkman-Lawrence paper appeared in the same journal and year. Thus each of the surviving founders had a thesis or preprints at roughly the same time, and all three had their major papers appear at about the same time. Each of these originators made substantial contributions to the subject beyond those made by the others.

There are some much earlier works which may be regarded as predecessors of oriented matroids. An ancient ancestor of allowable sequences is Perrin (1881/82);

early axiomatic studies can be found in Sperner (1949), see also Karzel (1969). L. Gutierrez Novoa in 1965 studied what amounts to signed bases, and R. Buchi studied generalized convex closures, with some of his ideas appearing in the thesis of P. Mei, his student, in 1971. There were also several other early papers that pointed in the direction of oriented matroid theory. These include Minty (1966), Fulkerson (1968), and Rockafellar (1969). Minty's paper, which gave axioms for the regular or unimodular case, very likely inspired the other two. Rockafellar's paper actually proposed that an axiom system for general oriented matroids should be developed.

Several people later rediscovered what amounts to an axiom system for oriented matroids (or some special case thereof), without realizing that their work overlapped with already published papers.

"If you are not too ambitious, it can be a pleasure to realize that you have rediscovered something previously known, because at least then you know that you were on the right track." (I.M. Gel'fand)

Exercises

3.1 Let $\mathcal{C}$ be the set of circuits of an oriented matroid $\mathcal{M}$ and $_{-A}\mathcal{C} = \{_{-A}X : X \in \mathcal{C}\}$ some reorientation of it.

Verify that $_{-A}\mathcal{C}$ also satisfies the axioms in Definition 3.2.1.

3.2 Write a computer program for checking whether a given collection of sign vectors is (resp. can be augmented to) the set of circuits (resp. set of vectors) of an oriented matroid.

3.3 Compute all circuits and cocircuits of the one-element deletion CUBE\8 of the CUBE defined in Section 3.2.

3.4 Consider the oriented matroid corresponding to a directed graph D. Explain how to compute its circuits, cocircuits, maximal vectors, maximal covectors and chirotope directly from D. Interpret the axiom systems in terms of D.

3.5 Interpret Corollary 3.4.6 ("every element is contained either in a positive circuit, or in a positive cocircuit, but not both") for the oriented matroid of a digraph. Give a constructive proof.

3.6 Give a geometric interpretation for the modular elimination axiom for circuits (C3″) in Theorem 3.6.1.

3.7 Show that cocircuit elimination according to (C3) is unique *if and only if* X and Y form a modular pair. In this case

$(X^+ \cap Y^+) \subseteq (X^+\backslash Y^-) \cup (Y^+\backslash X^-) \subseteq Z^+$ and
$(X^- \cap Y^-) \subseteq (X^-\backslash Y^+) \cup (Y^-\backslash X^+) \subseteq Z^-$,

so that the strong elimination axiom (C3′) yields no extra choice.

3.8 Instead of signed sets or sign vectors, one can use subsets of a set E with involution $*: E \longrightarrow E$, $e \longmapsto e^*$ (such that $e^* \neq e$, but $e^{**} = e$ for all $e \in E$) to define symmetrically signed oriented matroids.

Formulate the circuit axioms in this framework. Discuss the variations that arise from allowing circuits of the form $\{e, e^*\}$.

(Folkman and Lawrence 1978)

3.9 Let $\mathcal{C} \subseteq \{+, -, 0\}^E$ be the set of circuits of an acyclic simple oriented matroid. Define:

$$\text{conv}(A) := A \cup \{e \in E \backslash A : X^+ \subseteq A \text{ and } X^- = \{e\} \text{ for some } X \in \mathcal{C}\}$$

for all $A \subseteq E$. Show that

(a) conv $: 2^E \longrightarrow 2^E$ is a closure operator (that is, $A \subseteq \text{conv}(A) = \text{conv}(\text{conv}(A))$ and $A \subseteq B \implies \text{conv}(A) \subseteq \text{conv}(B)$).

(b) $e \notin \text{conv}(E \backslash e)$ if and only if $Y^0 = \{e\}$, $Y^+ = E \backslash e$ for some covector Y.

(c) $\text{conv}(A)$ is equal to the intersection of all supersets of A of the form $Y^+ \cup Y^0$ (geometrically: closed halfspaces) for cocircuits $Y \in \mathcal{C}^*$.

(d) $\text{conv}(A) = \{e \in E : A \subseteq T^+ \text{ implies } e \in T^+ \text{ for all topes } T \in \mathcal{T}\}$.

(Las Vergnas 1975a, 1980a)

3.10 Let $\mathcal{L}$ be the set of covectors of an acyclic simple oriented matroid $\mathcal{M}$, let conv $: 2^E \longrightarrow 2^E$ be its convex closure (defined in Exercise 3.9) and let $\overline{A} = \cap \{X^0 : A \subseteq X^0, X \in \mathcal{L}\}$ be the closure operator of the underlying matroid. Show that

(a) $A \subseteq \text{conv}(A) \subseteq \overline{A}$, for all $A \subseteq E$, (in particular, all the flats of the underlying matroid are conv-closed)

(b) if $Y \in \mathcal{L}$ and $Y^- = \emptyset$, then $Y^0 = \text{conv}(Y^0 \cap \{e \in E : e \notin \text{conv}(E \backslash e)\})$,

(c) conv satisfies the anti-exchange property: If $e, f \in E \backslash \text{conv}(A)$, $e \neq f$ and $e \in \text{conv}(A \cup f)$, then $f \notin \text{conv}(A \cup e)$.   (Edelman 1982)

3.11 Let E be a finite set with involution $*$. Show that the following axioms characterize oriented matroids in terms of the *convex closure operator* conv $: 2^E \longrightarrow 2^E$.

(CV0) $\text{conv}(\emptyset) = \emptyset$.

(CV1) conv *is a closure operator*
       (that is $A \subseteq \text{conv}(A) = \text{conv}(\text{conv}(A))$, and $A \subseteq B$ implies $\text{conv}(A) \subseteq \text{conv}(B)$.)

(CV2) $\text{conv}(A^*) = \text{conv}(A)^*$ for $A \subseteq E$
       (where $A^* := \{e^* : e \in A\}$.)

(CV3) If $e \in \text{conv}(A \cup e^*)$, then $e \in \text{conv}(A)$.

(CV4) $e \in \text{conv}(A \cup f^*)$ and $e \notin \text{conv}(A)$ imply $f \in \text{conv}(A \cup e^*) \backslash f$.

(convex closure axioms)

(Folkman and Lawrence 1978) (Buchi and Fenton 1988)

3.12 From the previous exercise, derive an axiomatization for the convex-closure operators of simple acyclic oriented matroids $\mathcal{M} = (E, \mathcal{C})$.

3.13 Define and investigate *infinite* oriented matroids.

(a) Show how to modify the axiom systems for

– circuits

– convex closure

in this case, to get equivalent axiomatizations.   (Buchi and Fenton 1988)

(b) Show how to adopt the axiom systems for

– cocircuits

– chirotopes

in the infinite case of finite rank.

(c)* Develop the framework to deal with geometry, realization, topological representation etc. for infinite oriented matroids.

An interesting case to treat with this theory is given by the set of all vectors (and dense subsets) of Euclidean unit spheres.

3.14 Show that $(_{-A}\mathcal{M})^* =_{-A} (\mathcal{M}^*)$, that is, dualization and reorientation commute.

3.15* Characterize the signed basis graphs of oriented matroids as a class of 2-edge-colored graphs.   (Las Vergnas 1978a)

3.16 Is it true that the ideal of algebraic relations among the $r \times r$-minors of a generic $r \times n$-matrix $(n \geq r)$ is generated by the 3-term Grassmann-Plücker relations

$$\det(x_1, x_2, x_3, \ldots, x_r) \cdot \det(y_1, y_2, x_3, \ldots, x_r)$$
$$- \quad \det(x_1, y_1, x_3, \ldots, x_r) \cdot \det(x_2, y_2, x_3, \ldots, x_r)$$
$$+ \quad \det(x_1, y_2, x_3, \ldots, x_r) \cdot \det(x_2, y_1, x_3, \ldots, x_r) \quad ?$$

3.17 Let $\mathcal{M}$ be a connected oriented matroid and e be any element of $\mathcal{M}$.

(a) Show that $\mathcal{M}$ is completely determined by the set of signed circuits of $\mathcal{M}$ containing e.   (Hamidoune and Las Vergnas 1986)

(b*) Find a specific axiomatization for the collection of circuits containing a given element in an oriented matroid.

3.18 (a) Let M be a matroid on an ordered set E.

Let $B_1, B_2, \ldots$ be the sequence of bases of M in the lexicographic ordering of $\binom{E}{r}$ induced by the ordering of E.

Show that for all $i \geq 2$, there is $j < i$ such that $|B_i \Delta B_j| = 2$ (a consequence of lexicographic shelling of matroid bases, see Chapter 7 of White 1992).

Deduce an algorithm to construct a base orientation from signed circuits.

(b) Show that a uniform oriented matroid $\mathcal{M}$ on $E = \{1, 2, \ldots, n\}$ is determined by the set of circuits whose largest two elements are consecutive. (Las Vergnas)

3.19 Let $\mathcal{M}$ be a uniform oriented matroid of rank r on n elements.

(a) Show that the subsets of circuits defined in Exercises 3.17 and 3.18 contain the same number of circuits, namely $\binom{n-1}{r}$.

(b) Is this number minimal? In other words, is this the smallest number of circuits that is sufficient to determine every uniform matroid with these parameters?   (The answer is "no": Forge and Ramírez Alfonsín 1998a)

(c)* Similarly, what is the smallest number of signs that suffices to reconstruct an arbitrary uniform chirotope from it?

3.20 Let χ be an acyclic chirotope of rank r. Define its λ-function

$$\lambda : E^{r-1} \longrightarrow \mathbb{N}$$

$$(x_1, \ldots, x_{r-1}) \longmapsto |\{x \in E : \chi(x_1, \ldots, x_{r-1}, x) = +\}|.$$

(a) Show that λ determines χ. Show that this fails if χ is not required to be acyclic.

(b) Conclude that the number of rank r oriented matroids on n points is at most

$$2^{O(n^{r-1} \log n)}.$$

(c)* Characterize the λ-functions of acyclic oriented matroids.

(Goodman and Pollack 1983)

3.21 In Theorem 3.6.2, show that (B1$'$) can be weakened to

> (B1$''$) χ is alternating, and let $\mathcal{B}$ be the set of r-subsets B of E such that $\chi(x_1, \ldots, x_r) \neq 0$ for some ordering $(x_1, \ldots, x_r)$ of B.
>
> Then there is $B \in \mathcal{B}$ with the property that for all $B' \in \mathcal{B}$ and $x \in B \backslash B'$ there is $y \in B' \backslash B$ such that $B \backslash \{x\} \cup \{y\} \in \mathcal{B}$.

(Guedes de Oliveira 1988)

3.22 Given the set $\mathcal{L} \subseteq \{+, -, 0\}^E$ of covectors of an oriented matroid, and element $g \in E$, let $\mathcal{A} := \{X \in \{+, -, 0\}^{E \backslash g} : (X, +) \in \mathcal{L}\}$ be the set of affine covectors of $\mathcal{M}$ (see Section 4.5 for a motivation for this model of arrangements in affine space).

(a) Show that

(i) If $X, Y \in \mathcal{A}$, then $X \circ Y \in \mathcal{A}$ and $X \circ (-Y) \in \mathcal{A}$.

(ii) If $X, Y \in \mathcal{A}$ and $X_e = -Y_e \neq 0$, then there is a $Z \in \mathcal{A}$ such that $Z_e = 0$ and $Z_f = (X \circ Y)_f$ for all $f \in E - g$ unless $X_f = -Y_f \neq 0$.

(b) Characterize the sets of "affine covectors" $\mathcal{A}$ coming from an oriented matroid $\mathcal{L}$ in the described fashion.   (Karlander 1992)

(c)* Generalize the development of "affine oriented matroids" (b) to the (discrete) infinite case. (The hyperplane arrangements of affine Coxeter groups should be covered by this.)

3.23 Following Section 2.2, we can view a set $\mathcal{V} \subseteq \{+, -, 0\}^n$ as a collection of faces in the boundary complex of the n-cube. Interpret the vector axioms (V1), (V2), (V3), (V3$'$) and (V3$''$) from this perspective.

3.24 Let $\mathcal{M}$ be an oriented matroid of rank r. For every ordered basis $B =$

$(b_1, b_2, \ldots, b_r)$ and any sign vector $\underline{\alpha} = (\alpha_1, \alpha_2, \ldots, \alpha_r) \in \{+, -\}^r$, define $T_{B,\underline{\alpha}} \in \{+, -, 0\}^E$ by

$$T_{B,\underline{\alpha}} := \alpha_1 c^*(b_1, B) \circ \alpha_2 c^*(b_2, B) \circ \ldots \circ \alpha_r c^*(b_r, B)$$

using the basic cocircuits associated with B (see definition preceding Proposition 3.4.1).

(a) Show that $T_{B,\underline{\alpha}}$ always is a tope (maximal covector) of $\mathcal{M}$.

(b) Show that every tope T of $\mathcal{M}$ equals $T_{B,\underline{\alpha}}$ for some B and $\underline{\alpha}$.

3.25 Using the previous exercise, give an algorithm for constructing all covectors of an oriented matroid from its chirotope.

3.26 Determine the f-vector and combinatorial type of all topes in the arrangement of eight hyperplanes associated with CUBE.

3.27 Let $\mathcal{M}$ be an oriented matroid without loops or coloops. Then
  (a) $\{+, -\}^E = \mathcal{T} \cup \mathcal{W} \cup \{U \in \{+, -\}^E : X \leq U$ and $Y \leq U$
  $\qquad\qquad\qquad$ for some $X \in \mathcal{C}$, $Y \in \mathcal{C}^*$, $\underline{X} \cap \underline{Y} = \emptyset\}$

  (b) Conclude that $\{+, -\}^E = \mathcal{T} \cup \mathcal{W}$ if and only if $\mathcal{M}$ is uniform.

3.28 Show that $\mathcal{T} \subseteq \{+, -\}^E$ is the system of topes of a uniform oriented matroid if and only if the following axioms hold:

(T0) $\quad \mathcal{T} \neq \emptyset$

(T1) $\quad \mathcal{T} = -\mathcal{T}$

(TU2) *If $X \in \{+, -, 0\}^E$, $X \neq 0$, is a restriction of some $T \in \mathcal{T}$, then*
  either there is a $T \in \mathcal{T}$ with $X \circ T \in \mathcal{T}$, $X \circ (-T) \notin \mathcal{T}$,
  or $X \circ U \in \mathcal{T}$ for every $U \in \{+, -\}^E$. $\qquad$ (Lawrence's axioms)

$\qquad\qquad\qquad\qquad\qquad\qquad\qquad\qquad\qquad\qquad\qquad$ (Lawrence 1983)

3.29 Show that if $\mathcal{W}$ is the set of maximal vectors of an oriented matroid $\mathcal{M}$, then the closure operator of $\underline{\mathcal{M}}$ is given by $\mathrm{cl}(A) = E \backslash \{\underline{Y} : Y \in \{+, -, 0\}^E$, $\underline{Y} \cap A = \emptyset$, $Y \perp W$ for all $W \in \mathcal{W}\}$
  Derive from this that $\mathrm{cl}(\{e\}) = \bar{e} := \{f \in E : W_e = W_f$ for all $W \in \mathcal{W}$, or $W_e = -W_f$ for all $W \in \mathcal{W}\}$ is the closure of e (the set of elements that are parallel or antiparallel to e).

3.30 Show that if $\mathcal{T}$ is the set of topes of $\mathcal{M}$, then the closure operator of $\underline{\mathcal{M}}$ is given by

  $\mathrm{cl}(A) = A \cup \{e \in E \backslash A :$ there is $B \subseteq A$ such that
  $$\mathcal{T}(B) = \{+, -\}^B \quad \text{and} \quad \mathcal{T}(B \cup e) \neq \{+, -\}^{B \cup e}\}.$$

  In particular,

  $\mathrm{cl}(\{e\}) = \{f \in E : T_e = T_f$ for all $T \in \mathcal{T}$, or $T_e = -T_f$ for all $T \in \mathcal{T}\}$

  is the closure of e.

3.31 It is a recently solved problem to give a non-recursive axiomatization of oriented matroids in terms of topes.

(a) Prove that the following axiom holds for the family $\mathcal{T} \subseteq \{+, -\}^E$ of topes of every loop-free oriented matroid:

(T2) *If $X \in \{+, -, 0\}^E$, is a restriction of some $T \in \mathcal{T}$, then*

      either there is a $T \in \mathcal{T}$ with $X \circ T \in \mathcal{T}$, $X \circ (-T) \notin \mathcal{T}$,

      or $X \circ T \in \mathcal{T}$ for every $T \in \mathcal{T}$.   (da Silva's axiom)

(b) The axiom (TU2) of Exercise 3.28 implies (T2) – but every non-uniform matroid shows that the converse is false.

(c) Show that the axioms (T0), (T1) and (T2) sufficient to characterize oriented matroids.   (da Silva 1987, 1991)

3.32 ("Acycloids")

    (a) Show that the axioms (T1) and (T2) together imply the following weaker axiom:

(T2′) *For $T^1, T^2 \in \mathcal{T}$ with $T^1 \neq T^2$, there is an element $e \in E$ with $T_e^1 = -T_e^2$ such that $_{\bar{e}}T^1 \in \mathcal{T}$.*

    (b) Conclude that the set of topes of any oriented matroid can be interpreted as a

    – centrally symmetric

    – isometric

    subgraph of a cube graph.

    (c) Show that (T0)+(T1)+(T2′) do not imply (T2).

    (d) Show that if $\mathcal{T} \subseteq \{+, -\}^E$ satisfies (T0)+(T1)+(T2′), and every *contraction* (to be defined suitably) also satisfies (T2′), then $\mathcal{T}$ is the tope system of an oriented matroid.

    (Families satisfying (T0)+(T1)+(T2′) have been studied as *acycloids*, see Handa (1990). The first example to show that not every acycloid is an oriented matroid – see (c) – was found by Fukuda. Note that (d) is the recursive characterization of topes of Proposition 3.8.3.) (See Handa 1985, 1987, 1990.)

3.33 A matroid with a signature $\mathcal{C}$ of its set of circuits is said to be *weakly orientable* if there is a cocircuit signature $\mathcal{C}^*$ of M^* such that $X \perp Y$ holds for all $X \in \mathcal{C}$ and all $Y \in \mathcal{C}^*$ with $|X \cap Y| = 2$.

    (a) Derive an axiomatic for the circuits of a weakly oriented matroid.

    (b) Show that a weak orientation of a matroid induces well-defined weak orientations on its minors.   (Bland and Jensen 1987)

3.34 Let $\mathcal{M}$ be an oriented matroid on a set E and A be a subset of E. Show that the properties (i) and (ii) are equivalent:

    (i) A meets every positive circuit of $\mathcal{M}$ and is inclusion minimal with this property.

    (ii) The reorientation $_{-A}\mathcal{M}$ is acyclic, and A is inclusion minimal with this property.   (Berge and Las Vergnas 1984)

4

From Face Lattices to Topology

A central concept for the combinatorial study of convex polytopes and hyperplane arrangements in $\mathbb{R}^d$ is that of a *face lattice*: the cells on the boundary of the polytope or the cells in the induced decomposition of $\mathbb{R}^d$ ordered by inclusion. These are precisely the kind of objects that will be studied in this chapter, except in a more general axiomatized version. A surprising amount of detailed information, much of it highly non-trivial also in the realizable case, can be obtained this way.

The lattices with which we will be concerned are formed by the covectors of an oriented matroid under a natural partial ordering. In this chapter oriented matroids will primarily be viewed as such ordered structures. The first two sections develop the elementary combinatorial properties, which are basic for an understanding of oriented matroid theory. The step to topology is taken in Section 4.3. There it is shown that the covector lattice of an oriented matroid uniquely determines a regular cell decomposition of a sphere.

The details of the constructions in Section 4.3 add up to a proof for one direction of the Topological Representation Theorem for oriented matroids (the more difficult one). This proof differs from the two previously known ones by Folkman and Lawrence (1978) and by Edmonds and Mandel (1982), although the mathematical underpinnings are similar to those of the Edmonds-Mandel proof. It makes systematic use of a general technique for poset shellability developed in Björner (1980, 1984a) and Björner and Wachs (1983), which has been applied to several classes of finite posets in connection with certain algebraic and topological questions. The application to oriented matroids presented in Sections 4.3 and 4.5 is due to Björner and Ziegler (1988), and appears here for the first time. The parts of this technique that are needed for the purposes of this book are developed in an Appendix (Section 4.7), together with a thumbnail sketch of regular cell complexes and a review of some topological generalities.

The work of proving the Topological Representation Theorem, begun in this chapter, is continued and completed in Chapter 5 (Sections 5.1 and 5.2). A reader who seeks the quickest path to this important theorem (including a complete proof) needs only read the following sections in this chapter: 4.1, 4.2 up to Lemma 4.2.12, 4.3 through Proposition 4.3.6.

In Sections 4.4–4.6 we discuss some special aspects of oriented matroids that are byproducts of or otherwise closely related to the topological analysis of the covector lattice given in Section 4.3.

4.1 The big face lattice

Our point of departure will be the covector axiomatization of oriented matroids. These axioms and connections with other axiomatizations were already discussed in Chapter 3. Nevertheless we begin with a quick review of this material.

Our reasons for making this review, in spite of the slight overlap with Chapter 3, are:

(1)   The discussion here is in terms of covectors rather than (as in Chapter 3) in terms of vectors and signed sets. This causes some differences in notation, concepts and geometric point of view. Also, order-theoretic aspects of covectors are emphasized here.

(2)   We want to lay the foundation for the discussion leading to the Topological Representation Theorem with extra care.

(3)   This review should facilitate access to the material in this chapter for someone who hasn't carefully studied Chapter 3.

Let E be a finite set and consider sign vectors $X, Y \in \{+, -, 0\}^E$. The *support* of a vector X is $\underline{X} = \{e \in E : X_e \neq 0\}$; its *zero set* is

$$z(X) = X^0 = E \setminus \underline{X} = \{e \in E : X_e = 0\}.$$

The *opposite* of a vector X is $-X$, defined by

$$(-X)_e = \begin{cases} -, & \text{if } X_e = +, \\ +, & \text{if } X_e = -, \\ 0, & \text{if } X_e = 0. \end{cases}$$

The *zero vector* is 0, with $0_e = 0$ for all $e \in E$. The *composition* of two vectors X and Y is $X \circ Y$, defined by

$$(X \circ Y)_e = \begin{cases} X_e, & \text{if } X_e \neq 0, \\ Y_e, & \text{otherwise .} \end{cases}$$

(Note that this product is associative, but not commutative.)

The *separation set* of X and Y is $S(X, Y) = \{e \in E : X_e = -Y_e \neq 0\}$. If $S(X, Y) = \emptyset$ we say that the vectors X and Y are *conformal*; this is the case precisely if $X \circ Y = Y \circ X$.

The following axiom system consists of vector axioms (V0), (V1), (V2) and (V3′) of Theorem 3.7.5 and Corollary 3.7.9, stated in dual form (i.e., for covectors rather than vectors).

4.1.1 Covector Axioms. *A set $\mathcal{L} \subseteq \{+, -, 0\}^E$ is the set of covectors of an oriented matroid if and only if it satisfies:*

(L0)	$0 \in \mathcal{L}$,
(L1)	$X \in \mathcal{L}$ implies $-X \in \mathcal{L}$,
(L2)	$X, Y \in \mathcal{L}$ implies $X \circ Y \in \mathcal{L}$,
(L3)	if $X, Y \in \mathcal{L}$ and $e \in S(X, Y)$ then there exists $Z \in \mathcal{L}$ such that $Z_e = 0$ and $Z_f = (X \circ Y)_f = (Y \circ X)_f$ for all $f \notin S(X, Y)$.

Since oriented matroids will in this chapter and the next usually be thought of in terms of their covectors, we will sometimes simply say that such a set $\mathcal{L}$ of sign vectors, or the pair $(E, \mathcal{L})$, *is* an oriented matroid.

The motivating model for this axiom system is the following. Suppose $\mathcal{A} = (H_e = \{x \in \mathbb{R}^d : \langle x, a_e \rangle = 0\})_{e \in E}$ is a central arrangement of oriented hyperplanes in $\mathbb{R}^d$, orthogonal to a family $(a_e)_{e \in E}$ of non-zero vectors. For each point $x \in \mathbb{R}^d$, let $\sigma(x) = (\sigma_e(x) = \mathrm{sign}\langle x, a_e \rangle)_{e \in E}$. This defines a mapping

$$\sigma : \mathbb{R}^d \to \{+, -, 0\}^E,$$

and clearly $\sigma(x) = \sigma(y)$ if and only if x and y belong to the same cell in the decomposition of $\mathbb{R}^d$ induced by $\mathcal{A}$. Therefore the sign vectors in the image $\sigma(\mathbb{R}^d)$ index these cells.

It is instructive to interpret the covector axioms geometrically for such a system $\mathcal{L} = \sigma(\mathbb{R}^d) \subseteq \{+, -, 0\}^E$. The first two axioms are immediate, since $\sigma(0) = 0$ and $\sigma(-x) = -\sigma(x)$. Suppose $\sigma(x) = X$ and $\sigma(y) = Y$. Then $\sigma(x + \epsilon(y - x)) = X \circ Y$, for sufficiently small $\epsilon > 0$, and if $\sigma_e(x) = -\sigma_e(y) \neq 0$ then choosing $z = x + \delta(y - x), 0 < \delta < 1$, so that $\sigma_e(z) = 0$, the vector $Z = \sigma(z)$ will satisfy (L3). See Figure 4.1.1.

The oriented matroids which come from arrangements of hyperplanes $\mathcal{A}$ in the described fashion are precisely the realizable ones.

In the following we will frequently need poset (partially ordered set) terminology, which we now review. All posets are assumed to be finite.

If a poset $P = (P, \leq)$ has a unique minimal (resp. maximal) element, we denote it by $\hat{0}$ (resp., by $\hat{1}$). A poset is *bounded* if it has $\hat{0}$ and $\hat{1}$. A poset P is *pure* if all maximal chains $x_0 < x_1 < \cdots < x_\ell$ have the same *length* ℓ, called the *length* of P. In this case the *rank* $\rho(x)$ of an element $x \in P$ is the length of the subposet $P_{\leq x} = \{y \in P : y \leq x\}$, and ρ is the *rank function* of P. The poset P is *graded* if it is pure and bounded, in which case its length equals $\rho(\hat{1})$.

For $x, y \in P$ we say that y *covers* x if $x < y$ and no $z \in P$ satisfies $x < z < y$. We denote this relation by $x \lessdot y$, and also say that y is a *cover* of x and that x is a *cocover* of y.

If P is bounded, the covers of $\hat{0}$ are the *atoms* and the cocovers of $\hat{1}$ are the *coatoms*. In particular, we may speak about the atoms and coatoms of an *interval* $[x, y] = \{z \in P : x \leq z \leq y\}$. The set of coatoms of $[x, y]$ will

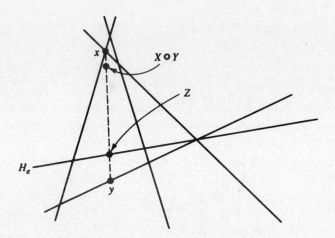

Figure 4.1.1: Motivating model for covector axioms.

be denoted coat$[x,y]$. In addition to the *closed intervals* $[x,y]$, there are *open intervals* $(x,y) = \{z \in P : x < z < y\}$.

A poset L is said to be a *lattice* if for all pairs x, $y \in L$ there exists a least upper bound (or, *join*) $x \vee y$ and a greatest lower bound (or, *meet*) $x \wedge y$. In a finite lattice L joins and meets exist for arbitrary subsets, hence L is bounded ($\hat{0} = \wedge L$, $\hat{1} = \vee L$). Here is a useful characterization: A finite poset is a lattice if and only if $\hat{0}$ exists and joins $x \vee y$ exist for all pairs x, $y \in L$.

The *opposite* (or, *order dual*) $P^{op} = (P, \leq^{op})$ of a poset $P = (P, \leq)$ is defined on the same set P by the opposite order relation: $x \leq^{op} y$ if and only if $y \leq x$. An *order ideal* of P is a subset $I \subseteq P$ such that $x \leq y$ and $y \in I$ imply $x \in I$. Dually, $F \subseteq P$ is a *filter* if $x \geq y$ and $y \in F$ imply $x \in F$. A *linear ordering* of a set E is a poset $(E, \leq)$ in which all pairs of elements are comparable, i.e., $x \leq y$ or $y \leq x$ holds for all x, $y \in E$. A linearly ordered set of size n is isomorphic to the set $E_n = \{1, 2, \cdots, n\}$ under the natural order of the integers.

Figure 4.1.2: Ordering of signs.

We now return to the discussion of covectors in an oriented matroid. Let "$\leq$"

be the partial order on the set $\{+, -, 0\}$ defined by $0 < +$ and $0 < -$, with $+$ and $-$ incomparable (see Figure 4.1.2). This induces the product partial order on $\{+, -, 0\}^E$, in which sign vectors are compared componentwise. Thus $Y \leq X$ if and only if $S(Y, X) = \emptyset$ and $\underline{Y} \subseteq \underline{X}$, or equivalently, if $Y_e \in \{0, X_e\}$ for all $e \in E$. This ordering is depicted for $|E| = 2$ in Figure 4.1.3.

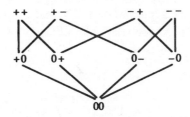

Figure 4.1.3: Ordering of sign vectors.

Now, let $\mathcal{L} \subseteq \{+, -, 0\}^E$ be the set of covectors of an oriented matroid. As a subset of $\{+, -, 0\}^E$ the set $\mathcal{L}$ has an induced partial order with bottom element 0. The poset $\widehat{\mathcal{L}} = \mathcal{L} \cup \{\hat{1}\}$, i.e., the poset $\mathcal{L}$ with a top element $\hat{1}$ adjoined, is a lattice. For, as is easy to see, the join in $\widehat{\mathcal{L}}$ of X and Y equals $X \circ Y = Y \circ X$ if $S(X, Y) = \emptyset$, and equals $\hat{1}$ otherwise.

4.1.2 Definition.

(i)   The lattice $\mathcal{F}_{big}(\mathcal{L}) = \widehat{\mathcal{L}} = (\widehat{\mathcal{L}}, \leq)$ is called the *face lattice* of the oriented matroid $\mathcal{L}$ (sometimes also the *"big" face lattice*). The maximal elements of $\mathcal{L}$ are called *topes* (or *regions*). Let $\mathcal{T}(\mathcal{L})$ denote the set of topes.

(ii)   Suppose that $\mathcal{L}$ is an acyclic oriented matroid with positive tope T. Then the interval $\mathcal{F}_{em}(\mathcal{L}) = [0, T]$ is called the *Edmonds-Mandel (face) lattice*, and its order dual $\mathcal{F}_{lv}(\mathcal{L}) = [0, T]^{op}$ is called the *Las Vergnas (face) lattice* of $\mathcal{L}$.

(iii) A tope is *simplicial* (or a *simplex*) if the interval $[0, T]$ is Boolean (i.e., isomorphic to the lattice of subsets of a finite set). The oriented matroid $\mathcal{L}$ is *simplicial* if all its topes are simplicial.

The face lattice $\mathcal{F}_{big}(\mathcal{L})$ was introduced by Folkman and Lawrence (1978) and then studied in greater detail by Edmonds and Mandel (1982). The Las Vergnas lattice $\mathcal{F}_{lv}(\mathcal{L})$ originated in Las Vergnas (1975a, 1980a).

Recall that an oriented matroid is *acyclic* if there is no positive circuit, or equivalently (Proposition 3.4.8) if there exists a *positive tope*, meaning a tope B such that $B_e \in \{+, 0\}$ for all $e \in E$. Given an oriented matroid $(E, \mathcal{L})$ and a tope $B \in \mathcal{T}(\mathcal{L})$ one may always via reorientation (Lemma 4.1.8) assume that $\mathcal{L}$ is acyclic with positive tope B.

Let us look at a few examples.

4.1.3 Example. From the discussion after Axioms 4.1.1 it follows that if $\mathcal{L} =$

$\sigma(\mathbb{R}^d)$ is a linear oriented matroid coming from a central arrangement of hyperplanes $\mathcal{A}$ in $\mathbb{R}^d$, then $\mathcal{F}_{big}(\mathcal{L})$ is isomorphic to the face poset $\mathcal{F}(\mathcal{A})$, and also to the face poset of the regular cell decomposition of the unit sphere S^{d-1} induced by $\mathcal{A}$ (cf. Section 1.2(c) and Section 2.1). In particular, the topes of $\mathcal{L}$ correspond to the maximal cells, i.e., to the regions (convex d-cones) of $\mathbb{R}^d \setminus \bigcup \mathcal{A}$.

4.1.4 Example. The oriented matroid $\mathcal{L} = \{+,-,0\}^{E_n}$ corresponds to the arrangement of coordinate hyperplanes in $\mathbb{R}^n$, and $\mathcal{F}_{big}(\mathcal{L})$ is the face lattice of the n-dimensional cross-polytope (the polar of the n-cube).

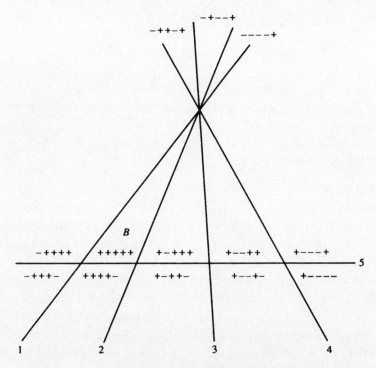

Figure 4.1.4: A central arrangement of five planes in $\mathbb{R}^3$.

4.1.5 Example. A central arrangement of 5 oriented planes in $\mathbb{R}^3$ is drawn in Figure 4.1.4. The picture shows the intersection of the arrangement with an affine plane away from the origin. The corresponding oriented matroid $\mathcal{L} = \sigma(\mathbb{R}^3) \subseteq \{+,-,0\}^{E_5}$ can be read off from the cell decomposition. For example, the 13 "visible" topes (sign vectors of regions) are named in the picture (the 3 topes not visible in this affine section are $(-----)$, $(-+---)$ and $(-++--)$). The face lattice $\mathcal{F}_{big}(\mathcal{L})$ is shown in Figure 4.1.5.

4.1.6 Example. Let P be a convex d-polytope. Embed P in the hyperplane $x_{d+1} = 1$ in $\mathbb{R}^{d+1}$, see Figure 4.1.6.

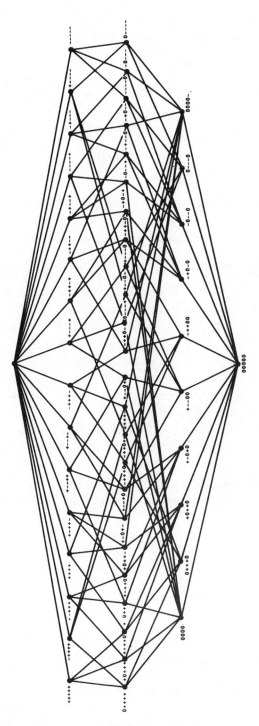

Figure 4.1.5: A big face lattice.

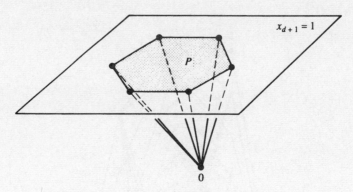

Figure 4.1.6: The two oriented matroids of a convex polytope.

Two oriented matroids naturally arise from this situation, and both encode P's face lattice $\mathcal{F}(P)$.

(1) Let E be the set of extreme points of P. As a set of vectors in $\mathbb{R}^{d+1}$ there is an induced structure of oriented matroid $\mathcal{M}_1 = (E, \mathcal{C})$. Since all vectors E are on the positive side of the plane $x_{d+1} = 0$, there can be no positive circuit in $\mathcal{C}$, so $\mathcal{M}_1$ is acyclic. The facets of P are precisely those subsets $F \subseteq E$ such that $F \cup \{0\}$ span a hyperplane in $\mathbb{R}^{d+1}$ having all of $E \backslash F$ on the same side. This is equivalent to saying that F is the zero-set of a positive cocircuit in $\mathcal{M}_1$. Hence, $\mathcal{F}(P) \cong [O, T]^{op} = \mathcal{F}_{lv}(\mathcal{M}_1)$, where T is the positive tope of $\mathcal{M}_1$. In other words, the face lattice of P is isomorphic to the Las Vergnas lattice of $\mathcal{M}_1$.

(2) Let E be the arrangement of hyperplanes in $\mathbb{R}^{d+1}$ spanned by sets $F \cup \{0\}$, where F is a facet of P. Orient these hyperplanes so that the interior of P is on the positive side of them all. The associated oriented matroid $\mathcal{M}_2 = (E, \mathcal{L})$ is acyclic, with positive tope T corresponding to the interior of the convex cone C_P spanned by $P \cup \{0\}$. Now, the boundaries of C_P and P are combinatorially isomorphic, and the face lattice of C_P is isomorphic to $[0, T]$ because of the inclusion-preserving bijection between the cells induced by the arrangement and $\mathcal{L}$. Therefore $\mathcal{F}(P) \cong [0, T] = \mathcal{F}_{em}(\mathcal{L})$, i.e., the face lattice of P is isomorphic to the Edmonds-Mandel lattice of $\mathcal{M}_2$.

Face lattices of convex polytopes will be further discussed from an oriented matroid point of view in Chapter 9.

4.1.7 Example. If $\mathcal{L}$ arises from a hyperplane arrangement $\mathcal{A}$ as in Example 4.1.3 then $\mathcal{F}_{big}(\mathcal{L})$ is the face lattice of a convex polytope, namely the polar of the zonotope associated with $\mathcal{A}$ (see Proposition 2.2.2). In general, however, $\mathcal{F}_{big}(\mathcal{L})$, $\mathcal{F}_{lv}(\mathcal{L})$ and $\mathcal{F}_{em}(\mathcal{L})$ are not polytope lattices. This follows from the fact, first shown by J. Lawrence, that acyclic oriented matroids $\mathcal{L}$ exist for which $\mathcal{F}_{lv}(\mathcal{L})$ is not polytopal. Lawrence's construction is discussed in Section 9.3.

Before proceeding we will review the concepts of *minors, reorientation, loops* and *parallel elements* from the covector point of view. These notions were discussed in terms of some of the other axiomatizations in Chapter 3. The *restriction* of a sign vector $X \in \{+,-,0\}^E$ to a subset $F \subseteq E$ is the sign vector $X|_F \in \{+,-,0\}^F$ defined by $(X|_F)_e = X_e$ for all $e \in F$.

4.1.8 Lemma. *Let $\mathcal{L} \subseteq \{+,-,0\}^E$ be the set of covectors of an oriented matroid $\mathcal{M}$ on E, and let $A \subseteq E$.*

(i) *The set of covectors of the deletion $\mathcal{M} \backslash A$ equals*
$$\mathcal{L} \setminus A = \{X|_{E \backslash A} : X \in \mathcal{L}\} \subseteq \{+,-,0\}^{E \backslash A}.$$

(ii) *The set of covectors of the contraction $\mathcal{M}/A$ equals*
$$\mathcal{L}/A = \{X|_{E \backslash A} : X \in \mathcal{L} \text{ and } A \subseteq X^0\} \subseteq \{+,-,0\}^{E \backslash A}.$$

(iii) *The set of covectors of the reorientation $_{-A}\mathcal{M}$ equals*
$$_{-A}\mathcal{L} = \{_{-A}X : X \in \mathcal{L}\}.$$

Proof. For (i) and (ii) dualize Proposition 3.7.11. For (iii) see Remark 3.2.3. ☐

Note that if $A = \underline{X}$ for some $X \in \mathcal{L}$ then the face lattice $\mathcal{F}_{big}(\mathcal{L} \backslash A)$ is isomorphic to the interval $[X, \hat{1}]$ in $\hat{\mathcal{L}}$. In particular, all upper intervals $[X, \hat{1}]$ in $\mathcal{F}_{big}(\mathcal{L}) = \hat{\mathcal{L}}$ are themselves oriented matroid lattices. Furthermore, if $A = z(Y)$ for some $Y \in \mathcal{L}$ then (after reorientation to get rid of all minuses in Y) the face lattice $\mathcal{F}_{em}(\mathcal{L}/A)$ is isomorphic to the interval $[0, Y]$ in $\mathcal{L}$. Hence, all lower intervals $[0, Y]$ in $\mathcal{L}$ are themselves tope lattices (or, Edmonds-Mandel face lattices). From this the following can be concluded.

4.1.9 Proposition. *Let $[X, Y]$ be an interval in $\hat{\mathcal{L}}$. Then there exists some acyclic oriented matroid $\mathcal{L}'$, obtained from $\mathcal{L}$ by deletion and contraction, such that*

(i) $[X, Y] \cong \mathcal{F}_{big}(\mathcal{L}')$, *if $Y = \hat{1}$,*

(ii) $[X, Y] \cong \mathcal{F}_{em}(\mathcal{L}')$, *if $Y \neq \hat{1}$.*

Note that the contraction $\mathcal{L}/A$ is not necessarily acyclic, even if $\mathcal{L}$ is, cf. Proposition 3.4.8(b).

From axiom (L2) it is clear that all topes T of an oriented matroid $(E, \mathcal{L})$ have the same support and the same zero set $E_o = z(T)$. The elements in E_o are called *loops*.

4.1.10 Lemma. *Let $e, f \in E \backslash E_o$. Then the following conditions are equivalent:*

(i) $X_e = X_f$ *for all $X \in \mathcal{L}$ or $X_e = -X_f$ for all $X \in \mathcal{L}$,*

(ii) $X_e = 0 \Longleftrightarrow X_f = 0$, *for all $X \in \mathcal{L}$,*

(ii') $X_e = 0 \Longrightarrow X_f = 0$, *for all $X \in \mathcal{L}$,*

(iii) $e \in S(T^1, T^2) \Longleftrightarrow f \in S(T^1, T^2)$, *for all $T^1, T^2 \in \mathcal{T}(\mathcal{L})$.*

Proof. The cycles of implications (i) $\Longrightarrow$ (iii) $\Longrightarrow$ (ii) $\Longrightarrow$ (i) and (ii) $\Longleftrightarrow$ (ii)' are straightforward using axioms (L1), (L2) and (L3). ☐

Two elements $e, f \in E \backslash E_\circ$ are said to be *parallel* (denoted $e \| f$) if they satisfy the conditions of Lemma 4.1.10. For $e \in E \backslash E_\circ$ let

$$\bar{e} = \{f \in E \backslash E_\circ : f \| e\}$$

denote its parallelism class. Also, let

$$\overline{E} = \{\bar{e} : e \in E \backslash E_\circ\}.$$

The oriented matroid is *simple* if it does not have loops or distinct parallel elements.

Starting with any oriented matroid it is possible by Lemma 4.1.10 (i) to reorient so that $X_e = X_f$ for all $X \in \mathcal{L}$ and all $e \| f$. Such a sign vector X on the set E determines a sign vector $\overline{X}$ on $\overline{E}$. This leads to the following conclusion.

4.1.11 Lemma. *For every oriented matroid $\mathcal{L}$ on E there is a simple oriented matroid $\overline{\mathcal{L}}$ on $\overline{E}$ and a poset isomorphism $\mathcal{L} \to \overline{\mathcal{L}}$, mapping $X \mapsto \overline{X}$, such that*

(i) $z(\overline{X}) = \{\bar{e} : e \in z(X) \backslash E_\circ\}$, *for all $X \in \mathcal{L}$, and*

(ii) $S(\overline{X}, \overline{Y}) = \{\bar{e} : e \in S(X, Y)\}$, *for all $X, Y \in \mathcal{L}$.*

Furthermore, $\overline{\mathcal{L}}$ is uniquely determined by $\mathcal{L}$ up to reorientation.

Consequently, for the study of face lattices of oriented matroids it would suffice to consider simple oriented matroids. However, the very useful contraction operation leads to the introduction of loops and parallel elements, and therefore we must consider general oriented matroids as well.

The following technical lemma of Edmonds and Mandel (1982) will prove useful.

4.1.12 Lemma. *Suppose that $X, Y \in \mathcal{L}$, $\underline{Y} \subseteq \underline{X}$ and $Y \not\leq X$. Then there exists $Z \in \mathcal{L}$ such that $Z \lessdot X$ and $Z_e = X_e$ for all $e \notin S(X, Y)$.*

Proof. This is a slightly sharper version of condition (V3″) in Proposition 3.7.10. We will give an independent proof. Assume that the assertion is false. Among all pairs $X, Y \in \mathcal{L}$ that violate the assertion choose one for which $|S(X, Y)|$ is minimal. By axiom (L3) there exists $V \in \mathcal{L}$ such that $V_e = X_e$ for all $e \notin S(X, Y)$ and $S(X, V) \subset S(X, Y)$. In case $S(X, V) = \emptyset$, then $V < X$ and any coatom Z in the interval $[V, X]$ would satisfy the assertion, contrary to the assumption that there is no such Z. In case $S(X, V) \neq \emptyset$, then the pair X, V would also violate the assertion, contrary to the minimality criterion for choosing X, Y. $\square$

The connection between an oriented matroid $\mathcal{L}$, as given by the Covector Axioms 4.1.1, and its underlying matroid (in the form of its geometric lattice L of flats) is provided by the zero map. The following is implicit in Folkman and Lawrence (1978) and Las Vergnas (1980a), and explicit in Edmonds and Mandel (1982).

4.1.13 Proposition. *Let $\mathcal{L} \subseteq \{+, -, 0\}^E$ be the set of covectors of an oriented matroid.*

(i)   *The set $L = \{z(X) : X \in \mathcal{L}\}$ is the collection of flats of the underlying matroid.*

(ii)   *The map $z : \mathcal{L} \to L$ is a cover-preserving, order-reversing surjection of $\mathcal{L}$ onto the geometric lattice L. It satisfies the algebraic property*

$$z(X \circ Y) = z(X) \cap z(Y) = z(X) \wedge z(Y).$$

Proof. (i) The covectors $X \in \mathcal{L}$ are precisely the compositions $C_1 \circ C_2 \circ \cdots \circ C_k$ of cocircuits C_i. Hence the zero sets $z(X)$ are precisely the intersections $\bigcap z(C_i)$. But $z(C_i)$ is the same thing as the set complement of the cocircuit $\underline{C_i}$ of the underlying matroid $\underline{M}$. Consequently, the zero sets $z(X)$ are precisely the intersections of hyperplanes of $\underline{M}$, i.e., the flats of $\underline{M}$.

(ii) The only non-trivial statement here is that z preserves covers. To see this, let $X, Y, Z \in \mathcal{L}$ such that $X < Y$ and $z(X) \supset z(Z) \supset z(Y)$. We will show that Y does not cover X.

Replacing Z by $X \circ Z$ we may assume $X < Z$. If $Z < Y$ we are done. Otherwise by Lemma 4.1.12 there exists $W \in \mathcal{L}$ such that $W < Y$ and $W_e = Y_e$ for all $e \notin S(Z, Y)$. Then $X \leq W$, since $X_e = Y_e = W_e$ for all $e \in \underline{X}$, and $X \neq W$, since $X_f = 0 \neq Y_f = W_f$ for any $f \in \underline{Y} \backslash \underline{Z}$. So, $X < W < Y$.   □

The following result of Folkman and Lawrence (1978), Las Vergnas (1980a) and Edmonds and Mandel (1982) gives two basic combinatorial properties of the face lattice $\widehat{\mathcal{L}} = \mathcal{F}_{big}(\mathcal{L})$ of an oriented matroid. Recall that the *rank* of an oriented matroid is defined to equal the rank of its underlying matroid.

4.1.14 Theorem. *Let $\mathcal{L}$ be the set of covectors of an oriented matroid of rank r.*

(i)   *$\widehat{\mathcal{L}}$ is a graded lattice of length $r + 1$.*

(ii)   *All intervals of length 2 in $\widehat{\mathcal{L}}$ have cardinality 4 (see Figure 4.1.7 (a)).*

Proof. From Proposition 4.1.13 (ii) it follows that maximal chains in $\mathcal{L}$ are mapped bijectively onto maximal chains in L. Since L is a graded lattice of rank r, this proves part (i).

Let $X \lessdot Z \lessdot Y$ be given in $\widehat{\mathcal{L}}$. To prove part (ii) we will show that there is a unique $Z' \in \mathcal{L}$ such that $X \lessdot Z' \lessdot Y$ and $Z' \neq Z$.

Passing to the deletion $\mathcal{L} \backslash \underline{X}$, we may assume without loss of generality that $X = 0$ (see Proposition 4.1.9). If $Y = \hat{1}$ it then comes down to proving that a rank 1 oriented matroid has exactly two topes, which is easy (if T and T' are topes and $T' \neq T, -T$ then axiom (L3) implies the existence of $Z \in \mathcal{L}$ such that $\emptyset \subset z(Z) \subset z(T)$). So we may also assume that $Y \neq \hat{1}$, i.e., that $Y \in \mathcal{L}$.

After these simplifications we now start from the situation $0 \lessdot Z \lessdot Y$. Applying Lemma 4.1.12 to $-Z$ and Y gives the existence of a vector $Z' \in \mathcal{L}$ such that $Z' \lessdot Y$ and $Z'_e = Y_e$ for all $e \in \underline{Y} \backslash \underline{Z}$. Hence, the open interval $(0, Y)$ contains at least the two distinct elements Z and Z'. It remains to show that it cannot contain any other element.

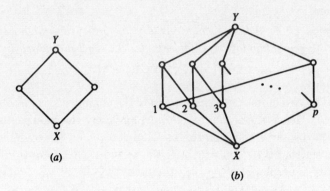

Figure 4.1.7: Intervals of length 2 and 3.

Suppose that $0 \lessdot W \lessdot Y$, for some $W \in \mathcal{L}, W \neq Z, Z'$. Then $Z \leq Z{\circ}W \leq Y$ and $Z' \leq Z'{\circ}W \leq Y$. If $Z = Z{\circ}W$ then $W \leq Z$, and similarly if $Z' = Z'{\circ}W$. So, assume that $Z{\circ}W = Y = Z'{\circ}W$. Let $A = \underline{Z} \cap \underline{Z'}, B = \underline{Z}\backslash\underline{Z'}$ and $C = \underline{Z'}\backslash\underline{Z}$. Then $B \cup C \subseteq \underline{W} \subset \underline{Y} = A \cup B \cup C$.

Now, choose two elements, $a \in A\backslash\underline{W}$ and $c \in C$, and apply axiom (L3) twice. Applied to the pair $Z, -Z'$ we get a vector $V \in \mathcal{L}$ such that $V_a = 0$, $V_B = Y_B$, $V_C = -Y_C$, and $\underline{V} \subseteq \underline{Y}$. Then, applied to the pair V, W we get a vector $U \in \mathcal{L}$ such that $U_c = U_a = 0, U_B = Y_B$, and $\underline{U} \subseteq \underline{Y}$.

If $U_C = 0$, then $\emptyset \neq \underline{U} \subset \underline{Z}$, which by Proposition 4.1.13 (ii) contradicts that $0 \lessdot Z$. Similarly, if $U_C \neq 0$, then $\underline{Z} \subset \underline{Z{\circ}U} \subset \underline{Y}$, which contradicts that $Z \lessdot Y$. $\qquad\square$

4.1.15 Corollary. *Let ρ denote the poset rank function of $\mathcal{L}$ and r the rank function of the underlying matroid.*

(i) $\rho(X) = \mathrm{rank}(\mathcal{L}) - r(X^0)$, *for all $X \in \mathcal{L}$.*

(ii) $\mathrm{rank}(\mathcal{L}/A) = \mathrm{rank}(\mathcal{L}) - r(A)$, *for all $A \subseteq E$.*

The lattice $\widehat{\mathcal{L}}$ is of course atomic: every covector is by Proposition 3.7.2 a join of atoms (conformal composition of cocircuits). It is also coatomic: every covector is a meet of topes. This is a consequence of the following lattice-theoretic property, which in turn follows from Theorem 4.1.14 (ii), see Exercise 4.4.

4.1.16 Corollary. *The lattice $\widehat{\mathcal{L}}$ is relatively complemented.*

That intervals of length 2 look like part (a) of Figure 4.1.7 ought to imply that intervals of length 3 look like part (b). This is almost true, but connectivity requires a separate proof. This will later follow from much stronger results, but we state already now the following characterization.

4.1.17. Proposition. *All intervals $[X, Y]$ of length 3 in $\widehat{\mathcal{L}}$ look as in Figure 4.1.7 (b), with $p \geq 3$. If $Y = \hat{1}$ then p is even.*

Proof. See Corollary 4.3.7 and Proposition 4.2.15. $\qquad\square$

We will return to the study of covector lattices in Section 4.3, where some deeper properties will be derived.

4.2 Topes I

Let $\mathcal{T} = \mathcal{T}(\mathcal{L})$ be the set of topes, i.e., maximal covectors, of an oriented matroid with covector set $\mathcal{L} \subseteq \{+, -, 0\}^E$. In linear oriented matroids coming from arrangements of hyperplanes $\mathcal{A}$ in $\mathbb{R}^d$ (as in Example 4.1.3) the topes correspond to the regions of $\mathbb{R}^d \backslash \bigcup \mathcal{A}$. Two regions are adjacent if they share a $(d-1)$-dimensional face, and the same notion of adjacency is easily transferred to the abstract setting of topes in a general oriented matroid. The relation of adjacency leads to some useful graph and poset structures on the set $\mathcal{T}$, which will be described in this section. At the end of the section we discuss how an oriented matroid is determined by its topes.

Let $\mathcal{L} \subseteq \{+, -, 0\}^E$ be the set of covectors of an oriented matroid of rank r. The topes are the elements of poset rank r in $\mathcal{L}$. The elements of poset rank $r-1$ in $\mathcal{L}$ will be called *subtopes*. By Theorem 4.1.14 (ii) every subtope is covered by exactly two topes, and it is therefore natural to think of the subtopes as representing edges connecting these pairs of topes.

4.2.1 Definition. Two topes are *adjacent* if some subtope is covered by both. The adjacent pairs are the edges of the *tope graph* $\mathcal{T} = \mathcal{T}(\mathcal{L})$. For topes T^1 and T^2, let $d(T^1, T^2)$ denote the *graph distance*, i.e., the length of a shortest path from T^1 to T^2 in the tope graph (which by Proposition 4.2.3 below is connected).

The tope graph of an oriented matroid was introduced by Edmonds and Mandel (1982). In the realizable case (arrangements of hyperplanes) the tope graph is a much older concept (although not by that name), since adjacency of regions is a very natural relation. For instance, it plays an important role in Tits' (1974) combinatorial study of reflection arrangements (Coxeter complexes) and was used by Deligne (1972) for simplicial arrangements. The tope graph of a hyperplane arrangement is the 1-skeleton of the associated zonotope, by Proposition 2.2.2.

The tope graph of the arrangement of 5 planes from Example 4.1.5 is shown in Figure 4.2.1. The tope graph of the oriented matroid

$$\{+, -, 0\}^E, \quad |E| = n,$$

is the graph of the n-cube (cf. Example 4.1.4).

For the rest of this section we assume that the oriented matroid $\mathcal{L} \subseteq \{+, -, 0\}^E$ is *simple*. By Lemma 4.1.11 this causes no loss of generality. It is easy for the reader to work out the changes in formulation for the following developments in the non-simple case (replacing E by the set $\overline{E}$ of parallelism classes of non-loops, replacing $S(X, Y)$ by the set of separating blocks $\overline{S}(X, Y) = \{\overline{e} \in \overline{E} : \overline{e} \subseteq S(X, Y)\}$, replacing $|E|$ by $|\overline{E}|$, etc.).

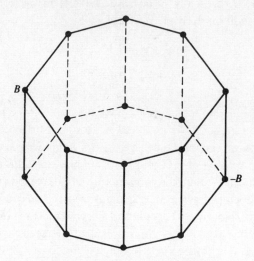

Figure 4.2.1: A tope graph.

4.2.2 Lemma.   *Let $X \in \mathcal{L}, T^1, T^2 \in \mathcal{T}$, and $e \in E$. Then:*

(i)   *X is a subtope $\Longleftrightarrow |X^0| = 1$.*

(ii)   *T^1 and T^2 are adjacent $\Longleftrightarrow |S(T^1, T^2)| = 1$.*

(iii) *There is a subtope Y such that $Y < T^1$ and $Y^0 = e$*

    $\Longleftrightarrow$ There is a tope T such that $S(T, T^1) = e$.

Proof. Suppose that X is a subtope covered by T^1 and T^2, $T^1 \neq T^2$. Then $\emptyset \neq S(T^1, T^2) \subseteq X^0$.

We want to show that $|X^0| = 1$. Suppose to the contrary that $e, f \in X^0, e \neq f$. By Lemma 4.1.10 (ii) there exists $Z \in \mathcal{L}$ such that $Z_e \neq 0$ and $Z_f = 0$, since e and f are not parallel. Then $\underline{X} \subset \underline{X \circ Z} \subset \underline{T^1} = E$, which by Proposition 4.1.13 (ii) contradicts the fact that $X \lessdot T^1$. This proves the forward direction of parts (i) and (ii).

Suppose $S(T^1, T^2) = e$. Axiom (L3) gives the existence of $Y \in \mathcal{L}$ such that $Y_e = 0$ and $Y_f = T_f^1 = T_f^2$ for all $f \neq e$, i.e., Y is a subtope covered by T^1 and T^2. This proves the backward direction of parts (ii) and (iii). The rest is obvious.   □

It is a consequence of Lemma 4.2.2(ii) that $d(T^1, T^2) \geq |S(T^1, T^2)|$, since at each step along a minimal path the size of the separation set can decrease by at most one. We will now show that equality holds, that is, a path (necessarily minimal) from T^1 to T^2 in the tope graph can be constructed by successively changing the signs of the elements of $S(T^1, T^2)$ in some order. An equivalent formulation is that the natural embedding of the tope graph $\mathcal{T}(\mathcal{L})$ into the graph of the n-cube is isometric. This was observed by Las Vergnas (1980a), see also Cordovil (1982a).

4.2.3 Proposition. *The graph distance between any pair of topes equals the size of their separation set:* $d(T^1, T^2) = |S(T^1, T^2)|$.

Proof. By Lemma 4.1.12 there exists $Z \in \mathcal{L}$ such that $Z \lessdot T^1$ and $Z_f = T_f^1 = T_f^2$ for all $f \notin S(T^1, T^2)$. The subtope Z has exactly one zero coordinate, say $Z_e = 0$. Then $e \in S(T^1, T^2)$, and the unique tope $T \neq T^1$ such that $Z \lessdot T$ satisfies $S(T, T^2) = S(T^1, T^2) \backslash e$. By induction, T^2 can be reached in $|S(T, T^2)|$ steps from T, and since T is adjacent to T^1 we are done. $\qquad \square$

From the set equation $S(T^1, T^2) = S(T^1, T) \triangle S(T, T^2)$, clearly valid for any three topes $T, T^1, T^2 \in \mathcal{T}$, we deduce the following formula.

4.2.4 Lemma.
$$d(T^1, T^2) = d(T^1, T) + d(T, T^2) - 2|S(T^1, T) \cap S(T, T^2)|.$$

Some special metric properties of the tope graph are captured in the following concept of convexity, which for the case of reflection arrangements appeared in Tits (1974).

4.2.5 Definition. A subset $\mathcal{Q} \subseteq \mathcal{T}$ is *T-convex* if it contains every shortest path between any two of its members. In other words, if $T^1, T^2 \in \mathcal{Q}$, $T \in \mathcal{T}$, and $d(T^1, T^2) = d(T^1, T) + d(T, T^2)$, then $T \in \mathcal{Q}$.

Any intersection of T-convex sets is clearly T-convex. Hence, every subset $\mathcal{Q} \subseteq \mathcal{T}$ is contained in a least T-convex superset. This *T-convex hull*, denoted $\mathrm{conv}_T(\mathcal{Q})$, is the intersection of all T-convex supersets of $\mathcal{Q}$.

Each $e \in E$ determines a partition of $\mathcal{T}$ into two *halfspaces* $\mathcal{T}_e^+ = \{T \in \mathcal{T} : T_e = +\}$ and $\mathcal{T}_e^- = \{T \in \mathcal{T} : T_e = -\}$. Every halfspace is T-convex (this is easily seen from Proposition 4.2.3), hence so is every intersection of halfspaces. Intersections of halfspaces are called "supertopes" by Edmonds and Mandel (1982), and we will now show that this notion coincides with that of T-convex sets.

4.2.6 Proposition. *A set of topes is T-convex if and only if it is an intersection of halfspaces.*

Proof. Assume that $\emptyset \neq \mathcal{Q} \subseteq \mathcal{T}$ is T-convex, and let $\mathcal{H}$ be the intersection of the halfspaces that contain $\mathcal{Q}$. Suppose that $\mathcal{H} \backslash \mathcal{Q} \neq \emptyset$. Then we can find a pair of adjacent topes T^1, T^2 such that $T^1 \in \mathcal{H} \backslash \mathcal{Q}, T^2 \in \mathcal{Q}$. Let $S(T^1, T^2) = e$, and assume that $T^1 \in \mathcal{T}_e^-, T^2 \in \mathcal{T}_e^+$. If $\mathcal{Q} \cap \mathcal{T}_e^- \neq \emptyset$, say $T \in \mathcal{Q} \cap \mathcal{T}_e^-$, then the T-convexity of $\mathcal{Q}$ and $S(T, T^1) \cap S(T^1, T^2) = \emptyset$ implies (via Lemma 4.2.4) that $T^1 \in \mathcal{Q}$, a contradiction. Hence $\mathcal{Q} \cap \mathcal{T}_e^- = \emptyset$, and therefore $\mathcal{Q} \subseteq \mathcal{H} \cap \mathcal{T}_e^+ \neq \mathcal{H}$, another contradiction. Consequently, $\mathcal{Q} = \mathcal{H}$. $\qquad \square$

A reformulation of this result is that the T-convex hull of $\mathcal{Q} \subseteq \mathcal{T}$ equals the intersection of all halfspaces that contain $\mathcal{Q}$. In other words:

4.2.7 Corollary. *For any fixed $B \in T$ and any subset $\mathcal{Q} \subseteq T$,*
$$\operatorname{conv}_T(\mathcal{Q}) = \{T \in T : \bigcap_{Q \in \mathcal{Q}} S(B,Q) \subseteq S(B,T) \subseteq \bigcup_{Q \in \mathcal{Q}} S(B,Q)\}.$$

4.2.8 Corollary. *For $T^1, T^2 \in T$,*
$$\operatorname{conv}_T(\{T^1, T^2\}) = \{T \in T : S(T^1, T) \subseteq S(T^1, T^2)\}$$
$$= \{T \in T : d(T^1, T) + d(T, T^2) = d(T^1, T^2)\}.$$

The last corollary, a special case of the first, means that the T-convex hull of any pair of topes equals the union of all minimal paths between them.

By choosing a base tope $B \in T$ and directing the edges of the tope graph away from B we induce a useful poset structure on the set of topes.

4.2.9 Definition. Let $B \in T$. Then a partial order on the set T is defined by:
$$T^1 \preceq T^2 \iff S(B, T^1) \subseteq S(B, T^2).$$
Denote by $T(\mathcal{L}, B)$ this *tope poset* based at B.

The tope poset of an oriented matroid was introduced by Edmonds and Mandel (1982), and in the realizable case independently by Edelman (1984b) who initiated a detailed study of this class of posets. For reflection arrangements the tope poset is identical to weak Bruhat order (Section 2.3(b)).

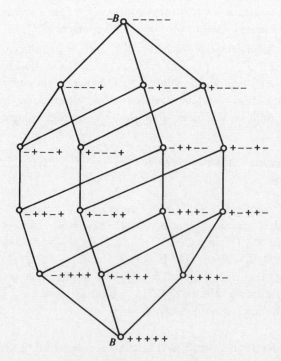

Figure 4.2.2: A tope poset.

Figure 4.2.2 shows the tope poset based at the positive tope of the arrangement in Example 4.1.5 (see also Figure 4.2.1 for comparison).

4.2.10 Proposition. $\mathcal{T}(\mathcal{L}, B)$ *is a graded poset with bottom element B, top element $-B$, and poset rank function $\rho_B(T) = d(B, T) = |S(B, T)|$. Furthermore, for $T^1, T^2 \in \mathcal{T}$ the following conditions are equivalent:*

(i) $T^1 \preceq T^2$,

(ii) $d(B, T^1) + d(T^1, T^2) = d(B, T^2)$,

(iii) $T^1 \in \text{conv}_T(\{B, T^2\})$.

Proof. Directly from Proposition 4.2.3 and Corollary 4.2.8. $\square$

4.2.11 Corollary.

(i)   *Intervals $[T^1, T^2]$ in $\mathcal{T}(\mathcal{L}, B)$ are T-convex.*

(ii)   *The interval $[T^1, T^2]$ has the same structure in $\mathcal{T}(\mathcal{L}, T^1)$ as in $\mathcal{T}(\mathcal{L}, B)$.*

For $X \in \mathcal{L}$ let $\text{star}(X) = \{T \in \mathcal{T} : X \leq T\}$. The following observation is a direct consequence of the definitions.

4.2.12 Lemma. *For all $X \in \mathcal{L}$ and $B \in \mathcal{T}$:*

$$\text{star}(X) = [X \circ B, X \circ (-B)],$$

the right hand side denoting an interval in the tope poset $\mathcal{T}(\mathcal{L}, B)$. In particular, $\text{star}(X)$ is T-convex.

From knowledge of the set of topes $\mathcal{T}(\mathcal{L})$ one can deduce complete knowledge of the entire oriented matroid $\mathcal{L}$. This was first observed by A. Mandel (unpublished), see Cordovil (1985) and da Silva (1987).

4.2.13 Theorem (Mandel). *The set $\mathcal{T}$ determines $\mathcal{L}$ via*

$$\mathcal{L} = \{X \in \{+, -, 0\}^E : X \circ T \in \mathcal{T} \text{ for all } T \in \mathcal{T}\}.$$

Proof. The forward inclusion is clear. For the backward inclusion we will argue by induction on $\text{rank}(\mathcal{L})$, the rank 1 case being clear. Also, we assume here that $\mathcal{L}$ is simple (the non-simple case requires some notational adjustment only).

Suppose X is a sign vector such that $X \circ T \in \mathcal{T}$ for all $T \in \mathcal{T}$. If $\underline{X} = E$, then $X = X \circ T \in \mathcal{T} \subseteq \mathcal{L}$. If $\underline{X} \neq E$, then pick an element $e \in X^0$ and let Y be a tope in $\mathcal{L}/e$. Then $Y^0 = e$, and Y is covered in $\mathcal{L}$ by two topes T^1 and T^2 such that $S(T^1, T^2) = e$. From this and Axiom 4.1.1 (L3) applied to $X \circ T^1$ and $X \circ T^2$ we conclude that $X \circ Y \in \mathcal{L}$. Hence, $X \circ Y \in \mathcal{L}/e$ for all $Y \in \mathcal{T}(\mathcal{L}/e)$, so by induction $X \in \mathcal{L}/e \subseteq \mathcal{L}$. $\square$

It is possible to go one step further. If all information about the topes is erased, except for the relation of adjacency, then the set $\mathcal{L}$ can still be reconstructed, as we will now show.

4.2.14 Theorem (Björner, Edelman and Ziegler 1990). *A simple oriented matroid is uniquely determined (up to reorientation) by its unlabeled tope graph.*

Proof. Let $(E, \mathcal{L})$ be a simple oriented matroid with tope graph $\mathcal{T}$. Think of $\mathcal{T}$ first as a labeled graph: each vertex is labeled by the sign vector of that tope and each edge is labeled by the element of E that separates the two topes at its endpoints. With all this information $\mathcal{L}$ can be reconstructed. For this the labels of the vertices suffice, by Theorem 4.2.13.

Now erase the labels of the vertices but keep the labels on the edges. Mark an arbitrary vertex with the all-plus sign vector, and then search the graph starting at that vertex. When coming to an unmarked node u along an e-labeled edge from a marked node v, then switch the e-coordinate of the sign vector at v and assign the new vector to u. In this way the sign vectors of the topes are reconstructed up to reorientation, so we are back in business.

Finally, erase all labels, both of the vertices and of the edges. As we have seen it is enough to reconstruct the labels of the edges (up to isomorphism). For this we have to be able to tell for each edge precisely which other ones have the same label. This can be done with a simple metric test: take an edge (T^1, T^2); then another edge (T^3, T^4) has the same label as (T^1, T^2) if and only if $(d(T^3, T^1) - d(T^3, T^2)) \cdot (d(T^4, T^1) - d(T^4, T^2)) < 0$.

The reason for this is as follows. Suppose that in the labeled tope graph T^1 and T^2 are connected by an e-labeled edge, i.e., $S(T^1, T^2) = \{e\}$. Let $\mathcal{T}_e^+$ and $\mathcal{T}_e^-$ denote the two halfspaces of topes, as in Proposition 4.2.6, and say that $T^1 \in \mathcal{T}_e^+$ and $T^2 \in \mathcal{T}_e^-$. Lemma 4.2.4 shows that for arbitrary $T \in \mathcal{T}$ we have: $T \in \mathcal{T}_e^+$ if and only if $d(T, T^2) = d(T, T^1) + 1$, and similarly for $\mathcal{T}_e^-$. Hence, whether $T \in \mathcal{T}_e^+$ can be decided from checking whether $d(T, T^1) < d(T, T^2)$, and this metric information is contained in the unlabeled tope graph. Having thus identified the sets $\mathcal{T}_e^+$ and $\mathcal{T}_e^-$ it is clear how to spot all the e-labeled edges: they are the edges connecting a node in $\mathcal{T}_e^+$ with a node in $\mathcal{T}_e^-$. $\qquad\square$

The proof of Theorem 4.2.14 shows how to graph-theoretically construct the classes of label-equivalent edges in the tope graph. In the realizable case these are precisely the *zones* of the corresponding zonotope (see Proposition 2.2.6, and also Proposition 2.2.9).

It is implied by the preceding that the tope graph determines the size $|E|$ of the ground set. This is also a consequence of another metric fact: $\mathrm{diam}(\mathcal{T}) = d(T, -T) = |E|$.

No graph-theoretic characterization is known for the class of (unlabeled) tope graphs of oriented matroids. Such a characterization would by Theorem 4.2.14 lead to yet another axiomatization of oriented matroids. A characterization of tope graphs for oriented matroids of rank at most 3 is given in Fukuda and Handa (1990).

We end by stating a few elementary graph-theoretic properties of tope graphs. All are simple consequences of the definitions.

4.2.15 Proposition. *The tope graph $\mathcal{T} = \mathcal{T}(\mathcal{L})$ of an oriented matroid $\mathcal{L} \subseteq \{+, -, 0\}^E$ has the following properties as an unlabeled graph:*

(i) *For each node $T \in \mathcal{T}$ there is a unique node $-T \in \mathcal{T}$ satisfying $d(T, -T) = \mathrm{diam}(\mathcal{T})$.*

(ii) *The mapping $T \mapsto -T$ is a fixed-point-free automorphism of $\mathcal{T}$.*

(iii) *All cycles in $\mathcal{T}$ are of even length. Equivalently, the graph $\mathcal{T}$ is bipartite.*

(iv) *The degree of each vertex is $\geq \mathrm{rank}(\mathcal{L})$, with equality if and only if the corresponding tope is simplicial (see Exercise 4.4).*

The discussion of tope graphs and tope posets will be continued in Section 4.4.

4.3 Shellability and sphericity

The study of covector lattices of oriented matroids will now be resumed. We start with the construction of recursive coatom orderings, which will make possible the interpretation of covector lattices as cell decompositions of spheres

From now on we will assume familiarity with regular cell complexes and with shellability and recursive coatom orderings of posets. These concepts and their topological ramifications are reviewed in Section 4.7.

In the following $(E, \mathcal{L})$ will be an oriented matroid with set of covectors $\mathcal{L} \subseteq \{+, -, 0\}^E$. We will from now on use the simplified notation $[X]$ for lower intervals $[0, X]$ in the covector lattice $\mathcal{L}$.

Let $R \in \mathcal{T}(\mathcal{L})$ and fix a maximal chain

$$\mathbf{m} : R = R^0 \prec R^1 \prec \ldots \prec R^m = -R$$

in the tope poset $\mathcal{T}(\mathcal{L}, R)$. Then letting $\bar{e}_i = S(R^{i-1}, R^i)$, for $1 \leq i \leq m$, we get a linear order on the set $\overline{E} = \{\bar{e}_1, \bar{e}_2, \ldots, \bar{e}_m\}$ of parallelism classes by putting $\bar{e}_1 < \bar{e}_2 < \ldots < \bar{e}_m$.

This induces a linear order on the set of coatoms of the interval $[R]$ by restriction:

$$X^1 \prec_{\mathbf{m}} X^2 \iff z(X^1) < z(X^2),$$

for $X^1, X^2 \in \mathrm{coat}[R]$. (Recall from Lemma 4.2.2 (i) that $z(X) \in \overline{E}$ for all $X \in \mathrm{coat}[R]$.) This order " $\prec_{\mathbf{m}}$ " we will call *the linear ordering on $\mathrm{coat}[R]$ induced by* $\mathbf{m}$.

4.3.1 Proposition (Edmonds and Mandel 1978, 1982). *Let $(E, \mathcal{L})$ be an oriented matroid and R one of its topes. Then via the construction above every maximal chain $\mathbf{m}$ in the poset $\mathcal{T}(\mathcal{L}, R)$ induces a recursive coatom ordering $\prec_{\mathbf{m}}$ of the interval $[R]$ in $\mathcal{L}$.*

Proof. If $\mathrm{rank}(\mathcal{L}) \leq 2$ the length of the interval $[R]$ is at most 2 and every coatom ordering is recursive. We proceed by induction on $\mathrm{rank}(\mathcal{L})$. Without loss of generality we may assume that $\mathcal{L}$ is simple (cf. Lemma 4.1.11).

Suppose that

$$\mathbf{m} : R = R^0 \prec R^1 \prec \ldots \prec R^m = -R$$

is a maximal chain in $\mathcal{T}(\mathcal{L}, R)$, and let $E = \{e_1, e_2, \ldots, e_m\}$ with $z(R^{i-1} \wedge R^i) = \{e_i\}$ for $1 \le i \le m$. We claim that the linear ordering $\prec_\mathbf{m}$ of $\mathrm{coat}[R]$ induced by $\mathbf{m}$ is recursive.

Let $X \in \mathrm{coat}[R]$ with $z(X) = \{e_i\}$. Also, let $X^i = R^{i-1} \wedge R^i$. Then both X and X^i are topes in the contracted oriented matroid $\mathcal{L}/e_i$. Clearly, $S(X, X^i) = S(R, R^{i-1}) = \{e_1, e_2, \ldots, e_{i-1}\}$.

For each $Y \in \mathrm{coat}[X]$ we have $z(Y) = \{e_i\} \cup A$, where A is a parallelism class of $\mathcal{L}/e_i$. Since $S(X, X^i)$ is a union of such classes we conclude that the coatoms Y of $[X]$ are of two distinct types: those for which $z(Y) \subseteq \{e_1, \ldots, e_i\}$ and those for which $z(Y) \subseteq \{e_i, \ldots, e_m\}$. Let

$$Q_X = \{Y \in \mathrm{coat}[X] : z(Y) \subseteq \{e_1, \ldots, e_i\}\}.$$

The set Q_X is related to the ordering $\prec_\mathbf{m}$ as follows. Each $Y \in \mathrm{coat}[X]$ satisfies $Y \lessdot X' \lessdot R$ for exactly one $X' \ne X$ (by Theorem 4.1.14 (ii)). Clearly, $X' \prec_\mathbf{m} X$ (i.e., $z(X') \in \{e_1, \ldots, e_{i-1}\}$) if and only if $Y \in Q_X$.

According to Definition 4.7.17 there are two properties to verify:

(i) Suppose that $X' \prec_\mathbf{m} X$ and $Z \lessdot X, X'$. This implies that

$$D := z(Z) \cap \{e_1, \ldots, e_{i-1}\} \ne \emptyset$$

and $e_i \in z(Z)$. Hence, $\underline{Z \circ X^i} = E \backslash \{e_i\} = \underline{X}$ and $S(Z \circ X^i, X) = D$. By Lemma 4.1.12 we deduce the existence of $Y \in \mathcal{L}$ such that $Y \lessdot X$ and $Y_e = X_e$ for all $e \notin D$. Consequently, $Y \in Q_X$ and $Z \le Y$.

We have shown that

$$[X] \cap \left(\bigcup_{X' \prec_\mathbf{m} X} [X'] \right) \subseteq \bigcup_{Y \in Q_X} [Y].$$

The reverse inclusion follows from the observation (already made) that $Y \in Q_X$ if and only if $Y \lessdot X'$ for some $X' \in \mathrm{coat}[R]$ such that $X' \prec_\mathbf{m} X$.

(ii) By the induction hypothesis every maximal chain in $\mathcal{T}(\mathcal{L}/e_i, X)$ induces a recursive ordering of $\mathrm{coat}[X]$. Let $\mathbf{n}$ be such a chain which contains X^i. Since $S(X, X^i) = \{e_1, e_2, \ldots, e_{i-1}\}$ it is apparent that the elements of Q_X come first in the recursive ordering $\prec_\mathbf{n}$ of $\mathrm{coat}[X]$. $\square$

A *linear extension* of a poset P is a linear ordering $<_L$ of P which is compatible with the given partial order, that is, such that $x < y$ implies $x <_L y$, for all $x, y \in P$.

4.3.2 Proposition (Lawrence 1984a). *Let $(E, \mathcal{L})$ be an oriented matroid and B one of its topes. Then every linear extension of $\mathcal{T}(\mathcal{L}, B)$ is a recursive coatom ordering of $\mathcal{F}_{big}(\mathcal{L}) = \widehat{\mathcal{L}}$.*

Proof. Again, we may without loss of generality assume that $\mathcal{L}$ is simple. Suppose that $R_0 = B, R_1, R_2, \ldots, R_t = -B$ is a linear extension of $\mathcal{T}(\mathcal{L}, B)$, i.e., $S(B, R_i) \subseteq S(B, R_k)$ implies $i \le k$. For tope R_k define

$$Q_k = \{X \in \mathrm{coat}[R_k] : z(X) \subseteq S(B, R_k)\}.$$

Equivalently, if $X \lessdot R_k$ then $X \in Q_k$ if and only if $R \prec R_k$ for the unique tope $R \neq R_k$ such that $X \lessdot R$.

There are again two properties to verify.

(i)   The following inclusions are clear from what has been said:

$$\bigcup_{X \in Q_k} [X] \subseteq [R_k] \cap \left(\bigcup_{R \prec R_k} [R] \right) \subseteq [R_k] \cap \left(\bigcup_{i<k} [R_i] \right).$$

The three sets are in fact equal. Suppose that $Y < R_i, R_k$ and $i < k$. The set $\mathrm{star}(Y) = [Y \circ B, Y \circ (-B)]$ is an interval in $\mathcal{T}(\mathcal{L}, B)$ by Lemma 4.2.12. If $Y \circ B = R_k$, then $R_k \prec R_i$, which contradicts $i < k$. Hence $Y \circ B \prec R_k$, and any coatom R in the interval $[Y \circ B, R_k] \subseteq \mathrm{star}(Y)$ gives $Y \le X = R \wedge R_k$ and $X \in Q_k$.

(ii)   Let **m** be a maximal chain in $\mathcal{T}(\mathcal{L}, R_k)$ that contains the base tope B. By Proposition 4.3.1 **m** induces a recursive ordering of $\mathrm{coat}[R_k]$. The construction of this ordering is such that the coatoms X for which $z(X) \in S(B, R_k)$, i.e., the coatoms in Q_k, come first. □

The point has now been reached where the step from combinatorics to topology can be taken. We apply the results from Section 4.7 to derive the basic Sphericity Theorems 4.3.3 and 4.3.5.

4.3.3 Theorem. *Let $(E, \mathcal{L})$ be an oriented matroid of rank r. Then $\mathcal{F}_{big}$ $(\mathcal{L})$ is isomorphic to the face lattice of a shellable regular cell decomposition $\Delta(\mathcal{L})$ of the $(r-1)$-sphere, unique up to cellular homeomorphism. Furthermore, every ordering of the maximal cells corresponding to a linear extension of $\mathcal{T}(\mathcal{L}, B)$ is a shelling of $\Delta(\mathcal{L})$.*

Proof. The lattice $\mathcal{F}_{big}(\mathcal{L}) = \widehat{\mathcal{L}}$ has been shown to be graded and thin (Theorem 4 .1.14) and to admit a recursive coatom ordering (Proposition 4.3.2). By Theorem 4.7.24 these properties imply (in fact, are equivalent to) the statement in Theorem 4.3.3. The remark on shellings also follows from Proposition 4.3.2, by Lemma 4.7.18. □

Via Proposition 4.7.26(iv) we deduce:

4.3.4 Corollary. *$\mathcal{F}_{big}(\mathcal{L})^{op}$ is isomorphic to the face lattice of a PL regular cell decomposition $\Delta^{op}(\mathcal{L})$ of the $(r-1)$-sphere.*

In a similar way, Proposition 4.3.1 implies the following.

4.3.5 Theorem. *Let $(E, \mathcal{L})$ be an acyclic oriented matroid of rank r.*

(i)   *The Edmonds-Mandel lattice $\mathcal{F}_{em}$ $(\mathcal{L})$ is isomorphic to the face lattice of a shellable regular cell decomposition of the $(r-2)$-sphere.*

(ii)   *The Las Vergnas lattice $\mathcal{F}_{lv}(\mathcal{L})$ is isomorphic to the face lattice of a PL regular cell decomposition of the $(r-2)$-sphere.*

Theorem 4.3.3 and Theorem 4.3.5(i) are due to Folkman and Lawrence (1978), with the extra ingredient of "PL" and shellability added by Edmonds and Mandel (1982) and Lawrence (1984a). Corollary 4.3.4 and Theorem 4.3.5(ii) are due to Edmonds and Mandel (1982).

The Las Vergnas cell complex and the Edmonds-Mandel cell complex are opposite to each other, in the sense that there is a one-to-one inclusion-reversing correspondence between their cells. It is not known whether the Las Vergnas PL spheres or the spheres $\Delta^{op}(\mathcal{L})$ are shellable. All regular cell complexes referred to in the theorems above have the intersection property, since their face posets are lattices (Exercise 4.47).

Propositions 4.3.1 and 4.3.2 contain in great concentration the combinatorics that make the topological interpretation of oriented matroids possible. Knowing, as we now do, that the face lattices of oriented matroids encode existing cell complexes, these theorems can be recast as shellability results for such complexes (cf. Lemma 4.7.18). Such shellability results are non-trivial and of interest even in the realizable case. For instance, as is remarked in Example 4.7.15 the shellability theorem of Bruggesser and Mani (1971) for boundary complexes of convex polytopes is a special case of Proposition 4.3.1. Also, a shellability result of Björner (1984c) for Coxeter complexes (reflection arrangements) is in the finite case equivalent to that instance of Proposition 4.3.2. Lawrence's proof of Proposition 4.3.2 appeared in a 1984 preprint which was unfortunately never published and therefore not widely known. The result was since independently rediscovered by Björner and Ziegler (1988) with the proof presented here, and by Hochstättler (1990).

The "big" oriented matroid sphere $\Delta(\mathcal{L})$ constructed in Theorem 4.3.3 has some combinatorially induced subcomplexes that are of crucial importance for the Topological Representation Theorem 5.2.1. These subcomplexes will now be investigated.

As we know, the cells of $\Delta(\mathcal{L})$ may be identified with the sign vectors of $\mathcal{L}\setminus\{0\}$ in inclusion-preserving manner, so without further discussion order ideals in $\mathcal{L}$ can be used to designate subcomplexes of $\Delta(\mathcal{L})$.

For each $e \in E, A \subseteq E$, and $i \in \{+, -\}$, let

$$
\begin{aligned}
\Delta_e^0 &= \{X \in \mathcal{L} : X_e = 0\}, \\
\Delta_e^i &= \{X \in \mathcal{L} : X_e \in \{0, i\}\}, \\
\Delta_A^0 &= \bigcap_{e \in A} \Delta_e^0, \text{ and } \Delta_A^i = \bigcap_{e \in A} \Delta_e^i.
\end{aligned}
$$

(4.3.1)

We observe that $\Delta_A^0 \cong \Delta(\mathcal{L}/A)$, cf. Lemma 4.1.8, so that these subcomplexes correspond to matroid contraction.

In the following, $\mathcal{L} \subseteq \{+, -, 0\}^E$ will be a loop-free oriented matroid of rank $r \geq 2$, and "cl" and "r" will denote the closure operator and rank function of the underlying matroid.

4.3.6 Proposition.   *The system of subcomplexes (4.3.1) of the regular cell complex $\Delta(\mathcal{L})$ has the following properties:*

(a) $\Delta_A^0 = \Delta_{clA}^0$ is a shellable $(r - r(A) - 1)$-sphere, for all $A \subseteq E$.

(b) If $e \in E \setminus clA$, then $\Delta_A^0 \cap \Delta_e^i$, for $i \in \{+, -\}$, are shellable $(r - r(A) - 1)$-balls, each with boundary $\Delta_{A \cup e}^0$.

(c) Every non-empty intersection $\Delta_A^+ \cap \Delta_B^-$, for $A, B \subseteq E$, is a shellable sphere or a shellable ball.

Proof.

(a) We have that
$$\{X \in \mathcal{L} : A \subseteq z(X)\} = \{X \in \mathcal{L} : clA \subseteq z(X)\},$$
by Proposition 4.1.13(i), which means that $\Delta_A^0 = \Delta_{clA}^0$. Furthermore, $\Delta_A^0 \cong \Delta(\mathcal{L}/A)$ as was already observed, and rank $(\mathcal{L}/A) = r - r(A)$ by Corollary 4.1.15. Hence, by Theorem 4.3.3, Δ_A^0 is a shellable $(r - r(A) - 1)$-sphere.

(b) By passing to the contraction $\mathcal{L}/A$, we may without loss of generality assume that $clA = \emptyset$. So, it suffices to prove that Δ_e^+ and Δ_e^- are shellable $(r - 1)$-balls, each with boundary Δ_e^0. Let
$$B \in \mathcal{T}_e^+ := \{R \in \mathcal{T}(\mathcal{L}) : R_e = +\}.$$
Since $\mathcal{T}_e^+$ is an order ideal in $\mathcal{T}(\mathcal{L}, B)$ there are linear extensions of $\mathcal{T}(\mathcal{L}, B)$ in which $\mathcal{T}_e^+$ comes first. This means by Theorem 4.3.3 that there are shellings of $\Delta(\mathcal{L})$ in which the maximal cells of Δ_e^+ come first, which by Proposition 4.7.26(ii) shows that Δ_e^+ is a shellable ball. Its boundary is clearly Δ_e^0.

(c) This is proved like part (b) by passing to a suitable contraction $\mathcal{L}/C$ (this gives the shellable spheres) and by finding suitable initial segments of shellings from linear extensions of $\mathcal{T}(\mathcal{L}/C, D)$ (this gives the shellable balls). We leave the details to the reader. $\qquad\square$

In the terminology of Chapter 5, Proposition 4.3.6 shows that the family of subcomplexes $(\Delta_e^0)_{e \in E}$ forms an "arrangement of pseudospheres" in the oriented matroid sphere $\Delta(\mathcal{L})$, and this proves one direction of the Topological Representation Theorem 5.2.1.

Before ending we will deduce some additional information from the constructions in this section. First of all, for the record let us state the following, cf. Proposition 4.1.9.

4.3.7 Corollary. Let $[X, Y]$ be an interval in $\mathcal{F}_{big}$ $(\mathcal{L})$. Then $[X, Y]$ is isomorphic to the face lattice of a shellable regular cell decomposition of the $(\rho(Y) - \rho(X) - 2)$-sphere.

4.3.8 Corollary. The Möbius function of $\mathcal{F}_{big}$ $(\mathcal{L})$ is given by
$$\mu(X, Y) = (-1)^{\rho(Y) - \rho(X)},$$
for $X \leq Y$, with ρ denoting the poset rank function.

Let E be a set of size n. It was pointed out in Example 4.1.4 that the set $\{+, -, 0\}^E$ of *all* sign vectors on E with their usual ordering form a simplicial complex triangulating the boundary of the n-dimensional cross-polytope. The topes of any loop-free oriented matroid $\mathcal{L}$ on E therefore form a subcollection of the maximal simplices of this polytope boundary, and we will now compute the homotopy type of the subcomplex $\Gamma(\mathcal{L})$ that they generate. This complex, properly defined by

$$\Gamma(\mathcal{L}) = \{X \in \{+, -, 0\}^E : X \leq Y \text{ for some } Y \in \mathcal{L}\},$$

was studied by Edelman (1984a) in a slightly different guise (see Exercise 4.22). The motivation for interest in $\Gamma(\mathcal{L})$ is to reach a deeper understanding of the Euler characteristic formula of Corollary 4.3.10, which via a Möbius inversion is equivalent to the important Las Vergnas–Zaslavsky Theorem 4.6.1.

4.3.9 Proposition (Edelman 1984a). *Let $(E, \mathcal{L})$ be an oriented matroid of rank r. Then the complex $\Gamma(\mathcal{L})$ is homotopy equivalent to the $(r - 1)$-sphere.*

4.3.10 Corollary.

$$\sum_{X \in \Gamma(\mathcal{L})} (-1)^{|X|} = (-1)^r.$$

Proof. Consider the natural embedding $f : \mathcal{L} \backslash \{0\} \longrightarrow \Gamma(\mathcal{L}) \backslash \{0\}$. This map is order-preserving, and we know from Theorem 4.3.3 that the order complex of $\mathcal{L} \backslash \{0\}$ triangulates the $(r - 1)$-sphere. Hence, by Lemma 4.7.29 we only need to verify that the filter $F_X = \{Y \in \mathcal{L} : Y \geq X\}$ is contractible for all $X \in \Gamma(\mathcal{L}) \backslash \{0\}$.

Suppose that $0 \neq X \leq R \in \mathcal{T}(\mathcal{L})$, and let

$$J_X = \mathcal{L} \backslash F_X = \{Y \in \mathcal{L} : Y \not\geq X\}.$$

If $Y \in J_X$, then clearly $Y \circ (-R) \in J_X$, so the order ideal J_X is generated by the set of topes $J_X \cap \mathcal{T}(\mathcal{L})$. This set, in turn, forms an order ideal in the tope poset $\mathcal{T}(\mathcal{L}, -R)$, so by Proposition 4.3.2 there exists some recursive coatom ordering of $\widehat{\mathcal{L}}$ in which the topes in J_X come first. Hence, by Lemma 4.7.28, F_X is contractible. $\qquad \square$

The following technical lemma will be of use in the next section.

4.3.11 Lemma. *Let $B, R \in \mathcal{T}(\mathcal{L})$, and consider the set*

$$F_R = \{X \in (0, B) : z(X) \subseteq S(B, R)\},$$

which is a filter in the open interval $(0, B)$ of $\mathcal{L}$. If $R \neq B, -B$, then the poset F_R is contractible.

Proof. Observe that F_R is a non-empty proper subset of $(0, B)$. Let

$$J_R = (0, B) \backslash F_R.$$

If $X \in J_R$ then $X \circ (-R) \neq B$ and

$$S(B, R) \cap S(B, X \circ (-R)) = \emptyset.$$

By Lemma 4.1.12 applied to $X \circ (-R)$ and B there exists $Z \in \mathcal{L}$ such that $Z \lessdot B$ and $Z_e = B_e$ for all $e \notin S(B, X \circ (-R))$. In particular, $X \leq Z$ and $Z \in J_R$. Hence, the order ideal J_R is generated by the set of coatoms $J_R \cap \text{coat}[0, B]$. By

Proposition 4.3.1 any maximal chain in $\mathcal{T}(\mathcal{L}, B)$ containing the tope $-R$ induces a recursive coatom ordering of $[0, B]$ in which the elements of $J_R \cap \text{coat}[0, B]$ come first. Therefore, by Lemma 4.7.28, F_R is contractible. $\qquad\square$

4.4 Topes II

In this section we continue the study of tope graphs and tope posets, taking advantage of the topological perspective developed in the previous section. For tope posets the Möbius function and homotopy type of intervals will be discussed, and also the question when a tope poset is a lattice. For tope graphs we will determine the degree of connectivity and also present two homotopy results.

We begin with the following observation about the topological nature of tope graphs, which follows directly from the definitions and Corollary 4.3.4. Throughout this section $\mathcal{L}$ is the set of covectors of a rank r oriented matroid.

4.4.1 Lemma. *The tope graph $\mathcal{T}(\mathcal{L})$ is the 1-skeleton of the PL $(r-1)$-sphere $\Delta^{op}(\mathcal{L})$.*

The following result was proved in the realizable case (arrangements of hyperplanes) by Edelman and Walker (1985). The case of reflection arrangements also follows from the results on weak Bruhat order of Björner (1984b). Another special case appears in Cordovil (1983b). Our proof follows that of Edelman and Walker.

4.4.2 Theorem. *Let (T^1, T^2) be a non-empty open interval in the tope poset $\mathcal{T}(\mathcal{L}, B)$ of an oriented matroid $\mathcal{L}$ of rank r. Let $\rho(X)$ denote the poset rank function in $\mathcal{L}$. Then:*

(i) (T^1, T^2) *is homotopy equivalent to the $(r - \rho(X) - 2)$-sphere, if $[T^1, T^2] = \text{star}(X)$ for some $X \in \mathcal{L}$.*

(ii) (T^1, T^2) *is contractible, otherwise.*

Since the Möbius function $\mu(T^1, T^2)$ is the reduced Euler characteristic of the open interval (T^1, T^2) we deduce:

4.4.3 Corollary (Edelman 1984b). *The Möbius function of $\mathcal{T}(\mathcal{L}, B)$ is given by*

$$\mu(T^1, T^2) = \begin{cases} (-1)^{r-\rho(X)}, & \text{if } [T^1, T^2] = \text{star}(X), \ X \in \mathcal{L}, \\ 0, & \text{otherwise.} \end{cases}$$

Proof. We may without loss of generality assume that $T^1 = B$ (cf. Corollary 4.2.11). Put also $T^2 = T$.

The mapping $f : X \mapsto X \circ (-B)$ is an order-reversing embedding of the interval $(0, B)_{\mathcal{L}}$ into the interval $(B, -B)_{\mathcal{T}}$, where we have denoted by subscripts $\mathcal{L}$ resp. $\mathcal{T} = \mathcal{T}(\mathcal{L}, B)$ the ambient posets of these intervals. For each $R \in (B, -B)_{\mathcal{T}}$ let

$$F_R = f^{-1}(\mathcal{T}_{\leq R}) = \{X \in (0, B)_{\mathcal{L}} : z(X) \subseteq S(B, R)\}.$$

We know from Lemma 4.3.11 that the filter F_R is contractible for all topes $R \neq B, -B$.

Lemma 4.7.29 applied to the restriction of f to

$$f^{-1}(\mathcal{T}_{<T}) = F_T \backslash f^{-1}(T)$$

now gives that $(B,T)_{\mathcal{T}}$ and $f^{-1}(\mathcal{T}_{<T})$ are homotopy equivalent. Since $f(X) = T$ if and only if $[B,T]_{\mathcal{T}} = \text{star}(X)$, by Lemma 4.2.12, we can analyze the two cases of the theorem:

Case (ii) means that $f^{-1}(T) = \emptyset$, so $f^{-1}(\mathcal{T}_{<T}) = F_T$ which we know is contractible, hence so is $(B,T)_{\mathcal{T}}$.

In case (i), when $T = f(X) = X \circ (-B)$ for some $X \in (0,B)_{\mathcal{L}}$, we get that $(B,T)_{\mathcal{T}}$ and $f^{-1}(\mathcal{T}_{<T}) = (X,B)_{\mathcal{L}}$ are homotopy equivalent. But the order complex of $(X,B)_{\mathcal{L}}$ triangulates a $(\rho(B) - \rho(X) - 2)$-dimensional sphere by Corollary 4.3.7, so the proof is complete.

The Möbius function formula of Corollary 4.4.3 is, to be pedantic, implied by the topology only for intervals $[T^1, T^2]$ of length ≥ 2. It is easily checked directly in the trivial cases of length ≤ 1. $\square$

Call an interval $[T^1, T^2]$ in $\mathcal{T}(\mathcal{L},B)$ *elementary* if it satisfies any (and hence all) of the following equivalent conditions:

(i)   the open interval (T^1, T^2) consists of two disjoint chains,

(ii) (T^1, T^2) is homotopy equivalent to the 0-sphere,

(iii) $[T^1, T^2] = \text{star}(X) = [X \circ B, X \circ (-B)]$, for some $X \in \mathcal{L}$ with $\rho(X) = r-2$.

The forward implications are clear, the second one from Theorem 4.4.2 and Lemma 4.2.12. For (iii) $\Longrightarrow$ (i) use that the interval $[X, \hat{1}]$ is the face lattice of an even n-gon (Proposition 4.1.17).

The following technical lemma, implicit in Salvetti (1987) and Cordovil and Moreira (1990), will be needed. The simplicial case goes back to Deligne (1972) and Edelman (1984b).

4.4.4 Lemma.   *Let B, A, A' and R be topes, and suppose that A and A' are atoms in the interval $[B, R]$ of $\mathcal{T}(\mathcal{L},B)$. Then there exists a sequence of atoms $A = A_0, A_1, \ldots, A_k = A'$ and a sequence of other elements $T_1, T_2, \ldots, T_k$ in $[B, R]$ such that $[B, T_i]$ is elementary and contains A_{i-1} and A_i, for all $1 \leq i \leq k$. Furthermore, if the tope B is simplicial then $k = 1$. (See Figure 4.4.1.)*

Proof. Let $F_R = \{X \in (0,B)_{\mathcal{L}} : z(X) \subseteq S(B,R)\}$. F_R is an order ideal in $(0,B)_{\mathcal{L}}^{op}$, and can therefore be thought of as a subcomplex of the Las Vergnas sphere of the tope B. For $R \neq -B$ Lemma 4.3.11 shows that F_R is contractible. If $R = -B$ then $F_R = (0,B)_{\mathcal{L}}$, and hence F_R is an $(r-2)$-sphere (Theorem 4.3.5). Hence, in either case the regular cell complex F_R is connected (we may assume that $r \geq 3$, the $r \leq 2$ case being trivial), from which it follows that its 1-skeleton is connected.

Let X and X' be subtopes such that $X = B \wedge A$ and $X' = B \wedge A'$. Then

X and X' are rank $r-1$ elements of F_R, and by connectivity of F_R there exist
rank $r-1$ elements (vertices) $X = X_0, X_1, \ldots, X_k = X'$ and rank $r-2$ elements
(edges) $Y_1, Y_2, \ldots, Y_k$ of F_R such that $Y_i < X_{i-1}, X_i$ for all $1 \le i \le k$. The
argument is now completed with letting $A_i = X_i \circ (-B)$ and $T_i = Y_i \circ (-B)$
for $1 \le i \le k$.

If B is simplicial, or equivalently: if the interval $[0, B]$ is Boolean, then we can
set $Y_1 = X \wedge X'$. $\qquad\square$

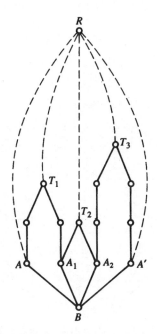

Figure 4.4.1: Sketch for Lemma 4.4.4.

The tope poset $T(\mathcal{L}, B)$ is not always a lattice. For examples of non-lattices
coming from arrangements of hyperplanes, see Edelman (1984b). No characteri-
zation is known for the pairs $(\mathcal{L}, B)$ giving rise to *lattices* $T(\mathcal{L}, B)$. See Exercise
4.13 for some information in this connection. The following gives a sufficient
condition.

4.4.5 Proposition (Björner, Edelman and Ziegler 1990). *If $\mathcal{L}$ is simplicial (all topes are simplicial) then $T(\mathcal{L}, B)$ is a lattice for any choice of base
tope B.*

Proof. The proof boils down to a combinatorial lemma: A bounded poset P is a
lattice if whenever two elements a and a' cover a third then the join $a \vee a'$ exists.
This condition is in our case verified by the simplicial case of Lemma 4.4.4, in
view of Corollary 4.2.11 (ii). $\qquad\square$

We will now describe a notion of homotopy for paths in the tope graph which is

natural from both a combinatorial and a topological point of view. This concept
of homotopy has since long been used in the theory of reflection arrangements.
It was developed for simplicial arrangements by Deligne (1972) and its basic
properties were extended to general real arrangements by Salvetti (1987). The
final extension to oriented matroids was made by Cordovil and Moreira (1990).

Let $\mathcal{P} = T_1 T_2 \ldots T_k$ and $\mathcal{Q} = T_k T_{k+1} \ldots T_s$ be paths in the tope graph $\mathcal{T}(\mathcal{L})$
of an oriented matroid $\mathcal{L}$ of rank r, and denote by $\mathcal{PQ} = T_1 T_2 \ldots T_k T_{k+1} \ldots T_s$
their concatenation. Say that paths $\mathcal{PR}^1\mathcal{Q}$ and $\mathcal{PR}^2\mathcal{Q}$ differ by an *elementary
homotopy* (a symmetric relation) if the paths $\mathcal{R}^1$ and $\mathcal{R}^2$ are related in either of
the following two ways:

(i) $\mathcal{R}^1 = ABA$, $\mathcal{R}^2 = A$ (the empty path at A),

(ii) $\mathcal{R}^1 = A_1 A_2 \ldots A_c$, $\mathcal{R}^2 = B_1 B_2 \ldots B_d$, where

$$A_1, A_2, \ldots, A_c = B_d, B_{d-1}, \ldots, B_1 = A_1$$

are the topes of $\mathrm{star}(X)$ in cyclic order, for some $X \in \mathcal{L}$ of rank $r - 2$.

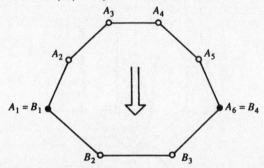

Figure 4.4.2: Elementary homotopy of type (ii)

The second condition is illustrated in Figure 4.4.2, which should be seen in
the light of Proposition 4.1.17. Note that the number $c + d$ must be even.

4.4.6 Proposition.   *Let $\mathcal{P}$ and $\mathcal{Q}$ be two paths in $\mathcal{T}(\mathcal{L})$ with the same endpoints. Then $\mathcal{P}$ can be transformed into $\mathcal{Q}$ via a sequence of elementary homotopies. Equivalently, any closed path in $\mathcal{T}(\mathcal{L})$ can be deformed into the empty
path via elementary homotopies.*

Proof. Attach 2-cells into the circuits $\mathrm{star}(X)$ of the tope graph that define
elementary homotopies of type (ii). This creates a 2-dimensional regular cell
complex Δ, and the theorem is equivalent to the statement that this complex
is simply connected (in the topological sense). But we observe that Δ is the
2-skeleton of the complex $\Delta^{op}(\mathcal{L})$ of Corollary 4.3.4, which is simply connected
since it is an $(r - 1)$-sphere. (We may assume that $r \geq 3$, the $r = 2$ case
being trivial.) Hence, the result follows from the general fact that a regular cell
complex is simply connected if and only if its 2-skeleton is.   □

Some variations on this theme appear in Exercise 4.16. The case of reflection arrangements (Section 2.3) is of special interest, due to the one-to-one correspondence between paths in the tope graph (starting at a base region B) and expressions for the elements of the reflection group as words in the Coxeter generators (the reflections through the walls of B). The simple-connectedness expressed by Proposition 4.4.6 is in that case equivalent to the fact that all relations in the reflection group are generated by the Coxeter relations (2.3.5). This connection of Coxeter relations with simple-connectedness has long been known, see e.g. Sections 5.3 and 11.1 of Coxeter (1973). It is used for "reflection oriented matroids" in Björner and Ziegler (1992b).

4.4.7 Proposition (Cordovil and Moreira 1990). *Let $\mathcal{P}$ and $\mathcal{Q}$ be two minimal paths in $\mathcal{T}(\mathcal{L})$ with the same endpoints. Then $\mathcal{P}$ can be transformed into $\mathcal{Q}$ via a sequence of elementary homotopies in such a way that all intermediate paths are also minimal.*

Proof. Call the endpoints B and R and argue by induction on their distance $|S(B,R)|$. For distances ≤ 2 there is no problem.

Suppose that A and A' are the first topes along the paths $\mathcal{P}$ and $\mathcal{Q}$ after B, and that $A \neq A'$. We are then in the situation of Lemma 4.4.4, and use of the induction hypothesis for the shorter intervals $[A_i, R]$, $0 \leq i \leq k$, completes the proof. $\qquad\square$

In the transformations of Proposition 4.4.7 no elementary homotopies of type (i) are used, and the ones of type (ii) involve paths of equal length: $c = d$. For reflection arrangements the result is equivalent to the finite case of the solution to the word problem for reduced expressions in a Coxeter group of Tits (1968). If R is the region corresponding to a group element w (see Section 2.3), then the minimal paths in the tope graph from the base region B to R correspond to the reduced decompositions of w, and the elementary homotopies of paths correspond to the Coxeter relations (2.3.5) with $s \neq s'$. An application of this to pseudoline arrangements is given in the proof of Theorem 6.4.1.

We will end this section with some results about the connectivity of tope graphs and some related graphs.

A graph is said to be *k-connected* if every pair of non-adjacent vertices can be connected by k vertex-disjoint paths. A well-known theorem of Balinski (1961) states that the graph of a convex d-polytope is d-connected. This result was generalized by Barnette (1973b) to 1-skeleta of a rather general class of combinatorial $(d-1)$-pseudomanifolds. Because of Lemma 4.4.1, Barnette's theorem applies to oriented matroids, as was observed by Cordovil and Fukuda (1987).

4.4.8 Proposition (Barnette, Cordovil and Fukuda). *The tope graph $\mathcal{T}(\mathcal{L})$ is r-connected, $r = \text{rank}(\mathcal{L})$.*

If $\mathcal{T}(\mathcal{L})$ has a vertex of degree r, or equivalently if $\mathcal{L}$ has a simplicial tope (see

Exercise 4.4), then $\mathcal{T}(\mathcal{L})$ cannot be $(r+1)$-connected. Hence by Corollary 2.1.7 r-connectivity is best possible for all *realizable* oriented matroids.

The same reasoning applies to the cell decompositions of spheres considered in Theorem 4.3.5, from which the following can be concluded.

4.4.9 Proposition (Barnette, Cordovil and Fukuda).   *The 1-skeleta of the face lattices $\mathcal{F}_{em}$ $(\mathcal{L})$ and $\mathcal{F}_{lv}$ $(\mathcal{L})$ are $(r-1)$-connected graphs, $r = \operatorname{rank}(\mathcal{L})$.*

In the realizable case this gives the theorem of Balinski (1961) about polytope graphs. By the same method one can conclude that the 1-skeleton of $\mathcal{F}_{big}$ $(\mathcal{L})$, called the *cocircuit graph*, is r-connected. However, in this case a sharper result is known.

4.4.10 Proposition (Cordovil and Fukuda 1987).   *The cocircuit graph of $\mathcal{L}$ is $2(r-1)$-connected, where $r = \operatorname{rank}(\mathcal{L})$.*

The degree of connectivity stated in the last two propositions is best possible. See Cordovil, Fukuda and Guedes de Oliveira (1991) for more information about cocircuit graphs, in particular about the question to which extent is $\mathcal{L}$ determined by its cocircuit graph.

4.5 The affine face lattice

Throughout this section we will consider triples $(E, \mathcal{L}, g)$, called *affine oriented matroids*, where $(E, \mathcal{L})$ is an oriented matroid and $g \in E$ is a distinguished element, which is not a loop. As usual, $\mathcal{L}$ denotes the set of covectors.

4.5.1 Definition. For such a triple $(E, \mathcal{L}, g)$ let
$$\mathcal{L}^+ = \mathcal{L}_g^+ = \{X \in \mathcal{L} : X_g = +\} \text{ and } \widehat{\mathcal{L}}^+ = \mathcal{L}^+ \cup \{0, \hat{1}\}.$$
With the induced order as a subset of $\widehat{\mathcal{L}}$, we call $\widehat{\mathcal{L}}^+$ (and abusively also $\mathcal{L}^+$) the *affine face lattice* of $(E, \mathcal{L}, g)$. The *bounded complex* of $(E, \mathcal{L}, g)$ is
$$\mathcal{L}^{++} = \{X \in \mathcal{L}^+ : \mathcal{L}_{\leq X} \subseteq \widehat{\mathcal{L}}^+\}.$$

The reason for interest in affine oriented matroids and their face lattices comes from affine or non-central hyperplane arrangements affine hyperplane arrangements in $\mathbb{R}^d$ (Section 2.1), e.g. in connection with linear programming (Section 10.1). Suppose that $\mathcal{A} = (H_e)_{e \in E}$ is an arrangement of affine hyperplanes in $\mathbb{R}^d$, and for each hyperplane choose a positive and a negative side. Then we get a mapping
$$\sigma : \mathbb{R}^d \longrightarrow \{+, -, 0\}^E,$$
where the sign vector $\sigma(x)$ is a position vector for the point x with respect to each hyperplane H_e (on the positive side, on the negative side, or right on). This setup is identical to the one described in Section 4.1, and again the system of sign vectors $\sigma(\mathbb{R}^d) \subseteq \{+, -, 0\}^E$ serves as an indexing set for the cells in the decomposition of $\mathbb{R}^d$ induced by $\mathcal{A}$.

The sign vectors obtained from an affine arrangement $\mathcal{A} = (H_e)_{e \in E}$ can be embedded into those of a larger oriented matroid as follows.

Take an affine embedding of $\mathbb{R}^d$ into $\mathbb{R}^{d+1}$, say $\mathbb{R}^d = (\mathbb{R}^d, 1) \hookrightarrow \mathbb{R}^{d+1}$, and let $\mathcal{A}' = (H'_e)_{e \in E \cup g}$ be the central arrangement in $\mathbb{R}^{d+1}$ consisting of $H'_e = \mathrm{span}(H_e)$, for all $e \in E$, and the "new" hyperplane $H'_g = \{x \in \mathbb{R}^{d+1} : x_{d+1} = 0\}$, see Figure 4.6.2. Then the original set of sign vectors $\sigma(\mathbb{R}^d)$ appears as $\mathcal{L}^+$ for the oriented matroid $(E \cup g, \mathcal{L})$ of the arrangement $\mathcal{A}'$, except that in $\mathcal{L}^+$ they have been embellished by a dummy "+" in position g. Notice that the order relation in $\mathcal{L}^+$ correctly describes the inclusion of cells in the decomposition of $\mathbb{R}^d$ induced by $\mathcal{A}$. The bounded complex $\mathcal{L}^{++}$, finally, corresponds to the system of cells of $\mathcal{A}$ which are bounded in the usual metric sense.

Affine sign vector systems, i.e. sets of the form $\mathcal{L}^+$, have been given an intrinsic characterization not depending on an ambient oriented matroid by Karlander (1992).

4.5.2 Example. Let $\mathcal{A}$ be the affine arrangement of 4 lines in $\mathbb{R}^2$ depicted in Figure 4.5.1. The sign vectors corresponding to regions are indicated in Figure 4.5.1 and the bounded complex consists of the two triangles and their faces (shaded). Figure 4.5.2 shows a view of the upper hemisphere of the central arrangement $\mathcal{A}'$ in $\mathbb{R}^3$ into which $\mathcal{A}$ is embedded. The new hyperplane g is "at the horizon".

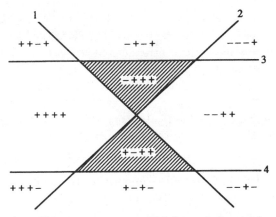

Figure 4.5.1: An affine arrangement. The bounded complex is shaded.

The face lattice $\mathcal{L}^+$ is in this example a pure poset of length 2 with rank levels of size 5, 14 and 10, from bottom to top. Its order complex $\Delta_{ord}(\mathcal{L}^+)$ is shaded in Figure 4.5.2 and shown as a subcomplex of the barycentric subdivision $\Delta_{ord}(\mathcal{L})$.

The main objective in this section is to prove that $\Delta_{ord}(\mathcal{L}^+)$ is a shellable ball and that the bounded complex is contractible, see Theorem 4.5.7 below. The first part was shown in the realizable case by Ziegler (1988), however, his proof does not generalize to the non-realizable case. The question about the homotopy type of the bounded complex was raised by Zaslavsky (1975a, 1977). It seems

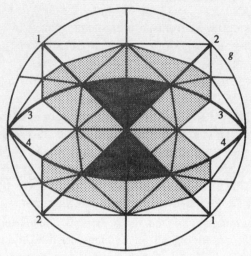

Figure 4.5.2. The order complex $\Delta_{ord}(\mathcal{L}^+)$ is a ball
(shaded); the bounded complex is contractible (dark).

intuitively obvious that the bounded complex is contractible, and probably in most cases a full-dimensional ball, but surprisingly there seems to be no simple proof for this fact, even in the realizable case. The method of proof here, due to Björner and Ziegler (1988), will be via recursive coatom orderings, extending the ideas used in Section 4.3.

We begin with a few elementary observations concerning the combinatorial structure of the poset $\mathcal{L}^+$. Put $r = \text{rank}(\mathcal{L})$. Note that the following result implies that an affine hyperplane arrangement is essential (Definition 2.1.1) if and only if every minimal cell is a vertex.

4.5.3 Proposition.

(i) $\widehat{\mathcal{L}}^+$ *is a coatomic graded lattice of length $r + 1$; it is a join-subsemilattice of $\widehat{\mathcal{L}}$.*

(ii) $\mathcal{L}^+$ *is a pure poset of length $r - 1$; it is a filter in $\mathcal{L}$.*

Proof. Each $X \in \mathcal{L}^+$ is conformally composed of cocircuits:
$$X = C^1 \circ C^2 \circ \ldots \circ C^k.$$
If $C_g^i = 0$ for all $1 \leq i \leq k$, then also $X_g = 0$. Hence, $C^i \in \mathcal{L}^+$ for some $1 \leq i \leq k$. It follows that all minimal elements of $\mathcal{L}^+$ are cocircuits, and hence (from Theorem 4.1.14) that $\mathcal{L}^+$ is pure and of length $r - 1$. Since $\widehat{\mathcal{L}}^+$ has joins (inherited from $\widehat{\mathcal{L}}$) and a least element 0 it is a lattice. Finally, $\widehat{\mathcal{L}}^+$ is coatomic since $\widehat{\mathcal{L}}$ is coatomic (Corollary 4.1.16).   □

A few more definitions are needed. For any subset $A \subseteq \mathcal{L}$, let
$$A^+ = A \cap \mathcal{L}^+ = \{X \in A : X_g = +\}$$
and
$$A^* = A \cap (\mathcal{L}^+ \cup \{0\}).$$

Let

$$\mathcal{L}^{\infty} = \{X \in \mathcal{L} : X_g = 0\},$$

which geometrically is the "horizon" of the upper hemisphere $\mathcal{L}^+$. For $X \in \mathcal{L}^+$ let $X^{\infty} = \vee\{Y \in \mathcal{L}^{\infty} : Y < X\}$, i.e., the unique maximal face of X at infinity (the join is taken in $\widehat{\mathcal{L}}$). Note that $X^{\infty} = 0$ if and only if $X \in \mathcal{L}^{++}$. Recall the notation

$$\mathrm{star}(X) = \{T \in \mathcal{T} : X \le T\}$$

from Section 4.2.

4.5.4 Proposition. *Let $(E, \mathcal{L}, g)$ be an affine oriented matroid and $R \in \mathcal{T}^+$. Then every maximal chain $\mathbf{m}$ in the poset $\mathcal{T}(\mathcal{L}, R)$ that includes a tope from $\mathrm{star}^+(-R^{\infty})$ induces a recursive coatom ordering $\prec_{\mathbf{m}}$ of the interval $[R]^*$ in $\widehat{\mathcal{L}}^+$.*

The intended construction here is of course the same as for Proposition 4.3.1. Note that if $R^{\infty} = 0$, i.e., if the tope R is bounded, then $\mathrm{star}^+(-R^{\infty}) = \mathcal{T}^+$ and $[R]^+ = [R]$. Hence, in this case there are no new restrictions and the result reduces to Proposition 4.3.1.

4.5.5 Lemma. *$\mathcal{T}^+$ is an order ideal in the tope poset $\mathcal{T}(\mathcal{L}, R)$ and all its maximal elements belong to $\mathrm{star}^+(-R^{\infty})$.*

Proof. That $\mathcal{T}^+$ is an order ideal is clear from its $\mathcal{T}$-convexity. If $B \in \mathcal{T}^+$ then

$$S(B, R) \subseteq \underline{R^{\infty}} \cup S(B, R) = \underline{R^{\infty}} \cup [S(B, R) \cap z(R^{\infty})] = S((-R^{\infty}) \circ B, R),$$

which means that $B \preceq (-R^{\infty}) \circ B \in \mathrm{star}^+(-R^{\infty})$. ☐

Proof of Proposition 4.5.4. The proof runs parallel to the proof of Proposition 4.3.1, although some new difficulties must be dealt with.

We use induction on $\mathrm{rank}(\mathcal{L})$, the $\mathrm{rank}(\mathcal{L}) \le 2$ case being trivial. We may without loss of generality assume that $\mathcal{L}$ is simple. We will also assume that $R^{\infty} \ne 0$, since (as was already observed) the $R^{\infty} = 0$ case is subsumed by Proposition 4.3.1.

Choose a maximal chain in $\mathcal{T}(\mathcal{L}, R)$

$$\mathbf{m} : R = R^0 \prec R^1 \prec \ldots \prec R^m = -R,$$

that passes through $\mathrm{star}^+(-R^{\infty})$. Let $E = \{e_1, e_2, \ldots, e_m\}$ with

$$z(R^{i-1} \wedge R^i) = \{e_i\}$$

for $1 \le i \le m$. Then order the set of coatoms of $[R]^*$ by: $X \prec_{\mathbf{m}} Y$ if and only if $z(X) < z(Y)$, for $X, Y \in \mathrm{coat}[R]^*$. The claim is that this is a recursive coatom ordering.

Let $X \in \mathrm{coat}[R]^*$ with $z(X) = \{e_i\}$. Then let $X^i = R^{i-1} \wedge R^i$. Note that $X^i \notin \mathcal{L}^+$ is a possibility. We have that X and X^i are topes in $\mathcal{L}/e_i$ and that $S(X, X^i) = S(R, R^{i-1}) = \{e_1, e_2, \ldots, e_{i-1}\}$.

For $Y \in \text{coat}[X]^*$, either

$$z(Y) \subseteq \{e_1, \ldots, e_i\}$$

or

$$z(Y) \subseteq \{e_i, \ldots, e_m\}$$

for the same reason as in the proof of Proposition 4.3.1. Let

$$Q_X^+ = \{Y \in \text{coat}[X]^* : z(Y) \subseteq \{e_1, \ldots, e_i\}\}.$$

Then $Q_X^+ = Q_X \cap \mathcal{L}^+$ for the set Q_X defined in that proof.

We now come to the two properties of Definition 4.7.17 that must be verified:

(i) $[X]^* \cap \left(\bigcup_{\substack{X' \in \mathcal{L}^+ \\ X' \prec_m X}} [X']^*\right) = \bigcup_{Y \in Q_X^+} [Y]^*$, and

(ii) there is a recursive ordering of $\text{coat}[X]^*$ in which the elements of Q_X^+ come first.

Property (i). Applying the $(\cdot)^*$ operator to the corresponding set equation from the proof of Proposition 4.3.1 we get

$$\text{(i)}' \qquad [X']^* \cap \left(\bigcup_{X' \prec_m X} [X']^*\right) = \bigcup_{Y \in Q_X} [Y]^*.$$

If $Z \in \mathcal{L}^+$ then $[Z]^*$ is the interval $\{Z' \in \mathcal{L}^+ \cup \{0\} : 0 \leq Z' \leq Z\}$ in $\widehat{\mathcal{L}}^+$ that we want, whereas if $Z \notin \mathcal{L}^+$ then $[Z]^* = \{0\}$. Therefore, if X *is not* first in the $\prec_m$ ordering of $\text{coat}[R]^+$, and hence $Q_X^+ \neq \emptyset$ by Claim B below, we obtain formula (i) from (i)' as desired. Suppose on the other hand that X *is* first in $\text{coat}[R]^+$. Then we must have that $X = R \wedge R^1$ and $z(X) = \{e_1\}$, since $R^1 \in \mathcal{L}^+$ follows from $R^\infty \neq 0$ and the condition that the chain **m** passes through $\text{star}^+(-R^\infty)$. Hence, X is then first also in the $\prec_m$ ordering of $\text{coat}[R]$, hence $Q_X = \emptyset$, hence $Q_X^+ = \emptyset$. Consequently formula (i) is valid also in this case (both sides being empty).

The verification of property (i) will be complete with the proof of Claim B below, whose proof in turn requires Claim A.

Claim A. If $X^i \notin \mathcal{L}^+$, then $(-X^\infty) \circ X \preceq X^i$ in $T(\mathcal{L}/e_i, X)$.

Claim B. If X is not first in the $\prec_m$ ordering of $\text{coat}[R]^+$, then $Q_X^+ \neq \emptyset$.

Proof of Claim A. Choose j so that $R^j \in \text{star}^+(-R^\infty)$. It was required of **m** that such j exists. Since $X^i = R^{i-1} \wedge R^i \notin \mathcal{L}^+$ it follows that $R^i \notin \mathcal{L}^+$, and hence that $i > j$. Since $\text{star}(-R^\infty)$ is T-convex (Lemma 4.2.12) and contains R^j and $R^m = -R$ it must contain all R^k with $j \leq k \leq m$, and in particular R^{i-1} and R^i. Hence, $X^i = R^{i-1} \wedge R^i \geq -R^\infty \geq -X^\infty$.

Moving to the contraction $\mathcal{L}/e_i$ where X and X^i are topes, we now have:

$$X^i \in \text{star}_{\mathcal{L}/e_i}(-X^\infty) = [(-X^\infty) \circ X, -X],$$

the latter being an interval in $T(\mathcal{L}/e_i, X)$ by Lemma 4.2.12. This implies Claim A.

Proof of Claim B. Case 1: $X^i \in \mathcal{L}^+$. We know that

$$S(X, X^i) = \{e_1, \ldots, e_{i-1}\},$$

and this set is not empty since X is not first. By Lemma 4.1.12 there exists $Y \lessdot X$ such that $Y_e = X_e$ for all $e \notin S(X, X^i)$. Then $Y_g = +$ and $z(Y) \subseteq \{e_1, \ldots, e_i\}$, hence $Y \in Q_X^+$.

Case 2: $X^i \notin \mathcal{L}^+$. We may assume that $X^\infty \neq 0$. Otherwise all coatoms of $[X]$ are in $\mathcal{L}^+$ (the subtope X being bounded) and $Q_X^+ = Q_X$, and $Q_X \neq \emptyset$ since X is not first in the recursive coatom ordering $\prec_m$ of $[R]$.

Lemma 4.1.12 applied to $-X^\infty$ and X gives the existence of $Y \lessdot X$ such that $Y_e = X_e$ for all $e \notin S(-X^\infty, X) = \underline{X^\infty}$. Hence, $Y_g = +$. We have that $\underline{X^\infty} = S(X, (-X^\infty) \circ X) \subseteq S(X, X^i) = \{e_1, \ldots, e_{i-1}\}$, with the middle inclusion provided by Claim A. Therefore

$$z(Y) \subseteq \{e_1, \ldots, e_i\},$$

and consequently $Y \in Q_X^+$.

Property (ii). By the induction hypothesis every maximal chain **n** in the poset $\mathcal{T}(\mathcal{L}/e_i, X)$ that intersects $\mathrm{star}^+_{\mathcal{L}/e_i}(-X^\infty)$ induces a recursive coatom ordering of the interval $[X]^*$ in $\widehat{\mathcal{L}}^+$. If such a chain **n** can be found which contains X^i, then the elements of Q_X^+ will come first in the recursive ordering $\prec_n$ of $\mathrm{coat}[X]^*$, just as in the proof of Proposition 4.3.1. Hence, we must show that X^i is comparable in $\mathcal{T}(\mathcal{L}/e_i, X)$ to some $Z \in \mathrm{star}^+_{\mathcal{L}/e_i}(-X^\infty)$, so that a maximal chain **n** through X^i and Z can be chosen.

Case 1: $X^i \in \mathcal{L}^+$. Then X^i belongs to the order ideal $\mathcal{T}^+(\mathcal{L}/e_i)$ in $\mathcal{T}(\mathcal{L}/e_i, X)$. Let $Z \succeq X^i$ be a maximal element of that ideal. Then $Z \in \mathrm{star}^+_{\mathcal{L}/e_i}(-X^\infty)$, by Lemma 4.5.5.

Case 2: $X^i \notin \mathcal{L}^+$. This is settled by Claim A above. $\qquad\square$

Let $B \in \mathcal{T}^+ = \{T \in \mathcal{T} : T_g = +\}$. Then $\mathcal{T}^+$ is an order ideal in the tope poset $\mathcal{T}(\mathcal{L}, B)$. Let $\mathcal{T}^+(\mathcal{L}, B)$ denote this order ideal as a poset in its own right.

4.5.6 Proposition. *Let $(E, \mathcal{L}, g)$ be an affine oriented matroid and $B \in \mathcal{T}^+$. Then every linear extension of $\mathcal{T}^+(\mathcal{L}, B)$ is a recursive coatom ordering of $\widehat{\mathcal{L}}^+$.*

Proof. Suppose that $B = R_0, R_1, \ldots, R_s$ is a linear extension of $\mathcal{T}^+(\mathcal{L}, B)$, i.e., $S(B, R_i) \subseteq S(B, R_k)$ implies $i \leq k$. This can be continued, $R_{s+1}, \ldots, R_t = -B$, to a linear extension of $\mathcal{T}(\mathcal{L}, B)$, and hence we may fall back on the proof of Proposition 4.3.2. Define

$$Q_k = \{X \in \mathrm{coat}[R_k] : z(X) \subseteq S(B, R_k)\},$$

as was done there. Note that $Q_k \subseteq \mathcal{L}^+$, for $1 \leq k \leq s$.

Property (i). We directly obtain

$$\bigcup_{X \in Q_k} [X]^* = [R_k]^* \cap \left(\bigcup_{i < k} [R_i]^* \right)$$

by applying the $(\cdot)^*$ operator to the corresponding set equation in the proof of Proposition 4.3.2. The kind of difficulties that were encountered in the proof of Proposition 4.5.4 at this point do not occur, since $Q_k^+ = Q_k$ and $R_i \in \mathcal{L}^+$ for all $i < k$.

Property (ii). Lemma 4.5.5 shows that it is possible to choose a maximal chain m in $\mathcal{T}(\mathcal{L}, R_k)$ that contains both B and some element from $\text{star}^+(-R_k^\infty)$. By Proposition 4.5.4 such a chain induces a recursive ordering of $\text{coat}[R_k]^*$, and the proof of Proposition 4.3.2 showed that in this ordering the elements of Q_k come first. $\square$

We can now draw the topological conclusions. Since $\mathcal{L}^{++}$ is an order ideal in $\mathcal{L}\backslash\{0\}$ it can via the cellular interpretation of $\mathcal{L}$ in Section 4.3 be considered a subcomplex $\Delta(\mathcal{L}^{++})$ of the regular cell complex $\Delta(\mathcal{L})$. The poset $\mathcal{L}^+$ does not have any cellular interpretation, however its opposite does, see Corollary 4.5.8 below. The following result is illustrated by Example 4.5.2.

4.5.7 Theorem (Björner and Ziegler).   *Let $(E, \mathcal{L}, g)$ be an affine oriented matroid of rank r. Then:*

(i)   *The order complex $\Delta_{ord}(\mathcal{L}^+)$ is a shellable $(r-1)$-ball.*

(ii)   *The bounded complex $\Delta(\mathcal{L}^{++})$ is contractible.*

Proof.

(i) Since $\widehat{\mathcal{L}}^+$ admits recursive coatom orderings, it follows that $\Delta_{ord}(\mathcal{L}^+)$ is shellable (Proposition 4.7.19), and hence is a PL $(r-1)$-ball (Proposition 4.7.22).

(ii) Let $\Delta = \Delta_{ord}(\mathcal{L}^+)$. We claim that the boundary $\partial\Delta$ of the $(r-1)$-ball Δ is the subcomplex Δ_B induced on the vertex set $B = \mathcal{L}^+\backslash\mathcal{L}^{++}$. Since $\partial\Delta$ is generated by the $(r-2)$-simplices that face only one $(r-1)$-simplex, we must check that a general simplex $\sigma : X_0 < X_1 < \ldots < X_k$ in Δ can be extended to such an $(r-2)$-simplex if and only if $\sigma \in \Delta_B$. This follows from the easy observation that $\mathcal{L}^+\backslash\mathcal{L}^{++}$ as a filter in $\mathcal{L}^+$ is generated by its members of poset rank 1 (in $\mathcal{L}^+$). Geometrically this last statement means that every cell in the positive hemisphere which touches the horizon has an edge on its boundary that also touches the horizon.

By Lemma 4.7.27, $\|\Delta_{ord}(\mathcal{L}^{++})\|$ is a strong deformation retract of $\|\Delta_{ord}(\mathcal{L}^+)\|\backslash\|\Delta_B\|$, which as we have seen is an open ball. Hence, the regular cell complex $\Delta(\mathcal{L}^{++})$ is contractible. $\square$

As a poset with the opposite ordering, $(\mathcal{L}^+)^{op}$ is an order ideal in $\mathcal{L}^{op}$, and therefore $\mathcal{L}^+$ can be interpreted as a face poset "upside down" of a certain subcomplex of the PL sphere $\Delta^{op}(\mathcal{L})$ opposite to $\Delta(\mathcal{L})$, see Corollary 4.3.4.

4.5.8 Corollary. $(\mathcal{L}^+)^{op}$ *is the face poset of a PL regular cell decomposition of the $(r-1)$-ball.*

4.5.9 Corollary.   *Let $\rho(X)$ denote the poset rank function in $\mathcal{L}$. Then*

(i) $\sum\limits_{X \in \mathcal{L}^+} (-1)^{r-\rho(X)} = 1.$

(ii) $\sum\limits_{X \in \mathcal{L}^{++}} (-1)^{\rho(X)-1} = 1.$

Proof. These are the Euler characteristics.   □

Some generalizations of the results in this section are indicated in Exercises 4.24–4.27.

4.6 Enumeration of cells

Let Δ be a d-dimensional regular cell complex, and for $0 \le k \le d$ let f_k be the number of k-dimensional cells in Δ. The sequence

$$f(\Delta) = (f_0, f_1, \ldots, f_d)$$

is called the *f-vector* of Δ.

More generally, for $J = \{j_1, j_2, \ldots, j_k\}$, $0 \le j_1 < j_2 < \ldots < j_k \le d$, let f_J be the number of chains $\sigma_1 \subset \sigma_2 \subset \ldots \subset \sigma_k$ in the face poset $\mathcal{F}(\Delta)$ such that $\dim(\sigma_i) = j_i$ for $1 \le i \le k$. The sequence (or array) $\tilde{f}(\Delta) = (f_J)_{J \subseteq \{0,\ldots,d\}}$ is called the *flag vector* (or, *fine f-vector*) of Δ.

This section is devoted to some basic properties of f-vectors and flag vectors of oriented matroid complexes. Let $\mathcal{L} \subseteq \{+,-,0\}^E$ be the set of covectors of a simple oriented matroid of rank r, and let $\Delta(\mathcal{L})$ be the corresponding regular cell complex (Theorem 4.3.3). Then

$$f_k(\Delta(\mathcal{L})) = \mathrm{card}\{X \in \mathcal{L} : \rho(X) = k+1\},$$

for $0 \le k \le r-1$, and the f-vectors $f(\mathcal{L}) := f(\Delta(\mathcal{L}))$ will be the main object of study. The f-vector $f(\mathcal{L}^{++})$ of the bounded complex of an affine oriented matroid will also be considered. Notice that $f_{r-1}(\mathcal{L})$ is the number of topes and $f_{r-1}(\mathcal{L}^{++})$ is the number of "bounded" topes.

The questions treated here have obvious relevance for the problem of counting the number of regions and k-cells induced by a central or affine hyperplane arrangement in $\mathbb{R}^r$, see Corollary 4.6.8 below.

We will assume familiarity with the notions of *Möbius function* $\mu_L(A, B)$, *characteristic polynomial* $p(L; \lambda)$ and *beta invariant* $\beta(L)$ of a geometric lattice L, see Chapter 7 of White (1987) for an expository treatment. Also, recall from Proposition 4.1.13 that $L(\mathcal{L}) = \{z(X) : X \in \mathcal{L}\}$ is the geometric lattice of flats of the underlying matroid of $\mathcal{L}$. It is our convention to let $\rho(X)$ denote the poset rank function for $X \in \mathcal{L}$ and $r(A)$ the matroid rank function for $A \subseteq E$. The two notions of rank are related as in Corollary 4.1.15.

The following result was independently discovered by Las Vergnas (1975a, 1980a) and Zaslavsky (1975a,b, 1977). For the case of hyperplane arrangements it had earlier been obtained in an equivalent form by Winder (1966).

4.6.1 Theorem (Las Vergnas–Zaslavsky). *Let $(E, \mathcal{L})$ be a simple oriented matroid of rank r, and let L be the geometric lattice of its underlying matroid. Then the numbers of topes of $\mathcal{L}$ equals*

$$(-1)^r p(L; -1) = \sum_{A \in L} |\mu_L(\emptyset, A)|.$$

It follows that the number of regions induced by a central hyperplane arrangement depends only on the underlying matroid structure and not on the combinatorial type of the induced cell decomposition (which is not determined by L, cf. Exercise 5.5). It will soon appear that the same is true for all entries of the flag vector (see Corollary 4.6.3 below).

Theorem 4.6.1 is usually proved with an inductive deletion-contraction argument (the Tutte-Grothendieck method), see Zaslavsky (1975a) and Las Vergnas (1980a). Zaslavsky (1975a, 1977) also gave a "topological" proof, which via Möbius inversion on a geometric lattice leads to evaluating an Euler characteristic. We will take this second approach, also used by Edelman (1984a).

Proof. For $A \subseteq E$, let $\alpha(A)$ denote the number of topes of $\mathcal{L} \backslash (E \backslash A)$, the restriction of $\mathcal{L}$ to A (cf. Lemma 4.1.8). Put also $\alpha(\emptyset) = 1$. Corollary 4.3.10 shows that

$$\sum_{A \subseteq B} (-1)^{|A|} \alpha(A) = (-1)^{r(B)},$$

for all $B \subseteq E$. A Möbius inversion gives

$$\sum_{A \subseteq B} (-1)^{|B| - |A|} (-1)^{r(A)} = (-1)^{|B|} \alpha(B),$$

for all $B \subseteq E$, and in particular

$$\alpha(E) = \sum_{A \subseteq E} (-1)^{|A| - r(A)} = (-1)^r p(L; -1).$$

The last equality follows from the Boolean expansion

$$p(L; \lambda) = \sum_{A \subseteq E} (-1)^{|A|} \lambda^{r - r(A)},$$

see White (1987, Proposition 7.2.1).   □

Theorem 4.6.1 is equivalent to the $A_1 = \emptyset, k = 1$ case of the following generalization. The hypotheses are the same.

4.6.2 Proposition. *Let $A_1 \subset A_2 \subset \ldots \subset A_k$ be a chain of flats in L. Put $A_{k+1} = E$. Then*

$$|z^{-1}(A_1, A_2, \ldots, A_k)| = \prod_{i=1}^{k} \sum_{\substack{A_i \leq B \leq A_{i+1} \\ B \in L}} |\mu_L(A_i, B)|.$$

Proof. Theorem 4.6.1 makes it possible to determine the size of inverse images of singletons under the surjective zero set map $z : \mathcal{L} \longrightarrow L$. Namely, for $A \in L$

the set $z^{-1}(A)$ consists of the topes of the contraction $\mathcal{L}/A$, and since $L(\mathcal{L}/A)$ equals the interval $[A, E]$ in L their number is $\sum_{A \leq B \leq E} |\mu_L(A, B)|$.

We can now count the chains $X_1 > X_2 > \ldots > X_k$ in $\mathcal{L}$ such that $z(X_i) = A_i$ for $1 \leq i \leq k$. First, choose X_k. This can be done in $\sum_{A_k \leq B \leq E} |\mu_L(A_k, B)|$ ways. Then, choose X_{k-1}. This means choosing a covector in

$$[X_k, \hat{1}] \cong \mathcal{F}_{big}(\mathcal{L} \backslash (E \backslash A_k))$$

with zero set A_{k-1}, which can be done in $\sum_{A_{k-1} \leq B \leq A_k} |\mu_L(A_{k-1}, B)|$ ways. Here we have used Proposition 4.1.9 and the fact that the interval $[\emptyset, A_k]$ in L is the geometric lattice of the restriction $\mathcal{L} \backslash (E \backslash A_k)$. Having chosen X_k and X_{k-1} we next choose X_{k-2}, which by the same reasoning can be done in $\sum_{A_{k-2} \leq B \leq A_{k-1}} |\mu_L(A_{k-2}, B)|$ ways, and so on. $\square$

4.6.3 Corollary (Bayer and Sturmfels 1990). *The flag vector of an oriented matroid depends only on the underlying matroid.*

Proof. If $0 \leq j_1 < j_2 < \ldots < j_k \leq r - 1$, then

$$f_{\{j_1, j_2, \ldots, j_k\}}(\mathcal{L}) = \sum |z^{-1}(A_1, A_2, \ldots, A_k)|,$$

the sum extending over all chains $A_1 \supset A_2 \supset \ldots \supset A_k$ in L such that $r(A_i) = r - j_i - 1, 1 \leq i \leq k$. $\square$

Using Proposition 4.6.2, Billera, Ehrenborg and Readdy (1997) expressed the flag vector of an oriented matroid explicitly in terms of the flag vector of the geometric lattice L of its underlying matroid.

4.6.4 Corollary (Zaslavsky 1975a). *For $1 \leq k \leq r$:*

$$f_{k-1}(\mathcal{L}) = \text{card}\{X \in \mathcal{L} : \rho(X) = k\} = \sum |\mu_L(A, B)|,$$

the sum extending over all $A, B \in L$ such that $A \leq B$ and $r(A) = r - k$.

We now turn to the f-vector of the bounded complex $\mathcal{L}^{++}$ of an affine oriented matroid $(E, \mathcal{L}, g)$, see Definition 4.5.1. The following result is due to Zaslavsky (1975a). He at first considered only the realizable case; the general case appeared in Greene and Zaslavsky (1983) (announced in Greene (1977)) and Las Vergnas (1977).

4.6.5 Theorem (Zaslavsky). *The number of bounded topes of an affine oriented matroid $(E, \mathcal{L}, g)$ equals the beta invariant of the underlying geometric lattice, $\beta(L(\mathcal{L}))$.*

Again we find that the number in question depends only on the underlying matroid. Surprisingly, it does not depend on the choice of $g \in E$, either.

Proof. Let $L^g = \{A \in L(\mathcal{L}) : g \notin A\}$, and for each $A \in L^g$ let $\gamma(A)$ be the number of $\mathcal{L}/A$-topes in $\mathcal{L}^{++}$. Equivalently,

$$\gamma(A) = \text{card}\{X \in \mathcal{L}^{++} : z(X) = A\}.$$

Then for all $A \in L^g$:

$$\sum_{\substack{B \in L^g \\ A \subseteq B}} (-1)^{r-r(B)-1} \gamma(B) = \sum_{X \in (\mathcal{L}/A)^{++}} (-1)^{\rho(X)-1} = 1,$$

where the second equality comes from Corollary 4.5.9. Möbius inversion then gives

$$\sum_{\substack{B \in L^g \\ A \subseteq B}} \mu_L(A, B) = (-1)^{r - r(A) - 1} \gamma(A),$$

for all $A \in L^g$, and in particular

$$\gamma(\emptyset) = (-1)^{r-1} \sum_{B \in L^g} \mu_L(\emptyset, B) = \beta(L).$$

For the last equality see White (1987, Proposition 7.3.1).   □

We define a *pointed matroid* (E, M, g) to be a matroid (E, M) together with a choice of base point $g \in E$. Two pointed matroids are isomorphic if some isomorphism of the unpointed matroids sends one base point to the other. The underlying matroid of an affine oriented matroid $(E, \mathcal{L}, g)$ is naturally pointed. For $L = L(\mathcal{L})$, let $L^g = \{A \in L : g \notin A\}$.

4.6.6 Corollary. *For $1 \le k \le r$:*

$$f_{k-1}\,(\mathcal{L}^{++}) = \operatorname{card}\{X \in \mathcal{L}^{++} : \rho(X) = k\} = (-1)^{k-1} \sum \mu_L(A, B),$$

the sum extending over all $A, B \in L^g$ such that $A \le B$ and $r(A) = r - k$.

Consequently, the f-vector $f(\mathcal{L}^{++})$ depends only on the underlying pointed matroid.

Proof. In the proof of Theorem 4.6.5 it was shown that

$$\gamma(A) = (-1)^{r - r(A) - 1} \sum_{B \in L^g} \mu_L(A, B).$$

The result follows by summing over all $A \in L^g$ such that $r(A) = r - k$.   □

It is surprising, in view of Corollary 4.6.6, that the flag vector $\tilde{f}(\mathcal{L}^{++})$ is *not* an invariant of the underlying pointed matroid. Also, for fixed oriented matroid $(E, \mathcal{L})$ the f-vector $f(\mathcal{L}^{++})$ depends on the choice of $g \in E$.

4.6.7 Example. Let $\mathcal{A}'$ be the arrangement of 5 planes in $\mathbb{R}^3$ discussed in Example 4.5.2. The bounded complex $\mathcal{L}_5^{++}$ (i.e., for the choice $g = 5$) is shown in Figure 4.5.2, and the complexes $\mathcal{L}_3^{++}$ and $\mathcal{L}_1^{++}$ appear in Figure 4.6.1.

For $g = 5$ and $g = 3$ there is the same underlying pointed matroid. We find that $f(\mathcal{L}_5^{++}) = f(\mathcal{L}_3^{++}) = (5, 6, 2)$, but $\tilde{f}(\mathcal{L}_5^{++}) \neq \tilde{f}(\mathcal{L}_3^{++})$ since $f_{\{0,2\}}\,(\mathcal{L}_5^{++}) = 6$ but $f_{\{0,2\}}\,(\mathcal{L}_3^{++}) = 7$.

For $g = 5$ and $g = 1$ there is the same underlying matroid but different pointed matroids. The f-vectors are different, since $f(\mathcal{L}_1^{++}) = (4, 5, 2)$.

The preceding results are directly applicable to the problem of counting the number of regions determined by an arrangement of affine hyperplanes (translates of $(d-1)$-dimensional subspaces) in $\mathbb{R}^d$. Suppose that $\mathcal{A} = (H_e)_{e \in E}$ is an essential affine hyperplane arrangement (Definition 2.1.1). Extend $\mathcal{A}$ to a central arrangement $\mathcal{A}'$ in $\mathbb{R}^{d+1}$ by placing a copy of $\mathcal{A}$ on the d-plane $x_{d+1} = 1$ in $\mathbb{R}^{d+1}$

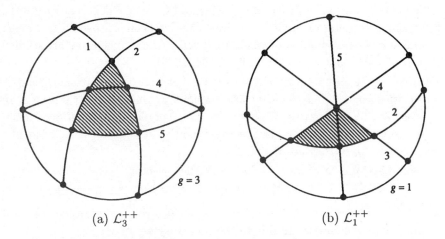

Figure 4.6.1: The bounded complex for different choices of horizon g.

and putting into $\mathcal{A}'$ all hyperplanes $H'_e = \text{span}\,(H_e \cup \{0\})$ for $e \in E$ and also the hyperplane $x_{d+1} = 0$ (Figure 4.6.2). Let M be the ordinary matroid of linear dependences of the hyperplanes in $\mathcal{A}'$. The geometric picture together with the preceding theorems immediately gives the following.

Corollary 4.6.8 (Zaslavsky 1975a). *The total number of regions determined by $\mathcal{A}$ is $\frac{1}{2}(-1)^{d+1}p(M;-1)$. The number of bounded regions is $\beta(M)$.*

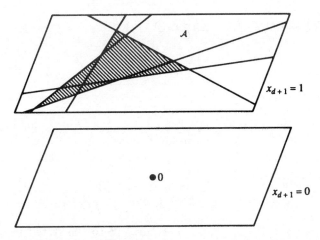

Figure 4.6.2: Counting the number of regions.

There are many variants and refinements to the formulas in Corollary 4.6.8 for counting regions, and cells of arbitrary dimensions, in partitionings of space by affine hyperplanes. See Zaslavsky (1975a) and Greene and Zaslavsky (1983), and for some particular examples Exercise 4.35. Extremal arrangements, having the smallest or largest number of k-cells for given size and dimension, are studied in Shannon (1976) and Zaslavsky (1981, 1983, 1985).

Theorems 4.6.1 and 4.6.5 have been further generalized and strengthened to pairs of oriented matroids on the same ground set related by a strong map (Definition 7.7.2) in a series of papers by Las Vergnas (1977, 1978c, 1980b, 1984b), see Exercise 4.41. The last of these papers contains a survey of results of this type and how they are inter-related.

We will mention a few more results about the f-vector

$$f(\mathcal{L}) = (f_0, f_1, \ldots, f_{r-1})$$

of a rank r oriented matroid $\mathcal{L}$, with no attempt at a comprehensive survey. All numbers f_k are of course even, due to the central symmetry.

The following interesting result has been proved in varying degrees of generality by Schneider (1987), Varchenko (1989), and Fukuda, Saito, Tamura and Tokuyama (1991). The proof is by induction on size.

4.6.9 Proposition.   For $0 \leq j < k \leq r - 1$, $(j,k) \neq (0,1)$, we have:

$$f_{\{j,k\}} < 2^{k-j} \binom{k}{j} f_k.$$

Equivalently, the average number of j-faces of a k-cell is less than the number of j-faces of the k-cube. Furthermore, $f_{\{0,1\}} = 2f_1$.

It follows, for instance, that the average number of edges of a 2-cell is less than 4. This strengthens the fact that an oriented matroid complex must always have a triangular 2-cell, which also follows from Theorem 6.5.2 via contraction to rank 3.

4.6.10 Corollary (Varchenko 1989).   For $0 \leq j < k \leq r-1$, $(j,k) \neq (0,1)$, we have:

$$\binom{r-1}{k} f_j < \binom{r-1}{j} f_k.$$

Furthermore, $(r-1)f_0 \leq f_1$.

Proof. Suppose that $X \in \mathcal{L}$, $\rho(X) = j+1$. Then $[X, \hat{1}]$ is the big face lattice of a rank $r - j - 1$ oriented matroid (Proposition 4.1.9), which has at least $2^{k-j} \binom{r-j-1}{k-j}$ covectors of rank $k - j$ (Exercise 4.37). This means that every j-cell is below at least that many k-cells, i.e.,

$$f_{\{j,k\}} \geq 2^{k-j} \binom{r-1-j}{k-j} f_j.$$

Thus, $\binom{r-1-j}{k-j} f_j < \binom{k}{j} f_k$, which is equivalent to the stated inequality. The $(j,k) = (0,1)$ case is elementary. $\square$

The following special cases of these inequalities were observed by Fukuda, Saito, Tamura and Tokuyama (1991):

(i) $f_k < f_{r-1-k}$,   for $0 \le k \le \left[\frac{r-2}{2}\right]$.

(ii) $f_k < \binom{r-1}{k} f_{r-1}$,   for $0 \le k \le r-2$.

(iii) $f_k < \frac{k+1}{r-k-1} f_{k+1}$,   for $1 \le k \le r-2$.

From the fact that the oriented matroid complex is a cell decomposition of the $(r-1)$-sphere, we deduce the *Euler-Poincaré formula*:

4.6.11 Corollary. $f_0 - f_1 + f_2 - \ldots + (-1)^{r-1} f_{r-1} = 1 + (-1)^{r-1}$.

This can also be proved with Möbius function techniques starting from the expressions given in Corollary 4.6.4, see Cordovil, Las Vergnas and Mandel (1982).

Some other results on the same level of generality as the Euler-Poincaré formula can be cited. For this we need the following arithmetic function. For positive integers n and k there exists a unique expansion

$$n = \binom{a_k}{k} + \binom{a_{k-1}}{k-1} + \ldots + \binom{a_i}{i},$$

such that $a_k > a_{k-1} > \ldots > a_i \ge i \ge 1$. Then define

$$\partial_{k-1}(n) = \binom{a_k}{k-1} + \binom{a_{k-1}}{k-2} + \ldots + \binom{a_i}{i-1}.$$

Also, let $\partial_{k-1}(0) = 0$. For instance, $39 = \binom{7}{3} + \binom{3}{2} + \binom{1}{1}$, so $\partial_2(39) = 25$.

4.6.12 Proposition.   *Let* $f(\mathcal{L}) = (f_0, f_1, \ldots, f_{r-1})$ *be the f-vector of an oriented matroid of rank* r, *and put* $f_r = 1$. *Then,*

(1) (Wegner 1984) $\partial_k(f_k) \le f_{k-1}, \text{ for } 1 \le k \le r-1$,

(2) (Björner and Kalai 1991)

$$\partial_k(f_{k+1} - f_{k+2} + f_{k+3} - \ldots) \le f_k - f_{k+1} + f_{k+2} - \ldots,$$

for $1 \le k \le r-2$.

Part (1), the so-called *Kruskal-Katona relations*, were shown by Wegner to hold for *any* regular cell complex with the intersection property (Definition 4.7.7). Similarly, part (2) holds for any such complex whose space has the homology of a sphere. Actually, part (2) implies part (1), but the implication is not very straightforward. Since oriented matroids have so much more combinatorial and topological structure these relations are quite weak, but nevertheless non-trivial in the present context.

It would be very interesting to have a characterization of the set of vectors $(n, f_0, f_1, \ldots, f_{r-1})$ of rank r oriented matroids with n elements. Such a characterization seems to be outside reach even in the rank 3 case, where however some interesting information exists about pairs (n, f_i), see Grünbaum (1972).

In closing let us mention that the f-vectors of Las Vergnas and Edmonds-Mandel face lattices (matroid polytopes) are an interesting area of study. The distinctions between such lattices and face lattices of convex polytopes are subtle, and it is not yet clear whether the induced classes of f-vectors differ or not. See Section 9.3 for further discussion of these topics.

There is an enormous literature on the f-vectors of convex polytopes (see e.g.

Bayer and Lee, 1992, or Klee and Kleinschmidt, 1995), and some results have been non-trivially generalized to matroid polytopes. Noteworthy is the Upper Bound Theorem for topes, due to Edmonds and Mandel (1978, 1982).

4.7 Appendix: Regular cell complexes, posets and shellability

The combinatorial method in topology consists in decomposing a topological space into cells governed by some combinatorial data, and then deducing topological information about the space from the combinatorial encoding. In this chapter we are faced with the opposite situation: given a combinatorial encoding (the covectors of an oriented matroid) we want to synthetically construct a topological space with a corresponding decomposition. The step from combinatorics to topology is taken (in Section 4.3) via a "cellular interpretation theorem" for posets (Theorem 4.7.24). It is the purpose of this section to review the background material needed for a proper understanding of this tool. Along the way we will also review some general topological material needed elsewhere in the book.

(a) Generalities

We will assume familiarity with basic topological concepts such as *homeomorphisms* of spaces (denoted $T \cong T'$), abstract and geometric *simplicial complexes*, the *geometric realization* $\|\Delta\|$ of an abstract simplicial complex Δ, etc. As general references we suggest the books by Munkres (1984) and Spanier (1966). Topological methods in combinatorics are surveyed, together with a condensed review of basic material, in Björner (1995).

With any poset (partially ordered set) P we associate its *order complex* $\Delta_{ord}(P)$. This is a simplicial complex whose vertices are the elements of P and whose simplices are the chains $x_0 < x_1 < \cdots < x_k$ in P. The geometric realization $\|\Delta_{ord}(P)\|$ will at times be denoted by $\|P\|$, or even just by P if no confusion can arise.

The following simple rule for gluing homeomorphisms will be of use.

4.7.1 Lemma. *Let $T = T_1 \cup T_2$ and $U = U_1 \cup U_2$ be topological spaces with T_1, T_2, U_1 and U_2 closed as subsets of T and U. Let $h_1 : T_1 \to U_1$ and $h_2 : T_2 \to U_2$ be homeomorphisms that restrict to the same homeomorphism $T_1 \cap T_2 \to U_1 \cap U_2$. Then the function*

$$h_1 \cup h_2 : T \to U$$

defined by $(h_1 \cup h_2)(x) = h_i(x)$ for all $x \in T_i$, $i = 1, 2$, is a homeomorphism.

The two most important spaces here will be the *standard d-ball*

$$B^d = \{x \in \mathbb{R}^d : \|x\| \leq 1\}$$

and its boundary, the *standard $(d-1)$-sphere*

$$S^{d-1} = \partial B^d = \{x \in \mathbb{R}^d : \|x\| = 1\},$$

for $d \geq 0$. Note that $S^{-1} = \emptyset$, $B^0 = \{\text{point}\}$ and $S^0 = \{\text{two points}\}$. When saying that a space T "is a sphere", we mean that T is homeomorphic to S^d for some d, and similarly for balls. Boundaries of convex d-polytopes are important examples of $(d-1)$-spheres. Prefixes of type "d-..." will always denote dimension.

4.7.2 Lemma. *Suppose $h : \partial B_1 \rightarrow \partial B_2$ is a homeomorphism of the boundaries of two d-balls B_1 and B_2. Then h can be extended to a homeomorphism $\widehat{h} : B_1 \rightarrow B_2$ of the entire balls.*

Proof. Via some choice of homeomorphisms we may assume that $B_1 = B_2 = B^d$. Then use radial extension, i.e., define

$$\widehat{h}(x) = \begin{cases} \|x\| \cdot h\left(\frac{x}{\|x\|}\right), & \text{if } x \in B^d \backslash \{0\}, \\ 0, & \text{if } x = 0. \end{cases}$$

$\square$

The d-ball can be embedded as a proper subspace of itself, but this is not true for the d-sphere.

4.7.3 Lemma. *The d-sphere is not homeomorphic to any proper subspace of itself.*

This follows from the Borsuk-Ulam theorem, or from basic homology theory (e.g. Alexander duality).

(b) Regular cell complexes

The cell decompositions with which we will be concerned are of the following kind. By a *ball* in a topological space T we mean a subspace $\sigma \subseteq T$ which is homeomorphic to the standard d-ball, for some d. The (relative) interior $\overset{\circ}{\sigma}$ and boundary $\partial \sigma = \sigma \backslash \overset{\circ}{\sigma}$ of a ball σ have the established meanings. If $\dim \sigma = 0$, then $\overset{\circ}{\sigma} = \sigma = \{\text{point }\}$.

4.7.4 Definition.

(1)   A *regular cell complex* Δ is a finite collection of balls σ in a Hausdorff space $\|\Delta\| = \bigcup_{\sigma \in \Delta} \sigma$ such that
      (i) the interiors $\overset{\circ}{\sigma}$ partition $\|\Delta\|$, (i.e., every $x \in \|\Delta\|$ lies in exactly one $\overset{\circ}{\sigma}$), and
      (ii) the boundary $\partial \sigma$ is a union of some members of Δ, for all $\sigma \in \Delta$.

(2)   The balls $\sigma \in \Delta$ are called the *closed cells* of Δ, their interiors $\overset{\circ}{\sigma}$ are the *open cells*. The space $\|\Delta\|$ is called the *underlying space* of Δ.

(3)   If $T \cong \|\Delta\|$, then Δ is said to provide (via the homeomorphism) a *regular cell decomposition* of the space T.

(4)   The *face poset* $\mathcal{F}(\Delta) = (\Delta, \leq)$ is the set of closed cells ordered by containment. The *augmented face poset* $\widehat{\mathcal{F}}(\Delta) = \mathcal{F}(\Delta) \cup \{\hat{0}, \hat{1}\}$ is the face poset enlarged by new elements such that $\hat{0} < \sigma < \hat{1}$ for all $\sigma \in \Delta$.

Regular cell complexes have been thoroughly studied in topology, where they are called "regular finite CW complexes". See Cooke and Finney (1967) and Lundell and Weingram (1969). Edmonds and Mandel (1982) call them simply "ball complexes".

Actually, the usual definition of a regular CW complex in the topological literature appears to be weaker in that condition (ii) is replaced by "the boundary $\partial\sigma$ is contained in a union of open cells of dimension less than $\dim \sigma$, for all $\sigma \in \Delta$ ". The equivalence of the two definitions follows from the *normality* of regular CW complexes: If $\sigma, \tau \in \Delta, \dim(\tau) < \dim(\sigma)$ and $\overset{\circ}{\tau} \cap \sigma \neq \emptyset$, then $\tau \subseteq \partial\sigma$. Proofs for this non-trivial fact appear in both of the cited books. Since we will not consider non-regular CW complexes we take the seemingly stronger Definition 4.7.4 (1) as our point of departure.

Note that the normality of regular cell complexes implies that every intersection of closed cells, if non-empty, is a union of closed cells (a subcomplex).

4.7.5 Example. A *polyhedral complex* Γ is a finite collection of convex polytopes in $\mathbb{R}^d$ such that (i) if $\pi \in \Gamma$ and σ is a face of π then $\sigma \in \Gamma$, and (ii) if $\pi, \sigma \in \Gamma$ have non-empty intersection then $\pi \cap \sigma$ is a face both of π and of σ. A polyhedral complex is clearly a regular cell complex, whose closed cells are the participating polytopes, and whose underlying space is the union of these polytopes (a subspace of $\mathbb{R}^d$).

A *geometric simplicial complex* is a polyhedral complex Γ such that every $\pi \in \Gamma$ is a *simplex*, that is, the convex hull of an affinely independent set of points.

4.7.6 Example. To illustrate the generality of regular cell complexes, as compared to polyhedral complexes, consider the regular cell decomposition of the 2-sphere shown in Figure 4.7.1 (a). It has two vertices: α and β, two edges: γ and δ, and two 2-cells: ϵ and η. The face poset is shown in Figure 4.7.1 (b).

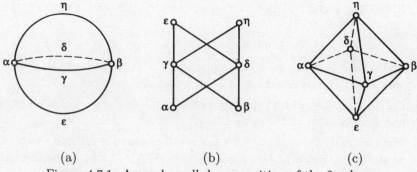

(a)   (b)   (c)

Figure 4.7.1: A regular cell decomposition of the 2-sphere.

The dimension of a cell is well-defined by the ball to which it is homeomorphic, so we may speak about k-cells (k-dimensional cells). Here is some more terminology.

4.7.7 Definition. Let Δ be a regular cell complex.

(1)   The 0-cells and 1-cells are called *vertices* and *edges*, respectively.

(2)   If $\sigma, \tau \in \Delta$ and $\sigma \subseteq \tau$ then σ is said to be a *face* of τ.

(3) $\Gamma \subseteq \Delta$ is a *subcomplex* of Δ if $\tau \in \Gamma$ implies that every face of τ also belongs to Γ. (Equivalently: if $\mathcal{F}(\Gamma)$ is an order ideal in $\mathcal{F}(\Delta)$.)

(4) $\dim \Delta = \max_{\sigma \in \Delta} \dim \sigma$.

(5) Δ is *pure* if all maximal cells have the same dimension (i.e., every cell is contained in a $(\dim \Delta)$-dimensional cell).

(6)   The k-*skeleton* of Δ is the subcomplex $\mathrm{skel}_k(\Delta) = \{\sigma \in \Delta : \dim \sigma \le k\}$.

(7) Δ has the *intersection property* if whenever $\sigma, \tau \in \Delta$ have non-empty intersection then $\sigma \cap \tau \in \Delta$.

The 1-skeleton of a regular cell complex is a loop-free graph. Conversely, every loop-free graph (parallel edges allowed) is a 1-dimensional regular cell complex.

Graphs with parallel edges, and also Figure 4.7.1 (a), give examples of regular cell complexes not having the intersection property. The class of regular cell complexes with the intersection property, which strictly contains the class of polyhedral complexes, is of particular interest combinatorially. See Proposition 4.7.13 below.

The following result, showing that a regular cell complex is homeomorphic to the order complex of its face poset, is of crucial importance. This property fails for more general types of cell complexes (e.g., for non-regular CW complexes) and is the main reason why the class of regular complexes is tractable from a combinatorial point of view.

4.7.8 Proposition.   *Let Δ be a regular cell complex.   Then*
$$\|\Delta\| \cong \|\Delta_{ord}(\mathcal{F}(\Delta))\|.$$
Furthermore, this homeomorphism can be chosen so that it restricts to a homeomorphism between σ and $\|\Delta_{ord}(\mathcal{F}_{\le\sigma})\|$, where $\mathcal{F}_{\le\sigma}$ is the subposet $\{\tau \in \mathcal{F}(\Delta) : \tau \le \sigma\}$, for all $\sigma \in \Delta$.

Proof. We will use induction on the number of cells in Δ, the result being obviously true if $\dim (\Delta) = 0$.

Let σ be a maximal cell in Δ of dimension $d \ge 1$, and consider the subcomplexes $\Delta' = \Delta \backslash \{\sigma\}$ and $\Delta'' = \{\tau \in \Delta : \tau < \sigma\}$. Then $\|\Delta''\| = \mathring{\sigma} \cong S^{d-1}$ and $\mathcal{F}(\Delta'') = \mathcal{F}_{<\sigma} = \mathcal{F}_{\le\sigma} \backslash \{\sigma\}$. By the induction assumption there exists a compatible homeomorphism $h' : \|\Delta'\| \to \|\mathcal{F}(\Delta')\|$. Being compatible in the sense of the proposition implies that h' restricts to a homeomorphism $h'' : \|\Delta''\| \to \|\mathcal{F}_{<\sigma}\|$. In particular, $\|\mathcal{F}_{<\sigma}\| \cong S^{d-1}$, and it follows that $\|\mathcal{F}_{\le\sigma}\| \cong B^d$, since $\Delta_{ord}(\mathcal{F}_{\le\sigma})$ is a cone with base $\Delta_{ord}(\mathcal{F}_{<\sigma})$.

Now choose a homeomorphism $h : \sigma \to \|\mathcal{F}_{\leq\sigma}\|$ that extends the homeomorphism h'' of their boundaries (Lemma 4.7.2). By Lemma 4.7.1 the homeomorphisms h and h' can be glued to give a compatible homeomorphism $h \cup h' : \|\Delta\| \to \|\mathcal{F}(\Delta)\|$. $\qquad\square$

We have seen that the space of a regular cell complex is completely determined (up to homeomorphism) by its face poset. Hence, it is permissible to think of regular cell complexes as a class of finite posets. Part (c) of Figure 4.7.1 shows the order complex $\Delta_{ord}(\mathcal{F}(\Delta))$ of the regular cell complex in part (a). A comparison of the two pictures illustrates Proposition 4.7.8.

4.7.9 Corollary.   *Let Δ and Δ' be two regular cell complexes, and assume there is a poset isomorphism $f : \mathcal{F}(\Delta) \to \mathcal{F}(\Delta')$. Then there exists a homeomorphism $h : \|\Delta\| \to \|\Delta'\|$ such that $h(\sigma) = f(\sigma)$ for all $\sigma \in \Delta$.*

Proof.   Choose compatible homeomorphisms, according to Proposition 4.7.8, and compose: $\|\Delta\| \cong \|\mathcal{F}(\Delta)\| \cong \|\mathcal{F}(\Delta')\| \cong \|\Delta'\|$. $\qquad\square$

Proposition 4.7.8 shows that a regular cell complex can be subdivided into a simplicial complex via one "barycentric subdivision". Since a d-dimensional simplicial complex has a rectilinear geometric realization in $\mathbb{R}^{2d+1}$, we draw the following important conclusion.

4.7.10 Corollary.   *Every d-dimensional regular cell complex can be embedded into $\mathbb{R}^{2d+1}$ so that its barycentric subdivision is a geometric simplicial complex.*

The following special case of Proposition 4.7.8 was also an ingredient in its proof.

4.7.11 Corollary.   *If $\mathcal{F} = \mathcal{F}(\Delta)$ is the face poset of a regular cell complex then $\|\mathcal{F}_{<\sigma}\|$ is homeomorphic to the $(\dim\sigma - 1)$-dimensional sphere, for all $\sigma \in \Delta$.*

Since any triangulation of the d-sphere is pure and every $(d-1)$-simplex of such a triangulation is contained in exactly two d-simplices (these are well-known properties of the triangulation of any closed manifold), the following combinatorial facts about the face poset $\mathcal{F} = \mathcal{F}(\Delta)$ follow from Corollary 4.7.11.

4.7.12 Corollary.   (i) *All maximal chains in the subposet $\mathcal{F}_{\leq\sigma}$ have the same length, equal to $\dim\sigma$, for all $\sigma \in \mathcal{F}$.*
(ii)   *Every closed interval of length 2 in $\mathcal{F} \cup \{\hat{0}\}$ has cardinality 4.*

A regular cell complex Δ has the intersection property if and only if its augmented face poset $\widehat{\mathcal{F}}(\Delta)$ is a lattice, as is easy to see. This leads to the useful fact that such complexes can be encoded as set systems. Let $\text{vert}(\Delta) = \{\text{vertices in } \Delta\}$, and for any cell $\sigma \in \Delta$ let

$$\text{vert}(\sigma) = \text{vert}(\Delta) \cap \sigma.$$

4.7.13 Proposition. *If Δ is a regular cell complex with the intersection property, and $\sigma, \tau \in \Delta$, then*

(i) $\mathrm{vert}(\sigma) = \mathrm{vert}(\tau)$ *if and only if $\sigma = \tau$,*

(ii) $\mathrm{vert}(\sigma) \subseteq \mathrm{vert}(\tau)$ *if and only if $\sigma \subseteq \tau$.*

Consequently, the face poset $\mathcal{F}(\Delta)$ is isomorphic to the set family $\{\mathrm{vert}(\sigma) : \sigma \in \Delta\}$ ordered by inclusion.

We conclude that in this situation the set family $\{\mathrm{vert}(\sigma) : \sigma \in \Delta\}$ gives a complete encoding of the complex Δ, and hence of the space $\|\Delta\|$ up to homeomorphism. Example 4.7.6 shows that such vertex-set encoding fails for regular cell complexes without the intersection property.

Proof of Proposition 4.7.13. Fix $\sigma \in \Delta$, and consider the interval $[\hat{0}, \sigma]$ in $\widehat{\mathcal{F}}(\Delta)$. Every subinterval of length 2 has cardinality 4 by Corollary 4.7.12. Hence, since $\widehat{\mathcal{F}}(\Delta)$ is a lattice it follows that $[\hat{0}, \sigma]$ is in fact an *atomic* lattice (see Exercise 4.4), meaning that every element is the join of some set of atoms. In particular, $\sigma = \bigvee \mathrm{vert}(\sigma)$. Everything follows at once from this fact. $\square$

(c) Shellability and recursive coatom orderings

For each cell σ in a regular cell complex Δ let $\delta\sigma$ denote the subcomplex consisting of all proper faces of σ. Thus, $\delta\sigma$ is a pure $(\dim\sigma - 1)$-dimensional complex giving a regular cell decomposition of the boundary of σ: $\|\delta\sigma\| = \partial\sigma$.

4.7.14 Definition. Let Δ be a pure d-dimensional regular cell complex. A linear ordering $\sigma_1, \sigma_2, \cdots, \sigma_t$ of its maximal cells is called a *shelling* if either $d = 0$, or if $d \geq 1$ and the following conditions are satisfied:

(i) $\delta\sigma_j \cap (\bigcup_{i=1}^{j-1}\delta\sigma_i)$ is pure and $(d-1)$-dimensional, for $2 \leq j \leq t$,

(ii) $\delta\sigma_j$ has a shelling in which the $(d-1)$-cells of $\delta\sigma_j \cap (\bigcup_{i=1}^{j-1}\delta\sigma_i)$ come first, for $2 \leq j \leq t$,

(iii) $\delta\sigma_1$ has a shelling.

A complex which admits a shelling is said to be *shellable*.

4.7.15 Example. The boundary complex of a convex polytope is shellable in this sense. This was shown by Bruggesser and Mani (1971) with the following geometric construction.

Suppose that $(F_e)_{e \in E}$ is the set of facets (maximal faces) of a convex d-polytope $\mathcal{P} \subseteq \mathbb{R}^d$. Let H_e be the affine hyperplane spanned by F_e. Now choose a line L that passes through the interior of $\mathcal{P}$ and such that the intersection points $x_e = L \cap H_e, e \in E$, are distinct. Relabel the indexing set E so that these intersection points come in order

$$x_1 < \cdots < x_k < y < x_{k+1} < \cdots < x_t$$

along the line L, for some interior point y in $\mathcal{P}$. Then

$$F_{k+1}, F_{k+2}, \cdots, F_t, F_1, F_2, \cdots, F_k$$

is a shelling of $\mathcal{P}'$ s boundary.

An intuitive way to see this argument is to think that we leave the surface of the "planet" $\mathcal{P}$ in a rocket that travels through space along the straight line L, passes through the point of infinity, and then returns to $\mathcal{P}$ along L from the opposite side. The order in which facets appear over the horizon as we leave $\mathcal{P}$, followed by the order in which they disappear (after passing through ∞), gives the shelling. Such shellings of convex polytopes are called *line shellings*, see Figure 4.7.2.

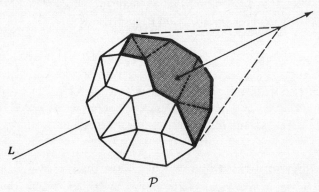

$\mathcal{P}$

Figure 4.7.2: Line shelling.

Proposition 4.3.1, due to Edmonds and Mandel, gives a generalization of the Bruggesser-Mani result to arbitrary topes of oriented matroids. The role of the line L is there played by a maximal chain in the tope poset. The geometric meaning of this for a convex polytope is that the line L need not be straight, there are many permissible curved trajectories through space that produce "line" shellings.

The Bruggesser-Mani result has the consequence that condition (iii) in Definition 4.7.14 can be deleted in the case of polyhedral complexes. Three weaker notions of shellability for polyhedral complexes are discussed in Danaraj and Klee (1974).

4.7.16 Example. In the case of a simplicial complex both conditions (ii) and (iii) in Definition 4.7.14 are superfluous, since every permutation of the facets of a simplex is a shelling (Exercise 4.48). The definition of a shelling therefore loses its recursive character; all that is required is that the j-th simplex intersects the union of the earlier ones in a pure $(d-1)$-dimensional subcomplex of its boundary. This can be reformulated in purely set-theoretic terms for abstract simplicial complexes as follows:

Let $\Delta \subseteq 2^E$ be a pure d-dimensional abstract simplicial complex. A linear ordering $F_1, F_2, \cdots, F_t$ of its maximal simplices (which are $(d+1)$-element subsets of E) is a shelling if for all $1 \le i < j \le t$ there exists $1 \le k < j$ and $z \in F_j$ such that $F_i \cap F_j \subseteq F_k \cap F_j = F_j \backslash z$.

Shellability of simplicial complexes has several applications in the theory of ordinary matroids. For this, see Chapter 7 of White (1992).

The following concept of recursive coatom orderings for posets is closely related to shellability. It was introduced in Björner and Wachs (1983), where it is shown that the existence of a recursive coatom ordering in a poset is equivalent to a special kind of shellability for its order complex, called "lexicographic shellability". Recall from the review of poset terminology in Section 4.1 the definition of "graded poset" and that the set of coatoms in such a poset P is denoted coat(P). We will here use the simplified notation $[x] := [\hat{0}, x]$ for lower intervals in a graded poset.

4.7.17 Definition. Let P be a graded poset. A linear ordering x_1, $x_2, \cdots, x_t$ of its coatoms is a *recursive coatom ordering* if either length $(P) \leq 2$, or if length $(P) > 2$ and for all $1 \leq j \leq t$ there is a distinguished subset $Q_j \subseteq \text{coat}[x_j]$ such that

(i) $[x_j] \cap (\bigcup_{i<j}[x_i]) = \bigcup_{y \in Q_j}[y]$, and

(ii) $[x_j]$ has a recursive coatom ordering in which the elements of Q_j come first.

The distinguished subsets Q_j are uniquely determined by the coatom ordering, since $Q_j = \{y \in \text{coat}[x_j] : y < x_i \text{ for some } i < j\}$ by (i). It follows that $Q_1 = \emptyset$ and (since $\hat{0} \in [x_j] \cap [x_1]$) that $Q_j \neq \emptyset$ for all $j > 1$.

Let $\widehat{\mathcal{F}}(\Delta)$ be the face poset of a pure regular cell complex Δ. Comparison of Definitions 4.7.14 and 4.7.17 shows that a shelling of the maximal cells of Δ is a recursive coatom ordering of $\widehat{\mathcal{F}}(\Delta)$, and vice versa. Hence, we get:

4.7.18 Lemma.   *A pure regular cell complex is shellable if and only if its face poset admits a recursive coatom ordering.*

In view of this lemma it could at this point seem unnecessary to introduce the two distinct concepts. Let us therefore say a few words of motivation. Whereas shellability is a property of *cell complexes* (with an interesting history in convex geometry and *PL* topology), recursive coatom orderings are applicable to very general classes of *abstract posets* that may have little in common with face posets of cell complexes. In the particular context of this chapter we need to work with an abstractly given poset (the covector lattice of an oriented matroid in Section 4.3 and of an affine oriented matroid in Section 4.5), about which we do not yet know whether it is the face lattice of a cell complex. In fact, in Section 4.3 *this is precisely what we are trying to establish*, whereas in Section 4.5 the poset at hand is not the face lattice of any cell complex. In this situation recursive coatom orderings provide a neutral combinatorial tool without geometric-topological overtones, and we think that by this conceptual separation of the combinatorics and the geometry greater clarity is achieved.

We shall say that a pure poset is *shellable* if its order complex is shellable. The following result is from Björner and Wachs (1983).

4.7.19 Proposition.   *Suppose that the graded poset P admits a recursive coatom ordering. Then every open interval (x, y) in P is shellable.*

Proof. Fix a recursive coatom ordering of P. Then for each coatom x_j fix a compatible recursive coatom ordering of $[x_j]$, according to condition (ii) of Definition 4.7.17. By successive recursive calls on this condition it is possible to choose a compatible recursive coatom ordering of $[y_k]$ for each unrefinable upper chain $y_k \lessdot y_{k-1} \lessdot \ldots \lessdot y_1 \lessdot \hat{1}$ in P.

We will first show that the order complex of the open interval $(\hat{0}, \hat{1})$ is shellable. Its maximal simplices are the maximal chains of $P \backslash \{\hat{0}, \hat{1}\}$, which we denote here by $(y_1, y_2, \ldots, y_r)$ as an abbreviation for

$$\hat{0} \lessdot y_r \lessdot y_{r-1} \lessdot \ldots \lessdot y_1 \lessdot \hat{1}.$$

Define a relation "$\prec$" on the set of maximal chains by

$$(y_1, y_2, \ldots, y_r) \prec (z_1, z_2, \ldots, z_r)$$

if $y_i = z_i$ for $1 \le i \le k$, $y_{k+1} \ne z_{k+1}$ and y_{k+1} precedes z_{k+1} in the compatible recursive coatom ordering of $[y_k]$ chosen for $(y_1, y_2, \ldots, y_k)$. This relation is antisymmetric and transitive, and it determines a linear ordering of the maximal chains. We claim that this is a shelling.

Suppose that $(y_1, y_2, \ldots, y_r) \prec (z_1, z_2, \ldots, z_r)$. More precisely, put $y_0 = z_0 = \hat{1}$ and $y_{r+1} = z_{r+1} = \hat{0}$, say that $y_i = z_i$ for $0 \le i \le k$, $y_{k+1} \ne z_{k+1}$, and let g be the least integer such that $g > k$ and $y_g = z_g$. If for some $k < j < g$ we can find an element z'_j such that $z_{j-1} > z'_j > z_{j+1}$ and z'_j precedes z_j in the recursive coatom ordering of $[z_{j-1}]$ chosen for $(z_1, z_2, \ldots, z_{j-1})$, then we are done. Because, letting $\mathbf{y} = (y_1, y_2, \ldots, y_r)$, $\mathbf{z} = (z_1, z_2, \ldots, z_r)$ and $\mathbf{z}' = (\mathbf{z} \backslash z_j) \cup z'_j$, we then have that $\mathbf{z}' \prec \mathbf{z}$ and $\mathbf{y} \cap \mathbf{z} \subseteq \mathbf{z}' \cap \mathbf{z} = \mathbf{z} \backslash z_j$, which agrees with the simplicial formulation of shellability in Example 4.7.16.

The problem has thus been reduced to finding an element z'_j as explained in the preceding paragraph. Let us for $q \ge p+2$ say that "z_{p+1} is first in $[z_q, z_p]$" as an abbreviation for the statement "in the compatible recursive coatom ordering of $[z_p]$ chosen for $(z_1, z_2, \ldots, z_p)$, restricted to $\mathrm{coat}[z_q, z_p]$, the earliest element is z_{p+1}".

Claim: If $k \le p \le g - 3$ and z_{p+1} is not first in $[z_g, z_p]$, then either z_{p+1} is not first in $[z_{p+2}, z_p]$ or z_{p+2} is not first in $[z_g, z_{p+1}]$.

Assuming the claim, we can quickly finish the proof: since z_{k+1} is not first in $[z_g, z_k]$ (y_{k+1} being earlier) it follows from the claim that z_j is not first in $[z_{j+1}, z_{j-1}]$ for some $k < j < g$, and then choosing z'_j to be earlier than z_j we are done.

It remains to prove the claim, see Figure 4.7.3. Suppose that z_{p+1} is preceded by u_{p+1} in $[z_g, z_p]$, and that z_{p+1} is first in $[z_{p+2}, z_p]$. Then since $z_g \in [z_{p+1}] \cap [u_{p+1}]$ condition (i) of Definition 4.7.17 shows that there exists an element u_{p+2} in $[z_g, z_p]$ which is covered by z_{p+1} and also by some element that precedes z_{p+1} in $\mathrm{coat}[z_g, z_p]$. Since z_{p+1} is first in $[z_{p+2}, z_p]$ we have that z_{p+2} is not covered by any element that precedes z_{p+1} in $\mathrm{coat}[z_g, z_p]$. Hence condition (ii) shows that

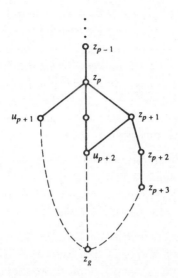

Figure 4.7.3: Sketch for the proof of Proposition 4.7.19.

u_{p+2} precedes z_{p+2} in $\mathrm{coat}[z_g, z_{p+1}]$, in particular z_{p+2} is not first, as was to be shown.

Shellability has now been established for the open interval $(\hat{0}, \hat{1})$. Since all lower intervals $[y]$ by definition have recursive coatom orderings the result immediately follows for all lower open intervals $(\hat{0}, y), y \in P \backslash \{\hat{0}\}$. Since this is all we will need for the purposes of this book, we end the proof at this point. The case of general intervals is easily dealt with along the same lines (see Exercise 4.49). □

(d) PL topology

Topological constructions with spheres and balls that intuitively "ought to" produce other spheres and balls often fail to do so. However, with the magic letters "*PL*" much of the trouble disappears. For this reason some basic *PL* (= piecewise linear) topology is important for us. Basic references are Zeeman (1963) and Hudson (1969).

Edmonds and Mandel (1982) developed *PL* topology in parallel to oriented matroid theory, with a concept of "constructibility" playing an important role in both connections. We will instead use the traditional approach to *PL* balls and spheres (as in Chapter 1 of Hudson 1969) together with shellability.

Let Γ and Δ be two geometric simplicial complexes (or regular cell complexes). Then Γ is a *subdivision* of Δ if $\|\Gamma\| = \|\Delta\|$ and every closed cell of Γ is a subset of some closed cell of Δ.

Let Δ be a geometric simplicial complex with underlying space $\|\Delta\|$. Each point $x \in \|\Delta\|$ can be uniquely expressed in terms of *barycentric coordinates*

ξ_i with respect to the vertices v_i. These are real numbers $\xi_i \geq 0$ such that $x = \sum \xi_i v_i$, $\sum \xi_i = 1$, and $\{v_i : \xi_i > 0\} \in \Delta$. A mapping $f : \|\Delta\| \longrightarrow \mathbb{R}^d$ is said to be *linear* if $f(\sum \xi_i v_i) = \sum \xi_i f(v_i)$ for each point $\sum \xi_i v_i \in \|\Delta\|$. The mapping f is *piecewise linear* if it is linear with respect to some simplicial subdivision of Δ. Two complexes Δ and Δ' are *PL homeomorphic* if there exists a piecewise linear map $\|\Delta\| \longrightarrow \|\Delta'\|$ which is also a homeomorphism. The simplest triangulations of the d-ball and the d-sphere are given by the d-simplex and the boundary of the $(d+1)$-simplex, respectively, and these complexes serve as the "standard" d-ball and d-sphere in PL theory.

4.7.20 Definition. A (geometric) simplicial complex Δ is a *PL d-ball* if it is PL homeomorphic to the d-simplex. It is a *PL d-sphere* if it is PL homeomorphic to the boundary of the $(d+1)$-simplex. (*Remark*: The definition can be extended to regular cell complexes, see Lemma 4.7.25.)

A triangulation of the topological d-sphere (or d-ball) may or may not be PL. Being PL is in fact a combinatorial property, that is, it depends only on the abstract simplicial complex Δ and not on its particular geometric realization. It is known that every triangulation of the d-sphere is PL for $d \leq 3$, but that non-PL triangulations exist for $d \geq 5$.

The following characterization is easily seen to be equivalent to Definition 4.7.20: A simplicial complex Δ is a PL d-ball (resp., d-sphere) if and only if it admits a simplicial subdivision combinatorially isomorphic to some subdivision of the d-simplex (resp., boundary of the $(d+1)$-simplex).

In the following theorem we gather the basic technical properties of PL balls and spheres that will be needed. Proofs and more precise formulations can be found in Chapter 1 of Hudson (1969). The *link* of a simplex σ in a simplicial complex Δ is by definition the subcomplex $lk_\Delta(\sigma) = \{\tau \in \Delta : \tau \cup \sigma \in \Delta, \tau \cap \sigma = \emptyset\}$.

4.7.21 Theorem.

(i)   *The union of two PL d-balls, whose intersection is a PL $(d-1)$-ball lying in the boundary of each, is a PL d-ball.*

(ii)   *The union of two PL d-balls, which intersect along their entire boundaries, is a PL d-sphere.*

(iii) *(Newman's Theorem) The closure of the complement of a PL d-ball embedded in a PL d-sphere is itself a PL d-ball.*

(iv)   *If Δ is a PL d-sphere and $\sigma \in \Delta$, then $lk_\Delta(\sigma)$ is a PL $(d - \dim \sigma - 1)$-sphere.*

(v)   *The cone over a PL d-sphere is a PL $(d+1)$-ball.*

Except for (ii) and (v), these statements with "PL" removed are false. Counterexamples to the non-PL versions of (i) and (iii) appear on p.69 of Rushing (1973), and it follows from R.D. Edwards' double-suspension theorem (see Daverman 1986) that S^5 has (necessarily non-PL) triangulations for which some links are not homeomorphic to spheres.

The following result appears implicitly in Bing (1964) and explicitly in Danaraj and Klee (1974).

4.7.22 Proposition. *Let Δ be a shellable d-dimensional simplicial complex.*

(i) *If every $(d-1)$-cell is a face of exactly two d-cells, then Δ is a PL d-sphere.*

(ii) *If every $(d-1)$-cell is a face of one or two d-cells, the first case occurring, then Δ is a PL d-ball.*

Proof. Let $F_1, F_2, \ldots, F_t$ be the d-simplices of Δ arranged in a shelling order, cf. Example 4.7.16, and let Δ_j be the subcomplex of Δ generated by $\{F_1, F_2, \ldots, F_j\}$, for $1 \le j \le t$. Being a shelling order means that $\Gamma_j := \delta F_j \cap \Delta_{j-1}$ is a pure $(d-1)$-dimensional subcomplex of the simplex boundary δF_j, for $2 \le j \le t$. In fact, Γ_j is a PL $(d-1)$-ball if $\Gamma_j \neq \delta F_j$, and Γ_j is a PL $(d-1)$-sphere if $\Gamma_j = \delta F_j$. These facts are easy to see directly (particularly the second one), or can otherwise be gotten by induction since Γ_j is shellable of lower dimension and satisfies the conditions of the proposition.

Assume now that every $(d-1)$-simplex of Δ is a face of at most two d-simplices. We will show that if Δ_{j-1} is a PL d-ball then either Δ_j is a PL d-ball, or Δ_j is a PL d-sphere and $j = t$. Since Δ_1 is a PL d-ball, the result readily follows from this.

Suppose that Δ_{j-1} is a PL d-ball. Its boundary is the $(d-1)$-subcomplex generated by all $(d-1)$-simplices contained in only one d-simplex. So, Γ_j lies in the boundary of Δ_{j-1}.

Case 1. $\Gamma_j \neq \delta F_j$. Then Δ_j is also a PL d-ball by part (i) of Theorem 4.7.21.

Case 2. $\Gamma_j = \delta F_j$. Then Γ_j must equal the boundary of Δ_{j-1}, by Lemma 4.7.3, so Δ_j is a PL d-sphere by part (ii) of Theorem 4.7.21. Furthermore, in this case we must have that $j = t$, since no more d-simplex F_{j+1} can be added to Δ_j without forcing some $(d-1)$-simplex into three d-simplices. $\square$

The usefulness of Proposition 4.7.22 lies in being able to conclude from a combinatorial test (shellability) the topological type of the complex (sphere or ball). The PL property is what makes this possible. It should be mentioned that the implication "shellable" $\implies$ "PL" for simplicial spheres and balls is strict. See Lickorish (1991) for a construction of non-shellable PL d-spheres, $d \ge 3$, and Rudin (1958) for a non-shellable PL 3-ball. In this connection it is an interesting fact that every simplicial PL sphere is the boundary of a shellable ball (Pachner 1986).

(e) Cellular interpretation of posets

Given a poset P, we would like to decide: Does there exist a regular cell complex Δ whose face poset is isomorphic to P? In other words: Is it possible to interpret the elements of P as topological cells and the order relation of P as inclusion of cells in a topologically realizable way? The motivating example in this book is of course when P is the covector lattice of an oriented matroid, and the cellular interpretation was in this case worked out by Folkman and Lawrence (1978) and Edmonds and Mandel (1982) using quite different tools. We will here follow a third approach, which was developed by Björner (1984a) for the cellular interpretation of Bruhat order of Coxeter groups, but which suits oriented matroid theory equally well.

A first answer to our question will be given in terms of the topology of the order complex of P, or rather that of some of its subcomplexes. Note that if $P \cong \mathcal{F}(\Delta)$ then the order complex of P is a kind of derivative of Δ (its first barycentric subdivision), so from a general point of view we are dealing with an integration problem: Which order complexes are derivatives? Recall the notation $P_{<x} = \{y \in P : y < x\}$, and $\|P\| = \|\Delta_{ord}(P)\|$ for the geometric realization of the order complex $\Delta_{ord}(P)$.

4.7.23 Proposition.   *Let P be a poset. Then $P \cong \mathcal{F}(\Delta)$ for some regular cell complex Δ if and only if $\|P_{<x}\|$ is homeomorphic to a sphere for all $x \in P$. Furthermore, then Δ is uniquely determined by P up to cellular homeomorphism.*

Proof. If $\mathcal{F}(\Delta) \cong P \cong \mathcal{F}(\Delta')$, for regular cell complexes Δ and Δ', then everything follows from Corollaries 4.7.9 and 4.7.11.

Suppose that $\|P_{<x}\|$ is homeomorphic to a sphere, for all $x \in P$. We will show by induction on the size of P that $P \cong \mathcal{F}(\Delta)$ for some regular cell complex Δ. The result is clearly true if P is an antichain (in which case $P_{<x} = \emptyset \cong S^{-1}$, for all $x \in P$).

Let y be a maximal element of P such that $P_{<y} \neq \emptyset$. Then by induction there exists some regular cell complex Δ' such that

$$\mathcal{F}(\Delta') \cong P' := P - y.$$

The subcomplex Δ'' corresponding to $P_{<y}$ is homeomorphic to the d-sphere for some $d \geq 0$, since by Proposition 4.7.8 and assumption we have

$$\|\Delta''\| \cong \|\mathcal{F}(\Delta'')\| \cong \|P_{<y}\| \cong S^d.$$

Now, enlarge the complex Δ' with a new $(d+1)$-cell σ that is glued into the spherical "hole" determined by Δ'' via some arbitrary homeomorphism $\overset{\circ}{\sigma} = S^d \cong \|\Delta''\|$. This creates a regular cell complex $\Delta = \Delta' \cup \{\sigma\}$ with the newly attached cell σ corresponding to the poset element y, such that $\mathcal{F}(\Delta') \cong P'$ can be enlarged to $\mathcal{F}(\Delta) \cong P$. $\square$

From an algorithmic point of view the answer given by Proposition 4.7.23 is unsatisfactory, since it is known (see Volodin, Kuznetsov and Fomenko 1974) that no algorithm can exist for deciding when a complex is homeomorphic to the 5-sphere. One might ask also for a characterization of the face lattices of polyhedral complexes. This leads to the much studied *Steinitz problem*, which is decidable but of high complexity, see Section 9.5.

We now come to the main tool, which is applied to oriented matroids in Section 4.3 and for which much of the preceding material was developed. Let us call a graded poset P *thin* if all intervals $[x, y]$ of length 2 have cardinality 4, see Figure 4.7.4 (a). We will say that P is *subthin* if all intervals $[x, y]$ of length 2 with $y \neq \hat{1}$ have cardinality 4 and all such intervals with $y = \hat{1}$ have cardinality 3 or 4, with at least one of cardinality 3, see Figure 4.7.4 (b).

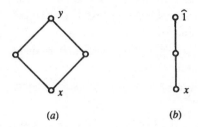

(a)   (b)

Figure 4.7.4: Thin and subthin posets.

4.7.24 Theorem (Björner 1984a). *Let P be a graded poset of length $d + 2$. Then:*

(i) $P \cong \widehat{\mathcal{F}}(\Delta)$ *for some shellable regular cell decomposition Δ of the d-sphere*
   $\Longleftrightarrow$ *P is thin and admits a recursive coatom ordering.*

(ii) $P \cong \widehat{\mathcal{F}}(\Delta)$ *for some shellable regular cell decomposition Δ of the d-ball*
   $\Longleftrightarrow$ *P is subthin and admits a recursive coatom ordering.*

Furthermore, in both cases Δ is uniquely determined by P up to cellular homeomorphism, and recursive coatom orderings in P correspond via the isomorphism to shellings of Δ.

Proof. Suppose first that $P \cong \widehat{\mathcal{F}}(\Delta)$ for a shellable regular cell complex Δ. Then P admits a recursive coatom ordering and in fact recursive coatom orderings correspond to shellings, by Lemma 4.7.18. Furthermore, all intervals $[x, y]$ in P of length 2 with $y \neq \hat{1}$ have cardinality 4, by Corollary 4.7.12 (ii). If $\|\Delta\| \cong S^d$ then this also extends to the case $y = \hat{1}$, for the same reason (every $(d-1)$-simplex in a triangulation of the d-sphere is a face of exactly two d-simplices). Hence, P is then thin. If $\|\Delta\| \cong B^d$, then P is subthin for similar reasons (every $(d-1)$-simplex in a triangulation of the d-ball is a face of one or two d-simplices, the first case occurring exactly at the boundary). The triangulations referred to in these parenthetical remarks are those provided by the order complex $\Delta_{ord}(P \setminus \{\hat{0}, \hat{1}\}) \cong \Delta_{ord}(\mathcal{F}(\Delta))$, cf. Proposition 4.7.8.

Conversely, suppose that P admits a recursive coatom ordering and that all intervals $[z, y]$ of length 2 with $y \neq \hat{1}$ have cardinality 4. Let $x \in P \setminus \{\hat{0}, \hat{1}\}$. By Proposition 4.7.19 the order complex of the open interval $(\hat{0}, x)$ is shellable, and by Proposition 4.7.22 it is a sphere. Hence, by Proposition 4.7.23 P is isomorphic to the augmented face lattice $\widehat{\mathcal{F}}(\Delta)$ of some regular cell complex Δ, uniquely determined by P up to cellular homeomorphism. Lemma 4.7.18 shows that Δ is shellable. It remains only to conclude that $\|\Delta\| \cong S^d$ if P is thin, and $\|\Delta\| \cong B^d$ if P is subthin. But this also follows from Proposition 4.7.22, using Propositions 4.7.8 and 4.7.19 (for the "big" interval $(\hat{0}, \hat{1})$). $\qquad\square$

The theorem can be summarized by saying that, as classes of mathematical objects, shellable regular cell decompositions of spheres and graded thin posets having a recursive coatom ordering are canonically equivalent.

Definition 4.7.20 of *PL* spheres and balls was formulated only for simplicial complexes. However, the definition is applicable to any space that can be the domain of a *PL* homeomorphism, i.e., to any subspace of $\mathbb{R}^d$ admitting triangulation by a finite geometric simplicial complex. In particular, the definition is applicable to regular cell complexes via their realization in $\mathbb{R}^{2d+1}$ (see Corollary 4.7.10). Since it is known that any two geometric simplicial complexes with the same underlying space have a common subdivision (Zeeman 1963, p.8), the following can be deduced.

4.7.25 Lemma. *A regular cell complex Δ is a PL d-sphere if and only if its simplicial subdivision $\Delta_{ord}(\mathcal{F}(\Delta))$ (the order complex of its face poset) is a PL d-sphere, and similarly for PL d-balls.*

Only the simplicial case of *PL* balls and spheres have been dealt with previously, in the interest of simplicity, and now the step to the general case can be taken with ease via Lemma 4.7.25. Here are a few facts about cellular *PL* spheres that we shall need.

4.7.26 Proposition. *Let Δ be a regular cell decomposition of the d-sphere.*

(i) *If $\mathcal{F}(\Delta)$ is shellable (in particular if Δ is shellable), then Δ is a PL sphere.*

(ii) *If $\sigma_1, \sigma_2, \ldots, \sigma_t$ is a shelling order of the maximal cells of Δ, then the subcomplex generated by $\{\sigma_1, \sigma_2, \ldots, \sigma_k\}$ for $1 \le k < t$ is a PL d-ball.*

(iii) *If Δ is a PL sphere, then every closed cell $\sigma \in \Delta$ is a PL ball.*

(iv) *If Δ is a PL sphere, then there exists a PL regular cell decomposition Δ^{op} of the d-sphere with anti-isomorphic face poset:*

$$\mathcal{F}(\Delta^{op}) \cong \mathcal{F}(\Delta)^{op}.$$

Proof. We have the chain of implications:

$$\Delta \text{ shellable} \implies \mathcal{F}(\Delta) \text{ shellable} \implies \Delta_{ord}(\mathcal{F}(\Delta)) \text{ a } PL\text{sphere}$$
$$\implies \Delta \text{ a } PL \text{ sphere.}$$

The first arrow follows from Lemma 4.7.18 and Proposition 4.7.19, the second from Proposition 4.7.22 (i), and the third is the obvious direction of Lemma 4.7.25. This proves part (i).

Part (ii) is proved in the same way, just using part (ii) of Proposition 4.7.22 for the second arrow.

For part (iii) we observe that $\Delta_{ord}(\mathcal{F}_{<\sigma})$ is the link of a simplex in the simplicial *PL* sphere $\Delta_{ord}(\mathcal{F})$, where $\mathcal{F} = \mathcal{F}(\Delta)$. Hence, using Theorem 4.7.21 we find that $\Delta_{ord}(\mathcal{F}_{<\sigma})$ is a *PL* sphere, and that $\Delta_{ord}(\mathcal{F}_{\le\sigma})$, being a cone over $\Delta_{ord}(\mathcal{F}_{<\sigma})$, is a *PL* ball. Hence, σ is a *PL* ball.

For part (iv), let $\mathcal{F} = \mathcal{F}(\Delta)$ and $P = \mathcal{F}^{op}$. Then

$$\Delta_{ord}(P_{<\sigma}) = \Delta_{ord}(\mathcal{F}_{>\sigma})$$

for all $\sigma \in P$, and hence we conclude that $\Delta_{ord}(P_{<\sigma})$ is a PL sphere, being the link of a simplex in the PL sphere $\Delta_{ord}(\mathcal{F})$. It follows from Proposition 4.7.23 that $P \cong \mathcal{F}(\Delta^{op})$ for some regular cell complex Δ^{op}. Furthermore, Δ^{op} is a PL sphere since

$$\Delta_{ord}(\mathcal{F}(\Delta^{op})) = \Delta_{ord}(\mathcal{F}(\Delta)).$$

$\square$

(f) Homotopy type

The notion of homotopy type of spaces appears in several places in the book. We will here review the basic definitions and some useful lemmas.

Two mappings $f_0, f_1 : T_1 \longrightarrow T_2$ of topological spaces are said to be *homotopic* (written $f_0 \sim f_1$) if there exists a mapping (called a *homotopy*) $F : T_1 \times [0, 1] \longrightarrow T_2$ such that $F(x, 0) = f_0(x)$ and $F(x, 1) = f_1(x)$ for all $x \in T_1$ (all mappings are of course assumed to be continuous). The spaces are *homotopy equivalent* (or, have the *same homotopy type*) if there exist mappings $f_1 : T_1 \longrightarrow T_2$ and $f_2 : T_2 \longrightarrow T_1$ such that $f_2 \circ f_1 \sim id_{T_1}$ and $f_1 \circ f_2 \sim id_{T_2}$. A space which is homotopy equivalent to a point (e.g. the d-ball) is called *contractible*. If $T_2 \subseteq T_1$ and there exists a homotopy $F : T_1 \times [0, 1] \longrightarrow T_1$ such that $F(x, 0) \in T_2$ and $F(x, 1) = x$ for all $x \in T_1$, and $F(y, t) = y$ for all $y \in T_2$ and all $t \in [0, 1]$, then T_2 is called a *strong deformation retract* of T_1, and T_1 and T_2 are clearly homotopy equivalent. See the general references cited in Section 4.7(a) for more information.

We will occasionally use that if two spaces are homotopy equivalent then they have the same Euler characteristic (in fact, isomorphic homology groups). This is relevant for the Möbius function $\mu(x, y)$, $x \leq y$, of a poset, since $\mu(x, y)$ is one less than the Euler characteristic of the open interval (x, y).

4.7.27 Lemma. *Let Δ be a simplicial complex, and $V = A \cup B$ a non-trivial partition of its vertices into two classes. Let $\Delta_A = \{F \in \Delta : F \subseteq A\}$ be the induced subcomplex on vertices A, and similarly for Δ_B. Then $\|\Delta_A\|$ is a strong deformation retract of $\|\Delta\| \backslash \|\Delta_B\|$.*

Proof. Express all points $x \in \|\Delta\|$ in barycentric coordinates:

$$x = \sum_{a \in A} \xi_a a + \sum_{b \in B} \xi_b b, \qquad \xi_a, \xi_b \geq 0, \ \sum \xi = 1$$

(see the discussion preceding Definition 4.7.20). Then $x \in \|\Delta_B\|$ if and only if $\xi_a = 0$ for all $a \in A$. We can therefore define a homotopy $F : (\|\Delta\| \backslash \|\Delta_B\|) \times [0, 1] \longrightarrow (\|\Delta\| \backslash \|\Delta_B\|)$ by

$$F(x, t) = t \sum_{b \in B} \xi_b b + \frac{1 - t \sum_{b \in B} \xi_b}{\sum_{a \in A} \xi_a} \sum_{a \in A} \xi_a a,$$

which gives a strong deformation retraction of $\|\Delta\| \backslash \|\Delta_B\|$ onto the subspace $\|\Delta_A\|$. $\square$

4.7.28 Lemma (Edelman and Walker 1985). *Let P be a thin graded poset having a recursive coatom ordering $x_1, x_2, \ldots, x_t$. Let J_k be the order ideal of $\overline{P} = P \backslash \{\hat{0}, \hat{1}\}$ generated by $\{x_1, x_2, \ldots, x_k\}$, where $1 \leq k < t$. Then the filter $\overline{P} \backslash J_k$ is contractible.*

Proof. We know from Theorem 4.7.24 that $\|\overline{P}\|$ is a PL d-sphere, and from Proposition 4.7.26 (ii) that the subcomplex $\|J_k\|$ is a PL d-ball. Therefore, by Theorem 4.7.21 (iii) we have that $\|\overline{P}\| \backslash \|J_k\|$ is an open ball (the interior of a PL d-ball), which by Lemma 4.7.27 deformation retracts onto $\|\overline{P} \backslash J_k\|$. $\square$

4.7.29 Lemma (Quillen 1978). *Let $f : P \longrightarrow Q$ be an order-preserving or order-reversing mapping of posets. Suppose that $f^{-1}(Q_{\leq x})$ is contractible for all $x \in Q$. Then f induces homotopy equivalence of $\|P\|$ and $\|Q\|$.*

A simple proof for this lemma appears in Walker (1981); see also Björner (1995).

Putting together some facts from homotopy theory (see Spanier 1966, especially pp. 377 and 405) one can deduce the following analogue of Lemma 4.7.29: If $p : E \longrightarrow B$ is a map of topological spaces with suitable properties (a fibration of CW complexes) and the fiber $p^{-1}(b)$ is contractible for all $b \in B$ (or, for some $b \in B$, if B is path-connected), then p induces homotopy equivalence of E and B. This is applied to realization spaces of oriented matroids in Lemma 8.2.1.

Exercises

4.1 Let $\mathcal{L} \subseteq \{+, -, 0\}^E$ be an oriented matroid. Show that the following conditions are equivalent for $X, Y \in \mathcal{L}$:

(i) X and Y are conformal (i.e., $S(X, Y) = \emptyset$),

(ii) $X \circ Y = Y \circ X$,

(iii) $Y \leq X \circ Y$,

(iv) the join $X \vee Y$ exists in $\mathcal{L}$, and $X \vee Y = X \circ Y$.

4.2 Prove the equivalence of the various characterizations of parallel elements in Lemma 4.1.10.

4.3 Show that $(\widehat{\mathcal{L}})^{op}$ is in general not isomorphic to the big face lattice of any oriented matroid.

4.4 A lattice L of finite length is called *relatively complemented* if for all $x < y < z$ in L there exists $y' \in L$ such that $y \wedge y' = x$ and $y \vee y' = z$. It is *atomic* if every element is a join of atoms and *coatomic* if every element is a meet of coatoms.

(a) Show that L is relatively complemented if and only if every interval of length 2 has cardinality ≥ 4. (Björner 1981)

(b) Assume that L is a relatively complemented graded lattice of length r, and for $0 \le k \le r$ let $L_k = \{$elements in L of rank $k\}$. Show that:

(i) L is atomic and coatomic,

(ii) $|L_k| \ge \binom{r}{k}$,

(iii) $|L_1| = r \iff |L_{r-1}| = r \iff L$ is Boolean.

(c) Let $\mathcal{L}$ be an oriented matroid of rank r with tope graph $\mathcal{T} = \mathcal{T}(\mathcal{L})$. Show that:

(i) every tope has at least r neighbors in $\mathcal{T}$,

(ii) $\mathcal{T}$ has exactly r neighbors in $\mathcal{T}$ if and only if the tope T is simplicial,

(iii) $\mathcal{T}$ is an r-regular graph $\iff \mathcal{L}$ is a simplicial oriented matroid.

4.5* Does every oriented matroid have a simplicial tope? (See Conjecture 7.3.10.)

4.6 Let $T^1, T^2 \in \mathcal{T}$. Suppose that for every subtope X covered by T^1 and for all $e \in X^0$ we have $T_e^1 = T_e^2$. Show that $T^1 = T^2$.

4.7 Show that the set $\mathcal{L}$ of all covectors is completely determined by the subset of all covectors of rank k, for any $1 \le k \le r$.

4.8 Show that all cycles in the tope graph $\mathcal{T}(\mathcal{L})$ are of even length.

4.9 Show that $\mathcal{Q} \subseteq \mathcal{T}$ is T-convex if and only if $\mathcal{Q}$ is an order ideal of the tope poset $\mathcal{T}(\mathcal{L}, B)$ for all $B \in \mathcal{T}$.

4.10 A subset $\mathcal{G} \subseteq \mathcal{T}$ is called *gated* if for every $T \in \mathcal{T}$ there is a $G_T \in \mathcal{G}$ (a *gate*) such that $d(T, G) = d(T, G_T) + d(G_T, G)$ for all $G \in \mathcal{G}$. (This terminology is due to Dress and Scharlau 1987.) Show the following:

(a) The tope G_T, if it exists, is unique.

(b) Every gated set is T-convex.

(c) A subset $\mathcal{G} \subseteq \mathcal{T}$ is gated if and only if $\mathcal{G} = \text{star}(X)$ for some $X \in \mathcal{L}$.

(d) A subset $\mathcal{G} \subseteq \mathcal{T}$ is gated if and only if $\mathcal{G}$ is an interval of the tope poset $\mathcal{T}(\mathcal{L}, B)$ for all $B \in \mathcal{T}$.

4.11 Let $\mathcal{T}(\mathcal{L}, B)$ be the tope poset of a simple oriented matroid $\mathcal{L} \subseteq \{+, -, 0\}^E$, based at B.

(a) Show that $\mathcal{T}(\mathcal{L}, B) \cong \mathcal{T}(\mathcal{L}, B)^{op} = \mathcal{T}(\mathcal{L}, -B)$.

(b) Deduce that the number of elements of rank k in $\mathcal{T}(\mathcal{L}, B)$ equals the number of elements of rank $|E| - k$.

4.12* Characterize those pairs $\mathcal{L}, B$ for which $\mathcal{T}(\mathcal{L}, B)$ is a lattice.

4.13 Let $\mathcal{T} \subseteq \{+, -\}^E$ be the set of topes of an acyclic simple oriented matroid with positive tope B. Say that $A \subseteq E$ is *convex* if $\text{conv}(A) = A$, in the sense of Exercise 3.9 and Chapter 9.

Show the following.

(a) $S(B, T)$ is convex for each $T \in \mathcal{T}$.

For parts (b)–(d) suppose also that the tope poset $\mathcal{T}(\mathcal{L}, B)$ is a lattice.

(b) A subset $A \subseteq E$ is of the form $A = S(B, T)$ for some tope $T \in \mathcal{T}$ if and only if A and $E \backslash A$ are convex.

(c) For topes $T, T^1, T^2 \in \mathcal{T}$:

$$\mathrm{conv}(S(B, T^1) \cup S(B, T^2)) = S(B, T) \iff T^1 \vee T^2 = T.$$

(d) The base tope B is simplicial.

(Björner, Edelman and Ziegler 1990)

4.14 Let $\mathcal{L} \subseteq \{+, -, 0\}^E$ be an oriented matroid and $L = \{X^0 : X \in \mathcal{L}\}$ the underlying geometric lattice. Suppose L is supersolvable (i.e., has a maximal chain of modular elements) with characteristic polynomial

$$p(L; t) = \prod_{i=1}^{r} (t - e_i)$$

(see White 1987, p. 128). Show that there exists a tope $B \in \mathcal{T}$ such that

$$\sum_{T \in \mathcal{T}} q^{d(B,T)} = \prod_{i=1}^{r} (1 + q + q^2 + \ldots + q^{e_i}).$$

(Björner, Edelman and Ziegler 1990)

4.15 Deduce Proposition 4.4.6 from Proposition 4.4.7 using only elementary arguments.

4.16 Let $\mathcal{Q} \subseteq \mathcal{T}$ and consider the subgraph of $\mathcal{T}(\mathcal{L})$ induced on the vertex set $\mathcal{Q}$. Show that for the following two choices of $\mathcal{Q}$ this subgraph has the homotopy property expressed by Proposition 4.4.6.

(a) $\mathcal{Q} =$ all topes in a T-convex set.

(b) $\mathcal{Q} =$ all topes in the complement of a T-convex set.

(c) Draw from part (a) the weaker conclusion for Proposition 4.4.7 that the intermediate paths stay within the union of all minimal paths (but are not necessarily minimal themselves). (*Hint*: Exercises 4.25 and 4.26 are relevant.)

4.17 Let the notation be as in the proof of Proposition 4.3.2. Show that

$$\bigcup_{X \in Q_k} [X] = [R_k] \cap \left(\bigcup_{R \lessdot R_k} [R] \right).$$

4.18* (a) Are the spheres $\Delta^{op}(\mathcal{L})$ shellable? Equivalently, does $\widehat{\mathcal{L}}$ admit a recursive *atom* ordering?

(b) Are Las Vergnas spheres shellable? (Note that this would be implied by a positive answer to part (a). By *Las Vergnas sphere* we mean any regular cell decomposition of the d-sphere whose face lattice is isomorphic to the Las Vergnas lattice of an acyclic oriented matroid, and similarly for *Edmonds-Mandel spheres*.)

4.19* Let $\mathcal{F}$ be a collection of convex polyhedral cones in $\mathbb{R}^d$ satisfying

(i) the origin is a vertex of each cone in $\mathcal{F}$,

(ii) if C is a face of $C' \in \mathcal{F}$, then $C \in \mathcal{F}$,

(iii) $C \cap C'$ is a face of C and of C', for all $C, C' \in \mathcal{F}$,

(iv) $\mathbb{R}^d = \bigcup_{C \in \mathcal{F}} C$.

Such an $\mathcal{F}$ is called a *fan*. Let $\Delta(\mathcal{F}) = \{C \cap S^{d-1} : C \in \mathcal{F}\backslash\{0\}\}$. Then $\Delta(\mathcal{F})$ is a regular cell decomposition of the $(d-1)$-sphere.

(a) Is $\Delta(\mathcal{F})$ an Edmonds-Mandel sphere?

(b) Is $\Delta(\mathcal{F})$ shellable?

4.20 Let $\mathcal{B}$ be the 3-dimensional Barnette sphere on 8 vertices, described in Proposition 9.5.3.

(a) Show that there exists a simplicial fan $\mathcal{F}$ in $\mathbb{R}^4$ such that $\Delta(\mathcal{F}) \cong \mathcal{B}$.

(b) Conclude from this and Proposition 9.5.3 the reason why the question "Is $\Delta(\mathcal{F})$ a Las Vergnas sphere?" does not appear in the previous problem.

4.21 Let $\text{graph}(\mathcal{L})$ denote the 1-skeleton of $\mathcal{F}_{big}(\mathcal{L})$. Show the following:

(a) If $r = \text{rank}(\mathcal{L}) \leq 3$, then $\text{graph}(\mathcal{L})$ is not $(2r-1)$-connected.

(b) Let $\mathcal{L}$ be the rank 4 oriented matroid given by 2 skew lines in $\mathbb{R}^4$ with 3 points on each. Then $\text{graph}(\mathcal{L})$ is 8-connected.

4.22 Define the complex $\Gamma(\mathcal{L})$ as in Proposition 4.3.9, and let $\mathcal{C} \subseteq \{+,-,0\}^E$ be the set of circuits of $\mathcal{L}$.

(a) Show that $\Gamma(\mathcal{L}) = \{X \in \{+,-,0\}^E : X \not\geq C \text{ for all } C \in \mathcal{C}\}$.

(b) Let $\mathcal{A} = (S_e)_{e \in E}$ be a signed and essential arrangement of pseudospheres corresponding to $\mathcal{L}$ (Theorem 5.2.1). Show that

$$\Gamma(\mathcal{L}) = \{X \in \{+,-,0\}^E : \bigcap_{e \in X} S_e^{X_e} \neq \emptyset\},$$

where S_e^+, S_e^- are the open sides of S_e. (In other words, $\Gamma(\mathcal{L})$ is the *nerve* of the covering of the $(r-1)$-sphere by open sides of $\mathcal{A}$.) (Edelman 1984a)

4.23 Derive Corollary 4.3.10 from Theorem 4.6.1.

4.24 Let $\mathcal{Q}$ be a T-convex set of topes, $\mathcal{Q} \neq \emptyset, T$, and let $\Delta(\mathcal{Q})$ denote the subcomplex of $\Delta(\mathcal{L})$ generated by $\mathcal{Q}$. Show the following:

(a) $\Delta(\mathcal{Q})$ is a shellable $(r-1)$-ball.

(b) The subcomplex of $\Delta(\mathcal{L})$ consisting of all cells that are disjoint from $\Delta(\mathcal{Q})$ is contractible.

4.25 For $e \in E$, let $\mathcal{L}_e^+ = \{X \in \mathcal{L} : X_e = +\}$. Show that the poset $\bigcap_{e \in A} \mathcal{L}_e^+$ is contractible, if non-empty, for $\emptyset \neq A \subseteq E$.

4.26 Let $\mathcal{L}_A^+ = \bigcup_{e \in A} \mathcal{L}_e^+$, for $\emptyset \neq A \subseteq E$ such that $T \not\subseteq \mathcal{L}_A^+$. Show the following:

(a) $\mathcal{L}_A^+$ admits a recursive coatom ordering.

(b) $\mathcal{L}_A^+$ is a PL $(r-1)$-ball.

(c) $\mathcal{L}_A^{++} = \{X \in \mathcal{L}_A^+ : \mathcal{L}_{\leq X}\backslash\{0\} \subseteq \mathcal{L}_A^+\}$ is contractible.

4.27　Let σ be a proper face of a tope (maximal cell) in $\Delta(\mathcal{L})$.

(a) Show that the subcomplex of the tope boundary consisting of all cells that are disjoint from σ is contractible.

(b) Give an example showing that such a subcomplex needs not be pure.

4.28　(a) Is the bounded complex $\mathcal{L}^{++}$ always pure?　　　　　(Zaslavsky 1975a)

(b)* Assuming that $\mathcal{L}^{++}$ is full-dimensional and that $\mathcal{L}$ is uniform or simplicial, does it follow that $\mathcal{L}^{++}$ is homeomorphic to the $(r-1)$-ball?

4.29　Give an example of an affine hyperplane arrangement whose bounded complex is not star-convex, that is, which has no "center point" from which all other points in the bounded complex can be seen. (Hint: there are examples consisting of six lines in the plane.)

4.30　(a) Let G be a simple connected graph with vertex set $\{1, 2, \ldots, n\}$ and edge set E. For $e = (i, j) \in E$, let $H_e = \{x \in \mathbb{R}^n : x_i = x_j\}$. Show that the regions of the arrangement $\mathcal{A}_G = (H_e)_{e \in E}$ are in one-to-one correspondence with the acyclic orientations of G. (An acyclic orientation is a choice of direction for each edge such that no directed cycles are induced.)

(Greene 1977, Greene and Zaslavsky 1983)

(b) Show that the number of acyclic orientations of G is equal to $(-1)^n \chi_G(-1)$, where χ_G is the chromatic polynomial of G.

(Stanley 1973)

(c) Show that for any fixed edge $(i, j) \in E$ the number of acyclic orientations of G such that i is the unique source and j is the unique sink equals $\beta(M)$, where M is the circuit matroid of G. (Greene and Zaslavsky 1983)

(d) Let G be the complete graph K_n. Interpret the previous results in terms of permutations, as in Example 2.3.3.

4.31　Let L be the geometric lattice of flats of the r-uniform matroid on n elements. Show that:

(a) $\mu_L(\hat{0}, \hat{1}) = (-1)^r \binom{n-1}{r-1}$,

(b) $\sum_{x \in L} |\mu_L(\hat{0}, x)| = 2 \sum_{i=0}^{r-1} \binom{n-1}{i}$.

4.32　Let $(E, \mathcal{L})$ be a uniform oriented matroid of rank r, $|E| = n$. Furthermore, let $1 \le j_1 < j_2 < \ldots < j_k \le r$ and $J = \{j_1 - 1, j_2 - 1, \ldots, j_k - 1\}$. Show that

$$f_J(\mathcal{L}) = \frac{n! 2^{j_k - j_1 + 1} \sum_{i=0}^{j_1 - 1} \binom{n - r + j_1 - 1}{i}}{(n - r + j_1)! (j_2 - j_1)! \ldots (j_k - j_{k-1})! (r - j_k)!}$$

4.33　Let $(E, \mathcal{L}, g)$ be an affine oriented matroid.

(a) Show that the f-vector $f(\mathcal{L}^+)$ depends only on the underlying pointed matroid, but may depend on the choice of g.　　　　(Zaslavsky 1975a)

(b)* What about the flag vector $\tilde{f}(\mathcal{L}^+)$?

4.34　Determine the f-vectors $f(\mathcal{L}^+)$ and $f(\mathcal{L}^{++})$ for $(E, \mathcal{L}, g)$ such that $\mathcal{L}$ is uniform.　　　　(Zaslavsky 1975a)

4.35 Let $\mathcal{A}$ be an arrangement of affine hyperplanes in $\mathbb{R}^d$, and let $L = L(\mathcal{A})$ be the intersection poset (i.e., the meet-semilattice of non-empty intersections of subfamilies of $\mathcal{A}$ ordered by reverse inclusion, see Definition 2.1.3). Prove the following:

(a) The number of regions into which $\mathcal{A}$ partitions $\mathbb{R}^d$ is $\sum_{s \in L} |\mu(\hat{0}, s)|$,

(b) The number of bounded regions is $|\sum_{s \in L} \mu(\hat{0}, s)|$.

(c) If $\mathcal{A}$ is in general position (i.e., any i hyperplanes meet in a $(d-i)$-dimensional affine subspace, for $2 \leq i \leq d+1$) and $|\mathcal{A}| = n$, then the number of regions is

$$\binom{n}{0} + \binom{n}{1} + \ldots + \binom{n}{d},$$

and the number of bounded regions is $\binom{n-1}{d}$.

(d) If $\mathcal{A}$ is in general position and $|\mathcal{A}| = n$, then the number of k-cells and the number of bounded k-cells are

$$\sum_{i=d-k}^{d} \binom{n}{i}\binom{i}{d-k} \quad \text{and} \quad \sum_{i=d-k}^{d} (-1)^{d-i}\binom{n}{i}\binom{i}{d-k},$$

respectively, $0 \leq k \leq d$. (Buck 1943, Zaslavsky 1975a)

4.36 For a zonotope $\mathcal{Z}$ with associated matroid M (Section 2.2) show that the number of vertices outside any fixed zone is $2\beta(M)$.

(Greene and Zaslavsky 1983)

4.37 Let $f_{k-1}(\mathcal{L})$ be the number of covectors of rank k in a simple rank r oriented matroid $(E, \mathcal{L})$. Show that:

(a) $f_{k-1}(\mathcal{L}) \geq 2^k \binom{r}{k}$, for $1 \leq k \leq r$,

(b) $f_{k-1}(\mathcal{L}) \geq 2^k |E|$, for $1 \leq k \leq r-1$.

(*Hint*: For part (b) use the "hyperplane theorem" for matroids, see White 1987, p. 158.)

4.38 (a) Show that $|\mathcal{L}| \leq 2^{r-1} \cdot |\mathcal{T}| + 1$.

(b) Discuss the algorithmic complexity of determining the full set $\mathcal{L}$ of covectors from the set $\mathcal{T}$ of topes, as in Theorems 4.2.13 and 4.2.14.

(Fukuda, Saito and Tamura 1991)

4.39 Let $(f_{r-1}, f_{r-2}, \ldots, f_0)$ be the reversed f-vector of an oriented matroid. Show that the relations of Proposition 4.6.12 have dual versions that are valid for this sequence.

4.40* Does there exist some f-vector of a matroid polytope (Las Vergnas or Edmonds-Mandel face lattice) which is not the f-vector of some convex polytope?

4.41 Let $\mathcal{M}$ and $\mathcal{M}'$ be two oriented matroids on a set E such that every covector of $\mathcal{M}$ is a covector of $\mathcal{M}'$, see Proposition 7.7.1. Let $M = \underline{\mathcal{M}}$ and $M' = \underline{\mathcal{M}'}$.

(a) Show that the number of subsets $A \subseteq E$ such that $_{-A}\mathcal{M}$ is acyclic and $_{-A}\mathcal{M}'$ is totally cyclic is equal to

$$\sum_{A \subseteq E} (-1)^{r(M') - r_{M'}(A) + |A| - r_M(A)}.$$

(b) Show that Theorem 4.6.1 is the special case $r(M) = r(M')$ of (a), and that Theorem 4.6.5 is a straightforward consequence of the special case $r(M) = r(M') + 1$.

(Las Vergnas 1977)

4.42 Let $\mathcal{M}$ be an oriented matroid on an ordered set E, and $M = \underline{\mathcal{M}}$. Let $o(\mathcal{M})$ be the number of elements $e \in E$ such that e is the smallest element of some positive circuit of $\mathcal{M}$. For $A \subseteq E$ set $o(A) = o(_{-A}\mathcal{M})$ and $o^*(A) = o(_{-A}\mathcal{M}^*)$.

Recall that the *Tutte polynomial* of M is the polynomial

$$t(M; x, y) = \sum_{A \subseteq E} (x - 1)^{r(M) - r_M(A)} (y - 1)^{|A| - r_M(A)},$$

see White (1992).

(a) Show that

$$t(M; x, y) = \sum_{A \subseteq E} \left(\frac{x}{2}\right)^{o^*(A)} \left(\frac{y}{2}\right)^{o(A)}$$

(b) Show that Theorem 4.6.1 is a specialization of (a).

(c) Give a common generalization of (a) and 4.41(a).

(Las Vergnas 1984b)

4.43 Let M be an ordinary matroid on E with a signature $\mathcal{C}$ of its circuits. Suppose that $\mathcal{C}$ is *minorable*, that is, it induces a well-defined circuit signature on every minor of M.

(a) Show that the number of subsets $A \subseteq E$ such that $_{-A}\mathcal{C}$ contains no positive circuit is at most $t(M; 2, 0)$. (See Exercise 4.42 for the definition of $t(M; x, y)$.)

(b) Show that if M has no loops, we have equality in (a) if and only if $\mathcal{C}$ is the set of circuits of an oriented matroid. (*Hint*: use Theorem 3.3.7.)

(Las Vergnas 1990)

4.44 (a)* Given an oriented matroid $(E, \mathcal{L})$ of rank r, does there exist an oriented matroid $(E', \mathcal{L}')$ of rank $r + 1$ such that $\mathcal{F}_{big}(\mathcal{L}) \cong \mathcal{F}_{em}(\mathcal{L}')$?

(b) Show that this is true in the realizable case.

4.45 The incidence matrix $I(\mathcal{S}, E)$ of a family $\mathcal{S}$ of signed subsets of a set E is a $\{0, +1, -1\}$-matrix whose columns are indexed by the elements of E and whose rows are the signed incidence vectors of the members of $\mathcal{S}$.

(a) Show that rank $I(\mathcal{T}, E) = |E|$, if $\mathcal{T}$ is the set of topes of a simple oriented matroid on E.

(b) Show that rank $I(\mathcal{C}, E) \geq |E| - r$, if $\mathcal{C}$ is the set of circuits of a rank r oriented matroid on E.

(c) Show that equality holds in (b) if the matroid is graphic.

(d)* Determine rank $I(\mathcal{C}, E)$ for general oriented matroids.

4.46 Let $\mathcal{T} \subseteq \{+, -\}^E$ be the set of topes of a simple oriented matroid whose underlying matroid is M. Let ω be a linear ordering of the set E, and consider the broken circuit complex $BC_\omega(M)$. (See Chapter 7 of White 1992 for definitions relating to $BC_\omega(M)$.)

Describe a direct and geometrically motivated bijection $\mathcal{T} \longleftrightarrow BC_\omega(M)$. The bijection should preferably be such that the topes of one halfspace $\{T \in \mathcal{T} : T_g = -\}$ correspond to the reduced broken circuit complex, and the "bounded" topes of the opposite halfspace, whose number is $\beta(M)$, correspond to the bases of internal activity 1 and external activity 0.

(Las Vergnas 1984c; for graphic matroids also Blass and Sagan 1986)

4.47 Let Δ be a regular cell complex with augmented face poset $\widehat{\mathcal{F}}(\Delta)$. Show that Δ has the intersection property if and only if $\widehat{\mathcal{F}}(\Delta)$ is a lattice.

4.48 Show the following:

(a) Every permutation of the facets of a d-simplex is a shelling of its boundary complex.

(b) No other d-polytope has this property.

(c) No other regular cell complex with the intersection property whose underlying space is the $(d-1)$-sphere has this property.

4.49 Let $[x, y]$ be an interval in a graded poset P. Show that:

(a) If P admits a recursive coatom ordering, then so does $[x, y]$.

(b) If P is shellable, then so is $[x, y]$.

(c) Complete the proof of Proposition 4.7.19.

(Björner and Wachs 1983)

4.50 Show that Proposition 4.7.22 can be generalized to regular cell complexes Δ. (Björner 1984a)

4.51 Let $\widehat{\mathcal{F}}(\Delta)$ be the augmented face poset of a regular cell complex, and let ρ denote poset rank.

(a) Show that the Möbius function of $\widehat{\mathcal{F}}(\Delta)$ is given by
$$\mu(\sigma, \tau) = (-1)^{\rho(\tau) - \rho(\sigma)},$$
for all $\sigma \leq \tau \neq \hat{1}$ in $\widehat{\mathcal{F}}(\Delta)$.

(b) If $\|\Delta\|$ is a topological manifold without boundary then this formula is valid also for all $\hat{0} \neq \sigma \leq \tau$.

(c) If $\|\Delta\|$ is a sphere then the formula extends to include also $\sigma = \hat{0}, \tau = \hat{1}$. (*Remark:* For (b) use the fact that the link of any simplex in a triangulation of a manifold has the homology of a sphere.)

4.52 Show that Corollary 4.3.4 with "*PL*" removed can be deduced from Theorem 4.3.3 with "shellable" removed, that is, from the Folkman and Lawrence (1978) version of the sphericity theorem.   (G. Rybnikov)

4.53 Let L be a geometric lattice of rank r, and for $0 \leq i \leq r-1$ define

$$f_i := \sum_{\substack{x \leq y \\ r(x)=i}} |\mu_L(x,y)|.$$

Show the following:

(a) $f_0 - f_1 + f_2 - \ldots + (-1)^{r-1} f_{r-1} = 1 + (-1)^{r-1}$.

(b) $f_i \geq 2^{r-i} \binom{r}{i}$, for all $1 \leq i \leq r-1$.

(c) Equality holds in (b) for some i (equivalently, for all i) if and only if L is Boolean.

(d) If L is the lattice of flats of an orientable matroid, then $r \cdot f_0 \leq 2 \cdot f_1$.

(e) Use part (d) to show that the Fano matroid F_7 and the affine plane $AG(3,3)$ are not orientable (cf. Example 6.6.2). Can the non-orientability of the MacLane matroid $\mathtt{ML}_8$ be deduced this way?

(f) Using the underlying idea of part (d) and the results of Section 4.6, state some other f-vector and flag vector criteria for non-orientability.

5

Topological Models for Oriented Matroids

It was described already in Chapter 1 how *realizable* oriented matroids correspond to arrangements of hyperplanes in $\mathbb{R}^d$. The content of the Topological Representation Theorem is that *general* oriented matroids similarly correspond to arrangements of generalized hyperplanes, each obtained from a flat hyperplane by tame topological deformation. This important result, the proof of which was begun already in the preceding chapter, is the main concern of this chapter. In the last section it is discussed to what extent points rather than hyperplanes can be used to represent the elements of an oriented matroid in a topologically deformed arrangement. It turns out that in the non-realizable case there are severe limitations to such an analogue of projective polarity.

5.1 Arrangements of pseudospheres

A linear oriented matroid can be described as an arrangement $\mathcal{A}$ of oriented hyperplanes through the origin in $\mathbb{R}^{d+1}$. This information is equivalently represented by the arrangement $\mathcal{A}' = \{H \cap S^d : H \in \mathcal{A}\}$ of linear $(d-1)$-subspheres in S^d together with a choice of orientation (positive and negative hemisphere) for each $H \cap S^d$. If these subspheres are topologically deformed in a tame way that preserves their intersection pattern, the arrangement will still represent an oriented matroid. This idea leads to "arrangements of pseudospheres", which will be discussed in this section.

We start with a discussion of codimension 1 subspheres of S^d. If S is a $(d-1)$-sphere embedded in S^d (i.e., $S \subseteq S^d$ and S is homeomorphic to S^{d-1}), then by the Jordan-Brouwer separation theorem $S^d \backslash S$ consists of two connected components such that S is the boundary of each. These components can have non-trivial topology for $d \geq 3$, e.g., if $d = 3$ and S is the "Alexander horned sphere" then one of the components is not simply connected. We will single out the class of $(d-1)$-subspheres that partition S^d in the nicest possible way.

Say that two $(d-1)$-subspheres S, S' of S^d are *equivalent* if $h(S) = S'$ for some homeomorphism $h : S^d \to S^d$.

5.1.1 Lemma. *For a $(d-1)$-subsphere S of S^d the following conditions are equivalent:*

(i)   *S is equivalent to the equator $S^{d-1} = \{x \in S^d : x_{d+1} = 0\}$,*

(ii) *S is equivalent to some piecewise-linearly embedded $(d-1)$-subsphere,*

(iii) *the closure of each connected component of $S^d \backslash S$ is homeomorphic to the d-ball.*

Proof. That (i) implies the two other conditions is obvious. For (ii)$\Longrightarrow$(i), see Theorems 1.7.2 and 1.8.2 in Rushing (1973). This implication will not be used here. We will sketch a proof of (iii)$\Longrightarrow$(i).

Let B_1 and B_2 be the closures of the connected components of $S^d \backslash S$, and assume that B_1 and B_2 are d-balls. Let $C_1 = \{x \in S^d : x_{d+1} \geq 0\}$ and $C_2 = -C_1$. By assumption there exists a homeomorphism $f_1 : B_1 \to C_1$. The restriction of f_1 to $S = B_1 \cap B_2$ gives a homeomorphism $S \to S^{d-1} = C_1 \cap C_2$, which by Lemma 4.7.2 can be extended to a homeomorphism $f_2 : B_2 \to C_2$. Since f_1 and f_2 agree on the intersection of their domains they can be glued together (Lemma 4.7.1) to give a homeomorphism $h = (f_1 \cup f_2) : S^d \to S^d$ for which $h(S) = S^{d-1}$. ☐

The equivalence relation on $(d-1)$-subspheres of S^d defined above has been much studied, see Rushing (1973). The equivalence class singled out in the lemma consists of the *tame* subspheres, all other subspheres are *wild*. It is known for instance that there are uncountably many equivalence classes of wild 2-spheres in S^3. On the other hand, every 1-sphere in S^2 is tame (the Schönflies theorem). (*Remark:* In codimension 2 one must also distinguish between knotted and unknotted tame subspheres, but this distinction disappears in codimension 1, as well as in codimensions greater than 2.)

5.1.2 Definition. A $(d-1)$-subsphere S in S^d (or in any d-sphere) satisfying the conditions in Lemma 5.1.1 will be called a *pseudosphere in S^d*. The two connected components of $S^d \backslash S$ are its *sides*. The closures of the sides are called *closed sides* (or *pseudohemispheres*).

Each side of a pseudosphere in S^d is homeomorphic to an open d-ball. However, contrary to intuition this property does not characterize pseudospheres: see Rushing (1973, p. 68), for an example of a wild sphere whose complement consists of two open balls. Each closed side of a pseudosphere is, by Lemma 5.1.1, homeomorphic to a closed d-ball.

A few words about terminology are called for. A pseudosphere in S^d is by definition a tame codimension 1 subsphere. Why then use the word "pseudo-sphere" when referring to a particularly nice kind of sphere? The reason is mainly historical: the "pseudo" terminology has become established in the matroid literature ("pseudoline", "pseudohyperplane", etc.), and we have chosen to keep this convention. By a *linear* pseudosphere we mean one of the form $S^d \cap H$, for a hyperplane H through the origin of $\mathbb{R}^{d+1}$.

The following notion of arrangement will play a key role.

5.1.3 Definition. A finite multiset $\mathcal{A} = (S_e)_{e \in E}$ of pseudospheres in S^d is called an *arrangement of pseudospheres* if the following conditions hold:

(A1) $S_A = \bigcap_{e \in A} S_e$ is a sphere, for all $A \subseteq E$.

(A2) If $S_A \not\subseteq S_e$, for $A \subseteq E, e \in E$, and S_e^+ and S_e^- are the two sides of S_e, then $S_A \cap S_e$ is a pseudosphere in S_A with sides $S_A \cap S_e^+$ and $S_A \cap S_e^-$.
  (If $S_A \cap S_e = S^{-1} = \emptyset$ is the empty sphere in a zero sphere $S_A \cong S^0$, then the sides of this empty sphere are the two points of S_A.)

(A3) The intersection of an arbitrary collection of closed sides is either a sphere or a ball. (*Note:* This axiom is redundant, see comment below.)

By a *signed arrangement of pseudospheres* we mean an arrangement $\mathcal{A}$ together with a fixed choice of *positive side*, denoted S_e^+, and *negative side*, denoted S_e^-, for every $S_e \in \mathcal{A}$.

Axiom (A1) allows in particular that $S_A = \emptyset$ (remember that $\emptyset$ is the (-1)-sphere). If $S_E = \emptyset$ the arrangement is called *essential*.

The notion of an arrangement of pseudospheres is due to Folkman and Lawrence (1978), although much simplification and clarification was achieved by Edmonds and Mandel (1982). In particular, Edmonds and Mandel showed that the *ball axiom* (A3) is implied by (A1) and (A2), and can therefore be deleted from the definition. Unfortunately there seems to be no easily accessible proof of this fact. The proof of Edmonds and Mandel uses the generalized Schönflies theorem and some technical results in the literature concerning "local flatness". We will keep axiom (A3) in our definition of arrangements in order to simplify the proofs in this section.

Our definition of an arrangement allows repeated copies of the same pseudosphere. An arrangement in which each pseudosphere occurs only once is called *repetition-free*. A signed arrangement in which each pseudosphere occurs exactly twice, and then with opposite orientations (choice of positive side), is said to be *symmetrically signed.* There is an obvious bijective correspondence between unsigned repetition-free arrangements and symmetrically signed arrangements.

It is very instructive, particularly upon first encounter, to work out the details of Definition 5.1.3 in low dimensions. For this see Section 6.1, where arrangements of pseudospheres in S^1 and S^2 are discussed.

Observe the following recursive property of Definition 5.1.3: If $A, B \subseteq E, A \cap B = \emptyset$, and if $S_A \not\subseteq S_e$ for all $e \in B$, then $\mathcal{A}_A^B = (S_A \cap S_e)_{e \in B}$ is an arrangement of pseudospheres in S_A. If $\mathcal{A}$ is signed then by axiom (A2) there is a naturally induced signature of $\mathcal{A}_A^B$. The $A = \emptyset$ case just says that the restriction of an arrangement to a subset of the pseudospheres is also an arrangement.

The position of each point $x \in S^d$ with respect to a signed arrangement $(S_e)_{e \in E}$ of pseudospheres is given by a sign vector $\sigma(x) \in \{+, -, 0\}^E$, defined by

$$\sigma(x)_e = \begin{cases} +, & \text{if } x \in S_e^+ \\ -, & \text{if } x \in S_e^- \\ 0, & \text{if } x \in S_e. \end{cases}$$

The collection $\sigma(S^d)$ of all such sign vectors serves as an indexing set for a partition of S^d into "cells" $\sigma^{-1}(X)$, for $X \in \sigma(S^d)$.

Our aim in this section is to prove three basic results about pseudosphere arrangements:

(1)   they give rise to oriented matroids (Theorem 5.1.4),

(2)   they decompose S^d into a regular cell complex (Proposition 5.1.5), and

(3)   their topological equivalence classes are combinatorially determined (Theorem 5.1.6).

These results are due to Folkman and Lawrence (1978).

5.1.4 Theorem.   *Let $\mathcal{A} = (S_e)_{e \in E}$ be a signed arrangement of pseudospheres in S^d. Then*

$$\mathcal{L}(\mathcal{A}) := \{\sigma(x) : x \in S^d\} \cup \{0\} \subseteq \{+, -, 0\}^E$$

is the set of covectors of an oriented matroid. If $\dim S_E = k$, then $\operatorname{rank} \mathcal{L}(\mathcal{A}) = d - k$. (In particular, if $\mathcal{A}$ is essential then $\operatorname{rank} \mathcal{L}(\mathcal{A}) = d + 1$.)

We define the *rank* of an arrangement rank of an arrangement $\mathcal{A} = (S_e)_{e \in E}$ to be the rank of the associated oriented matroid $\mathcal{L}(\mathcal{A})$. Also the *intersection lattice* $L(\mathcal{A})$ is the geometric lattice of the underlying matroid. Thus, by Proposition 4.1.13, $L(\mathcal{A}) = \{z(X) : X \in \mathcal{L}(\mathcal{A})\}$. Alternatively, $L(\mathcal{A})$ is the family of spheres S_A (intersections of subspheres S_e), $A \subseteq E$, ordered by reverse inclusion (Exercise 5.3).

An essential arrangement of pseudospheres naturally partitions S^d into "cells", which form a regular cell decomposition of S^d. This is the content of the next result, which for ease of notation we have formulated in the signed case.

5.1.5 Proposition.   *Let $\mathcal{A} = (S_e)_{e \in E}$ be a signed and essential arrangement of pseudospheres. Then the sets $\sigma^{-1}(X)$, for $X \in \mathcal{L}(\mathcal{A}) \backslash \{0\}$, are the open cells of a regular cell decomposition $\Delta(\mathcal{A})$ of S^d. Furthermore, the mapping $X \mapsto \{y \in S^d : \sigma(y) \leq X\}$ gives an isomorphism*

$$\widehat{\mathcal{L}}(\mathcal{A}) \cong \widehat{\mathcal{F}}(\Delta(\mathcal{A}))$$

of the big face lattice of the oriented matriod $\mathcal{L}(\mathcal{A})$ and the face lattice of the regular cell complex $\Delta(\mathcal{A})$.

The complex $\Delta(\mathcal{A})$ is a PL sphere (in fact, it is shellable). This strengthening of Proposition 5.1.5, first shown by Edmonds and Mandel (1982), will follow as a consequence of Theorem 5.2.1. Notice that the proposition implies that the closure of an open cell $\sigma^{-1}(X)$ of $\Delta(\mathcal{A})$ is the set $\{y \in S^d : \sigma(y) \leq X\}$.

To make the relationship between arrangements and oriented matroids more precise we define the following equivalence relation: Two (signed) arrangements $\mathcal{A} = (S_e)_{e \in E}$ and $\mathcal{A}' = (S'_e)_{e \in E'}$ of pseudospheres in S^d are *topologically equivalent* (written $\mathcal{A} \sim \mathcal{A}'$) if there exists some homeomorphism $h : S^d \to S^d$ and some bijection $g : E \to E'$ such that $h(S_e) = S'_{g(e)}$ for all $e \in E$ (and also $h(S_e^+) = (S'_{g(e)})^+$ for all $e \in E$ in the signed case).

5.1.6 Theorem. *Two signed arrangements $\mathcal{A}$ and $\mathcal{A}'$ in S^d are topologically equivalent if and only if $\mathcal{L}(\mathcal{A}) \cong \mathcal{L}(\mathcal{A}')$.*

5.1.7 Corollary. *Let $\mathcal{A} = (S_e)_{e \in E}$ be a signed arrangement of pseudospheres in S^d. The oriented matroid $\mathcal{L}(\mathcal{A})$ is realizable (in the sense of Chapter 8) if and only if there exists a homeomorphism $h : S^d \to S^d$ such that $h(S_e)$ is a linear subsphere for all $e \in E$.*

The proof of these theorems will follow two technical lemmas. The first one makes the ball axiom more precise. The assumptions are as in Theorem 5.1.4.

5.1.8 Lemma. *Let $X \in \mathcal{L}(\mathcal{A}) \setminus \{0\}$, and $A = z(X)$. Then*
$$B_X = \{y \in S^d : \sigma(y) \le X\}$$
is a $(\dim S_A)$-dimensional ball. Furthermore, its relative boundary, as a subset of the sphere S_A, is
$$\partial B_X = \{y \in S^d : \sigma(y) < X\}.$$

Proof. Assume first that $A = \emptyset$, so that $S_A = S^d$. Choose a point $x \in S^d$ such that $\sigma(x) = X$, and for each $e \in E$ let S_e^x be the side of S_e that contains x. Then $B_X = \bigcap_{e \in E} \overline{S_e^x}$, so from Axiom (A3) of Definition 5.1.3 we have that B_X is either a sphere or a ball. We will show that it is a ball of dimension d.

The set-theoretic formula $\mathrm{Int}(A \cap B) = \mathrm{Int}(A) \cap \mathrm{Int}(B)$ for interiors implies that

$$(5.1.1) \qquad \mathrm{Int}(B_X) = \mathrm{Int}(\bigcap_{e \in E} \overline{S_e^x}) = \bigcap_{e \in E} \mathrm{Int}(\overline{S_e^x}) = \bigcap_{e \in E} S_e^x.$$

Hence, $x \in \mathrm{Int}(B_X)$. Therefore B_X is a k-dimensional ball or sphere, $k \le d$, which lies embedded in S^d with non-empty interior. By the invariance of domain theorem (see e.g. Spanier 1966) this cannot happen unless $k = d$. But being a proper subset of S^d, B_X is not a d-sphere (see Lemma 4.7.3). Hence, B_X is a d-ball.

We have already established in equation (5.1.1) that
$$\mathrm{Int}(B_X) = \{y \in S^d : \sigma(y) = X\}.$$
The expression for $\partial B_X = B_X \setminus \mathrm{Int}(B_X)$ then follows.

The case when $A \ne \emptyset$ can be reduced to the previous situation by passing to the arrangement $\mathcal{A}_A^{E \setminus A}$ induced in the sphere S_A. $\qquad \square$

Two of the covector axioms, to be verified in the proof of Theorem 5.1.4, require a different version of Lemma 5.1.8 that depends on two sign vectors X and Y. The idea is similar.

5.1.9 Lemma. *Let $X, Y \in \mathcal{L}(\mathcal{A})\backslash\{0\}$, and $A = z(X) \cap z(Y)$. Suppose that $\emptyset \neq B \subseteq (E\backslash A)\backslash S(X, Y)$, and for each $e \in B$ let S_e^* be that side of S_e which is determined by X_e and Y_e (this is well-defined since $\{X_e, Y_e\}$ by choice of B contains either $+1$ or -1, but not both). Then $S_A \cap (\bigcap_{e\in B} \overline{S_e^*})$ is a $(\dim S_A)$-dimensional ball.*

Proof. Consider the arrangement $\mathcal{A}_A^B = (S_A \cap S_e)_{e\in B}$. There are points $x, y \in S_A$ such that $\sigma(x) = X$ and $\sigma(y) = Y$. If $\underline{X} \cap B = B$, then the claim follows directly from Lemma 5.1.8 applied to the arrangement $\mathcal{A}_A^B$. Similarly if $\underline{Y} \cap B = B$, which also covers the case when $\underline{X} \cap B = \emptyset$.

We are left to deal with the situation when $\emptyset \neq \underline{X} \cap B \neq B$. Let $C = B\backslash \underline{X}$. Then $C \subseteq \underline{Y}$. Lemma 5.1.8 applied to the arrangement $(S_A \cap S_e)_{e\in C}$ shows that $S_A \cap (\bigcap_{e\in C} \overline{S_e^{Y_e}})$ is a $(\dim S_A)$-ball, in which y is an interior point and x is on the boundary. Then every neighborhood of x contains interior points of $S_A \cap (\bigcap_{e\in C} \overline{S_e^{Y_e}})$, and picking one such point z_1 sufficiently close to x we find for $\sigma(z_1) = Z$ that $Z_e = X_e$ for all $e \in \underline{X} \cap B$, and $Z_e = Y_e$ for all $e \in \underline{Y} \cap B$. Therefore, Lemma 5.1.8 applied to the vector Z and the arrangement $\mathcal{A}_A^B$ gives the desired result.

The proof is illustrated in Figure 5.1.1. For example, let $B = \{2, 3, 4, 6\}$ and $C = \{2, 4\}$. $\square$

We have now assembled the tools needed to prove Theorem 5.1.4. For geometric intuition the proof should be compared to that in the linear case, which is discussed in connection with Figure 4.1.1. Observe the similar features of Figures 4.1.1 and 5.1.1, that emphasize the parallel lines of reasoning.

Proof of Theorem 5.1.4. We will verify the Covector Axioms 4.1.1.

Axiom (L0). Clear.

Axiom (L2). Let $x, y \in S^d$ with $\sigma(x) = X$, $\sigma(y) = Y$, $A = z(X) \cap z(Y)$, and $B = \underline{Y} \cap z(X)$. Assume that $X \neq 0$, $Y \neq 0$ and $B \neq \emptyset$ (otherwise there is nothing non-trivial to prove), and let $R = \bigcap_{e\in B} \overline{S_e^{Y_e}}$. Then from Lemma 5.1.9 it follows that $S_A \cap R$ is a full-dimensional ball inside the sphere S_A. Now, $x, y \in S_A \cap R$, and being a ball it is possible to connect x to y with a path π that lies in the relative interior of $S_A \cap R$ except possibly for its endpoints. (Actually, x lies on the boundary of $S_A \cap R$ and y in its relative interior, but we don't need this.) If we move along π some sufficiently small distance away from x we reach a point z_1 which by construction satisfies $\sigma(z_1) = X \circ Y$. See Figure 5.1.1.

Axiom (L3). The argument here is similar. Let $x, y \in S^d$ with $\sigma(x) = X$ and $\sigma(y) = Y$, and suppose that $e \in S(X, Y)$. Let $A = z(X) \cap z(Y)$, $B_1 = \underline{X}\backslash S(X, Y)$, $B_2 = \underline{Y}\backslash S(X, Y)$, and

$$R = \left(\bigcap_{e\in B_1} \overline{S_e^{X_e}}\right) \cap \left(\bigcap_{e\in B_2} \overline{S_e^{Y_e}}\right).$$

We may assume that $B_1 \neq \emptyset$ or $B_2 \neq \emptyset$, since otherwise $X = -Y$ and $Z = 0$ will satisfy axiom (L3). Assume that $B_1 \neq \emptyset$. Then Lemma 5.1.9 shows that $S_A \cap R$ is a full-dimensional ball inside the sphere S_A. Connect x to y with a path π that lies in the relative interior of $S_A \cap R$ except possibly for its endpoints. Since x and y are in different connected components of $S^d \backslash S_e$, the path π must intersect S_e at some interior point, say $z_2 \in \pi \cap S_e$. Then $Z = \sigma(z_2)$ has the required properties: $Z_e = 0$ and $Z_f = (X \circ Y)_f$ for all $f \notin S(X, Y)$. See Figure 5.1.1.

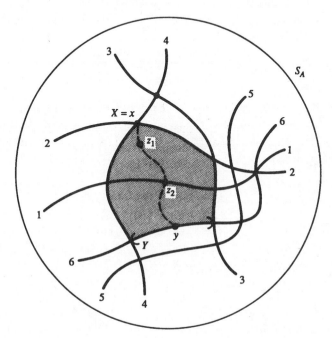

Figure 5.1.1: Sketch for the proofs of Lemma 5.1.9 and of Theorem 5.1.4.

Axiom (L1). This will be proved by induction on the number of pseudo-spheres, using that axiom (L2) is already verified. For arrangements of only one pseudosphere it is trivial.

Let $X \in \mathcal{L}(\mathcal{A}) \backslash \{0\}$, and $A = z(X)$. We know from Lemma 5.1.8 that $B_X = \{y \in S^d : \sigma(y) \leq X\}$ is a full-dimensional ball embedded in the sphere S_A.

Case 1. $\dim S_A = \dim S_E + 1$. Then S_E is a pseudosphere in S_A, and $S_A \cap S_e = S_E$ for all $e \in E \backslash A$. Clearly, $\sigma(x) = X$ for all x lying on one side of S_E in S_A, and by axiom (A2) $\sigma(y) = -X$ for all y lying on the other side. Hence, $-X \in \mathcal{L}(\mathcal{A})$.

Case 2. $\dim S_A > \dim S_E + 1$. Let $f \in \underline{X}$. From the induction assumption applied to the arrangement $\mathcal{A} \backslash \{S_f\}$ we have that there exists $Z \in \mathcal{L}(\mathcal{A})$ such that $Z_e = -X_e$ for all $e \in E \backslash f$.

The boundary of the $(\dim S_A)$-dimensional ball B_X cannot be properly contained in the $(\dim S_A - 1)$-dimensional sphere $S_A \cap S_f$ (Lemma 4.7.3). Also, $\partial B_X \neq S_A \cap S_f$, since otherwise $S_A \cap S_f = S_A \cap S_e$ for all $e \in \underline{X}$ which would imply that $S_A \cap S_f = S_E$, contradicting the codimension > 1 assumption. Hence, $\partial B_X \backslash S_f \neq \emptyset$. Pick $y \in \partial B_X \backslash S_f$ and let $Y = \sigma(y)$. Then $Y_f = X_f$, and from

Lemma 5.1.8 we see that $Y < X$, say $Y_g = 0 \neq X_g$. The induction assumption applied to the smaller arrangement $(S_g \cap S_e)_{e \in E \setminus g}$ in S_g shows that $-Y \in \mathcal{L}(\mathcal{A})$. But then, using axiom (L2), $-X = (-Y) \circ Z \in \mathcal{L}(\mathcal{A})$.

All oriented matroid axioms 4.1.1 have now been verified, and it remains only to prove the statement about rank. This will be done by induction on $d - k$, where $k = \dim S_E$. If $k = d - 1$, then clearly rank $\mathcal{L}(\mathcal{A}) = 1$.

Pick one pseudosphere $S_f, f \in E$, and let $A = \{e \in E : S_e = S_f\}$. The induced arrangement $\mathcal{A}_A^{E \setminus A} = (S_f \cap S_e)_{e \in E \setminus A}$ is $(d-1)$-dimensional. Hence, by induction the oriented matroid

$$\mathcal{L}(\mathcal{A}_A^{E \setminus A}) \cong \mathcal{L}(\mathcal{A})/A = \mathcal{L}(\mathcal{A})/f$$

has rank $d - k - 1$. It follows that rank $\mathcal{L}(\mathcal{A}) = d - k$. □

The preceding proof can be carried out without making full use of the fact that the sets B_X are balls. This way Theorem 5.1.4 can be proved using only axioms (A1) and (A2) in the definition of pseudosphere arrangement, and without first deducing the ball axiom (A3) in full force. See Exercise 5.6.

Proof of Proposition 5.1.5. By Lemma 5.1.8 the family of sets

$$\{B_X : X \in \mathcal{L}(\mathcal{A}) \setminus \{0\}\}$$

gives a covering of S^d by closed balls, and the boundary ∂B_X of each ball of dimension $\geq \dim S_E + 2$ is the union of lower-dimensional balls. Therefore, if $\mathcal{A}$ is essential (so $\dim S_E = -1$), the proposition follows directly from the Definition 4.7.4 of regular cell complexes. □

A non-essential arrangement $\mathcal{A} = (S_e)_{e \in E}$ of pseudospheres in S^d gives rise to a "relative cell-decomposition of S^d modulo S_E". This follows from Lemma 5.1.8 in just the same way. In this case we have a partition of S^d into sets $\sigma^{-1}(X)$, for $X \in \mathcal{L}(\mathcal{A})$, such that $\sigma^{-1}(0) = S_E$ is a non-empty sphere and $\sigma^{-1}(X)$, for $X \neq 0$, are open cells with closure $\sigma^{-1}([0, X])$. We will not have reason to further consider such relative cell decompositions.

Proof of Theorem 5.1.6. One direction is clear. For the other, assume that $g : E \to E'$ is a bijection that induces an isomorphism $\tilde{g} : \mathcal{L}(\mathcal{A}) \to \mathcal{L}(\mathcal{A}')$ of oriented matroids. The two induced cell decompositions $\Delta(\mathcal{A})$ and $\Delta(\mathcal{A}')$ of S^d have isomorphic face lattices. In fact, Proposition 5.1.5 shows that $\tilde{g}$ can be considered an isomorphism $\mathcal{F}(\Delta(\mathcal{A})) \to \mathcal{F}(\Delta(\mathcal{A}'))$. (We assume here that the arrangements are essential, the general case can be handled similarly.) By Corollary 4.7.9 the isomorphism $\tilde{g}$ of face lattices can be extended to a homeomorphism $h : S^d \to S^d$ inducing $\tilde{g}$. This means that $h(\sigma) = \tilde{g}(\sigma)$ for each closed cell $\sigma \in \Delta(\mathcal{A})$. This pair of maps h and g show that $\mathcal{A} \sim \mathcal{A}'$. □

5.2 The topological representation theorem

We have now come to the main result of this chapter, the Topological Representation Theorem of Folkman and Lawrence (1978). A proof based on *PL*

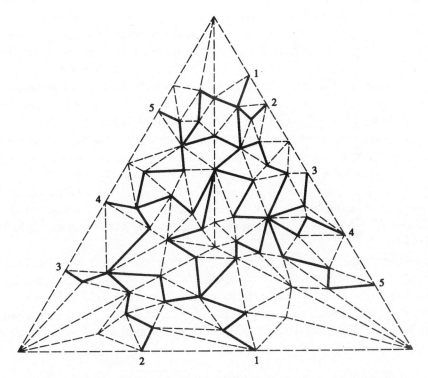

Figure 5.2.1: *PL* representation of a rank 3 oriented matroid
as an arrangement of five pseudocircles on the
boundary of a tetrahedron.

topology was given by Edmonds and Mandel (1982), leading to a stronger version
of the theorem, see the comment below.

5.2.1 Topological Representation Theorem.   *Let $\mathcal{L} \subseteq \{+,-,0\}^E$. Then
the following conditions are equivalent:*

(i)   *$\mathcal{L}$ is the set of covectors of a loop-free oriented matroid of rank $d + 1$.*

(ii)   *$\mathcal{L} = \mathcal{L}(\mathcal{A})$ for some signed arrangement $\mathcal{A} = (S_e)_{e \in E}$ of pseudospheres in
S^{d+1+k}, such that $\dim(\bigcap_{e \in E} S_e) = k$.*

(iii) *$\mathcal{L} = \mathcal{L}(\mathcal{A})$ for some signed arrangement $\mathcal{A}$ of pseudospheres in S^d, which
is essential and centrally symmetric and whose induced cell complex $\Delta(\mathcal{A})$
is shellable.*

By saying that an arrangement $\mathcal{A} = (S_e)_{e \in E}$ in S^d is *centrally symmetric* we
mean that $-S_e = S_e$ for all $e \in E$, i.e., that each individual pseudosphere is
invariant under the antipodal mapping $x \mapsto -x$ of S^d. In the signed case this is
equivalent to requiring that $-S_e^+ = S_e^-$ for all $e \in E$.

Recall that the *real projective d-space* $\mathbb{P}^d$ is the topological space obtained from S^d by identifying all pairs of antipodal points. The double covering map $\pi : S^d \to \mathbb{P}^d$, given by $\pi(x) = \{x, -x\}$, gives a natural identification of centrally symmetric subsets of S^d and general subsets of $\mathbb{P}^d$. This way centrally symmetric pseudospheres in S^d correspond to *pseudohyperplanes* in $\mathbb{P}^d$. Thus, the significance of central symmetry in Theorem 5.2.1 is that every oriented matroid can be represented (up to reorientation) by an *arrangement of pseudohyperplanes in projective space*. We will not in the text develop the details of this projective point of view other than in the 2-dimensional case, which is of special importance (see Section 6.2). See Exercise 5.8 for further information.

The significance of shellability in Theorem 5.2.1 is mainly that all aspects of the topological representation of oriented matroids can be carried out in PL topology. For the technical meaning of this see Section 4.7. The PL version of Theorem 5.2.1 is due to Edmonds and Mandel (1982). Figure 5.2.1 illustrates the PL representation of a rank 3 oriented matroid on the boundary of the 3-simplex. Three facets of the simplex are visible and the back facet is hidden from view.

It follows from Theorems 5.1.4 and 5.1.6 that the map $\mathcal{A} \mapsto \mathcal{L}(\mathcal{A})$ induces an injective map from equivalence classes of essential signed arrangements of pseudospheres in S^d to isomorphism classes of loop-free oriented matroids of rank $d + 1$. Theorem 5.2.1 shows that this mapping is surjective. This proves the first part of the following corollary. The second part is a direct consequence of the first.

5.2.2 Corollary.

(a)   *Loop-free oriented matroids of rank $d + 1$ (up to isomorphism) are in one-to-one correspondence with essential signed arrangements of pseudospheres in S^d (up to topological equivalence).*

(b)   *Loop-free oriented matroids of rank $d + 1$ (up to reorientation and isomorphism) are in one-to-one correspondence with essential arrangements of pseudospheres in S^d (up to topological equivalence).*

Clearly, Theorem 5.2.1 and Corollary 5.2.2 can be formulated for simple oriented matroids (no loops or parallel elements) if the corresponding arrangements are required to be repetition-free.

Proof. Most of Theorem 5.2.1 has already been proved. The implication (iii)$\Rightarrow$(ii) is trivial, and (ii)$\Rightarrow$(i) is given by Theorem 5.1.4.

The details of the implication (i)$\Rightarrow$(iii), except for central symmetry, are given in Section 4.3. Assume condition (i). By Theorem 4.3.3 there exists a shellable regular cell complex $\Delta(\mathcal{L})$ whose face lattice can be identified with $\hat{\mathcal{L}}$ and whose underlying space is homeomorphic to S^d. A comparison of Proposition 4.3.6 and Definition 5.1.3 shows that the system of subcomplexes $\mathcal{A} = \left(\Delta_e^0 \right)_{e \in E}$ forms an arrangement of pseudospheres in $\Delta(\mathcal{L}) \cong S^d$. (The fact that all relevant induced codimension 1 subspheres are tame follows from the fact that their closed sides are balls, using Proposition 4.3.6(b) and Lemma 5.1.1(iii).) A signature of the arrangement $\mathcal{A}$ is naturally induced by $\mathcal{L}$ in such a way that $\mathcal{L} = \mathcal{L}(\mathcal{A})$. The construction shows that $\Delta(\mathcal{A}) = \Delta(\mathcal{L})$, and $\mathcal{A}$ is essential since the cells are indexed by $\mathcal{L}\backslash\{0\}$.

It remains only to prove the statement about central symmetry. This will be a consequence of the following two claims. If $\mathcal{A} = (S_e)_{e \in E}$ is an arrangement of pseudospheres in S^d, a homeomorphism $\alpha : S^d \to S^d$ will be called a *compatible involution* if $\alpha(x) \neq x = \alpha^2(x)$ for all $x \in S^d$ and $\alpha(S_e) = S_e$ for all $e \in E$. Notice that then α must exchange the two sides of each S_e, otherwise Brouwer's fixed point theorem would be violated.

Claim 1. For every essential arrangement there is a compatible involution.

To prove this, assume that $\mathcal{A} = (S_e)_{e \in E}$ is an essential signed arrangement in S^d. Then the cells of the induced cell decomposition $\Delta(\mathcal{A})$ are indexed by the non-zero vectors of the associated oriented matroid $\mathcal{L}(\mathcal{A})$. The mapping $X \mapsto -X$ is a fixed-point-free order 2 automorphism of the face poset $\mathcal{L}(\mathcal{A})\backslash\{0\}$. By Corollary 4.7.9 it can be extended to a fixed-point-free order 2 homeomorphism $\alpha : \Delta(\mathcal{A}) \to \Delta(\mathcal{A})$ such that $\alpha(\sigma^{-1}(X)) = \sigma^{-1}(-X)$ for all $X \in \mathcal{L}(\mathcal{A})\backslash\{0\}$. Clearly, α is a compatible involution.

Claim 2. If $\mathcal{A} = (S_e)_{e \in E}$ is an essential arrangement in S^d and α a compatible involution, then there exists a homeomorphism $h : S^d \to S^d$ such that $h\alpha(x) = -h(x)$ for all $x \in S^d$.

This will be proved by induction on d, the $d = 1$ case being obvious. Suppose without loss of generality that $\mathcal{A}$ is repetition-free, and let $f \in E$. The induced arrangement $\mathcal{A}' = (S_f \cap S_e)_{e \in E\backslash f}$ in S_f is $(d-1)$-dimensional and the restriction of α to S_f is a compatible involution. Therefore by induction there exists a homeomorphism $g : S_f \to S^{d-1}$ such that $g\alpha(x) = -g(x)$ for all $x \in S_f$. See Figure 5.2.2.

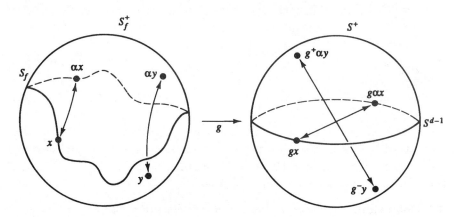

Figure 5.2.2: Sketch for the proof of Claim 2

Let S_f^+, S_f^- and S^+, S^- denote the (open) sides of the subspheres S_f and

S^{d-1}, respectively. The homeomorphism g of their boundaries can be extended across the d-balls $\overline{S_f^+}$ and $\overline{S^+}$ (Lemma 4.7.2). This gives a homeomorphism $g^+ : \overline{S_f^+} \to \overline{S^+}$ such that $g^+(x) = g(x)$ for all $x \in S_f$. Define a homeomorphism $g^- : \overline{S_f^-} \to \overline{S^-}$ by letting $g^-(y) = -g^+\alpha(y)$ for all $y \in \overline{S_f^-}$. Again, $g^-(x) = g(x)$ for all $x \in S_f$. Since g^+ and g^- agree on the intersection of their domains, these maps can be glued together (Lemma 4.7.1) to obtain a homeomorphism $h : S^d \to S^d$, defined by

$$h(x) = \begin{cases} g^+(x), & \text{if } x \in \overline{S_f^+}, \\ g^-(x), & \text{if } x \in \overline{S_f^-}. \end{cases}$$

By construction h satisfies $h\alpha(x) = -h(x)$ for all $x \in S^d$.

It is a consequence of these claims that *every essential arrangement of pseudospheres is topologically equivalent to some centrally symmetric arrangement.* For, given $\mathcal{A} = (S_e)_{e \in E}$ let α be a compatible involution, and then let h be a homeomorphism such as in Claim 2. Then the arrangement $h(\mathcal{A}) = (h(S_e))_{e \in E}$ is centrally symmetric. This concludes the proof of Theorem 5.2.1. $\qquad\square$

5.3 Pseudoconfigurations of points

The two "polar" geometric situations that are abstracted by oriented matroids – namely hyperplane arrangements and point configurations – motivate two alternative representations of oriented matroids by arrangements of pseudospheres. We will now formalize a generalization of the point configuration picture, which we call "pseudoconfigurations of points". In this context, *adjoints* of oriented matroids are introduced.

The material on point configurations in this section extends the work on "generalized configurations" of Goodman (1980), Cordovil (1982a) and Goodman and Pollack (1984b), which treats the case of rank 3. Also, very general representations of ordinary matroids by point configurations on surfaces, which do not relate to the combinatorics of pseudosphere arrangements, appear in the work of Mason (1971). The topological model that we use corresponds to the combinatorial framework of oriented matroids with adjoints, as introduced by Bachem and Kern (1986a).

For motivation of the following, consider a realizable oriented matroid $\mathcal{M}$ of rank r given by a point configuration P in real affine space $\mathbb{R}^{r-1}$. The whole oriented matroid $\mathcal{M}$ is determined by the cocircuits of $\mathcal{M}$, that is, by the partitions of P induced by the hyperplanes that are spanned by subsets of P. This suggests to try to represent an arbitrary oriented matroid by a point configuration together with the pseudohyperplanes it "spans".

5.3.1 Definition. A *pseudoconfiguration of points* of rank r is a pair $(\mathcal{A}, P)$, where $\mathcal{A} = (S_e)_{e \in E}$ is an essential arrangement of pseudospheres of rank r, and P is a collection of vertices of $\mathcal{A}$ with $|P| \geq r$, such that

(i)   every subset of P of size $r - 1$ is contained in a pseudosphere $S_e \in \mathcal{A}$, and

(ii)   every pseudosphere in $\mathcal{A}$ contains a subset of P that is not contained in any other pseudosphere of $\mathcal{A}$.

Here are a few initial remarks concerning this definition:

(a)   The fundamental requirement is (i): all the pseudospheres in $\mathcal{A}$ necessary to specify hyperplanes for a matroid on the point set P can be found in $\mathcal{A}$. Condition (ii) only deletes redundant pseudospheres that are not "spanned" by subsets of P. From (ii) it follows that P is not contained in a single pseudosphere S_e.

(b)   For a pseudoconfiguration of points it is geometrically irrelevant how $\mathcal{A}$ is signed. Therefore we assume that it is unsigned.

(c)   Condition (ii) implies that $\mathcal{A}$ is repetition-free: there cannot be multiple copies of pseudospheres. However, we do allow the possibility that P contains opposite pairs of vertices of $\mathcal{A}$. (In most cases this can be excluded with no loss of generality.)

(d)   By the *rank* of a pseudosphere arrangement we mean the rank of the associated oriented matroid, see Theorem 5.1.4. Hence, a pseudoconfiguration of points of rank r "lives" in an arrangement of pseudospheres on S^{r-1}.

Pseudoconfigurations of points can be viewed as objects that are the "projective polars" of pseudosphere arrangements. For example, the point configuration of Figure 5.3.2(b) is "projectively polar" to the arrangement of lines in Figure 5.3.2(a), having reversed point-line incidences.

Our first key result states that every pseudoconfiguration of points represents an oriented matroid on P – this is the "projective polar" of Theorem 5.1.4. In particular, the pseudospheres of a pseudoconfiguration of points describe the hyperplanes of a matroid on P, and the partition of P into positive, negative and zero parts with respect to any of the pseudospheres describes a cocircuit signature for this matroid.

We will use sign vectors $\bar{\sigma}(S_e) \in \{+,-,0\}^P$ to indicate the partition of the point set P by the oriented pseudosphere S_e. That is, for $p \in P$ we set

$$\bar{\sigma}(S_e)_p := \begin{cases} + & \text{if } p \in S_e^+, \\ - & \text{if } p \in S_e^-, \\ 0 & \text{if } p \in S_e. \end{cases}$$

Note that this $\bar{\sigma}$ describes the signs of the point configuration P with respect to a fixed pseudosphere S_e, whereas the σ used in the previous sections records the signs of a pseudosphere arrangement with respect to a single point p, such that $\bar{\sigma}(S_e)_p = \sigma(p)_e$ for a point p and a pseudosphere S_e.

5.3.2 Proposition.   *Every pseudoconfiguration of points* $(\mathcal{A}, P)$ *uniquely determines an oriented matroid* $\mathcal{M}_P = (P, \mathcal{C}^*)$ *on* P, *whose cocircuits are given by*

$$\mathcal{C}^* = \mathcal{C}^*(\mathcal{A}, P) := \{\bar{\sigma}(S_e) : e \in E\} \cup \{-\bar{\sigma}(S_e) : e \in E\}.$$

$\mathcal{M}_P$ *is called the oriented matroid of the pseudoconfiguration* $(\mathcal{A}, P)$.

In this proposition an orientation has been arbitrarily chosen for each pseudo-sphere $S_e \in \mathcal{A}$. Observe that $\mathcal{M}_P$ depends only on the reorientation class of $\mathcal{A}$. In fact, if S_e is reoriented, we get that $\bar{\sigma}(-S_e) = -\bar{\sigma}(S_e)$, so

$$C^* = \{\pm\bar{\sigma}(S_e) : e \in E\}$$

is not changed. Also, note that $\mathcal{M}_P$ need not be simple – it can have antiparallel elements, corresponding to opposite vertices in P.

The proof of Proposition 5.3.2 will be done in two parts. The first verifies that the set $\{S_e \cap P : e \in E\} = \{z(\bar{\sigma}(S_e)) : e \in E\}$ is the collection of hyperplanes of a matroid M_P of rank r on the set P. This requires a lattice theoretic argument within the intersection lattice of $\mathcal{A}$, which is isolated and deferred to Lemma 5.3.3 below.

The second half of the proof is geometric. It verifies that signed cocircuit elimination holds for every modular pair of vectors $\bar{\sigma}(e_1), \bar{\sigma}(e_2) \in C^*(\mathcal{A}, P)$, that is, for spheres $S_{e_1}, S_{e_2} \in \mathcal{A}$ such that $P \cap S_{e_1} \cap S_{e_2}$ has rank $r - 2$.

Proof of Proposition 5.3.2.

Part 1. $\mathcal{A}$ is an essential pseudosphere arrangement in S^{r-1}, so its poset of cells

$$\mathcal{L} := \{\sigma(p) : p \in S^{r-1}\}$$

is an oriented matroid of rank r. We consider P as a subset of the set of atoms of $\mathcal{L}$, that is, as a set of cocircuits of the oriented matroid.

For a while we will ignore orientation, which corresponds to taking supports. So we get a support map $X \longmapsto \underline{X}$, which is a surjective and order preserving map

$$\mathcal{L} \to \check{L} := \{\underline{X} : X \in \mathcal{L}\}.$$

We order $\check{L}$ by inclusion. By Proposition 4.1.13(i) it is the cocircuit span (set of complements of flats) of a matroid of rank r, so $\check{L}$ is the opposite (order dual) of a geometric lattice, which we refer to as an *inverted geometric lattice*.

The lattice $\check{L}$ can via the map $A \mapsto S_{E\backslash A} = \bigcap_{e \in E\backslash A} S_e$ also be viewed as the set of unsigned spheres in $\mathcal{A}$ (i.e., all spheres S_e and their intersections), ordered by inclusion. In particular, P is identified with a set of atoms of $\check{L}$, whereas the coatoms of $\check{L}$ are in bijection with the unsigned pseudospheres S_e in $\mathcal{A}$. (Here we assume, without loss of generality, that P does not contain opposite vertices.)

From Definition 5.3.1 we deduce that $|P| \geq r$ and

(i)   no subset P_0 of P with $|P_0| < r$ has join $\hat{1}$,

(ii)   every coatom of $\check{L}$ is the join of a subset of P.

Hence, Lemma 5.3.3 below implies that the subposet

$$L_P := \{\vee P_0 : P_0 \subseteq P\}$$

of $\check{L}$ is a geometric lattice of rank r, whose coatoms are all the coatoms of $\check{L}$. In other words, $\{P \cap S_e : e \in E\}$ is the set of hyperplanes of a rank r matroid $M_P = (P, C^*)$ on P, where $C^* := \{P\backslash S_e : e \in E\}$ is the corresponding family of (unsigned) cocircuits.

Part 2. We now want to prove that $\mathcal{C}^* = \{\bar{\sigma}(S_e) : e \in E\}$ is a valid cocircuit signature of M_P. For this we first note that $\overline{\bar{\sigma}(S_e)} = P \backslash S_e$, so that $\mathcal{C}^*$ is the set system underlying $\mathcal{C}^*$.

By Theorem 3.6.1* it suffices to prove signed cocircuit elimination for every *modular pair* of cocircuits $\bar{\sigma}(e_1)$ and $\bar{\sigma}(e_2)$ in $\mathcal{C}^*$. Now a pair of cocircuits is modular (as a pair of subsets of the dual underlying matroid $\underline{M}_P^*$) if and only if their complements intersect in a flat of rank $r - 2$ of $\underline{M}_P$.

Let $S_1 := S_{e_1}$ and $S_2 := S_{e_2}$ be the signed pseudospheres of $\mathcal{A}$ corresponding to the modular pair of cocircuits $\bar{\sigma}(e_1)$ and $\bar{\sigma}(e_2)$. Consider a point $p \in P$ in $S_1^+ \cap S_2^-$. Now modularity says that

$$\rho(P \cap (S_1 \cap S_2)) = r - 2,$$

which implies that $S_A := S_1 \cap S_2$ is a sphere of corank 2. So, there is an (up to sign) unique pseudosphere $S_h \in \mathcal{A}$ with $p \in S_h$ that contains S_A, namely the lattice join in L_P of S_A and p.

(*Remark:* The three spheres S_1, S_2 and S_h form a subarrangement of rank 2 in $\mathcal{A}$, so the rest of this argument will in fact amount to rather trivial statements about rank 2 arrangements, as treated in Section 6.1.)

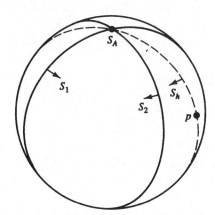

Figure 5.3.1: Signed elimination of a modular pair of cocircuits (sketch).

S_A is a sphere of codimension 1 in S_h, so it divides S_h into two connected halves, namely into

$$S_h \cap S_1^+ = S_h \cap S_2^- = S_h \cap (S_1^+ \cap S_2^-),$$

which contains p, and into

$$S_h \cap S_1^- = S_h \cap S_2^+ = S_h \cap (S_1^- \cap S_2^+).$$

Thus $S_h \cap (S_1^+ \cap S_2^+) = \emptyset$, which means that $S_1^+ \cap S_2^+$ is completely contained in one side of S_h. That is, $S_1^+ \cap S_2^+ \subseteq S_h^+$ with the right choice of positive side for S_h.

By symmetry, we now get $S_1^- \cap S_2^- \subseteq S_h^-$, and thus $\bar{\sigma}(h)$ is the (unique) signed cocircuit obtained by signed elimination of p between $\bar{\sigma}(e_1)$ and $\bar{\sigma}(e_2)$. $\qquad\square$

5.3.3 Lemma. *Let $\check{L}$ be an inverted geometric lattice of rank r, and let P be a set of atoms of $\check{L}$ with $|P| \geq r$, such that*

(i) $\vee P_0 < \hat{1}$ *for every subset $P_0 \subseteq P$ with $|P_0| = r - 1$, and*

(ii)   *every coatom of $\check{L}$ is the join of a subset of P.*

Then the subposet $L_P := \{\vee P_0 : P_0 \subseteq P\}$ of $\check{L}$ is a geometric lattice of rank r, and $\check{L}^{op}$ is an adjoint of L_P, (cf. Definition 5.3.5).

Proof. We first show that the family $\mathbf{B}$ of subsets $P_0 \subseteq P$ that are minimal (under inclusion) such that $\vee P_0 = \hat{1}$ are the bases of a rank r matroid on P. For this, first note that $\mathbf{B} \neq \emptyset$ from (ii).

Every $B \in \mathbf{B}$ satisfies $|B| = r$. To see this let $B = \{p_1, \ldots, p_s\}$ and consider the multichain

$$\hat{0} < p_1 \leq p_1 \vee p_2 \leq \ \ldots \ \leq p_1 \vee \ldots \vee p_{s-1} \leq p_1 \vee \ldots \vee p_s = \hat{1}.$$

If for some i ($2 \leq i \leq s$) we had $p_1 \vee \ldots \vee p_{i-1} = p_1 \vee \ldots \vee p_i$, then $\vee(B \backslash \{p_i\}) = \vee B$, so B would not be minimal. Hence we have a strict chain

$$\hat{0} < p_1 < p_1 \vee p_2 < \ \ldots \ < p_1 \vee \ldots \vee p_{s-1} < p_1 \vee \ldots \vee p_s = \hat{1},$$

which implies that $s \leq r$ because $\check{L}$ has rank r. However, from (i) and $\vee B = \hat{1}$ we have that $s \geq r$, so $s = r$.

Next consider two sets B_0 and B_1 in $\mathbf{B}$, and let $p_0 \in B_0 \backslash B_1$. Then $\vee(B_0 \backslash \{p_0\}) < \hat{1}$, but $\vee((B_0 \backslash \{p_0\}) \cup B_1) = \hat{1}$, which implies that B_1 contains an element p_1 such that $\vee((B_0 \backslash \{p_0\}) \cup \{p_1\}) = \hat{1}$. Hence $(B_0 \backslash \{p_0\}) \cup \{p_1\} \in \mathbf{B}$, and the basis exchange axiom has been verified.

Thus $M_P = (P, \mathbf{B})$ is a matroid of rank r given by its family of bases. Its hyperplanes are the maximal subsets of P that do not contain a basis, so they are the sets $\{p \in P : p \leq H\}$ for the coatoms H of $\check{L}$. So the hyperplanes of M_P correspond to the coatoms of $\check{L}$, which are the coatoms of L_P by (ii). Consequently the flats of M_P, being the intersections of the hyperplanes, correspond to the elements of L_P, so that L_P is the lattice of flats of M_P. $\qquad\square$

Pseudoconfigurations of points can by Proposition 5.3.2 be used to represent oriented matroids. To see the analogy with and the differences from the representations used in Section 5.2, let us reformulate the Topological Representation Theorem: its one direction states that every loop-free rank r oriented matroid $\mathcal{M} = (E, \mathcal{C}^*)$ has a TYPE I *representation*, by which we mean an essential pseudosphere arrangement $\mathcal{A} = (S_e)_{e \in E}$ in S^{r-1} indexed by the elements of E, such that $\mathcal{C}^*$ is in bijection with the vertex set P of $\mathcal{A}$ via $\mathcal{C}^* = \{\sigma(p) : p \in P\}$.

5.3.4 Definition. Let $\mathcal{M} = (P, \mathcal{C}^*)$ be an oriented matroid of rank r. A TYPE II *representation* of $\mathcal{M}$ is a pseudoconfiguration of points $(\mathcal{A}, P)$ of rank r such that $\mathcal{M}$ is the oriented matroid of $(\mathcal{A}, P)$.

Thus a TYPE II representation of $\mathcal{M}$ amounts to a pseudosphere arrangement $\mathcal{A} = (S_y)_{y \in F}$ in which P is (identified with) a subset of the vertices, and the pseudospheres S_y are in bijection with the pairs of opposite cocircuits of $\mathcal{M}$, via $\mathcal{C}^* = \{\pm \bar{\sigma}(S_y) : y \in F\}$.

Proposition 5.3.2 leads us to ask whether *every* simple oriented matroid has a TYPE II representation. The answer will turn out to be negative – for non-trivial reasons.

For this, we have to understand the relationship between the two oriented matroids associated with a pseudoconfiguration of points $(\mathcal{A}, P)$: namely the oriented matroid $\mathcal{M}(\mathcal{A})$ TYPE I-represented by $\mathcal{A}$, and the oriented matroid $\mathcal{M}_P$ that is TYPE II-represented by $(\mathcal{A}, P)$. There is a purely combinatorial description of the relationship between $\mathcal{M}(\mathcal{A})$ and $\mathcal{M}_P$, using the notion of an "oriented adjoint". We will therefore give a review of adjoints, starting with the unoriented case.

An *adjoint* of a matroid M can be defined as a matroid M^{ad} of the same rank, such that there is an order preserving embedding

$$L^{op} \hookrightarrow L^{ad}$$

which identifies the coatoms of L (hyperplanes of M) with the atoms of L^{ad} (points of M^{ad}). Here we think of M and M^{ad} as simple matroids with geometric lattices L and L^{ad}.

We will use that every geometric lattice is completely determined by its subposet of atoms and coatoms: namely, this subposet identifies every coatom (hyperplane) with the set of atoms (points) below it, and thus the flats of L are given as the meets of coatoms (intersections of hyperplanes). Therefore an embedding $L^{op} \hookrightarrow L^{ad}$ is equivalent to an embedding of the corresponding subposets of atoms and coatoms. In particular, one can see that every embedding of L^{op} into L^{ad} with $r(L^{op}) = r(L^{ad})$ preserves not only ranks and cover relations, but also meets. This will turn out to be important later.

The bijection between the hyperplanes of M and the points of M^{ad} allows us to interpret the adjoint as a matroid on the set of hyperplanes of M, or, equivalently, as a matroid on the set of cocircuits of M. This, and the characterization as an embedding of the subposet of atoms and coatoms, suggests the right generalization to oriented matroids.

5.3.5 Definition (Cheung 1974, Bachem and Kern 1986a).

(1) Let $M = M(E)$ be a matroid of rank r. An *adjoint* of M is a matroid M^{ad} of the same rank r on $E^{ad} := C^*(M)$ such that the sets $Z[e] = \{Y \in E^{ad} : e \in Y\}$, for all $e \in E$, are cocircuits of M^{ad}.

(2) Let $\mathcal{M} = \mathcal{M}(E)$ be an oriented matroid of rank r. An *adjoint* of $\mathcal{M}$ is an oriented matroid $\mathcal{M}^{ad}$ of rank r on a subset $E^{ad} \subseteq C^*(\mathcal{M})$ with $C^*(\mathcal{M}) = E^{ad} \cup (-E^{ad})$, such that the sign vectors $Z[e] = (Y_e)_{Y \in E^{ad}}$, for all $e \in E$, are cocircuits of $\mathcal{M}^{ad}$.

Equivalently, consider the cocircuit incidence matrix (cf. Problem 4.45), whose *rows*

- in the case of unoriented matroids, are all the cocircuit vectors,
- for oriented matroids contain exactly one member from every pair of opposite cocircuit vectors.

Then an adjoint is an (oriented) matroid *some* of whose cocircuits are given by the *columns* of the cocircuit matrix. This matrix representation of adjoints, due to Billera and Munson (1984a), is sometimes very handy. We will use it in connection with the Lawrence construction in Section 9.3.

Note that the underlying matroid of an oriented adjoint $\mathcal{M}^{ad}$ yields an adjoint M^{ad} of $M = \underline{\mathcal{M}}$.

Adjoints are not unique (in the oriented as well as in the unoriented version) and they may also fail to exist. For example, in Sections 7.4 and 7.5 we will show that the Vámos matroid $V'(8)$ is orientable, but it does not have an adjoint, and thus no orientation of $V'(8)$ can have an oriented adjoint. Furthermore there is no unique way to recover the original $\mathcal{M}$ from an adjoint (Exercise 5.13(a)). Also there are uniform oriented matroids that do not have an adjoint, as we will see later (Section 10.4).

However, every realizable oriented matroid has a (realizable) adjoint. To see this, consider a TYPE I representation of an oriented matroid $\mathcal{M}$ by an arrangement of linear pseudospheres $\mathcal{A}$. Let $P = E^{ad}$ be the set of vertices of $\mathcal{A}$. Now consider the arrangement of *all* linear pseudospheres spanned by subsets of P. This yields a new, larger arrangement $\mathcal{A}' \supseteq \mathcal{A}$, such that $(\mathcal{A}', P)$ is a TYPE II representation of an adjoint of $\mathcal{M}$. Thus every realized oriented matroid has a unique adjoint associated with it – however, the combinatorial type of that adjoint can depend on the particular realization (Exercise 5.13(b)).

The significance of adjoints for our current discussion comes from the following simple but basic observation, which amounts to "a topological representation theorem for oriented matroids with adjoint".

5.3.6 Theorem.   *A simple oriented matroid has a representation by a pseudoconfiguration of points (a TYPE II representation) if and only if it has an adjoint.*

Proof. If $(\mathcal{A}, P)$ is a TYPE II representation of an oriented matroid $\mathcal{M}$, then $\mathcal{A}$ gives a TYPE I representation of an adjoint $\mathcal{M}^{ad}$. Conversely, from a TYPE I representation $\mathcal{A}$ of an adjoint $\mathcal{M}^{ad}$ it is clear how to select vertices P giving a TYPE II representation $(\mathcal{A}, P)$ of $\mathcal{M}$. The result therefore follows from the Topological Representation Theorem 5.2.1.   □

This gives us a nice geometric interpretation of oriented adjoints, which will turn out useful later. It is illustrated in Figure 5.3.2 for a small example. For this example, $E = \{1, 2, 3, 4, 5\}$ and

$$\mathcal{C}^* = \{\pm(00{+}{+}{+}), \pm(0{-}0{+}0), \pm(0{-}{-}0{-}), \pm({+}00{+}{+}), \pm({+}0{-}00), \pm({+}{+}00{+})\}.$$

To test this geometric point of view, the reader could check that if $\mathcal{M}$ has an adjoint, then also every minor of $\mathcal{M}$ has an adjoint (Exercise 5.15) – this is hard to see from the algebraic definition of adjoints, but (nearly) obvious from the TYPE I/II representations.

We have seen adjoints arise in connection with the representation of oriented

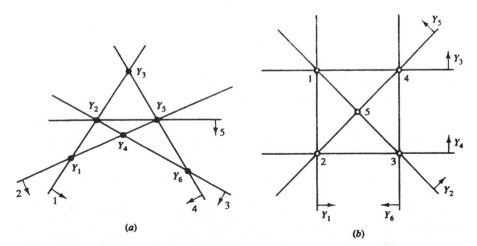

Figure 5.3.2: Topological representations of an oriented matroid $(E,\mathcal{C}^*)$:
(a) Type I representation. (b) Type II representation.

matroids by pseudoconfigurations of points. This set-up might obscure the fact that the existence of an adjoint is a very strong property. This becomes clear when one studies the intersection properties of oriented matroids (see Section 7.5.) Each intersection property requires the existence of certain single element extensions. All these properties are implied by the existence of an adjoint, which suggests to consider the existence of an adjoint itself as a strong intersection property.

We have now reduced the (topological) question

"Which oriented matroids have TYPE II representations?"

to the (combinatorial) problem

"Which oriented matroids have adjoints?".

An incomplete answer will be given in Chapter 7, using different tools. As a preview, and to collect the results, we note the following:

(i)   every realizable oriented matroid has a realizable adjoint (see above),

(ii)   every rank 3 oriented matroid has an adjoint (Proposition 6.3.6),

(iii) if $\mathcal{M}$ has an adjoint, then it has strong intersection properties (Section 7.5), implying that certain generalizations of classical convexity theorems hold for $\mathcal{M}$ (Section 9.2),

(iv)   if $\mathcal{M}$ has an adjoint, then so does $\underline{\mathcal{M}}$ (see above),

(v)   the Vámos matroid $V'(8)$ has no adjoint, so its oriented version has no oriented adjoint (Sections 7.4 and 7.5),

(vi) the Edmonds-Fukuda-Mandel oriented matroid EFM(8) has no adjoint (Section 10.4), although the underlying matroid $\underline{EFM(8)} = U_{4,8}$ is realizable.

Exercises

5.1 Show that the words "with sides $S_A \cap S_e^+$ and $S_A \cap S_e^-$" cannot be dropped from axiom (A2) in Definition 5.1.3.

5.2 Let $\mathcal{A}$ be an arrangement of pseudospheres in S^d. Show that every non-empty intersection of open sides of these pseudospheres is homeomorphic to the open d-ball.

(Folkman and Lawrence 1978, Edmonds and Mandel 1982)

5.3 Show that the two descriptions of the intersection lattice $L(\mathcal{A})$ given after Theorem 5.1.4 are equivalent.

5.4 In Theorem 5.1.6 it suffices to assume that $\mathcal{L}(\mathcal{A}) \cong \mathcal{L}(\mathcal{A}')$ is an isomorphism of unlabeled posets. Prove this.

5.5 Let $\mathcal{A}$ and $\mathcal{A}'$ be two essential arrangements of pseudospheres in S^d, and let M and M' be their underlying (unoriented) matroids. Show that if the cell complexes $\Delta(\mathcal{A})$ and $\Delta(\mathcal{A}')$ are combinatorially isomorphic then also $M \cong M'$, but not conversely.

5.6 (a) Assuming only the axioms (A1) and (A2) in Definition 5.1.3, prove the following weaker form of Lemma 5.1.8:

Let $\mathcal{A}$ be a signed arrangement of pseudospheres, and let $X \in \mathcal{L}(\mathcal{A}) \setminus \{0\}$. Then the set $B_X = \{y \in S^d : \sigma(y) \leq X\}$ has the following properties as a subset of the sphere $S_{z(X)}$:

   (i) B_X is a closed set with relative boundary $\partial B_X = \{y \in S^d : \sigma(y) < X\}$,
   (ii) $B_X = \overline{IntB_X}$, where $IntB_X = B_X \setminus \partial B_X$,
   (iii) $IntB_X$ is path connected.

(b) Prove Theorem 5.1.4 assuming only the axioms (A1) and (A2) in Definition 5.1.3.   (Edmonds and Mandel 1982)

(c) Let "arrangement" mean a finite collection of codimension 1 subspheres in S^d satisfying axioms (A1) and (A2) with the word "pseudosphere" in (A2) replaced by "codimension 1 subsphere". Can Theorem 5.1.4 be proved for some non-trivial class of such arrangements?

(In other words: Which arrangements of *wild* spheres give rise to oriented matroids?)   (Hochstättler 1995)

5.7 Let $\mathcal{A} = (S_e)_{e \in E}$ be an arrangement of pseudospheres in S^d, with $\dim S_E = k$. Let $\mathcal{M} = \mathcal{M}(\mathcal{A})$ be the corresponding oriented matroid of rank $d - k$. Let $\mathcal{M}' = \mathcal{M} \oplus B^{k+1}$ be the (unique) oriented matroid obtained by adding $k + 1$ coloops to $\mathcal{M}$.

(a) Show that $\mathcal{A}$ can be extended to a pseudosphere arrangement $\mathcal{A}'$ in S^d representing $\mathcal{M}'$.

(b) Let S be the intersection of the $k+1$ pseudospheres which in $\mathcal{A}'$ represent the "new" coloops of $\mathcal{M}'$. Show that S is a $(d-k-1)$-sphere, and that $(S \cap S_e)_{e \in E}$ is an essential arrangement in S whose oriented matroid is $\mathcal{M}$.

(This is the *essential arrangement associated with* $\mathcal{M}$.)

5.8   Let $\mathbb{P}^d = S^d/(x \sim -x)$ denote d-dimensional real projective space, here modeled by the standard d-sphere modulo identification of antipodal points. The canonical inclusion $S^k \subseteq S^d$, obtained by adding $d-k$ zero coordinates, induces an inclusion $\mathbb{P}^k \subseteq \mathbb{P}^d$, for $0 \leq k \leq d$. A k-*pseudoflat* in $\mathbb{P}^d$ is the image of $\mathbb{P}^k$ by a self-homeomorphism of $\mathbb{P}^d$. A finite collection $\mathcal{P} = (H_e)_{e \in E}$ of *pseudohyperplanes* (i.e., $(d-1)$-pseudoflats) in $\mathbb{P}^d$ is called a *(projective)* *arrangement of pseudohyperplanes* if the intersection of every subcollection is a pseudoflat or empty. Let $\pi : S^d \longrightarrow \mathbb{P}^d$ be the double-covering map given by $\pi(x) = \{x, -x\}$.

(a) Show that $H \subseteq \mathbb{P}^d$ is a pseudohyperplane if and only if $\pi^{-1}(H)$ is a centrally symmetric pseudosphere.

(b) Show that $\mathcal{P} = (H_e)_{e \in E}$ is an arrangement of pseudohyperplanes if and only if $(\pi^{-1}(H_e))_{e \in E}$ is a centrally symmetric arrangement of pseudospheres.

(c) Deduce the following projective version of the Topological Representation Theorem: the reorientation classes of simple oriented matroids of rank $d+1$ are in one-to-one correspondence with the topological equivalence classes of essential arrangements of pseudohyperplanes in $\mathbb{P}^d$. (*Remark*: see Section 6.2 for the case $d = 2$.)

(Folkman and Lawrence 1978, Edmonds and Mandel 1982)

5.9   Show the following:

(a) Every essential arrangement $\mathcal{P}$ of pseudohyperplanes in $\mathbb{P}^d$ induces a regular cell decomposition $\Delta(\mathcal{P})$ of the space $\mathbb{P}^d$.

(b) The cell complex $\Delta(\mathcal{P})$ is shellable if and only if $d \leq 1$.

5.10   (a) Show that an oriented matroid has a restriction of size k that is realizable if and only if it can be represented by a pseudosphere arrangement in which k of the pseudospheres are linear.

(b) Show that if $\mathcal{M}$ can be represented by an arrangement of smooth, transversally intersecting pseudospheres, then every restriction of $\mathcal{M}$ of rank $r(\mathcal{M}) - 1$ is realizable.

(c)* Disprove the converse of (b).

5.11   Define a *Miller arrangement* to be an arrangement of pseudospheres such that one of the pseudospheres S_g is linear and all other pseudospheres S_e are linear in an open neighborhood of S_g. (This models affine arrangements of pseudohyperplanes, all of whose pseudohyperplanes are linear outside a bounded region.)   (Miller 1987)

(a) Show that every oriented matroid of rank 3 is Miller representable (that is, can be represented by a Miller arrangement). (Compare with Section 6.3.)

(b) Show that an oriented matroid $\mathcal{M} = (E, \mathcal{C})$ is Miller representable if and only if $\mathcal{M}^* \backslash e$ is realizable, for some $e \in E$.

(c) Conclude that for $r \geq 4$ there are oriented matroids of rank r that are not Miller representable.

(d) Show that every oriented matroid on at most 9 points is Miller representable.

(e)* What is the minimum cardinality of an oriented matroid that is not Miller representable?

5.12 Show that in a pseudoconfiguration of points $(\mathcal{A}, P)$ of rank r (Definition 5.3.1) every pseudosphere contains a subset of P *of size* $r - 1$ that is not contained in any other pseudosphere of $\mathcal{A}$.

5.13 (a) Construct two non-isomorphic simple oriented matroids of rank 3 that have isomorphic adjoints.

(b) Construct an oriented matroid of rank 3 with two realizations that generate non-isomorphic adjoints.   (Bachem and Kern 1986a,b)

5.14 (a) Show that if P(9) is the non-Pappus oriented matroid (Figure 1.3.3), then every adjoint of P(9) contains a submatroid isomorphic to P(9).

(b) Does RS(8) have an adjoint?

5.15 Show that if $\mathcal{M}$ has an adjoint $\mathcal{M}^{ad}$, then every minor of $\mathcal{M}$ has an adjoint that is a minor of $\mathcal{M}^{ad}$.   (Cordovil 1987)

5.16 Prove that if $\mathcal{M}$ is not realizable, then $\mathcal{M}^{ad}$ cannot be realizable.

5.17 (a) Show that every adjoint is a simple (oriented) matroid.

(b) Show that if $\mathcal{M}$ is a simple oriented matroid and $\mathcal{M}^{ad}$ an adjoint, then $|E^{ad}| \geq |E|$, with equality if and only if $\mathcal{M}$ is the free oriented matroid with $|E| = r(\mathcal{M})$.

5.18 Show that if $\mathcal{M}^{adad}$ is an adjoint of an adjoint of $\mathcal{M}$, then $\mathcal{M}$ is (isomorphic to) a submatroid of $\mathcal{M}^{adad}$.

6

Arrangements of Pseudolines

Oriented matroids differ from rectilinear geometry by a tame topological deformation. This statement is given precise meaning by the Topological Representation Theorem 5.2.1. Sometimes this deviation allows for phenomena that cannot otherwise occur, and in this way oriented matroids provide a precise language for discussing the question: Which properties in space are truly geometrical, and which are essentially combinatorial?

To understand the Topological Representation Theorem it is very instructive to study the rank 3 case (the first non-trivial case), where visualization is easy. In the projective version, this identifies rank 3 oriented matroids with *arrangements of pseudolines*, a topic studied long before the advent of oriented matroid theory.

We will not attempt to give a comprehensive treatment of the basic results concerning arrangements of pseudolines. Grünbaum (1972) gives an excellent exposition of this material with many interesting examples, and his monograph is still the best place to enter the subject. This chapter is instead devoted to a rather detailed discussion of those aspects of the subject which are important from an oriented matroid point of view. Also, some general results that update the information in Grünbaum (1972) are given.

6.1 Arrangements of pseudospheres in low dimensions

The Topological Representation Theorem 5.2.1 assumes much simpler form when stated for oriented matroids of rank at most 3. Let us look at these cases in turn.

Rank 1. Since $S^0 = \{-1, +1\}$, and $S^{-1} = \emptyset$ with sides $\{-1\}$ and $\{+1\}$ as a subsphere of S^0, a signed arrangement $\mathcal{A} = (S_e)_{e \in E}$ of (-1)-spheres in S^0 amounts to a choice of one of the two elements -1 and $+1$ for each $e \in E$.

$$-1 \circ \qquad \circ +1$$

Figure 6.1.1: Rank 1 oriented matroid.

It follows that $|E| = 1$ and $|\mathcal{C}^*| = 2$ for every simple rank 1 oriented matroid $\mathcal{M} = (E, \mathcal{C})$.

Rank 2. This case is a little more interesting. The pseudospheres in S^1 are pairs of points, and their sides are the two semicircles into which they partition S^1. The requirement for an arrangement $\mathcal{A} = (S_e)_{e \in E}$ of 0-spheres in S^1 that $S_e \cap S_f$ is a (-1)-sphere with sides $S_e \cap S_f^+$ and $S_e \cap S_f^-$ (Definition 5.1.3(ii)), means that any two distinct 0-spheres $S_e = \{a, a'\}$ and $S_f = \{b, b'\}$ are in a crossing position on S^1, i.e., come in circular order $aba'b'$ or the reverse. Hence, a typical repetition-free arrangement of four pseudospheres in S^1 looks as in Figure 6.1.2.

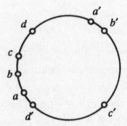

Figure 6.1.2: Rank 2 oriented matroid.

One sees that up to relabeling of E and reorientation there exists exactly one simple rank 2 oriented matroid on every ground set E, for $|E| \geq 2$.

It is obvious that every one-dimensional arrangement, such as the one in Figure 6.1.2, can be "straightened", meaning that some self-homeomorphism of the circle will move each related pair of points into antipodal position. Equivalently, every rank 2 oriented matroid is realizable.

Rank 3. Let us call a tamely embedded copy of S^1 in S^2 a *pseudocircle* (cf. the discussion in connection with Lemma 5.1.1). It is a non-trivial topological fact that *every* simple closed curve (i.e., homeomorphic image of S^1) in S^2 is a pseudocircle (there are no "wild" circles). This is the content of the Schönflies theorem, see Rushing (1973). By an *arrangement of pseudocircles* $\mathcal{A} = (S_e)_{e \in E}$ on S^2 we shall mean a collection of pseudocircles $S_e \subseteq S^2$ such that

(i) $\bigcap_{e \in E} S_e = \emptyset$,

(ii) if $e \neq f$, then S_e and S_f intersect at exactly two points where they cross,

(iii) if $S_e \cap S_f \not\subseteq S_g$, then the two points in $S_e \cap S_f$ are separated by S_g (i.e., lie in different components of its complement).

It is easy to verify that this concept is equivalent to the $d = 2$ (and repetition-free) case of Definition 5.1.3; hence there is a one-to-one correspondence between

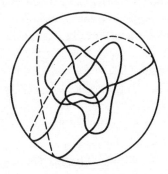

Figure 6.1.3: Rank 3 oriented matroid.

equivalence classes of arrangements $\mathcal{A} = (S_e)_{e \in E}$ of pseudocircles on S^2 and reorientation classes of rank 3 simple oriented matroids on E (Corollary 5.2.2).

Figures 6.1.1, 6.1.2, and 6.1.3 show unsigned arrangements. To get specific oriented matroids from them one must choose positive and negative sides of each pseudosphere.

It seems intuitively clear that there could be difficulty with trying to "straighten" an arrangement of pseudocircles on S^2, i.e., finding a self-homeomorphism of S^2 which moves each pseudocircle to a great circle. Indeed, there exist arrangements of 9 pseudocircles for which this cannot be done (see Figures 1.3.3 and 8.3.1). Starting in the next section we will continue the discussion of this and other phenomena of 2-dimensional arrangements in the projective version: "arrangements of pseudolines".

The representation of rank 3 oriented matroids as arrangements of pseudocircles can sometimes provide a good point of view for quick proofs, as exemplified in the following.

6.1.1. Proposition.   *Let M be an orientable simple matroid of rank 3. Then:*
(i)   *M has a 2-point line,*
(ii)   *if the points of M are 2-colored there exists a monochromatic line.*

Part (i), the "Sylvester-Gallai theorem" for oriented matroids, will receive a sharper statement and the result will be discussed in its proper context with attributions in Section 6.5. For part (ii), see Chakerian (1970) and Edmonds, Lovász and Mandel (1980). Part (ii) was generalized to higher-dimensional hyperplane arrangements by Shannon (1974).

We will need the following auxiliary result.

6.1.2. Lemma.   *For every planarly embedded simple graph:*
(i)   *some vertex has degree at most 5,*
(ii)   *if the edges are 2-colored then there exists a vertex around which the edges of each color class are consecutive in the cyclic ordering induced by the embedding.*

Proof. Part (i) is a well-known consequence of Euler's formula $f_0 - f_1 + f_2 = 2$, for the number of vertices, edges and faces of a planarly embedded graph. Part (ii), also a consequence of Euler's formula, was used by A. Cauchy in 1812 as a key idea in the proof of his rigidity theorem for 3-dimensional polytopes (see Chakerian, 1970, and Fenchel, 1983). We will give the simple proof.

Let p_i be the number of i-gonal faces, so that $f_2 = \sum_{i \geq 3} p_i$. Also, $2f_1 = \sum_{i \geq 3} i p_i$. Hence, $4f_1 - 4f_2 = \sum_{i \geq 3} 2(i-2)p_i$. Let c be the number of red-blue corners (i.e., corners of a face where two differently colored edges meet).

Suppose that the conclusion in part (ii) is false. Then $c \geq 4f_0$, since each vertex must have at least 4 red-blue corners. On the other hand, each i-gonal face can have at most i red-blue corners, and at most $i-1$ ones for odd i. Hence,

$$4f_0 \leq c \leq 2p_3 + \sum_{i \geq 4} i p_i \leq \sum_{i \geq 3} 2(i-2)p_i = 4f_1 - 4f_2,$$

which contradicts Euler's formula. □

Proof of Proposition 6.1.1. Represent M by an arrangement of pseudocircles in S^2. Then the lines of the matroid M are maximal collections of pseudocircles with non-empty intersection. The arrangement determines a simple graph G whose vertices are the points of intersection and whose edges are the segments of pseudocircles between such points (this is the 1-skeleton of the induced cell complex). Now, if all lines in M have at least 3 points, then every vertex in G will have degree at least 6, in violation of part (i) of the lemma. On the other hand, if the pseudocircles are 2-colored and through every intersection point there is at least one pseudocircle of each color, then the induced edge-coloring of G will violate part (ii). □

6.2 Arrangements of pseudolines

Let $\pi : S^2 \to \mathbb{P}^2$, $\pi(x) = \{x, -x\}$, be the double covering of the real projective plane by the 2-sphere. $\mathbb{P}^2$ is here as usual represented by S^2 modulo identification of antipodal points, and $\mathbb{P}^1$ is the embedded copy of the projective line corresponding to the equator of S^2. The images under π of centrally symmetric pseudocircles have several characterizations.

6.2.1 Lemma. *Let L be a simple closed curve embedded in $\mathbb{P}^2$. The following conditions are equivalent:*
(1) $L = h(\mathbb{P}^1)$, *for some self-homeomorphism h of $\mathbb{P}^2$,*
(2) $\mathbb{P}^2 \setminus L$ *is homeomorphic to an open 2-ball,*
(3) $\mathbb{P}^2 \setminus L$ *has one connected component,*
(4) $\pi^{-1}(L)$ *is a pseudocircle in S^2,*
(5) $\pi^{-1}(L)$ *is connected.*

Furthermore, two curves that both satisfy these conditions have non-empty intersection, and if they meet in exactly one point they cross each other at that point.

Proof. Parametrize L by a continuous mapping $f : [0,1] \rightarrow \mathbb{P}^2$, i.e., choose a mapping f such that $f([0,1]) = L, f$ is injective on $(0,1)$, and $f(0) = f(1)$. Pick a point $a \in S^2$ such that $\pi(a) = f(0)$. Then f can be uniquely lifted to a continuous map $\hat{f} : [0,1] \rightarrow S^2$ such that $\pi \circ \hat{f} = f$ and $\hat{f}(0) = a$. There are two cases: either $\hat{f}(1) = a$ or $\hat{f}(1) = -a$. In the first case $\pi^{-1}(L)$ consists of the two disjoint simple closed curves $\hat{f}([0,1])$ and $-\hat{f}([0,1]), S^2 - \pi^{-1}(L)$ has three connected components (two open 2-balls and one open annulus), and $\mathbb{P}^2 \backslash L$ has two components (one open 2-ball and one copy of the Möbius strip). In the second case $\pi^{-1}(L)$ equals the centrally symmetric simple closed curve $\hat{f}([0,1]) \cup -\hat{f}([0,1]), S^2 \backslash \pi^{-1}(L)$ has two components (both open 2-balls), and $\mathbb{P}^2 \backslash L$ has one component (an open 2-ball).

We have seen that simple closed curves in $\mathbb{P}^2$ are of two kinds, and that conditions (2), (3), (4) and (5) describe curves of the second kind. Hence these conditions are equivalent. The implication (1) $\Rightarrow$ (2) is obvious.

The proof of (4)$\Rightarrow$(1), and also the statement about intersections, contains the same ingredients that were used to discuss central symmetry in the proof of Theorem 5.2.1. We will restate these arguments in the present context, starting with the following intuitively obvious fact.

Claim 1. Let S be a centrally symmetric pseudocircle in S^2 and $x \in S^2 \backslash S$. Then x and $-x$ belong to different components of $S^2 \backslash S$.

To see this, let $y \in S^2 \backslash S$. The implication (4) $\Rightarrow$ (2) shows that there exists a path in $\mathbb{P}^2 \backslash \pi(S)$ from $\pi(y)$ to $\pi(x) = \pi(-x)$. This path can be uniquely lifted to a path in $S^2 \backslash S$ from y to either x or $-x$. Hence, every point in $S^2 \backslash S$ can be reached by a path from either x or $-x$, so if these points were located in the same path-component then $S^2 \backslash S$ would be connected, in violation of the Jordan curve theorem. This proves Claim 1.

It follows immediately that two centrally symmetric pseudocircles on S^2 must intersect in at least two points (antipodally situated), and must cross each other there unless they meet somewhere else. The corresponding properties then hold also for their projections under π.

Claim 2. Every centrally symmetric pseudocircle in S^2 is the image of the equator S^1 under an equivariant homeomorphism $h : S^2 \rightarrow S^2$. (Equivariant means that $h(-x) = -h(x)$ for all $x \in S^2$.)

Let $x \in S \subseteq S^2$, where S is a centrally symmetric pseudocircle. Then $\{x, -x\}$ partitions S into two open arcs A^+ and A^-, and S partitions S^2 into two open 2-balls D^+ and D^-. Clearly $-A^+ = A^-$, and $-D^+ = D^-$ by Claim 1. Let $y = (1,0,0)$. Then, similarly, $\{y, -y\}$ partitions S^1 into arcs B^+ and B^-, and S^1 partitions S^2 into 2-balls E^+ and E^-.

Pick arbitrarily a homeomorphism of closed arcs $g : \overline{B^+} \rightarrow \overline{A^+}$. Extend this to an equivariant homeomorphism $g : S^1 \rightarrow S$ by setting $g(z) = -g(-z)$ for all $z \in B^-$. At this stage g is a homeomorphism of the boundaries of the two closed 2-balls $\overline{E^+}$ and $\overline{D^+}$ (here we use the Schönflies theorem, which guarantees that $\overline{D^+}$ is indeed a closed 2-ball). Extend g arbitrarily to a homeomorphism $h : \overline{E^+} \to \overline{D^+}$ (see Lemma 4.7.2). Finally, extend this to $h : S^2 \to S^2$ by setting $h(z) = -h(-z)$ for all $z \in E^-$. Then h is an equivariant homeomorphism such that $h(S^1) = g(S^1) = S$. This proves Claim 2.

An equivariant homeomorphism of S^2 obviously induces a homeomorphism of $\mathbb{P}^2$. The implication $(4) \Rightarrow (1)$ is therefore a direct consequence of Claim 2.   □

As shown by the preceding proof, the simple closed curves in $\mathbb{P}^2$ are of two kinds. One way to state the distinction is that there are two homotopy classes of such curves (the fundamental group has order 2). We will here be interested only in the homotopically non-trivial curves, and in arrangements of such.

6.2.2 Definition.   A simple closed curve in $\mathbb{P}^2$ is called a *pseudoline* if it satisfies the equivalent conditions of Lemma 6.2.1. A collection of pseudolines $\mathcal{A} = (L_e)_{e \in E}$ is called an *arrangement of pseudolines* if $\bigcap \mathcal{A} = \emptyset$ and every pair of pseudolines L_e and L_f in $\mathcal{A}$, $e \neq f$, intersect in exactly one point.

It is clear from the preceding discussion that arrangements of pseudolines are precisely the images of arrangements of centrally symmetric pseudocircles on S^2 (as defined in Section 6.1) under the projection map $\pi : S^2 \to \mathbb{P}^2$. Hence, the following is a reformulation of the Topological Representation Theorem (or, rather of its Corollary 5.2.2) for rank 3 oriented matroids. See Exercise 5.8 for the generalization to higher ranks.

6.2.3 Theorem.   *Equivalence classes of arrangements of pseudolines $\mathcal{A} = (L_e)_{e \in E}$ correspond in one-to-one fashion to reorientation classes of rank 3 simple oriented matroids on E.*

The following terminology will be used in connection with a pseudoline arrangement $\mathcal{A}$. There is an induced regular cell decomposition $\widetilde{\Delta}(\mathcal{A})$ of $\mathbb{P}^2$, obtained as the projection $\pi(\Delta)$ of the centrally symmetric cell decomposition $\Delta = \Delta(\pi^{-1}(\mathcal{A}))$ of S^2 (Proposition 5.1.5). The 0-cells, 1-cells and 2-cells of $\widetilde{\Delta}(\mathcal{A})$ will be called *vertices, edges* and *2-cells* of $\mathcal{A}$, respectively. $\mathcal{A}$ is *simplicial* if every 2-cell is a triangle (has three edges). $\mathcal{A}$ is *simple* if exactly two pseudolines meet at every vertex. Note that simple pseudoline arrangements correspond to *uniform* rank 3 oriented matroids. (*Warning*: The word "simple" has different meanings for arrangements and for matroids. A simple (oriented) matroid is one that lacks loops and parallel elements.)

An early study of arrangements of lines and hyperplanes appeared in the work of Steiner (1826). A century later arrangements of pseudolines were introduced into mathematics by Levi (1926). As mentioned in the introduction, Grünbaum (1972) gives an excellent account of the developments up to 1972 concerning arrangements of lines and pseudolines. With the proof of the Topological Representation Theorem by Folkman and Lawrence (1978) it was realized that arrangements of pseudolines are equivalent to rank 3 oriented matroids. This connection between seemingly disparate mathematical objects has stimulated much fruitful work.

In the rest of this section we will discuss the correspondence between pseudo-line arrangements and rank 3 oriented matroids in microscopic detail. Although very elementary, the details of this correspondence are very instructive and definitely important for an understanding of oriented matroids.

The real projective plane will be represented by a circular diagram, where opposite points on the boundary are identified.

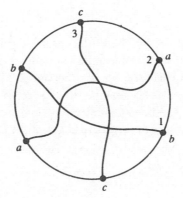

Figure 6.2.1.

Figure 6.2.1 shows an arrangement of three pseudolines. In the standard circular diagram model of $\mathbb{P}^2$ we may assume that "the line at infinity" (the line represented by the boundary) meets every pseudoline of the arrangement in exactly one point (on condition that the line at infinity is not itself a member of the arrangement). To see this for a given arrangement, add a pseudoline in general position using Proposition 6.3.4 below, then move it to "infinity" with a homeomorphism h as in Part (1) of Lemma 6.2.1, then delete it from the arrangement, and – *voilà*!

To get an oriented matroid out of an arrangement of pseudolines, such as the one in Figure 6.2.1, we must "orient" the pseudolines, i.e., assign positive and negative signs to their "sides". This can, strictly speaking, only be done at the level of the arrangement of pseudocircles on S^2 to which the given projective arrangement lifts, since pseudolines have only *one* side and the projective plane is non-orientable. For instance, if we should decide that the central triangle in Figure 6.2.1 is on the positive side of each of the three lines, we would get the seemingly paradoxical labeling of the three adjacent regions in Figure 6.2.2.

To stay clear of this confusion, we must realize that a pseudoline splits the projective plane only modulo another pseudoline. Hence a reasonable thing to do is to choose one of the lines in the arrangement as a reference line and label the sides of the other pseudolines modulo this one. (A different approach would

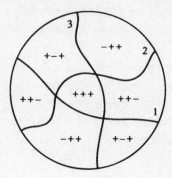

Figure 6.2.2.

be to orient the pseudolines by choosing a preferred direction of traversal for each. We leave it to the reader to formulate how the sign vectors of the oriented matroid are produced in this case, see also Section 1.2.) Via a homeomorphism we may without loss of generality assume that the reference line is the boundary of our circular diagram (the line at infinity).

For instance, if in Figure 6.2.1 (which now has four lines) we again decide that the central triangle is on the positive side of the three original lines we get the labeling shown in Figure 6.2.3.

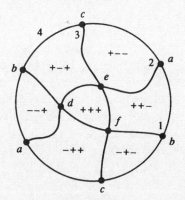

Figure 6.2.3.

Having the labels of the regions, as in Figure 6.2.3, we must add a plus at the end and then symmetrize with respect to sign to get the topes of the associated oriented matroid (again, think of the situation when lifted back to the sphere). In this case we read off the set of topes:

$$\mathcal{T} = \{++++, +-++, +--+, ++-+, -+-+, -+++, --++,$$
$$----, -+--, -++-, --+-, +-+-, +---, ++--\}.$$

The cocircuits correspond to vertices of the induced cell decomposition of $\mathbb{P}^2$, so again from Figure 6.2.3 we read off:

$$C^* = \{-0+0,\ 0-+0,\ +-00,\ 00++,\ +00+, 0+0+,$$
$$+0-0,\ 0+-0,\ -+00,\ 00--,\ -00-,\ 0-0-\}.$$

For vertices on the reference line (here: a, b and c) their two appearances in the diagram give the two cocircuits which they represent. For interior vertices (such as d, e and f) we have to remember to symmetrize with respect to sign to get both cocircuits.

In the opposite direction, suppose that a rank 3 oriented matroid is handed to us in the form of a list of sign vectors. How can we most effectively construct a corresponding arrangement of pseudolines? In principle, one could compute the signed cocircuit span $\mathcal{L}$ and from it read off the two-dimensional cell complex $\Delta(\mathcal{L})$ and the arrangement of lines on it. In practise, a step-wise extension procedure is quicker.

Suppose for instance that

(6.2.1)

$$
\begin{array}{rcccccc}
a & = & 0 & 0 & + & + & + \\
b & = & 0 & + & - & 0 & - \\
c & = & + & - & 0 & 0 & - \\
d & = & + & - & 0 & + & 0 \\
e & = & 0 & + & 0 & + & + \\
f & = & + & 0 & 0 & + & + \\
g & = & + & 0 & - & 0 & - \\
h & = & + & 0 & - & + & 0 \\
i & = & + & - & + & 0 & 0 \\
j & = & 0 & + & - & + & 0 \\
\end{array}
$$

and that the set of cocircuits of a given oriented matroid $\mathcal{M} = (E, C^*)$ is

$$C^* = \{\pm a,\ \pm b,\ \pm c,\ \pm d,\ \pm e,\ \pm f,\ \pm g,\ \pm h,\ \pm i,\ \pm j\}.$$

We see that $|E| = 5$, and we may choose the line at infinity to represent the first element. The first three lines are not concurrent (since the first three columns of the sign matrix (6.2.1) have no zero row), so we may represent them as in Figure 6.2.4.

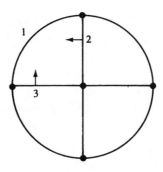

Figure 6.2.4.

Here orientations of lines 2 and 3 have also been chosen (indicated by small arrows pointing to the positive side), which of course are relative to line 1 and at this stage arbitrary.

Now permute the rows of (6.2.1) to obtain

$$
\begin{array}{rcccc|c|c}
a &=& 0 & 0 & + & + & + \\
e &=& 0 & + & 0 & + & + \\
f &=& + & 0 & 0 & + & + \\
b &=& 0 & + & - & 0 & - \\
c &=& + & - & 0 & 0 & - \\
g &=& + & 0 & - & 0 & - \\
d &=& + & - & 0 & + & 0 \\
h &=& + & 0 & - & + & 0 \\
i &=& + & - & + & 0 & 0 \\
j &=& 0 & + & - & + & 0 \\
\end{array}
$$

(6.2.2)

We have singled out the points of intersection of lines 1, 2 and 3 (first three rows), those intersection points that will be added when line 4 appears (next three rows), and finally those that will come from line 5.

From the information in the first six rows of (6.2.2), Figure 6.2.4 can be augmented to Figure 6.2.5.

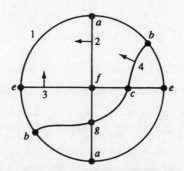

Figure 6.2.5.

To determine the position of line 5 relative to the earlier ones, we can read off from the last four rows of (6.2.2) how its intersection points are located with respect to these. For instance, the point where line 5 cuts line 3 (point "d") is on the positive side of lines 1 and 4 and on the negative side of line 2, hence in the representation of Figure 6.2.5 it must be located on line 3 between f and c. Put a small "x" at each point of intersection thus determined. It will then be possible, as in Figure 6.2.6, to trace a correct position for the fifth line and to determine its orientation.

Extending an oriented matroid one element at a time is a very useful technique, which is treated in detail in Section 7.1.

The projective version of arrangements of pseudolines that we have discussed is the correct one from the oriented matroid point of view. However, it would

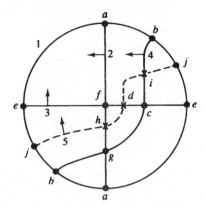

Figure 6.2.6.

seem intuitively appealing to be able to consider also arrangements of pseudo-lines in the affine plane, suitably defined (and, of course, higher-dimensional analogues). To keep in touch with oriented matroid theory we will only consider affine arrangements that have projective continuations. A more general kind of affine arrangement is considered in Assman and Kleitman (1984).

Use as a topological model of the affine plane the interior of the unit circle S^1, which we here call A^2. Define an *affine pseudoline* to be the image $L = f((0, 1))$ of a continuous injective mapping $f : [0, 1] \to B^2 = A^2 \cup S^1$ such that $f^{-1}(S^1) = \{0, 1\}$.

Figure 6.2.7: Affine pseudoline.

An affine pseudoline has the following properties:

(1) $A^2 \backslash L$ has two connected components. The closure of each component is homeomorphic to B^2, hence each component is homeomorphic to A^2.

(2) $L = h(\{(x, y) \in A^2 | y = 0\})$ for some self-homeomorphism $h : A^2 \to A^2$.

Since L forms a simple closed curve together with each of the two segments into which $f(0)$ and $f(1)$ separate S^1, part (1) follows from the Schönflies theorem. Part (2) follows from part (1) by gluing together suitable homeomorphisms between the closures of the components in $A^2 \backslash L$ and the upper and lower parts of A^2. Such homeomorphisms can be constructed by radial extension from homeomorphisms of the boundaries of these regions, which are circles.

Figure 6.2.8: Affine arrangement of pseudolines.

By an *affine arrangement of pseudolines* $\mathcal{A} = (L_e)_{e \in E}$ we will understand a collection of affine pseudolines L_e in A^2 such that

(i)   every pair meets in at most one point,

(ii) if two pseudolines meet at a point they cross each other there,

(iii) being disjoint (or equal) is an equivalence relation (mimicking parallelism), and

(iv) for any two disjoint pseudolines there is a third one that intersects both (and hence the whole parallelism class).

See Figure 6.2.8 for an illustration.

As mentioned, an affine arrangement can be extended to a projective arrangement, and hence it corresponds to a certain rank 3 oriented matroid. For instance, by giving the affine diagram in Figure 6.2.8 a collar in the form of an annulus whose outer boundary is the line at infinity, the affine arrangement is converted to a projective arrangement. See Figure 6.2.9.

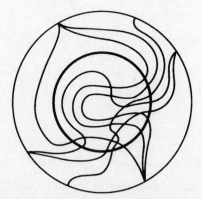

Figure 6.2.9: Projective continuation.

There will be no further mention of affine arrangements of pseudolines. However, a different concept of affine oriented matroids is discussed in Sections 4.5 and 10.1. Also, from now on we will not necessarily draw projective arrangements in the circular fashion of Figures 6.2.1–6.2.6, the line at infinity (whether included in the arrangement or not) usually being left to the reader's imagination.

6.3 How far can things be stretched?

An arrangement $\mathcal{A} = (L_e)_{e \in E}$ of pseudolines is called *stretchable* if it satisfies any one of the following equivalent conditions:

(1) Some self-homeomorphism of the projective plane moves all lines L_e into straight lines.

(2) The cell decomposition of $\mathbb{P}^2$ induced by $\mathcal{A}$ is combinatorially isomorphic to the cell decomposition induced by some arrangement of straight lines.

(3) The oriented matroid corresponding to $\mathcal{A}$ is realizable.

The equivalence of these three statements follows from the $r = 3$ case of Theorem 5.1.6 and its Corollary 5.1.7. Let us say that two pseudoline arrangements are *isomorphic* if the induced cell complexes are combinatorially isomorphic (have the same face lattice). By Theorem 5.1.6 two arrangements are isomorphic if and only if some homeomorphism of $\mathbb{P}^2$ transforms one into the other, and a pseudoline arrangement is stretchable if and only if it is isomorphic to a straight-line arrangement. (*Remark*: This notion of "isomorphic" is by Theorem 5.1.6 identical to the notion of "topologically equivalent" defined in Section 5.1 and used elsewhere in the book. We use the word "isomorphic" here only to conform with standard usage in the literature on pseudoline arrangements, as established by Grünbaum (1972).)

It has been known since the pioneering paper of Levi (1926), that non-stretchable arrangements exist. Figure 1.3.3 shows one such arrangement of nine pseudolines, whose non-stretchability derives from Pappus's theorem. Another non-stretchable arrangement of nine lines which is *simple*, i.e., more than two lines never cross at any point, was constructed by Ringel (1956), see Figure 8.3.1. These examples are minimal in the sense of the following result, which was conjectured by Grünbaum (1972) and proved by Goodman and Pollack (1980b) using allowable sequences. See also Theorem 8.2.4.

6.3.1 Theorem.   *Every arrangement of at most eight pseudolines is stretchable. Equivalently, every rank 3 oriented matroid of size at most eight is realizable.*

The smallest known non-stretchable *simplicial* arrangement (every 2-cell in the cell complex is a triangle) has 15 pseudolines, and it appears to be undecided whether smaller non-stretchable simplicial arrangements exist. See Grünbaum (1972) for information about these examples and much related material.

It follows from the work of Mnëv (1988) that the problem of determining whether a pseudoline arrangement is stretchable is equivalent to the "existential theory of the reals". In particular, the problem is NP-hard. See Theorem 8.7.2 and Corollary 8.7.3.

In this section we will discuss different ways of dealing with non-stretchability. We are interested in "nice" representations of pseudoline arrangements, where the lines while not straight are still manageable, individually and together. At

the end there will also be a few remarks about adjoints of rank 3 oriented matroids and the related polarity principle for pseudoline arrangements.

One early suggestion for line representation is by Ringel (1956), whose pseudolines are obtained from straight lines by replacing a bounded segment with a
"wiggly" path. It is clear that every pseudoline arrangement can be realized with
such lines; just replace a small enough part of each pseudoline in a neighborhood
of infinity (assumed *not* in the arrangement) by a straight segment. Ringel's idea
was generalized to higher dimensions by Miller (1987), whose pseudohyperplanes
are linear hyperplanes outside a bounded region. However, in higher dimensions
such pseudohyperplanes are not general enough to represent all oriented matroids
(Exercise 5.11).

Several possibilities for representations with piecewise linear pseudolines exist.
The following result, proved by induction on the number of lines, is due to
Grünbaum (1972). Recall that the words *"vertex"*, *"edge"* and *"2-cell"* of an
arrangement always refer to the induced cell decomposition of $\mathbb{P}^2$.

6.3.2 Proposition.   *Every arrangement of pseudolines is isomorphic to an
arrangement whose edges are linear and whose 2-cells are convex polygons.*

The most useful kind of representation has turned out to be the *"wiring diagrams"* of Goodman (1980). These are drawn as in Figure 6.3.1, usually with
piecewise linear pseudolines, although the lines could certainly be smooth (more
about this later).

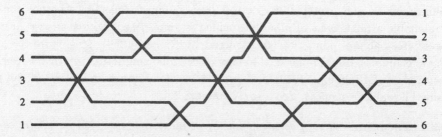

Figure 6.3.1: Wiring diagram.

It is a useful convention to let the lines of a wiring diagram be labeled from 1
to n in the up-going direction on the left side (and, of course, oppositely on the
right side). By extending each "wire" by the straight line through its endpoints
in both directions towards infinity a wiring diagram can be thought of as an
arrangement of pseudolines (in fact, Ringel pseudolines).

6.3.3 Theorem (Goodman and Pollack 1984b).   *Every arrangement of
pseudolines is isomorphic to a wiring diagram arrangement.*

To prove this we will need the following result of Levi (1926). This fundamental
tool for working with pseudoline arrangements guarantees that new pseudolines
can be added through not-yet-collinear pairs of points, just as in the linear case.
For a proof see Grünbaum (1972).

6.3.4 Proposition (Levi's Enlargement Lemma). *Suppose that $\mathcal{A} = (L_e)_{e \in E}$ is an arrangement of pseudolines and $x, y \in \mathbb{P}^2$ two points which do not both lie on any of the pseudolines $L_e, e \in E$. Then there exists a pseudoline L' through x and y that intersects each pseudoline L_e in exactly one point. In other words, there exists an enlarged arrangement $\mathcal{A}' = \mathcal{A} \cup \{L'\}$ for which $x, y \in L'$. Furthermore, L' can be chosen so that $L' \backslash \{x, y\}$ does not contain any vertices of $\mathcal{A}$.*

It is an interesting fact that such an enlargement lemma fails in higher dimensions. Examples can be constructed of an arrangement of pseudoplanes in $\mathbb{P}^3$ (a rank 4 oriented matroid) and three given points, such that no pseudoplane through these points can be properly added to the arrangement. See Proposition 10.4.5 for the details. Richter-Gebert (1993b) has constructed examples showing that "three given points" can be lowered to "two given points" in the previous statement, see Proposition 7.7.4.

Proof of Theorem 6.3.3. Pick a free point p on the line at infinity L_∞, which is assumed not to be in the arrangement. As remarked in connection with Figure 6.2.1 we may assume that L_∞ meets every pseudoline of the arrangement in exactly one point. Enlarge the arrangement by adding for each vertex x of the original arrangement a pseudoline L_x through p and x, but avoiding all vertices other than x. See Figure 6.3.2, where p is the north-south pole, and the new lines L_x are dashed.

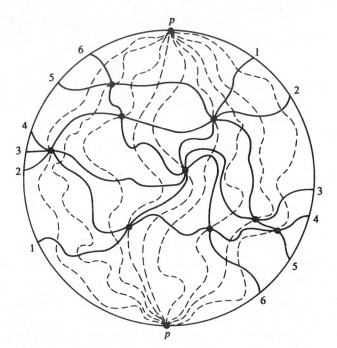

Figure 6.3.2: Sketch 1 for the proof of Theorem 6.3.3.

The new lines L_x divide the affine plane $\mathbb{P}^2 \backslash L_\infty$ into strips, and there are no lines crossing in a strip. Furthermore, these strips are naturally ordered from left to right. One can now describe how the given arrangement is transformed into a wiring diagram arrangement. First observe that the arrangement $\{L_x\}$ of new lines is combinatorially isomorphic to a pencil of straight lines through p. Therefore a homeomorphism exists that will straighten the lines L_x. After such a transformation the situation looks like that in Figure 6.3.3. It is easy to visualize (for the technical details see Goodman 1980, or Goodman and Pollack 1984b) how from such a position the original arrangement is further transformed into a wiring diagram arrangement, by a continuous deformation of $\mathbb{P}^2$ that maps each straightened line L_x and each straightened strip onto themselves. For instance, compressing the arrangement in Figure 6.3.3 vertically into a band around "the equator", and then adjusting the lines appropriately, we get the wiring diagram of Figure 6.3.1. It is clear that the combinatorial type of the original arrangement is unaffected by this sequence of transformations.

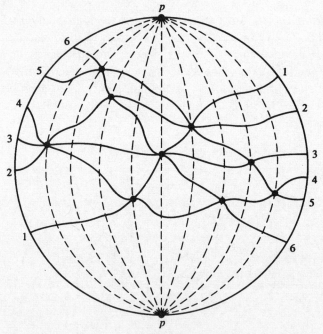

Figure 6.3.3: Sketch 2 for the proof of Theorem 6.3.3.

□

Theorem 6.3.3 has a certain resemblance to the correspondence between links and braids, studied in knot theory. (This was pointed out to us by A.M. Vershik.) More precisely, it resembles the theorem of J.W. Alexander that every knot or link is isotopic to a closed braid. Is there more than formal similarity? One can prove a result (reminiscent of A.A. Markov's theorem in knot theory) characterizing the equivalence relation on wiring diagrams of being isomorphic as pseudoline arrangements, see Exercise 6.12. For the theorems of Alexander and Markov, and information about links and braids generally, see Hansen (1989).

Each individual pseudoline in a wiring diagram can be thought of as a piecewise linear function graph $y = f(x)$ on the unit interval, $0 \leq x \leq 1$. By taking other kinds of continuous functions one gets alternative concepts of wiring diagrams, and to measure the intrinsic distance from linearity of a non-stretchable arrangement one idea is to use polynomial curves of lowest possible degree.

Call an arrangement of pseudolines *d-stretchable* if it is isomorphic to an arrangement of wiring diagram type for which each pseudoline $y = p_i(x)$, $0 \leq x \leq 1$, is given by a polynomial p_i of degree at most d. So, the meaning of 1-stretchable coincides with stretchable. This concept was introduced by Goodman and Pollack (1985c), who proved the following.

6.3.5 Proposition. *Let d_n be the smallest d such that every simple arrangement of n pseudolines is d-stretchable. Then there exist positive constants c_1 and c_2 such that*

$$c_1\sqrt{n} \leq d_n \leq c_2 n^2.$$

The lower bound is deduced using Sturm sequences and a theorem of Milnor on Betti numbers of real semialgebraic sets. The upper bound comes from a theorem of Newman on polynomial co-monotone approximation applied to wiring diagrams.

In view of this result it is natural to ask for higher-dimensional analogues. We know from the Topological Representation Theorem that an arrangement of pseudospheres in S^d has a piecewise linear representation on the boundary of a $(d+1)$-simplex. By projection (and some hand-waving) we get a realization as an arrangement of piecewise linear pseudohyperplanes in projective d-space. To what extent can these be smoothed, or what degrees would be required from polynomial hypersurfaces realizing the arrangement? It can be shown that smooth realizations with transversal tangent planes at all intersections is too much to hope for (Exercise 5.10(b)).

Every rank 3 oriented matroid $\mathcal{M} = (E, \mathcal{C})$ has a TYPE I (or pseudoline) representation, by Theorem 6.2.3. Does it also have a TYPE II representation, in the sense of Section 5.3? A TYPE II representation would require an arrangement of pseudolines $\mathcal{A} = (L_Y)_{Y \in \mathcal{C}^*}$, indexed by the set of cocircuits of $\mathcal{M}$, and such that E is identified with a subset of the vertices of $\mathcal{A}$ so that $Y_e = +, -, 0$ for $Y \in \mathcal{C}^*$ precisely when $e \in E$ lies on the positive side of, negative side of, respectively, on the pseudoline L_Y. As was discussed in Section 5.3, the existence of a TYPE II representation is equivalent to the existence of an adjoint oriented matroid.

6.3.6 Proposition (Goodman 1980, Cordovil 1982a, Goodman and Pollack 1984b). *Every rank 3 oriented matroid has an adjoint.*

Proof. Take a TYPE I representation $\mathcal{A}$ of $\mathcal{M}$. Enlarge the pseudoline arrange-

ment $\mathcal{A}$ to an arrangement $\mathcal{A}'$ by adding pseudolines through all pairs of intersection points of $\mathcal{A}$ that are not collinear (Proposition 6.3.4). Then $\mathcal{A}'$ together
with the original intersection points (i.e., vertices) of $\mathcal{A}$ form a pseudoconfiguration of points. The oriented matroid $\mathcal{M}^{ad}$ determined by it (Proposition 5.3.2)
is clearly an adjoint to $\mathcal{M}$. See Figure 6.3.4 for an illustration of the proof.   □

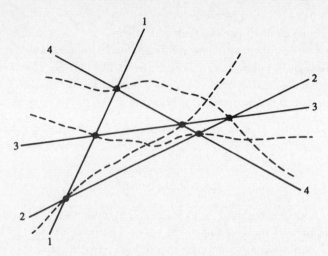

Figure 6.3.4: Construction of an adjoint.

We conclude that for rank 3 oriented matroids there are always two kinds of
topological representations available: as an arrangement of pseudolines, or as a
pseudoconfiguration of points. It was noted in Section 5.3 that only the former
kind of representation is universal for ranks $r \geq 4$.

The passage from a pseudoline arrangement $\mathcal{A}$ to the adjoint pseudoline ar-
rangement $\mathcal{A}^{ad}$, via $\mathcal{A} \rightarrow \mathcal{M} \rightarrow \mathcal{M}^{ad} \rightarrow \mathcal{A}^{ad}$, is the *polarity principle* of Goodman (1980), see Exercise 6.4. Note that in the linear case this is just the polarity
principle of projective planes, which makes it possible to always exchange the
roles of points and of lines.

Remark. In these connections it is customary to speak of "duality", but to be
consistent with our usage elsewhere in this book we prefer to call this polarity.
See Exercise 1.21.

6.4 Allowable sequences, wiring diagrams and homotopy

Allowable sequences of permutations associated to point configurations were
discussed in Section 1.10. Here we will continue this discussion in the polar
setting of (pseudo)line arrangements. As an application a homotopy theorem of
Ringel will be proved.

Recall the definition: A sequence Σ of permutations is an *allowable sequence*
if

(1) Σ begins with the identity permutation $12 \ldots n$ and ends with the reverse permutation $n(n-1) \ldots 1$,

(2)   the move between consecutive terms of Σ consists of reversing one or more non-overlapping increasing substrings.

By "substring" we mean what could more precisely be called "adjacent substring", i.e., the positions occupied must be consecutive.

Any wiring diagram determines an allowable sequence. For instance, the successive permutations of the wires in Figure 6.3.1 gives the allowable sequence:

$$123456 \xrightarrow{234} 143256 \xrightarrow{56} 143265 \xrightarrow{26} 143625 \xrightarrow{14} 413625 \xrightarrow{136}$$
$$463125 \xrightarrow{125} 463521 \xrightarrow{46} 643521 \xrightarrow{35} 645321 \xrightarrow{45} 654321.$$

Conversely, any allowable sequence uniquely determines a wiring diagram. So, there is a one-to-one correspondence

(6.4.1)   allowable sequences $\longleftrightarrow$ wiring diagrams.

In view of Theorem 6.3.3 this implies that every arrangement of pseudolines determines an allowable sequence, but not uniquely. In the straight-line case this is projectively polar to the construction in Section 1.10 of an allowable sequence associated to a point configuration.

In the above correspondence, if in some move of the allowable sequence more than one increasing substring is reversed then at the corresponding horizontal coordinate of the wiring diagram more than one crossing takes place. In the following we will for simplicity assume that there are no such multiple flips and crossings, and leave to the reader to check that all works out also in the general case.

Let Σ be an allowable sequence of permutations. The rank 3 oriented matroid of the wiring diagram associated to Σ is the *little* oriented matroid determined by Σ. Let $\mathcal{A}$ be the arrangement of this wiring diagram, let p be the north-south pole in $\mathbb{P}^2$ as in Figure 6.3.3. Then enlarge $\mathcal{A}$ to an arrangement $\mathcal{A}'$ by adding for each vertex x of $\mathcal{A}$ a new pseudoline L_x through p and x, which avoids all other vertices of $\mathcal{A}$ and which partitions this set of vertices according to which moves come before and after move x in Σ. This is similar to what was done in the proof of Theorem 6.3.3, and the enlargement leads to an arrangement $\mathcal{A}'$ like the one depicted in Figure 6.3.3. The rank 3 oriented matroid associated to $\mathcal{A}'$ is the *big* oriented matroid determined by Σ.

The little and the big oriented matroid are determined by Σ up to reorientation. These oriented matroids were associated to Σ in projectively polar fashion in Section 1.10.

Let K denote the family of "new" pseudolines L_x of the arrangement $\mathcal{A}'$ constructed above. The pseudolines in K all contain p, every vertex of $\mathcal{A}'$ is contained in some pseudoline L_x, and every L_x contains exactly one point of intersection of pseudolines of $\mathcal{A}$. This is equivalent to saying that in the big oriented matroid associated to $\mathcal{A}'$ the line K is modular and every element in K is on exactly one line spanned by elements not in K. We will now show that conversely every oriented matroid with such special structure arises as the big oriented matroid of an allowable sequence.

Assume that $\mathcal{M}(E)$ is a rank 3 oriented matroid having a modular line K such that every element in K is on exactly one line spanned by elements of $E \backslash K$. Let $\mathcal{A}'$ be an arrangement of pseudolines realizing $\mathcal{M}$. In this model K is a family of pseudolines in $\mathcal{A}'$, and being a line in $\mathcal{M}$ means that K is the pencil of pseudolines through some point p. Taking for the line at infinity any line L_∞ through p not containing any other vertex of $\mathcal{A}'$ we find ourselves once more in the situation of Figure 6.3.3. The fact that K is modular in M means that every vertex of $\mathcal{A}'$ lies on one of the pseudolines in K, and the other hypothesis we made about K as a line in M means that every pseudoline in K contains exactly one intersection point of pseudolines of the subarrangement $\mathcal{A} = \mathcal{A}' \backslash K$. Now, numbering the lines of $\mathcal{A}$ from 1 to n in the order that their intersection points appear on L_∞, and letting Σ be the allowable sequence corresponding to the wiring diagram $\mathcal{A}$, the construction shows that M is the big oriented matroid determined by Σ.

As already pointed out, the correspondence between allowable sequences and rank 3 oriented matroids is not in any way bijective, since several choices are made during the constructions (see also Theorem 6.6.4). By introducing additional normalization the correspondence can be made tighter, but we leave this aside. Instead we will exemplify the use of allowable sequences in the topological theory of rank 3 oriented matroids.

The papers by Ringel (1956, 1957) are to a large extent motivated by the question of understanding the relation of combinatorial equivalence (i.e., isomorphism) between arrangements of pseudolines. A characterization of such equivalence for simple arrangements is given in terms of sign-patterns, and the possible sign-patterns that arise are also characterized. In modern terminology this amounts to an axiomatization of maximal covectors (topes) of uniform rank 3 oriented matroids (cf. Section 3.8). An important role in Ringel's work is played by the local changes in an arrangement resulting from perturbing one of the pseudolines, and we shall now discuss some of his results in this direction.

Let $\mathcal{A} = (L_e)_{e \in E}$ be a simple arrangement of pseudolines in $\mathbb{P}^2$. By a *move* we will mean the continuous deformation of one of the lines, say L_0, keeping all the other lines fixed. More precisely, we have in mind a parametrized family of pseudoline arrangements

$$\mathcal{A}_t = (L_e)_{e \in E} \cup \{L_t\}, 0 \leq t \leq 1,$$

such that for some parametrizations $f_0 : S^1 \to L_0$ and $f_1 : S^1 \to L_1$ there exists a continuous mapping (homotopy) $F : S^1 \times [0,1] \to \mathbb{P}^2$ such that $F(x,0) = f_0(x)$ and $F(x,1) = f_1(x)$ for all $x \in S^1$.

Following Ringel's terminology we will call a move for which no vertex is crossed (i.e., no vertex of $\mathcal{A}_0 \backslash \{L_0\}$ belongs to $\mathrm{Im}F$) a 0-*move*. Combinatorial isomorphism type is unaffected by a 0-move: $\mathcal{A}_0 \cong \mathcal{A}_1$. A move for which exactly one vertex is crossed is called a 1-*move*. Formally, a 1-move is a move

$\mathcal{A}_t, 0 \le t \le 1$, such that $F(x, t)$ is a vertex of the subarrangement $\mathcal{A}_0 - \{L_0\}$ for exactly one value (x', t') of the domain, and all points $F(x', t'')$ and $F(x', t''')$ with $t'' < t' < t'''$ and $t''' - t''$ sufficiently small lie on opposite sides of the two pseudolines that intersect at the vertex $v = F(x', t')$. Intuitively, 1-moves correspond to *triangle-switches*, as illustrated in Figure 6.4.1.

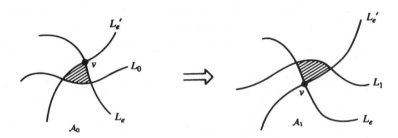

Figure 6.4.1: Triangle-switch.

6.4.1 Theorem (Ringel 1956, 1957). *Every pair of simple arrangements of n pseudolines is connected by a sequence of 0-moves and 1-moves.*

For the case of straight-line arrangements the sequence of moves can be chosen so that every move is a rigid motion of the line being moved (i.e., no deformation). This is actually not so hard to prove (Exercise 6.5). The proof of the "homotopy theorem" in the general case is a bit more demanding. Ringel's original proof uses sign-matrices. Another proof in the language of pseudoconfigurations of points was given by Roudneff (1988b). We shall sketch a proof that via allowable sequences directly relates the question to Coxeter relations in the symmetric group. In preparation for this proof we will review some definitions and facts. See Section 2.3 for more background.

The symmetric group S_n of permutations of $E_n = \{1, 2, \ldots, n\}$ is generated by the adjacent transpositions $s_i = (i, i+1), 1 \le i \le n-1$. These generators satisfy the *Coxeter relations*

$$(6.4.2) \quad \begin{cases} \text{(i)} & s_i^2 = id, & 1 \le i \le n-1, \\ \text{(ii)} & s_i s_j = s_j s_i, & |i - j| \ge 2, \\ \text{(iii)} & s_i s_{i+1} s_i = s_{i+1} s_i s_{i+1}, & 1 \le i \le n-2. \end{cases}$$

In fact, all relations among the s_i in S_n are implied by the Coxeter relations, see Example 2.3.3. (This is actually a consequence of Proposition 4.4.7, as was remarked in that connection.) A decomposition

$$\pi = s_{i_1} s_{i_2} \ldots s_{i_t}, \quad 1 \le i_1, \ldots, i_t \le n-1,$$

expressing a permutation π as a word of minimal length t in these generators, is called a *reduced decomposition* of π. The length of the reduced decompositions of π equals the number of inversions in π. For example, $\pi = s_2 s_4 s_3 s_4 s_1 s_2$ is a reduced decomposition of $\pi = 35142$. Hence, the reduced decompositions of maximal length, here to be called *maximal reduced decompositions*, are the decompositions of the permutation $n(n-1)\ldots 1$, and have length $\binom{n}{2}$.

Let us call an allowable sequence of permutations in S_n *simple* if each move consists in reversing exactly one substring and this is a substring of size two. Algebraically each such move consists in multiplying on the right by a Coxeter generator s_i so that length (number of inversions) increases. Therefore a simple allowable sequence determines a maximal reduced decomposition, as illustrated by the following example:

$$1234 = id$$
$$1324 = s_2$$
$$3124 = s_2 s_1$$
$$3142 = s_2 s_1 s_3$$
$$3412 = s_2 s_1 s_3 s_2$$
$$3421 = s_2 s_1 s_3 s_2 s_3$$
$$4321 = s_2 s_1 s_3 s_2 s_3 s_1$$

Conversely, by reading prefixes of increasing length in a maximal reduced decomposition, a simple allowable sequence is determined. Thus we get a one-to-one correspondence

(6.4.3)   simple allowable sequences $\longleftrightarrow$ maximal reduced decompositions.

Simple allowable sequences are also identical to maximal chains in weak Bruhat order on S_n (see Section 2.3, particularly Figure 2.3.8).

A wiring diagram will be called *simple* if exactly two wires meet at every intersection point and if (as usual) the intersection points have distinct horizontal coordinates. Clearly, (6.4.1) specializes to a one-to-one correspondence

(6.4.4)   simple allowable sequences $\longleftrightarrow$ simple wiring diagrams.

Proof of Theorem 6.4.1. Suppose given two simple arrangements $\mathcal{A}$ and $\mathcal{B}$, each with n pseudolines. First we normalize these arrangements into simple wiring diagrams using a sequence of 0-moves. We leave to the reader to check that this can be done (Exercise 6.7). From now on we may then assume that $\mathcal{A}$ and $\mathcal{B}$ are simple wiring diagrams.

Via (6.4.3) and (6.4.4) there are maximal reduced decompositions $mrd(\mathcal{A})$ and $mrd(\mathcal{B})$ canonically associated with $\mathcal{A}$ and $\mathcal{B}$. We now use the following fact from Coxeter group theory (Tits, 1968): *Every pair of reduced decompositions of the same group element is connected by a sequence of Coxeter relations* (6.4.2) *of types* (ii) *and* (iii). This is implied by Proposition 4.4.7, as was remarked in that connection. This fact applies in particular to $mrd(\mathcal{A})$ and $mrd(\mathcal{B})$. The effect of a Coxeter relation of type (iii) on the corresponding simple wiring diagram is illustrated in Figure 6.4.2.

One sees that a Coxeter relation of type (iii) corresponds to a certain kind of

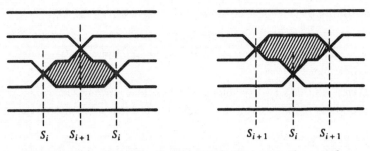

Figure 6.4.2: Coxeter relation producing triangle-switch.

triangle-switch. It can be realized by a 1-move followed by some 0-moves. On the other hand, a Coxeter relation of type (ii) corresponds to a trivial modification of the simple wiring diagram which can be realized by a sequence of 0-moves. Changing $mrd(\mathcal{A})$ into $mrd(\mathcal{B})$ via a sequence of Coxeter relations therefore leads to changing $\mathcal{A}$ into $\mathcal{B}$ via a sequence of 0-moves and 1-moves, and the proof is complete. □

The "homotopies" considered in Theorem 6.4.1 are special in that only one line is moved at a time. It is natural to consider also continuous deformation of the whole arrangement such that combinatorial type is preserved throughout, called *isotopy*. The following result is according to Ringel "nicht schwer zu zeigen".

6.4.2 Proposition (Ringel 1956). *Given two combinatorially equivalent simple arrangements of pseudolines $\mathcal{A}$ and $\mathcal{A}'$, there exists an isotopy that changes $\mathcal{A}$ either into $\mathcal{A}'$ or into the mirror image of $\mathcal{A}'$.*

Ringel raised the question whether for straight-line arrangements $\mathcal{A}$ and $\mathcal{A}'$ such an isotopy can be found which stays in the realizable straight-line case throughout. From research in realization spaces of oriented matroids it is now known that the answer to this question is negative. The smallest known counterexamples involve simple arrangements of 14 lines. See Section 8.6 for more about this.

Triangle-switches (or, 1-moves) in simple pseudoline arrangements correspond to "mutations" in rank 3 uniform oriented matroids. Mutations are defined and discussed in Section 7.3. From this point of view Theorem 6.4.1 can be reformulated as saying that the "mutation graph" of uniform oriented matroids of rank 3 is connected. Whether the same is true for rank n uniform oriented matroids, giving a higher-dimensional analogue of Ringel's Homotopy Theorem 6.4.1, is an open problem. Roudneff and Sturmfels (1988) have given a positive answer in the realizable case. See their paper and also Section 7.3 for further discussion of this topic.

6.5 Three enumerative questions

In this section we will report what is known about the following problems:

(1)   How many distinct arrangements of n pseudolines are there?

(2)   How many triangles must there be in a pseudoline arrangement?

(3)   How many points where exactly two pseudolines meet must exist in an arrangement?

Notice that the last two problems concern the number of 2-cells and 0-cells with the smallest possible number of incidences in the induced cell decomposition of $\mathbb{P}^2$.

Let a_n denote the number of isomorphism classes of arrangements of pseudolines. Equivalently, a_n is the number of reorientation classes of rank 3 oriented matroids. What is the asymptotic order of magnitude of a_n? Let us also consider the number s_n of isomorphism classes of simple arrangements of pseudolines.

6.5.1 Proposition.   *For sufficiently large n,*

$$\frac{1}{8}n^2 \;<\; \log_2 s_n \;<\; \log_2 a_n \;<\; 1.0850\, n^2$$

and

$$\log_2 s_n \;<\; 0.6988\, n^2.$$

The lower bound for the number of simple arrangements was proved by Goodman and Pollack (1983), see also Example 7.4.1. Quadratic upper bounds follow from the zone theorem for pseudoline arrangements of Edelsbrunner, O'Rourke and Seidel (1986). The constant 1.0850 above follows (by induction on n) from the sharper version of Bern, Eppstein, Plassmann and Yao (1991). This was pointed out to us by M. Bern and R. Seidel. The upper bound on $\log s_n$ is by Felsner (1996, 1997). It has been conjectured by Knuth (1992) that in fact $\log_2 s_n \leq \binom{n}{2}$. See Section 7.4 for more about enumeration of oriented matroids.

Via the connection with allowable sequences an asymptotically slightly weaker upper bound can be obtained for s_n. Select one representative from each isomorphism class of simple pseudoline arrangements. Without loss of generality these may be taken to be simple wiring diagrams (Theorem 6.3.3). Therefore by (6.4.3) and (6.4.4): $s_n \leq r_n$, where r_n is the number of maximal length reduced decompositions in the symmetric group S_n. The following explicit expression for r_n was obtained by Stanley (1984a), see also Section 4 of Edelman and Greene (1987):

$$r_n = \frac{\binom{n}{2}!}{(2n-3)(2n-5)^2(2n-7)^3 \cdot \ldots \cdot 5^{n-3} \cdot 3^{n-2}}.$$

From this expression an $O(n^2 \log n)$ upper bound follows for $\log s_n$.

We know from Theorem 2.1.5 that every *realizable* oriented matroid of cardinality n has at least n simplicial topes. (Actually, at least $2n$ simplicial topes, but we take the projective point of view and identify opposite pairs.) On the other hand, as mentioned in Sections 1.5 and 7.2, there exists a rank 4 oriented matroid on 8 elements having only 7 simplicial topes. Our ignorance about simplicial topes in general oriented matroids of rank ≥ 4 is almost complete: we don't even know whether they must exist (Conjecture 7.3.10/Problem 4.5).

In rank 3 much more is known, since the existence and properties of triangles in pseudoline arrangements have long been studied. Simplicial topes are interesting for several reasons, one being that they give positions within an oriented matroid where the simplest possible local changes can take place. For rank 3 this was discussed in the previous section (triangle-switches), for higher ranks see Section 7.3 (mutations).

The following theorem collects the main facts known about the number p_3 of triangles in the cell complex determined by an arrangement of pseudolines in $\mathbb{P}^2$.

6.5.2 Theorem. *For every arrangement $\mathcal{A}$ of n pseudolines:*

(i) $p_3 \geq n$,

(ii) *for every pseudoline L in $\mathcal{A}$ there are at least three triangles having an edge on L,*

(iii) *if $p_3 = n$ and $\mathcal{A}$ is stretchable, then $\mathcal{A}$ is simple,*

(iv) $p_3 \leq \frac{1}{3}n(n-1)$, *for $n \geq 10$,*

(v)   *equality holds in* (iv) *for infinitely many simple arrangements, and hence for infinitely many values of n.*

Part (ii) and its corollary (i) are due to Levi (1926). Shannon's Theorem 2.1.5 is the direct generalization to higher dimensions, but as was mentioned it is only valid there in the "straight" case. This shows that the geometric complexity of oriented matroids is greater in higher ranks, and that intuition gained in the rank 3 case may be too optimistic and simplistic.

Part (iii) is due to Shannon (1974) and Roudneff (1988a), in response to a conjecture by Grünbaum (1972). It is known that the stretchability assumption cannot be dropped. Part (iv) is due to Roudneff (1989b). Part (v) was proved independently by Harborth (1984) and Roudneff (1986).

The fact that $\frac{1}{3}n(n-1)$ is the best polynomial upper bound for the number of triangles is interesting from the point of view of geometric complexity. Consider the question: How many values of a chirotope function $\chi : \binom{E_n}{3} \longrightarrow \{+, -\}$ are needed (in the worst case) for χ to be completely determined? Translated into a question about pseudoline arrangements we see that $2^{\frac{1}{3}n(n-1)}$ is a lower bound for this amount of information, since a triangle-switch (Figure 6.4.1) will change the orientation only at that triangle leaving the orientation of all other triples unaffected. Unfortunately such triangle-switches are not independent (switching one triangle can destroy an adjacent one), so it seems difficult to relate this $2^{\frac{1}{3}n(n-1)}$ lower bound to the $2^{n^2/8}$ lower bound of Proposition 6.5.1.

A special question concerning triangles in line arrangements is discussed in Section 1.3.

Let us call a vertex (intersection point) of a pseudoline arrangement *ordinary* if exactly two pseudolines meet there. Proposition 6.1.1 (i) can be reformulated as saying that ordinary vertices always exist. However, a considerably stronger result is known.

6.5.3. Theorem (Csima and Sawyer 1992). *Every arrangement of $n \neq 7$ pseudolines has at least $\frac{6n}{13}$ ordinary vertices.*

Equivalently, every orientable simple rank 3 matroid of size $n \neq 7$ has at least $\frac{6n}{13}$ 2-point lines.

The existence of 2-point lines for arbitrary finite non-collinear point configurations in the real affine or projective plane is the well-known *Sylvester-Gallai Theorem*, whose history was touched upon in Section 1.11. The topological proof for it given in Section 6.1 goes back to Melchior (1940) and has been rediscovered several times, see Chakerian (1970). The Sylvester-Gallai Theorem has stimulated a lot of research on generalizations, one of the strongest known being Theorem 6.5.3. (Twenty years earlier, Kelly and Rottenberg (1972) had established that every arrangement of n pseudolines has at least $\frac{3n}{7}$ ordinary vertices.) See Grünbaum (1972) and Erdős and Purdy (1993) for surveys, and Bokowski and Richter-Gebert (1992) for a recent related result. Grünbaum (1972) conjectured (for the straight-line case) the following strengthening of Theorem 6.5.3, which is still open:

Every arrangement of n pseudolines has at least $f(n)$ ordinary vertices, where $f(n) = \frac{n}{2}$ if n is even $(n \neq 4)$ and $f(n) = 3[\frac{n}{4}]$ if n is odd $(n \neq 3, 5, 13)$.

If true, these bounds are known to be best possible. The following result extends the Sylvester-Gallai Theorem to higher dimensions.

6.5.4 Proposition (Hansen 1965). *Let C be a finite spanning configuration of points in d-dimensional real affine or projective space. Then among the hyperplanes spanned by subsets of C there is at least one with the property that the points of C which it contains, with the exception of precisely one of them, lie in a $(d-2)$-dimensional subspace.*

Another proof of Hansen's result was given by Shannon (1976). Proposition 6.5.4 was generalized by Edmonds and Mandel (1982) to the class of all oriented matroids $\mathcal{M}$ that can be extended to an Euclidean oriented matroid $(\mathcal{M}, g)$ by the addition of an element g in general position. The generalization to *all* oriented matroids is still open (Problem 6.1).

6.6 Orientable matroids of rank 3

Quite little is known about the problem of how to determine whether a given matroid is orientable. In this section we have gathered the available information in the rank 3 case, where the close links to pseudoline arrangements and allowable sequences improve the situation somewhat. We will return to the general case in Section 7.9.

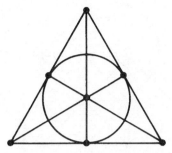

Figure 6.6.1: The minimal non-orientable Fano plane F_7.

Every property of oriented matroids which depends only on the underlying matroid (i.e., does not require a specified orientation or reorientation class) can of course be turned into a non-orientability criterion by negation. This way we obtain the following, see also Exercise 4.53.

6.6.1 Proposition. *Let M be a simple rank 3 matroid on n points.*

(i) *If M has fewer than $\frac{3n}{7}$ 2-point lines, then M is not orientable.*

(ii) *If M has a 2-coloring of the points without any monochromatic line, then M is not orientable.*

(iii) *If M has a modular line that contains k points, the remaining $n - k$ points are not all on a line, and if $k < 2\lfloor \frac{n-k}{2} \rfloor$, then M is not orientable.*

Proof. The first two parts are reformulations of Proposition 6.1.1 and Theorem 6.5.3. For part (iii), suppose that M is orientable and consider a representation of M by a pseudoconfiguration of points such that the modular line becomes the line at infinity. The result then follows from Theorem 1.11.1. □

These sufficient conditions for non-orientability are far from being necessary: the existence of sufficiently many 2-point lines, or of only sufficiently large modular lines, etc., does not imply orientability.

6.6.2 Examples.

(1) *Finite projective planes are never orientable.* A projective plane of order q has $q^2 + q + 1$ points and every line contains $q + 1$ points and is modular, so either of conditions (i) or (iii) above applies. The non-orientability of the Fano plane $F_7 = PG(3, 2)$, shown in Figure 6.6.1, was proved by Bland and Las Vergnas (1978) using exhaustive enumeration of possibilities. Since every contraction of F_7 has smaller rank, whereas deletion of any point produces the orientable matroid $M(K_4)$ of Section 1.1, F_7 is in fact (minor-) minimal non-orientable. This minimality property is not shared by any other projective plane.

(2) *Affine planes of order $q > 2$ are never orientable,* since every line has q points. The special case of $AG(3, 3)$ is shown in Figure 6.6.2. Affine planes are never minimal non-orientable.

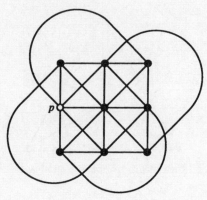

Figure 6.6.2: The affine plane AG(3, 3), and the minimal
non-orientable MacLane matroid ML$_8$.

(3)   *The MacLane matroid* ML$_8$ *is minimal non-orientable.* This matroid, in-
      troduced by MacLane (1936), is obtained from AG(3, 3) by deleting any
      point p. Its non-orientability does not follow from Proposition 6.6.1, in-
      stead it can be checked via enumeration of possibilities. Alternatively, one
      can observe that if ML$_8$ were orientable then it would by Theorem 6.3.1 be
      representable over $\mathbb{R}$. However, MacLane showed that ML$_8$ is representable
      over $\mathbb{C}$ but not over $\mathbb{R}$. Deletion of any point from ML$_8$ gives a matroid
      which is isomorphic to the affine matroid of a configuration of 7 points in
      $\mathbb{R}^2$, and hence is orientable.

We have encountered two examples, F$_7$ and ML$_8$, of minimal non-orientable
matroids of rank 3. An infinite family of minimal non-orientable matroids of
rank 3, which includes ML$_8$, was constructed by Ziegler (1991), see Exercise 6.15.
(See also Theorem 7.9.2.)

Using the existence of adjoints, Proposition 6.6.1 produces the following com-
panion result. Part (i) is a version of the "polar" of the Kelly–Rottenberg Theorem 6.5.3, due to Goodman (1980).

6.6.3 Proposition.   *Let M be a simple rank 3 matroid with n points and l
lines.*

(i)   *If M is connected, $n \geq 6$, and M has fewer than $\frac{3l}{7}$ pairs of non-intersecting
       lines, then M is not orientable.*

(ii)   *If M has a 2-coloring of the lines without any monochromatic point and
       such that no two disjoint lines have the same color, then M is not ori-
       entable.*

(iii) *If M has a point p contained in k lines, among the remaining $l - k$ lines
       there is a subset L whose members intersect pairwise but have no vertex
       common to them all, and if $k < 2\lfloor \frac{|L|}{2} \rfloor$, then M is not orientable.*

Proof. Let M^{ad} be the adjoint of M, whose points are the lines of M and whose
lines are the pencils of lines of M (corresponding to the points of M) together

with the pairs of disjoint lines of M. If M is orientable, then so is M^{ad}. This follows from Proposition 6.3.6.

Now apply Proposition 6.6.1 to M^{ad}.

For part (i) it is necessary to exclude the possibility that M has a point that lies on only two lines L_1 and L_2. Suppose this were the case and let $n_1 = |L_1|, n_2 = |L_2|$, so that $n = n_1 + n_2 - 1$. Since M is connected, $n_1 \geq 3$ and $n_2 \geq 3$, and equality in both cases is impossible since $n \geq 6$. For such a matroid there are $l = 2 + (n_1 - 1)(n_2 - 1)$ lines and $m = 2 \cdot \binom{n_1-1}{2}\binom{n_2-1}{2}$ pairs of disjoint lines, so that $m > \frac{3l}{7}$ follows, contradicting the hypothesis. Hence, the 2-point lines of M^{ad} are precisely the pairs of disjoint lines of M.

Part (ii) is straightforward.

For part (iii) we apply Proposition 6.6.1 (iii) to the submatroid M' of M^{ad} generated by the lines of M containing p together with the lines in L. For, if M' is not orientable, then neither is M^{ad}, and hence not M. $\qquad\square$

In view of these criteria, which may sound pretty "special" or "weak", we note that we *cannot* expect that there is *any* efficient algorithm for deciding whether a given rank 3 matroid is orientable or not. Specifically: The Problem ORIENTABILITY$_3$ is NP-complete, as was shown in Richter-Gebert (1998).

Here we only note the quite trivial fact that ORIENTABILITY$_3$ is in NP: if M is orientable, then we can list all basis orientations in $O(n^3)$ space and the 3-term Grassmann–Plücker relations (Theorem 3.6.2) can be checked in $O(n^5)$ time. Furthermore, the equivalence (1) $\Longleftrightarrow$ (4) of the following theorem shows that there is an even more efficient procedure for proving orientability.

6.6.4 Theorem (Folkman and Lawrence 1978, Goodman 1980, Cordovil 1982a). *Let $M = M(E)$ be a simple matroid of rank 3. Then the following conditions are equivalent:*

(1)   *M is orientable.*

(2)   *M is the matroid of a pseudoline arrangement $\mathcal{A} = (L_e)_{e \in E}$ (i.e., the lines of M are exactly the maximal subcollections of $\mathcal{A}$ with non-empty intersection).*

(3)   *M is the (little) matroid of an allowable sequence Σ (i.e., for some identification $E = E_n$ the lines of M are exactly the increasing substrings reversed in the moves of Σ).*

(4)   *There exist compatible linear orderings of the set E of points and of the set L of lines of M such that, if $e_1 < e_2 < e_3$ for three non-collinear points, then the three lines spanned have either the order $\overline{e_1e_2} < \overline{e_1e_3} < \overline{e_2e_3}$ or the order $\overline{e_2e_3} < \overline{e_1e_3} < \overline{e_1e_2}$.*

Proof. The equivalence (1) $\Longleftrightarrow$ (2) follows from the Topological Representation Theorem in rank 3 (Theorem 6.2.3), whereas (2) $\Longleftrightarrow$ (3) was shown in Section 6.4.

For $(3) \Rightarrow (4)$, let Σ be an allowable sequence with matroid $M = M(E_n)$, and with corresponding wiring diagram $\Sigma^{\approx}$ as in (6.4.1). Order the points of M naturally: $1 < 2 < \ldots < n$, and order the lines according to the sequence of moves of Σ (equivalently, according to increasing horizontal coordinate of the corresponding intersection point of $\Sigma^{\approx}$). For example, the allowable sequence of Figure 6.3.1 has the following ordering of its lines: $234 < 56 < 26 < \ldots < 35 < 45$. If several lines belong to the same move their internal ordering can be arbitrarily assigned.

Now, let $e_1 < e_2 < e_3$ be three non-collinear points in M, i.e., three wires in $\Sigma^{\approx}$ not through a point. Erase all other wires from the diagram $\Sigma^{\approx}$. There are (up to equivalence) only two possible types for the remaining diagram, as shown in Figure 6.6.3. These lead to the orders $\overline{e_1 e_2} < \overline{e_1 e_3} < \overline{e_2 e_3}$ and $\overline{e_2 e_3} < \overline{e_1 e_3} < \overline{e_1 e_2}$, respectively.

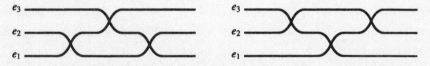

Figure 6.6.3: Sketch for the proof of Theorem 6.6.4.

To prove $(4) \Longrightarrow (3)$, label the elements of E by $1, 2, \ldots, n$ so that the compatible orderings of E and L are given by $1 < 2 < \ldots < n$ and (say) $l_1 < l_2 < \ldots < l_m$. Now, put $\sigma_0 = 12 \ldots n$, and construct σ_i from σ_{i-1} inductively by reversing the substring corresponding to the points of l_i, for $1 \leq i \leq m$. We claim that l_i is an adjacent, increasing substring of σ_{i-1}. It then follows that $\Sigma = (\sigma_0, \sigma_1, \ldots, \sigma_m)$ is an allowable sequence with matroid M. Note that $\sigma_m = n(n-1) \ldots 1$, since each pair of points has been transposed exactly once, namely when reversing the line that they span.

That l_i is an adjacent, increasing substring of $\sigma_{i-1}, 1 \leq i \leq m$, will be proved by induction.

To see that l_i appears as an increasing substring in σ_{i-1}, we observe that l_i appears increasingly in σ_0 and that the reversal of adjacent substrings $l_1, \ldots, l_{i-1}$ only reverses the order of the points in these substrings. Since every pair of points determines a unique line, no pair of points in l_i can therefore appear transposed in σ_{i-1}.

Suppose that l_i is not an adjacent substring of σ_{i-1}. Then some three points e, e_1 and e_2, such that $l_i = \overline{e_1 e_2}$ and $e \notin l_i$, appear in σ_{i-1} in the order $\ldots e_1 \ldots e \ldots e_2 \ldots$. Since l_i appears increasingly we know that $e_1 < e_2$. There are therefore three cases to check.

(i) $e_1 < e < e_2$. Then the pairs (e_1, e) and (e, e_2) have not yet been reversed, so $\overline{e_1 e_2} < \overline{e_1 e}$ and $\overline{e_1 e_2} < \overline{e e_2}$, which contradicts the compatibility condition (4).

(ii) $e < e_1 < e_2$. This means that (e, e_1) has been reversed but (e, e_2) has not, so $\overline{ee_1} < \overline{e_1 e_2} < \overline{ee_2}$, which contradicts (4).

(iii) $e_1 < e_2 < e$. This case is symmetric to (ii).   □

Exercises

6.1* Is it true that every orientable simple matroid of rank ≥ 3 has a pair of cocircuits C_1 and C_2 such that $|C_1 \backslash C_2| = 1$?

(*Remark:* This is true in the realizable case and also in rank 3. The statement is in these cases equivalent to Hansen's Proposition 6.5.4 and to the generalized Sylvester–Gallai Theorem 6.1.1 (i), respectively. A generalization of these two cases, due to Edmonds and Mandel (1982), is mentioned after Proposition 6.5.4.)

6.2 Show that part (ii) of Lemma 6.1.2 is true for any simple graph embedded into the real projective plane, but is false for embeddings into the torus.

6.3 Give a reasonably direct proof of the rank 3 case of the Topological Representation Theorem (see Theorem 6.2.3).

(*Remark:* What we have in mind is a proof that takes advantage of the topological simplicity and additional structure of the planar case, not a mere translation of the general proof. For instance, is there a simple direct proof of the implication (1) $\implies$ (4) in Theorem 6.6.4? Since (4) $\implies$ (3) $\implies$ (2) $\implies$ (1) is fairly elementary, this might be useful. One solution to this problem was given by Bokowski, Mock and Streinu (1999).)

6.4 Prove the following "polarity principle": If $\mathcal{A}$ is an arrangement of pseudolines and $\mathcal{C}$ a set of points in $\mathbb{P}^2$, and if I is the set of all true statements of the form "$p(\in \mathcal{C})$ is incident to $L(\in \mathcal{A})$", then there is an arrangement $\hat{\mathcal{C}}$ of pseudolines and a set $\hat{\mathcal{A}}$ of points, such that the set of incidences holding between members of $\hat{\mathcal{A}}$ and members of $\hat{\mathcal{C}}$ is the dual $\hat{I}$ of I.

(*Hint:* See the proof of Proposition 6.3.6.)   (Goodman 1980)

6.5 (a) Prove Theorem 6.4.1 in the realizable case with a geometric argument using straight-line arrangements at every stage.   (Ringel 1957)

(b) Prove Theorem 6.4.1 in the realizable case with an algebraic–geometric argument for the realization space of point configurations.

6.6 Show that Theorem 6.4.1, as well as the versions for the realizable case in the previous exercise, are true for *labeled* arrangements (line i goes to line i).

6.7 Show that the isomorphism in Theorem 6.3.3 can be realized by a sequence of 0-moves (defined in Section 6.4).

6.8* Does there exist any useful combinatorial representation, such as allowable sequences and wiring diagrams, for oriented matroids of rank ≥ 4?

6.9   Show that two labeled arrangements of $n \geq 5$ pseudolines $(L_1, \ldots, L_n)$ and $(L'_1, \ldots, L'_n)$ in $\mathbb{P}^2$ are isomorphic if and only if $(L_{i_1}, \ldots, L_{i_5})$ and $(L'_{i_1}, \ldots, L'_{i_5})$ are isomorphic for all $1 \leq i_1 < \ldots < i_5 \leq n$. (Roudneff 1988c)

6.10 Let (*) denote the inequality (iv) of Theorem 6.5.2.

      (a) Prove (*) by an elementary counting argument for all *simple* arrangements of at least 4 pseudolines, and show that equality then holds if and only if every edge is adjacent to exactly one triangle.

      (b) Prove that a simple arrangement of at least 5 pseudolines achieving equality in (*) cannot have a 4-sided cell.   (Roudneff 1986)

6.11* Let $\mathcal{A} = (L_e)_{e \in E}$ be an arrangement of pseudolines and T the set of its triangles. Say that pseudoline L_e and triangle t are *adjacent* if one edge of t lies on L_e. This defines a bipartite graph $G_{\mathcal{A}} \subseteq \mathcal{A} \times T$.

      (a) Is the graph $G_{\mathcal{A}}$ connected? (*Remark:* This would imply a positive answer to Problem 6.27.)

      (b) If for two arrangements $\mathcal{A}$ and $\mathcal{A}'$ the graphs $G_{\mathcal{A}}$ and $G_{\mathcal{A}'}$ are isomorphic, must $\mathcal{A}$ and $\mathcal{A}'$ be isomorphic?

6.12 Characterize the equivalence classes of wiring diagrams that correspond to the same (labeled) pseudoline arrangement. That is, describe "local moves" between wiring diagrams that generate the equivalence classes.

6.13 Find an orientable matroid of rank 3 to which the (non-orientable) Fano plane $\mathbf{F}_7$ is an adjoint.

6.14 (a) If a matroid M has at most 7 points and is not realizable over $\mathbb{R}$, then $M \cong \mathbf{F}_7$ or $M \cong \mathbf{F}_7^*$.

      (b) Show that if M is a minimal non-orientable matroid of rank 3 on at most 8 points, then $M \cong \mathbf{F}_7$ or $M \cong \mathtt{ML}_8$.

6.15 Let H_n (for $n \geq 3$) be the rank 3 matroid on $3n + 2$ points given by the Figure below, and let G_n be the matroid on $3n - 1$ points obtained by deleting a 3-point line from H_n.

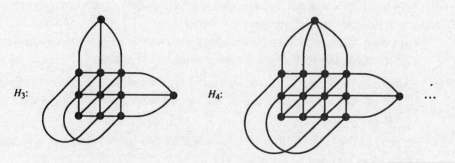

H_3: H_4: $\ldots$

      (a) Show that H_n is representable over $\mathbb{C}$ for all $n \geq 3$.

      (b) Show that $G_3 \cong \mathtt{ML}_8$ is the MacLane matroid.

(c) Show that every proper minor of G_n is orientable, but G_n is not, for all $n \geq 3$. (Ziegler 1991)

6.16* Decide the independence of the six criteria for non-orientability in Propositions 6.6.1 and 6.6.3: for any two conditions, is there a matroid satisfying one but not the other?

6.17* Find an upper bound on the number of orientable matroids (possibly those of rank 3) showing that asymptotically most matroids (of rank 3) are not orientable.

The following list of open problems on pseudoline arrangements has been contributed by **J.-P. Roudneff**. Unless otherwise indicated the problems and remarks are due to him. Throughout $\mathcal{A}$ will denote an arrangement of n pseudolines in $\mathbb{P}^2$. The number of k-sided 2-cells is denoted by p_k, so as usual p_3 is the number of triangles. $\mathcal{A}$ is a *near-pencil* if $n-1$ of the pseudolines have a vertex in common.

6.18* Classify simplicial arrangements. In particular, is it true that, except for a finite number of sporadic examples, every simplicial arrangement has a non-trivial group of symmetries? (B. Grünbaum)

6.19* Let V be a finite set of points in $\mathbb{P}^2$, with $|V| \geq 10$. Consider the arrangement $\mathcal{A}$ obtained by drawing all the straight lines connecting pairs of points in V. Show that if $\mathcal{A}$ is simplicial then $\mathcal{A}$ is a near-pencil. (B. Grünbaum)

6.20 (a) Show that a necessary condition for the existence of a simple arrangement $\mathcal{A}$ of n pseudolines with $p_3 = \frac{1}{3}n(n-1)$ is: n even, $n \not\equiv 2 \pmod{3}$, and $n \neq 12$.

(b)* Show that these conditions are also sufficient. (Roudneff 1986)

6.21 (a) Are there infinitely many arrangements of *straight lines* that have $p_3 = \frac{1}{3}n(n-1)$? (The answer is "yes": Forge and Ramírez Alfonsín 1998a)

(b)* Decide the same problem for arrangements of pseudolines with *odd* n. (B. Grünbaum)

6.22 If $\mathcal{A}$ is an arrangement of straight lines with $p_3 = n$ (necessarily simple by Theorem 6.5.2 (iii)), must there then be two triangles of $\mathcal{A}$ that have a vertex in common?

(The answer is "no": Ramírez Alfonsín (1998) has a counterexample for $n = 12$.)

6.23* Suppose that $\mathcal{A}$ is not a near-pencil, and let L be a given pseudoline in $\mathcal{A}$. Prove that at least $n-3$ triangles of $\mathcal{A}$ have no edge in common with L.

(A positive answer has been given for arrangements of straight lines by R. Shannon, using a metrical argument and duality.) (B. Grünbaum)

6.24* What is the maximum number of k-sided 2-cells in arrangements of $n \geq k$ pseudolines? Show that the maximum is attained only for simple arrangements.

(The problem has been completely solved for $k = 4$ by Grünbaum and Roudneff.)   (B. Grünbaum)

6.25* Prove that $\Sigma_{k \geq 4}(k - 6)p_k \leq 0$ holds for every arrangement.

(Strommer 1977)

6.26* A *proper k-coloring* of $\mathcal{A}$ is a coloring with k colors of the pseudolines of $\dot{\mathcal{A}}$ such that no vertex of $\mathcal{A}$ incident with at least 3 pseudolines of $\mathcal{A}$ is monochromatic. Prove that every arrangement is properly 3-colorable.

(B. Grünbaum)

6.27* Consider a coloring of the pseudolines of $\mathcal{A}$ with 2 colors, no color-class being empty. Show that there is a triangle of $\mathcal{A}$ which is supported by a pseudoline of each color.

6.28* Any arrangement $\mathcal{A}$ of n straight lines can be represented by n great circles on the 2-sphere. Suppose that $\mathcal{A}$ is simple and let $a_1, a_2, \ldots, a_m$ (with $m = n(n-1)+2$) denote the areas of the spherical polygons on the sphere. Show that $(\min a_i) / (\max a_i) \leq \epsilon(n)$, with $\epsilon(n) \longrightarrow 0$ as $n \longrightarrow \infty$.

(L. Fejes-Tóth)

7

Constructions

This chapter gives an exposition of some important constructions. It also demonstrates various ways to handle and manipulate oriented matroids. It discusses how oriented matroids can be extended, deformed, locally perturbed, flipped, glued together, and how the old and the newly obtained oriented matroids are related. This will lead to new insight into the structure of oriented matroids and their important features.

7.1 Single element extensions

For the construction and analysis of oriented matroids it is of interest to describe the set of all single element extensions of a given oriented matroid. This is also useful for proofs by induction on the size of the ground set. We note that every oriented matroid can be obtained by a sequence of single element extensions. By dualization, all results in this section can be translated into results about *single element liftings*, which will be applied to triangulations in Section 9.6. The material is drawn from Las Vergnas (1978b).

7.1.1 Definition. Let $\mathcal{M}$ be an oriented matroid on E. An *extension* of $\mathcal{M}$ is an oriented matroid $\widetilde{\mathcal{M}}$ on a ground set $\widetilde{E}$ that contains E, such that the restriction of $\widetilde{\mathcal{M}}$ to E is $\mathcal{M}$.

$\widetilde{\mathcal{M}}$ is a *single element extension* if $|\widetilde{E}\backslash E| = 1$, that is, $\widetilde{E} = E \cup p$ for some $p \notin E$.

In the remainder of this book, we will exclude the *trivial* case where p is a coloop of $\widetilde{\mathcal{M}}$ (equivalently, if $r(\widetilde{\mathcal{M}}) = r(\mathcal{M}) + 1$), and thus all single element extensions considered here are *non-trivial*.

Prototypical for the following is the characterization of single element extensions of ordinary matroids by Crapo (1965). We will now briefly review this

material. Further information can be found in Crapo and Rota (1970) and in
Section 7.3 of White (1986).

Various matroid cryptomorphisms allow us to reformulate Crapo's result in
different axiom systems. The formulation in terms of modular cuts of geometric
lattices is geometrically appealing, but does not generalize to the oriented case.
However, by taking complements we get a description of point extensions in
terms of "linear subclasses of cocircuits", which does generalize. In order to
stress the analogy with our subsequent treatment for oriented matroids, we will
give a formulation in vector notation.

We now review the concept of modular pairs and modular triples (see White
1986, Section 3.2; White 1987, Section 2.7). If these are applied to oriented
matroids, this always refers to the underlying matroid.

Two subsets $A, B \subseteq E$ form a *modular pair in M* if

$$r(A) + r(B) = r(A \cup B) + r(A \cap B).$$

Thus two different hyperplanes H^1, H^2 of a matroid form a modular pair if and
only if $r(H^1 \cap H^2) = r - 2$, that is, if they intersect in a coline.

A routine computation shows that A, B form a modular pair for M exactly if
$E \backslash A, E \backslash B$ form a modular pair for M^*. Thus we will call two cocircuits Y^1, Y^2
a *modular pair of cocircuits* in M if they form a modular pair for M^*, and hence
if and only if their complements form a modular pair of hyperplanes of M (cf.
Section 3.6).

By a *modular triple of hyperplanes* we mean three hyperplanes that intersect
in a coline ($r(H^1 \cap H^2 \cap H^3) = r - 2$), that is, distinct hyperplanes H^1, H^2, H^3
such that H^1, H^2 form a modular pair and $H^3 \supseteq H^1 \cap H^2$. Note that the order
of the H^i does not matter. A *modular triple of cocircuits* is a triple Y^1, Y^2, Y^3
such that $\{E \backslash Y^1, E \backslash Y^2, E \backslash Y^3\}$ is a modular triple of hyperplanes.

The following proposition shows that a single element extension $\widetilde{M}$ is completely determined if for every hyperplane we know whether it contains the new
point p.

7.1.2 Proposition (Crapo 1965).

(i) Let $M = M(E)$ be a matroid, $C^* \subseteq \{0,1\}^E$ its collection of cocircuits, given
     by the incidence vectors. Let $\widetilde{M} = M \cup p$ be a single element extension of
     M by a new element p, with cocircuits $\widetilde{C}^*$. Then for every cocircuit $Y \in C^*$
     there is a unique cocircuit of $\widetilde{M}$ contained in $Y \cup p$, that is, there is a unique
     function

$$\sigma : C^* \longrightarrow \{0, 1\}$$

     such that

$$\{(Y, \sigma(Y)) : Y \in C^*\} \subseteq \widetilde{C}^*,$$

     where the last component of the extended vector $(Y, \sigma(Y))$ corresponds to p.

(ii) Furthermore, all the cocircuits of $\widetilde{M}$ are given by

$$\widetilde{C}^* = \{(Y, \sigma(Y)) : Y \in C^*\} \cup$$

(7.1.1)
$$\{(Y^1 \circ Y^2, 0) : \{Y^1, Y^2\} \text{ is a modular pair of cocircuits}$$
$$\text{such that } \sigma(Y) = 1 \text{ for all } Y \subseteq Y^1 \circ Y^2\},$$

where $Y^1 \circ Y^2$ denotes the composition of the two $0/1$-vectors, that is, the union of the corresponding sets. Thus $\widetilde{M}$ is uniquely determined by σ.

Proof. The hyperplanes of a single element extension $\widetilde{M}$ of a matroid M are given by

- either H or $H \cup p$, for every hyperplane H of M, and
- the sets $G \cup p$, where $G = H_1 \cap H_2$ is an intersection of two hyperplanes of M for which there is no hyperplane $H \supset G$ of M such that $H \cup p$ is a hyperplane of $\widetilde{M}$.

Now taking complements, and putting

$$\sigma(E \backslash H) = 1 \text{ if } H \text{ is a hyperplane of } \widetilde{M}, \text{ and}$$
$$\sigma(E \backslash H) = 0 \text{ if } H \cup p \text{ is a hyperplane of } \widetilde{M},$$

the proposition follows. □

7.1.3 Theorem (Crapo 1965). *Let M be a matroid. Then, for a function $\sigma : C^* \longrightarrow \{0, 1\}$, which to every $Y \in C^*$ assigns a signature $\sigma(Y) \in \{0, 1\}$, the following are equivalent:*

(1)   *There exists a single element extension $\widetilde{M}$ of M such that*
$$\{(Y, \sigma(Y)) : Y \in C^*\} \subseteq \widetilde{C}^*.$$

(2) σ *defines a single element extension on every rank 2 contraction of M.*

(3)   *If two cocircuits in a modular triple have signature 0, then so has the third. (Equivalently, if a modular pair of cocircuits has signature 0, then so has every other cocircuit contained in their union.)*

Proof. For the implication (1)$\Longrightarrow$(2) observe that for every $E_0 \subseteq E$, the contraction $\widetilde{M}/E_0$ is a single element extension of M/E_0, with

$$C^*(M/E_0) = \{Y \in C^*(M) : Y \subseteq E \backslash E_0\}$$

and thus

$$\{(Y, \sigma(Y)) : Y \in C^*(M/E_0)\}$$
$$= \{(Y, \sigma(Y)) : Y \in C^*(M) \text{ and } Y \subseteq E \backslash E_0\}$$
$$\subseteq \{\widetilde{Y} \in C^*(\widetilde{M}) : \widetilde{Y} \subseteq (E \cup p) \backslash E_0\}$$
$$= C^*(\widetilde{M}/E_0).$$

For (2)$\Longrightarrow$(3), note that $\sigma(Y^1) = \sigma(Y^2) = 0$ implies that M is extended by an element p contained in $H_1 \cap H_2 = (E \backslash Y_1) \cap (E \backslash Y_2)$. The hyperplanes H_i have rank 1 in the contraction $M/(H_1 \cap H_2)$, thus $p \in H_1 \cap H_2$ is a loop, hence contained in every hyperplane $H \supseteq H_1 \cap H_2$. Thus $\sigma(E \backslash H) = 0$.

Finally, for (3)$\Longrightarrow$(1), we have to verify that under condition (3), the set $\widetilde{C}^*$

given by formula (7.1.1) satisfies cocircuit exchange. (It is clear that $\widetilde{C^*}$ is an antichain.) For details, consult Chapter 10 of Crapo and Rota (1970) or Section 7.3 of White (1986).   □

A set of cocircuits of a matroid M is defined to be a *linear subclass of cocircuits* if its characteristic function σ satisfies the equivalent conditions of Theorem 7.1.3. The condition obtained by taking complements of the cocircuits in (3) requires a collection of hyperplanes such that if any two intersect in a coline, then every hyperplane containing this coline is in the collection. This describes a *linear subclass of hyperplanes* (Crapo 1965). Our description of single element extensions of oriented matroids will be a direct generalization of the formulation of Theorem 7.1.3.

We will for the following assume that $\mathcal{M}$ is a fixed oriented matroid on $E = \{1, \dots, n\}$, and $\widetilde{\mathcal{M}}$ is a single element extension of $\mathcal{M}$ on $\widetilde{E} = E_{n+1} = \{1, \dots, n+1\}$ with $p := n + 1$. Recall that we only consider *non-trivial* single element extensions, with $r(\widetilde{\mathcal{M}}) = r(\mathcal{M})$. Also the case where $\mathcal{M}$ has rank 0 or 1 will be excluded for simplicity. Some crucial first observations are collected in the following proposition.

7.1.4 Proposition (Las Vergnas 1978b).

(i)   Let $\widetilde{\mathcal{M}}$ be a single element extension of $\mathcal{M}$. Then for every cocircuit $Y \in C^*$, there is a unique way to extend Y to a cocircuit of $\widetilde{\mathcal{M}}$: there is a unique function

$$\sigma : C^* \longrightarrow \{+, -, 0\}$$

such that

$$\{(Y, \sigma(Y)) : Y \in C^*\} \subseteq \widetilde{C^*},$$

that is, $(Y, \sigma(Y))$ is a cocircuit of $\widetilde{\mathcal{M}}$ for every cocircuit Y of $\mathcal{M}$. Furthermore, this σ satisfies $\sigma(-Y) = -\sigma(Y)$ for all $Y \in C^*$.

(ii) $\widetilde{\mathcal{M}}$ is uniquely determined by σ, with
$$\begin{aligned} \widetilde{C^*} = &\{(Y, \sigma(Y)) : Y \in C^*\} \cup \\ &\{(Y^1 \circ Y^2, 0) : Y^1, Y^2 \in C^*, \ \sigma(Y^1) = -\sigma(Y^2) \neq 0, \\ &S(Y^1, Y^2) = \emptyset, \ \rho(Y^1 \circ Y^2) = 2\}. \end{aligned} \tag{7.1.2}$$

(Here "$\circ$" denotes the (conformal) composition of covectors, and ρ denotes the rank function on $\mathcal{L}$, which can be derived from the rank function on the geometric lattice $\mathcal{L}$ by $\rho(Z) = r(\hat{1}) - r(z(Z))$.)

Proof. Both parts follow from "orienting" Proposition 7.1.2, observing that a single element extension of $\mathcal{M}$ also defines a single element extension of the underlying matroid. An alternative way to see them is to translate into the pseudosphere arrangement picture, where cocircuits are vertices.   □

Part (i) of this proposition will allow us to characterize single element extensions by describing the functions σ on C^* that give rise to single element extensions.

7.1.5 Definition (Las Vergnas 1978b). The functions $\sigma : C^* \longrightarrow \{+, -, 0\}$ that correspond to single element extensions (via Proposition 7.1.4) are called *localizations*.

In the point configuration picture, the function σ indicates for every hyperplane $H \subseteq E$ determined by points of $\mathcal{M}$ whether the new point p is supposed to lie on the hyperplane, on the positive side of H, or on its negative side.

In the pseudosphere arrangement picture, σ signifies for every vertex whether it is supposed to lie on the positive side or on the negative side of S_{n+1} or on S_{n+1} itself.

We want to characterize the functions σ that are localizations, i.e., that determine single element extensions. Recalling that oriented matroids can be axiomatized as geometries that are consistently oriented in codimension 2 (Section 3.6), one is led to study the case of rank 2 oriented matroids in detail.

7.1.6 Remark. For every $n \geq 2$ there is, up to reorientation and permutation of the elements, exactly one simple rank 2 oriented matroid on n elements.

This is clear from the Topological Representation Theorem (compare this to our discussion in Section 6.1 of the rank 2 case), but it is also easy to verify directly from any axiom system for oriented matroids.

Thus, if we delete multiple points and loops, relabel and reorient, every rank 2 oriented matroid is isomorphic to a "standard" one, given by n consecutive points on the (affine) real line. Equivalently, this oriented matroid is given by a central line arrangement in $\mathbb{R}^2$, where the choice of a "positive side" for every line specifies an orientation.

7.1.7 Example. Every oriented matroid of rank 2 can be represented by a central line arrangement in the plane, as indicated by an example in Figure 7.1.1. Its cocircuits correspond to the rays emanating from the origin. For the example of Figure 7.1.1, we show the cocircuits (in vector notation) in Figure 7.1.2.

Now there are *three types of cocircuit signatures* $\sigma : C^* \longrightarrow \{+, -, 0\}$ *that define a single element extension of a rank 2 oriented matroid*. Of course every such cocircuit signature has to satisfy $\sigma(-Y) = -\sigma(Y)$ for all $Y \in C^*$.

Type I: The cocircuit signature given by $\sigma(Y) = 0$ for all $Y \in C^*$ defines the extension of $\mathcal{M}$ by a loop.

Type II: If there is an $e \in E$ and a sign $\alpha_e \in \{+, -\}$ such that $\sigma(Y) = \alpha_e Y_e$ for all $Y \in C^*$, then this defines the extension of $\mathcal{M}$ by an element p that is parallel (if $\alpha_e = +$) or antiparallel (if $\alpha_e = -$) to e. We assume for this that e is not a loop, to exclude the situation of Type I.

For the example of Figure 7.1.1, the cocircuit signature in Figure 7.1.3 produces an extension by an element antiparallel to 3.

Such a cocircuit signature has always exactly one pair $Y, -Y$ of signed cociruits with signature $\sigma(Y) = \sigma(-Y) = 0$.

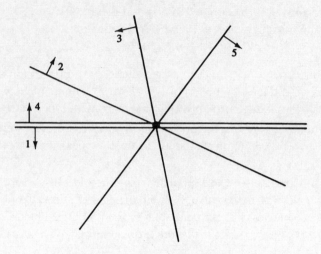

Figure 7.1.1: Example of rank 2.

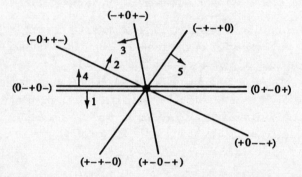

Figure 7.1.2: Cocircuits for the Example of Figure 7.1.1.

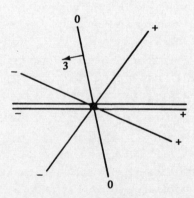

Figure 7.1.3: Cocircuit signature of Type II.

Type III: If there are $e, f \in E$ and a pair of signs $[\alpha_e, \alpha_f] \in \{+, -\}^2$ such that

$$\sigma(Y) = \begin{cases} \alpha_e Y_e & \text{for } Y_e \neq 0, \\ \alpha_f Y_f & \text{otherwise,} \end{cases}$$

then this also defines an extension of $\mathcal{M}$. If we assume that $\{e, f\}$ is independent (otherwise we get a signature of Type I or II) then $\sigma(Y) \neq 0$ for all $Y \in C^*$. The corresponding single element extension adds to $\mathcal{M}$ a line that is obtained by "perturbing e in direction f", after the sign reversals indicated by α_e and α_f.

Our example, for $e = 3$, $f = 5$ and $\alpha_e = \alpha_f = +$, produces the situation of Figure 7.1.4.

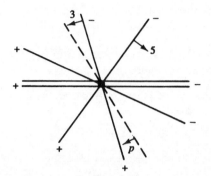

Figure 7.1.4: Cocircuit signature of Type III.

These three types of possible cocircuit signatures describe all single element extensions of a matroid of rank 2, and in all cases the extension is determined uniquely. In fact, in the first two cases we have

$$\widetilde{C^*} = \{(Y, \sigma(Y)) : Y \in C^*\},$$

whereas in the third case we have to add two "new" cocircuits of the form $(Y^1 \circ Y^2, 0)$, where Y^1 and Y^2 can be chosen to be conformal cocircuits with $\sigma(Y^1) = -\sigma(Y^2) \neq 0$. In this case $\rho(Y^1 \circ Y^2) = 2$ because $Y^1 \circ Y^2 = Y^2 \circ Y^1 = Y^1 \vee Y^2$ lies in the signed cocircuit span $\mathcal{L}$. (Here ρ denotes the rank function on $\mathcal{L}$.)

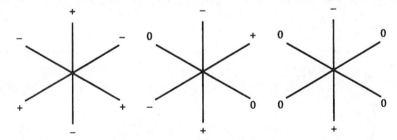

Figure 7.1.5: Forbidden subconfigurations for cocircuit signatures.

Finally, observe that these three types of admissible cocircuit signatures (localizations) can also be described as follows: a symmetric cocircuit signature of an oriented matroid of rank 2 defines an extension if and only if it does not have a three element restriction that is signed according to one of the three "forbidden configurations" of Figure 7.1.5.

The detailed discussion of the rank 2 case will be useful for illustrating the structure of "lexicographic extensions". It will turn out that *all* extensions of rank 2 oriented matroids in fact are lexicographic extensions (Remark 7.2.5).

Now we use the analysis of the rank 2 case to state the characterization theorem for single element extensions. Observe the close analogy to the "unoriented" Theorem 7.1.3.

7.1.8 Theorem (Las Vergnas 1978b). *Let $\mathcal{M}$ be an oriented matroid, and*
$$\sigma : \mathcal{C}^* \longrightarrow \{+,-,0\}$$
a cocircuit signature, satisfying $\sigma(-Y) = -\sigma(Y)$ for all $Y \in \mathcal{C}^*$. Then the following statements are equivalent.

(1) σ *is a localization: there exists a single element extension $\widetilde{\mathcal{M}}$ of $\mathcal{M}$ such that*
$$\{(Y,\sigma(Y)) : Y \in \mathcal{C}^*\} \subseteq \widetilde{\mathcal{C}^*}.$$

(2) σ *defines a single element extension on every contraction of $\mathcal{M}$ of rank 2. That is, the signature on every rank 2 contraction is of one of the Types I, II and III described in Example 7.1.7.*

(3)   *The signature σ produces none of the three excluded subconfigurations (minors) of rank 2 on three elements, as given by Figure 7.1.5.*

Before proving this theorem, a few comments are in order.

(i)   If $S \subseteq E$ is a subset of rank $r(\mathcal{M}) - 2$, then $\mathcal{M}/S$ is an oriented matroid of rank 2, whose cocircuits are the cocircuits contained in $E \backslash S$. Thus the cocircuit signature on $\mathcal{C}^*(\mathcal{M}/S)$ referred to in condition (2) of the theorem is canonically given by the cocircuit signature σ on $\mathcal{C}^*$.

(ii)   If $\mathcal{M}$ is given by a pseudosphere arrangement, then the cocircuits are the vertices. The theorem then states that a labeling σ of the vertices by $+, -, 0$ is a localization if it is a localization on every pseudoline, that is, if on every 1-sphere in the arrangement it is of one of the three types indicated by Figure 7.1.6.

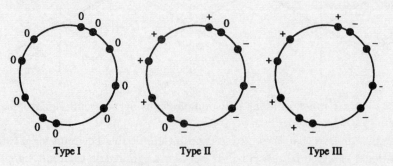

Figure 7.1.6: Localizations of the three possible Types.

This will be used for the explicit construction of examples in Section 10.4.

(iii)   The theorem simplifies for uniform oriented matroids: if $\mathcal{M}$ is uniform, then a localization σ defines a uniform single element extension if and only if it only takes values $+$ and $-$. Cocircuit signatures defining uniform extensions of uniform oriented matroids can therefore be characterized by the condition that the signature induced on a contraction of rank 2 is always of Type III.

Proof of Theorem 7.1.8. For $(1) \Longrightarrow (2)$, let $\mathcal{M}_1 = \mathcal{M}/E_0$ be a contraction of $\mathcal{M}$. Then (in set notation) every cocircuit of $\mathcal{M}_1$ is also a cocircuit of $\mathcal{M}$, that is,

$$\check{\mathcal{C}}^* = \{Y|_{E \backslash E_0} : Y \in \mathcal{C}^*, \ \underline{Y} \subseteq E \backslash E_0\}.$$

Thus the same proof as for the implication $(1) \Longrightarrow (2)$ of Theorem 7.1.3 applies.

The equivalence $(2) \Longleftrightarrow (3)$ follows from our analysis of the rank 2 case in Example 7.1.7.

For the implication $(2) \Longrightarrow (1)$ we have to verify that under the condition (2) the signed set system (7.1.2) from Proposition 7.1.4, i.e.,

$$\widetilde{\mathcal{C}}^* = \{(Y, \sigma(Y)) : Y \in \mathcal{C}^*\} \cup$$
(7.1.2) $$\{(Y^1 \circ Y^2, 0) \ : Y^1, Y^2 \in \mathcal{C}^*, \ \sigma(Y^1) = -\sigma(Y^2) \neq 0,$$
$$S(Y^1, Y^2) = \emptyset, \ \rho(Y^1 \circ Y^2) = 2\},$$

satisfies signed cocircuit exchange. However, by Proposition 7.1.2 we know that σ defines a single element extension $\widetilde{M}$ of the underlying matroid $M := \underline{\mathcal{M}}$, whose cocircuits are given by $C^*(\widetilde{M}) = \underline{\widetilde{\mathcal{C}}^*}$. This allows us to apply Theorem 3.6.2*: we only have to check oriented cocircuit exchange for modular pairs.

So, let $Y^1, Y^2 \in \widetilde{\mathcal{C}}^*$ form a modular pair, that is,

$$E_0 = \widetilde{E} \backslash (\underline{Y}^1 \cup \underline{Y}^2) = \{e \in \widetilde{E} : Y_e^1 = Y_e^2 = 0\}$$

satisfies $r(E_0) = r - 2$. We now distinguish three cases.

Case 1: If $p \notin E_0$, then we consider the subsystem

$$C^*(\widetilde{\mathcal{M}}/E_0) := \{Y \in \widetilde{\mathcal{C}}^* : E_0 \cap \underline{Y} = \emptyset\},$$

which contains Y^1 and Y^2. This is the system of cocircuits of the extension that σ defines on the rank 2 contraction $\mathcal{M}/E_0$, so oriented cocircuit exchange can be performed.

Case 2: If $p \in E_0$ but $r(E_0 \backslash p) = r - 2$, then we similarly pass to $C^*(\widetilde{\mathcal{M}}/(E_0 \backslash p))$, where oriented cocircuit exchange works by assumption.

Case 3: If $p \in E_0$ and $r(E_0 \backslash p) = r - 3$, then we also pass to the system $C^*(\widetilde{\mathcal{M}}/(E_0 \backslash p))$. This is an orientation $\mathcal{M}_1$ of the rank 3 matroid $M_1 := \widetilde{M}/(E_0 \backslash p)$ on the ground set $E_1 := (E \backslash E_0) \cup \{p\}$.

We reorient $\mathcal{M}_1$ such that $Y_e^1 \geq 0$ for $e \in \underline{Y}^1 \backslash \underline{Y}^2$, and $Y_e^2 \geq 0$ for $e \in \underline{Y}^2 \backslash \underline{Y}^1$. Furthermore, we fix some $e_1 \in \underline{Y}^2 \backslash \underline{Y}^1$ and $e_2 \in \underline{Y}^1 \backslash \underline{Y}^2$.

Now let $f \in E$ with $Y_f^1 = -Y_f^2 = +$. Then $f \notin E_0$, hence $\{p, f\}$ has rank 2 in M_1, so $\overline{p, f}$ is a hyperplane, and there is a unique cocircuit Z such that $Z_p = Z_f = 0$ and $Z_{e_1} = +$. We claim that Z correctly eliminates f from Y^1 and Y^2, that is,

$$Z^+ \subseteq (Y^1)^+ \cup (Y^2)^+ \quad Z^- \subseteq (Y^1)^- \cup (Y^2)^-.$$

Note that if this fails, then it already fails in a restriction of M_1 to a five element subset $\{e_1, e_2, f, p, q\} \subseteq E_1$. This leads to the analysis of a number of cases corresponding to the possible isomorphism types of these restrictions, for which we omit the details. $\qquad \Box$

Proposition 7.1.4 and Theorem 7.1.8 together form an abstract description of *all* single element extensions of a given oriented matroid, by showing that they are equivalent to localizations, of which several characterizations of varying effectiveness are given: (2) gives a manageable tool to handle examples, as will be seen in Section 10.4. Correspondingly, (3) means that there is always a "short proof" (a small bad minor) if a cocircuit signature σ is *not* a localization. The following alternative characterization will not be used later.

7.1.9 Corollary (Edmonds and Mandel 1982). *A symmetric cocircuit signature* $\sigma : \mathcal{C}^* \longrightarrow \{+, -, 0\}$ *is a localization if and only if*

$$\mathcal{S} := \sigma^{-1}(\{+, 0\}) = \{Y \in \mathcal{C}^* : \sigma(Y) \geq 0\}$$

satisfies the (weak) signed cocircuit exchange axiom (C3).

Proof. By Theorem 7.1.8 (2)$\Longrightarrow$(1), the "if"-part reduces to the case of a modular pair of cocircuits and thus to a rank 2 contraction. Now in rank 2, elimination between two cocircuits $Y^1 \neq -Y^2$ yields a cocircuit *between* the two (in the obvious geometric sense): for $e \in S(Y^1, Y^2)$ elimination yields the unique cocircuit Y *between* Y^1 and Y^2. (Geometric intuition comes from the Figures of Example 7.1.7 or from Figure 7.1.6.)

Thus $\mathcal{S} \cap \mathcal{C}^*(\mathcal{M}/E_0)$ satisfies cocircuit exchange if and only if it is *convex* (that is, if it contains Y^1 and Y^2 with $Y^1 \neq -Y^2$, then it contains all the cocircuits Y which satisfy $Y^1 \circ Y = Y^1 \circ Y^2$ and $Y^2 \circ Y = Y^2 \circ Y^1$). Together with the symmetry condition $\sigma(-Y) = -\sigma(Y)$, this means that σ satisfies condition (2) of Theorem 7.1.8.

For the "only if"-part let $Y^1, Y^2 \in \mathcal{S}$ and $e \in S(Y^1, Y^2)$. If $\sigma(Y^1) = \sigma(Y^2) = +$, then $(Y^1, +)$ and $(Y^2, +)$ are cocircuits of $\widetilde{\mathcal{M}}$, and by strong cocircuit exchange we produce a cocircuit $\widetilde{Y}$ of $\widetilde{\mathcal{M}}$ with $\widetilde{Y}_p = +$. Thus if we write $\widetilde{Y} = (Y, +)$, then $Y \in \mathcal{S}$ follows from formula (7.1.2), and Y is a cocircuit resulting from (weak) cocircuit exchange in $\mathcal{M}$. The same argument applies if $\sigma(Y^1) = +$ and $\sigma(Y^2) = 0$.

Finally, if $\sigma(Y^1) = \sigma(Y^2) = 0$, then Y^1 and Y^2 are cocircuits of $\widetilde{\mathcal{M}}/p$. Cocircuit elimination results in $Y \in \mathcal{C}^*(\widetilde{\mathcal{M}}/p)$, such that $(Y, 0)$ is a cocircuit of $\widetilde{\mathcal{M}}$. Now by (7.1.2), either $Y \in \mathcal{C}^*$, or $Y = Y^3 \vee Y^4$, where $\sigma(Y^3) = +$, $\sigma(Y^4) = -$. Thus either Y or Y^3 satisfies the conditions. $\qquad \Box$

Remarks.

(1) Corollary 7.1.9 does *not* have an unoriented counterpart. In fact, simple examples show that neither $S_0 = \{Y : \sigma(Y) = 0\}$ nor $S_1 = C^* \backslash S_0$ satisfy cocircuit exchange in general.

(2) Our proof shows that if σ is a localization, then also $\mathcal{S}_+ := \sigma^{-1}(\{+\}) = \{Y \in C^* : \sigma(Y) = +\}$ satisfies cocircuit exchange. The converse is false – see the first configuration of Figure 7.1.5.

7.2 Lexicographic extensions and the extension lattice

In this section we will describe the construction of "lexicographic extensions", a class of explicitly given single element extensions which has various uses and applications.

We start with a definition of the extension lattice, which organizes and encodes the complete set of all single element extensions of an oriented matroid. Its relevance for the analysis of oriented matroids will be illustrated in Section 7.5. The idea is to put a partial order on the set of single element extensions, by declaring that *a single element extension is "smaller" than another one if it is obtained by "moving the extension point into a more special position"*. This makes sense for matroids as well as for oriented matroids.

More formally, the set of single element extensions of a matroid is given by the linear subclasses of cocircuits, as described by Theorem 7.1.3(3). We will partially order them by inclusion, resulting in the *extension lattice $E(M)$* of a matroid (Cheung 1974). For this lattice, note that $M \cup p^1 \leq M \cup p^2$ holds exactly if $\sigma^1 \leq \sigma^2$, where the σ-function is the characteristic vector of the *complement* of the linear subclass of cocircuits: $\sigma(Y) = 1$ if and only if Y is not in the subclass.

The poset $E(M)$ has a unique minimal element (corresponding to the extension by a loop, $\sigma \equiv 0$) and a unique maximal element (corresponding to the extension by a point in general position, $\sigma \equiv 1$). Now since the intersection of linear subclasses of cocircuits is again linear, we get that for $\sigma^1, \sigma^2 \in E(M)$, the union $\sigma^1 \circ \sigma^2$ is again in $E(M)$, and thus $E(M)$ is a lattice. (Here we, interchangeably, interpret σ as a set of cocircuits, as a 0/1-vector or as a function $C^* \longrightarrow \{0, 1\}$. Thus $\sigma^1 \circ \sigma^2(Y) = 1$ unless $\sigma^1(Y) = \sigma^2(Y) = 0$.)

Furthermore, for every flat $F \in L$ the corresponding principal extension is given by $\sigma^F = \{Y \in C^* : Y \cap F = \emptyset\}$, which leads to an inclusion map $L^{op} \hookrightarrow E(M)$. Here L^{op} denotes the order dual (inverted lattice) of L.

Consult Bachem and Kern (1986a,b) for more material on the extension lattice of a matroid. The analogous construction for oriented matroids yields the following.

7.2.1 Definition (Bachem and Kern 1986a,b). Let $\mathcal{M}$ be an oriented matroid. The *extension poset* of $\mathcal{M}$ is the set $\mathcal{E}(\mathcal{M})$ of all single element extensions, ordered by

$$M \cup p^1 \leq M \cup p^2 \quad \text{if and only if} \quad \sigma^1 \leq \sigma^2.$$

The natural ordering on the set of localizations is given by regarding them as $\{+,-,0\}$-vectors. Equivalently, we can define $\sigma^1 \leq \sigma^2$ by any of the following equivalent conditions:

(i) $\sigma^1(Y) \in \{0, \sigma^2(Y)\}$ for all Y,

(ii) $\mathcal{S}^1 \subseteq \mathcal{S}^2$, using the sets $\mathcal{S}(\mathcal{M})$ of Corollary 7.1.9,

(iii) $M \cup p^2 \rightsquigarrow M \cup p^1$ is a weak map, as defined in Section 7.7.

For the rest of this section, it will be convenient to consider the localizations σ as sign vectors $\sigma = (\sigma(Y))_{Y \in \mathcal{C}^*}$ on the set of all cocircuits of $\mathcal{M}$, where the ordering $\leq$ on localizations coincides with the usual order on signed covectors.

In the following proposition, we collect the basic properties of $\mathcal{E}(\mathcal{M})$.

7.2.2 Proposition (Las Vergnas 1978b, Bachem and Kern 1986a).

(1)   *$\mathcal{E}(\mathcal{M})$ has a unique minimal element $\sigma \equiv 0$, and it is closed under composition: $\sigma^1 \circ \sigma^2 \in \mathcal{E}(\mathcal{M})$ for $\sigma^1, \sigma^2 \in \mathcal{E}(\mathcal{M})$.*

(2)   *The maximal elements of $\mathcal{E}(\mathcal{M})$ are the localizations that define extensions of $\mathcal{M}$ by elements in general position, that is, they satisfy $\sigma(Y) \neq 0$ for all $Y \in \mathcal{C}^*$.*

(3)   *Adjoining a maximal element to $\mathcal{E}(\mathcal{M})$ one obtains a lattice, the extension lattice $\mathcal{E}(\mathcal{M}) \cup \hat{1}$ of $\mathcal{M}$.*

Proof.

(1)   follows directly from Theorem 7.1.8(1)$\Longleftrightarrow$(2) and the analysis of the rank 2 case, where we only have to check that $\sigma^1 \circ \sigma^2 \in \mathcal{E}(\mathcal{M})$ for the three types of localizations as given by Figure 7.1.6.

(2)   is clear from (1). For

(3)   observe that there is a join (smallest common upper bound) for any $\sigma^1, \sigma^2 \in \mathcal{E}(\mathcal{M})$: it is $\sigma^1 \circ \sigma^2$ if the two extensions are conformal (such that $\sigma^1 \circ \sigma^2 = \sigma^2 \circ \sigma^1 = \sigma^1 \vee \sigma^2$), and is $\sigma^1 \circ \sigma^2 = \hat{1}$ otherwise, if they do not have an upper bound in $\mathcal{E}(\mathcal{M})$.   □

The geometric interpretation is that the extension $\sigma^1 \circ \sigma^2$ *is obtained by taking the point extension corresponding to σ^1 and perturbing it in direction of the point described by σ^2.*

Proposition 7.2.2(2) states that *every single element extension can be perturbed to an extension by an element in general position,* that is, "every localization is incident to a uniform localization" (Las Vergnas 1978b).

In fact, Proposition 7.2.2(1) yields a very useful *composition operation* on single element extensions. (The analogy to the composition of covectors is not completely coincidental, as we will soon see. Namely, we derive the (canonical) embedding of adjoints into the extension lattice, which implies "good intersection properties" for all oriented matroids with adjoint – see Section 7.5.)

To put the composition operation to use, let us recall as "building blocks" the simplest single element extensions. First, $\sigma \equiv 0$ describes the extension of $\mathcal{M} = \mathcal{M}(E)$ by a new loop. Secondly, for $e \in E$, the signature $\sigma[e^+]$ defined by

$$\sigma[e^+](Y) := Y_e \quad \text{for all} \quad Y \in \mathcal{C}^*(\mathcal{M})$$

describes the extension of $\mathcal{M}$ by an element that is parallel to e. Similarly, $\sigma[e^-]$ with

$$\sigma[e^-](Y) := -Y_e \quad \text{for all} \quad Y \in \mathcal{C}^*(\mathcal{M})$$

is the extension by an element that is antiparallel to e. We will use a combined notation $\sigma[e^\alpha]$ for $\alpha \in \{+, -\}$ to cover both cases.

All this is quite trivial, but by forming compositions of these "elementary" (anti)parallel extensions we get a useful class of single element extensions, which have explicit descriptions.

7.2.3 Definition (Las Vergnas 1978b). Let $\mathcal{M}$ be an oriented matroid, $I = [e_1, \ldots, e_k]$ an ordered subset of E, and

$$\alpha = [\alpha_1, \ldots, \alpha_k] \in \{+, -\}^k.$$

Then the *lexicographic extension of $\mathcal{M}$ by* $p := [I^\alpha] = [e_1^{\alpha_1}, \ldots, e_k^{\alpha_k}]$ is the single element extension $\widetilde{\mathcal{M}} = \mathcal{M} \cup p = \mathcal{M}[I^\alpha] = \mathcal{M}[e_1^{\alpha_1}, \ldots, e_k^{\alpha_k}]$ of $\mathcal{M}$ given by $\sigma : \mathcal{C}^* \longrightarrow \{+, -, 0\}$ with

$$\sigma[e_1^{\alpha_1}, \ldots, e_k^{\alpha_k}] = \sigma[e_1^{\alpha_1}] \circ \ldots \circ \sigma[e_k^{\alpha_k}].$$

Thus the lexicographic extensions are exactly those extensions which arise as compositions of extensions by parallel or antiparallel elements.

Lexicographic extensions are a particular way to construct orientations of principal matroid extensions. Recall that principal matroid extensions correspond to placing a new point "in general position" on a given flat F. In oriented matroids, we need more information to determine the position of the point on the flat. The solution is to specify an ordered basis $[e_1, \ldots, e_k]$ of the flat F, together with a string of signs $[\alpha_1, \ldots, \alpha_k] \in \{+, -\}^k$.

In the case of a realizable oriented matroid, given by any vector configuration in $\mathbb{R}^n$, the relative position of the new point p can then be completely determined as

$$p = \alpha_1 e_1 + \epsilon \alpha_2 e_2 + \epsilon^2 \alpha_3 e_3 + \ldots + \epsilon^{k-1} \alpha_k e_k$$

for some small $\epsilon > 0$. Combinatorially, and thus also covering the non-realizable case, this corresponds to the following inductive procedure: put $p_0 = \mathbf{0}$ and get p_i by moving p_{i-1} within the flat

$$F_i = \overline{\{e_1, \ldots, e_i\}}$$

off the hyperplane $F_{i-1} = \overline{\{e_1, \ldots, e_{i-1}\}}$ in F_i to the side of F_{i-1} on which $\alpha_i e_i$ lies. Formally, the preceding idea amounts to the following (see also Billera and Munson 1984b,c).

7.2.4 Proposition (Las Vergnas 1978b). *Let $\mathcal{M}$ be an oriented matroid, $I = [e_1, \ldots, e_k]$ an ordered subset of E, and $\alpha = [\alpha_1, \ldots, \alpha_k] \in \{+, -\}^k$.*

Then the lexicographic extension $M[I^\alpha] = M[e_1^{\alpha_1}, \ldots, e_k^{\alpha_k}]$ of M is given by $\sigma : C^* \longrightarrow \{+, -, 0\}$ with

$$\sigma(Y) = \begin{cases} \alpha_i Y_{e_i} & \text{if } i \text{ is minimal such that } Y_{e_i} \neq 0, \\ 0 & \text{if } Y_{e_i} = 0 \text{ for all } 1 \leq i \leq k. \end{cases}$$

Since i can be minimal with $Y_{e_i} \neq 0$ only if $e_i \notin \{e_1, \ldots, e_{i-1}\}$, the lexicographic extension $M[I^\alpha]$ is determined by the lexicographically first ("greedy") basis of I together with the corresponding signs. It is therefore customary to require (without loss of generality) that I is an independent set.

Let us record how the chirotope of a lexicographic extension is determined: it is given by

$$\widetilde{\chi}(B) \quad = \chi(B) \quad \text{for } p \notin B,$$
$$\widetilde{\chi}(f_1, \ldots, f_{r-1}, p) = \alpha_i \chi(f_1, \ldots, f_{r-1}, e_i) \quad \text{for the smallest } i$$
$$\text{such that this is not zero}$$
$$= \alpha_1 \chi(f_1, \ldots, f_{r-1}, e_1) \circ \alpha_2 \chi(f_1, \ldots, f_{r-1}, e_2)$$
$$\circ \ldots \circ \alpha_k \chi(f_1, \ldots, f_{r-1}, e_k)$$

As special cases we can recover for $I = \emptyset$ the extension of M by a loop, and for $I = \{e\}$ the extension of M by an element that is parallel or antiparallel to e, depending on whether $\alpha_e = +$ or $\alpha_e = -$.

7.2.5 Remark. All single element extensions of an oriented matroid of rank 2 are lexicographic.

In fact, in the classification of Example 7.1.7, "Type I" is the extension by a loop, "Type II" is the extension by a parallel or antiparallel element, and "Type III" is the lexicographic extension corresponding to a basis of M.

If M is uniform, then $\widetilde{M}$ is uniform again if and only if I contains a basis. More generally, we can observe the following facts.

7.2.6 Lemma (Todd 1984). Let $\widetilde{M} = M[I^\alpha]$ arise from M by lexicographic extension by a point p, where $I \subseteq E$ is independent of rank k. Then the underlying matroid of $\widetilde{M}$ is the principal extension of $\underline{M}$ in the flat spanned by I. Furthermore $X \in \{+, -, 0\}^{E \cup p}$, defined by

$$X_e = \begin{cases} \alpha_e & \text{for } e \in I, \\ - & \text{for } e = p, \\ 0 & \text{otherwise,} \end{cases}$$

is a circuit of $\widetilde{M}$ of size $k + 1$, and every circuit of $\widetilde{M}$ other than $\pm X$ that contains p has at least $k + 2$ elements.

Proof. By Proposition 7.2.4 $\underline{\widetilde{M}}$ is the specified principal extension. This already determines the supports of the circuits of $\widetilde{M}$. Now it follows that X has the specified signs from the fact that they are orthogonal to the cocircuits of $\widetilde{M}$, as given by Proposition 7.2.4.   □

See Todd (1984, 1985) for a more detailed description of circuits in lexicographic extensions.

If $\widetilde{\mathcal{M}}$ is the lexicographic extension $\widetilde{\mathcal{M}} = \mathcal{M}[e_1^{\alpha_1}, \ldots, e_k^{\alpha_k}]$ of $\mathcal{M}$ by p, then e_1 and p are inseparable. (Inseparability and the inseparability graph are studied in Section 7.8.) In fact, $\widetilde{\mathcal{M}}$ can be characterized as the unique single element extension of $\mathcal{M}$ such that

(i) e_1 and p are inseparable in $\widetilde{\mathcal{M}}$, covariant if $\alpha_1 = +$, contravariant if $\alpha_1 = -$, and

(ii) $\widetilde{\mathcal{M}}/e_1$ is the lexicographic extension $(\mathcal{M}/e_1)[e_2^{\alpha_2}, \ldots, e_k^{\alpha_k}]$, where $\mathcal{M}[\emptyset]$ is the extension of $\mathcal{M}$ by a loop.

In particular, this means that any extension of $\mathcal{M}$ by a non-loop that is not in an inseparable pair cannot be lexicographic. Thus there are many oriented matroid extensions (in fact, principal ones) that do not arise by lexicographic extensions.

The smallest example of this arises in rank 3 for $|E| = 5$. It is sketched in Figure 7.2.1.

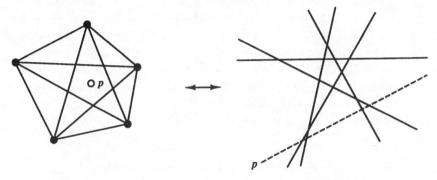

Figure 7.2.1: An extension that is not lexicographic.

It is not hard to reformulate lexicographic extensions in the language of pseudosphere arrangements. This yields a description of $\widetilde{\mathcal{A}} = \mathcal{A}[e_1^{\alpha_1}, \ldots, e_k^{\alpha_k}]$ as an arrangement to which we add a pseudosphere S_p (anti)parallel to S_{e_1}, which is then in $k-1$ steps deformed to more general position, so that after those steps $S_{e_1} \cap S_p = S_{e_1} \cap S_{e_2} \cap \ldots \cap S_{e_k}$. In particular, S_p is the "deformation" of a second copy of S_{e_1} to a nearby pseudosphere in more general position, as in Figure 7.2.2. Assuming $\alpha_1 = +$ for the moment, we can (with $p \leftrightarrow e_1$) identify the ground sets of the oriented matroids $\mathcal{M}$ and $\widetilde{\mathcal{M}} \backslash e_1$. We then say that $\widetilde{\mathcal{M}} \backslash e_1$ arises as a *global perturbation* of $\mathcal{M}$, because S_{e_1} is perturbed (almost) everywhere to obtain S_p. In particular, every tope (region) of $\mathcal{M}$ is also a tope of $\widetilde{\mathcal{M}} \backslash e_1$, so "moving S_p back to the special position S_{e_1}" corresponds to a weak map $\widetilde{\mathcal{M}} \backslash e_1 \rightsquigarrow \mathcal{M}$, see Section 7.7.

Very recently, the extension poset $\mathcal{E}(\mathcal{M})$ has received considerable attention: the *extension space conjecture* states that if $\mathcal{M}$ is a realizable oriented matroid, then the simplicial complex $\Delta(\mathcal{E}(\mathcal{M}) \backslash 0)$ is homotopy equivalent to an $(r-1)$-sphere. This was proved in Sturmfels and Ziegler (1993) for $r \leq 3$ and in

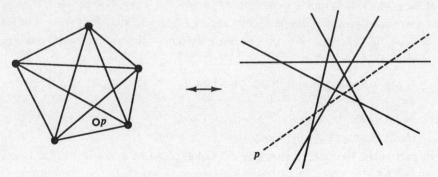

Figure 7.2.2: A lexicographic extension.

some other cases. It was shown by Mnëv and Richter-Gebert (1993) that the realizability assumption on $\mathcal{M}$ is essential: $\mathcal{E}(\mathcal{M})$ can be disconnected, even have isolated points in general.

The extension space conjecture is a special case of the *Generalized Baues conjecture* of Billera, Kapranov and Sturmfels (1994), via the Bohne-Dress Tiling Theorem 2.2.13. It also arises in connection with the space determined by the collection of all oriented matroids of rank r on $\{1, \ldots, n\}$ ordered by weak maps, see Section 7.7. This space, the *MacPhersonian* (Gel'fand and MacPherson 1992, Mnëv and Ziegler 1993, Babson 1992), is of great interest as the classifying space of vector bundles on *combinatorial differential manifolds*, as defined by MacPherson (1993). We refer to Sections A.2 and A.3 for further discussion.

7.3 Local perturbations and mutations

There are several interesting ways in which oriented matroids can be locally changed, or "perturbed", in order to get new oriented matroids. The first example of such a local transformation was given in Las Vergnas (1975a), where it was used to construct, for example, orientations of the non-Desargues and the non-Pappus matroids, and of the Vámos cube, see Example 7.4.1. These ideas were developed extensively by Edmonds and Fukuda (1982), Edmonds and Mandel (1982) and Fukuda and Tamura (1988a,b).

In this section, we will discuss local perturbations and some related operations in their simplest form. Our intuition will once more be guided by the Topological Representation Theorem, which suggests that we can deform a pseudosphere locally, "in the neighborhood of a single vertex", without affecting the rest of the pseudosphere arrangement.

The operations we discuss in this context are the *local perturbation* ("pushing a pseudosphere off a given vertex"), its *reverse perturbation* ("pushing a pseudosphere onto a nearby vertex"), and the *flipping* ("pushing a pseudosphere over a given vertex"), which can be obtained as a combination of first a reverse perturbation, then a local perturbation. A more general set-up, allowing deformation "along a flat", can be found in Fukuda and Tamura (1988b).

We then treat *mutations*, which correspond to the reversal of one base sign in a chirotope. In the uniform case, this is the same as a flipping. Mutations are interesting because they correspond to simplicial regions (topes) in the corresponding pseudosphere arrangement. This correspondence is also established here.

We start with a discussion of local perturbations. For this, consider a pseudosphere arrangement $\mathcal{A}$, in which the pseudosphere S_f goes through a vertex $\mathbf{w}$ of $\mathcal{A}_0 := \mathcal{A}\backslash\{S_f\}$. As in Figure 7.3.1, we want to "deform" S_f locally (in a small neighborhood of $\mathbf{w}$ that does not contain any other vertex of $\mathcal{A}_0$) so that it is "pushed off" $\mathbf{w}$.

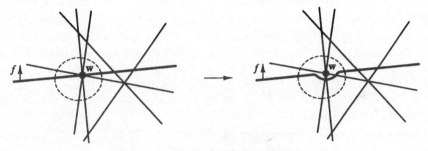

Figure 7.3.1: A local perturbation.

In the case of a pseudoline arrangement without parallel elements this is always possible, but in higher ranks there are restrictions. However, these are easily identified if we consider the localization (signature) σ_f that S_f induces on the vertices of $\mathcal{A}_0$. We are trying to change the sign $\sigma_f(\mathbf{w})$ from 0 to $+$ (and $\sigma_f(-\mathbf{w})$ from 0 to $-$). Now from Theorem 7.1.8 we get that the new σ_{LP} on the vertices of $\mathcal{A}_0$ has to have one of the three Types of Figure 7.1.6. Now consider any pseudoline which contains $\mathbf{w}$, and note that on it the signs of $\pm\mathbf{w}$ are changed from 0 to $\pm$. This yields a valid signature unless the pseudoline is of Type I and contains more than two pairs of opposite vertices $\pm\mathbf{w}$ and $\pm\mathbf{w}'$.

Translating this description into the language of cocircuit signatures, we have just obtained the following "local perturbation theorem".

7.3.1 Theorem (Edmonds and Fukuda 1982, Edmonds and Mandel 1982). *Let $\mathcal{M} = \mathcal{M}(E)$ be a simple oriented matroid of rank $r \geq 2$. If $f \in E$ is not a coloop, then $\mathcal{M}_0 := \mathcal{M}\backslash f$ has rank r, and $\mathcal{M}$ is a single element extension of $\mathcal{M}_0$ whose localization we denote by σ_f.*

Now let $W^0 \in \mathcal{C}^*(\mathcal{M}_0)$ be a cocircuit with $\sigma_f(W^0) = 0$, such that $W := (W^0, 0)$ is a cocircuit of $\mathcal{M}$. Then the following three conditions are equivalent:

(1) *The cocircuit signature $\sigma_{LP} : \mathcal{C}^*(\mathcal{M}) \longrightarrow \{+,-,0\}$ defined by*

$$\sigma_{LP}(Y^0) = \begin{cases} + & \text{if } Y^0 = W^0, \\ - & \text{if } Y^0 = -W^0, \\ \sigma_f(Y^0) & \text{otherwise} \end{cases}$$

is a localization.

(2)   If W' is another cocircuit, such that $W \circ W'$ is an edge in $\mathcal{M}_0$ (that is, $W' \in \mathcal{C}^*(\mathcal{M}_0)$ such that W and W' form a modular pair in $\underline{\mathcal{M}}$) and if $\underline{W} \cap \underline{W'} \neq \emptyset$, then $\sigma_f(W \circ W') \neq 0$.

(3)   If H, H', H'' form a modular triple of hyperplanes for the underlying matroid $M = \underline{\mathcal{M}}$ (so $r_M(H \cap H') = r - 2$) and if $f \in H \cap H'$, then $r_M((H \cap H') \backslash f) < r - 2$.

In the case that the conditions (1)-(3) hold, the oriented matroid

$$\mathcal{M}_{LP} := \mathcal{M} \cup p$$

defined by σ_{LP} is called a local perturbation of $\mathcal{M}$.

The local perturbation construction is very useful for creating new oriented matroids from old ones. An important feature of it is that the condition (3) only depends on the underlying matroid, which yields a matroid construction that preserves orientability. We formulate here only one special case which will be used in Section 7.4.

7.3.2 Corollary.   Let $M = M(E)$ be an orientable matroid of rank r, and let **B** be its family of bases.

Assume that M has a hyperplane (flat of rank $r - 1$) $H \subseteq E$ that is a circuit (thus $|H| = r$). Then $\mathbf{B} \cup \{H\}$ is again the family of bases of an orientable matroid.

Proof. Choose $f \in H$ arbitrarily, and then apply the local perturbation operation of Theorem 7.3.1 to f and one of the two opposite cocircuits corresponding to $E \backslash H$, to construct the new oriented matroid. Geometrically, this describes the special case where **w** is a "simple" vertex of $\mathcal{A}_0 = \mathcal{A} \backslash \{S_f\}$, determined by the $r - 1$ pseudospheres in $\{S_e : e \in H\} \backslash \{S_f\}$.   □

Another "obvious" corollary of the local perturbation theorem is that if S_f can be pushed off **w** to the one side, then it can also be pushed to the other one. This is also a consequence of the following more general lemma, which will be useful later.

7.3.3 Lemma.   Let $\sigma : \mathcal{C}^*(\mathcal{M}) \longrightarrow \{+, -, 0\}$ be a cocircuit signature, and choose $W \in \mathcal{C}^*(\mathcal{M})$.

For $\alpha = +, -, 0$, let σ^α be the cocircuit signature given by

$$\sigma^\alpha(Y^0) = \begin{cases} \alpha & \text{if } Y^0 = W^0, \\ -\alpha & \text{if } Y^0 = -W^0, \\ \sigma_f(Y^0) & \text{otherwise.} \end{cases}$$

If σ^α is a localization for two values of α, then it is also a localization for the third value.

Proof. Theorem 7.1.8 reduces this to rank 2, where it is easily checked in the three cases given by Figure 7.1.6.   □

What about the reverse operation of the local perturbation? It would correspond to "pushing a pseudosphere S_f onto a vertex **w** by a local deformation". For this, the vertex **w** has to be "close" to S_f – meaning "nothing separates S_f from **w**".

To make this precise, we can again rely on Theorem 7.1.8 to reduce to the case of rank 2, where the correct conditions are clear and the proofs are trivial.

7.3.4 Definition.

(1) Let $\mathcal{M}$ be an oriented matroid on E of rank r, and fix an element $f \in E$ (not a coloop) and a cocircuit $W \in \mathcal{C}^*(\mathcal{M})$ with $W_f \neq 0$. Partition E into $R := \{e \in E : W_e = 0\} \cup \{f\}$ and $S := E \backslash R$. Then $\mathcal{M}(R)$ is a restriction of $\mathcal{M}$ of rank r. Denote by $\mathcal{U} := \{U \in \mathcal{C}^*(\mathcal{M}) : U(R) \in \mathcal{C}^*(\mathcal{M}(R))\}$ the set of cocircuits of $\mathcal{M}$ that are extensions of cocircuits of $\mathcal{M}(R)$.

We will say that W is *near f* if for every $U \in \mathcal{U}$, one has either $U(S) = W(S)$ or $U(S) = -W(S)$.

(2) Let $\mathcal{A} = (S_e)_{e \in E}$ be a pseudosphere representation of $\mathcal{M}$, where **w** is the vertex corresponding to W, and S_f is a pseudosphere with $\mathbf{w} \notin S_f$. Putting again $R := \{e \in E : W_e = 0\} \cup \{f\}$ and $S := E \backslash R$, we get that $\mathcal{A}(R)$ is the rank r subarrangement of $\mathcal{A}$ that consists of S_f and all the pseudospheres through **w**. (The subarrangement $\mathcal{A}(S)$ can have smaller rank.)

Now **w** is *near S_f* if all the vertices of $\mathcal{A}(R)$ lie inside the two regions of $\mathcal{A}(S)$ that contain **w** and $-\mathbf{w}$. See Figure 7.3.2.

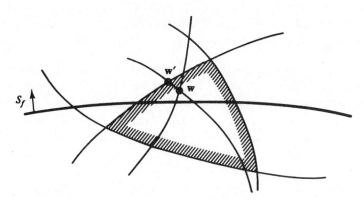

Figure 7.3.2: Here **w** is near S_f, but **w**' is not.

7.3.5 Proposition (Fukuda and Tamura 1988a). *Let $\mathcal{M}$ be an oriented matroid of rank r, let $f \in E$ be not a coloop, and let $W \in \mathcal{C}^*(\mathcal{M})$ be a cocircuit that is near f. Then*

$$(\mathcal{C}^*(\mathcal{M}) \backslash \mathcal{U}) \cup \{W^0, -W^0\}$$

(with the notation of Definition 7.3.4, where W^0 is the cocircuit of $\mathcal{M} \backslash f$ obtained by deleting f from W) is the collection of cocircuits of an oriented matroid $\mathcal{M}_{RP}$. This $\mathcal{M}_{RP}$ is called a reverse perturbation of $\mathcal{M}$.

A pseudosphere representation of $\mathcal{M}_{RP}$ is obtained by "pushing S_f onto $\mathbf{w}$" in an arrangement $\mathcal{A}$ that represents $\mathcal{M}$.

Proof. Let σ_f be the localization of f on $\mathcal{C}^*(\mathcal{M}\backslash f)$. Then

$$\sigma_{RP} : \mathcal{C}^*(\mathcal{M}\backslash f) \longrightarrow \{+,-,0\}$$

is defined by

$$\sigma_{RP}(Y^0) = \begin{cases} 0 & \text{if } Y^0 = \pm W^0, \\ \sigma_f(Y^0) & \text{otherwise.} \end{cases}$$

Now σ_{RP} is a localization: for this we again use Theorem 7.1.8, and observe that on every line through W, the localization has Type III, which is changed into Type II for σ_{RP}. $\qquad\qquad\qquad\qquad\qquad\qquad\qquad\qquad\qquad\qquad\qquad$ $\square$

Now we invoke Lemma 7.3.3, to see that whenever we can perform a reverse perturbation, we can follow it by a local perturbation. In other words, if by a reverse perturbation we can "push S_f onto $\mathbf{w}$", then we can also "push it across $\mathbf{w}$". The combined operation is called a *flipping* by Fukuda and Tamura (1988a), and we get the following "flipping theorem" with the same proof as for Proposition 7.3.5.

7.3.6 Corollary (Fukuda and Tamura 1988a). *In the situation of Proposition 7.3.5,*

$$(\mathcal{C}^*(\mathcal{M})\backslash\mathcal{U}) \uplus \{_{-R}U : U \in \mathcal{U}\}$$

is the set of cocircuits of a new oriented matroid $\mathcal{M}_F$. This oriented matroid $\mathcal{M}_F$ is called a flipping of $\mathcal{M}$. Its representation by a pseudosphere arrangement is obtained from a representation of $\mathcal{M}$ by "pushing S_f across $\mathbf{w}$".

A key observation is that the flipping operation preserves the underlying matroid. Thus in particular, if $\mathcal{M}$ is uniform, then so is $\mathcal{M}_F$. In the context of uniform oriented matroids, the flippings of $\mathcal{M}$ correspond to reversals of single basis sign ("mutations"), and to simplicial regions in the corresponding pseudosphere arrangement. Our goal for the rest of this section is to explain and prove these correspondences.

In the non-uniform case, we will see that simplicial regions do not correspond to flippings or mutations, but still a lot can be said. The aim of the following theorem is to provide the links and identifications that can be made in this context between different aspects of oriented matroid theory. Let us for this review, collect and introduce a few notations and concepts.

o An oriented matroid $\mathcal{M}$ is *acyclic* if it has no positive circuit. This means that $T^+ = (++\ldots+)$ is a tope, which is identified with the *positive region* $\bigcap_{e \in E} S_e^+$ of the pseudosphere arrangement $\mathcal{A}$ that represents $\mathcal{M}$.

o For an acyclic oriented matroid $\mathcal{M}$, we define the *convex closure*

$$\text{conv}_{\mathcal{M}}(A) := A \cup \{e \in E : X^+ = \{e\}, \ X^- \subseteq A \text{ for some } X \in \mathcal{C}(\mathcal{M})\}$$

(1)   The positive tope $T^+ = (++\ldots+) \in \mathcal{T}$ is a simplicial region in the arrangement $\mathcal{A}$ that is bounded by the pseudospheres $S_{b_1}, \ldots, S_{b_r}$. Equivalently, the intersection $S_{b_1}^+ \cap \ldots \cap S_{b_r}^+$ is a single region (namely, the positive one).

(2)   The elements $b_1, \ldots, b_r$ are the only extreme points of $\mathcal{M}$: $\mathrm{conv}_{\mathcal{M}}(B) = E$.

(3)   The basic cocircuits are all non-negative: $c^*(b_i, B) \geq 0$ for all $i = 1, \ldots, r$.

(4)   The tableau of B is non-negative: $T^*[B] \geq 0$.

Of course, a similar characterization can be given for any simplicial region of $\mathcal{A}$ – this is obtained from the above case after an acyclic reorientation.

Let us now consider the uniform case, where more can be said. Thus, from now on we consider a uniform oriented matroid $\mathcal{M}$ of rank $r \geq 2$. In this case, every simplicial region corresponds to a flipping, which pushes one of its boundary pseudospheres over the opposite vertex. (It is easily seen that all the possible flippings at a simplicial region produce the same result.)

In the chirotope picture, flippings correspond to the reversal of a single base sign. These reversals have become known as *mutations* (Roudneff and Sturmfels 1988, Bienia and da Silva 1990).

Of course, it is not always "legal" to reverse a base sign. The obstructions are in the chirotope picture imposed by the Grassmann-Plücker relations: these are quite restrictive. However, we may observe that if it is "legal" to set a certain base sign to zero, then this describes a reverse perturbation. Similarly, reversing the sign of one basis in a chirotope corresponds to a flipping if it produces an oriented matroid. The converse ("every flipping reverses a single base sign") is false in the non-uniform case – see the example of Figure 7.3.1.

7.3.8 Definition.   Let $\mathcal{M}$ be a uniform oriented matroid of rank r, given by its chirotope $\chi : \Lambda(n, r) \longrightarrow \{+, -\}$, where $\Lambda(n, r)$ denotes the set of r-subset of E, written as increasing r-tuples.

The r-tuple $\lambda \in \Lambda(n, r)$ is a *mutation* of $\mathcal{M}$ if the mapping $_{\overline{\lambda}}\chi$ given by

$$_{\overline{\lambda}}\chi(\mu) := \begin{cases} -\chi(\mu) & \text{if } \mu = \lambda \\ \ \ \chi(\mu) & \text{otherwise} \end{cases}$$

also defines an oriented matroid – that is, if it also satisfies (B2″).

The set of mutations of χ is denoted by $\mathrm{Mut}(\chi) \subseteq \Lambda(n, r)$.

In other words, the mutations are those bases whose signs are not determined by the orientations of the other bases. A computational geometry approach to oriented matroids leads one to consider *reduced systems* of χ, that is, minimal subsets $\mathcal{R} \subseteq \Lambda(n, r)$ that completely determine χ, such that $\chi'|_{\mathcal{R}} = \chi|_{\mathcal{R}}$ together with (B2″) imply $\chi' = \chi$. A mutation is a basis that is contained in every reduced system, which implies $\mathrm{Mut}(\chi) \subseteq \mathcal{R}$ for every reduced system.

A naive approach to test whether a given λ is a mutation would be to test whether λ is determined by any 3-term Grassmann-Plücker condition. A more efficient method is provided by the following theorem, which identifies mutations and simplicial regions; see Roudneff and Sturmfels (1988).

for every subset $A \subseteq E$. An element $e \in E$ is an *extreme poin* $\text{conv}_{\mathcal{M}}(E \backslash e)$, that is, if there is no $X \in \mathcal{C}$ with $X^+ = \{e\}$. (Sec discusses convexity in oriented matroids.) Equivalently, e is an extrem if S_e defines a facet of the positive region.

- If $B = \{b_1, \ldots, b_r\}$ is a basis, then the corresponding pseudosphere. $S_{b_2}, \ldots, S_{b_r}$ subdivide the sphere S^{r-1} into 2^r simplicial cells that are divided by the pseudospheres S_e ($e \in E \backslash B$) of $\mathcal{A}$.

- In particular, the intersection of hemispheres $S_{b_1}^+ \cap S_{b_2}^+ \cap \ldots \cap S_{b_r}^+$ is su subdivided simplicial cell T^B whose vertices are given by the basic cocirc $v_1 := c^*(b_1, B), \ldots, v_r := c^*(b_r, B)$.

Note that for $e \in N := E \backslash B$, the pseudosphere S_e dissects the cell T^B if a only if two of these vertices lie on different sides of S_e. Assuming that $\mathcal{M}$ acyclic, we know $T^B \cap S_e^+ \neq \emptyset$, so this is equivalent to the condition that son v_i lies on the negative side of S_e, which means that $c^*(b_i, B)_e = -$ for some

- In linear programming language (see Section 10.2) the $(r \times (n-r))$-matrix

$$T^*[B] := (c^*(b, B)_e)_{b \in B, e \in N} =: (\tau_{be})_{b,e}$$

is called the *(reduced dual) tableau* or the *standard representative matrix* of $\mathcal{M}$ with respect to B. (We will in this Section just talk about *the tableau* $T^*[B]$ of B.) Its rows correspond to the *basic elements* $b_i \in B$, the columns are indexed by the *non-basic elements* $n_i \in N := E \backslash B$. The rows of $T^*[B]$ are the "interesting" part of the basic cocircuits of B.

Note that $c^*(b_1, B) \circ \ldots \circ c^*(b_r, B)$ is a tope associated with the basis B, as in Exercise 3.24. Also, the basic circuits $c(e, B)$ ($e \in N$) can be read off from the columns of the tableau as

$$c(e, B)_f = \begin{cases} \delta_{fe} & \text{for } f \in N, \\ -\tau_{fe} & \text{for } f \in B. \end{cases}$$

So if $T^*[B]$ is non-negative, then $\mathcal{M}$ is acyclic and all points $e \in E$ are in the convex hull of B.

- The tableau $T^*[B]$ can be read off from the chirotope as follows. Let the basis B be given by an ordered r-tuple $\lambda = (\lambda_1, \ldots, \lambda_r) \in E^r$, and the non-basic N correspond to the $(n-r)$-tuple $\mu = (\mu_1, \ldots, \mu_{n-r}) \in E^{n-r}$ with $|E| = n$. For $i = 1, \ldots, r$ and $j = 1, \ldots, n-r$ define

$$\lambda[i \to j] := (\lambda_1, \ldots, \lambda_{i-1}, \mu_j, \lambda_{i+1}, \ldots, \lambda_r) \in E^r.$$

Then

$$T^*[B] = (\chi(\lambda[i \to j]))_{i=1,\ldots,r; \ j=1,\ldots,n-r}$$

expresses the tableau in terms of the chirotope.

With these explanations, we get the following equivalent formulations of "simplicial topes", apparently first observed in Edmonds and Mandel (1982).

7.3.7 Proposition. *Let $\mathcal{M}$ be a simple acyclic oriented matroid of rank r represented by the pseudoarrangement $\mathcal{A}$, and let $B = \{b_1, \ldots, b_r\}$ be a basis of $\mathcal{M}$. Then the following are equivalent.*

7.3.9 Theorem. *Let $\mathcal{M}$ be a uniform oriented matroid of rank r, represented by the arrangement $\mathcal{A}$ of pseudospheres. Let $\lambda = (\lambda_1, \ldots, \lambda_r) \in \Lambda(n, r)$ be the ordered r-tuple corresponding to the basis $B = \{b_1, \ldots, b_r\}$. We assume that $n > r$ ($\mathcal{M}$ is not the free matroid). Then the following conditions are equivalent.*

(1) *The pseudospheres $S_{b_1}, S_{b_2}, \ldots, S_{b_r}$ bound a simplicial region of $\mathcal{A}$.*

(2) *$\mathcal{M}$ has an acyclic reorientation in which B is the set of extreme points.*

(3) *There is a flipping of S_{b_1} over the two vertices $\pm\mathbf{w}$ in $S_{b_2} \cap \ldots \cap S_{b_r}$.*

(4) *The tableau $T^*[B]$ has rank 1 as a real (± 1)-matrix.*

(5) *λ is a mutation of $\mathcal{M}$.*

Proof.

(a) The equivalence of (1), (2) and (4) follows as a special case from Proposition 7.3.7. To see this, note that the tableau of (4) is a $\{+1, -1\}$-matrix in the uniform case; its rank is 1 if and only if we can obtain the matrix with all entries $+1$ by multiplying some of its rows and columns by -1. This operation is equivalent to forming an acyclic reorientation of $\mathcal{M}$ which satisfies the equivalent conditions of Proposition 7.3.7 for B.

(b) For (1) $\Longrightarrow$ (3) reorient so that the simplicial region is positive. Then observe that the vertices of the subarrangement $\mathcal{A}(B)$ are exactly the vertices of the positive region and its negatives, so by Corollary 7.3.6 (with $f = b_1$ and $R = B$) we can flip. For the reverse implication (3) $\Longrightarrow$ (1), let us suppose that we can flip S_{b_1} over $c^*(b_1, B)$. The nearness condition of Definition 7.3.4 implies that the cocircuits $c^*(b_i, B)$ are all conformal. Hence they determine a simplicial region.

(c) To see (4) $\Longleftrightarrow$ (5), observe that λ is a mutation if and only if $\chi(\lambda)$ is not determined by a 3-term Grassmann-Plücker condition of the form

$$\{\lambda - \{\lambda_{i_1}, \lambda_{i_2}\} | \{\lambda_{i_1}, \lambda_{i_2}, \mu_{j_1}, \mu_{j_2}\}\},$$

which states that the three products

$$\chi(\lambda[i_1 \to j_1]) \cdot \chi(\lambda[i_2 \to j_2])$$
$$-\chi(\lambda[i_1 \to j_2]) \cdot \chi(\lambda[i_2 \to j_1])$$
$$\chi(\lambda) \cdot \chi(\lambda[i_1 \to j_1][i_2 \to j_2])$$

are not all equal. Thus $\chi(\lambda)$ is *not* determined by this Grassmann-Plücker relation if and only if the first two of these three products are not equal. This is the case if and only if the $\mathbb{R}$-determinant

$$\det \begin{pmatrix} \chi(\lambda[i_1 \to j_1]) & \chi(\lambda[i_1 \to j_2]) \\ \chi(\lambda[i_2 \to j_1]) & \chi(\lambda[i_2 \to j_2]) \end{pmatrix}$$

is zero. But this is a 2×2-minor of the tableau – given by the row indices $\lambda_{i_1}, \lambda_{i_2}$ and the column indices μ_{j_1}, μ_{j_2}. In this way the 2×2-minors of $T^*[B]$ are identified with the 3-term Grassmann-Plücker relations that involve $\chi(\lambda)$.

We conclude that λ is a mutation (that is, not determined by any 3-term Grassmann-Plücker relation) if and only if all 2×2-minors of $T^*[B]$ vanish. This in turn is the case if and only if $T^*[B]$ has rank 1. $\qquad\square$

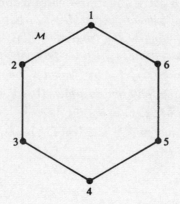

Figure 7.3.3: The mutations of the hexagon:
$$\text{Mut}(\mathcal{M}_6) = \{123, 234, 345, 456, 156, 126\}$$

In Figure 7.3.3 we sketch a rank 3 example to illustrate our analysis of mutations, flippings and simplicial regions. For this hexagon, we compute

$$T^*[156] = \begin{pmatrix} \chi(256) & \chi(356) & \chi(456) \\ \chi(125) & \chi(136) & \chi(146) \\ \chi(152) & \chi(153) & \chi(154) \end{pmatrix} = \begin{pmatrix} + & + & + \\ + & + & + \\ - & - & - \end{pmatrix},$$

thus 156 is a mutation of $\mathcal{M}_6$. Similarly,

$$T^*[146] = \begin{pmatrix} \chi(246) & \chi(346) & \chi(546) \\ \chi(124) & \chi(136) & \chi(156) \\ \chi(142) & \chi(143) & \chi(145) \end{pmatrix} = \begin{pmatrix} + & + & - \\ + & + & + \\ - & - & + \end{pmatrix}$$

shows that $\chi(356)$ is determined by the Grassmann-Plücker relation $\{6|1435\}$. In fact,

$$\chi(146) \cdot \chi(356) = +\chi(146) = +$$
$$-\chi(136) \cdot \chi(456) = -(+) \cdot (+) = -$$
$$\chi(156) \cdot \chi(436) = (+) \cdot (-) = -,$$

thus 146 is not a mutation of $\mathcal{M}_6$.

One question we have not discussed yet is the *existence* of mutations. For this recall Shannon's Theorem 2.1.5 which states that every *realizable* uniform oriented matroid on n points has at least n mutations. The same conclusion holds for rank 3 oriented matroids also if they are not realizable – see Theorem 6.5.2(i). However, the example RS(8) constructed in Section 1.5 proves that this statement cannot be extended: this uniform oriented matroid on 8 points (rank 4) has only 7 mutations. Starting with the Roudneff-Sturmfels example, and using a new *connected sum* operation, Richter-Gebert (1992, 1993b) recently constructed, for every $n \geq 2$, a uniform oriented matroid of rank 4 on $4n$ elements with only $3n+1$ mutations. Furthermore, he obtained an oriented matroid of rank 4 on 20 elements such that one element is not contained in any mutation, i.e., one pseudo-hyperplane is not adjacent to any simplicial region. Thus Shannon's theorem fails badly in the non-realizable case. We are left with the following conjecture, which was motivated by the results of Camion (1968).

7.3.10 Conjecture (Las Vergnas 1980a). *Every simple oriented matroid of rank r has a reorientation with exactly r extreme points.*

In view of Theorem 7.3.9, there are several equivalent ways to formulate this conjecture. For instance, it asks whether every arrangement of pseudospheres has a simplicial region (Problem 4.5). In the uniform case, it is equivalent to the existence of mutations.

The correctness of Conjecture 7.3.10 is known for the realizable case, for rank $r \leq 3$, and for $n - r \leq 3$ by duality, using Exercise 7.9. It has also been shown by Edmonds and Mandel (1982) under the condition that $\mathcal{M}$ has an extension by an element g in general position so that $(\mathcal{M}, g)$ is Euclidean (cf. Section 10.5).

7.4 Many oriented matroids

Here we will present two applications of the local perturbation operation introduced in Section 7.3. Mainly, we will show that the local perturbation operation is so general that it can be used to give good estimates on the number of oriented matroids of rank r on a ground set of n elements. Before that, we show that the Vámos matroid (see Ingleton 1971, Welsh 1976, or White 1987) and some variants, which we call V'-matroids, are orientable. This yields an explicit class of oriented matroids that is easily (see Section 7.5) shown to be non-realizable.

7.4.1 Example (The V'-matroids). Consider any configuration of eight points in $\mathbb{R}^3$ of the form

$$\begin{array}{llll}
\mathbf{a} = (0,0,\alpha) & \mathbf{b} = (1,0,\beta) & \mathbf{c} = (1,1,\gamma) & \mathbf{d} = (0,1,\delta) \\
\mathbf{a}' = (0,0,\alpha') & \mathbf{b}' = (1,0,\beta') & \mathbf{c}' = (1,1,\gamma') & \mathbf{d}' = (0,1,\delta')
\end{array}$$

with $\alpha \neq \alpha'$, $\beta \neq \beta'$, $\gamma \neq \gamma'$ and $\delta \neq \delta'$. It determines a matroid V of rank 4 on eight points, which of course depends on the specific choice of $\alpha, \alpha', \beta, \ldots$ A V-*matroid* is a matroid of this type, of rank 4, on 8 points. V-matroids are by definition realizable and orientable. A V'-*matroid* is a matroid that is obtained from a V-matroid by making $\{\mathbf{a}, \mathbf{a}', \mathbf{c}, \mathbf{c}'\}$ into a basis (see Figure 7.4.1). This includes the *Vámos cube* $\mathsf{V}'(8)$, which is the V'-matroid for generic choices of $\alpha, \alpha', \beta, \ldots$, as well as several weak map images of it. A specific orientation of $\mathsf{V}'(8)$ is given in Bland and Las Vergnas (1978). ·

From Corollary 7.3.2 of the local perturbation theorem we get that *every V'-matroid is orientable*. In fact, a V'-matroid arises from the corresponding V-matroid "by breaking the plane $\mathbf{aa'cc'}$ to become a basis". So, if we represent the V-matroid by a (linear) arrangement of pseudospheres, then a representation of the V'-matroid is given, for example, by deforming $S_\mathbf{a}$ off the two opposite vertices given by $S_{\mathbf{a}'} \cap S_\mathbf{c} \cap S_{\mathbf{c}'}$.

We will now construct a very large class of oriented matroids by local perturbations, large enough to get good asymptotic estimates of the number of all oriented matroids for given n and r.

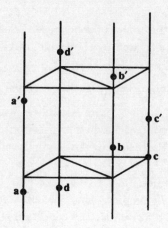

Figure 7.4.1: A V'-matroid.

Theorem 7.4.2. (Ziegler) *For $n \geq r \geq 3$ let $f(n,r)$ denote the number of uniform oriented matroids of rank r on the set $E = \{1, 2, \ldots, n\}$. Then*

$$f(n,r) \geq 2^{\left(\frac{n-r}{2(r-1)}\right)^{r-1}} = 2^{c_r(n-r)^{r-1}}$$

for $c_r = \left(\frac{1}{2(r-1)}\right)^{r-1} > 0$.

Proof. Let $n' \leq n$ be the largest integer that can be written as

$$n' = (2m+1)(r-1) + 1$$

with $m \in \mathbb{N}_0$. Thus we have

$$(2m+1)(r-1) + 1 = n' \leq n < n' + 2(r-1).$$

We will now give an explicit description of a class of "many" oriented matroids that are derived from certain affine hyperplane arrangements of size n' in $\mathbb{R}^{r-1}$.

Consider the affine arrangement given by the $(m+1)(r-1)$ hyperplanes

$$H_i^j := \{\mathbf{x} \in \mathbb{R}^{r-1} : x_i = j\} \qquad i = 1, 2, \ldots, r-1; \ j = 0, 1, \ldots, m.$$

This defines a "grid" to which we add the $(r-1)m+1$ "diagonal" hyperplanes

$$H_k := \{\mathbf{x} \in \mathbb{R}^{r-1} : \sum_{i=1}^{r-1} x_i = k\} \qquad k = 0, 1, \ldots, (r-1)m.$$

Through each of the $(m+1)^{r-1}$ vertices of this "grid" goes exactly one of the "diagonal" hyperplanes H_k, which yields two choices for a local perturbation at this vertex, by the local perturbation theorem.

This yields $2^{(m+1)^{r-1}}$ different oriented matroids, which can be perturbed "at infinity" to get the same number of different uniform oriented matroids of rank r on n points. (This perturbation at infinity can be done with the global perturbations of Section 7.2, see also Corollary 7.7.9.)

Now we estimate this number with $2m+1 = \frac{n'-1}{r-1}$ and $n' > n - 2(r-1)$ as

$$m+1 = \frac{n'+r-2}{2(r-1)} > \frac{n-r}{2(r-1)},$$

which proves the theorem.   □

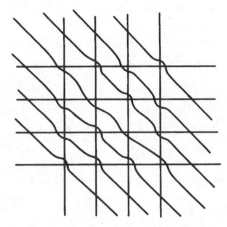

Figure 7.4.2: A typical perturbed arrangement as in
the proof of Theorem 7.4.2, for $r = 3$, $m = 3$, $n' = 15$.

Remarks.

(1) For rank $r = 2$, there are $2^{n-1}(n-1)! = 2^{O(n \log n)}$ different uniform oriented matroids. The two factors come from reorientation and permutation of elements: thus there is exactly one *reorientation class* of *unlabeled* oriented matroids of rank 2, see Remark 7.1.6.

However, for $r \geq 3$, as assumed in Theorem 7.4.2, factors like $n!$ for permutation of elements and 2^n from reorientations can be neglected for the asymptotic study of $f(n, r)$, because they are dominated by the rapid growth of this function.

(2) The constant c_r of our construction can be improved (increased) by simple modifications of our argument. Also, it increases by a factor of $\frac{\log 3}{\log 2}$ if we count non-uniform oriented matroids as well.

(3) For the case $r = 3$, the above result with a better constant was achieved by Goodman and Pollack (1983), who showed that there are at least $2^{n^2/8}$ different oriented matroids of rank 3 on n points, see Proposition 6.5.1.

(4) For fixed $r \geq 3$, the bound of Theorem 7.4.2 is *best possible* except for the constant. In fact, by Edelsbrunner, O'Rourke and Seidel (1986), an arrangement of pseudolines can be constructed in $O(n^2)$ time, which implies that there is a constant $d_3 > 0$ such that there are less than $2^{d_3 n^2}$ oriented matroids of rank 3. For a given oriented matroid $\mathcal{M}$ of higher rank one can compute each of the $\binom{n}{r-3} = O(n^{r-3})$ contractions of rank 3 in $O(n^2)$ time each, and those determine $\mathcal{M}$. So there are constants $d_r > 0$ such that

$$f(n, r) \leq 2^{d_r n^{r-1}} = 2^{O(n^{r-1})}$$

for $r \geq 3$. (This argument is due to H. Edelsbrunner and a similar one to R. Seidel, personal communications, 1990.)

An alternative method for proving such an upper bound is to use the algorithm for constructing arrangements by Edelsbrunner, O'Rourke and Seidel (1986). For this we need the corrected version of the "zone theorem" in Edelsbrunner,

Seidel and Sharir (1991), which also works for pseudoarrangements and produces explicit constants d_r.

(5) "Most" of the oriented matroids of Theorem 7.4.2 are *not realizable*: there are too many of them, see Theorem 8.7.5.

7.4.3 Corollary. *There are constants $d'_r > c'_r > 0$ such that for $n \geq r \geq 3$*

(i)   *the number $f(n,r)$ of uniform oriented matroids of rank r on n points and*

(ii)   *the number $g(n,r)$ of all oriented matroids of rank r on n points satisfy*

$$2^{c'_r n^{r-1}} < f(n,r) \leq g(n,r) < 2^{d'_r n^{r-1}}.$$

In both the uniform and the general case, most of the oriented matroids are not realizable.

7.5 Intersection properties and adjoints

In this section we will study the hierarchy of intersection properties of matroids and oriented matroids, of which the existence of an adjoint (see Section 5.3) is the strongest.

As a motivation for this, we will in Proposition 7.5.1 review the usual argument for showing that the Vámos matroid and the V'-matroids of Section 7.4 are not realizable: a certain line cannot be intersected with a certain (hyper)plane of the matroid. This argument suggests generalizations to flats of various dimensions, and also to oriented matroids. We will find that the failure of a certain intersection property on an oriented matroid can sometimes, but not always, be traced to the underlying matroid.

It will be shown that non-realizability is linked to the existence of certain single element extensions. This suggests linking the structure of an (oriented) matroid with that of the extension lattice, which we will achieve in this section.

7.5.1 Proposition. *The V'-matroids of Example 7.4.1 are not realizable.*

Proof. Assume that a V'-matroid is represented as a configuration of 8 points $\mathbf{a}, \mathbf{a}', \mathbf{b}, \mathbf{b}', \mathbf{c}, \mathbf{c}', \mathbf{d}, \mathbf{d}'$ in projective 3-space. Then we can extend it by the point $\mathbf{p}$ in which the line $\mathbf{aa}'$ intersects the plane $\mathbf{bb'cc}'$. This $\mathbf{p}$ lies on the planes $\mathbf{aa'bb}'$ and $\mathbf{aa'dd}'$, and therefore also on the lines $\mathbf{bb}' = \mathbf{aa'bb}' \cap \mathbf{bb'dd}'$ and $\mathbf{dd}' = \mathbf{aa'dd}' \cap \mathbf{bb'dd}'$. Thus $\mathbf{p}$ lies on the planes $\mathbf{bb'cc}'$ and $\mathbf{dd'cc}'$, and thus also on the intersection line $\mathbf{cc}' = \mathbf{bb'cc}' \cap \mathbf{dd'cc}'$ of these two planes. But this means that the lines $\mathbf{aa}'$ and $\mathbf{cc}'$ intersect in $\mathbf{p}$, and thus $\mathbf{aa'cc}'$ is a plane, contradicting the definition of the V'-matroid.   □

This argument in fact shows that no V'-matroid can be realized over any field. The important point for us is that non-realizability over $\mathbb{R}$ implies that no orientation of such a matroid can be realizable.

The V'-matroids, and other examples, motivate the following definitions.

7.5.2 Definition (Bachem and Wanka 1985, 1989a). Let $\mathcal{M}$ be an oriented matroid of rank r.

(0) $\mathcal{M}$ satisfies the *intersection property* IP_0 if it has an adjoint.

(1) $\mathcal{M}$ has *Levi's intersection property* IP_1 if for every collection of $r-1$ hyperplanes $\{H_1,\ldots,H_{r-1}\}$ there is a single element extension $\widetilde{\mathcal{M}} = \mathcal{M} \cup p$ such that p is not a loop and $H_1 \cup p,\ldots,H_{r-1} \cup p$ are hyperplanes of $\widetilde{\mathcal{M}}$ (that is, "$p \in \overline{H}_i$").

(2) $\mathcal{M}$ has the *Generalized Euclidean intersection property* IP_2 if for every pair of flats F and G with $r(F \wedge G) < r(F) + r(G) - r(F \vee G)$ there is a nontrivial single element extension $\widetilde{\mathcal{M}} = \mathcal{M} \cup p$ such that p is not a loop and $F \cup p$ and $G \cup p$ are flats of $\widetilde{\mathcal{M}}$ (that is, "$p \in \overline{F} \cap \overline{G}$").

(3) $\mathcal{M}$ has the *Euclidean intersection property* IP_3 if for every hyperplane H and line ℓ there is a single element extension $\widetilde{\mathcal{M}} = \mathcal{M} \cup p$ such that p is not a loop, $\ell \cup p$ is a line of $\widetilde{\mathcal{M}}$, and $H \cup p$ is a hyperplane of $\widetilde{\mathcal{M}}$ (that is, "$p \in \overline{\ell} \cap \overline{H}$").

As always one can switch here from hyperplanes and flats to cocircuits and covectors by taking complements in E. Thus an oriented matroid of rank r has the property IP_1 if and only if for every set $Y^1,\ldots,Y^{r-1}$ of cocircuits, there is a non-zero localization σ with

$$\sigma(Y^1) = \ldots = \sigma(Y^{r-1}) = 0.$$

Similar reformulations exist for IP_2 and IP_3.

As a rule, it is in practise hard to verify and deal with the various intersection properties. Part of the problem lies in the fact that their definition refers to single element extensions, which may be considered as structure that lies outside the oriented matroid. Also, in contrast to IP_3 (see Corollary 10.5.9), not all intersection properties are preserved under duality. For instance, Alfter, Kern and Wanka (1990) have shown that the (orientable, rank 3, 10-point) non-Desargues matroid does have an adjoint (IP_0), but its dual does not.

The following implications relate the various intersection properties.

7.5.3 Proposition (Bachem and Wanka 1985, 1989a, Bachem and Kern 1986a). *Between the intersection properties of Definition 7.5.2, one has the implications*

$$\text{realizable} \implies (\mathrm{IP}_0) \implies (\mathrm{IP}_1) \implies (\mathrm{IP}_2) \implies (\mathrm{IP}_3).$$

Proof. The implications $(\mathrm{IP}_1) \implies (\mathrm{IP}_2) \implies (\mathrm{IP}_3)$ are quite straightforward. For $(\mathrm{IP}_0) \implies (\mathrm{IP}_1)$, let $\mathcal{M}$ have an adjoint, and let $\mathcal{A}$ be a Type II representation of $\mathcal{M}$ (see Section 5.3). Thus the hyperplanes $H_1,\ldots,H_{r-1}$ of $\mathcal{M}$ correspond to pseudospheres $S_1,\ldots,S_{r-1}$ in $\mathcal{A}$.

Now the intersection $S_1 \cap \ldots \cap S_{r-1}$ is a subsphere of codimension at most $r-1$, and therefore non-empty. Thus we can choose a vertex p with $p \in S_i$ for all i. By the criterion (2) of Theorem 7.1.8, this vertex specifies a single element extension of $\mathcal{M}$ with the required properties: we only have to show that p consistently describes the location of a new pseudosphere in every subarrangement of rank 2, which is obvious. □

For example, the V'-matroids (Example 7.4.1) do not have the Euclidean intersection property (IP$_3$), by the proof of Proposition 7.5.1.

7.5.4 Corollary. *The V'-matroids do not have adjoints.*

The intersection properties for oriented matroids of Definition 7.5.2 are formulated solely in terms of flats and single element extensions. Thus they apply to ordinary matroids with the analogous definitions. However, stating results and proofs concerning intersection properties, one has to carefully distinguish between the oriented and the unoriented case. In fact, as Bachem and Wanka (1989a) note, there are

– matroids that violate some of the intersection properties (so that their orientations violate them, too): for example, the Vámos matroid and its orientations violate IP$_3$,

– matroids that satisfy an intersection property, with orientations that fail it: for example, the uniform matroid $U_{4,8}$ satisfies IP$_3$, but its orientation EFM(8) does not (see Section 10.4),

– orientable matroids with intersection properties that are not satisfied by any orientation. Furthermore, Remark 7.5.7 will show an example of a matroid result whose oriented matroid counterpart is not proved (although probably true).

Note, however, that Definition 7.5.2 and Proposition 7.5.3 generalize to the unoriented case.

7.5.5 Proposition (Cordovil 1987). *The intersection properties IP$_i$ are minor closed, that is, if an oriented matroid $\mathcal{M}$ has property IP$_i$ for some $i \in \{0, 1, 2, 3\}$, then so does every minor of $\mathcal{M}$.*

This easy proposition suggests that it might be interesting to seek an excluded minor characterizations of the intersection properties. The following proposition and remark represent what is known. Note that all oriented matroids of rank at most 3 satisfy property (IP$_0$), and thus all intersection properties, by Proposition 6.3.6.

7.5.6 Proposition (Bachem and Wanka 1989a). *The following equivalences are valid for all oriented matroids of rank 4:*

$$(\text{IP}_1) \Longleftrightarrow (\text{IP}_2) \Longleftrightarrow (\text{IP}_3).$$

7.5.7 Remark. Bachem and Wanka (1989a) have shown that none of the implications between the *matroid* properties (IP$_i$) of Proposition 7.5.3 can be characterized by a finite list of excluded minors. However, it is open whether the matroids constructed by Bachem and Wanka are orientable. Thus it is not clear whether the implications of 7.5.3 for *oriented matroids* have characterizations by a finite list of excluded minors.

For some oriented matroid analogues of classical convexity theorems (Section 9.2), we will have to assume sufficiently strong intersection properties. For example, Edmonds and Mandel (1982) observed that the "Hahn-Banach Theorem" is not in general true for oriented matroids. However, (IP_2) implies the Hahn-Banach property (Theorem 9.2.1(6)), while for example the Vámos matroid shows that the converse fails. It is not clear whether (IP_3) is a necessary or sufficient condition for the Hahn-Banach property to hold.

The proof of Proposition 7.5.3 allows us to explain the connection between the existence of an adjoint and the existence of single element extensions more precisely. For this, let $\mathcal{M}$ be represented by a pseudoconfiguration of points $(\mathcal{A}, P)$. Then

○ the k-flats of $\mathcal{M}$ are represented as spheres of codimension k in the arrangement $\mathcal{A}$,

○ these spheres intersect according to their dimensions, as required by the sphere axioms for arrangements, and

○ every cell of $\mathcal{A}$ (every covector of $\mathcal{M}^{ad}$) specifies a *consistent* choice of a single element extension of $\mathcal{M}$, by "placing a new point of $\mathcal{M}$ on this cell".

So, every covector of $\mathcal{M}^{ad}$ corresponds to a single element extension of $\mathcal{M}$. This yields an embedding

$$\mathcal{L}^{ad} \hookrightarrow \mathcal{E}(\mathcal{M}),$$

for which the following theorem also provides the unoriented analogue.

At this point note that there is *no* direct oriented matroid analogue of the embedding $L^{op} \hookrightarrow E(M)$: there is no natural embedding $\hat{\mathcal{L}}^{op} \hookrightarrow \mathcal{E}(\mathcal{M})$ of the inverted big face lattice of an oriented matroid into its extension lattice. However, the embedding of the cocircuit span of an adjoint into the extension lattice is very natural. For this we can, in fact, interpret both objects as subsets of $\{+, -, 0\}^{\mathcal{C}^*}$ (respectively, $\{0, 1\}^{\mathcal{C}^*}$ in the unoriented case).

7.5.8 Theorem (Cheung 1974, Bachem and Kern 1986a).

(a) *Let M be a matroid of rank r, and $L^{ad} \subseteq \{0, 1\}^{C^*}$ the lattice of flats of an adjoint of M, given by the collection of covectors (characteristic vectors of flat complements).*
Let $E(L) \subseteq \{0, 1\}^{C^*}$ be the extension lattice, given by the characteristic vectors of complements of linear subclasses of cocircuits. Then

$$L^{ad} \subseteq E(L).$$

(b) *Let $\mathcal{M}$ be an oriented matroid of rank r, and $\mathcal{L}^{ad} \subseteq \{+, -, 0\}^{E^{ad}}$ the lattice of flats of an adjoint of $\mathcal{M}$, given by the collection of covectors (cocircuit span).*
Let $\mathcal{E}(\mathcal{M}) \subseteq \{+, -, 0\}^{\mathcal{C}^*}$ be the extension lattice, given by the cocircuit signatures. Then

$$\mathcal{L}^{ad} \subseteq \mathcal{E}(\mathcal{M}).$$

Proof. In both cases (a) and (b) the vector families considered are closed under composition and endowed with the obvious partial orders. So the only part we have to show is that the cocircuits of an adjoint lie in the extension lattice.

(a) For ordinary matroids, this follows from Crapo's characterization 7.1.2: every single cocircuit of M forms a linear subclass.

(b) For oriented matroids, let the pseudosphere arrangement $\mathcal{A}$ be a Type II representation of M and thus a Type I representation of M^{ad}. Let W be a cocircuit of M^{ad}, given by a vertex $\mathbf{w}$ of $\mathcal{A}$. It corresponds to the sign $\sigma_w(f) = W_f = +, -$ or 0 for every sphere S_f ($f \in E^{ad}$) in $\mathcal{A}$, depending on whether $\mathbf{w} \in S_f^+$, $\mathbf{w} \in S_f^-$ or $\mathbf{w} \in S_f$.

To see that $\sigma_w : E^{ad} \longrightarrow \{+, -, 0\}$ is a localization, we use the criterion of Theorem 7.1.8(2). It requires that σ_w describes the consistent location of a new pseudosphere with respect to a subarrangement of pseudospheres in $\mathcal{A}$ of rank 2. One easily checks that it does. $\square$

7.6 Direct sum and union

We now develop several techniques for constructing a new, more complicated oriented matroid from two smaller ones. This starts with the (quite trivial) direct sum construction, and leads to the powerful oriented matroid union, which produces large classes of oriented matroids with interesting properties.

Let $\mathcal{M}_1$ and $\mathcal{M}_2$ be oriented matroids on sets E_1 and E_2 which are assumed to be disjoint. For $i = 1, 2$, let $\mathcal{C}_i$ be the set of circuits, $\mathcal{C}_i^*$ the set of cocircuits, χ_i the chirotope, $\mathcal{V}_i$ the set of vectors, $\mathcal{L}_i$ the set of covectors, and r_i the rank of $\mathcal{M}_i$.

7.6.1 Proposition. *There exists an oriented matroid $\mathcal{M}$ of rank $r_1 + r_2$ on the disjoint union $E_1 \uplus E_2$ such that*

(a)   *the set of circuits of $\mathcal{M}$ equals $\mathcal{C} = \mathcal{C}_1 \cup \mathcal{C}_2$;*

(b)   *the set of cocircuits of $\mathcal{M}$ equals $\mathcal{C}^* = \mathcal{C}_1^* \cup \mathcal{C}_2^*$;*

(c)   *the chirotope χ of $\mathcal{M}$ satisfies*

$$\chi(e_1, \ldots, e_{r_1}, f_1, \ldots, f_{r_2}) = \chi_1(e_1, \ldots, e_{r_1}) \cdot \chi_2(f_1, \ldots, f_{r_2})$$

   for all

$$e_1, \ldots, e_{r_1} \in E_1, \quad f_1, \ldots, f_{r_2} \in E_2;$$

(d)   *the set $\mathcal{V}$ of $\mathcal{M}$ equals $\mathcal{V} = \{ X_1 \cup X_2 \mid X_1 \in \mathcal{V}_1 \ and \ X_2 \in \mathcal{V}_2 \};$*

(e)   *the set $\mathcal{L}$ of $\mathcal{M}$ equals $\mathcal{L} = \{ Y_1 \cup Y_2 \mid Y_1 \in \mathcal{L}_1 \ and \ Y_2 \in \mathcal{L}_2 \}.$*

The oriented matroid $\mathcal{M}$ described in Proposition 7.6.1 is called the *direct sum* of $\mathcal{M}_1$ and $\mathcal{M}_2$, and it is denoted by $\mathcal{M}_1 \oplus \mathcal{M}_2$. We can see from part (c) that the bases of the direct sum $\mathcal{M}_1 \oplus \mathcal{M}_2$ are the unions of bases of $\mathcal{M}_1$ and bases of $\mathcal{M}_2$. Consequently the underlying matroid $\underline{\mathcal{M}_1 \oplus \mathcal{M}_2}$ equals the direct sum $\mathcal{M}_1 \oplus \mathcal{M}_2$ of the respective underlying matroids. In part (d) note that the empty signed set (= the zero vector) is contained in $\mathcal{V}_1$ and in $\mathcal{V}_2$, and therefore $\mathcal{V}_1 \cup \mathcal{V}_2 \subset \mathcal{V}$. Similarly, we have $\mathcal{L}_1 \cup \mathcal{L}_2 \subset \mathcal{L}$ in part (e).

In the realizable case, the vector configuration $\mathcal{M}_1 \oplus \mathcal{M}_2$ is obtained by placing the vector configurations $\mathcal{M}_1$ and $\mathcal{M}_2$ into orthogonal subspaces of dimensions r_1 and r_2 in $\mathbb{R}^{r_1 + r_2}$.

While the matroid operation of direct sum has a *unique* analogue for oriented matroids, this is not the case for the union operation. In order to define the union of two oriented matroids on a set E we will have to make several choices along the way. In particular, we shall assume that the set E is linearly ordered, and the result of the construction will depend on this ordering.

The following construction is due to Las Vergnas (1978b). Let $E = \{e_1 < e_2 < \ldots < e_n\}$ and $E' = \{e_1' < e_2' < \ldots < e_m'\}$ be disjoint linearly ordered sets, and let G be any bipartite graph on $E \cup E'$. For each $e_i' \in E'$ we consider the set $G_i = \{e_{i1} < e_{i2} < \ldots < e_{id_i}\}$ consisting of all $e \in E$ with $(e, e_i') \in G$. Now suppose that $\mathcal{M}$ is any oriented matroid on E. Then the bipartite graph G *induces* an oriented matroid $\mathcal{M}'$ on E' as follows. Let $\mathcal{M}_{m+1} := \mathcal{M}$, and for $i = m, m-1, \ldots, 2, 1$ define $\mathcal{M}_i$ to be the oriented matroid on $E \cup \{e_m', e_{m-1}', \ldots, e_i'\}$ which is obtained from $\mathcal{M}_{i+1}$ by the lexicographic extension $e_i' := [e_{i1}^+, e_{i2}^+, \ldots, e_{id_i}^+]$. (Here we are making a choice; for the purposes of constructing a union of oriented matroids any other lexicographic extension of the form $[e_{i1}^{\alpha_1}, e_{i2}^{\alpha_2}, \ldots, e_{id_i}^{\alpha_{d_i}}]$, or a different principal extension on the same flat, would also work.) Let $\mathcal{M}'$ be the oriented matroid obtained from $\mathcal{M}_1$ by deleting the set E. Note that the rank of $\mathcal{M}'$ is less or equal to the rank of $\mathcal{M}$. The underlying matroid $\underline{\mathcal{M}'}$ is the matroid induced by the bipartite graph G from $\underline{\mathcal{M}}$ (see Las Vergnas 1975c, White 1986).

We now define the *union* of two oriented matroids $\mathcal{M}_1$ and $\mathcal{M}_2$ on the same ground set E as follows. First identify E with disjoint sets E' and E'' respectively, and consider the direct sum $\mathcal{M}_1 \oplus \mathcal{M}_2$, an oriented matroid on the disjoint union $E' \cup E''$. Now let G be the bipartite graph on $(E' \cup E'') \cup E$ whose edges are (e', e) and (e'', e) for all $e \in E$. The *union* $\mathcal{M}_1 \cup \mathcal{M}_2$ is defined to be the oriented matroid induced from $\mathcal{M}_1 \oplus \mathcal{M}_2$ by the bipartite graph G.

7.6.2 Theorem (Las Vergnas 1978b). *The union of two orientable matroids on the same ground set is again orientable.*

In fact, it follows directly from the construction that the underlying matroid $\underline{\mathcal{M}_1 \cup \mathcal{M}_2}$ of the union as defined above equals the union $\underline{\mathcal{M}_1} \cup \underline{\mathcal{M}_2}$ of the respective underlying matroids. This implies that

$$\text{rank}(\mathcal{M}_1 \cup \mathcal{M}_2) \leq \text{rank}(\mathcal{M}_1) + \text{rank}(\mathcal{M}_2).$$

Here equality holds if and only if there exist bases B_1 of $\mathcal{M}_1$ and B_2 of $\mathcal{M}_2$ such that $B_1 \cap B_2 = \emptyset$. In that case $B_1 \cup B_2$ is a basis of $\mathcal{M}_1 \cup \mathcal{M}_2$.

7.6.3 Example. The union of oriented matroids is not a commutative operation. Consider the rank 2 oriented matroids $\mathcal{M}_1$ and $\mathcal{M}_2$ on $E = \{1, 2, 3, 4, 5\}$

which are defined by their chirotopes χ_1 and χ_2 as

$$\chi_1: \qquad \begin{array}{ccccc} 12+ & 13+ & 14+ & 15+ & 23+ \\ 24+ & 25+ & 34+ & 35+ & 45+ \end{array}$$

$$\chi_2: \qquad \begin{array}{ccccc} 12+ & 13+ & 14+ & 15+ & 23- \\ 24+ & 25+ & 34+ & 35+ & 45+ \end{array}$$

Geometrically, we can think of $\mathcal{M}_1$ and $\mathcal{M}_2$ as configurations of five points on the projective line, ordered $1,2,3,4,5$ and $1,3,2,4,5$ respectively. We consider the direct sum $\mathcal{M}_1 \oplus \mathcal{M}_2$ which is a rank 4 oriented matroid on

$$\{1',2',3',4',5',1'',2'',3'',4'',5''\}.$$

Geometrically, this corresponds to placing the configurations $\mathcal{M}_1$ and $\mathcal{M}_2$ onto two skew lines in 3-space.

In order to compute the union $\mathcal{M}_1 \cup \mathcal{M}_2$ from the direct sum $\mathcal{M}_1 \oplus \mathcal{M}_2$ we successively perform the lexicographic extensions $5 = [5'^+, 5''^+]$, $4 = [4'^+, 4''^+]$, $3 = [3'^+, 3''^+]$, $2 = [2'^+, 2''^+]$ and $1 = [1'^+, 1''^+]$. Let $\widetilde{\mathcal{M}}$ be the resulting rank 4 oriented matroid on

$$\{1',2',3',4',5',1'',2'',3'',4'',5'',1,2,3,4,5\},$$

and let $\widetilde{\chi}$ be its chirotope. From the definitions of lexicographic extensions we find

$$\widetilde{\chi}(1,2,3,4) = \widetilde{\chi}(1',2,3,4) = \widetilde{\chi}(1',2',3,4)$$
$$= \widetilde{\chi}(1',2',3'',4) = \widetilde{\chi}(1',2',3'',4'').$$

Here we are using the fact that $3'$ and $4'$ are contained in the flat spanned by $\{1',2'\}$. Using Proposition 7.6.1, we obtain

$$\widetilde{\chi}(1,2,3,4) = \widetilde{\chi}(1',2',3'',4'') = \chi_1(1,2)\cdot\chi_2(3,4) = (+)\cdot(+) = +.$$

Now, $\mathcal{M}_1 \cup \mathcal{M}_2$ is obtained from $\widetilde{\mathcal{M}}$ by deleting

$$\{1',2',3',4',5',1'',2'',3'',4'',5''\},$$

and so the chirotope χ_{12} of $\mathcal{M}_1 \cup \mathcal{M}_2$ satisfies

$$
\begin{array}{rclcl}
\chi_{12}(1,2,3,4) & = & \chi_1(1,2)\cdot\chi_2(3,4) & = & + \\
\chi_{12}(1,2,3,5) & = & \chi_1(1,2)\cdot\chi_2(3,5) & = & + \\
\chi_{12}(1,2,4,5) & = & \chi_1(1,2)\cdot\chi_2(4,5) & = & + \\
\chi_{12}(1,3,4,5) & = & \chi_1(1,3)\cdot\chi_2(4,5) & = & + \\
\chi_{12}(2,3,4,5) & = & \chi_1(2,3)\cdot\chi_2(4,5) & = & +.
\end{array}
$$

Geometrically, we obtain $\mathcal{M}_1 \cup \mathcal{M}_2$ from $\mathcal{M}_1 \oplus \mathcal{M}_2$ by first placing the point 5 on the segment $\overline{5'5''}$ very close to $5'$, then placing 4 on the segment $\overline{4'4''}$ even closer to $4'$, then placing 3 on the segment $\overline{3'3''}$ even closer to $3'$, etc. The resulting point configuration in 3-space has the unique circuit (or Radon partition) $1\overline{2}3\overline{4}5$.

On the other hand, the chirotope χ_{21} of the union $\mathcal{M}_2 \cup \mathcal{M}_1$ is given by

$$\chi_{21}(1,2,3,4) = \chi_2(1,2) \cdot \chi_1(3,4) = +$$
$$\chi_{21}(1,2,3,5) = \chi_2(1,2) \cdot \chi_1(3,5) = +$$
$$\chi_{21}(1,2,4,5) = \chi_2(1,2) \cdot \chi_1(4,5) = +$$
$$\chi_{21}(1,3,4,5) = \chi_2(1,3) \cdot \chi_1(4,5) = +$$
$$\chi_{21}(2,3,4,5) = \chi_2(2,3) \cdot \chi_1(4,5) = -.$$

This oriented matroid has the unique circuit $1\overline{2}3\overline{4}5$. So, $\mathcal{M}_1 \cup \mathcal{M}_2 \neq \mathcal{M}_2 \cup \mathcal{M}_1$, showing that the union of oriented matroids is not commutative.

In Example 7.6.3 we computed the chirotope of the union directly from the two chirotopes. The following result shows that this works in general.

7.6.4 Proposition (Lawrence and Weinberg 1980). *Let $\mathcal{M}_1$ and $\mathcal{M}_2$ be oriented matroids of rank r_1 and r_2 on E_n such that $\mathcal{M}_1 \cup \mathcal{M}_2$ has rank $r_1 + r_2$. Suppose $\{e_1 < e_2 < \ldots < e_{r_1}\}$ is a basis of $\mathcal{M}_1$, and $\{f_1 < f_2 < \ldots < f_{r_2}\}$ is a basis of $\mathcal{M}_2$ such that for all lexicographic earlier permutations $(\tilde{e}_1, \ldots, \tilde{e}_{r_1}, \tilde{f}_1, \ldots, \tilde{f}_{r_2})$ of $(e_1, \ldots, e_{r_1}, f_1, \ldots, f_{r_2})$ either $\{\tilde{e}_1, \ldots, \tilde{e}_{r_1}\}$ is not a basis of $\mathcal{M}_1$ or $\{\tilde{f}_1, \ldots, \tilde{f}_{r_2}\}$ is not a basis of $\mathcal{M}_2$. Then*

$$\chi_{12}(e_1, \ldots, e_{r_1}, f_1, \ldots, f_{r_2}) = \chi_1(e_1, \ldots, e_{r_1}) \cdot \chi_2(f_1, \ldots, f_{r_2}),$$

where χ_{12} is the chirotope of $\mathcal{M}_1 \cup \mathcal{M}_2$.

Proof. Consider the direct sum $\mathcal{M}_1 \oplus \mathcal{M}_2$, a rank $r_1 + r_2$ oriented matroid on $\{1', \ldots, n', 1'', \ldots, n''\}$. Then define the oriented matroid $\widetilde{\mathcal{M}}$ with chirotope $\widetilde{\chi}$ which is obtained from $\mathcal{M}_1 \oplus \mathcal{M}_2$ by the successive lexicographic extensions

$$n = [n'^+, n''^+], \ n-1 = [(n-1)'^+, (n-1)''^+], \ \ldots, \ 1 = [1'^+, 1''^+].$$

Consider the basis $\{b_1 < b_2 < \ldots < b_{r_1+r_2}\} = \{e_1, \ldots, e_{r_1}, f_1, \ldots, f_{r_2}\}$ with the properties above. In order to compute $\widetilde{\chi}(b_1, b_2, \ldots, b_{r_1+r_2})$, we proceed as follows:

$$\widetilde{\chi}(b_1, b_2, \ldots, b_{r_1+r_2}) = \widetilde{\chi}(b_1', b_2, \ldots, b_{r_1+r_2}) \circ \widetilde{\chi}(b_1'', b_2, \ldots, b_{r_1+r_2})$$
$$= \widetilde{\chi}(b_1', b_2', \ldots, b_{r_1+r_2}) \circ \widetilde{\chi}(b_1', b_2'', \ldots, b_{r_1+r_2}) \circ \widetilde{\chi}(b_1'', b_2', \ldots, b_{r_1+r_2})$$
$$\circ \widetilde{\chi}(b_1'', b_2'', \ldots, b_{r_1+r_2})$$
$$= \widetilde{\chi}(b_1', b_2', b_3' \ldots, b_{r_1+r_2}) \circ \widetilde{\chi}(b_1', b_2', b_3'' \ldots, b_{r_1+r_2}) \circ \cdots \cdots \quad \cdots\cdots$$

Thus $\widetilde{\chi}(b_1, b_2, \ldots, b_{r_1+r_2})$ is equal to the first non-zero entry in the lexicographic list

$$\widetilde{\chi}(b_1', \ldots, b_{r_1+r_2-1}', b_{r_1+r_2}'), \ \widetilde{\chi}(b_1', \ldots, b_{r_1+r_2-1}', b_{r_1+r_2}''),$$
$$\widetilde{\chi}(b_1', \ldots, b_{r_1+r_2-1}'', b_{r_1+r_2}'), \ \ldots$$

In this first non-zero entry precisely the elements $e_1, e_2, \ldots, e_{r_1}$ are primed and the elements $f_1, f_2, \ldots, f_{r_2}$ are double-primed. Using Proposition 7.6.1 (c) we find

$$\chi_{12}(e_1, \ldots, e_{r_1}, f_1, \ldots, f_{r_2}) = \widetilde{\chi}(e_1, \ldots, e_{r_1}, f_1, \ldots, f_{r_2})$$
$$= \widetilde{\chi}(e_1', \ldots, e_{r_1}', f_1'', \ldots, f_{r_2}'') = \chi_1(e_1', \ldots, e_{r_1}') \cdot \chi_2(f_1', \ldots, f_{r_2}').$$

This completes the proof of Proposition 7.6.4. $\qquad\square$

7.6.5 Example. Let $\mathcal{M}_1$ and $\mathcal{M}_2$ be the rank 2 oriented matroids on $\{1, 2, 3, 4, 5\}$ whose chirotopes are given by

$$
\begin{array}{llllll}
\chi_1: & 12\ 0 & 13\ + & 14\ + & 15\ + & 23\ + \\
        & 24\ + & 25\ + & 34\ 0 & 35\ + & 45\ +
\end{array}
$$

$$
\begin{array}{llllll}
\chi_2: & 12\ + & 13\ + & 14\ + & 15\ + & 23\ - \\
        & 24\ - & 25\ - & 34\ 0 & 35\ 0 & 45\ 0
\end{array}
$$

We compute the chirotope χ_{12} of the union $\mathcal{M}_1 \cup \mathcal{M}_2$ as follows:

$$
\begin{aligned}
\chi_{12}(1, 2, 3, 4) &= -\chi_1(1, 3) \cdot \chi_2(2, 4) &=& \quad + \\
\chi_{12}(1, 2, 3, 5) &= -\chi_1(1, 3) \cdot \chi_2(2, 5) &=& \quad + \\
\chi_{12}(1, 2, 4, 5) &= -\chi_1(1, 4) \cdot \chi_2(2, 5) &=& \quad + \\
\chi_{12}(1, 3, 4, 5) &= -\chi_1(3, 5) \cdot \chi_2(1, 4) &=& \quad - \\
\chi_{12}(2, 3, 4, 5) &= -\chi_1(3, 5) \cdot \chi_2(2, 4) &=& \quad +.
\end{aligned}
$$

We note that the union of two realizable oriented matroids is again realizable because the operations of direct sum, lexicographic extension and deletion preserve realizability. For details on the realization space of the union of two realizable oriented matroids see Section 8.2.

We will discuss a specific class of uniform oriented matroids which was introduced by Lawrence (1984b). It has interesting applications to the study of mutations of oriented matroids (Roudneff and Sturmfels 1988), see Exercise 7.24.

A rank r uniform oriented matroid $\mathcal{M}$ on E_n is in the class Γ if there exist rank 1 uniform oriented matroids $\mathcal{M}_1, \mathcal{M}_2, \ldots, \mathcal{M}_r$ on E_n such that $\mathcal{M} = \mathcal{M}_1 \cup \mathcal{M}_2 \cup \ldots \cup \mathcal{M}_r$. Here it is important that the oriented matroids $\mathcal{M}_i$ have a common ground set E_n with a fixed linear ordering of its elements. The oriented matroids in the class Γ are by definition unions of realizable matroids and hence realizable.

From Proposition 7.6.4 we conclude that

$$
\chi(i_1, i_2, \ldots, i_r) \quad = \quad \chi_1(i_1) \cdot \chi_2(i_2) \cdot \ldots \cdot \chi_r(i_r)
$$

if $i_1 < i_2 < \ldots < i_r$ and $\chi, \chi_1, \ldots, \chi_r$ are the chirotopes of $\mathcal{M}, \mathcal{M}_1, \ldots, \mathcal{M}_r$ respectively. Thus we can represent $\mathcal{M}$ by the $r \times n$-matrix $A = (a_{ij})$ where $a_{ij} = \chi_i(j)$. The sign $\chi(i_1, i_2, \ldots, i_r)$ of the ordered basis $(i_1, i_2, \ldots, i_r)$ equals the product of the main diagonal along the columns of A indexed by $i_1, i_2, \ldots, i_r$.

7.6.6 Example. Consider the 3×6 sign-matrix

$$
A \quad = \quad \begin{pmatrix} - & + & + & + & + & - \\ + & + & + & + & + & - \\ + & + & - & + & + & + \end{pmatrix}.
$$

The three rows of this matrix stand for rank 1 uniform oriented matroids $\mathcal{M}_1$, $\mathcal{M}_2$, $\mathcal{M}_3$; for instance $\mathcal{M}_1$ is the unique uniform rank 1 oriented matroid $\overline{1}2345\overline{6}$. Let $\mathcal{M} = \mathcal{M}_1 \cup \mathcal{M}_2 \cup \mathcal{M}_3$ be their union, which is a rank 3 ori-

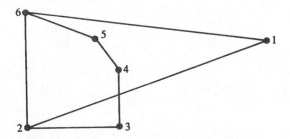

Figure 7.6.1: The rank 3 oriented matroid in Γ
obtained by taking the union of $(-++++-)$,
$(+++++-)$ and $(++-+++)$.

ented matroid in Γ shown in Figure 7.6.1. Its chirotope equals

123+	124−	125−	126−	134−
135−	136−	145−	146−	156−
234+	235+	236+	245+	246+
256+	345+	346+	356+	456+

We close this section with the following result of Lawrence (1984b).

7.6.7 Proposition. *The class Γ is closed under minors and duality.*

Sketch of proof: Let $\mathcal{M} = \mathcal{M}_1 \cup \mathcal{M}_2 \cup \ldots \cup \mathcal{M}_r$ be a rank r oriented matroid on E_n in the class Γ, and let A be the representative sign-matrix whose columns are the rank 1 uniform oriented matroids $\mathcal{M}_i$. For all $e \in E_n$ we have

$$\mathcal{M} \setminus e \;=\; (\mathcal{M}_1 \setminus e) \cup (\mathcal{M}_2 \setminus e) \cup \ldots \cup (\mathcal{M}_r \setminus e),$$

and so all deletions of $\mathcal{M}$ are in Γ. By the duality of contraction and deletion, it suffices to show that $\mathcal{M}^* \in \Gamma$.

Given the $r \times n$ sign-matrix A which represents $\mathcal{M}$, it is possible to explicitly construct an $(n-r) \times n$ sign-matrix B (depending on A) which represents the dual $\mathcal{M}^*$. This representation proves that $\mathcal{M}^*$ is also in Γ.

For simplicity we will explain here the construction of the dual matrix B only for the special case $r = 3$, $n = 7$. The description and verification of this construction for general r and n is left as an exercise to the reader.

Let $\mathcal{M}$ be a rank 3 oriented matroid in Γ and let χ denote its chirotope. If its representation matrix equals

$$A \;=\; \begin{pmatrix} a_{11} & a_{12} & a_{13} & a_{14} & a_{15} & * & * \\ * & a_{22} & a_{23} & a_{24} & a_{25} & a_{26} & * \\ * & * & a_{33} & .a_{34} & a_{35} & a_{36} & a_{37} \end{pmatrix}$$

with $a_{ij} \in \{-1, +1\}$, then $\chi(i, j, k) = a_{1i} a_{2j} a_{3j}$ for all $1 \le i < j < k \le 7$.

(Here the stars denote irrelevant entries.) Consider the 4×7 sign-matrix $B \quad =$

$$\begin{pmatrix} 1 & -a_{11}a_{12} & a_{11}a_{12}a_{22}a_{23} & -a_{11}a_{12}a_{22}a_{23}a_{33}a_{34} \\ * & 1 & -a_{12}a_{13} & a_{12}a_{13}a_{23}a_{24} \\ * & * & 1 & -a_{13}a_{14} \\ * & * & * & 1 \end{pmatrix}$$

$$\left. \begin{matrix} * & * & * \\ -a_{12}a_{13}a_{23}a_{24}a_{34}a_{35} & * & * \\ a_{13}a_{14}a_{24}a_{25} & -a_{13}a_{14}a_{24}a_{25}a_{35}a_{36} & * \\ -a_{14}a_{15} & a_{14}a_{15}a_{25}a_{26} & -a_{14}a_{15}a_{25}a_{26}a_{36}a_{37} \end{matrix} \right)$$

and the chirotope $\widetilde{\chi}$ of the induced rank 4 oriented matroid in Γ. We compute

$$\widetilde{\chi}(1,2,3,4) \;=\; 1 \;=\; \chi(5,6,7) \cdot \chi(5,6,7)$$

$$\widetilde{\chi}(1,2,3,5) \;=\; -a_{14}a_{15} \;=\; -a_{14}a_{26}a_{37}\,a_{15}a_{26}a_{37}$$
$$=\; -\chi(4,6,7)\,\chi(5,6,7)$$

$$\widetilde{\chi}(1,2,3,6) \;=\; a_{14}a_{15}a_{25}a_{26} \;=\; a_{14}a_{25}a_{37}\,a_{15}a_{26}a_{37}$$
$$=\; \chi(4,5,7)\,\chi(5,6,7)$$

$$\widetilde{\chi}(1,2,3,7) \;=\; -a_{14}a_{15}a_{25}a_{26}a_{36}a_{37} \;=\; -a_{14}a_{25}a_{36}\,a_{15}a_{26}a_{37}$$
$$=\; \chi(4,5,6)\,\chi(5,6,7)$$

$$\widetilde{\chi}(1,2,4,5) \;=\; a_{13}a_{14}^2 a_{15} \;=\; a_{13}a_{26}a_{37}\,a_{15}a_{26}a_{37}$$
$$=\; \chi(3,6,7)\,\chi(5,6,7)$$

$$\widetilde{\chi}(1,2,4,6) \;=\; -a_{13}a_{14}^2 a_{15}a_{25}a_{26} \;=\; -a_{13}a_{25}a_{37}\,a_{15}a_{26}a_{37}$$
$$=\; -\chi(3,5,7)\,\chi(5,6,7)$$

$$\widetilde{\chi}(1,2,4,7) \;=\; a_{13}a_{14}^2 a_{15}a_{25}a_{26}a_{36}a_{37} \;=\; a_{13}a_{25}a_{36}\,a_{15}a_{26}a_{37}$$
$$=\; \chi(3,5,6)\,\chi(5,6,7)$$

$$\widetilde{\chi}(1,2,5,6) \;=\; a_{13}a_{14}^2 a_{24}a_{25}^2 a_{15}a_{26} \;=\; a_{13}a_{24}a_{37}\,a_{15}a_{26}a_{37}$$
$$=\; \chi(3,4,7)\,\chi(5,6,7)$$

$$\widetilde{\chi}(1,2,5,7) \;=\; -a_{13}a_{14}^2 a_{24}a_{25}^2 a_{15}a_{26}a_{36}a_{37} \;=\; -a_{13}a_{24}a_{36}a_{15}a_{26}a_{37}$$
$$=\; -\chi(3,4,6)\chi(5,6,7)$$

$$\widetilde{\chi}(1,2,6,7) \;=\; \ldots\ \ldots\ \ldots\ \ldots\ \ldots$$

This shows that the map $\widetilde{\chi}$ is equal, up to the global sign $\overset{\bullet}{\chi}(5,6,7)$, to the chirotope χ^* of the dual oriented matroid $\mathcal{M}^*$. Thus the rank 4 uniform oriented matroid $\mathcal{M}^*$ is obtained by taking the union of the four rank 1 oriented matroid defined by the rows of B. $\square$

7.7 Strong maps and weak maps

There are two kinds of morphisms between ordinary matroids, known as "strong maps" and "weak maps". Strong maps are a generalization of the linear algebra concept of a linear map, whereas weak maps are a way to make precise the notion of going from more general position to more special position. See also

Propositions 2.4.6 and 2.4.7 for a discussion of weak maps in the context of the matroid stratification of the Grassmannian.

Oriented matroid maps have been studied considerably less than their ordinary matroid counterparts – see Chapters 8 and 9 of White (1986) for a full exposition of the latter. The case of oriented matroids presents, however, more technical difficulties. The current state of knowledge provides more questions than answers.

Maps of oriented matroids have recently gained attention from the work of Gel'fand and MacPherson (1992), see also MacPherson (1993). Oriented matroid maps are the key ingredient in their combinatorial construction of Pontrjagin classes. In Proposition 7.7.5 we show that their definition of oriented matroid weak maps is equivalent to the straightforward generalizations of the ordinary matroid definitions in terms of bases and circuits.

We start with a discussion of strong maps. They were first defined by Las Vergnas (1975b). The following proposition describes some basic facts.

7.7.1 Proposition. *Let $\mathcal{M}_1$ and $\mathcal{M}_2$ be two oriented matroids on the same ground set E. Then the following are equivalent:*

(1) *Every covector of $\mathcal{M}_2$ is a covector of $\mathcal{M}_1$.*

(2) *Every cocircuit of $\mathcal{M}_2$ is a covector of $\mathcal{M}_1$.*

(3) *Every vector of $\mathcal{M}_1$ is a vector of $\mathcal{M}_2$.*

(4) *Every circuit of $\mathcal{M}_1$ is a vector of $\mathcal{M}_2$.*

Proof. (1) $\Longrightarrow$ (2) and (3) $\Longrightarrow$ (4) are trivial. To see (2) $\Longrightarrow$ (1), let Z be a covector of $\mathcal{M}_2$. Then $Z = Y^1 \circ Y^2 \circ \ldots \circ Y^m$, where $Y^1, Y^2 \ldots, Y^m$ are cocircuits of $\mathcal{M}_2$. By assumption, $Y^1, Y^2 \ldots, Y^m$ must also be covectors of $\mathcal{M}_1$, hence Z is also. By orthogonality, we also have (4) $\Longrightarrow$ (3).

Let $\mathcal{V}_i$ be the family of vectors and $\mathcal{L}_i$ the family of covectors of $\mathcal{M}_i$, for $i = 1, 2$. Then (1) is $\mathcal{L}_2 \subseteq \mathcal{L}_1$, but this is equivalent to $\mathcal{V}_2 = \mathcal{L}_2^\perp \supseteq \mathcal{L}_1^\perp = \mathcal{V}_1$, which is (3). $\square$

Figure 7.7.1 illustrates the conditions of Proposition 7.7.1.

7.7.2 Definition. In the situation described by the equivalent conditions of Proposition 7.7.1, we say that there is a *quotient map*, or a *strong map* (induced by the identity map on E) from $\mathcal{M}_1$ to $\mathcal{M}_2$. We denote this situation by "$\mathcal{M}_1 \longrightarrow \mathcal{M}_2$".

In analogy with the case of ordinary matroids, we could have defined strong maps more generally between two oriented matroids on different sets, where the map on the underlying sets need be neither surjective nor injective, and may send some elements to "zero". We leave the details to Exercise 7.27.

7.7.3 Corollary.

(i) *A strong map $\mathcal{M}_1 \longrightarrow \mathcal{M}_2$ of oriented matroids induces a strong map $\underline{\mathcal{M}_1} \longrightarrow \underline{\mathcal{M}_2}$ of their underlying matroids.*

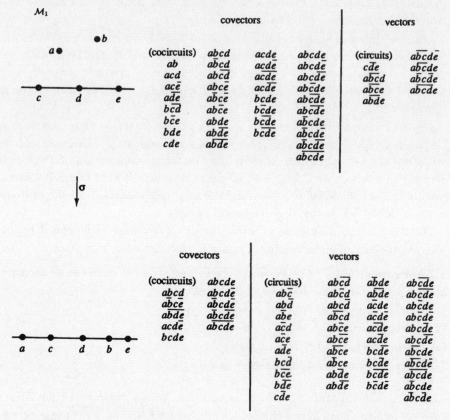

Figure 7.7.1: Example of a strong map of oriented matroids.

(ii) If there is a strong map $\mathcal{M}_1 \longrightarrow \mathcal{M}_2$, then $r(\mathcal{M}_1) \geq r(\mathcal{M}_2)$.

(iii) If there is a strong map $\mathcal{M}_1 \longrightarrow \mathcal{M}_2$, then $\mathcal{M}_2^* \longrightarrow \mathcal{M}_1^*$ also is a strong map.

Proof. This follows from the above Proposition 7.7.1, together with Proposition 8.1.6 of White (1986).   □

Much of the utility of strong maps for ordinary matroids, as well as the notion that they generalize linear projections, arise from the factorization theorem, which states that a strong map can always be factored as an extension followed by a contraction. Obviously, it would be very desirable to have a corresponding theorem for strong maps of oriented matroids.

Thus it was conjectured by Las Vergnas, see (Allys and Las Vergnas 1992), that such a factorization always exists. This conjecture is easy to verify in the cases of strong maps $\mathcal{M}^1 \longrightarrow \mathcal{M}^2$ with $r(\mathcal{M}^1) \leq 3$, with $r(\mathcal{M}_2) \leq 1$ or with $r(\mathcal{M}^1) - r(\mathcal{M}^2) \leq 1$; see Exercise 7.30. The general case has very recently been *disproved*:

7.7.4 Proposition   (Richter-Gebert 1993b).   *There is a strong map from a uniform matroid* R(12) *of rank 4 on 12 points to a uniform matroid of rank 2 that does not admit a factorization into an extension followed by a contraction.*

It remains an attractive question whether a strong map can always be factored if $\mathcal{M}^1$ is realizable. This has a certain plausibility in the general framework of extension spaces and OM-Grassmannians presented in Mněv and Ziegler (1993). In Figure 7.7.2 we indicate a single element extension of the matroid $\mathcal{M}_1$ of Figure 7.7.1, such that the contraction by that element gives $\mathcal{M}_2$. This is the desired factorization for this example.

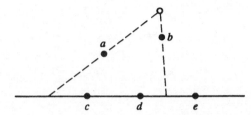

Figure 7.7.2: A factorization of the strong map of Figure 7.7.1.

Now we turn our attention to weak maps, which we will characterize by several equivalent conditions. The possibly most intuitive one, in terms of the chirotopes, only works in the rank-preserving case. The following result links the definitions suggested by matroid theory to the condition (3) used by Gel'fand and MacPherson (1992).

7.7.5 Proposition (Ziegler and White).   *Let $\mathcal{M}_1$ and $\mathcal{M}_2$ be two oriented matroids on the same ground set E. Then the following conditions are equivalent:*

(1)   *For every $X^1 \in \mathcal{C}(\mathcal{M}_1)$, there exists $X^2 \in \mathcal{C}(\mathcal{M}_2)$ with $X^2 \leq X^1$*

   ("*every signed circuit of $\mathcal{M}_1$ contains a signed circuit of $\mathcal{M}_2$*").

(2)   *For every $X^1 \in \mathcal{V}(\mathcal{M}_1) \setminus \{0\}$, there exists $X^2 \in \mathcal{V}(\mathcal{M}_2) \setminus \{0\}$ with $X^2 \leq X^1$*

   ("*every vector of $\mathcal{M}_1$ contains a vector of $\mathcal{M}_2$*").

(3)   *For every $Y^2 \in \mathcal{L}(\mathcal{M}_2)$, there exists $Y^1 \in \mathcal{L}(\mathcal{M}_1)$ with $Y^2 \leq Y^1$*

   ("*every covector of $\mathcal{M}_2$ is contained in a covector of $\mathcal{M}_1$*").

(4)   *For every $T^2 \in \mathcal{T}(\mathcal{M}_2)$, there exists $T^1 \in \mathcal{T}(\mathcal{M}_1)$ with $T^2 \leq T^1$*

   ("*every tope of $\mathcal{M}_2$ is contained in a tope of $\mathcal{M}_1$*", or $\mathcal{T}(\mathcal{M}_2 \setminus L_2) \subseteq \mathcal{T}(\mathcal{M}_1 \setminus L_2)$, where L_2 is the set of loops of $\mathcal{M}_2$).

Now assume, furthermore, $\mathcal{M}_1$ and $\mathcal{M}_2$ have the same rank $r(\mathcal{M}_1) = r(\mathcal{M}_2)$, and let χ_1 and χ_2 be chirotopes for $\mathcal{M}_1$ and $\mathcal{M}_2$, respectively. Then these conditions are also equivalent to

(5)   *either $\chi_2 \leq \chi_1$, or $\chi_2 \leq -\chi_1$*

   ("*χ_2 is obtained from χ_1 by setting some non-zeroes to zero*").

(5*)   *either $\chi_2^* \leq \chi_1^*$, or $\chi_2^* \leq -\chi_1^*$*

   ("*χ_2^* is obtained from χ_1^* by setting some non-zeroes to zero*").

Proof. First observe that we may assume that neither $\mathcal{M}_1$ nor $\mathcal{M}_2$ have any loops. In fact, any of the conditions imply that $L_1 \subseteq L_2$ holds for the respective sets of loops, and the conditions hold for $\mathcal{M}_1$ and $\mathcal{M}_2$ if and only if they hold for $\mathcal{M}_1\backslash L_2$ and $\mathcal{M}_2\backslash L_2$. With this, we may in particular replace (4) by (4')
$\mathcal{T}(\mathcal{M}_2) \subseteq \mathcal{T}(\mathcal{M}_1)$.

(1)$\Longleftrightarrow$(2) is clear, using that every vector contains a circuit, and that every circuit is a vector. (3)$\Longleftrightarrow$(4) is also clear, using that every covector is contained in a tope, and that every tope is a covector.

(1)$\Longrightarrow$(4'): for this let $T \in \{+,-\}^E\backslash\mathcal{T}(\mathcal{M}_1)$, then there is a circuit $X^1 \in \mathcal{C}(\mathcal{M}_1)$ with $X^1 \le T$. By (1) we find $X^2 \in \mathcal{C}(\mathcal{M}_2)$ with $X^2 \le X^1$. So we get $X^2 \le T$, and thus $T \notin \mathcal{T}(\mathcal{M}_2)$.

(4')$\Longrightarrow$(1): let X^1 be a circuit of $\mathcal{M}_1$. We consider the restriction $\mathcal{M}_1(\underline{X}^1)$, whose topes are the restrictions of topes of $\mathcal{M}_1$ to $\underline{X}^1$. By construction, we get that $\mathcal{T}(\mathcal{M}_1(\underline{X}^1)) = \{+,-\}^{\underline{X}^1}\backslash\{X^1, -X^1\}$.

From (4') we get $X^1 \notin \mathcal{T}(\mathcal{M}_1(\underline{X}^1)) \supseteq \mathcal{T}(\mathcal{M}_2(\underline{X}^1))$, so X^1 is not a tope of $\mathcal{M}_2(\underline{X}^1)$, either. We conclude that X^1 contains a circuit $X^2 \le X^1$ of $\mathcal{M}_2(\underline{X}^1)$, where $X^2 \in \mathcal{C}(\mathcal{M}_2(\underline{X}^1)) \subseteq \mathcal{C}(\mathcal{M}_2))$.

(1)$\Longrightarrow$(5): for this, it suffices to show that χ_1 and χ_2 are consistent (that is, either $\chi_1 \equiv \chi_2$, or $\chi_1 \equiv -\chi_2$) on any two adjacent bases $B = \{b, b_2, \ldots, b_r\}$ and $B' = \{b', b_2, \ldots, b_r\}$ of $\mathcal{M}_2$.

By (1), neither B nor B' contains a circuit of $\mathcal{M}_1$, so they are bases of $\mathcal{M}_1$, as well. Thus we can restrict both $\mathcal{M}_1$ and $\mathcal{M}_2$ to the ground set $B \cup B'$ of size $r + 1$. There we know $\mathcal{C}(\mathcal{M}_1) = \{X^1, -X^1\}$ for $X^1 = c_1(b', B)$, and similarly $\mathcal{C}(\mathcal{M}_2) = \{X^2, -X^2\}$ for $X^2 = c_2(b', B)$.

From (1) we now get $X^1 \ge X^2$, with $\sigma := X^2(b) = X^1(b) \ne 0$ and $X^2(b') = X^1(b') = +$. Hence the construction rule of circuit signatures from chirotopes in Section 3.5 yields

$$\chi_1(b, b_2, \ldots, b_r) = -X^1(b)\chi_1(b', b_2, \ldots, b_r)$$

and

$$\chi_2(b, b_2, \ldots, b_r) = -X^2(b)\chi_2(b', b_2, \ldots, b_r).$$

This means that $\chi_1(B) = -\sigma\chi_1(B')$ and $\chi_2(B) = -\sigma\chi_2(B')$, which implies that χ_1 and χ_2 either agree on both B and B', or they disagree on both B and B', depending on whether $\sigma = -$ or $\sigma = +$.

(5)$\Longrightarrow$(1). Here consider a counter-example, given by two oriented matroids $\mathcal{M}_1$ and $\mathcal{M}_2$ and a circuit $X^1 \in \mathcal{C}(\mathcal{M}_1)$, such that $|E|$ is minimal.

First assume that $\mathcal{M}_2$ has a coloop e_0. Then $X^1\backslash e_0 \ne \emptyset$, because $\mathcal{M}_1$ has no loops. From the definition of the chirotope of a contraction we see that $\mathcal{M}_1/e_0$ and $\mathcal{M}_2/e_0$ again satisfy (5), and $X^1\backslash e_0$ contains a circuit M^1/e_0, using Proposition 3.3.2 and Theorem 3.2.5. By minimality of the example, we can find $X^2 \in \mathcal{C}(\mathcal{M}_2/e_0)$ with $X^2 \le X^1\backslash e_0$. But e_0 is a coloop of $\mathcal{M}_2$, so $\mathcal{C}(\mathcal{M}_2/e_0) = \mathcal{C}(\mathcal{M}_2)$. Hence $X^2 \le X^1$ for $X^2 \in \mathcal{C}(\mathcal{M}_2)$, so (1) is not violated for X^1. Therefore $\mathcal{M}_2$ cannot have a coloop.

Now choose a basis B^2 of $\mathcal{M}_2$ such that $|B^2 \cap X^1| = r_2(X^1)$, that is, such that B^2 contains an $\mathcal{M}_2$-basis of $\underline{X}^1$. Then the restrictions of $\mathcal{M}_1$ and $\mathcal{M}_2$ to $B^2 \cup \underline{X}^1$ still satisfy (5) (note that B^2 is a basis of $\mathcal{M}_1$, too), but they still violate (1).

Thus, from the minimality of our counter-example, we can conclude $E = B^2 \cup \underline{X}^1$. Also, since $\mathcal{M}_2$ does not have coloops, we get $B^2 \subseteq \underline{X}^1$, that is, $\underline{X}^1 = E$ is a circuit in $\mathcal{M}_2$. This means that $\mathcal{M}_1 = \mathcal{M}_2$, so from (5) we get that $\chi_1 = \pm\chi_2$, hence $\mathcal{M}_1 = \mathcal{M}_2$. Thus (1) holds trivially, and there is no counter-example.

(5)$\Longleftrightarrow$(5*), finally, is clear from the definition of the dual chirotope in Section 3.5.   □

7.7.6 Definition. In the situation of Proposition 7.7.5(1-4), we say that there is a *weak map* from $\mathcal{M}_1$ to $\mathcal{M}_2$, and denote this situation by $\mathcal{M}_1 \rightsquigarrow \mathcal{M}_2$. There is a *rank preserving weak map* from $\mathcal{M}_1$ to $\mathcal{M}_2$, if additionally $\mathcal{M}_1$ and $\mathcal{M}_2$ have the same rank, that is, in the situation in Proposition 7.7.5(5).

7.7.7 Corollary.

(i)   *A weak map $\mathcal{M}_1 \rightsquigarrow \mathcal{M}_2$ of oriented matroids induces a weak map $\mathcal{M}_1 \rightsquigarrow \mathcal{M}_2$ of their underlying matroids.*

(ii)   *If there is a weak map $\mathcal{M}_1 \rightsquigarrow \mathcal{M}_2$, then $r(\mathcal{M}_1) \geq r(\mathcal{M}_2)$.*

(iii) *If there is a rank preserving weak map $\mathcal{M}_1 \rightsquigarrow \mathcal{M}_2$, then $\mathcal{M}_1^* \rightsquigarrow \mathcal{M}_2^*$ is a (rank preserving) weak map as well.*

(iv) *Every strong map of oriented matroids $\mathcal{M}_1 \longrightarrow \mathcal{M}_2$ is a weak map as well.*

Note that in the case where the rank is not preserved, even for ordinary matroids, weak maps are *not* a self-dual concept as in the case of strong maps: this is clear from Corollary 7.7.7(ii,iii).

The idea of weak maps is that $\mathcal{M}_1 \rightsquigarrow \mathcal{M}_2$ is not only a rank preserving weak map of the underlying matroids, but that in the resulting special position of $\mathcal{M}_2$, the orientation of $\mathcal{M}_1$ is not strictly violated. For example, in Figure 7.7.3, $\mathcal{M}_1 \rightsquigarrow \mathcal{M}_2$ is a weak map of oriented matroids, but $\mathcal{M}_1 \rightsquigarrow \mathcal{M}_3$ is not, because b is allowed to move onto the line ae, but not to cross it. In particular, if χ_1 and χ_3 are normalized to agree on abc, then they disagree on abe. An explicit geometric/topological characterization of weak maps was obtained by Anderson (1999c).

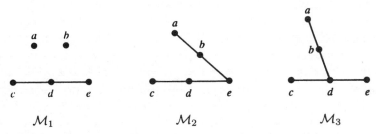

Figure 7.7.3: $\mathcal{M}_1 \rightsquigarrow \mathcal{M}_2$ is a weak map, but $\mathcal{M}_1 \rightsquigarrow \mathcal{M}_3$ is not.

Unlike the case of strong maps, where every rank preserving strong map is the identity (Exercise 7.25), rank preserving weak maps capture most of the complexity of general weak maps. There is a (quite trivial) matroid theorem that says that every weak map of matroids is a truncation followed by a rank preserving weak map – again, it would be very nice to have an analogue for weak maps of oriented matroids (Problem 7.33).

We close this section with the observation that every oriented matroid is the weak map image of a uniform oriented matroid of the same rank. We need a lemma that was already mentioned in Section 7.2.

7.7.8 Lemma.   *Let $\mathcal{M}$ be an oriented matroid, and let σ_1, σ_2 be cocircuit signatures that represent two single element extensions $\widetilde{\mathcal{M}}^1$ and $\widetilde{\mathcal{M}}^2$ of $\mathcal{M}$.*

Then $\sigma_1 \geq \sigma_2$ (in the extension lattice) if and only if $\widetilde{\mathcal{M}}^1 \rightsquigarrow \widetilde{\mathcal{M}}^2$ is a weak map.

Proof. This is clear from the definitions of $\widetilde{\mathcal{M}}^1 \rightsquigarrow \widetilde{\mathcal{M}}^2$ (7.7.6) and $\sigma_1 \geq \sigma_2$ (7.2.1) together with the description of all the cocircuits of a single element extension in Proposition 7.1.4.   □

So, in particular every reverse perturbation of an element corresponds to a weak map.

7.7.9 Corollary.   *Every oriented matroid $\mathcal{M}$ is the weak map image of a uniform oriented matroid of the same rank.*

Proof. We argue by induction on the number of elements $e \in E$ such that $e \in \underline{X}$ for some circuit $X \in \mathcal{C}(\mathcal{M})$ with $|\underline{X}| \leq r(\mathcal{M})$.

By Proposition 7.2.2 (2), every single element can be perturbed to general position, and the result follows.   □

7.8 Inseparability graphs

The inseparability graph of an oriented matroid is an important invariant of its reorientation class. It was introduced in Bland and Las Vergnas (1978) and later studied in detail in Roudneff (1989) and Cordovil and Duchet (1990).

Let $\mathcal{M}$ be a rank r oriented matroid without loops on E, $|E| = n$. Two elements $x, y \in E$ are *covariant* in $\mathcal{M}$ if x and y have the same sign in all circuits containing them, or, equivalently, if they have opposite signs in all cocircuits containing them. They are called *contravariant* if they have the same sign in all cocircuits containing them, or, equivalently (Exercise 7.36), if they have opposite signs in all circuits containing them. Two elements x, y are *inseparable* in $\mathcal{M}$ if they are either covariant or contravariant. The *inseparability graph* of $\mathcal{M}$ is the graph $\mathrm{IG}(\mathcal{M})$ on E whose edges are the inseparable pairs. The following observation follows directly from the definitions.

7.8.1 Observation. *The inseparability graph* $\text{IG}(\mathcal{M})$ *is invariant under duality and reorientations, i.e.,* $\text{IG}(\mathcal{M}) = \text{IG}(\mathcal{M}^*) = \text{IG}({}_{\overline{A}}\mathcal{M})$ *for all* $A \subseteq E$.

The graph $\text{IG}(\mathcal{M})$ has the following geometric meaning when $\mathcal{M}$ is the oriented matroid of n points $x_1, x_2, \ldots, x_n$ in projective $(r-1)$-space $\mathbb{P}^{r-1}$. Two points x_i and x_j are inseparable if and only if they lie in the same connected component of $\mathbb{P}^{r-1} \setminus (\bigcup \mathcal{A})$, where $\mathcal{A}$ is the arrangement of all hyperplanes spanned by $\{x_1, x_2, \ldots, x_n\}$ but not passing through x_i or x_j. In the polar picture, suppose $\mathcal{M}$ is the oriented matroid of a hyperplane arrangement $\mathcal{H} = \{H_1, H_2, \ldots, H_n\}$ in $\mathbb{P}^{r-1}$. Then H_i and H_j are inseparable if and only if all vertices of the cell complex $\Delta(\mathcal{H})$ not on H_i or H_j lie in the same connected component of $\mathbb{P}^{r-1} \setminus (H_i \cup H_j)$.

Note that every lexicographic extension of an oriented matroid introduces an inseparable pair. Suppose that $\mathcal{M} \cup p$ is obtained from $\mathcal{M}$ by the lexicographic extension $p = [e_1^{\sigma_1}, e_2^{\sigma_2}, \ldots, e_k^{\sigma_k}]$. If $\sigma_1 = +$, then (p, e_1) is contravariant in $\mathcal{M} \cup p$, and if $\sigma_1 = -$, then (p, e_1) is covariant in $\mathcal{M} \cup p$.

It is the objective of this section to give an explicit description of the inseparability graphs of all *uniform* oriented matroids.

7.8.2 Proposition (Cordovil and Duchet 1990). *Let* $\mathcal{M}$ *be a rank* r *uniform oriented matroid on* E, *where* $|E| = n$.

(a) *If* $r = 1$, *then* $\text{IG}(\mathcal{M})$ *is the complete graph on* E;

(b) *if* $r = 2$, *then* $\text{IG}(\mathcal{M})$ *is an* n-cycle;

(c) *if* $2 \le r \le n - 2$, *then* $\text{IG}(\mathcal{M})$ *is either an* n-cycle or a disjoint union of $k \ge 2$ paths.

Proof.

(a) Here $\mathcal{M}$ has precisely two cocircuits X and $-X$ whose zero set is empty. Any two $x, y \in E$ either have the same sign in both X and $-X$ and are contravariant in $\mathcal{M}$, or they have opposite signs in both X and $-X$ and so are covariant in $\mathcal{M}$.

(b) By Observation 7.8.1, we may assume that $\mathcal{M}$ is the oriented matroid of n points $x_1, x_2, \ldots, x_n$ in increasing order on the affine line. For all $i = 1, 2, \ldots, n-1$, the pair (x_i, x_{i+1}) is contravariant because x_i and x_{i+1} lie on the same side of any other point x_j, thus they have the same sign in every cocircuit containing them. The pair (x_n, x_1) is covariant because x_1 and x_n lie on different sides of all other points x_j.

Now consider any other pair (x_i, x_j) with $j - i \ge 2$ and either $i > 1$ or $j < n$. This pair is not contravariant because of the cocircuit complementary to the hyperplane $\{x_{i+1}\}$. It is not covariant because of the hyperplane $\{x_1\}$ in case $i > 1$ or because of the hyperplane $\{x_n\}$ in case $j < n$.

(c) Let us show that $\text{IG}(\mathcal{M})$ has no vertices of degree ≥ 3. Suppose on the contrary that there are distinct $e_1, e_2, e_3, e_4 \in E$ such that (e_1, e_2), (e_1, e_3), $(e_1, e_4) \in \text{IG}(\mathcal{M})$. Pick distinct elements $e_5, e_6, \ldots, e_{r+1}, e_{r+2}$ in E and consider the restriction $\widetilde{M}$ of M to $\{e_1, \ldots, e_{r+1}, e_{r+2}\}$. The dual $\widetilde{M}^*$ of $\widetilde{M}$ is a rank 2 uniform oriented matroid. Clearly, $(e_1, e_2), (e_1, e_3), (e_1, e_4) \in \text{IG}(\widetilde{M}) = \text{IG}(\widetilde{M}^*)$ which is a contradiction to (b).

The proof of Proposition 7.8.2 is completed by establishing the following two facts.

– If $\text{IG}(M)$ contains a path $e_1, e_2, \ldots, e_n$ of all n points, then it is the cycle $e_1, \ldots, e_n, e_1$.

– Any cycle contained in $\text{IG}(M)$ has length n.

For details see Cordovil and Duchet (1990, Assertions 2.6.1 and 2.6.2).   □

The inseparability graph is an invariant of oriented matroids which can easily be computed and which can be used to test isomorphism of small examples. (A more powerful invariant is introduced in Exercise 7.39: the weighted inseparability graph.)

7.8.3 Example. We compute the inseparability graphs of all uniform rank 3 oriented matroids on 6 elements. Up to reorientation, there are four such oriented matroids, and we list their affine diagrams in Figure 7.8.1.

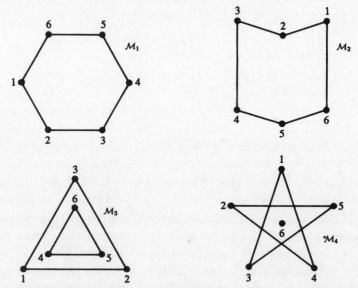

Figure 7.8.1: The rank 3 uniform oriented matroids on 6 elements.

The oriented matroid M_1 has the inseparability graph

$$\text{IG}(M_1) = \{\,(1,2), (2,3), (3,4), (4,5), (5,6), (6,1)\,\}.$$

All six pairs are contravariant. The pair $(1,3)$, for example, is neither contravariant nor covariant because of the cocircuits $1\overline{3}46$ and 1234.

The oriented matroid M_2 has the inseparability graph

$$\text{IG}(M_2) = \{\,(1,2), (2,3), (4,5), (5,6)\,\}.$$

As before, no pair is covariant, but now $(3,4)$ and $(1,6)$ are not contravariant any longer because of the newly created cocircuits $\overline{3}456$ and $\overline{1}456$.

For $\mathcal{M}_3$ the inseparability graph is $\mathtt{IG}(\mathcal{M}_3) = \{(1,4),(2,5),(3,6)\}$, where all three pairs are contravariant. All other pairs are separated by one of the lines $\overline{14}$, $\overline{25}$ or $\overline{36}$ and thus are not contravariant.

The oriented matroid $\mathcal{M}_4$ has no inseparable pairs. Any pair from $\{1,2,3,4,5\}$ is separated by a line through the center point 6. The pair $(1,6)$ is not inseparable because of the cocircuits $\overline{1}346$ and $12\overline{4}6$.

The following necessary condition for inseparability graphs of uniform oriented matroids is due to Roudneff (1989). In the remainder of this section we will abbreviate a *path* (resp. a *cycle*) with n vertices with P_n (resp. C_n). For any two graphs G and G', $G + G'$ denotes the disjoint union of G and G'. So, the inseparability graphs of uniform rank 3 oriented matroids on 6 elements are C_6, $P_3 + P_3$, $P_2 + P_2 + P_2$ and $\emptyset = P_1 + P_1 + P_1 + P_1 + P_1 + P_1$.

7.8.4 Proposition. *Let $\mathcal{M}$ be a uniform rank r oriented matroid on $E = \{e_1, e_2, \ldots, e_n\}$ such that the subgraph of $\mathtt{IG}(\mathcal{M})$ induced by $\{e_5, e_6, \ldots, e_n\}$ is an $(n-4)$-path. Then the subgraph of $\mathtt{IG}(\mathcal{M})$ induced by $\{e_1, e_2, e_3, e_4\}$ has at least two edges.*

Proof. The result is true for $r \leq 2$ or $n \leq r + 2$ by Proposition 7.8.2 and for $(r,n) = (3,6)$ by Example 7.8.3. Assume now that $n \geq 2r \geq 6$ but $(r,n) \neq (3,6)$. Suppose that $(e_5, e_6, \ldots, e_n)$ in this order is the $(n-4)$-path in $\mathtt{IG}(\mathcal{M})$.

Let $5 \leq i < j \leq n$. We will first prove that e_1 and e_2 are inseparable in $\mathcal{M} \setminus e_i$ if and only if e_1 and e_2 are inseparable in $\mathcal{M} \setminus e_j$. It suffices to show this for consecutive indices i and $j = i+1$. Without loss of generality we assume that the pair (e_1, e_2) is covariant in $\mathcal{M} \setminus e_i$.

Suppose e_1 and e_2 are separable in $\mathcal{M} \setminus e_{i+1}$. Then there exists a circuit C in $\mathcal{M} \setminus e_{i+1}$ such that $e_1 \in C^+$ and $e_2 \in C^-$. Such a circuit C cannot be a circuit of $\mathcal{M} \setminus e_i$ and therefore $e_i \in \underline{C}$. Let C' be the circuit of $\mathcal{M}$ such that $\underline{C'} = \underline{C} \cup e_{i+1} \setminus e_i$ and $\{e_1, e_2\} \subset C'^+$. By eliminating e_1 between C and $-C'$ and e_2 between C and C', we find that e_i and e_{i+1} are separable in $\mathcal{M}$. This is a contradiction to the hypothesis $(e_i, e_{i+1}) \in \mathtt{IG}(\mathcal{M})$.

We now proceed by induction and apply the statement of Proposition 7.8.4 to $\mathcal{M} \setminus e_n$. There are at least two edges in the subgraph of $\mathtt{IG}(\mathcal{M} \setminus e_n)$ induced by $\{e_1, e_2, e_3, e_4\}$. Suppose for instance that (e_1, e_2) is one of them and that (e_1, e_2) is covariant in $\mathcal{M} \setminus e_n$. By the previous paragraph, the pair (e_1, e_2) is inseparable in $\mathcal{M} \setminus e_i$ for $i = 5, 6, \ldots, n$, and after suitable reorientation, we get that (e_1, e_2) is covariant in $\mathcal{M} \setminus e_i$ for $i = 5, 6, \ldots, n$.

If e_1 and e_2 were separable in $\mathcal{M}$, then e_1 and e_2 have opposite signs in some circuit C of $\mathcal{M}$. Thus C contains $e_1, e_2, e_5, e_6, \ldots, e_n$, and so $|C| \geq n - 2 \geq 2r - 2$. As $|C| = r + 1$, we get $r \leq 2$, or $r = 3$ and $n = 6$, a contradiction.

Thus we have proved Proposition 7.8.4 for all cases with $n \geq 2r$. If $\mathcal{M}$ is a rank r uniform oriented matroid on n elements with $n < 2r$, then we can replace $\mathcal{M}$ by $\mathcal{M}^*$ which is a rank $n - r$ uniform oriented matroid on $n \geq 2(n - r)$ elements. Thus Proposition 7.8.4 holds for $\texttt{IG}(\mathcal{M}) = \texttt{IG}(\mathcal{M}^*)$.   □

By applying Proposition 7.8.4 to suitably chosen $(n - 4)$-paths in the graphs in Figure 7.8.2, we obtain the following.

7.8.5 Corollary. *The inseparability graph* $\texttt{IG}(\mathcal{M})$ *of a rank r uniform oriented matroid $\mathcal{M}$ on n elements is none of the seven forbidden graphs listed in Figure 7.8.2.*

We call a graph with n vertices *admissible* if it is an n-cycle or a disjoint union of $k \geq 2$ paths and it is none of the seven graphs listed in Figure 7.8.2. The main theorem of this section states that the necessary conditions given in Proposition 7.8.2 and Corollary 7.8.5 are in fact sufficient.

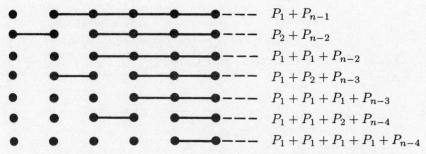

Figure 7.8.2: The seven forbidden graphs for admissibility.

7.8.6 Theorem (Roudneff 1989). *A graph G is the inseparability graph of a uniform oriented matroid if and only if it is admissible. If $3 \leq r \leq n - 3$ and G is an admissible graph on n vertices, then there exists a realizable rank r uniform oriented matroid $\mathcal{M}$ such that* $\texttt{IG}(\mathcal{M}) = G$. *Moreover, if $r \geq 4$, we can take $\mathcal{M}$ to be the oriented matroid defined by the vertex set of a simplicial $(r - 1)$-polytope.*

The proof of Theorem 7.8.6 is based on a case-by-case analysis and an explicit construction for each case. Here we illustrate this technique only for two cases. For the complete proof we refer to Roudneff (1989).

7.8.7 Example. Consider the admissible graph $G = P_1 + P_1 + P_3 + P_3$. A uniform rank 3 oriented matroid $\mathcal{M}_1$ with $G = \texttt{IG}(\mathcal{M}_1)$ is depicted in Figure 7.8.3.

We now construct a uniform rank 4 oriented matroid $\mathcal{M}_2$ with $G = \texttt{IG}(\mathcal{M}_2)$. Let $\mathcal{M}_2'$ be the rank 2 uniform oriented matroid of the ordered point configuration $1, 2, 3, 6, 4, 5, 7, 8$ on the affine line, and let $\mathcal{M}_2''$ be the rank 2 uniform oriented matroid defined by $1, 2, 4, 3, 5, 6, 7, 8$. Now define $\mathcal{M}_2 = \mathcal{M}_2' \cup \mathcal{M}_2''$ to be the union of the two rank 2 oriented matroids. We denote the respective chirotopes by χ_2, χ_2' and χ_2''.

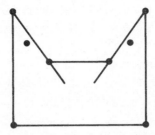

Figure 7.8.3: A uniform rank 3 oriented matroid $\mathcal{M}_1$ with
$$\text{IG}(\mathcal{M}_1) = P_1 + P_1 + P_3 + P_3.$$

The pair $(1,2)$ is contravariant in $\mathcal{M}_2$ because
$$\chi_2(1,i,j,k) = \chi_2'(1,i)\chi_2''(j,k) = \chi_2''(j,k) = \chi_2'(2,i)\chi_2''(j,k) = \chi_2(2,i,j,k)$$
for all $i,j,k \geq 3$. The pair $(2,3)$ is contravariant in $\mathcal{M}_2$ because $\chi_2(1,2,j,k) = \chi_2(1,3,j,k)$ and $\chi_2(2,i,j,k) = \chi_2(3,i,j,k)$ for all i, j, $k \geq 4$. Similarly, we find that both $(6,7)$ and $(7,8)$ are contravariant in $\mathcal{M}_2$.

We will show that these four are the only inseparable pairs of $\mathcal{M}_2$. First note that the pair $(1,8)$ is not contravariant since $\chi_2(1,4,5,7) = \chi_2'(1,4)\chi_2''(5,7) = +$ while $\chi_2(8,4,5,7) = -\chi_2'(4,5)\chi_2''(7,8) = -$. It is not covariant because $\chi_2(1,4,6,7) = \chi_2'(1,4)\chi_2'(6,7) = +$ and $\chi_2(8,4,6,7) = -\chi_2'(4,6)\chi_2''(7,8) = +$.

Now suppose $(i,j) \in \text{IG}(\mathcal{M}_2) \setminus \{(1,2),(2,3),(6,7),(7,8)\}$ with $i < j$. Using Proposition 7.8.2, we see that $(i,j) \notin \{(1,7),(1,8),(2,7),(2,8)\}$. Thus $\{i,j\} \subset \{1,2,3,4,5,6\}$ or $\{i,j\} \subset \{3,4,5,6,7,8\}$. In the first case we have
$$(i,j) \in \text{IG}(\mathcal{M}_2/\{7,8\}) = \{(1,2),(2,3),(3,6),(4,6),(4,5),(1,5)\}$$
and in the second case we have
$$(i,j) \in \text{IG}(\mathcal{M}_2/\{1,2\}) = \{(3,4),(3,5),(5,6),(6,7),(7,8),(4,8)\}.$$
This implies $(i,j) \in \{(1,5),(4,8)\}$.

The pair $(1,5)$ is not covariant because we have $\chi_2(1,2,3,4) = -$ and $\chi_2(5,2,3,4) = -\chi_2(2,3,4,5) = -$. It is not contravariant because $\chi_2(1,6,7,8) = +$, $\chi_2(5,6,7,8) = -$. The pair $(4,8)$ is not covariant because $\chi_2(4,5,6,7) = +$, $\chi_2(8,5,6,7) = +$. It is not contravariant since $\chi_2(1,2,3,4) = -$, $\chi_2(1,2,3,8) = +$.

In closing let us mention an interesting reconstruction theorem for uniform oriented matroids in terms of the inseparable pairs of all restrictions.

7.8.8 Theorem (Roudneff 1988c). *The reorientation class of a uniform oriented matroid $\mathcal{M}(E)$ is uniquely determined by the labeled inseparability graphs* $\text{IG}(\mathcal{M}(A))$, *for all $A \subseteq E$.*

7.9 Orientability

Not every matroid is orientable, and orientability is in fact a quite restrictive structural property. This leads one to ask some fundamental matroid theory questions, most of which are not well studied yet and to which only some partial answers are known:

(I)   Which matroids are orientable?

(II)   Which matroid constructions preserve orientability?

(III)   Under which condition is the orientation unique (up to reorientation)?

(IV)   How can the number of reorientation classes of a matroid be computed?

We have already discussed orientability for the case of rank 3 matroids in Section 6.6. We will now collect what is known in the general case. A complete answer can be given only for the binary case, see Proposition 7.9.3 below. The reader is also referred to Dress and Wenzel (1989) and to Guedes de Oliveira (1989) for two algebraic approaches to orientability.

We start with some simple cases of orientable matroids and orientability-preserving constructions. As mentioned the theory is not very well developed: there are many basic open problems and few satisfactory answers. We will touch upon some of them in the following discussion and the related exercises. Recall that principal matroid extensions place a point "in general position" on a flat, see White (1986).

7.9.1 Proposition.

(1)   *A matroid is orientable if and only if the associated simple matroid is orientable.*

(2)   *All matroids realizable over $\mathbb{R}$ (in particular all matroids of rank 2) are orientable.*

(3)   *Direct sums of orientable matroids are orientable.*

(4)   *Every minor of an orientable matroid is orientable.*

(5)   *The duals of orientable matroids are orientable.*

(6)   *Every principal extension of an orientable matroid is orientable.*

(7)   *Truncations of orientable matroids are orientable.*

(8)   *Unions of orientable matroids are orientable.*

(9)   *"Breaking a hyperplane that is a circuit" preserves orientability.*

Proof. Parts (1) to (5) are clear from our previous discussions. In particular every orientation of a matroid M canonically induces an orientation of its dual and of every minor.

For part (6), we have the "lexicographic extensions" of Section 7.1. We note, however, that there are several choices involved, and there is no canonical choice for the orientation of a principal extension.

Part (7) follows from combining (4) and (6). For (8) we refer to Theorem 7.6.2, and for (9) to Corollary 7.3.2.   □

Proposition 7.9.1 already describes reasonably large classes of orientable matroids. We will soon also see examples of non-orientable matroids. Proposition

7.9.1(4) suggests a characterization of orientable matroids by excluded minors, that is, by the collection of all minor-minimal non-orientable matroids. However, such a characterization cannot be achieved by a finite list. This was shown by Bland and Las Vergnas (1978), whose construction we now describe.

For $r \geq 3$ let M_r be the rank r matroid on the ground set

$$E = \{e_1, \ldots, e_r, e'_1, \ldots, e'_r\}$$

whose circuits are $\{e_i, e_j, e'_i, e'_j\}$ for $1 \leq i < j \leq r$, $\{e_1, \ldots, e_{i-1}, e'_i, e_{i+1}, \ldots, e_r\}$ for $1 \leq i \leq r$, and all $(r+1)$-sets that do not contain any of these circuits. Bland and Las Vergnas (1978) have shown that for $r \geq 4$ the matroid M_r is not orientable, whereas all of its proper minors are orientable. Thus orientability cannot be characterized by a finite list of excluded minors.

The Bland–Las Vergnas result was sharpened by Ziegler (1991), who constructed an infinite family of minimal non-orientable matroids of rank 3, see Exercise 6.15. Note that for these every contraction has rank at most 2 and is thus orientable. Therefore, it suffices to consider submatroids.

7.9.2 Theorem (Bland and Las Vergnas 1978, Ziegler 1991). *There is an infinite family of non-orientable matroids (of rank 3) all whose submatroids are orientable. Thus orientability of (rank 3) matroids cannot be characterized by a finite list of excluded minors.*

In this context we suggest that a "good" theorem on the orientability of Dilworth truncations (see Problem 7.43) might allow us to reduce some orientability problems to rank 3 – in analogy to the realizability problem in fixed characteristic, see White (1986). In rank 3, special tools are available, as we have seen in Section 6.6.

We proceed with a characterization of orientable binary matroids, with two different proofs. Recall for this (from White 1987, for example) that a matroid M is *binary* if it is representable over $\mathrm{GF}(2)$, and that it is *regular* if and only if it satisfies any of the following equivalent conditions:

(i) M is realizable over every field,

(ii) M is binary and realizable over $\mathbb{R}$,

(iii) M is real realizable with all basis determinants equal to ± 1,

(iv) M is real realizable with a representation matrix all of whose minors are ± 1 or 0.

7.9.3 Proposition (Bland and Las Vergnas 1978). *A binary matroid is orientable if and only if it is regular.*

Proof 1. In Example 6.6.2(1) we have seen that the Fano plane $F_7 = \mathrm{PG}(3, 2)$ is not orientable. With this, Proposition 7.9.1(4) implies that F_7^* is not orientable either. Thus if M is binary and orientable, then by 7.9.1(5) it cannot have an F_7 or F_7^* minor. Thus, by Tutte's characterization of regular matroids (see White 1987), M is regular. The converse is clear. $\square$

Proof 2. Let $\chi : \binom{E}{r} \longrightarrow \{0, +, -\}$ be an orientation of M. Because M is binary, at least one term in every 3-term Grassmann-Plücker relation is 0. This means that if we interpret χ as a map to $\{0, +1, -1\}$, then the 3-term Grassmann-Plücker relations (cf. Exercise 3.16) are in fact satisfied over $\mathbb{R}$, hence χ gives regular Grassmann-Plücker coordinates of M.   □

7.9.4 Corollary (Bland and Las Vergnas 1978). *If M is a regular matroid, then all orientations of M are realizable. They differ only by reorientation.*

Proof. This is clear from the second proof of Proposition 7.9.3, together with the projective uniqueness of regular matroids over $\mathbb{R}$ (see White 1987, Proposition 1.2.5).   □

Corollary 7.9.4 gives an exact answer to question (IV) for regular matroids. Some similar enumeration results have been obtained for matroids defined by affine dependencies on the vertex sets of regular polytopes: the number of reorientation classes is 2 for the icosahedron and the dodecahedron, 1 for the cube, the hypercube (in $\mathbb{R}^4$) and the 24-cell, see Las Vergnas, Roudneff and Salaün (1991). It is conjectured there that the affine matroid of the d-cube has a unique reorientation class for all d (Problem 7.12). An asymptotically exact answer to question (IV) was given in Corollary 7.4.3 for uniform matroids.

Exercises

7.1   Generalize Proposition 7.1.2(i): Let $\widetilde{\mathcal{M}} = \widetilde{\mathcal{M}}(\widetilde{E})$ be an extension of $\mathcal{M}(E)$ to some larger ground set. Show that if $r(\widetilde{\mathcal{M}}) = r(\mathcal{M})$, then there is a unique function

$$\sigma : \mathcal{C}^* \longrightarrow \{+, -, 0\}^{\widetilde{E} \setminus E} \text{ such that } \{(Y, \sigma(Y)) : Y \in \mathcal{C}^*\} \subseteq \widetilde{\mathcal{C}^*}.$$

Also show that uniqueness fails if the condition on the rank is dropped, although it holds for all single element extensions.

7.2   Consider an oriented matroid $\mathcal{M}$ of rank r, represented by an arrangement $\mathcal{A}$ of pseudospheres. Extend $\mathcal{M}$ lexicographically by $p = [b_1^{\alpha_1}, b_2^{\alpha_2}, \ldots, b_r^{\alpha_r}]$. Show that the "new" pseudosphere S_p dissects the tope $T_{B,\alpha}$ that is associated with the basis $B = (b_1, b_2, \ldots, b_r)$ and with $\alpha = (\alpha_1, \ldots, \alpha_r)$ according to Exercise 3.24.

7.3   Show that if e is not a loop, $\alpha \in \{+, -\}$, then the localization $\sigma[e^\alpha]$ is an atom of the extension lattice $\mathcal{E}(\mathcal{M})$.

Also show that, in general, this does not describe all atoms of $\mathcal{E}(\mathcal{M})$, and verify that not every coatom of an adjoint corresponds to an atom of $\mathcal{E}(\mathcal{M})$.

7.4   Let $\mathcal{M}$ have rank r.

(a) Show that the length of the lattice $\mathcal{E}(\mathcal{M})$ is at least $r + 1$: there are chains $\mathbf{c}$ of length $\ell(\mathbf{c}) = r+1$, i.e., of the form $\hat{0} = \sigma_0 < \sigma_1 < \ldots < \sigma_r < \hat{1}$ in $\hat{\mathcal{E}}(\mathcal{M})$.

(b) Show that $\mathcal{E}(\mathcal{M})$ can contain longer chains as well.

(c) Does every maximal chain have length at least $r+1$?

(d) Is it possible that every maximal chain has length $\ell(\mathbf{c}) > r+1$?

7.5 Choose $\mu \in \Lambda(n,d)$ and let $\chi : \Lambda(n,d) \longrightarrow \{+,-,0\}$ be any function. Show that if
$$\chi'(\lambda) := \begin{cases} \alpha & \text{for } \lambda = \mu, \\ \chi(\lambda) & \text{for } \lambda \neq \mu \end{cases}$$
is a chirotope for two values of $\alpha \in \{+,-,0\}$, then it is also a chirotope for the third. (This is the chirotope analogue of Lemma 7.3.3.)

7.6 Show that there is no "orientability version" of reverse perturbation construction. That is, give an example of an orientable matroid of rank 3, such that collapsing a basis yields a non-orientable matroid.

7.7 Describe explicitly the set of all cocircuits of a local perturbation, a reverse perturbation, and of a flipping. (Fukuda and Tamura 1988a)

7.8 Give an example of an oriented matroid $\mathcal{M}$ such that no single base sign can be reversed in its chirotope.

7.9 Show that $\mathcal{M}$ and $\mathcal{M}^*$ have the same number of simplicial regions. In particular, in the uniform case χ and χ^* have the same number of mutations.

7.10 The bound $|\mathrm{Mut}(\mathcal{M})| \geq n$ for realizable uniform oriented matroids is sharp for all $n > r$. For this, consider the oriented matroid $C^{n,r}$ given by the rows of
$$\begin{pmatrix} 1 & t_1 & \cdots & t_1^{r-1} \\ 1 & t_2 & \cdots & t_2^{r-1} \\ \vdots & \vdots & & \vdots \\ 1 & t_n & \cdots & t_n^{r-1} \end{pmatrix}$$
where $t_1 < t_2 < \ldots < t_n$, and show

(i) the chirotope of this oriented matroid is $\chi \equiv +$,

(ii) $\mathrm{Mut}(C^{n,r}) = \{12...r, 23...r+1, \ldots, n1...r-1\}$.

7.11 For every matroid of a regular 3-polytope, compute the number of reorientation classes. Explain the geometry.

7.12* Does the matroid of the d-dimensional cube have a unique reorientation class? (This has been shown for $d \leq 4$.)
(Las Vergnas, Roudneff and Salaün 1991)

7.13 Among the "many oriented matroids" constructed in Theorem 7.4.2, identify a small non-realizable (uniform) one of rank 3.

7.14 Estimate the constants d'_r of Corollary 7.4.3, for which the number of all oriented matroids of rank r on n points satisfies $g(n,r) \leq 2^{d'_r n^{r-1}}$.
(Edelsbrunner, Seidel and Sharir 1991)

7.15 Give an example

(a) of a non-realizable oriented matroid $\mathcal{M}$, $r(\mathcal{M}) \geq 4$, such that $\mathcal{M}^{ad}$ exists;

(b) of an oriented matroid $\mathcal{M}$ such that $\mathcal{M}^{ad}$ exists, but $(\mathcal{M}^{ad})^{ad}$ does not: $\mathcal{M}$ itself has an adjoint, but none of its adjoints has one.

(Kromberg 1995, Hochstättler and Kromberg 1996)

7.16 Show that $(\mathcal{M}^{ad})^{ad}$ contains $\mathcal{M}$ as a submatroid in a canonical way, but $\mathcal{M}^{ad}$ does not.

7.17 Show that if $\mathcal{M}^{ad}$ is a connected oriented matroid of rank $r(\mathcal{M}) \geq 4$, and if $\mathcal{M}$ admits a sequence $(\mathcal{M}_k)_{k \geq 0}$ with $\mathcal{M}_0 = \mathcal{M}$ and $\mathcal{M}_{k+1}$ is an adjoint of $\mathcal{M}_k$ for all $k \geq 0$, then $\mathcal{M}$ is realizable.

(Hint: Construct an infinite oriented matroid $\mathcal{M}_\infty$ as a "direct limit" $\lim_{i \to \infty} \mathcal{M}_{2i}$, and prove that $\mathcal{M}_\infty$ is a projective space. Now use that every irreducible projective space of rank at least 4 is coordinatizable, and show that the field is ordered. Now derive that $\mathcal{M}$ is canonically embedded in this projective space. Thus it is coordinatized over an ordered field, which implies that it is $\mathbb{R}$-realizable.)

7.18 The two unions $\mathcal{M}_1 \cup \mathcal{M}_2$ and $\mathcal{M}_2 \cup \mathcal{M}_1$ in Example 7.6.3 are distinct but isomorphic oriented matroids. Find two oriented matroids $\mathcal{N}_1$ and $\mathcal{N}_2$ such that $\mathcal{N}_1 \cup \mathcal{N}_2$ and $\mathcal{N}_2 \cup \mathcal{N}_1$ are not isomorphic and not reorientation equivalent.

7.19 Give a description of the union of two oriented matroids in terms of signed circuits.

7.20 Give a description in terms of signed circuits of the union of r uniform rank 1 oriented matroids.

7.21 Estimate the number $g_\Gamma(n, r)$ of oriented matroids of rank r on n points in the class Γ.

7.22 Complete the proof of Proposition 7.6.7, i.e., describe and verify the construction of the dual sign-matrix B for general r and n.

7.23 For any $n \geq 5$, give an example of a rank 3 uniform oriented matroid $\mathcal{M}$ on n points which is not contained in the class Γ, but all proper minors of $\mathcal{M}$ are in Γ.

7.24 Let $\mathcal{H}$ be a simple arrangement of n hyperplanes whose corresponding oriented matroid is in the class Γ. Prove that $\mathcal{H}$ has exactly n simplicial topes.   (Roudneff and Sturmfels 1988).

7.25 Let $\mathcal{M}_1 \longrightarrow \mathcal{M}_2$ be a strong map between two oriented matroids on the same ground set. Show that if $r(\mathcal{M}_1) = r(\mathcal{M}_2)$, then $\mathcal{M}_1 = \mathcal{M}_2$.

7.26 Let $\mathcal{M}_1$ and $\mathcal{M}_2$ be oriented matroids on the same ground set E. Show that if $\mathcal{M}_1 \rightsquigarrow \mathcal{M}_2$ and $\mathcal{M}_2 \rightsquigarrow \mathcal{M}_1$ are both weak maps, then $\mathcal{M}_1 = \mathcal{M}_2$.

7.27 Define strong and weak maps between oriented matroids on different sets, including an element "zero". Show that an arbitrary such map is equivalent, in a precise sense, to one which is the identity on the underlying sets (as in the unoriented case, see p.228 and p.255 of White 1986).

7.28* Characterize arbitrary strong and weak maps in terms of their chirotopes.

7.29 If $\mathcal{M}(E)$ is an oriented matroid and if $\widetilde{\mathcal{M}}(E \cup A)$ is an extension of $\mathcal{M}$, show that $\mathcal{M} \longrightarrow \widetilde{\mathcal{M}}/A$ ("an embedding followed by a contraction") is a strong map.

7.30 Show that every strong map $\mathcal{M}_1 \longrightarrow \mathcal{M}_2$ with $r(\mathcal{M}_1) \leq r(\mathcal{M}_2)+1$ or with $r(\mathcal{M}_2) \leq 1$ can be written as an extension followed by a contraction.

7.31 Let $\mathcal{M}_1$ and $\mathcal{M}_2$ be oriented matroids on the same ground set E. Show that if $r(\mathcal{M}_1) = r(\mathcal{M}_2)$, then the condition:

  (1′) For every $Y^1 \in \mathcal{C}^*(\mathcal{M}_1)$, there exists a $Y^2 \in \mathcal{C}^*(\mathcal{M}_2)$ with $Y^2 \leq Y^1$, ("every signed cocircuit of $\mathcal{M}_1$ contains a signed cocircuit of $\mathcal{M}_2$")

  is necessary and sufficient for a weak map $\mathcal{M}_1 \rightsquigarrow \mathcal{M}_2$.

  Prove that, in general, the condition (1′) is necessary, but not sufficient.

7.32 A *truncation* of $\mathcal{M}$ is an oriented matroid $\mathcal{M}^t := \mathcal{M}[p]/p$, where $\mathcal{M}[p]$ is a principal extension of $\mathcal{M}$ by an element in general position (such that $\underline{\mathcal{M}^t}$ is the first truncation of $\underline{\mathcal{M}}$). Show that every *truncation map* $\mathcal{M} \rightsquigarrow \mathcal{M}^t$ is a weak map.

7.33* Decide whether every weak map can be written as a multiple truncation map, followed by a rank preserving weak map. (If not, suitably extend the definition of a truncation in Exercise 7.32.)

7.34 Let $\mathcal{M}_1$ and $\mathcal{M}_2$ be oriented matroids of the same rank r, on the same ground set E. We say that $X^1 \in \mathcal{C}(\mathcal{M}_1)$ and $X^2 \in \mathcal{C}(\mathcal{M}_2)$ form a *stable pair* if $\underline{X}^1 \subseteq \underline{X}^2$ and $\underline{X}^1$ has the same rank in $\mathcal{M}_1$ and $\mathcal{M}_2$. (Thus $\underline{X}^2$ is the unique circuit of $\mathcal{M}_2$ that is contained in $\underline{X}^1$.) Show that $\mathcal{M}_1 \rightsquigarrow \mathcal{M}_2$ is a weak map if and only if $\underline{\mathcal{M}}_1 \rightsquigarrow \underline{\mathcal{M}}_2$ is a weak map, and for every stable pair either $X^2 \leq X^1$, or $X^2 \leq -X^1$.

  Also formulate the corresponding dual condition that characterizes rank preserving weak maps in terms of *stable pairs of cocircuits*.

7.35 Let $M_1 \rightsquigarrow M_2$ be a weak map of ordinary matroids, and let $\mathcal{M}_1$ be an orientation of M_1.

  (a) Show that there need *not* be an orientation $\mathcal{M}_2$ of M_2 such that $\mathcal{M}_1 \rightsquigarrow \mathcal{M}_2$ is a weak map.

  (b) Show that if M_1 and M_2 are regular (unimodular), then M_2 has an orientation $\mathcal{M}_2$ such that $\mathcal{M}_1 \rightsquigarrow \mathcal{M}_2$.

7.36 Let $x, y \in E$ be any two elements of an oriented matroid $\mathcal{M}$. Show that x and y have the same sign in all circuits of $\mathcal{M}$ containing both of them if and only if they have opposite signs in all cocircuits of $\mathcal{M}$ containing both of them.

7.37 Give definitions of "contravariant" and "covariant" in terms of chirotopes.

7.38 (a) Determine the inseparability graphs of all uniform rank 3 oriented matroids on 7 elements.

(b) Determine the inseparability graphs of all (not necessarily uniform) rank 3 oriented matroids on 6 elements.

7.39 Let M be an oriented matroid on a set E. For $x, y \in E$ let $equ(x,y)$ denote the number of circuits of M containing x and y with the same sign, and $diff(x,y)$ the number of circuits of M containing x and y with opposite signs. Set

$$w_{\max}(x,y) = \max\{equ(x,y), \ diff(x,y)\}$$
$$w_{\min}(x,y) = \min\{equ(x,y), \ diff(x,y)\}$$

and $w(x,y) := (w_{\max}(x,y), w_{\min}(x,y))$. Then the *weighted inseparability graph* $\mathrm{WIG}(M)$ is the graph with vertex set E such that every edge (x,y) has weight $w(x,y)$. Clearly $\mathrm{WIG}(M)$ is invariant under reorientation. Note that $\mathrm{IG}(M)$ is the subgraph of edges with $w_{\min}(x,y) = 0$. Any invariant of $\mathrm{WIG}(M)$ under isomorphism is an invariant of M under reorientation and isomorphism.

(a) In general, $\mathrm{WIG}(M) \neq \mathrm{WIG}(M^*)$. However, show that if M is uniform, then we have $\mathrm{WIG}(M) = \mathrm{WIG}(M^*)$.

(b) Suppose M is uniform of rank r on $n = |E|$ elements. Observe that for any $x \neq y$ we have $equ(x,y) + diff(x,y) = \binom{n-2}{r-1}$. Define $v(M) = (v_0(M), \dots, v_\ell(M))$, where $v_i(M)$ is the number of pairs of distinct elements $x, y \in E$ such that $w_{\min}(x,y) = i$, for $1 \leq i \leq \ell = \frac{1}{2}\binom{n-2}{r-1}$. Then $v(M)$ is invariant under reorientation and isomorphism.

As a test for isomorphism, compute $v(M)$ for the 24 non-realizable reorientation classes of $U_{4,8}$ (Bokowski and Richter-Gebert 1990b). (Result: 22 different vectors, only two pairs have the same vector.)

(c) In the two undecided cases of (b) the $\mathrm{WIG}(M)$ are isomorphic. Can these be decided by using triples instead of pairs? (There are 4 sign patterns to consider, in one-to-one correspondence with the unlabeled undirected graphs on 3 elements.)

(d)* Study the invariants $\mathrm{WIG}(M)$ and $v(M)$.

7.40 For any $n \geq r \geq 2$, give an example of a rank r uniform oriented matroid on n elements whose inseparability graph is the n-cycle C_n.

7.41 For any $n \geq r+3 \geq 6$ give an example of a rank r uniform oriented matroid on n elements whose inseparability graph is empty.

7.42 Show that if M is an orientable matroid, then so is its k-truncation $M^{[k]}$ (whose bases are the independent sets of M of size k) for $1 \leq k \leq r(M)$.

7.43* Let M be a matroid of rank $r \geq 4$, and $D^1(M)$ its first Dilworth truncation.
(a) Is it true that if M is orientable, then $D^1(M)$ is also orientable?
(b) Is the converse true?

(See White (1986) and Aigner (1979) for interpretation and details of the Dilworth construction.)

7.44* Give an example of an orientable matroid with finite characteristic set, that is, of an orientable matroid M that can be coordinatized over some finite field, but not in characteristic 0.

(Brylawski conjectures that this does not exist; see McNulty 1994)

7.45* Is the modular join of orientable matroids always orientable?

(See White (1986) for details. This problem might be a key to a decomposition theory for oriented matroids.)

7.46* Decide whether the matroids M_n and R_n given by Bachem and Wanka (1989a) are orientable.

7.47* For every prime power q, determine a minimal non-orientable submatroid of the projective plane of order q. (Ziegler 1991)

8

Realizability

This chapter deals with realizability of oriented matroids, an issue which was already touched upon in several earlier chapters. Here we are aiming for a deeper topological understanding of realizability, and that is the reason why we introduce the space $\mathcal{R}(\mathcal{M})$ of *all* vector realizations of a *fixed* oriented matroid $\mathcal{M}$. The realizability problem for $\mathcal{M}$ now becomes the question whether the semialgebraic variety $\mathcal{R}(\mathcal{M})$ is empty or not.

This global topological point of view furnishes us with an adequate language for studying an exciting class of non-trivial algebraic problems in discrete and computational geometry. In Section 8.2 we discuss specific oriented matroid constructions which preserve realizability. In Section 8.3 we show that realizable (uniform) oriented matroids cannot be characterized by excluding a finite collection of forbidden minors. Sections 8.4 and 8.5 deal with practical algorithms, complexity upper bounds and Bokowski's final polynomial method for compactly encoding non-realizability proofs. In Section 8.6 we present examples of oriented matroids $\mathcal{M}$ whose realization space $\mathcal{R}(\mathcal{M})$ is disconnected, and we describe how the underlying techniques are used to prove Mnëv's universality theorem *"Every semialgebraic variety is homotopy equivalent to some $\mathcal{R}(\mathcal{M})$"*. Applications of these results to computational geometry are discussed in Section 8.7. Throughout this chapter we will assume familiarity with the basics on coordinatizations of (unoriented) matroids. Here the reader is referred to Chapter 1 of White (1987).

8.1 The realization space of an oriented matroid

Let $\mathcal{M} = (E, \chi)$ be a rank r oriented matroid on a totally ordered n-element set E. We assume $E = \{1, 2, \ldots, n\}$ unless otherwise stated. $\mathcal{M}$ will be identified with its *chirotope* $\chi : E^r \to \{-, 0, +\}$ which maps non-bases to 0 and bases to their orientation ± 1. Recall that the chirotopes χ and $-\chi$ define the same oriented matroid. A *realization* of $\mathcal{M}$ is a mapping $\phi : E \to \mathbb{R}^r$ such that

$$\chi(e_1, e_2, \ldots, e_r) \quad = \quad \text{sign} \det(\phi(e_1), \phi(e_2), \ldots, \phi(e_r))$$

for all $e_1, e_2, \ldots, e_r \in E$. The oriented matroid $\mathcal{M}$ is called *realizable* if and only if such a realization ϕ exists. We remark that the terms *realizable, linear, coordinatizable, representable* and *stretchable* (for arrangements) are used synonymously throughout the literature.

As was explained in Chapters 1 and 2, we can think of realizable oriented matroids in many different ways. Depending on the situation, a realization ϕ of $\mathcal{M}$ can be interpreted

(i)   as a real $n \times r$-matrix,

(ii) as a (labeled) configuration of n vectors in $\mathbb{R}^r$,

(iii) as a (labeled) configuration of n points in real projective $(r-1)$-space,

(iv) as a (labeled) central arrangement of n hyperplanes in the vector space $\mathbb{R}^r$.

(v)   as a (labeled) arrangement of n hyperplanes in real projective $(r-1)$-space, or

(vi) as a point on the *Grassmann variety* of r-dimensional linear subspaces in $\mathbb{R}^n$.

Two realizations ϕ and ϕ' are *linearly equivalent* if there exists an invertible linear transformation $A \in GL(\mathbb{R}^r)$ such that $\phi' = A \circ \phi$. The set $\mathcal{R}(\mathcal{M})$ of linear equivalence classes of realizations of $\mathcal{M}$, in the quotient topology induced from $\mathbb{R}^{rn}$, is called the *realization space* of $\mathcal{M}$.

We next describe two alternative algebraic representations of the realization space. We will see in particular that $\mathcal{R}(\mathcal{M})$ is a *semialgebraic variety* (cf. Becker 1986, Benedetti and Risler 1990). This means that $\mathcal{R}(\mathcal{M})$ can be expressed as the set of solutions of a finite system of polynomial equations and inequalities in finitely many real variables.

The realization space $\mathcal{R}(\mathcal{M})$ of an oriented matroid $\mathcal{M}$ has been considered already in the Grassmannian set-up of Section 2.4, where it was also called the *oriented matroid stratum*. For completeness we recall this definition of the realization space as a subset of the exterior algebra. In the next paragraph a more elementary (but less conceptual) third description of the space $\mathcal{R}(\mathcal{M})$ will be given. Let $\wedge_r \mathbb{R}^n$ denote the r-fold exterior product of $\mathbb{R}^n$. This is the $\binom{n}{r}$-dimensional real vector space of alternating r-forms on $\mathbb{R}^n$. An element Ξ in $\wedge_r \mathbb{R}^n$ is said to be *decomposable* if $\Xi = \mathbf{v}_1 \wedge \mathbf{v}_2 \wedge \ldots \wedge \mathbf{v}_r$ for some $\mathbf{v}_1, \mathbf{v}_2, \ldots, \mathbf{v}_r \in \mathbb{R}^n$. The $(n-r)r$-dimensional projective variety of decomposable forms in $\wedge_r \mathbb{R}^n / \mathbb{R}^*$ is isomorphic to the Grassmann variety of r-dimensional linear subspaces in $\mathbb{R}^n$. This isomorphism is defined by assigning *Plücker coordinates* to each linear subspace of $\mathbb{R}^n$. The realization space of a rank r oriented matroid $\mathcal{M}$ on E equals the intersection in $\wedge_r \mathbb{R}^n$ of the orthant defined by the chirotope χ with the embedded Grassmann variety, i.e.,

$$\mathcal{R}(\mathcal{M}) = \{ \ \Xi \in \wedge_r \mathbb{R}^n / \mathbb{R}^* \ | \ \Xi \text{ decomposable and sign } \Xi = \chi \ \}.$$

The following equivalent description of $\mathcal{R}(\mathcal{M})$ is often more practical to work with, but it has the obvious disadvantage that it is dependent upon the choice of a basis of $\mathcal{M}$. Pick any basis $\{e_1, e_2, \ldots, e_r\}$ of the oriented matroid $\mathcal{M}$, and let $(e_1, e_2, \ldots, e_r)$ be the standard basis of $\mathbb{R}^r$. Then $\mathcal{R}(\mathcal{M})$ is isomorphic to the semialgebraic set of realizations $\phi : E \to \mathbb{R}^r$ of $\mathcal{M}$ for which $\phi(e_i) = \mathbf{e}_i$ for $i = 1, 2, \ldots, r$. This implies that $\mathcal{R}(\mathcal{M})$ is a set of $r \times (n - r)$-matrices all of whose minors have their signs prescribed. This description was used in Section 1.5 to give an algebraic proof for the non-realizability of $\mathrm{RS}(8)$. Here is another example:

8.1.1 Example (See Figure 8.1.1 (a).) Let $n = 7, r = 3$, and let $\mathcal{M}_3^7 = (E, \chi)$ be the uniform oriented matroid defined by $\chi : (1, 5, 7), (1, 6, 7), (2, 5, 7), (2, 6, 7),$ $\mapsto -1$ and $(i, j, k) \mapsto +1$ for all other increasing triples (i, j, k). Abbreviate with $[ijk]$ the 3×3-minor with column indices (i, j, k) of the matrix

$$\mathbf{A} \quad := \quad \begin{pmatrix} 1 & 0 & 0 & x_{41} & x_{51} & x_{61} & x_{71} \\ 0 & 1 & 0 & x_{42} & x_{52} & x_{62} & x_{72} \\ 0 & 0 & 1 & x_{43} & x_{53} & x_{63} & x_{73} \end{pmatrix}.$$

For example, $[257] = -x_{51}x_{73} + x_{53}x_{71}$. Then we have

$$\mathcal{R}(\mathcal{M}_3^7) \quad = \quad \{\, (x_{41}, x_{42}, \ldots, x_{73}) \in \mathbb{R}^{12} \mid \mathrm{sign}\,[ijk] \;=\; \chi(i, j, k) \text{ for all } i, j, k\,\}.$$

In other words, $\mathcal{R}(\mathcal{M}_3^7)$ is a semialgebraic variety in $\mathbb{R}^{12}$, defined by 34 sharp inequalities (12 1×1-determinants, 18 2×2-determinants, and four 3×3-determinants).

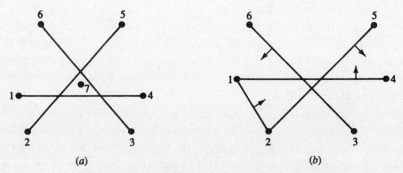

Figure 8.1.1: (a) The oriented matroid $\mathcal{M}_3^7$.
(b) $\mathcal{M}_3^7$ is not reducible by the point 7.

Note that the realization space of a simplex (an oriented matroid with $n = r$) is just a point. The realization space $\mathcal{R}(\mathcal{M})$ of any oriented matroid $\mathcal{M}$ is equal to the realization space $\mathcal{R}(\mathcal{M}^*)$ of the dual oriented matroid $\mathcal{M}^*$ under the canonical isomorphism between $\wedge_r \mathbb{R}^n$ and $\wedge_{n-r} \mathbb{R}^n$. This isomorphism is defined for decomposable r-forms, corresponding to r-flats, by taking the orthogonal complement in $\mathbb{R}^n$, and it is extended to all of $\wedge_r \mathbb{R}^n$ by linearity. This means that duality of oriented matroids is the combinatorial analogue of duality in the exterior algebra.

8.1.2 Example (Duality map for realization spaces). Let $e_1, e_2, \ldots, e_5$ denote the canonical basis of $\mathbb{R}^5$. The duality map between $\wedge_2 \mathbb{R}^5$ and $\wedge_3 \mathbb{R}^5$ maps an element

$$\Xi \;=\; a\,e_1 \wedge e_2 + b\,e_1 \wedge e_3 + c\,e_1 \wedge e_4 + d\,e_1 \wedge e_5 + e\,e_2 \wedge e_3$$
$$+ f\,e_2 \wedge e_4 + g\,e_2 \wedge e_5 + h\,e_3 \wedge e_4 + i\,e_3 \wedge e_5 + j\,e_4 \wedge e_5$$

of $\wedge_2 \mathbb{R}^5$ to the element

$$\Xi^* = a\,e_3 \wedge e_4 \wedge e_5 - b\,e_2 \wedge e_4 \wedge e_5 + c\,e_2 \wedge e_3 \wedge e_5$$
$$- d\,e_2 \wedge e_3 \wedge e_4 + e\,e_1 \wedge e_4 \wedge e_5 - f\,e_1 \wedge e_3 \wedge e_5$$
$$+ g\,e_1 \wedge e_3 \wedge e_4 + h\,e_1 \wedge e_2 \wedge e_5 - i\,e_1 \wedge e_2 \wedge e_4 + j\,e_1 \wedge e_2 \wedge e_3$$

of $\wedge_3 \mathbb{R}^5$. The Grassmann-Plücker relations (Theorem 2.4.3) state conditions in the indeterminates $a, b, c, \ldots$ for the 2-form Ξ and hence also for the 3-form Ξ^* to be decomposable. In our situation the Grassmann-Plücker relations reduce to

$$
\begin{aligned}
ah - bf + ce &= ai - bg + de = aj - cg + df \\
&= bj - ci + dh = ej - fi + gh = 0,
\end{aligned}
$$

(8.1.1)

and they are satisfied if and only if the coordinates of Ξ are the 2×2-minors of some 2×5-matrix. Taking the columns of that matrix, the 2-form Ξ corresponds to a configuration of 5 vectors in $\mathbb{R}^2$. Similarly, Ξ^* corresponds to a configuration of 5 vectors in $\mathbb{R}^3$. These vector configurations are obtained geometrically by projecting e_1, e_2, e_3, e_4, e_5 onto orthogonal pairs of subspaces in $\mathbb{R}^5$. The realization space of the rank 2 oriented matroid $\mathcal{M}$ of Ξ is isomorphic to the semialgebraic subset of $\mathbb{R}^{10}$ consisting of all vectors $(a', b', c', \ldots, j')$ satisfying the equations (8.1.1) and having componentwise the same sign as $(a, b, c, \ldots, j)$. Similarly, this semialgebraic set is isomorphic to the realization space of the rank 3 oriented matroid $\mathcal{M}^*$ of Ξ^*.

The deletion of a point $e \in E$ in any oriented matroid $\mathcal{M}$ induces a continuous map $\delta_e : \mathcal{R}(\mathcal{M}) \to \mathcal{R}(\mathcal{M} \setminus e)$. In terms of vector configurations, the *deletion map* δ_e is just defined by deleting the vector corresponding to e. Similarly, by duality, we can define a *contraction map* $\kappa_e : \mathcal{R}(\mathcal{M}) \to \mathcal{R}(\mathcal{M}/e)$ for every $e \in E$. For vector configurations the contraction map κ_e is the orthogonal projection onto the hyperplane perpendicular to the vector corresponding to e.

An oriented matroid $\mathcal{M}$ is said to be *reducible* by a point $e \in E$ if the deletion map $\delta_e : \mathcal{R}(\mathcal{M}) \to \mathcal{R}(\mathcal{M} \setminus e)$ is surjective. In other words, $\mathcal{M}$ is reducible by e if *every* realization of $\mathcal{M} \setminus e$ extends to a realization of $\mathcal{M}$.

8.1.1 Example (continued). Consider the deletion

$$\delta_7 : \mathcal{R}(\mathcal{M}_3^7) \to \mathcal{R}(\mathcal{M}_3^7 \setminus 7)$$

of the point $e = 7$ in Figure 8.1.1. This deletion map is induced from the projection

$$\mathbb{R}^{12} \to \mathbb{R}^9 \,, (x_{41}, \ldots, x_{62}, x_{63}, x_{71}, x_{72}, x_{73}) \mapsto (x_{41}, \ldots, x_{62}, x_{63})$$

onto the first nine coordinates. Figure 8.1.1 (b) shows a realization of $\mathcal{M}_3^7 \setminus 7$ which cannot be extended to a realization of $\mathcal{M}_3^7$. The arrows perpendicular to the lines $\overline{12}, \overline{14}, \overline{25}$ and $\overline{36}$ stand for open halfspaces into which the point 7 would have to be placed. As the intersection of these four halfspaces is empty, this specific realization of $\mathcal{M}_3^7 \setminus 7$ shows that $\mathcal{M}_3^7$ is *not* reducible by 7.

It is easy to see that $\mathcal{M}_3^7$ is reducible by the point 6, that $\mathcal{M}_3^7 \setminus \{6\}$ is reducible by 5, that $\mathcal{M}_3^7 \setminus \{6,5\}$ is reducible by 4, and that $\mathcal{M}_3^7 \setminus \{6,5,4\}$ is reducible by 7. Now, $\mathcal{M}_3^7 \setminus \{6,5,4,7\}$ is a simplex and hence $\mathcal{R}(\mathcal{M}_3^7 \setminus \{6,5,4,7\})$ is a point. Using the results of Section 8.2, this implies that the space $\mathcal{R}(\mathcal{M}_3^7)$ is non-empty and contractible.

8.1.2 Example. Consider the arrangement $\mathcal{A}$ of six lines $L_1, \ldots, L_6$ in the affine plane given in Figure 1.11.2, and let L_7 denote the line at infinity. Let $\mathcal{M}$ be the rank 3 oriented matroid on E_7 defined by (any reorientation of) $\mathcal{B} = \{L_1, \ldots, L_6, L_7\}$. Up to a trivial fibration, the realization space $\mathcal{R}(\mathcal{M})$ consists of all labeled arrangements combinatorially isomorphic to $\mathcal{B}$. Now consider the contraction map $\kappa_7 : \mathcal{R}(\mathcal{M}) \longrightarrow \mathcal{R}(\mathcal{M}/7)$ which assigns to an arrangement such as $\mathcal{B}$ its restriction $\mathcal{B} \cap L_7 = \{L_1 \cap L_7, \ldots, L_6 \cap L_7\}$ to the line at infinity. Equivalently, $\kappa_7(\mathcal{B})$ is the list of slopes $b_1, \ldots, b_6$ as in Example 1.11.2. Thus the counter-example to Ringel's conjecture presented in Example 1.11.2 is equivalent to saying that the contraction map $\kappa_7 : \mathcal{R}(\mathcal{M}) \to \mathcal{R}(\mathcal{M}/7)$ is not surjective.

Originally this example was constructed by Richter and Sturmfels (1991) using oriented matroid duality. The dual $\mathcal{M}^*$ is a rank 4 oriented matroid on E_7 such that the deletion map $\delta_7 : \mathcal{R}(\mathcal{M}^*) \to \mathcal{R}(\mathcal{M}^* \setminus 7)$ is not surjective. After reorientation, $\mathcal{M}^* \setminus 7$ is a configuration of generic points on the six lines spanned by the edges of a tetrahedron in affine 3-space. $\mathcal{M}^*$ is the single element extension of $\mathcal{M}^* \setminus 7$, where the point 7 is to be inserted into the "inverted" tetrahedron (i.e, the orientation of the tetrahedron has been reversed, similarly to the center triangle in Figure 8.1.1).

8.2 Constructions and realizability results

In order to derive interesting results on the realizability of matroids or oriented matroids a fundamental approach is to study constructions which preserve realizability. This line of thought will be pursued here from the realization space point of view, and we ask the more general question: *Which constructions preserve the homotopy type of the realization space of an oriented matroid ?* The following lemma shows that reducibility is a crucial issue for this question.

8.2.1 Lemma. *Let $\mathcal{M}$ be an oriented matroid and suppose that $\mathcal{M}$ is reducible by $e \in E$. Then $\mathcal{R}(\mathcal{M})$ is homotopy equivalent to $\mathcal{R}(\mathcal{M} \setminus e)$.*

Proof. We will show that the deletion map $\delta_e : \mathcal{R}(\mathcal{M}) \to \mathcal{R}(\mathcal{M} \setminus e)$ is a homotopy equivalence between both realization spaces. More precisely, we shall construct a continuous extension map $\gamma_e : \mathcal{R}(\mathcal{M} \setminus e) \to \mathcal{R}(\mathcal{M})$ with the property that the compositions $\gamma_e \circ \delta_e$ and $\delta_e \circ \gamma_e$ are homotopic to the identity maps on $\mathcal{R}(\mathcal{M})$ and $\mathcal{R}(\mathcal{M} \setminus e)$ respectively. Readers with a background in topology will observe that our explicit construction simply amounts to showing that δ_e defines a fibration with contractible fiber (see the remark at the end of Section 4.7(f)).

Fix an arbitrary realization Ξ of $\mathcal{M}$. The *residence* of a point e of $\mathcal{M}$ in Ξ is defined as the fiber $R_{e,\Xi} := \delta_e^{-1}(\delta_e(\Xi))$ of the deletion map. Without loss of generality we may assume $E = E_n$ and $e = n$. We now use the embedding of $\mathcal{R}(\mathcal{M})$ into $\mathbb{R}^{(n-r)r}$ which was defined in Section 8.1, and we write $\Xi = (x_{r+1}, \ldots, x_{n-1}, x_n)$ where $x_i \in \mathbb{R}^r$. Similarly we have $\delta_e(\Xi) = (x_{r+1}, \ldots, x_{n-1}) \in \mathcal{R}(\mathcal{M} \setminus e) \subset \mathbb{R}^{(n-r-1)r}$.

Let $\mathcal{C}^*$ denote the set of signed cocircuits of $\mathcal{M} \setminus e$, and let $\sigma : \mathcal{C}^* \to \{-, 0, +\}$ be the localization corresponding to its single element extension $\mathcal{M}$. There exists a linear hyperplane $H^0(C, \Xi)$ in $\mathbb{R}^r$ for each signed cocircuit $C \in \mathcal{C}^*$. We write $H^+(C, \Xi)$ and $H^-(C, \Xi)$ for the open halfspaces in $\mathbb{R}^r$ bounded by $H^0(C, \Xi)$. By the definition of localizations in Section 7.1, the point x_n is contained in $H^{\sigma(C)}(C, \Xi)$. The residence of e in Ξ equals the intersection of these cells with respect to all cocircuits of $\mathcal{M} \setminus e$:

$$R_{e,\Xi} = \bigcap_{C \in \mathcal{C}^*} H^{\sigma(C)}(C, \Xi).$$

Using the chirotope χ associated with the oriented matroid $\mathcal{M}$ we can write equivalently

$$R_{e,\Xi} = \left\{ x \in \mathbb{R}^r \mid \text{ sign } \det(x_{\lambda_1}, \ldots, x_{\lambda_{r-1}}, x) = \chi(\lambda_1, \ldots, \lambda_{r-1}, n) \right.$$
$$\left. \text{ for all } \lambda_i \in E \right\}.$$

Thus the residence $R_{e,\Xi}$ of e in Ξ is a relatively open convex cone in $\mathbb{R}^r$. As an example consider the oriented matroid $\mathcal{M}_3^7$ in Figure 8.1.1 (a), where we view all seven points as vectors in $\mathbb{R}^3$. The residence of the vector 7 is the open triangular cone bounded by the three flats 14, 25 and 36.

Next observe that the sets $H^{\sigma(C)}(C, \Xi)$ depend continuously on the parameter vector $\delta_e(\Xi) = (x_{r+1}, \ldots, x_{n-1}) \in \mathcal{R}(\mathcal{M} \setminus e)$ and so does their always nonempty intersection $R_{e,\Xi} = \delta_e^{-1}(\delta_e(\Xi))$. Consequently, there exists a continuous section $\gamma_e : \mathcal{R}(\mathcal{M} \setminus e) \to \mathcal{R}(\mathcal{M})$. This means that $\delta_e \circ \gamma_e$ equals the identity map on $\mathcal{R}(\mathcal{M} \setminus e)$.

In order to prove Lemma 8.2.1 it is sufficient to see that the map $\gamma_e \circ \delta_e$ is homotopic to the identity on $\mathcal{R}(\mathcal{M})$. We can interpret $\gamma_e \circ \delta_e$ geometrically as a normalization map which (for every realization of the oriented matroid $\mathcal{M}$) moves the point e into a certain special position within its residence. The fibers $R_{e,\Xi}$ being convex sets, we obtain a homotopy from $\gamma_e \circ \delta_e$ to $\text{id}_{\mathcal{R}(\mathcal{M})}$ by simply taking the convex combination :

$$[0,1] \times \mathcal{R}(\mathcal{M}) \quad \to \quad \mathcal{R}(\mathcal{M})$$
$$(t; x_{r+1}, \ldots, x_{n-1}, x_n) \quad \mapsto$$
$$\left(x_{r+1}, \ldots, x_{n-1}, t \cdot x_n + (1-t) \cdot \gamma_e(x_{r+1}, \ldots, x_{n-1}) \right).$$

This completes the proof of Lemma 8.2.1. $\qquad \square$

An ordering $(e_1, e_2, \ldots, e_n)$ of E is said to be a *reduction sequence* of the oriented matroid $\mathcal{M}$ if either $n = r$, or if $\mathcal{M}$ is reducible by e_n and $(e_1, e_2, \ldots, e_{n-1})$

is a reduction sequence for $\mathcal{M} \setminus e_n$. If $\mathcal{M}$ has a reduction sequence, then Lemma 8.2.1 implies that $\mathcal{M}$ is realizable and that its realization space $\mathcal{R}(\mathcal{M})$ is contractible. In this case the oriented matroid $\mathcal{M}$ satisfies the *isotopy property* which means that $\mathcal{R}(\mathcal{M})$ is connected. Our next result shows that oriented matroids which are built up by a sequence of lexicographic extensions are realizable and satisfy the isotopy property.

8.2.2 Proposition.   *Let $\mathcal{M} \cup e$ be a lexicographic extension of an oriented matroid $\mathcal{M}$. Then $\mathcal{M} \cup e$ is reducible by e.*

Proof. Let $\epsilon = \epsilon(\Xi)$ be a positive parameter which is sufficiently small, continuously depending on the realizations Ξ of $\mathcal{M}$. Using the notation of Definition 7.2.3, we define the extension map

$$\gamma_e : \mathcal{R}(\mathcal{M}) \to \mathcal{R}(\mathcal{M} \cup e)$$

$$\Xi = (\ldots, e_1, e_2, \ldots, e_k, \ldots) \mapsto (\Xi, \alpha_1 e_1 + \epsilon \alpha_2 e_2 + \ldots + \epsilon^{k-1} \alpha_k e_k).$$

The composition $\delta_e \circ \gamma_e$ with the deletion map $\mathcal{R}(\mathcal{M} \cup e) \to \mathcal{R}(\mathcal{M})$ equals the identity on $\mathcal{R}(\mathcal{M})$. Therefore δ_e is surjective. $\square$

In Remark 7.2.5 it has been observed that all extensions of a rank 2 oriented matroid are lexicographic. Hence Proposition 8.2.2 implies that in this case the realization spaces are well-behaved.

8.2.3 Corollary.   *Let $\mathcal{M} = (E, \chi)$ be an oriented matroid of rank 1 or 2. Then every ordering of E is a reduction sequence for $\mathcal{M}$. The realization space $\mathcal{R}(\mathcal{M})$ is non-empty and contractible.*

This raises the question whether all rank 3 oriented matroids admit a reduction sequence.

8.2.4 Theorem (Goodman and Pollack 1980b, Richter 1988).

(1)   *Every rank 3 oriented matroid on $n \leq 8$ points is realizable.*

(2)   *Every uniform rank 3 oriented matroid on $n \leq 8$ points has a reduction sequence.*

(3)   *There exists a realizable uniform rank 3 oriented matroid* Ric(9) *on 9 points which does not admit a reduction sequence.*

The oriented matroid Ric(9) without reduction sequence which proves part (3) of Theorem 8.2.4 is depicted in Figure 8.2.1 (a). By symmetry, it is sufficient to find a realization of Ric(9) \ 1 and a realization of Ric(9) \ 8, both of which do not extend to a realization of Ric(9). Such realizations are depicted in Figure 8.2.1 (b) and (c).

Richter proved part (2) of Theorem 8.2.4 by introducing a purely combinatorial criterion sufficient for reducibility in rank 3. He then proceeded with a case-by-case analysis which shows that all 135 combinatorially distinct simple arrangements of 8 pseudolines do satisfy his combinatorial condition. This result

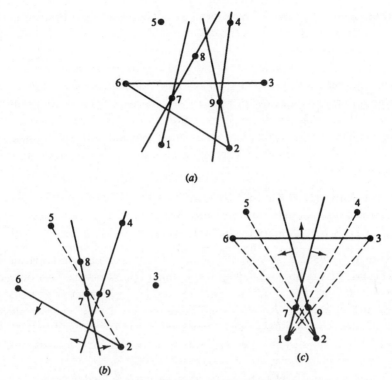

(a)

(b)

(c)

Figure 8.2.1: The realizable uniform oriented matroid
Ric(9) without reduction sequence.

implies in particular that every uniform rank 3 oriented matroid with up to 8 points is realizable. Part (1), which is an earlier result of Goodman and Pollack (1980b), states that in this assertion the requirement "uniform" can be dropped. Their proof uses the technique of allowable sequences (cf. Sections 1.10, 6.3 and 6.4) to show that all arrangements of eight pseudolines in the projective plane are stretchable. It appears that either method can be generalized to prove Theorem 8.2.4 (2) also for the non-uniform case, but this has not yet been verified.

Recall from Section 7.8 that two distinct points e and f of an oriented matroid $\mathcal{M}$ are *sign-invariant* if e and f have always the same sign or always an opposite sign in all signed circuits of $\mathcal{M}$ containing them. The graph $\mathrm{IG}(\mathcal{M})$ on E whose edges are the sign-invariant pairs of $\mathcal{M}$ is called the *inseparability graph* of $\mathcal{M}$.

8.2.5 Corollary (Roudneff 1987a). *Let $\{e, f\}$ be a sign-invariant pair in a rank 3 oriented matroid $\mathcal{M}$. Then $\mathcal{M}$ is reducible by e.*

Proof. We have seen in Section 7.8 that every lexicographic extension induces a sign-invariant pair. It turns out that the converse is true in rank 3. Consider the extension of $(\mathcal{M}/f)\backslash e$ by e. This extension in rank 2 is clearly lexicographic, say, $e = [e_1^{\sigma_1}, e_2^{\sigma_2}]$. If $\{e, f\}$ is contravariant, then $e = [f^+, e_1^{\sigma_1}, e_2^{\sigma_2}]$ is the (rank 3) extension of $\mathcal{M}\backslash e$ by e, and, if $\{e, f\}$ is covariant, then $e = [f^-, e_1^{\sigma_1}, e_2^{\sigma_2}]$ is the extension of $\mathcal{M}\setminus e$ by e. Thus Corollary 8.2.5 follows from Proposition 8.2.2. $\qquad\square$

Note that a sign-invariant pair of $\mathcal{M}$ is also sign-invariant in every minor of $\mathcal{M}$ that contains it. We proved in Proposition 7.8.2 that the inseparability graph of a uniform oriented matroid of rank $r \geq 2$ is either a cycle or a disjoint union of chains. Using this result and Theorem 8.2.4 we get a corollary which links certain combinatorial invariants of an oriented matroid with the topology of its realization space. It would be extremely interesting to find more and deeper results of this kind.

8.2.6 Corollary (Richter and Sturmfels 1991). *Let $\mathcal{M}$ be a rank 3 uniform oriented matroid with n points which has at least $n-8$ sign-invariant pairs. Then $\mathcal{R}(\mathcal{M})$ is a non-empty contractible space.*

The union of matroids is a typical realizability preserving construction in matroid theory. An oriented analogue to the union operation has been introduced in Section 7.6. Let $\mathcal{M}_1 = (E, \chi_1)$ and $\mathcal{M}_2 = (E, \chi_2)$ be oriented matroids of rank r_1 and r_2 on $E = E_n$ such that its union $\mathcal{M}_1 \cup \mathcal{M}_2 = (E, \chi)$ has rank $r := r_1 + r_2 \leq n$. Recall that an element $\alpha \in \binom{E}{r}$ is a basis of $\mathcal{M}_1 \cup \mathcal{M}_2$ if and only if it is the disjoint union of a basis β of $\mathcal{M}_1$ and a basis γ of $\mathcal{M}_2$. In this case we choose the lexicographically earliest basis β of $\mathcal{M}_1$ such that $\gamma = \alpha \setminus \beta$ is a basis of $\mathcal{M}_2$, and using Proposition 7.6.4 we find that

$$\chi(\alpha) \quad = \quad \mathrm{sign}(\beta,\gamma) \cdot \chi_1(\beta) \cdot \chi_2(\gamma),$$

where $\mathrm{sign}(\beta,\gamma)$ denotes the sign of the permutation $(\beta_1 \ldots \beta_{r_1} \gamma_1 \ldots \gamma_{r_2})$ of the sorted string $(\alpha_1, \alpha_2, \ldots, \alpha_r)$.

8.2.7 Proposition (Lawrence and Weinberg 1980). *The union $\mathcal{M}_1 \cup \mathcal{M}_2$ of two realizable oriented matroids $\mathcal{M}_1$ and $\mathcal{M}_2$ is realizable.*

Proof. We define a continuous *union map*

$$\omega \ : \ \mathcal{R}(\mathcal{M}_1) \times \mathcal{R}(\mathcal{M}_2) \quad \rightarrow \quad \mathcal{R}(\mathcal{M}_1 \cup \mathcal{M}_2)$$

as follows. Let Ξ_1 and Ξ_2 be realizations of $\mathcal{M}_1$ and $\mathcal{M}_2$ respectively. We think of Ξ_1 as a linear equivalence class of $n \times r_1$-matrices $(x_1, x_2, \ldots, x_n)$, and similarly Ξ_2 is represented by an $n \times r_2$-matrix $(y_1, y_2, \ldots, y_n)$.

Decompose $\mathbb{R}^r = \mathbb{R}^{r_1} \oplus \mathbb{R}^{r_2}$. The image of Ξ_1 and Ξ_2 under the union map is then defined by the $r \times n$-matrix

$$\omega(\Xi_1, \Xi_2) := (x_1 \oplus \epsilon^{n-1} y_1, \epsilon x_2 \oplus \epsilon^{n-2} y_2, \epsilon^2 x_3 \oplus \epsilon^{n-3} y_3, \ldots, \epsilon^{n-1} x_n \oplus y_n),$$

with ϵ yet to be specified. We can compute the maximal minors of $\omega(\Xi_1, \Xi_2)$ by Laplace expansion with respect to the first r_1 rows. For sufficiently small $\epsilon > 0$ the sign of each minor is given by the chirotope χ. We can now choose such a "sufficiently small" function $\epsilon = \epsilon(\Xi_1, \Xi_2)$ which depends continuously on the realizations Ξ_1 and Ξ_2. Moreover, we require that $\epsilon = \epsilon(\Xi_1, \Xi_2) = 1$ whenever this is small enough. Then ω is a well-defined continuous map into $\mathcal{R}(\mathcal{M}_1 \cup \mathcal{M}_2)$. $\qquad\square$

If the oriented matroids $\mathcal{M}_1 = (E, \chi_1)$ and $\mathcal{M}_2 = (E, \chi_2)$ are uniform, then their union $\mathcal{M}_1 \cup \mathcal{M}_2 = (E, \chi)$ is uniform too. The chirotope of the union is characterized by the equation

$$\chi(i_1, i_2, \ldots, i_{r_1}, j_1, j_2, \ldots, j_{r_2}) \quad = \quad \chi_1(i_1, i_2, \ldots, i_{r_1}) \cdot \chi_2(j_1, j_2, \ldots, j_{r_2})$$

for all $1 \leq i_1 < i_2 < \ldots < i_{r_1} < j_1 < j_2 < \ldots < j_{r_2} \leq n$. In the uniform case Proposition 8.2.7 can be strengthened substantially, and we obtain the following topological structure theorem for realization spaces under the union operation. We write $\overline{\beta} = \{1, 2, \ldots, r_1\}$ for the (lexicographically) first basis of $\mathcal{M}_1$ and $\overline{\gamma} = \{n - r_2 + 1, n - r_2 + 2, \ldots, n\}$ for the last basis of $\mathcal{M}_2$.

8.2.8 Proposition. *Suppose that $\mathcal{M}_1$ and $\mathcal{M}_2$ are uniform oriented matroids as above. Then we have the homotopy equivalence of realization spaces*

$$\mathcal{R}(\mathcal{M}_1 \cup \mathcal{M}_2) \quad \simeq \quad \mathcal{R}(\mathcal{M}_1 \setminus \overline{\gamma}) \times \mathcal{R}(\mathcal{M}_2 \setminus \overline{\beta}).$$

Proof. Consider the union map

$$\omega \; : \; \mathcal{R}(\mathcal{M}_1) \times \mathcal{R}(\mathcal{M}_2) \quad \to \quad \mathcal{R}(\mathcal{M}_1 \cup \mathcal{M}_2)$$

which was defined in the proof of Proposition 8.2.7. It follows from our construction that the image $\omega(\Xi_1, \Xi_2)$ does not depend on the vectors $\{x_i : i \in \overline{\gamma}\}$ or on the vectors $\{y_j : j \in \overline{\beta}\}$ at all. This means that the union map factors through a map

$$\widetilde{\omega} \; : \; \mathcal{R}(\mathcal{M}_1 \setminus \overline{\gamma}) \times \mathcal{R}(\mathcal{M}_2 \setminus \overline{\beta}) \quad \to \quad \mathcal{R}(\mathcal{M}_1 \cup \mathcal{M}_2),$$

i.e., we have $\omega = \widetilde{\omega} \circ (\delta_{\overline{\gamma}} \times \delta_{\overline{\beta}})$. It is our goal to prove that $\widetilde{\omega}$ is a homotopy equivalence.

To this end we first show that the oriented matroids $(\mathcal{M}_1 \cup \mathcal{M}_2)/\overline{\gamma}$ and $\mathcal{M}_1 \setminus \overline{\gamma}$ are equal. Let $\beta \in \binom{E_n}{r_1}$ such that $\beta \cap \overline{\gamma} = \emptyset$. Since $\overline{\gamma}$ is the last basis of $\mathcal{M}_2$, we have $\chi_1(\beta) = \chi(\beta \cup \overline{\gamma})/\chi_2(\overline{\gamma})$. The global ± 1 which is the orientation of the fixed basis $\overline{\gamma}$ of $\mathcal{M}_2$ may be disregarded, and we conclude that the chirotopes $\chi_1(\beta)$ and $\chi(\beta \cup \overline{\gamma})$, the latter viewed as a function of β, define the same oriented matroid. With a similar argument we find that the oriented matroids $(\mathcal{M}_1 \cup \mathcal{M}_2)/\overline{\beta}$ and $\mathcal{M}_2 \setminus \overline{\beta}$ are equal. This implies that the contraction map

$$\kappa_{\overline{\gamma}} \times \kappa_{\overline{\beta}} : \mathcal{R}(\mathcal{M}_1 \cup \mathcal{M}_2) \to \mathcal{R}((\mathcal{M}_1 \cup \mathcal{M}_2)/\overline{\gamma}) \times \mathcal{R}((\mathcal{M}_1 \cup \mathcal{M}_2)/\overline{\beta})$$

$$= \quad \mathcal{R}(\mathcal{M}_1 \setminus \overline{\gamma}) \times \mathcal{R}(\mathcal{M}_2 \setminus \overline{\beta})$$

takes us back to the product space we are interested in.

We now consider the composition $\widetilde{\omega} \circ (\kappa_{\overline{\gamma}} \times \kappa_{\overline{\beta}})$ which maps the realization space of $\mathcal{M}_1 \cup \mathcal{M}_2$ into itself. For every $\Xi \in \mathcal{R}(\mathcal{M}_1 \cup \mathcal{M}_2)$ the realizations $\Xi_1 = \kappa_{\overline{\gamma}}(\Xi)$ of $\mathcal{R}(\mathcal{M}_1 \setminus \overline{\gamma})$ and $\Xi_2 = \kappa_{\overline{\beta}}(\Xi)$ of $\mathcal{R}(\mathcal{M}_2 \setminus \overline{\beta})$ can be concatenated using the choice $\epsilon = 1$. We find $\widetilde{\omega}(\Xi_1, \Xi_2) = \Xi$ which means that $\widetilde{\omega} \circ (\kappa_{\overline{\gamma}} \times \kappa_{\overline{\beta}})$ is the identity on $\mathcal{R}(\mathcal{M}_1 \cup \mathcal{M}_2)$.

Conversely, the map $(\kappa_{\overline{\gamma}} \times \kappa_{\overline{\beta}}) \circ \widetilde{\omega}$ takes realizations $\Xi_1 = \{x_i\}$ of $\mathcal{R}(\mathcal{M}_1 \setminus \overline{\gamma})$ and $\Xi_2 = \{y_j\}$ of $\mathcal{R}(\mathcal{M}_2 \setminus \overline{\beta})$, and it only multiplies all coordinate vectors x_i and y_j by certain scalars ϵ^i which depend continuously on x_i and y_j. This scaling operation is clearly homotopic to the identity map on $\mathcal{R}(\mathcal{M}_1 \setminus \overline{\gamma}) \times \mathcal{R}(\mathcal{M}_2 \setminus \overline{\beta})$. This completes the proof of Proposition 8.2.8.   □

In Section 7.6 we considered the class Γ of oriented matroids which can be obtained by taking the union of uniform rank 1 oriented matroids on a fixed totally ordered ground set. By Proposition 7.6.7 due to Lawrence (1984b), the class Γ is closed under taking minors and forming duals. From Corollary 8.2.3 and Proposition 8.2.8 we obtain the following result.

8.2.9 Proposition.   *Let $\mathcal{M}$ be an oriented matroid in the class Γ. Then its realization space $\mathcal{R}(\mathcal{M})$ is non-empty and contractible.*

Every rank r oriented matroid $\mathcal{M} = (\chi, E_n)$ in the class Γ can be represented as an $n \times r$-matrix (a_{ij}) with entries from $\{-1, +1\}$. In this representation we have $\chi(i_1, i_2, \ldots, i_r) = a_{i_1 1} a_{i_2 2} \cdots a_{i_r r}$ for all $1 \leq i_1 < i_2 < \ldots < i_r \leq n$.

The constant $n \times r$-matrix $(a_{ij}) = (+1)$ corresponds to the *alternating matroid* $C^{n,r}$ all of whose sorted bases are positively oriented. This oriented matroid is of particular interest for the theory of convex polytopes because it corresponds to the cyclic $(r-1)$-polytope with n vertices (cf. Section 9.4).

8.2.10 Corollary.   *The realization space $\mathcal{R}(C^{n,r})$ of the alternating rank r oriented matroid $C^{n,r}$ on n elements is non-empty and contractible.*

8.3 The impossibility of a finite excluded minor characterization

This section deals with non-realizability results for oriented matroids. We start out with the observation that classical matroid theory can be used to construct certain basic examples of non-realizable oriented matroids. The non-Pappus matroid and the Vámos matroid are non-representable matroids which are orientable. Such an oriented version of the non-Pappus matroid is a non-realizable rank 3 oriented matroid on 9 points; similarly, we get a non-realizable rank 4 oriented matroid on 8 points from the Vámos matroid. These non-realizable oriented matroids can be extended to higher dimensions using duality and lexicographic extensions. Proposition 8.2.2 then implies the following.

8.3.1 Proposition.   *There exist non-realizable rank r oriented matroids on n points whenever $r = 3$ and $n \geq 9$, or $r \geq 4$ and $n \geq r + 4$.*

At this point we encounter an important difference between classical matroid theory and oriented matroid theory. The former is the *"theory of deviations from general position in geometry"* (A. Dress), and therefore uniform matroids are not particularly interesting in the classical theory. While uniform matroids are trivially coordinatizable over infinite fields, this is far from being true for uniform oriented matroids.

8.3.2 Proposition.   *There exist non-realizable* **uniform** *rank r oriented matroids on n points whenever $r = 3$ and $n \geq 9$, or $r \geq 4$ and $n \geq r + 4$.*

Proof. We take the oriented non-Pappus matroid, and we perturb it to a uniform oriented matroid $\mathrm{Rin}(9)$ which is still not realizable. The resulting uniform version of the non-Pappus matroid is depicted in Figure 8.3.1. It corresponds to the well-known non-stretchable arrangement due to Ringel (Ringel 1956, Grünbaum 1972).

We now present a very short algebraic non-realizability proof for the oriented matroid $\mathrm{Rin}(9)$. Let $[ijk]$ denote the 3×3-subdeterminant with column indices (i, j, k) of a generic 3×9-matrix $\mathbf{X} = (x_{ij})_{1 \leq i \leq 3, 1 \leq j \leq 9}$.

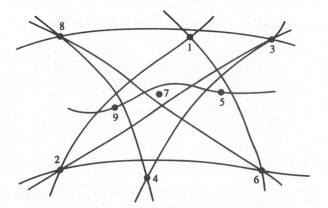

Figure 8.3.1: Ringel's non-realizable uniform
oriented matroid $\mathrm{Rin}(9)$.

The polynomials $[ijk]$ satisfy the following algebraic dependence

$$[246][184][175][437][197] \; + \; [129][184][175][437][467] \; +$$
$$[138][194][247][175][467] \; + \; [156][184][247][437][197] \; +$$
$$[345][184][247][176][197] \; + \; [489][247][175][176][143] \; +$$
$$[597][247][184][176][143] \; + \; [678][247][175][194][143] \; +$$
$$[237][194][184][175][467] \; = \; 0.$$

There are two computational methods for verifying that this bracket polynomial is indeed a *syzygy*. We can either plug in algebraically independent indeterminates x_{ij} and then check whether the resulting polynomial expands to zero, or we can use the *straightening algorithm* of classical invariant theory. For an exposition of the straightening algorithm and its connection with Buchberger's Gröbner basis method we refer to Sturmfels and White (1989).

The crucial point of the above syzygy is that all occurring ordered triples are positively oriented in Figure 8.3.1. For instance, we can see in this diagram that

$$\mathrm{Rin}(9): \; (2,4,6), (1,8,4), (1,7,5), (4,3,7), (1,9,7),$$
$$(1,2,9), (4,6,7), (1,3,8), \ldots \mapsto +1.$$

If we were able to specialize the generic matrix $\mathbf{X} = (x_{ij})$ to a realization of Rin(9), then all nine summands in the above syzygy would be positive. Since zero cannot be the sum of nine positive numbers, there exists no realization of Rin(9). This algebraic technique of proving non-realizability by exhibiting a single *final polynomial* will be explained in more detail in Section 8.5.

Applying a similar perturbation to the oriented Vámos matroid, we obtain the non-realizable oriented matroid RS(8) which is discussed and proved to be non-realizable in Section 1.5. Using duality and lexicographic extensions, Proposition 8.3.2 follows from the results of the previous section.   □

We summarize the known general results on realizability of oriented matroids of low dimension or low codimension.

8.3.3 Corollary.   *All oriented matroids $\mathcal{M}$ of rank r on E are realizable if and only if*

(i) $r \leq 2$,

(ii) $r = 3$ and $|E| \leq 8$,

(iii) $r = 4$ and $|E| \leq 7$,

(iv) $r = 5$ and $|E| \leq 8$,   or

(v) $r \geq 6$ and $|E| \leq r + 2$.

In all other cases, there exist non-realizable uniform oriented matroids.

Corollary 8.3.3 tells us that the smallest non-realizable uniform oriented matroids have the (r, n)-parameters $(3, 9)$ and $(4, 8)$. But there is a remarkable difference between these two parameters. For $(3, 9)$ Ringel's oriented matroid Rin(9) is the only (up to reorientation) non-realizable uniform oriented matroid, while for $(4, 8)$ there exist several non-realizable ones. A complete classification of all uniform oriented matroids with these parameters has been obtained by Richter and Bokowski. See page 486 of the appendix for the case $(3, 10)$.

8.3.4 Theorem (Richter 1988, Bokowski and Richter-Gebert 1990b).

(1)   *There are precisely 4382 (reorientation classes of) uniform rank 3 oriented matroids with 9 points. All these oriented matroids except Rin(9) are realizable.*

(2)   *There are precisely 2628 (reorientation classes of) uniform rank 4 oriented matroids with 8 points. Including RS(8) and EFM(8), precisely 24 of these oriented matroids are non-realizable.*

The class of realizable oriented matroids is clearly closed under taking minors. Whenever one encounters such a hereditary class of combinatorial objects, it is natural to ask whether this class has only a finite number of obstructions.

(∗)   *Are there oriented matroids $\mathcal{M}_1, \mathcal{M}_2, \ldots, \mathcal{M}_k$ such that any given oriented matroid $\mathcal{M}$ is non-realizable if and only if some $\mathcal{M}_i$ is a minor of $\mathcal{M}$?*

We recall that finite excluded minor characterizations are known for graphic, cographic, regular, binary and ternary matroids. These results, which are due to Tutte, Seymour, Reid and Bixby, rank among the outstanding achievements of classical matroid theory. Unfortunately these beautiful characterizations often do not apply to oriented matroids because an ordered field necessarily has characteristic 0. Under this perspective it becomes noteworthy that classical matroid theorists have almost exclusively been interested in finite characteristic p, and that only few matroid theory papers deal with coordinatizations over characteristic 0.

One important exception has the striking title *"The missing axiom of matroid theory is lost forever"*. In this paper Vámos (1978) shows that unlike in characteristic 2 and 3 there exists no finite excluded minor characterization for matroids which are realizable over fields of characteristic 0. One way of proving Vámos's theorem consists in constructing a suitable infinite sequence of plane incidence theorems, e.g. multiple copies of the Pappus or Desargues configuration. If the last incidence in each configuration is broken, then we obtain an infinite family of minor-minimally non-realizable matroids. In contrast to Vámos's original proof this approach generalizes immediately to oriented matroids, and hence the answer to our question (∗) is "no".

The finite excluded minor problem is more difficult for uniform oriented matroids. In fact, it had been conjectured by Grünbaum (1967) that stretchability of simple line arrangements in the projective plane can be characterized by excluding finitely many subarrangements. The following theorem shows that this conjecture is false.

8.3.5 Theorem (Haiman and Kahn; Bokowski and Sturmfels 1989a).
There does not exist a finite set $\{\mathcal{M}_1, \mathcal{M}_2, \ldots, \mathcal{M}_k\}$ of rank 3 uniform oriented matroids such that any given rank 3 uniform oriented matroid $\mathcal{M}$ is non-realizable if and only if some $\mathcal{M}_i$ is a minor of $\mathcal{M}$.

Theorem 8.3.5 was found by M. Haiman and J. Kahn (unpublished, around 1982/83; see Figure 8.3.3) and independently by Bokowski and Sturmfels (in 1984; see Figure 8.3.2), based on an example of Goodman and Pollack (1980a). An infinite family of minor-minimal non-realizable oriented matroids of increasing ranks had been constructed earlier by Edmonds and Mandel (1982).

8.3.6 Lemma. *For every $l \geq 5$ there exists a non-realizable uniform rank 3 oriented matroid B_3^{2l} on $2l$ points $E = \{1, 2, \ldots, l,\ 1', 2', \ldots, l'\}$ such that $B_3^{2l} \backslash e$ is realizable for every $e \in E$.*

The smallest member B_3^{10} of the family $\{B_3^{2l}\}_{l \geq 5}$ of minor-minimal non-realizable uniform oriented matroids is depicted in Figure 8.3.2.

In general, B_3^{2l} is defined as follows. Given a regular l-gon with counter-clockwise oriented vertices $1, 2, \ldots, l$ in the Euclidean plane, consider the parallel halflines $\overrightarrow{i, i+3}$ and $\overrightarrow{i+1, i+2}$ where $i = 1, \ldots, l$. (All labels are understood modulo l.) For every $i = 1, 2, \ldots, l$ we insert a new point i' with the following properties.

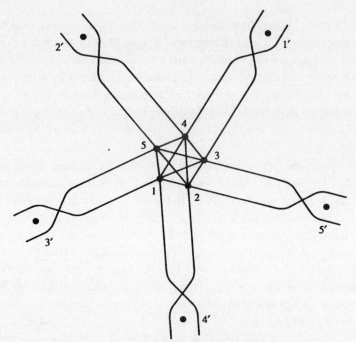

Figure 8.3.2: The non-realizable oriented matroid B_3^{10}.
Every minor of B_3^{10} is realizable.

(i) i' lies to the left of $\overrightarrow{i, i+3}$, i.e., $B_3^{2l}(i, i+3, i') := +1$.

(ii) i' lies to the right of $\overrightarrow{i+1, i+2}$, i.e., $B_3^{2l}(i+1, i+2, i') := -1$.

(iii) In all other triangle orientations i' is identified with the intersection point
       (at infinity) of $\overrightarrow{i, i+3}$ and $\overrightarrow{i+1, i+2}$.

These requirements uniquely define the oriented matroid B_3^{2l}.

We close this section with a description of the construction due to Haiman
and Kahn. Figure 8.3.3 shows a typical example of their infinite family. The
points are the elements of this rank 3 oriented matroid, and the curved lines
indicate the orientation of triples that would otherwise appear collinear. Triples
lying along one side of the lines A, B, or C are to be oriented "arbitrarily",
i.e., by inductively selecting general positions for these points very near the lines
in question. With these choices Figure 8.3.3 defines a uniform rank 3 oriented
matroid.

Now suppose the oriented matroid in Figure 8.3.3 were realizable. After a
suitable projective transformation, we may assume without loss of generality that
the lines A and C are parallel in the vertical direction, and that the lines D and E
are parallel in the horizontal direction. In this affine realization the line B must
have negative slope. Inductively we find that, for any pair of curved segments
such as U and V, the y-coordinate of the intersection point $U \cap C$ is larger than
the y-coordinate of the intersection point $V \cap A$. But this implies that the
line F has positive slope, which is impossible since E and D are horizontal.

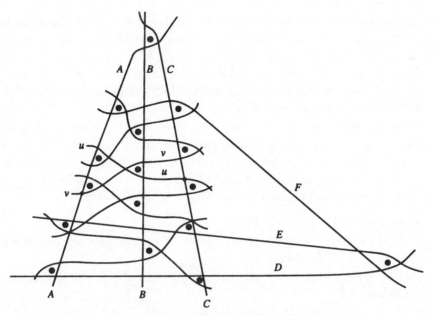

Figure 8.3.3: A minor-minimally non-realizable uniform
rank 3 oriented matroid.

Deleting any one point in this oriented matroid "frees up" the whole picture,
resulting in a realizable configuration.

8.4 Algorithms and complexity results

The results of the previous section imply that we cannot expect to find an
easy combinatorial characterization of realizable oriented matroids. This insight
suggests that we study realizability not only as a structural problem but also as
a computational problem.

In this section the computational realizability problem of oriented matroids is
considered for a general ordered field, with a particular emphasis on *real closed
fields* such as the real numbers. For an introduction to ordered fields and real
closed fields see e.g. (Jacobson 1964, Chapter VI). We will also sketch the *solvability sequence* method due to Bokowski, and we conclude with a complexity
upper bound for the realizability problem.

We have seen in Section 8.1 that the realization space $\mathcal{R}(\mathcal{M})$ of a rank r
oriented matroid on n points is the solution set of a system of $\binom{n}{r} - 1$ equations
and inequalities in $(n - r)r$ variables. Let K be an ordered subfield of the real
numbers. We write $\mathcal{R}_K(\mathcal{M})$ for the subset of K-rational points in $\mathcal{R}(\mathcal{M})$. It
is of course possible that an oriented matroid $\mathcal{M}$ is realizable (i.e. $\mathcal{R}(\mathcal{M}) \neq \emptyset$)
but not K-realizable (i.e. $\mathcal{R}_K(\mathcal{M}) = \emptyset$).

Our starting point is the question whether there exists at all an algorithm for
deciding whether a given oriented matroid $\mathcal{M}$ is realizable. The answer is "yes"
for realizability over the real numbers. By Tarski's famous theorem on quantifier

elimination for real closed fields (Tarski 1951), there exists a decision procedure
for systems of equations and inequalities over real closed fields. Tarski's theorem
also implies the following result on the real closure of an ordered field. Let $\mathcal{I}$ be
a system of polynomial equations and inequalities with coefficients in an ordered
field K. If $\mathcal{I}$ has a solution over some ordered field extension of K, then $\mathcal{I}$ has
a solution over the real closure of K.

As the rationals are contained in every ordered field, we get the following
universality property for the field $\mathbf{A}$ of real algebraic numbers which is the real
closure of $\mathbb{Q}$. If a system $\mathcal{I}$ of equations and inequalities with rational coefficients
has a solution in some arbitrary ordered field, then $\mathcal{I}$ has a solution in $\mathbf{A}$, too.
This has the following implications for oriented matroid theory.

Proposition 8.4.1 (Mandel 1978).

(1)   *There exists an algorithm which decides for an arbitrary oriented matroid
      $\mathcal{M}$ whether $\mathcal{M}$ is realizable.*

(2)   *Every realizable oriented matroid can be realized over the real algebraic
      numbers $\mathbf{A}$.*

The situation is entirely different if we turn our attention to solutions in
the field $\mathbb{Q}$ of *rational* numbers. In view of Matiyasevic's negative solution to
Hilbert's tenth problem in 1971, B. Grünbaum (1972) has conjectured that there
is no algorithm for enumerating all (isomorphism types of) line arrangements in
the *rational* projective plane. Matiyasevic's result "there exists no algorithm
to decide whether a system of diophantine equations has a solution among the
integers" does not imply Grünbaum's conjecture, and the corresponding problem
for *rational* numbers is still open (Mazur 1986). However, under the assumption that the analogue of Matiyasevic's theorem holds also for rational numbers,
we can conclude that rational realizability of oriented matroids is recursively
undecidable.

8.4.2 Proposition. (Sturmfels 1987e)   *The following three statements are
equivalent.*

(1)   *There exists an algorithm to decide for an arbitrary multivariate polynomial
      $f \in \mathbf{Z}[x_1, \ldots, x_n]$ whether f has zeros with coordinates in the field $\mathbb{Q}$ of
      rational numbers.*

(2)   *There exists an algorithm to decide for an arbitrary matroid G whether G
      is coordinatizable over $\mathbb{Q}$.*

(3)   *There exists an algorithm to decide for an arbitrary oriented matroid $\mathcal{M}$
      whether $\mathcal{M}$ is realizable over $\mathbb{Q}$.*

The proof of Proposition 8.4.2 is published in Bokowski and Sturmfels (1989b,
Chapter 2). Here the basic idea is to use the classical technique of projective
addition and multiplication. This technique is described and also applied to
matroid coordinatizations in Section 1.7 of White (1987). Extending a result of
MacLane (1936), White proves that every algebraic number, or, equivalently, every univariate polynomial over $\mathbf{Z}$, can be encoded into a suitable rank 3 matroid.

This method generalizes to oriented matroids and to polynomials in more than one variable. Hence arbitrary multivariate polynomials with integer coefficients can be encoded in suitable rank 3 oriented matroids. A careful analysis of this encoding procedure yields the proof of Proposition 8.4.2.

From the above argument we find that for every finite algebraic ordered field extension K of the rationals there exists an oriented matroid $\mathcal{M}$ such that $\mathcal{R}_K(\mathcal{M}) \neq \emptyset$ but $\mathcal{R}_\mathbb{Q}(\mathcal{M}) = \emptyset$. In particular, we get an infinite family of realizable oriented matroids which are not realizable over the rationals. A smallest such example, due to Grünbaum and Perles (see Grünbaum 1967, Fig. 5.5.2), is obtained as follows.

Consider a regular pentagon with counter-clockwise oriented vertices x_1, x_2, x_3, x_4, x_5 in the Euclidean plane. Apply a projective transformation such that the following four intersection points are moved from the line at infinity to a line in the affine plane: $x_6 := (x_1 \vee x_4) \wedge (x_2 \vee x_3)$, $x_7 := (x_2 \vee x_5) \wedge (x_3 \vee x_4)$, $x_8 := (x_3 \vee x_1) \wedge (x_4 \vee x_5)$, $x_9 := (x_4 \vee x_2) \wedge (x_5 \vee x_1)$. Let GP(9) denote the oriented matroid of affine dependences on $\{x_1, x_2, \ldots, x_9\}$. Using elementary algebra it can be seen that GP(9) (or its underlying matroid) is realizable over a subfield K of the reals if and only if $\sqrt{5} \in K$.

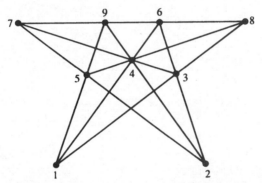

Figure 8.4.1: The non-rational oriented matroid GP(9).

A practical algorithm for solving the realizability problem for small parameters r, n and other special cases has been suggested by Bokowski. We outline the basic ideas of the method of *solvability sequences* as introduced in Bokowski and Sturmfels (1986a). Here we restrict ourselves to the case of uniform oriented matroids; however, most ideas generalize to all oriented matroids.

Let $\mathcal{M} = (E, \chi)$ be a rank r uniform oriented matroid on n points, and suppose that the basis $\beta := (\beta_1, \beta_2, \ldots, \beta_r) \in E^r$ is positively oriented in $\mathcal{M}$. The realization space of $\mathcal{M}$ equals the space of real $r \times n$-matrices $\mathbf{X} := (x_{kl})$ with $x_{i,\beta_j} = \delta_{ij}$ (Kronecker delta) and such that the maximal minors $[i_1 i_2 \ldots i_r]$ of $\mathbf{X}$ have their signs prescribed by χ. Using the identity $x_{ij} = [\beta_1 \ldots \beta_{i-1} j \beta_{i+1} \ldots \beta_r]$, the coordinate functions x_{ij} on the realization space $\mathcal{R}(\mathcal{M})$ can be thought of as brackets, too.

Let $(\bar{x}_1, \bar{x}_2, \ldots, \bar{x}_{r(n-r)})$ be a total order on the set $\{x_{ij}\}$ of variables, and consider all brackets as polynomial functions

$$[i_1 i_2 \ldots i_r](\bar{x}_1, \ldots, \bar{x}_{r(n-r)}).$$

Fix an index $t \in \{1, 2, \ldots, r(n-r)\}$ and consider the subset of those brackets which depend only on the first t variables $\bar{x}_1, \ldots, \bar{x}_t$. The corresponding open semialgebraic subset of $\mathbb{R}^t$

$$\{(\eta_1, \ldots, \eta_t) \in \mathbb{R}^t \mid$$
$$\forall [i_1 \ldots i_r] \in \mathbb{R}[\bar{x}_1, \ldots, \bar{x}_t] : \operatorname{sign}[i_1 \ldots i_r](\eta_1, \ldots, \eta_t) = \chi(i_1, \ldots, i_r)\}$$

is denoted by $\mathcal{R}_t$. The variable order $(\bar{x}_1, \bar{x}_2, \ldots, \bar{x}_{r(n-r)})$ is said to be a *solvability sequence* for $\mathcal{M}$ if every map in the sequence of projections

$$\mathcal{R}(\mathcal{M}) = \mathcal{R}_{r(n-r)} \rightarrow \mathcal{R}_{r(n-r)-1} \rightarrow \mathcal{R}_{r(n-r)-2}$$
$$\rightarrow \quad \ldots \quad \rightarrow \mathcal{R}_3 \rightarrow \mathcal{R}_2 \rightarrow \mathcal{R}_1$$

is surjective. The existence of a solvability sequence is connected with the topology of the realization space.

8.4.3 Proposition (Bokowski and Sturmfels 1986a). *Let $\mathcal{M}$ be a uniform oriented matroid which has a solvability sequence. Then its realization space $\mathcal{R}(\mathcal{M})$ is non-empty and contractible.*

The projection $\mathcal{R}_t \rightarrow \mathcal{R}_{t-1}$ corresponds algebraically to eliminating the variable $\bar{x}_t$. Since all defining polynomials are linear in $\bar{x}_t$, the fibers of the map $\mathcal{R}_t \rightarrow \mathcal{R}_{t-1}$ are either open line segments or empty. The proof of Proposition 8.4.3 follows with standard topological arguments similar to the ones used in the proof of Lemma 8.2.1.

Now the real problem consists in finding good combinatorial criteria which are sufficient to guarantee the existence of a solvability sequence for a uniform oriented matroid $\mathcal{M}$. Several such conditions, stated in the form of a "greedy-type" algorithm, have been found by Bokowski and Sturmfels (1986a). Their solvability sequence algorithm has been used to find coordinates for a large class of oriented matroids arising from the study of triangulated spheres (cf. Section 9.5). A subclass of oriented matroids for which this algorithm works has been named *max-realizable* in Bokowski and Sturmfels (1986a, Example 5.1). It can be seen that a uniform oriented matroid is max-realizable if it is in the class Γ, but not conversely. Therefore Proposition 8.2.9 can be considered a special case of Proposition 8.4.3.

We next discuss complexity upper bounds for the realizability problem of oriented matroids. Here a rank r oriented matroid $\mathcal{M} = (E_n, \chi)$ is assumed to be encoded by (lexicographically) listing the function values of its chirotope. This means that $\mathcal{M}$ can be encoded in $\lceil \log_2(3) \rceil \cdot \binom{n}{r} = O(n^r)$ bits.

Realizability problem for oriented matroids – decision version (ROM)

- **Input:** An oriented matroid $\mathcal{M}$, encoded as above.
- **Output:** "YES" if $\mathcal{M}$ is realizable over the real numbers, "NO" otherwise.

This problem reduces to the following one.

Existential theory of the reals (ETR)

- **Input:** Polynomials

$$f_1, f_2, \ldots, f_r, g_1, g_2, \ldots, g_s, h_1, h_2, \ldots, h_t \in \mathbf{Z}[x_1, x_2, \ldots, x_m]$$

whose total degree is bounded above by D and whose coefficients are bounded above (in their absolute value) by A.

- **Output:** "YES" if the polynomial system

$$f_1 = f_2 = \ldots = f_r = 0, g_1 \geq 0, g_2 \geq 0, \ldots, g_s \geq 0, h_1 > 0, h_2 > 0, \ldots, h_t > 0$$

has a solution over the real (algebraic) numbers, "NO" otherwise.

Tarski's original decision procedure for ETR based on quantifier elimination does not provide a useful complexity upper bound, and it is certainly not useful for any practical application. An interesting attempt to apply Tarski's procedure to the Steinitz, which is equivalent to ROM by the results of Section 9.5, can be found in Harbecke (1981).

In the 1970s Collins developed the first algorithm for ETR which has a reasonable complexity upper bound. The complexity of Collins's *cylindrical algebraic decomposition* algorithm is doubly exponential in the number of variables (cf. Collins 1975).

Over the years this complexity bound has been substantially improved by several authors. Grigor'ev and Vorobjov (1988) gave an algorithm for ETR which requires only *singly-exponential time*, and Canny (1988) gave an algorithm for ETR which requires only *polynomial space*. Renegar (1989) showed that the singly-exponential time bound and the polynomial space bound can be attained by the same algorithm, and Heintz, Roy and Solerno (1991, 1993) showed that ETR can be solved in *parallel polynomial time*. The following sharpening is due to Basu, Pollack and Roy (1997). See also Section A.5!

8.4.4 Proposition. *The existential theory of the reals (ETR) can be decided in single-exponential time and in polynomial space in the Turing machine model of complexity. More precisely, the number of bit operations can be bounded by* $\log(A) \binom{S}{m}^m S D^{O(m)}$, *for* $S := r + s + t$.

The question of lower bounds is more difficult. We know that ETR is NP-hard because quadratic programming, which is a special case of ETR, is known to be NP-hard (cf. Sahni 1974). On the other hand, it is an open question whether ETR is actually contained in NP. In view of the prominent open problem "$P = NP$?", even the possibility that ETR is decidable in polynomial time cannot be excluded at this moment. Proposition 8.4.4 implies the following result for oriented matroids.

8.4.5 Proposition. *The realizability problem for oriented matroids (ROM) can be decided in singly-exponential time and in polynomial space in the Turing machine model of complexity. More precisely, the number of bit operations is bounded by* $(n\,r)^{O((n-r)r)}$.

The question of lower bounds for ROM will be addressed in Section 8.7. Clearly, the best we can hope for is attaining the known lower bounds for ETR (i.e. NP-hardness). This hope will materialize a little later; we will see in Section 8.7 that the problems ETR and ROM are actually polynomially equivalent.

8.5 Final polynomials and the real Nullstellensatz

In the previous section we discussed the algorithmic version (ROM) of the realizability problem. From a certain practical point of view, this formulation as a decision problem and the resulting complexity bounds are somewhat misleading. Usually we are not only interested in the output "YES" or "NO"; what we really wish to find is either a realizability proof or a non-realizability proof.

Under this perspective one notices a peculiar asymmetry between the realizable and the non-realizable cases. (Complexity theorists will observe an analogy to the fact that CO-NP and NP are not equal, unless P = NP.) There exists a simple and systematic representation for a realizability proof, namely a sample point in the realization space $\mathcal{R}(\mathcal{M})$. Even the longest and most complicated derivation of a realizability proof can be discarded once we have found some coordinate matrix A. This means that we only need $(n - r)r$ real algebraic numbers to finally encode our proof. In fact, in the case where $\mathcal{M}$ is uniform, these numbers can be rationals because $\mathcal{R}_{\mathbb{Q}}(\mathcal{M})$ is dense in the open semialgebraic set $\mathcal{R}(\mathcal{M})$. Given a coordinate matrix A, it requires only polynomially many arithmetic operations to verify that the maximal minors of A have the correct signs and we really have a solution to the polynomial inequalities defining $\mathcal{R}(\mathcal{M})$.

Is there a similarly canonical way to represent the result of a non-realizability proof ? In linear programming we have a *dual solution* in order to witness the fact that no primal feasible solution exists. Is there there an analogous "dual solution" whenever the non-linear inequalities defining $\mathcal{R}(\mathcal{M})$ are inconsistent ?

The theory of *final polynomials* provides such an algebraic framework for compactly encoding non-realizability proofs. We have seen a first example of a final polynomial in our non-realizability proof for $\mathrm{Rin}(9)$ in Proposition 8.3.2. This method was first introduced by Bokowski for proving the non-existence of certain convex polytopes (cf. Altshuler, Bokowski and Steinberg 1980, Bokowski and Garms 1987, Bokowski and Sturmfels 1987). In this section we will see that final polynomials exist for all non-realizable oriented matroids. The key idea is to apply the duality expressed in the real Nullstellensatz of semialgebraic geometry to the specific semialgebraic sets $\mathcal{R}(\mathcal{M})$. We refer to Becker (1986) for an introduction to semialgebraic geometry. In the second half of this section we discuss a practical algorithm due to Bokowski and Richter (1990a) for computing "biquadratic" final polynomials for a large class of non-realizable oriented matroids.

In order to state the existence theorem for final polynomials of oriented matroids, we need to rephrase the realizability problem in a more abstract algebraic language. Let K be an ordered subfield of the real numbers, and let $K[x_{ij}]$

denote the polynomial ring over K freely generated by the rn entries x_{ij} of a generic $r \times n$-matrix $\mathbf{X} = (x_{ij})$. As before, we abbreviate

$$[i_1 i_2 \ldots i_r] \quad := \quad \det \begin{pmatrix} x_{1i_1} & \cdots & x_{1i_r} \\ x_{2i_1} & \cdots & x_{2i_r} \\ \vdots & \ddots & \vdots \\ x_{ri_1} & \cdots & x_{ri_r} \end{pmatrix},$$

and we write $K[\Lambda(n,r)]$ for the subring of $K[x_{ij}]$ which is generated by all *brackets* $[i_1 i_2 \ldots i_r]$. The ring $K[\Lambda(n,r)]$ is the coordinate ring of the Grassmann variety of r-dimensional linear subspaces in K^n. In other words, the decomposable r-vectors in $\wedge_r K^n$ are in one-to-one correspondence with the K-algebra homomorphisms $\Psi : K[\Lambda(n,r)] \to K$.

Now let $\mathcal{M} = (E, \chi)$ be a rank r oriented matroid on $E = \{1, 2, \ldots, n\}$.

8.5.1 Lemma. *Up to a multiplicative scalar, the K-rational points in the realization space $\mathcal{R}(\mathcal{M})$ are in one-to-one correspondence with the K-algebra homomorphisms $\Psi : K[\Lambda(n,r)] \to K$ such that*

$$\operatorname{sign} \Psi([i_1 i_2 \ldots i_r]) = \chi(i_1, i_2, \ldots, i_r)$$

for all $i_1, i_2, \ldots, i_r \in E$.

We will consider three sets $I_{\mathcal{M}}^K, N_{\mathcal{M}}^K$ and $P_{\mathcal{M}}^K$ of bracket polynomials which are respectively 0, non-negative and positive under every such realization map Ψ. Let $I_{\mathcal{M}}^K$ denote the ideal in $K[\Lambda(n,r)]$ which is generated by

$$\{ [i_1 i_2 \ldots i_r] \in \Lambda(n,r) \quad | \quad \chi(i_1, i_2, \ldots, i_r) = 0 \}.$$

The quotient ring $K[\Lambda(n,r)]/I_{\mathcal{M}}^K$ is the *bracket ring* of the matroid $M = \underline{\mathcal{M}}$ underlying $\mathcal{M}$. It is the coordinate ring of the algebraic variety of weak coordinatizations of $\underline{\mathcal{M}}$, embedded as a subvariety in the Grassmann variety (cf. Section 2.4).

Let $P_{\mathcal{M}}^K$ denote the multiplicative semigroup generated by the positive elements in the ordered field K and the set

$$\{ [i_1 i_2 \ldots i_r] \cdot \chi(i_1, i_2, \ldots, i_r) \quad | \quad 1 \leq i_1 < i_2 < \ldots < i_r \leq n \}$$

of positive brackets and negated negative brackets with respect to $\mathcal{M}$. Finally, define $N_{\mathcal{M}}^K$ to be the quadratic semiring in $K[\Lambda(n,r)]$ which is generated by $P_{\mathcal{M}}^K$ and the set $K[\Lambda(n,r)]^2$ of all squares in the polynomial ring $K[\Lambda(n,r)]$.

An explicit representation of $0 \in K[\Lambda(n,r)]$ as a sum of elements from $I_{\mathcal{M}}^K$, $N_{\mathcal{M}}^K$ and $P_{\mathcal{M}}^K$ is called a *final polynomial* for $\mathcal{M}$. With this definition we have the following result which was found independently by Dress (unpublished) and Sturmfels (1987b).

8.5.2 Theorem. *Let M be an oriented matroid and K an ordered field. Then one and only one of the following statements is true.*

(i) $\mathcal{M}$ *is realizable over some finite algebraic ordered field extension of K.*

(ii) *There exists a final polynomial for $\mathcal{M}$ with coefficients in K.*

As an example for a final polynomial recall our non-realizability proof of the

uniform non-Pappus matroid Rin(9) in Figure 8.3.1. We gave a final polynomial
which expressed

$$0 \in \underbrace{P^{\mathbb{R}}_{\text{Rin}(9)} + \cdots + P^{\mathbb{R}}_{\text{Rin}(9)}}_{9 \text{ times}} \quad \subset \quad P^{\mathbb{R}}_{\text{Rin}(9)} + N^{\mathbb{R}}_{\text{Rin}(9)}.$$

Theorem 8.5.2 states in other words that this proof technique can be applied
to every non-realizable oriented matroid. Using Theorem 8.5.2 we obtain the
following.

8.5.3 Corollary.   *An oriented matroid $\mathcal{M}$ is non-realizable (i.e. $\mathcal{R}(\mathcal{M}) = \emptyset$)
if and only if there exists a final polynomial for $\mathcal{M}$ with integer coefficients.*

The crucial question about final polynomials is how to find them. Unfortunately, the algorithms for solving inequality systems over the reals discussed in
Section 8.4 do not provide a final polynomial in the case of infeasibility. An algorithm which generates final polynomials, providing an effective version of the
real Nullstellensatz, was only recently found by Lombardi (1990).

An interesting and practical method for computing final polynomials for small
non-realizable oriented matroids has been suggested by Bokowski and Richter
(1990a). The basic idea underlying their *biquadratic final polynomials* is the
following. Let $\mathcal{R}(\mathcal{M}) \subset \wedge_r \mathbb{R}^n$ be the realization space of a rank r uniform
oriented matroid $\mathcal{M} = (E_n, \chi)$. With $\mathcal{M}$ we associate a convex polyhedron
$\mathcal{P}(\mathcal{M}) \subset \wedge_r \mathbb{R}^n$, defined by a system of linear inequalities in $\binom{n}{r}$ variables such
that

(i) $\mathcal{P}(\mathcal{M}) = \emptyset$ implies $\mathcal{R}(\mathcal{M}) = \emptyset$, and

(ii) every dual LP-solution, proving "$\mathcal{P}(\mathcal{M}) = \emptyset$", can be transformed into a
      final polynomial for $\mathcal{M}$.

In order to derive the linear programming feasibility problem which defines the
polyhedron $\mathcal{P}(\mathcal{M})$, we consider the orthant $(\wedge_r \mathbb{R}^n)_\chi$ of the chirotope χ. Using
bracket notation for the coordinate functions on $\wedge_r \mathbb{R}^n$, the orthant $(\wedge_r \mathbb{R}^n)_\chi$
consists of those r-vectors for which $[e_1 e_2 \ldots e_r] > 0$ for every positively ordered
basis $(e_1, e_2, \ldots, e_r)$ of $\mathcal{M}$. In the following we suppose that all occurring bases
and brackets are sorted to be positive in $\mathcal{M}$. Then every three-term Grassmann
Plücker relation can be written in the form

$$[\tau, \lambda_1, \lambda_2][\tau, \lambda_3, \lambda_4] + [\tau, \lambda_1, \lambda_4][\tau, \lambda_2, \lambda_3] \quad = \quad [\tau, \lambda_1, \lambda_3][\tau, \lambda_2, \lambda_4] \quad (1)$$

where τ is an ordered $(r-2)$-tuple and all six brackets are positive. Equation
(1) implies the following inequalities:

$$\begin{aligned}
[\tau, \lambda_1, \lambda_2][\tau, \lambda_3, \lambda_4] &< [\tau, \lambda_1, \lambda_3][\tau, \lambda_2, \lambda_4] \quad \text{and} \\
[\tau, \lambda_1, \lambda_4][\tau, \lambda_2, \lambda_3] &< [\tau, \lambda_1, \lambda_3][\tau, \lambda_2, \lambda_4]
\end{aligned} \quad (2)$$

Let $\mathcal{P}'(\mathcal{M})$ denote the solution set in $(\wedge_r \mathbb{R}^n)_\chi$ of the "biquadratic" inequality
system (2), where λ and τ range over all index sets. The realization space
$\mathcal{R}(\mathcal{M})$ is a subset of the semialgebraic variety $\mathcal{P}'(\mathcal{M})$. This implies that $\mathcal{P}'(\mathcal{M})$
being empty is a sufficient condition for $\mathcal{M}$ being non-realizable. Moreover,
using the implication "$(1) \Rightarrow (2)$", we can transform any final polynomial for (2)

into a final polynomial for (1). We will illustrate this transformation in Example 8.5.4.

In order to decide the solvability of (2), we now apply the bijective transformation

$$\log : (\wedge_r \mathbb{R}^n)_\chi \rightarrow \wedge_r \mathbb{R}^n, \quad [e_1, e_2, \ldots, e_r] \mapsto \log[e_1, e_2, \ldots, e_r]$$

and we define $\mathcal{P}(\mathcal{M}) := \log(\mathcal{P}'(\mathcal{M}))$. The convex polyhedron $\mathcal{P}(\mathcal{M})$ is the solution set of the linear system of inequalities

$$\begin{aligned}
[\tau, \lambda_1, \lambda_2] + [\tau, \lambda_3, \lambda_4] &< [\tau, \lambda_1, \lambda_3] + [\tau, \lambda_2, \lambda_4] \quad \text{and} \\
[\tau, \lambda_1, \lambda_4] + [\tau, \lambda_2, \lambda_3] &< [\tau, \lambda_1, \lambda_3] + [\tau, \lambda_2, \lambda_4]
\end{aligned} \tag{3}$$

At this point we can use linear programming to test the feasibility of the system (3). If $\mathcal{P}(\mathcal{M})$ is empty then we obtain a dual LP-solution, i.e., a positive integer linear combination of the left hand sides of (3) which equals the same linear combination of the right hand sides, resulting in the contradiction "$0 < 0$". By exponentiation we get a product of the left hand sides in (2) equal to the same product of the right hand sides. Using the syzygies (1), we transform this equation to an explicit representation, called a *biquadratic final polynomial* for $\mathcal{M}$, of 0 in $P_{\mathcal{M}}^{\mathbb{Q}} + N_{\mathcal{M}}^{\mathbb{Q}} + I_{\mathcal{M}}^{\mathbb{Q}}$.

8.5.4 Example. (A biquadratic final polynomial for RS(8)) Consider the non-realizable uniform rank 4 oriented matroid RS(8) which was discussed in Section 1.5. From the tables in Section 1.5 we can see that all brackets occurring in the following six syzygies are positive:

$$\begin{aligned}
[1234][1256] + [1245][1236] &= [1235][1246] \\
[1234][1357] + [1374][1235] &= [1354][1237] \\
[1234][1485] + [1354][1248] &= [1384][1245] \\
[1234][2376] + [2364][1237] &= [2374][1236] \\
[1234][2468] + [2384][1246] &= [2364][1248] \\
[1234][3478] + [2374][1384] &= [2384][1374]
\end{aligned} \tag{4}$$

This implies the six inequalities

$$\begin{aligned}
[1245][1236] &< [1235][1246] \\
[1374][1235] &< [1354][1237] \\
[1354][1248] &< [1384][1245] \\
[2364][1237] &< [2374][1236] \\
[2384][1246] &< [2364][1248] \\
[2374][1384] &< [2384][1374]
\end{aligned} \tag{5}$$

for the set $\mathcal{P}'(\text{RS}(8))$. Taking logarithms on both sides, we get the following system of linear inequalities for the polyhedron $\mathcal{P}(\mathcal{M}) \subset \wedge_4 \mathbb{R}^8$:

$$
\begin{array}{rcll}
L_1 & = & [1245] + [1236] - [1235] - [1246] & < \quad 0 \\
L_2 & = & [1374] + [1235] - [1354] - [1237] & < \quad 0 \\
L_3 & = & [1354] + [1248] - [1384] - [1245] & < \quad 0 \\
L_4 & = & [2364] + [1237] - [2374] - [1236] & < \quad 0 \\
L_5 & = & [2384] + [1246] - [2364] - [1248] & < \quad 0 \\
L_6 & = & [2374] + [1384] - [2384] - [1374] & < \quad 0
\end{array}
\tag{6}
$$

The system (6) is inconsistent, and by linear programming (or, in this example, by inspection) we get the dual solution

$$
L_1 + L_2 + L_3 + L_4 + L_5 + L_6 \quad = \quad 0. \tag{7}
$$

Exponentiating the identity (7), we find that the corresponding product of the right hand sides of (5) minus the product of the left hand sides of (5) equals 0. Now we replace all right hand sides by the left hand sides in (4). This yields the following polynomial

$$
\begin{aligned}
& ([1234][1256] + [1245][1236]) \cdot ([1234][1357] + [1374][1235]) \cdot \\
& ([1234][1485] + [1354][1248]) \cdot ([1234][2376] + [2364][1237]) \cdot \\
& ([1234][2468] + [2384][1246]) \cdot ([1234][3478] + [2374][1384]) \\
& - [1235][1246]\,[1354][1237]\,[1384][1245]\,[2374][1236]\,[2364][1248]\,[2384][1374]
\end{aligned}
\tag{8}
$$

which is 0 in the bracket ring $K[\Lambda(8,4)]$. Expanding the product cancels the only negative term, leaving a sum of 63 positive terms which, by construction, is 0 in $K[\Lambda(8,4)]$. By definition, this syzygy is a final polynomial for RS(8). This final polynomial can be simplified to a degree 4 final polynomial with 6 terms, isomorphic to the one given in the proof of Proposition 9.4.5.

The biquadratic method of Bokowski and Richter is particularly efficient when the non-realizable oriented matroid $\mathcal{M}$ has a non-trivial symmetry group $\Sigma_{\mathcal{M}}$. This is the subgroup of permutations of $E = \{1, 2, \ldots, n\}$ which maps $\mathcal{M}$ to itself. In that case all three semialgebraic varieties $\mathcal{R}(\mathcal{M}), \mathcal{P}'(\mathcal{M}), \mathcal{P}(\mathcal{M})$ are invariant under the canonical representation of $\Sigma_{\mathcal{M}}$ in $\wedge_r \mathbb{R}^n$, i.e., these three sets are symmetric with respect to the invariant linear subspace $(\wedge_r \mathbb{R}^n)^{\Sigma_{\mathcal{M}}}$. A recent construction of Shor (1991) shows that the subvariety $\mathcal{R}(\mathcal{M}) \cap (\wedge_r \mathbb{R}^n)^{\Sigma_{\mathcal{M}}}$ of symmetric realizations of $\mathcal{M}$ could be empty although $\mathcal{R}(\mathcal{M})$ is non-empty. However, such a behavior certainly does not happen with symmetric sets which are *convex*. The symmetric convex polyhedron $\mathcal{P}(\mathcal{M})$ is empty if and only if its subpolyhedron $\mathcal{P}(\mathcal{M}) \cap (\wedge_r \mathbb{R}^n)^{\Sigma_{\mathcal{M}}}$ of $\Sigma_{\mathcal{M}}$-stable points is empty. In the latter case there exists a dual linear programming solution which is symmetric with respect to $\Sigma_{\mathcal{M}}$, and it can be shown that this dual solution lifts to a symmetric biquadratic final polynomial for $\mathcal{M}$.

8.5.5 Proposition (Bokowski and Richter 1990a). *A non-realizable oriented matroid $\mathcal{M}$ has a biquadratic final polynomial if and only if it has a biquadratic final polynomial which is symmetric with respect to the symmetry group $\Sigma_{\mathcal{M}}$ of $\mathcal{M}$.*

8.6 The isotopy problem and Mnëv's universality theorem

All general constructions discussed so far in Chapter 8 generate oriented matroids whose realization spaces are contractible. Long before the dawn of oriented matroid theory it had been asked by G. Ringel whether the spaces $\mathcal{R}(\mathcal{M})$ are necessarily connected. He stated this problem for arrangements of lines in the plane.

Isotopy problem (Ringel 1956). *Can any two isomorphic line arrangements A_0 and A_1 in the projective plane be connected by a continuous path $(A_t)_{0 \le t \le 1}$ of line arrangements from the same isomorphism class ?*

Around 1980 Goodman and Pollack noted the chirotope version of Ringel's problem and its significance for discrete geometry, and in the years thereafter this problem became one of most prominent open questions in oriented matroid theory. White (1989) solved the isotopy problem by constructing a rank 3 oriented matroid on 42 points without the isotopy property, and subsequently Jaggi and Mani-Levitska (1988) found a uniform example with only 17 points. Independently from the combinatorics community in the West, the isotopy problem was studied and solved by the research groups of Vershik (1988) and Viro (1988) in Leningrad, now St. Petersburg. The highlight of these developments is the universality theorem due to Mnëv (1985a, 1988), which states that realization spaces of oriented matroids can be homotopy equivalent to an arbitrary semialgebraic variety.

In this section we give an elementary exposition of the solutions to Ringel's problem. For more details and a historical account on the above developments see also Bokowski and Sturmfels (1989b, Chapter VI). We first present the smallest known oriented matroid without the isotopy property. It is a simplified version of the line arrangement found by Suvorov (1988). (See Section A.1 for a new example with the same parameters.)

8.6.1 Proposition. *There exists a rank 3 oriented matroid $\mathrm{Suv}(14)$ on 14 points whose realization space $\mathcal{R}(\mathrm{Suv}(14))$ has two connected components.*

Proof. We first define the simple rank 3 matroid M on $E = \{1, 2, \ldots, 14\}$ underlying the oriented matroid $\mathrm{Suv}(14)$. The matroid M is defined by the following list of non-trivial lines

$$1\,3\,6\,8 \quad 1\,4\,5 \quad 1\,7\,10 \quad 1\,13\,14 \quad 2\,3\,5\,14 \quad 2\,4\,6\,9\,13 \quad 2\,7\,8 \quad 3\,4\,7$$
$$3\,12\,13 \quad 4\,8\,11 \quad 4\,12\,14 \quad 5\,6\,7 \quad 5\,9\,11 \quad 6\,10\,11\,12 \quad 8\,9\,10.$$

By a *projective operation* we mean any of the following three operations:

(i)   choosing a generic point in the projective plane,

(ii)   choosing a generic point on a line spanned by two earlier chosen points, or

(iii) choosing the intersection point of two lines which are spanned by earlier chosen points.

A sequence of these operations is called a *projective construction sequence* for a rank 3 matroid if all of its realizations can be obtained by this sequence. We find the following projective construction sequence for the above matroid M:

- Let $x_1 := (0,1,0)^T$, $x_2 := (1,0,0)^T$, $x_3 := (0,0,1)^T$, $x_4 := (1,1,1)^T$

- Let $x_5 := (x_2 \vee x_3) \wedge (x_1 \vee x_4)$

- Let $x_6 := (x_1 \vee x_3) \wedge (x_2 \vee x_4)$

- Let $x_7 := (x_5 \vee x_6) \wedge (x_3 \vee x_4)$

- Let $x_8 := (x_1 \vee x_3) \wedge (x_2 \vee x_7)$

- Let $x_9 := s \cdot x_4 + x_6$ where s is an indeterminate parameter.

- Let $x_{10} := (x_1 \vee x_7) \wedge (x_8 \vee x_9)$

- Let $x_{11} := (x_5 \vee x_9) \wedge (x_4 \vee x_8)$

- Let $x_{12} := t \cdot x_6 + x_{10}$ where t is an indeterminate parameter.

- Let $x_{13} := (x_3 \vee x_{12}) \wedge (x_2 \vee x_6)$

- Let $x_{14} := (x_2 \vee x_3) \wedge (x_4 \vee x_{12})$.

This construction results in the 3×14 coordinate matrix

$$\mathbf{X}(s,t) \quad = \quad (x_1, x_2, \ldots, x_{14}) \quad =$$

$$\begin{pmatrix} 0 & 1 & 0 & 1 & 1 & 0 & 1 & 0 & s & 2s & 2s+1 \\ 1 & 0 & 0 & 1 & 0 & 1 & 1 & 1 & s+1 & 3s+1 & 2s+2 \\ 0 & 0 & 1 & 1 & 1 & 1 & 2 & 2 & s+1 & 4s & 3+2s \end{pmatrix}$$

$$\begin{pmatrix} -2s & 2s & t-s-1 \\ t-3s-1 & 3s-t+1 & 0 \\ t-4s & 3s-t+1 & s-1 \end{pmatrix}$$

In order to obtain from this parametrized matrix a realization of the matroid M, we need to satisfy the additional requirements that point 14 is incident to the line $\{1,13\}$ and that point 11 is incident to the line $\{6,10,12\}$. These requirements imply the polynomial equations

$$\det(x_1, x_{13}, x_{14}) \quad = \quad -t^2 + 4st + 2t - 5s^2 - 2s - 1 \quad = \quad 0$$

$$\det(x_6, x_{10}, x_{11}) \quad = \quad 2s^2 - 3s - 1 \quad = \quad 0.$$

These two equations have four distinct roots over the complex numbers. Only two of these four roots are found to be real, namely,

$$(s_1, t_1) \quad = \quad (1.780776\ldots, 3.936742\ldots)$$

and

$$(s_2, t_2) \quad = \quad (1.780776\ldots, 5.186363\ldots).$$

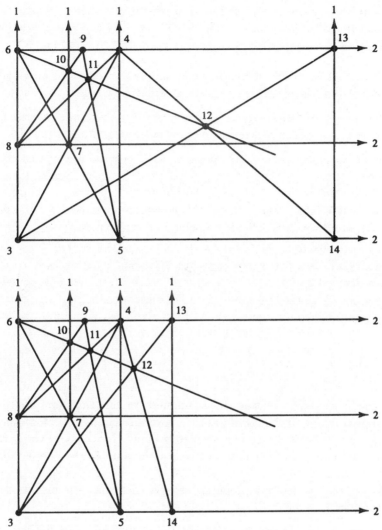

Figure 8.6.1: The two projectively distinct realizations of the
oriented matroid Suv(14).

Let $\mathbf{X}_i := \mathbf{X}(s_i, t_i)$ denote the corresponding specializations of the matrix $\mathbf{X}$. We have proved that, up to projective equivalence, $\mathbf{X}_1$ and $\mathbf{X}_2$ are the only two realizations of the matroid M. Projective diagrams of these two realizations are drawn in Figure 8.6.1. The points 1 and 2 are on the line at infinity, in the y-direction and x-direction respectively.

The crux of this configuration is that the matrices $\mathbf{X}_1$ and $\mathbf{X}_2$ also have the same *oriented* matroid Suv(14). This can be verified either algebraically or by checking that in Figure 8.6.1 all corresponding pairs of triangles have the same orientation. This implies that the realization space $\mathcal{R}(\text{Suv}(14))$ modulo projective transformations equals the disjoint union of two points. Here the projective group acts by positively scaling 11 of the 14 coordinate vectors, and we find that the realization space is the disjoint union of two contractible spaces: $\mathcal{R}(\mathrm{Suv}(14)) \cong \mathbb{R}^{11} \cup \mathbb{R}^{11}$. This completes the proof of Proposition 8.6.1.   □

In Suvorov (1988) the oriented matroid $\mathrm{Suv}(14)$ is perturbed into a uniform oriented matroid whose realization space is still disconnected.

8.6.2 Proposition (Suvorov).   *There exists a rank 3 uniform oriented matroid on 14 points whose realization space is disconnected.*

We omit Suvorov's proof of Proposition 8.6.2 because it is rather technical and specific. Instead we describe a general construction technique for uniform oriented matroids without the isotopy property given in Jaggi, Mani-Levitska, Sturmfels and White (1989). To get this construction started we first need an oriented matroid which is "closer" to being uniform than $\mathrm{Suv}(14)$.

A rank 3 oriented matroid $\mathcal{M}$ on $E = \{x_1, x_2, \ldots, x_n\}$ is said to be *constructible* if (x_1, x_2, x_3, x_4) is a projective basis and the point x_t is incident to at most two lines spanned by $\{x_1, x_2, \ldots, x_{t-1}\}$ for $t = 5, 6, \ldots, n$. We now define a constructible rank 3 oriented matroid $\mathcal{M}$ on 17 points with the property that $\mathcal{R}(\mathcal{M})$ has two connected components. Let $\mathcal{M}$ be the oriented matroid of the matrix

$$\mathbf{A}(t) := \begin{pmatrix} 1 & 0 & 0 & 1 & 0 & 1 & 1 & 0 & 1 & 1 & 1 & 1 & 5 & 5 & 0 & 1 & -1 \\ 0 & 1 & 0 & 1 & 1 & 0 & -1 & 2 & 2 & 2 & 4 & 6 & 0 & -1 & t & -t & t \\ 0 & 0 & 1 & 1 & 1 & 1 & 0 & 1 & 2 & 0 & 1 & 0 & 6 & 5 & 1 & 0 & t-1 \end{pmatrix}$$

where $\frac{1}{5} < t < \frac{1}{2}(1 - \frac{1}{\sqrt{5}})$ or $\frac{1}{2}(1 + \frac{1}{\sqrt{5}}) < t < \frac{4}{5}$. The oriented matroid $\mathcal{M}$ is well-defined because, evaluating all 3×3-minors, we find that $\mathbf{A}(t)$ has the same oriented matroid for all choices of t in the above specified range. Moreover, every other choice of t yields a different oriented matroid. It is easy to verify that $\mathcal{M}$ is constructible and that $\mathcal{M}$ is projectively unique up to the one parameter t. This shows that $\mathcal{R}(\mathcal{M}) \cong \mathbb{R}^{15} \cup \mathbb{R}^{15}$. Next we construct a uniform oriented matroid without the isotopy property, using a trick that was first used in Las Vergnas (1986a).

8.6.3 Proposition (Jaggi, Mani-Levitska, Sturmfels and White (1989).
Let $\mathcal{M}$ be a constructible rank 3 oriented matroid on n points. Then there exists a uniform rank 3 oriented matroid $\widetilde{\mathcal{M}}$ on at most $4(n-3)$ points such that $\mathcal{R}(\widetilde{\mathcal{M}})$ and $\mathcal{R}(\mathcal{M})$ are homotopy equivalent.

Proof. We define a sequence

$$\mathcal{M} =: \mathcal{M}_n, \mathcal{M}_{n-1}, \mathcal{M}_{n-2}, \ldots, \mathcal{M}_5, \mathcal{M}_4 =: \widetilde{\mathcal{M}}$$

of oriented matroids and maps between their realization spaces. Let $n \geq t \geq 5$. Then $\mathcal{M}_{t-1}$ is constructed from $\mathcal{M}_t$ as follows. First suppose that x_t is incident to exactly two lines $x_i \vee x_j$ and $x_k \vee x_l$ with $1 \leq i, j, k, l < t$. We let $\mathcal{M}'_t$ be the oriented matroid obtained from $\mathcal{M}_t$ by the four successive lexicographic extensions

(8.6.1) $\begin{aligned} x_{t,1} &:= [x_t^+, x_i^+, x_k^+], \quad x_{t,2} := [x_t^+, x_i^+, x_k^-], \\ x_{t,3} &:= [x_t^+, x_i^-, x_k^-], \quad x_{t,4} := [x_t^+, x_i^-, x_k^+]. \end{aligned}$

This means geometrically that the intersection point x_t is "caught" in the very small quadrangle $(x_{t,1}, x_{t,2}, x_{t,3}, x_{t,4})$. We define $\mathcal{M}_{t-1} := \mathcal{M}'_t \backslash x_t$ by deletion of that intersection point, and we let $\delta_t : \mathcal{R}(\mathcal{M}'_t) \to \mathcal{R}(\mathcal{M}_{t-1})$ denote the corresponding deletion map. The oriented matroids $\mathcal{M}_t, \mathcal{M}'_t, \mathcal{M}_{t-1}$ and the deletion maps between their realization spaces are illustrated in Figure 8.6.2.

The spaces $\mathcal{R}(\mathcal{M}_t)$ and $\mathcal{R}(\mathcal{M}'_t)$ are homotopy equivalent by Lemma 8.2.1 and Proposition 8.2.2. We will now show that also the map δ_t is a homotopy equivalence. Consider an arbitrary realization

$$X := (x_1, \ldots, x_{t-1}, x_{t,1}, x_{t,2}, x_{t,3}, x_{t,4}, x_{t+1,1}, \ldots, x_{n,4})$$

of $\mathcal{M}_{t-1}$. As a consequence of the lexicographic extension (8.6.1), $x_i \vee x_j$ and $x_k \vee x_l$ are the only lines spanned by $\{x_1, \ldots, x_{t-1}, x_{t+1,1}, \ldots, x_{n,4}\}$ which intersect the quadrangle $(x_{t,1}, x_{t,2}, x_{t,3}, x_{t,4})$. For any other such line the intersection point $x_t := (x_i \vee x_j) \wedge (x_k \vee x_l)$ is on the same side as $x_{t,1}, \ldots, x_{t,4}$. Therefore $(x_1, \ldots, x_t, x_{t+1,1}, \ldots, x_{n,4}) \in \mathcal{R}(\mathcal{M}_t)$ is a preimage of X under δ_t. This shows that δ_t is surjective, and with Lemma 8.2.1 we conclude that $\delta_t : \mathcal{R}(\mathcal{M}'_t) \to \mathcal{R}(\mathcal{M}_{t-1})$ is a homotopy equivalence. Hence the spaces $\mathcal{R}(\mathcal{M}_t)$ and $\mathcal{R}(\mathcal{M}_{t-1})$ are homotopy equivalent.

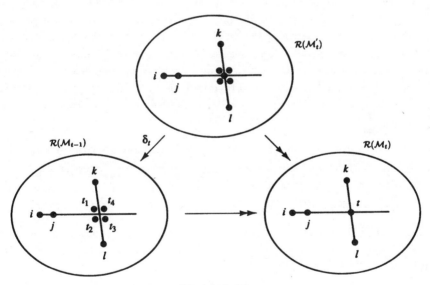

Figure 8.6.2:
The oriented matroids $\mathcal{M}_t, \mathcal{M}'_t$ and $\mathcal{M}_{t-1}$.

It remains to define $\mathcal{M}_{t-1}$ when x_t is incident to less than two lines in $\mathcal{M}_t$. If x_t is on no such line, then we simply set $\mathcal{M}_{t-1} := \mathcal{M}_t$. If x_t is incident to only one line $x_i \vee x_j$ with $1 \le i, j < t$, then we replace (8.6.1) by setting $x_{t,1} := [x_t^+, x_k^+], x_{t,2} := [x_t^+, x_i^+, x_k^-], x_{t,3} := [x_t^+, x_i^-, x_k^-]$ for some $x_k \notin x_i \vee x_j$. Now $x_t = (x_i \vee x_j) \wedge (x_k \vee x_{t,1})$ is "caught" in the triangle $(x_{t,1}, x_{t,2}, x_{t,3})$, and as before $\mathcal{R}(\mathcal{M}_{t-1})$ is homotopy equivalent to $\mathcal{R}(\mathcal{M}_t)$.

Iterating these constructions resolves all previous dependences, and we obtain a uniform oriented matroid $\widetilde{\mathcal{M}} := \mathcal{M}_4$ on $4(n-3)$ or fewer points whose realization space $\mathcal{R}(\widetilde{\mathcal{M}})$ is homotopy equivalent to $\mathcal{R}(\mathcal{M})$.   □

From Proposition 8.2.8 we get the following corollary.

8.6.4 Corollary.   *For every integer $C > 0$ there exists a uniform oriented matroid whose realization space has 2^C connected components.*

Suvorov's Proposition 8.6.2 together with Theorem 8.2.4 implies that the minimum number n_{min} of points in a rank 3 uniform oriented matroid without the isotopy property lies in the range $9 \leq n_{min} \leq 14$. The case $n_{min} = 9$ has also been excluded by the classification of Richter (1988).

8.6.5 Theorem (Richter 1988).   *The realization space of any realizable uniform rank 3 oriented matroid on 9 points is contractible.*

The remainder of this section is devoted to Mnëv's universality theorem for realization spaces of oriented matroids. Here we will only state this important result and briefly explain the key ideas for its proof. For more details the reader may wish to consult the note by Mnëv (1988). The complete proof is contained in Mnëv's 1986 dissertation. Recall that a *semialgebraic variety* is a subset of a finite dimensional real vector space which is defined by a finite set of polynomial equations and inequalities. Two semialgebraic varieties V and W are *stably equivalent* if they are related by a sequence of projections with contractible fibers and of homeomorphisms given by invertible rational maps; see Section A.1 for a precise definition. Stably equivalent varieties are in particular homotopy equivalent.

8.6.6 Universality Theorem (Mnëv 1988; see also Section A.1).

(1) *Let $V \subset \mathbb{R}^s$ be any semialgebraic variety. Then there exists a rank 3 oriented matroid $\mathcal{M}$ whose realization space $\mathcal{R}(\mathcal{M})$ is stably equivalent to V.*

(2) *If V is an open subset of $\mathbb{R}^s$, then $\mathcal{M}$ may be chosen to be uniform.*

The proof of Theorem 8.6.6 consists of two parts which are summarized in Lemmas 8.6.7 and 8.6.8. In order to state these lemmas we need to introduce some terminology. We define a language $\mathcal{A}$ over the alphabet $\{0, 1, x_1, x_2, \ldots, x_k\} \cup \{(,)\} \cup \{+, -, \times, :\}$ by the following recursive rule:

  – $0, 1, x_1, x_2, \ldots, x_k \in \mathcal{A}$;
  –   If $\Phi, \Psi \in \mathcal{A}$ and $\circ \in \{+, -, \times, :\}$, then $(\Phi \circ \Psi) \in \mathcal{A}$.

For all expressions $\Psi \in \mathcal{A}$ let $SW(\Psi)$ denote the set of those subwords of Ψ which belong to $\mathcal{A}$. Define $\sigma : \mathcal{A} \to \mathbf{Q}(x_1, x_2, \ldots, x_k) \cup \{\infty\}$ to be the map of "removing the parentheses", and abbreviate $SF(\Psi) := \sigma(SW(\Psi))$. A *computation* of a rational function $f \in \mathbb{Q}(x_1, x_2, \ldots, x_k)$ is a string in $\sigma^{-1}(f) \subset \mathcal{A}$. A *computation* of a rational vector function $\mathbf{f} = (f_1, f_2, \ldots, f_m) \in \mathbf{Q}(x_1, \ldots, x_k)^m$ is a vector of strings $\Phi = (\phi_1, \ldots, \phi_m) \in \mathcal{A}^m$ where ϕ_i is a computation for f_i.

The sets $SW(\Phi)$ and $SF(\Phi)$ are defined as the unions of the $SW(\phi_i)$ and $SF(\phi_i)$ respectively. If the map $\sigma : SW(\Phi) \to SF(\Phi)$ is a bijection, then the computation Φ is called a *formula* for the rational function $\mathbf{f}$.

Every computation Φ defines an equivalence relation $\sim_\Phi$ on $\mathbb{R}^k$ as follows. Given $\mathbf{x} \in \mathbb{R}^k$, we sort all subexpressions in the computation Φ according to their values at the point $\mathbf{x}$. Two points in $\mathbb{R}^k$ are equivalent if they induce the same ordering, i.e., we set

$$\mathbf{x} \sim_\Phi \mathbf{y} \quad : \Longleftrightarrow$$

$$\text{sign}(v(\mathbf{x}) - u(\mathbf{x})) = \text{sign}(v(\mathbf{y}) - u(\mathbf{y})) \quad \text{for all} \quad u, v \in SF(\Phi).$$

This equivalence relation defines a stratification $\sum(\mathbf{f}, \Phi)$ of $\mathbb{R}^k$ whose strata are semialgebraic varieties.

The crucial step in the argument of Mněv (1988) consists in showing that every semialgebraic variety occurs in this way.

8.6.7 Lemma.

(1) *Let $V \subset \mathbb{R}^s$ be any semialgebraic variety. Then there exist integers k, m, a rational function $\mathbf{f} \in \mathbb{Q}(x_1, \ldots, x_k)^m$, and a formula Φ for $\mathbf{f}$ such that the stratification $\sum(\mathbf{f}, \Phi)$ of $\mathbb{R}^k$ has a stratum W which is stably equivalent to V.*

(2) *If V is an open subset of $\mathbb{R}^s$, then the stratum W may be chosen to be open in $\mathbb{R}^k$.*

The first part of the universality theorem (Theorem 8.6.6) is now a direct consequence of the following lemma.

8.6.8 Lemma.

(1) *Let $\mathbf{f} \in \mathbb{Q}(x_1, \ldots, x_k)^m$, and let $\Phi \in \mathcal{A}$ be a formula for $\mathbf{f}$. For every stratum W in the stratification $\sum(\mathbf{f}, \Phi)$ of $\mathbb{R}^k$ there exists a rank 3 oriented matroid $\mathcal{M}$ whose realization space $\mathcal{R}(\mathcal{M})$ is stably equivalent to W.*

(2) *If W is open in $\mathbb{R}^k$ then $\mathcal{M}$ may be chosen to be constructible.*

The proof of Lemma 8.6.8 is based on the classical technique of projective addition and multiplication, which is referred to as the *van Staudt construction* by Mněv. As it is similar to the proof of Proposition 8.4.2 (but more complicated), we refer to our discussion and the additional references given in Section 8.4. Once the two lemmas are proved, then the second assertion of Theorem 8.6.6 follows directly from the perturbation result in Proposition 8.6.3.

8.7 Oriented matroids and robust computational geometry

Basic objects of study in computational geometry are configurations of points in $\mathbb{R}^d$ and their polars, arrangements of hyperplanes in $\mathbb{R}^d$. Most of the algorithms designed for these objects and the analysis of their complexity have assumed the use of a machine with infinite precision real arithmetic. The removal of this assumption leads to dramatically unpleasant results since we may get not only incorrect answers but nonsensical ones as well. For example, an incorrect decision about whether or not a point lies to the left of a directed line may lead to contradictory assumptions later in the algorithm. In this section we discuss results on oriented matroids which are related to computational geometry, and in particular to the robustness issue (cf. Goodman, Pollack and Sturmfels 1989).

Some of the difficulties that arise with non-robust algorithms have been illustrated by Hoffmann, Hopcroft and Karasick (1988). These authors analyze the problem of intersecting n non-convex polygons in the plane, and they show that arbitrary projective incidence theorems have to be taken into account when designing a correct and robust polygon intersection algorithm. They give an interesting example of an intersection of three convex polygons which, due to only locally correct choices, leads to a violation of the Pappus theorem.

If geometric operations are based on choices which are only locally correct, then this may result in global inconsistencies of the combinatorial specifications and the numerical data. Starting with a point configuration (i.e. a realizable oriented matroid), we may then end up with a non-realizable oriented matroid. In particular, a locally correct algorithm for the above polygon intersection problem can generate *arbitrary* non-realizable rank 3 oriented matroids.

In order to maintain consistency in its strongest form, the realizability problem of oriented matroids has to be taken into account. It is therefore of interest to find a good lower bound for the complexity of the realizability problem (ROM) stated in Section 8.4. We consider the following computational version of Mnëv's universality theorem.

Oriented matroid encoding of semialgebraic varieties. (OME)

– **Input:** Polynomials
$$f_1, f_2, \ldots, f_r, g_1, g_2, \ldots, g_s, h_1, h_2, \ldots, h_t \in \mathbf{Z}[x_1, x_2, \ldots, x_m]$$
whose total degree is bounded above by D and whose coefficients are bounded above (in their absolute value) by A.

– **Output:** A rank 3 oriented matroid $\mathcal{M}$ on n vertices whose realization space $\mathcal{R}(\mathcal{M})$ is stably equivalent to the semialgebraic variety
$$V = \{\mathbf{x} \in \mathbb{R}^m : f_1(x) = f_2(x) = \ldots = f_r(\mathbf{x}) = 0,$$
$$g_1(\mathbf{x}) \geq 0, \ldots, g_s(\mathbf{x}) \geq 0, \ h_1(\mathbf{x}) > 0, \ldots, h_t(\mathbf{x}) > 0\}.$$

Notice that here the set V is allowed to be empty. If V is empty then the output of OME will be a non-realizable oriented matroid $\mathcal{M}$. Closer inspection shows that Mnëv's proof of Theorem 8.6.6 can be modified to a polynomial time algorithm for OME. For details and a direct proof of Corollary 8.7.3 we refer to Shor (1991).

8.7.1 Lemma. *The problem OME can be solved in polynomial time in all parameters r, s, t, D, m and $\log(A)$. In particular, the size n of the constructed oriented matroid is bounded by a polynomial in r, s, t, D, m and $\log(A)$.*

This polynomial reduction together with the results of Section 8.4 implies the following equivalence.

8.7.2 Theorem. *The realizability problem for oriented matroids (ROM) is polynomially equivalent to the existential theory of the reals (ETR).*

It is well known that ETR is NP-hard because the NP-complete problem of quadratic programming can be polynomially reduced to ETR (cf. Sahni 1974). Consequently we get the same result for ROM. For earlier work on this subject, see also Huijari (1986).

8.7.3 Corollary. *The realizability problem for oriented matroids (ROM) is NP-hard.*

These results show that certain geometric algorithms which check the global consistency of all local choices cannot be expected to run in polynomial time. Thus oriented matroid theory provides a theoretical justification for weaker concepts of robustness which admit faster algorithms. Here we mention in particular the work of Milenkovic (1988) which circumvents the requirement of global realizability. He obtains fast robust algorithms which output locally realizable configurations (such as pseudoline arrangements). See Edelsbrunner and Mücke (1988), Guibas, Salesin and Stolfi (1989), and Milenkovic and Nackman (1990) for other approaches to robust geometric computations.

Closely related to robustness is the question of how much storage is needed for the coordinates of an n-point configuration in general position in the plane. This question is answered by the following "gap theorem" for realization spaces of oriented matroids. Let $\mathcal{M}$ be an acyclic rank 3 oriented matroid. A realization of $\mathcal{M}$ of the form $\{(1, x_i, y_i) : i = 1, 2, \ldots, n\}$, where x_i, y_i are integers, is called an *affine grid realization*. The *norm* of $\mathcal{M}$ is the integer

$$\nu(\mathcal{M}) \quad = \quad \min \quad \max\{|x_1|, \ldots, |x_n|, |y_1|, \ldots, |y_n|\}$$

where the minimum is taken over all affine grid realizations of $\mathcal{M}$. If $\mathcal{M}$ is not $\mathbb{Q}$-realizable then $\nu(\mathcal{M})$ is undefined.

8.7.4 Theorem (Goodman, Pollack and Sturmfels 1990).

(1) *There exists a constant c_1 such that for each $n \in \mathbb{N}$ there is a realizable acyclic uniform rank 3 oriented matroid $\mathcal{M}$ on n points with $\nu(\mathcal{M}) \geq 2^{2^{c_1 n}}$.*

(2) *There exists a constant c_2 such that for each realizable acyclic uniform rank 3 oriented matroid $\mathcal{M}$ on n points, $\nu(\mathcal{M}) \leq 2^{2^{c_2 n}}$.*

Proof. We sketch the proof. The upper bound in part (2) is derived from a general gap theorem for semialgebraic varieties due to Grigor'ev and Vorobjov (1988). The proof of the lower bound in part (1) is based on the following explicit construction in the projective plane. We start with the projective basis $(1, 1, 0), (1, 0, 1), (0, 1, 0), (0, 0, 1)$, and we first construct the points

$$
\begin{aligned}
(1, 0, 0) \quad &:= \quad \overline{(1,1,0)(0,1,0)} \;\wedge\; \overline{(1,0,1)(0,0,1)}, \\
(1, 1, 1) \quad &:= \quad \overline{(1,1,0)(0,0,1)} \;\wedge\; \overline{(1,0,1)(0,1,0)}, \\
(0, 1, -1) \quad &:= \quad \overline{(1,1,0)(1,0,1)} \;\wedge\; \overline{(0,1,0)(0,0,1)}, \\
(1, 2, 0) \quad &:= \quad \overline{(1,1,1)(0,1,-1)} \;\wedge\; \overline{(1,1,0)(0,1,0)}.
\end{aligned}
$$

Now notice that from $(1,1,0),(1,0,1),(0,1,0),(0,1,-1)$ and $(1,a,0)$ we can construct $(1,a^2,0)$ by the sequence

$$
\begin{aligned}
(1,0,a) &:= \overline{(1,a,0)(0,1,-1)} \wedge \overline{(1,0,1)(0,0,1)},\\
(0,a,-1) &:= \overline{(1,0,1)(1,a,0)} \wedge \overline{(0,1,0)(0,1,-1)},\\
(1,a^2,0) &:= \overline{(1,0,a)(0,a,-1)} \wedge \overline{(1,1,0)(0,1,0)}.
\end{aligned}
$$

By this process we can inductively construct a projectively unique oriented matroid $\mathcal{M}$ on $3r+8$ points which "computes" the number 2^{2^r}. It is not difficult to see that this non-uniform oriented matroid satisfies $\nu(\mathcal{M}) \geq 2^{2^r}$. We now apply the "scattering process" of Proposition 8.6.3 to the constructible oriented matroid $\mathcal{M}$, and we obtain a uniform oriented matroid $\widetilde{\mathcal{M}}$ on $12r+20$ points. The proof of Theorem 8.7.4 is then completed by showing that each step in the proof of Proposition 8.6.3 only increases the norm of the oriented matroid, that is, we get $\nu(\widetilde{\mathcal{M}}) \geq \nu(\mathcal{M}) \geq 2^{2^r}$. $\qquad\square$

An interesting example of an application of oriented matroid techniques in computational geometry can be found in Bienstock (1990), where Theorem 8.7.4 is used to derive lower bounds for certain graph layout problems.

Our next theorem concerns an asymptotic estimate for the number of realizable oriented matroids.

8.7.5 Theorem (Goodman and Pollack 1986a,b, Alon 1986). *The number of realizable rank r oriented matroids on E_n is bounded above by*

$$
\left(\frac{n}{r-1}\right)^{(r-1)^2 n\left(1+O\left(1/\log(n/r)\right)\right)} \qquad \text{for} \quad \frac{n}{r} \to \infty, \quad r \geq 3.
$$

This result is obtained from general upper bounds due to Milnor and Warren for the number of connected components of a semialgebraic variety. It is essentially best possible, see Exercise 8.26.

Theorem 8.7.5 implies that, in principle, realizable oriented matroids can be stored in as few as $O(r^2 n \log(n/r))$ bits. This number contrasts strongly with the fact that storing an oriented matroid requires as many as $O(n^{r-1})$ bits, see Corollary 7.4.3. In fact, storing the chirotope requires $O(n^r)$ bits. A more compact form of storing oriented matroids and point configurations is given by the λ-function of Goodman and Pollack (1983) which requires only $O(n^{r-1} \log n)$ bits – see also Exercise 3.20. Their approach is based on the observation that determining the oriented matroid of a point configuration can be interpreted as a *geometric sorting process*. Here the case $r=2$ is the usual sorting of a collection of numbers on the real line.

In the following let $\mathcal{M} = (E, \chi)$ be an acyclic rank r oriented matroid. (By reorientation, geometric sorting can be generalized to all oriented matroids.) The λ-function of $\mathcal{M}$ is the numerical function which counts for each hyperplane H of $\mathcal{M}$ the number of points in the open halfspace H^+. We define formally

$$
\begin{aligned}
\lambda_{\mathcal{M}} &: \quad E^{r-1} \quad \to \quad \mathbb{N}\\
(e_1,\ldots,e_{r-1}) &\mapsto \quad \#\{e \in E : \chi(e, e_1, \ldots, e_{r-1}) = +\}.
\end{aligned}
$$

The main theorem on geometric sorting states that the assignment $\mathcal{M} \mapsto \lambda_{\mathcal{M}}$ is injective.

8.7.6 Proposition (Goodman and Pollack 1983, Cordovil 1983c). *Let $\mathcal{M}$ be an acyclic rank r oriented matroid on E_n. Then $\mathcal{M}$ is uniquely determined by its λ-function $\lambda_{\mathcal{M}}$.*

Proof. We give a recursive algorithm for reconstructing $\mathcal{M}$ from $\lambda_{\mathcal{M}}$. For $r = 1$ and for $n - r = 0$ there exists only one acyclic oriented matroid $\mathcal{M}$ on E_n. Suppose now that $n \geq r + 1 \geq 3$.

A set of points $\{e_1, e_2, \ldots, e_{r-1}\}$ spans a facet of $\mathcal{M}$ if

$$\lambda_{\mathcal{M}}(e_1, e_2, \ldots, e_{r-1}) = 0$$

and

$$\lambda_{\mathcal{M}}(e_2, e_1, \ldots, e_{r-1}) > 0.$$

With this rule we can compute the Las Vergnas face lattice of $\mathcal{M}$ from $\lambda_{\mathcal{M}}$ (cf. Section 9.1). The set of extreme vertices of $\mathcal{M}$ is non-empty because $\mathcal{M}$ is acyclic. After relabeling we may assume that n is an extreme vertex of $\mathcal{M}$.

This implies that both the contraction $\mathcal{M}/n$ and the deletion $\mathcal{M} \backslash n$ are acyclic. We next show that the λ-functions $\lambda_{\mathcal{M} \backslash n}$ and $\lambda_{\mathcal{M}/n}$ of these two minors are determined by the given λ-function $\lambda_{\mathcal{M}}$. For the contraction this is obviously the case because

$$\lambda_{\mathcal{M}/n}(e_1, \ldots, e_{r-2}) = \lambda_{\mathcal{M}}(e_1, \ldots, e_{r-2}, n) \text{ for all } e_1, \ldots, e_{r-2} \in E_{n-1}.$$

By induction on r we may assume that $\mathcal{M}/n$ is determined by $\lambda_{\mathcal{M}/n}$. Then we have the following reconstruction rule for the λ-function of the deletion:

$$\lambda_{\mathcal{M} \backslash n}(e_1, \ldots, e_{r-1}) = \begin{cases} \lambda_{\mathcal{M}}(e_1, \ldots, e_{r-1}) - 1 & \text{if } (e_1, \ldots, e_{r-1}) \text{ is a positively} \\ & \text{oriented basis of } \mathcal{M}/n \\ \lambda_{\mathcal{M}}(e_1, \ldots, e_{r-1}) & \text{otherwise} \end{cases}$$

for all $e_1, \ldots, e_{r-1} \in E_{n-1}$. By induction on $n - r$ we may assume that $\mathcal{M}/n$ is determined from $\lambda_{\mathcal{M} \backslash n}$. The oriented matroid $\mathcal{M}$ is now determined from $\mathcal{M}/n$ and $\mathcal{M} \backslash n$. $\square$

Although computational geometry is still a comparatively young discipline, certain algorithms can already be considered as the "classics" in the field. One such classic due to Edelsbrunner, O'Rourke and Seidel (1986) is the optimal time construction of the face lattice $\mathcal{L}$ of a hyperplane arrangement which is presented by its coordinate matrix $\mathbf{X}$. The fact that this algorithm and complexity result constitute a basic building block for numerous other geometric algorithms suggests that oriented matroids might play an increasingly important role for computational geometry in the future.

Exercises

8.1   What is the dimension of the realization space of a uniform rank r oriented matroid on n vertices ? Characterize oriented matroids whose realization space has dimension 1.

8.2   Consider the 2-form Ξ in Example 8.1.2, and assume that its coordinates $a, b, \ldots, j$ are indeterminates which do satisfy the Grassmann-Plücker relations (8.1.1). Compute the corresponding configuration of 5 vectors in 2-space with coordinates in $\mathbb{Q}(a, b, \ldots, j)$. Compute also the dual configuration of 5 vectors in 3-space with coordinates in $\mathbb{Q}(a, b, \ldots, j)$.

8.3   Define the *oriented matroid category*, with a morphism between two oriented matroids $\mathcal{M}_1$ and $\mathcal{M}_2$ if and only if $\mathcal{M}_1$ is a minor of $\mathcal{M}_2$. Show that the map "$\mathcal{R}(\,\cdot\,)$" which assigns to each oriented matroid its realization space is a contravariant functor from the oriented matroid category into the category of topological spaces, where the morphisms are the continuous maps. How do non-realizable oriented matroids fit into this picture ?

8.4   Find an example of a realizable rank 3 oriented matroid $\mathcal{M}$ such that, for all points e of $\mathcal{M}$, the contraction map $\kappa_e : \mathcal{R}(\mathcal{M}) \to \mathcal{R}(\mathcal{M}/e)$ is not surjective. Can you find a uniform such example ?

8.5   Suppose that $\mathcal{M}$ is a realizable oriented matroid which is reducible by a point e. Express the proof of Lemma 8.2.1 in terms of coordinates, i.e., find an explicit continuous section $\gamma_e : \mathcal{R}(\mathcal{M} \setminus e) \to \mathcal{R}(\mathcal{M})$.

8.6* Show that every rank 3 oriented matroid on 8 points (not only the uniform ones) has a reduction sequence. Show that the realization space of every rank 3 oriented matroid on 9 points is contractible.

8.7   What is the maximum number n such that every uniform rank 4 oriented matroid on n points is obtained by taking the union of two rank 2 oriented matroids ?

8.8   Show that the oriented matroid B_3^{10} in Figure 8.3.2 is not realizable. Prove also that all proper minors of B_3^{10} are realizable.

8.9   Find an example of a non-realizable rank 4 oriented matroid $\mathcal{M}$ on 9 points such that all proper minors of $\mathcal{M}$ are realizable.

8.10 Does there exist a realizable oriented matroid $\mathcal{M}$ and a point e of $\mathcal{M}$ with the property $\dim(\mathcal{R}(\mathcal{M})) < \dim(\mathcal{R}(\mathcal{M} \setminus e))$?

8.11 A real $m \times n$-matrix A is called *totally positive* if all its subdeterminants are positive. Prove that the set of totally positive matrices is a contractible subset of $\mathbb{R}^{mn}$ in the usual topology.

8.12* Find a sharp upper and lower bound (in terms of m and n) for the smallest norm (i.e. maximal entry) among all totally positive $m \times n$-matrices with integer entries.

8.13 Give a rigorous proof that all proper minors of the Haiman-Kahn oriented matroids as in Figure 8.3.3 are realizable.

8.14 Show that the bracket ring $K[\Lambda(n,r)]$ is an integral domain. Show that $\mathcal{R}_K(\mathcal{M})$ can be identified with a certain set of maximal ideals in $K[\Lambda(n,r)]$.

8.15 Compute a final polynomial for the non-realizable oriented matroid EFM(8) in Section 10.4.

8.16* Let $\mathcal{M}$ be an oriented matroid which has a non-trivial symmetry group $\Sigma_\mathcal{M}$. Define the *unlabeled realization space* $\mathcal{R}(\mathcal{M})/\Sigma_\mathcal{M}$ whose points are the $\Sigma_\mathcal{M}$-orbits of realizations, and show that $\mathcal{R}(\mathcal{M})/\Sigma_\mathcal{M}$ has the structure of a semialgebraic variety. (*Hint*: Use invariant theory of finite permutation groups.)

8.17 Does there exist a simple arrangement of 8 hyperplanes in projective 4-space which does not have the isotopy property ?

8.18 Give an example of a rank 3 oriented matroid whose realization space $\mathcal{R}(\mathcal{M})$ has four connected components.

8.19* Does there exist a rank 3 oriented matroid on 10 points whose realization space is disconnected ?

8.20 Give an example of a realizable oriented matroid $\mathcal{M}$ and a point e of $\mathcal{M}$ such that $\mathcal{R}(\mathcal{M} \setminus e)$ has more connected components than $\mathcal{R}(\mathcal{M})$.

8.21 What is the number of labeled (resp. unlabeled) rank 3 oriented matroids on 6 points ? How many of them are uniform ?

8.22 What is the smallest grid $[-N, N] \times [-N, N]$ in the plane on which all uniform rank 3 oriented matroids on 6 points can be realized ?

8.23 Show that Proposition 8.7.6 is false without the hypothesis "acyclic".

8.24 Describe a *parallel* algorithm for computing all maximal minors of an $n \times r$-matrix for $n \gg r$. What can you say about the complexity of your algorithm ?

8.25* Describe the realization space of the union of two oriented matroids (generalizing Proposition 8.2.8 to the non-uniform case).

8.26 Show that the number of realizable oriented matroids on $E = \{1, 2, \ldots, n\}$ is bounded from below by
$$n^{(r-1)^2 n + O(n/\log n)}.$$

(Goodman and Pollack 1986a)

9

Convex Polytopes

The combinatorial theory of convex polytopes is an important area of application for oriented matroid theory. Several new results on polytopes as well as new simplified proofs for known results have been found, and it is fair to say that oriented matroids have significantly contributed to the progress of combinatorial convexity during the past decade. This chapter aims to be both an introduction to the basics and a survey on current research topics in this branch of discrete mathematics.

Section 9.1 is concerned with basic properties of matroid polytopes. We show that oriented matroid duality is essentially equivalent to the technique of Gale transforms. In Section 9.2 we discuss matroidal analogues to polytope constructions and some applications. Section 9.3 deals with the Lawrence construction, an important general method for encoding oriented matroid properties into polytopes. Cyclic and neighborly polytopes will be studied in Section 9.4, and triangulations of matroid polytopes in Section 9.6. In Section 9.5 we discuss an oriented matroid perspective on the Steinitz problem of characterizing face lattices of convex polytopes.

9.1 Introduction to matroid polytopes

Throughout this chapter we will interpret a rank r oriented matroid as a generalized point configuration in affine $(r-1)$-space. Using the language of oriented matroids, we can define the convex hull of such a configuration, and this allows us to study properties of convex polytopes in this purely combinatorial setting. The following basic definitions, due to Las Vergnas (1975a, 1980a), were already discussed in Exercise 3.9. In Exercise 3.11 we gave an axiomatization of oriented matroids in terms of their convex closure operators. A weaker notion of *abstract convexity* was developed independently by Edelman (1980, 1982) and

Jamison (1982). See Edelman and Jamison (1985) or Section 8.7 of White (1992), and also Exercise 3.10.

Let $\mathcal{M}$ be an acyclic oriented matroid on a set E. Given any cocircuit Y of $\mathcal{M}$, then the set Y^+ is said to be an *open halfspace* of $\mathcal{M}$, and $Y^+ \cup Y^0$ is a *closed halfspace* of $\mathcal{M}$. A *facet* of $\mathcal{M}$ is a hyperplane H such that $E \setminus H$ is an open halfspace; in other words, the facets of $\mathcal{M}$ are the complements of its positive cocircuits. Any intersection of facets is a *face* of $\mathcal{M}$; in other words, the faces of $\mathcal{M}$ are the complements of its positive covectors. In particular, a face of $\mathcal{M}$ which has rank 1 in the underlying matroid $\underline{\mathcal{M}}$ is called a *vertex* of $\mathcal{M}$. We say that $\mathcal{M}$ is a *matroid polytope* if all one-element subsets of E are vertices of $\mathcal{M}$. The *convex hull* in $\mathcal{M}$ of a subset A of E is defined as

$$\mathrm{conv}_{\mathcal{M}}(A) := A \cup \{x \in E \setminus A : \text{ there is a signed circuit } X \text{ of } \mathcal{M} \text{ with}$$

$$X^- = \{x\} \text{ and } X^+ \subseteq A\}.$$

A subset $K \subset E$ is *convex* if $\mathrm{conv}_{\mathcal{M}}(K) = K$.

As a good example consider the rank 4 matroid polytope $\mathrm{RS}(8)$ with eight vertices which was discussed in Section 1.5. Note that although $\mathrm{RS}(8)$ is not realizable as an oriented matroid, its face lattice can be realized as the face lattice of a simplicial 3-polytope with eight vertices.

The above definitions are consistent with the usual language of combinatorial convexity. As a recent update to Grünbaum's (1967) fundamental book on convex polytopes we recommend the survey of Klee and Kleinschmidt (1992). Suppose that P is a convex d-polytope with vertex set $V = \{v_1, v_2, \ldots, v_n\}$. Let $\mathcal{M}$ be the rank $d+1$ oriented matroid on E_n of affine dependences on V. Then $\mathcal{M}$ is a matroid polytope, and $\{e_1, \ldots, e_k\} \subset E_n$ is a facet (resp. face, vertex) of $\mathcal{M}$ if and only if $\mathrm{conv}(\{v_{e_1}, \ldots, v_{e_k}\})$ is a facet (resp. face, vertex) of P. Conversely, every *realizable* rank r matroid polytope $\mathcal{M}$ arises from some $(r-1)$-polytope P in this manner.

The following result is a consequence of the Topological Representation Theorem 5.2.1, see Theorem 4.3.5(ii).

9.1.1 Proposition. *The set $\mathcal{F}_{lv}(\mathcal{M})$ of faces of a rank r matroid polytope (ordered by inclusion) is a lattice which is isomorphic to the face lattice of a piecewise linear cell decomposition of the $(r-2)$-dimensional sphere.*

The lattice $\mathcal{F}_{lv}(\mathcal{M})$ is the *Las Vergnas face lattice* of the oriented matroid $\mathcal{M}$. Recall that the acyclic oriented matroid $\mathcal{M}$ has a unique positive tope or base region T in the big face lattice $\hat{\mathcal{L}}(\mathcal{M})$, see Definition 4.1.2. The lower interval $[0, T]$ of the positive tope T in $\hat{\mathcal{L}}(\mathcal{M})$ is the *Edmonds-Mandel face lattice* of $\mathcal{M}$ which is denoted by $\mathcal{F}_{em}(\mathcal{M})$. Since each positive covector of $\mathcal{M}$ is the complement of a face of $\mathcal{M}$, we see that the Edmonds-Mandel face lattice $\mathcal{F}_{em}(\mathcal{M})$ is anti-isomorphic to the Las Vergnas face lattice $\mathcal{F}_{lv}(\mathcal{M})$. It was proved in Chapter 4 that $\mathcal{F}_{em}(\mathcal{M})$ is the face lattice of a shellable $(r-2)$-dimensional PL-sphere, and this implies that $\mathcal{F}_{lv}(\mathcal{M})$ is the face lattice of an $(r-2)$-dimensional PL-sphere. It is not known in general whether $\mathcal{F}_{lv}(\mathcal{M})$ is shellable.

Throughout this chapter we are interested mainly in the Las Vergnas lattice, and so we will simply refer to $\mathcal{F}_{lv}(\mathcal{M})$ as the face lattice of $\mathcal{M}$. We begin with the observation that there are many equivalent ways to read off the faces of a matroid polytope.

9.1.2 Proposition (Las Vergnas 1975a, 1980a). *Let $\mathcal{M}$ be a matroid polytope on E. For any subset $F \subset E$ the following are equivalent:*

(a) *F is a face of the matroid polytope $\mathcal{M}$;*

(b) *$F \subset E$ is closed and $\mathcal{M}/F$ is acyclic;*

(c) *for every signed circuit X of $\mathcal{M}$, $X^+ \subset F$ implies $X^- \subset F$.*

Proof. (a) $\Leftrightarrow$ (b): Let F be a face of $\mathcal{M}$. By definition, this means that $E \setminus F$ is a positive covector of $\mathcal{M}$. Equivalently, $E \setminus F$ is a positive covector of $\mathcal{M}/F$, which means that $\mathcal{M}/F$ is acyclic.

(a) $\Rightarrow$ (c): Let F be a face and $Y = E \setminus F$ the corresponding positive covector of $\mathcal{M}$. Suppose that (c) is not true. Then there is a circuit X of $\mathcal{M}$ with $X^+ \subset F$ but $X^- \not\subset F$. This implies $X^+ \cap Y^+ = \emptyset$ and $X^- \cap Y^+ \neq \emptyset$, a contradiction to the orthogonality of circuits and covectors.

(c) $\Rightarrow$ (b): Suppose that $F \subset E$ is not closed in the underlying matroid of $\mathcal{M}$, say e lies in the closure of F but not in F. Then there exists a circuit X of $\mathcal{M}$, supported on $F \cup e$ and such that $e \in X^-$. Then the signed circuit X violates the condition (c).

Suppose now that $\mathcal{M}/F$ is not acyclic. Then there exists a positive circuit $\widetilde{X}$ of $\mathcal{M}/F$. We can write $\widetilde{X} = X \setminus F$ where X is a circuit of $\mathcal{M}$. Since $\widetilde{X}$ is positive, we have $X^- \subset F$ but $X^+ \not\subset F$. Hence the negated circuit $-X$ violates the condition (c). $\square$

As a corollary of Proposition 9.1.2 we obtain an easy rule for reading off the faces of a uniform matroid polytope. Note that the face lattice of a uniform matroid polytope is necessarily a simplicial (or triangulated) sphere. A regular octahedron shows that the converse fails: an oriented matroid whose face lattice is simplicial need not be uniform.

9.1.3 Corollary. *Let $\mathcal{M}$ be a uniform matroid polytope on a set E. A proper subset $F \subset E$ is a face of $\mathcal{M}$ if and only if $X^+ \not\subset F$ for all circuits X of $\mathcal{M}$.*

Proof. Since $\mathcal{M}$ is uniform, $\underline{M}$ has no proper flats which contain circuits. This implies that, for every proper subset $F \subset E$, the condition (c) of Proposition 9.1.2 is equivalent to $X^+ \not\subset F$ for all circuits X of $\mathcal{M}$. $\square$

We will next relate oriented matroid duality to the technique of Gale transforms in polytope theory. This duality enables us to draw planar diagrams to represent high-dimensional polytopes which have few more vertices than their dimension.

A *Gale transform* of a (labeled) point configuration

$$X = \{x_1, x_2, \ldots, x_n\}$$

in affine $(r-1)$-space is a (labeled) configuration of vectors

$$\overline{X} = \{\overline{x}_1, \overline{x}_2, \ldots, \overline{x}_n\}$$

in $\mathbb{R}^{n-r}$ such that the space of affine dependences on X is the orthogonal complement in $\mathbb{R}^n$ of the space of linear dependences on $\overline{X}$. The fact that orthogonality of subspaces in $\mathbb{R}^n$ corresponds to duality of realizable oriented matroids implies the following.

9.1.4 Proposition (McMullen 1979). *Let*

$$\overline{X} = \{\overline{x}_1, \overline{x}_2, \ldots, \overline{x}_n\} \subset \mathbb{R}^{n-r}$$

be a Gale transform of an affinely spanning point configuration

$$X = \{x_1, x_2, \ldots, x_n\} \subset \mathbb{R}^{r-1}.$$

Then the rank r oriented matroid $\mathcal{M}$ of affine dependences on X is dual to the rank $n - r$ oriented matroid $\mathcal{M}^*$ of linear dependences on $\overline{X}$.

As a consequence we obtain easy matroidal proofs for the fundamental properties of Gale transforms as stated in Grünbaum (1967). Given a subset $I \subset E_n$ of indices, we write $X(I) := \{x_i \in X : i \in I\}$ and $\overline{X}(I) := \{\overline{x}_i \in \overline{X} : i \in I\}$.

9.1.5 Proposition. *Let X and $\overline{X}$ as above, and consider the polytope $P = \operatorname{conv} X$.*

(1)   *conv $X(I)$ is a face of P if and only if $I = E_n$ or*

$$0 \in \operatorname{relint}(\operatorname{conv} \overline{X}(E_n \setminus I)).$$

(2)   *X is the set of vertices of P if and only if*

(2.1) *$\overline{X} = \{0\}$, in which case P is an $(r-1)$-simplex, or*

(2.2) *every open linear halfspace H^+ of $\mathbb{R}^{n-r}$ satisfies $|H^+ \cap \overline{X}| \geq 2$.*

(3)   *P is a simplicial polytope if and only if $\dim \overline{X}(E_n \setminus I) = n - r$ for every facet $X(I)$ of P.*

Proof. Let $\mathcal{M}$ be the oriented matroid associated with X. By Proposition 9.1.4, its dual $\mathcal{M}^*$ is the oriented matroid associated with $\overline{X}$.

(1) Let I be a proper subset of the index set E_n. Then conv $X(I)$ is a face of P if and only if I is a face of $\mathcal{M}$. Then $E \setminus I$ is positive covector of $\mathcal{M}$, and $E \setminus I$ is a positive vector of $\mathcal{M}^*$. This means that $E \setminus I$ supports a positive linear dependence $\sum_{e \in E \setminus I} \lambda_e \overline{x}_e = 0$, $\lambda_e > 0$. This is equivalent to $0 \in \operatorname{relint}(\operatorname{conv} \overline{X}(E \setminus I))$.

(2) If $\overline{X} = \{0\}$ then $\operatorname{rank}(\mathcal{M}^*) = 0$ and $\mathcal{M}$ is a simplex. Assume that $\operatorname{rank}(\mathcal{M}^*) \geq 1$ and that every open halfspace of $\mathcal{M}^*$ contains at least two points. Given any $e \in E$, then there is no cocircuit Y of $\mathcal{M}^*$ with $Y^+ = \{e\}$. Hence $\mathcal{M}^* \setminus e$ has no positive cocircuits and $\mathcal{M}/e$ has no positive circuits. Therefore $\mathcal{M}/e$ is acyclic which means that e is a vertex of $\mathcal{M}$.

Conversely, if $e \in E$ is a vertex of $\mathcal{M}$, then there is no circuit X of $\mathcal{M}$ with $X^+ \subseteq \{e\}$ by Proposition 9.1.2. If all $e \in E$ are vertices of $\mathcal{M}$, then $|X^+| \geq 2$ for every circuit X of $\mathcal{M}$, and thus $|Y^+| \geq 2$ for every cocircuit Y of $\mathcal{M}^*$. Hence every open halfspace of $\mathcal{M}^*$ contains at least two points.

(3) A facet F of $\mathcal{M}$ is a simplex if and only if it is contained in a basis of the underlying matroid $\underline{M}$. This means that $E \setminus F$ contains a basis of the dual $\mathcal{M}^*$, i.e., $E \setminus F$ has rank $n - r$ in the matroid $\mathcal{M}^*$.   □

9.1.6 Example (A matroid octahedron). We illustrate these concepts and results with a three-dimensional example. Let $\mathcal{M} = (E_6, \mathcal{C})$ be the uniform rank 4 oriented matroid on six points whose set of circuits equals

$$\mathcal{C} \quad = \quad \pm \{\, 12\overline{3}\overline{4}5, \, \overline{1}\overline{2}346, \, \overline{1}\overline{2}356, \, 124\overline{5}\overline{6}, \, 134\overline{5}\overline{6}, \, 2\overline{3}4\overline{5}6 \,\}.$$

Using the criterion of Corollary 9.1.3, we see that $F \subset E$ is a face of $\mathcal{M}$ if and only if F contains none of the sets 12, 34, and 56. This implies that the facets of $\mathcal{M}$ are given by

$$135, \ 136, \ 145, \ 146, \ 235, \ 236, \ 245, \ 246.$$

Hence $\mathcal{M}$ is a matroid polytope combinatorially equivalent to an octahedron. Any set of six vectors in $\mathbb{R}^2$ realizing the dual oriented matroid $\mathcal{M}^*$ is the **Gale transform** of an octahedron in general position. Using a suitable acyclic reorientation, we can represent $\mathcal{M}^*$ by six points in affine 1-space. See Figure 9.1.1 for such an affine Gale diagram of $\mathcal{M}^*$. We refer to Sturmfels (1988a) for details on affine Gale diagrams.

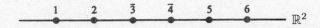

Figure 9.1.1: The dual $\mathcal{M}^*$ of the matroid octahedron $\mathcal{M}$.

The bar notation in Figure 9.1.1 is to be understood as follows. Let N be the acyclic rank 2 oriented matroid given by six points $1, 2, 3, 4, 5$ and 6 in increasing order on the real line. Then $\mathcal{M}^*$ equals the reorientation $_{-\{3,4\}}N$ of N. From this representation we can easily read off the above set $\mathcal{C}$ of circuits of $\mathcal{M}$ or cocircuits of $\mathcal{M}^*$.

Example 9.1.6 shows that, in general, there are many non-isomorphic matroid polytopes with the same face lattice. It can be seen from the affine Gale diagram of the octahedron $\mathcal{M}$ in Figure 9.1.1 that we can reverse the orientation of each of the three bases $\{1, 2\}$, $\{3, 4\}$, $\{5, 6\}$ and still get the same face lattice $\mathcal{F}_{lv}(\mathcal{M})$. The same holds true in the intermediate situation when one or more of these three sets becomes dependent. There is a total number of 27 matroid polytopes with the face lattice of an octahedron; and eight of these are uniform.

On the other hand, there also exist certain matroid polytopes which are uniquely determined by their face lattices. More precisely, a matroid polytope $\mathcal{M}$ is said to be *rigid* if $\mathcal{F}_{lv}(\mathcal{M}) = \mathcal{F}_{lv}(\widetilde{\mathcal{M}})$ implies $\mathcal{M} = \widetilde{\mathcal{M}}$ for every matroid polytope $\widetilde{\mathcal{M}}$ on the same set of vertices.

9.1.7 Example. As an example, we describe a rigid matroid polytope. Let $\mathcal{M}$ be the rank 5 matroid polytope on E_8 whose set of signed circuits is given by

$$\mathcal{C} \quad = \quad \pm \{ 12\bar{5}\bar{6}, \; 1\bar{3}\bar{5}7, \; 14\bar{5}\bar{8}, \; 23\bar{6}\bar{7}, \; 2\bar{4}\bar{6}8, \; 347\bar{8} \}.$$

Using Proposition 9.1.2, we find that $\mathcal{M}$ has six facets, namely

$$123567, \; 124568, \; 134578, \; 234678, \; 1368, \; 2457.$$

$\mathcal{M}$ is the oriented matroid corresponding to a prism over a tetrahedron. An affine Gale diagram for the matroid polytope $\mathcal{M}$, i.e. an acyclic reorientation of $\mathcal{M}^*$, is depicted in Figure 9.1.2.

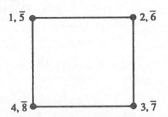

Figure 9.1.2: Affine Gale diagram of a prism over a tetrahedron.

We will show that $\mathcal{M}$ is a rigid matroid polytope. Let $\widetilde{\mathcal{M}}$ be any rank 5 matroid polytope on E_8 such that $\mathcal{F}_{lv}(\mathcal{M}) = \mathcal{F}_{lv}(\widetilde{\mathcal{M}})$. Then $\widetilde{\mathcal{M}}$ has the following positive cocircuits

$$15, \; 26, \; 37, \; 48, \; 1368, \; 2457,$$

which are the complements of the facets of $\widetilde{\mathcal{M}}$. Consequently these sets are positive circuits of the rank 3 oriented matroid $(\widetilde{\mathcal{M}})^*$. By circuit elimination, we find that $1\bar{2}3\bar{4}$ is a signed circuit of $(\widetilde{\mathcal{M}})^* \setminus \{5,6,7,8\}$. This is the unique signed circuit of that four point oriented matroid. But now note that $(\widetilde{\mathcal{M}})^* \setminus \{5,6,7,8\}$ uniquely determines $(\widetilde{\mathcal{M}})^*$ because 1 is antiparallel to 5, 2 is antiparallel to 6, 3 is antiparallel to 7, and 4 is antiparallel to 8. $\qquad\square$

In Section 9.3 we shall see a far-reaching generalization of this example. A matroid polytope whose set of vertices is partitioned into two-element positive cocircuits is necessarily rigid. This results in a fundamental construction technique due to Lawrence for matroid polytopes.

9.2 Convexity results and constructions

We have seen how some fundamental convexity concepts can be translated into the language of oriented matroids. These concepts will now be applied to study geometric results and constructions in this more general setting. As an illustration for the use of oriented matroid techniques in convexity, we will then discuss a result of Las Vergnas (1986b) on McMullen's problem concerning projective transformations of polytopes.

It turns out that some matroidal convexity results are surprisingly different from their geometric counterparts, and a generalization to all oriented matroids is not always possible. For example, oriented matroid analogues to the theorems of Carathéodory, Krein-Milman, Radon, Helly, and Hahn-Banach are given below. Most of these statements are restricted to oriented matroids with the generalized Euclidean intersection property IP_2 (see Section 7.5). While this class clearly includes the realizable oriented matroids, we will see in Section 10.4 that this intersection property fails for the rank 4 oriented matroid $EFM(8)$. As will be illustrated in Section 10.5, these intersection properties are also of considerable importance for linear programming in oriented matroids.

9.2.1 Theorem (Las Vergnas 1975a, 1980a, Bachem and Wanka 1988).
Let $\mathcal{M}$ be an acyclic oriented matroid of rank r on E.

(1) *"Carathéodory's Theorem":   Let $A \subseteq E$ and $e \in E$ such that $e \in \operatorname{conv}_{\mathcal{M}}(A)$. Then there exists a subset $B \subseteq A$ of cardinality at most r such that $e \in \operatorname{conv}_{\mathcal{M}}(B)$.*

(2) *"Weyl's Theorem": Let $A \subseteq E$. Then $\operatorname{conv}_{\mathcal{M}}(A)$ is equal to the intersection of the closed halfspaces of $\mathcal{M}$ containing A.*

(3) *"Krein-Milman Theorem":   Let $A \subseteq E$ and let S be the set of extreme points of $\mathcal{M}(A)$. Then $\operatorname{conv}_{\mathcal{M}}(A) = \operatorname{conv}_{\mathcal{M}}(S)$.*

Suppose furthermore that $\mathcal{M}$ has the generalized Euclidean intersection property IP_2.

(4) *"Radon's Theorem":   Let $A \subseteq E$ with $|A| \geq r + 1$. Then there exists a partition $A = A' \cup A''$ and a point extension $\widetilde{\mathcal{M}} = \mathcal{M} \cup p$ of $\mathcal{M}$ (where p is not a loop) such that $p \in \operatorname{conv}_{\widetilde{\mathcal{M}}}(A') \cap \operatorname{conv}_{\widetilde{\mathcal{M}}}(A'')$.*

(5) *"Helly's Theorem": Let $\{K_i\}_{i=1,\ldots,n}$ with $n \geq r + 1$ be a family of convex subsets of E such that every intersection of $r - 1$ of the sets K_i is non-empty. Then there exists an extension $\widetilde{\mathcal{M}} = \mathcal{M} \cup p$ (where p is not a loop) such that $p \in \bigcap_{i=1}^{n} \operatorname{conv}_{\widetilde{\mathcal{M}}}(K_i)$.*

(6) *"Hahn-Banach Theorem":   Two subsets $A, B \subseteq E$ are strongly separable (that is, $A \subseteq Y^+$ and $B \subseteq Y^-$ for some covector Y of $\mathcal{M}$) if and only if for every extension $\widetilde{\mathcal{M}}$ of $\mathcal{M}$ we have $\operatorname{conv}_{\widetilde{\mathcal{M}}}(A) \cap \operatorname{conv}_{\widetilde{\mathcal{M}}}(B) = \emptyset$.*

Proof.
(1) By definition, $e \in \operatorname{conv}_{\mathcal{M}}(A)$ means that there is a circuit X of $\mathcal{M}$ with $\underline{X} \subseteq A \cup \{e\}$ and $X^- = \{e\}$. Now, $|X| \leq r + 1$, so we can take $B := X^+ = \underline{X} \setminus \{e\}$.

(4) Since A has $\geq r + 1$ elements, it must contain a circuit X. We set $A' := X^+$ and $A'' := A \setminus A' \supset X^-$. Now the flats $\overline{X^+}$ and $\overline{X^-}$ satisfy $\operatorname{rank}_{\mathcal{M}}(X^+) + \operatorname{rank}_{\mathcal{M}}(X^-) = \operatorname{rank}_{\mathcal{M}}(\underline{X}) + 1$. By IP_2, $\mathcal{M}$ has a single element extension $\widetilde{\mathcal{M}} = \mathcal{M} \cup p$ such that p is not a loop, and

$$p \in \operatorname{conv}_{\widetilde{\mathcal{M}}}(X^+) \cap \operatorname{conv}_{\widetilde{\mathcal{M}}}(X^-) \subseteq \operatorname{conv}_{\widetilde{\mathcal{M}}}(A') \cap \operatorname{conv}_{\widetilde{\mathcal{M}}}(A'').$$

The assertion (2) can be derived from the 3-Painting Theorem 3.4.4, and then used to get (3), see Las Vergnas (1980a). Both (5) and (6) require more elaborate arguments, for which we refer to Bachem and Wanka (1988).   □

See also Goodman and Pollack (1982a,c) and Cordovil (1982b) for "Helly type theorems" in oriented matroids. We also note that Tverberg's Theorem, which is an important generalization of Helly's Theorem, has been shown to generalize to oriented matroids in the rank 3 case by Roudneff (1988b).

To stress the point that the convexity results (4) – (6) do not hold without the extra assumption IP_2, we observe that Radon's Theorem 9.2.1(4) holds for all reorientations of an oriented matroid $\mathcal{M}$ if and only if its reorientations all satisfy IP_2. (Geometrically speaking, an "affine point configuration" satisfies the intersection property IP_2 if and only if all its images under projective transformations satisfy Radon's Theorem.) To see this, consider any two flats F and G of $\mathcal{M}$ with $\mathrm{rank}(F) + \mathrm{rank}(G) > \mathrm{rank}(F \cup G)$, and choose bases F_0 and G_0 for these flats. Then $F_0 \cup G_0$ contains a circuit X, and after reorientation we may assume that $X^+ \subseteq F_0 \subseteq F$ and $X^- \subseteq G_0 \subseteq G$. Now the statement of Radon's Theorem can be applied to show IP_2.

We will illustrate this in more detail in Section 10.4, where specific rank 4 examples (reorientations of $\mathsf{EFM}(8)$) are constructed for which IP_2, Radon's Theorem 9.2.1(4) and the Hahn-Banach Theorem 9.2.1(6) fail. Note that these results do hold in rank 3 because all rank 3 oriented matroids satisfy IP_2.

Let $\mathcal{M}$ be a matroid polytope with vertex set E, and let $\widetilde{\mathcal{M}} = \mathcal{M} \cup p$ be a point extension of $\mathcal{M}$. Generalizing a classical result of Grünbaum (1967), we can give a description of the faces of $\widetilde{\mathcal{M}}$ in terms of the faces of $\mathcal{M}$. Let F be a facet of $\mathcal{M}$, and let Y be the unique cocircuit of the extension $\widetilde{\mathcal{M}}$ such that $F \subset Y^0$ and $E \subset Y^0 \cup Y^+$. The new point p is said to be *beneath* F if $p \in Y^+$, and it is *beyond* F if $p \in Y^-$.

9.2.2 Proposition (Munson 1981). *Let $\mathcal{M}$ and $\widetilde{\mathcal{M}} = \mathcal{M} \cup p$ be a matroid polytope. Then all faces of $\widetilde{\mathcal{M}}$ are obtained as follows.*

(1) *A face F of $\mathcal{M}$ is also a face of $\widetilde{\mathcal{M}}$ if and only if there exists a facet F' of $\mathcal{M}$ such that $F' \supset F$ and p is beneath F'.*

(2) *If F is a face of $\mathcal{M}$, then $F \cup \{p\}$ is a face of $\widetilde{\mathcal{M}}$ if and only if either*
   (a) *p is contained in the flat spanned by F in $\widetilde{\mathcal{M}}$, or*
   (b) *p is beyond some facet of $\mathcal{M}$ containing F and beneath some other facet of $\mathcal{M}$ containing F.*

Proof. Let σ be the localization of the point extension $\widetilde{\mathcal{M}} = \mathcal{M} \cup p$. From Proposition 7.1.4 we see that the positive cocircuits of $\widetilde{\mathcal{M}}$ are precisely the following subsets of E:

(1) $Y \cup \{p\}$, where Y is a positive cocircuit of $\mathcal{M}$ with $\sigma(Y) = +$;
(2a) Y, where Y is a positive cocircuit of $\mathcal{M}$ with $\sigma(Y) = 0$;
(2b) $Y^1 \cup Y^2$ where Y^1, Y^2 are positive cocircuits with
$$-\sigma(Y^1) = \sigma(Y^2) \neq 0.$$

The facets of $\widetilde{\mathcal{M}}$, the complements of positive cocircuits, are the following subsets of E:

(1) F, where F is a facet of $\mathcal{M}$ such that p is beneath F;

(2a) $F \cup \{p\}$, where F is a facet such that p is contained in the span of F in $\widetilde{\mathcal{M}}$;

(2b) $(F_1 \cap F_2) \cup \{p\}$, where F_1, F_2 are facets of $\mathcal{M}$ such that p is beneath F_1 and beyond F_2.

The proof of Proposition 9.2.2 can now be completed using the fact that the faces of a matroid polytope are the intersections its facets.   □

We will apply Proposition 9.2.2 to show that stellar subdivisions can be carried out on the level of matroid polytopes. Let $\mathcal{L}$ be the face poset of a regular cell complex with set of vertices E. Given a face $F \subset E$ of $\mathcal{L}$, then the *stellar subdivision* $\mathcal{L}' := \sigma(\mathcal{L}, F)$ of $\mathcal{L}$ at the face F is defined as follows. The poset $\mathcal{L}'$ consists of certain subsets of $E \cup p$, where p is a new point. If G is a face of a facet which does not contain F, then G is a face of $\mathcal{L}'$. If G is also a proper face of a facet of $\mathcal{L}$ which does contain F, then $G \cup \{p\}$ is a face of $\mathcal{L}'$. This describes all faces of the stellar subdivision $\mathcal{L}' = \sigma(\mathcal{L}, F)$.

9.2.3 Proposition.   *Let $\mathcal{M}$ be a matroid polytope of rank r on E, let $F \subset E$ be a face of $\mathcal{M}$, and let $\{e_1, e_2, \ldots, e_s\}$ be a basis of F. Consider the lexicographic extension $\widetilde{\mathcal{M}} = \mathcal{M} \cup p$ of $\mathcal{M}$ defined by $[e_1^+, \ldots, e_s^+, e_{s+1}^-, \ldots, e_r^-]$ where $\{e_1, \ldots, e_s, e_{s+1}, \ldots, e_r\}$ is a basis of $\mathcal{M}$. Then $\mathcal{F}_{lv}(\widetilde{\mathcal{M}}) = \sigma(\mathcal{F}_{lv}(\mathcal{M}), F)$, that is, the face lattice $\mathcal{F}_{lv}(\widetilde{\mathcal{M}})$ of $\widetilde{\mathcal{M}}$ is obtained by stellar subdivision of $\mathcal{F}_{lv}(\mathcal{M})$ at F.*

Proof. By the properties of lexicographic extensions, p is beyond a facet F' of $\mathcal{M}$ if and only if this facet contains the span F of $\{e_1, e_2, \ldots, e_s\}$. Indeed, if $F \not\subset F'$ then p has the same sign as the first $e_i \in F \setminus F'$ in the positive cocircuit complementary to F'. This sign being positive means that p is beneath the facet F'. Now Proposition 9.2.3 follows from Proposition 9.2.2 and the definition of stellar subdivisions.   □

The following two special cases of stellar subdivisions are of particular interest for polytope theory. If F is a facet of the sphere $\mathcal{L}$, then $\sigma(\mathcal{L}, F)$ is obtained by *stacking* the new point p onto the facet F. A *stacked polytope* is a simplicial polytope obtained from the simplex by successive stacking of points. Barnette's lower bound theorem (Barnette 1973a) states that among all simplicial d-polytopes with a given number of vertices, the stacked d-polytopes have the minimum number of i-dimensional faces for $i = 0, 1, \ldots, d - 1$. It can be shown that both this lower bound theorem and the upper bound theorem of McMullen (1971b) remain valid for face lattices of matroid polytopes.

If $F = \{v\}$ is a vertex of the lattice $\mathcal{L}$, then $\sigma(\mathcal{L}, \{v\})$ is obtained by *pulling* the vertex v. For more details we refer to the dissertation of Munson (1981) where pulling vertices, splitting facets, forming prisms and bipyramids, and several other important constructions are discussed from the matroidal point of view.

We close this section with an application of oriented matroid methods to a longstanding open problem from polytope theory. The following question has been introduced by McMullen; see Larman (1972):

(P1) Determine the largest integer $n = f(d)$ such that for any given n points in general position in affine d-space $\mathbb{R}^d$ there is an admissible projective transformation mapping these points onto the vertices of a convex polytope.

Larman (1972) has shown that $2d + 1 \le f(d) \le (d + 1)^2$ for any $d \ge 2$, and he conjectures that the lower bound is tight, i.e., $f(d) = 2d + 1$. The following improved bound has been obtained by Las Vergnas (1986b). The upper bound was much later lowered to $f(d) < \frac{5}{2}d$ by Ramírez Alfonsín (1999a).

9.2.4 Proposition (Las Vergnas 1986b). *For $d \ge 2$ we have $f(d) < (d + 1)(d + 2)/2$: There is a set of $(d + 1)(d + 2)/2$ points in general position in $\mathbb{R}^d$ which is not projectively equivalent to the set of vertices of a d-polytope.*

In order to prove this proposition, we need a reformulation of (P1) into the language of oriented matroids. The following generalization of McMullen's problem has been suggested by Cordovil and da Silva (1985):

(P2) Determine the largest integer $n = g(r)$ such that every uniform rank r oriented matroid $\mathcal{M}$ has an acyclic reorientation which is a matroid polytope.

McMullen's problem (P1) is the particular case of (P2) obtained by considering the rank $d + 1$ acyclic oriented matroid of the given n points in $\mathbb{R}^d$. Now, a projective transformation of $\mathbb{R}^d$ admissible on the n points corresponds to an acyclic reorientation of $\mathcal{M}$ and conversely. Hence we have $r = d + 1$ and $g(r) \le f(r + 1)$. Cordovil and da Silva have shown that $g(r) \ge 2r - 1$ for $r \ge 3$, thus generalizing Larman's lower bound. Notice furthermore that the acyclic reorientations of $\mathcal{M}$ correspond to the topes in an arrangement of pseudospheres representing the oriented matroid $\mathcal{M}$ (cf. Chapter 5). In this polar picture, we get the following reformulation of (P2):

(P3) Determine the largest integer $n = g(r)$ such that every arrangement of n pseudospheres of rank r has a tope that is bounded by all n pseudospheres.

The conjecture that $2d + 1 \le g(d + 1) \le f(d)$ holds with equality for all $d \ge 2$ is still open. For $d = 2, 3$ this is easy to see, and for $d = 4$ it was verified by Forge and Schuchert (1996).

Proof of Proposition 9.2.4. The idea is to consider a uniform oriented matroid "close" to the oriented cycle matroid of a tournament, and then to apply Rédei's classical theorem on Hamilton paths in tournaments.

Let T be a complete directed graph or *tournament* on $d + 1$ vertices. The oriented cycle matroid $\mathcal{M}_0$ of T is a rank d oriented matroid on $\binom{d+1}{2}$ elements. Note that $\mathcal{M}_0$ is not uniform for $d \ge 3$.

Consider any linear realization of $\mathcal{M}_0$ in $\mathbb{R}^d$ (for instance the standard one in the hyperplane $\sum x_k = 0$ of $\mathbb{R}^{d+1}$, where we associate with the directed edge $v_i v_j$ of T the point with i-th coordinate 1, j-th coordinate -1, and all other coordinates 0). Then we move these $d(d+1)/2$ points slightly in order to get general position in $\mathbb{R}^d$ without modifying any sign in any linear dependence corresponding to a $(d+1)$-element circuit of $\mathcal{M}_0$. By continuity this can be done by maintaining each point in a small neighborhood of its original position. Let $\mathcal{M}_1$ denote the resulting uniform oriented matroid.

Let $\widetilde{\mathcal{M}}_1$ be any acyclic reorientation of $\mathcal{M}_1$, and let $\widetilde{\mathcal{M}}_0$ be the corresponding acyclic reorientation of $\mathcal{M}_0$. By construction, all $(d+1)$-element circuits have the same signatures in $\widetilde{\mathcal{M}}_0$ and $\widetilde{\mathcal{M}}_1$, and hence $\widetilde{\mathcal{M}}_0$ is the oriented cycle matroid of a tournament $\widetilde{T}$ without Hamilton cycle. By Rédei's theorem (Rédei 1934), $\widetilde{T}$ contains a Hamilton path $v_1 v_2 \ldots, v_{d+1}$. Since $\widetilde{T}$ has no Hamilton cycle, the edge $e := v_1 v_{d+1}$ is in $\widetilde{T}$. In particular, $\widetilde{\mathcal{M}}_0$ has a signed circuit X with $|X^+| = d$ and $X^- = \{e\}$. By construction, X is also a signed circuit of $\widetilde{\mathcal{M}}_1$, hence e is an interior point of $\widetilde{\mathcal{M}}_1$. This proves that $f(d-1) < d(d+1)/2$ for $d \geq 3$, and consequently Proposition 9.2.4 follows. □

9.3 The Lawrence construction and its applications

For some time after convexity in oriented matroids had been introduced, it was not known whether the face lattices of oriented matroids were all polytopal (that is, isomorphic to face lattices of convex polytopes). Although it did not seem that this should be the case, many of the combinatorial properties of polytopes had been generalized to matroid polytopes, and no example of a non-polytopal lattice $\mathcal{F}_{lv}(\mathcal{M})$ or a polytopal property which failed for matroid polytopes had been exhibited.

In 1980 Lawrence found an oriented matroid construction which could be used to produce a class of matroid polytopes whose face lattices are not polytopal. A description of this construction appeared in Munson (1981) and Billera and Munson (1984a,b or c). Intuitively, the basic idea of this method is to construct, from any arrangement $\mathcal{H}$, a new, larger arrangement $\Lambda(\mathcal{H})$ of higher rank, such that the *entire face lattice* of $\mathcal{H}$ is encoded in the *positive tope* of $\Lambda(\mathcal{H})$.

Today the Lawrence construction is the most universal tool for transferring realizability results from oriented matroids (i.e. arrangements of hyperplanes or points) to the theory of convex polytopes; see Sturmfels (1988a), Bayer and Sturmfels (1990). Lawrence polytopes have been useful for a variety of combinatorial, geometric, topological and algebraic problems, and therefore we will give a detailed exposition of this technique. The Lawrence construction is fundamental also for our discussion of the Steinitz problem in Section 9.5.

Let $\mathcal{M}$ be a rank r oriented matroid on a set $E = \{e_1, e_2, \ldots, e_n\}$. Introduce n new elements $E' = \{e'_1, e'_2, \ldots, e'_n\}$, and write $A' := \{e' : e \in A\}$ for every subset $A \subset E$. We define a rank $n + r$ oriented matroid $\Lambda(\mathcal{M})$ on the set $E \cup E'$ by the sequence of lexicographic liftings as follows. The dual $\Lambda(\mathcal{M})^*$ is obtained from $\mathcal{M}^*$ by a sequence of lexicographic point extensions $e'_1 := [e_1^-], e'_2 := [e_2^-], \ldots, e'_n := [e_n^-]$. In other words, the new element e'_i is antiparallel to e_i in $\Lambda(\mathcal{M})^*$. We note that the Lawrence lifting $\Lambda(\mathcal{M})$ depends only on the reorientation class of $\mathcal{M}$, i.e., $\Lambda(\mathcal{M}) = \Lambda(_{-A}(\mathcal{M}))$ for all $A \subset E$.

9.3.1 Lemma. *Let $\mathcal{M}$ be a loopless oriented matroid on E, and let $\Lambda(\mathcal{M})$ be its Lawrence lifting as defined above.*

(a) *If $X = (X^+, X^-)$ is a circuit of $\mathcal{M}$, then $\left(X^+ \cup (X^-)', X^- \cup (X^+)'\right)$ is a circuit of $\Lambda(\mathcal{M})$. Moreover, all circuits of $\Lambda(\mathcal{M})$ are obtained in this way.*

(b) *If $e \in E$ is a coloop of $\mathcal{M}$, then both e and e' are coloops of $\Lambda(\mathcal{M})$. If $e \in E$ is not a coloop of $\mathcal{M}$, then $\{e, e'\}$ is a positive cocircuit of $\Lambda(\mathcal{M})$.*

(c) $\Lambda(\mathcal{M})$ *is a matroid polytope, called the Lawrence polytope associated with $\mathcal{M}$.*

Proof. Let N be any oriented matroid and suppose that $\widetilde{N} = N \cup e'$ is obtained from N by an antiparallel extension $e' := [e^-]$ for some element e of N. If $X = (X^+, X^-)$ is a cocircuit of N with $e \in X^+$, then $(X^+, X^- \cup \{e'\})$ is a cocircuit of $\widetilde{N}$, and if $e \in X^-$, then $(X^+ \cup \{e'\}, X^-)$ is a cocircuit of $\widetilde{N}$. All cocircuits of $\widetilde{N}$ are of this form. It follows from the orthogonality of circuits and cocircuits that $(\{e, e'\}, \emptyset)$ is in the signed circuit span of $\widetilde{N}$. If e is a loop of N, then both $\{e\}$ and $\{e'\}$ are loops of $\widetilde{N}$. Otherwise $\{e, e'\}$ is a positive circuit of $\widetilde{N}$. The assertions (a) and (b) now follow using duality and induction on the number of new points.

In order to see that $\Lambda(\mathcal{M})$ is a matroid polytope, we note that $\{e, e'\}$ is a face of $\Lambda(\mathcal{M})$ for every $e \in E$ by part (b). Since e is not a loop in $\mathcal{M}$, the set $E \setminus e$ spans $\mathcal{M}^*$ and therefore $(E \cup E') \setminus \{e, e'\}$ spans $\Lambda(\mathcal{M})^*$. Hence $\{e, e'\}$ is a simplicial face, and both e and e' are vertices of $\Lambda(\mathcal{M})$. □

It is not difficult to see that the conditions (a) and (b) characterize the Lawrence polytope $\Lambda(\mathcal{M})$. Before continuing our general discussion, we reformulate the Lawrence construction for the acyclic oriented matroid of a point configuration $X = \{x_1, x_2, \ldots, x_n\} \subset \mathbb{R}^{r-1}$. Let $\overline{X} = \{\overline{x}_1, \overline{x}_2, \ldots, \overline{x}_n\} \subset \mathbb{R}^{n-r}$ be a Gale transform of X. We centrally symmetrize $\overline{X}$, that is, we consider the set $\overline{X} \cup -\overline{X}$. By Proposition 9.1.5 (2), $\overline{X} \cup -\overline{X}$ is the Gale transform of an $(n + r - 1)$-dimensional polytope P, which is the Lawrence polytope associated with X.

9.3.2 Proposition (Bayer and Sturmfels 1990). *Let P be a d-polytope with $2n$ vertices $V = \{v_1, \ldots, v_n, v_1', \ldots, v_n'\} \subset \mathbb{R}^d$, and suppose that $r := d + 1 - n \geq 0$. Then the following statements are equivalent:*

(a) P *is a Lawrence polytope, i.e., the matroid polytope of P is of the form $\Lambda(\mathcal{M})$ where $\mathcal{M}$ is a realizable rank r oriented matroid on n elements;*

(b) P *has a centrally symmetric Gale diagram;*

(c) $\mathrm{conv}(V \setminus \{v_i, v_i'\})$ *is a face of P for all $i = 1, \ldots, n$;*

(d) *after a suitable projective transformation, homogeneous coordinates for V are given by the rows of a $2n \times (d+1)$-matrix $\begin{pmatrix} \mathbf{I} & \mathbf{0} \\ \mathbf{I} & \mathbf{B} \end{pmatrix}$, where $\mathbf{0}$ denotes the $n \times r$-zero matrix, $\mathbf{I}$ denotes the $n \times n$-unit matrix, and $\mathbf{B}$ is an $n \times r$-matrix of rank r.*

The proof of Proposition 9.3.2 is left as an exercise for the reader. We next give a complete description of all the cocircuits of $\Lambda(\mathcal{M})$ in terms of the cocircuits of $\mathcal{M}$. Let $\mathcal{M}$ be an oriented matroid on E with cocircuits $\mathcal{C}^*(\mathcal{M})$. For every $Y \in \mathcal{C}^*(\mathcal{M})$ and $A \subset E$, we define a sign vector Y^A on $E \cup E'$ by $Y^A :=$ $\left((Y^+ \setminus A) \cup (Y^- \cap A)', (Y^- \setminus A) \cup (Y^+ \cap A)' \right)$.

9.3.3 Proposition.   *The set of cocircuits of the Lawrence polytope $\Lambda(\mathcal{M})$ of $\mathcal{M}$ equals*

$$\mathcal{C}^*(\Lambda(\mathcal{M})) = \{ (\{e, e'\}, \emptyset), (\emptyset, \{e, e'\}) : e \in E \text{ is not a coloop of } \mathcal{M} \}$$
$$\cup \{ Y^A : Y \in \mathcal{C}^*(\mathcal{M}) \text{ and } A \subset \underline{Y} \}.$$

The proof of Proposition 9.3.3 is straightforward from Lemma 9.3.1. Let us take a look at some examples of Lawrence polytopes. It is clear from Proposition 9.3.2 that every prism over a simplex is a Lawrence polytope. To see this, pick the $n \times 1$-matrix $\mathbf{B} := (1, 1, 1, \ldots, 1)^T$ in order to obtain homogeneous coordinates as in Proposition 9.3.2(d) for the prism over an $(n-1)$-simplex. (This is an n-polytope with $2n$ vertices.)

In Example 9.1.7 we discussed the matroid polytope $\mathcal{M}$ corresponding to a prism over a 3-simplex. Using the identification $5 = 1'$, $6 = 2'$, $7 = 3'$, and $8 = 4'$, we see that $\mathcal{M}$ is the Lawrence polytope $\mathcal{M} = \Lambda(N)$ of the rank 1 uniform oriented matroid N on $\{1, 2, 3, 4\}$ with circuits $\mathcal{C}(N) = \pm\{12, 1\overline{3}, 14, 23, 2\overline{4}, 34\}$. If we apply Lemma 9.3.1(a) to $\mathcal{C}(N)$, then we see that $\mathcal{C}(\Lambda(N))$ equals the set of circuits of $\mathcal{M}$ listed in Example 9.1.7. Since rank 1 matroids have too little structure to exhibit all important features of their associated Lawrence polytopes, we will now consider a larger example.

9.3.4 Example. Let $\mathcal{M}$ be the rank 2 oriented matroid on E_5 defined by the matrix

$$\mathbf{A} = \begin{pmatrix} 1 & -1 & 0 & 1 & -2 \\ 0 & -1 & 1 & -2 & 1 \end{pmatrix}.$$

Its Lawrence polytope $\Lambda(\mathcal{M})$ is the rank 7 matroid polytope on E_{10} defined by the columns of the matrix

$$\Lambda(\mathbf{A}) = \begin{pmatrix} 1 & 0 & 0 & 0 & 0 & 1 & 0 & 0 & 0 & 0 \\ 0 & 1 & 0 & 0 & 0 & 0 & 1 & 0 & 0 & 0 \\ 0 & 0 & 1 & 0 & 0 & 0 & 0 & 1 & 0 & 0 \\ 0 & 0 & 0 & 1 & 0 & 0 & 0 & 0 & 1 & 0 \\ 0 & 0 & 0 & 0 & 1 & 0 & 0 & 0 & 0 & 1 \\ 0 & 0 & 0 & 0 & 0 & 1 & -1 & 0 & 1 & -2 \\ 0 & 0 & 0 & 0 & 0 & 0 & -1 & 1 & -2 & 1 \end{pmatrix}.$$

Similarly to the previous example, we identify $6 = 1'$, $7 = 2'$, $8 = 3'$, $9 = 4'$, and $0 = 5'$. It is our goal to determine the circuits, cocircuits and all facets of the 6-polytope $\Lambda(\mathcal{M})$ whose 10 vertices are homogeneously coordinatized by the columns of the matrix $\Lambda(\mathbf{A})$.

The duals of $\mathcal{M}$ and $\Lambda(\mathcal{M})$ are coordinatized by the matrices

$$\mathbf{B} \;=\; \begin{pmatrix} 1 & 1 & 1 & 0 & 0 \\ -1 & 0 & 2 & 1 & 0 \\ 2 & 0 & -1 & 0 & 1 \end{pmatrix}$$

and $(\mathbf{B}, -\mathbf{B})$ respectively. In Figure 9.3.1 we depict these realizations of the oriented matroids $\mathcal{M}, \mathcal{M}^*$ and $\Lambda(\mathcal{M})^*$.

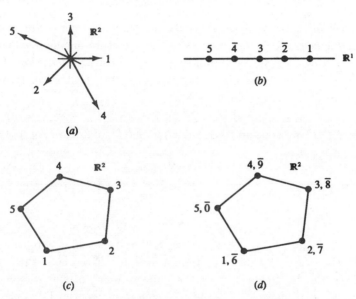

Figure 9.3.1: The oriented matroids $\mathcal{M}$ (a and b), $\mathcal{M}^*$ (c), and $\Lambda(\mathcal{M})^*$ (d).

Figure 9.3.1(a) shows the linear realization $\mathbf{A}$ of $\mathcal{M}$, while (b) shows an affine realization of $\mathcal{M}$ (using an acyclic reorientation). This is the affine Gale diagram of the matroid polytope $\mathcal{M}^*$, a pentagon, whose affine realization is depicted in (c). Figure 9.3.1(d) shows an affine reorientation (using acyclic reorientations on the new points) of $\Lambda(\mathcal{M})^*$, that is, (d) is an affine Gale diagram of the Lawrence polytope $\Lambda(\mathcal{M})$ in question. We see from Figure 9.3.1 that the circuits and cocircuits of $\mathcal{M}$ are given by

$$\mathcal{C}(\mathcal{M}) \;=\; \pm\{123, 12\overline{4}, 125, \overline{1}34, 1\overline{3}5, 145, 234, 23\overline{5}, \overline{2}45, 345\}$$

and

$$\mathcal{C}^*(\mathcal{M}) \;=\; \pm\{1\overline{2}34, 1\overline{2}35, 1\overline{2}45, 1\overline{3}45, 2\overline{3}45\}.$$

By Lemma 9.3.1, the circuits of the Lawrence extension $\Lambda(\mathcal{M})$ are given by

$$\mathcal{C}(\Lambda(\mathcal{M})) = \pm\{123\overline{678}, 12\overline{4}679, 125\overline{670}, \overline{1}34\overline{689}, 1\overline{3}5\overline{680}, 145690, 234789,$$
$$235\overline{780}, \overline{2}45\overline{790}, 345\overline{890}\}$$

Using Proposition 9.3.3, we can read off the cocircuits of $\Lambda(\mathcal{M})$. We obtain

$$\mathcal{C}^*(\Lambda(\mathcal{M})) =$$
$$\pm\{1\overline{2}3\overline{4}, 1\overline{2}3\overline{5}, 1\overline{2}3\overline{9}, 1\overline{2}3\overline{0}, 1\overline{2}4\overline{5}, \overline{1}2\overline{4}8, 1\overline{2}4\overline{0}, \overline{1}2\overline{5}8, \overline{1}2\overline{5}9, 1\overline{2}8\overline{9}, 1\overline{2}8\overline{0}, 1\overline{2}9\overline{0},$$
$$1\overline{3}4\overline{5}, 13\overline{4}\overline{7}, 1\overline{3}4\overline{0}, 13\overline{5}\overline{7}, \overline{1}3\overline{5}9, 1379, 1370, 1\overline{3}9\overline{0}, 14\overline{5}\overline{7}, 14\overline{5}\overline{8}, 1\overline{4}7\overline{8},$$
$$1470, 1480, 1\overline{5}7\overline{8}, 1\overline{5}7\overline{9}, 1\overline{5}8\overline{9}, 16, 17\overline{8}9, \ldots\}$$

As a result we obtain the following list of facets of $\Lambda(\mathcal{M})$:

    12346789, 12356780, 12456790, 124680, 124780, 13456890, 134780,

    134790, 135679, 135790, 23457890, 235679, 235689, 245689, 245680.

In the above example we have seen how the point and the vector realizations of an oriented matroid $\mathcal{M}$ are related to the structure and face lattice of the Lawrence polytope $\Lambda(\mathcal{M})$. We will now give an intrinsic description of the face lattice $\mathcal{F}_{lv}(\Lambda(\mathcal{M}))$ in terms of the big lattice $\widehat{\mathcal{L}}(\mathcal{M})$, that is, in terms of the (pseudo-)hyperplane arrangement of $\mathcal{M}$.

Let $\mathcal{M}$ be a rank r oriented matroid on E with set of covectors $\mathcal{L}(\mathcal{M})$. For any subset $I \subset E$, we consider the restriction $\mathcal{M}(I)$ of $\mathcal{M}$ to I, and we identify its set of covectors $\mathcal{L}(\mathcal{M}(I))$ with the image of $\mathcal{L}(\mathcal{M})$ under the canonical projection $\{-,0,+\}^E \to \{-,0,+\}^I$. We define the *cover poset* $\mathcal{U}(\mathcal{M})$ of $\mathcal{M}$ to be the set of pairs (F,I) where $I \subset E$ and $F \in \mathcal{L}(\mathcal{M}(I))$, partially ordered by the relation

$$(F,I) \preceq (G,J) \quad :\Longleftrightarrow \quad I \supseteq J, \quad F^+ \cap J \subseteq G^+, \quad \text{and} \quad F^- \cap J \subseteq G^-.$$

If we think of $\{\mathcal{L}(\mathcal{M}(I))\}_{I \subset E}$ as the "sheaf" of all posets of subarrangements, indexed by the poset 2^E and with the natural inclusion maps, then the cover poset $\mathcal{U}(\mathcal{M})$ can be thought of as the "direct limit" of this directed family of posets. We will show that $\mathcal{U}(\mathcal{M})$ is isomorphic to the Edmonds-Mandel face lattice $\mathcal{F}_{em}(\Lambda(\mathcal{M}))$ of the Lawrence polytope $\Lambda(\mathcal{M})$, and thus it is anti-isomorphic to $\mathcal{F}_{lv}(\Lambda(\mathcal{M}))$. In the following we abbreviate $I^c := E \setminus I$.

9.3.5 Proposition (Bayer and Sturmfels 1990). *For any oriented matroid $\mathcal{M}$ the map*

$$\phi : (F,I) \quad \mapsto \quad (F^+ \cup I^c) \cup (F^- \cup I^c)'$$

is an order-preserving isomorphism from the cover lattice $\mathcal{U}(\mathcal{M})$ to the Edmonds-Mandel face lattice $\mathcal{F}_{em}(\Lambda(\mathcal{M}))$ of the Lawrence polytope $\Lambda(\mathcal{M})$ associated with $\mathcal{M}$.

Proof. We first show that $\phi(F,I)$ is a positive covector of $\Lambda(\mathcal{M})$ for any $F \in \mathcal{L}(\mathcal{M}(I))$. The covector F of $\mathcal{M}(I)$ can be written as a conformal union $F = \cup_{j=1}^k F_j$ of cocircuits F_j. There exist cocircuits G_j of $\mathcal{M}$ such that $G_j \cap I = F_j$ for $j = 1, \ldots, k$. By Proposition 9.3.3, the sets $G_j^+ \cup (G_j^-)'$ are positive cocircuits of $\Lambda(\mathcal{M})$. Since $\{e, e'\}$ is a positive covector of $\Lambda(\mathcal{M})$ for every e, we conclude that

$$\phi(F,I) = \left(\cup_{j=1}^k F_j^+ \cup I^c \right) \cup \left(\cup_{j=1}^k F_j^- \cup I^c \right)'$$
$$= \cup_{j=1}^k (G_j^+ \cup (G_j^-)') \cup [\cup_{e \in I^c}\{e, e'\}]$$

is a positive covector of $\Lambda(\mathcal{M})$. The map ϕ is clearly order-preserving.

In order to show that ϕ is a lattice isomorphism, we will construct the inverse for ϕ. Given any positive covector C of $\Lambda(\mathcal{M})$, we write $S := C \cap E$ and $T := C \cap E'$, and we consider the sign vector

$$F := (S \setminus T', T' \setminus S).$$

We will show that F is a covector of $\mathcal{M}(I)$ where $I := E \setminus (S \cap T')$.

We consider a decomposition $C = \cup_{j=1}^k C_j$ into positive cocircuits C_j of $\Lambda(\mathcal{M})$, and we partition also each cocircuit $C_j = S_j \cup T_j$ as above. Suppose first that $S \cap T' = \emptyset$, or, equivalently $I = E$. By Proposition 9.3.3, the sign vectors $F_j = (S_j, T_j')$ are a conformal set of cocircuits of $\mathcal{M}$. So, their union $F = (S, T')$ is a covector of $\mathcal{M}(I) = \mathcal{M}$.

Now suppose that $S \cap T' \neq \emptyset$. In this case the sets

$$D_j = (S_j \setminus T') \cup (T_j \setminus S')$$

are positive cocircuits of $\Lambda(\mathcal{M}(I))$. Applying the argument in the previous paragraph to $D = \cup_{j=1}^k D_j = C \setminus [(S \cap T') \cup (S' \cap T)]$ (with $(D \cap E) \cap (D \cap E')' = \emptyset$), we see that the sign vector $F = (D \cap E, (D \cap E'))'$ is a covector of $\mathcal{M}(I)$. The proof is completed by checking that in both cases $\phi(F, I) = C$. □

As an easy corollary we obtain from Proposition 9.3.5 that Lawrence polytopes are rigid. From the realizability point of view of Section 9.5, this result may be viewed as the "main theorem on Lawrence polytopes". Proposition 9.3.6 is implicit in Munson (1981), and it appeared explicitly in Sturmfels (1985) and in Bokowski and Sturmfels (1987).

9.3.6 Proposition. *Every Lawrence polytope $\Lambda(\mathcal{M})$ is rigid, i.e., $\Lambda(\mathcal{M})$ is uniquely determined by its face lattice $\mathcal{F}_{lv}(\Lambda(\mathcal{M}))$.*

Proof. Observe that the big lattice

$$\mathcal{L}(\mathcal{M}) = \{ (F, E) \in \mathcal{U}(\mathcal{M}) : F \in \mathcal{L}(\mathcal{M}) \}$$

is a sublattice of $\mathcal{U}(\mathcal{M})$. Since $\mathcal{U}(\mathcal{M})$ and $\mathcal{F}_{em}(\Lambda(\mathcal{M}))$ are isomorphic by Proposition 9.3.5, we conclude that the big lattice $\mathcal{L}(\mathcal{M})$ is determined by $\mathcal{F}_{em}(\Lambda(\mathcal{M}))$. Hence the oriented matroid $\mathcal{M}$ is determined by the Las Vergnas face lattice $\mathcal{F}_{lv}(\Lambda(\mathcal{M}))$. This implies Proposition 9.3.6 because the Lawrence polytope $\Lambda(\mathcal{M})$ is determined by $\mathcal{M}$. □

9.3.7 Theorem (J. Lawrence, see Billera and Munson (1984a)).

(a) *The face lattice of $\Lambda(\mathcal{M})$ is polytopal if and only if $\mathcal{M}$ is realizable.*

(b) *There exists a rank 12 matroid polytope with 16 vertices whose face lattice is not polytopal.*

Proof. The statement (a) follows directly from Proposition 9.3.6. We obtain (b) by taking the non-realizable uniform rank 4 oriented matroid RS(8) discussed in Section 1.5. □

Theorem 9.3.7 shows that, using Lawrence polytopes, we can produce an infinite family of non-polytopal lattices arising from matroid polytopes. Stronger versions of this result will be given in Section 9.5. These results raise the important question of whether the class $\{\mathcal{F}_{lv}(\mathcal{M})\}$ of Las Vergnas face lattices equals the class $\{\mathcal{F}_{lv}(\mathcal{M})^{op}\} = \{\mathcal{F}_{em}(\mathcal{M})\}$ of Edmonds-Mandel face lattices of oriented matroids.

We define a *polar* of a matroid polytope $\mathcal{M}$ to be a matroid polytope $\mathcal{M}^\Delta$ such that $\mathcal{F}_{lv}(\mathcal{M}^\Delta) = \mathcal{F}_{lv}(\mathcal{M})^{op}$, i.e., $\mathcal{F}_{lv}(\mathcal{M})$ is anti-isomorphic to $\mathcal{F}_{lv}(\mathcal{M}^\Delta)$. Hence the above question is equivalent to asking whether every matroid polytope has a polar.

This issue is of interest as an analogue to the classical theorems of Weyl and Minkowski. These state that every convex polytope in $\mathbb{R}^d$ can be obtained both as the convex hull of finitely many points and as the bounded intersection of a finite number of closed halfspaces. While the Las Vergnas lattice $\mathcal{F}_{lv}(\mathcal{M})$ is the oriented matroid analogue of the convex hull of points, its polar, the Edmonds-Mandel lattice $\mathcal{F}_{em}(\mathcal{M})$, is the analogue of the intersection of a finite number of closed halfspaces. A matroid polytope $\mathcal{M}$ has the polar $\mathcal{M}^\Delta$ if and only if both $\mathcal{F}_{lv}(\mathcal{M}) = \mathcal{F}_{em}(\mathcal{M}^\Delta)$ and $\mathcal{F}_{lv}(\mathcal{M}^\Delta) = \mathcal{F}_{em}(\mathcal{M})$. In this case both $\mathcal{F}_{lv}(\mathcal{M})$ and $\mathcal{F}_{lv}(\mathcal{M}^\Delta)$ arise as the "convex hull of points" and as the "intersection of halfspaces". It has been proved by Billera and Munson (1984a) that not every matroid polytope admits a polar, and therefore not every lattice of the form $\mathcal{F}_{lv}(\mathcal{M})$ has this property. The key observation is the connection between polars and adjoints – as introduced in Section 7.5. We first observe that adjoints can be used to construct polars.

9.3.8 Proposition (Bachem and Kern 1986a, Cordovil 1987). *Let $\mathcal{M}$ be a matroid polytope with an adjoint $\mathcal{M}^{ad}$ on $E^{ad} = \mathcal{C}^*(\mathcal{M})$. Then the restriction $\mathcal{M}^{ad}(\{Y \in E^{ad} : Y \geq 0\})$ is a polar of $\mathcal{M}$.*

Proof. Let $(\mathcal{A}, E^{ad})$ be a Type II representation of $\mathcal{M}$ (cf. Theorem 5.3.6). That is, we construct a pseudosphere arrangement $\mathcal{A} = (S_Y)_{Y \in \mathcal{C}^*}$ of rank $r = \text{rank}(\mathcal{M})$ in which the points of E are identified with a certain set of vertices. Now we pass to the subarrangement $\mathcal{A}^0 = \{S_Y \in \mathcal{A} : Y \geq 0\}$ corresponding to the facets of $\mathcal{M}$. Then the elements of E are exactly the vertices of the positive tope T, and this implies the desired result

$$\mathcal{F}_{lv}(\mathcal{M}) = [\hat{0}, T] = \mathcal{F}_{em}(\mathcal{M}^{ad}(\{Y \geq 0\})). \qquad \square$$

The argument of Billera and Munson (1984a) now combines the Lawrence construction with polars and adjoints. They proved the following theorem using adjoints of the underlying matroid. Here we give the extension to oriented adjoints. For simplicity, we restrict ourselves to the case without coloops.

9.3.9 Proposition (Billera and Munson 1984a). *Let $\mathcal{M}$ be an oriented matroid without coloops such that its Lawrence lifting $\Lambda(\mathcal{M})$ has a polar. Then $\mathcal{M}$ has an adjoint.*

Proof. Let P be a polar of $\Lambda(\mathcal{M})$, with ground set

$$\{Y \in \mathcal{C}^*(\Lambda(\mathcal{M})) : Y \geq 0\},$$

of rank $r + n = \mathrm{rank}(\mathcal{M}) + |E|$. If

$$\mathcal{C}^*(\mathcal{M}) = \{W_1, W_2, \ldots, W_k\},$$

then the positive cocircuits of $\Lambda(\mathcal{M})$ are given as

$$\{Z_1, Z_2, \ldots, Z_n, W_1^p, \ldots, W_k^p\},$$

where $Z_i = \{e_i, e_i'\}$ and $W_j^p = (W_j^+) \cup (W_j^-)'$. We consider the matrix whose columns are the positive cocircuits of $\Lambda(\mathcal{M})$:

$$
\begin{array}{c}
\begin{array}{ccccccccc}
Z_1 & Z_2 & \cdots & Z_n & W_1^p & W_2^p & W_3^p & \cdots & W_k^p
\end{array} \\
\begin{array}{c}
Y_1 \\ Y_2 \\ \vdots \\ Y_n \\ Y_1' \\ Y_2' \\ \vdots \\ Y_n'
\end{array}
\left(
\begin{array}{ccccccccc}
+ & & & & \epsilon_{11} & \epsilon_{12} & \epsilon_{13} & \cdots & \epsilon_{1k} \\
& + & & & \epsilon_{21} & \epsilon_{22} & \epsilon_{23} & \cdots & \epsilon_{2k} \\
& & \ddots & & \vdots & \vdots & \vdots & \ddots & \vdots \\
& & & + & \epsilon_{n1} & \epsilon_{n2} & \epsilon_{n3} & \cdots & \epsilon_{nk} \\
+ & & & & \epsilon_{11}' & \epsilon_{12}' & \epsilon_{13}' & \cdots & \epsilon_{1k}' \\
& + & & & \epsilon_{21}' & \epsilon_{22}' & \epsilon_{23}' & \cdots & \epsilon_{2k}' \\
& & \ddots & & \vdots & \vdots & \vdots & \ddots & \vdots \\
& & & + & \epsilon_{n1}' & \epsilon_{n2}' & \epsilon_{n3}' & \cdots & \epsilon_{nk}'.
\end{array}
\right)
\end{array}
$$

In this matrix we have $\epsilon_{ij} = +$ if $e_i \in W_j^+$ and $\epsilon_{ij} = 0$ otherwise, and $\epsilon_{ij}' = +$ if $e_i \in W_j^-$ and $\epsilon_{ij}' = 0$ otherwise. By definition of a polar, the positive cocircuits of P are given by the rows $Y_1, \ldots, Y_n, Y_1', \ldots, Y_n'$ of this matrix, whose column indices give the ground set of P.

We will now show that the contraction $P_0 := P/\{Z_1, \ldots, Z_n\}$ is an adjoint of $\mathcal{M}$. To see this, we first observe that the set $\{Z_1, \ldots, Z_n\}$ is independent in P. Suppose it did contain a circuit X with $Z_i \in X$. Then we would get $|X \cap Y^i| = 1$, which is impossible since Y^i is a cocircuit of P. Thus P_0 has rank r, as required for an adjoint of $\mathcal{M}$. Its ground set is $\{W_1^p, W_2^p, \ldots, W_k^p\}$, which is naturally identified with the required ground set $\mathcal{C}^*(\mathcal{M}) = \{W_1, W_2, \ldots, W_k\} = E^{ad}$.

Applying the strong cocircuit elimination axiom to Y_i and Y_i', eliminating Z_i, we find that $F_i := (Y_i \setminus \{Z_i\}, Y_i' \setminus \{Z_i\})$ is a covector of P_0. It corresponds to the element $e_i \in E$, and in order to complete the proof we need to show that it is a cocircuit of P_0.

After relabeling, we may assume that $\{e_1, e_2, \ldots, e_r\}$ is a basis of $\mathcal{M}$, and that $W_1, W_2, \ldots, W_r$ are the basic cocircuits corresponding to this basis. For $1 \leq i < j \leq r$ we have $e_i \notin W_j$, hence $\epsilon_{ij} = \epsilon_{ij}' = 0$ and therefore $W_j \notin F_i$. This implies that

$$0 < F_1 < F_1 \circ F_2 < F_1 \circ F_2 \circ F_3 < \cdots < F_1 \circ F_2 \circ F_3 \circ \ldots \circ F_r$$

is a strictly increasing chain of covectors of P_0. Since the rank of P_0 equals r, this means that F_1 is a cocircuit. $\qquad\square$

As an immediate application of this result, we can construct oriented matroids without a polar. It is known that there are oriented matroids of rank 4 on eight points which do not have an adjoint. Any orientation of the Vámos matroid (see Example 7.4.1) or the oriented matroid EFM(8) in Section 10.4 fail to have adjoints. This implies the following.

9.3.10 Corollary (Billera and Munson 1984a).   *There exist rank 12 matroid polytopes with 16 vertices which do not have polars.*

The above results lead to the following apparent inconsistency. In the realizable case, the Farkas lemma is the key result in proving (and, in fact, is equivalent to) both the duality theorem of linear programming and the theorems of Weyl and Minkowski on polarity for polytopes and polyhedral cones. In Chapter 10 we show that the Farkas lemma and linear programming duality are valid for all oriented matroids. On the other hand, Corollary 9.3.10 shows that polarity fails for some oriented matroids. As Munson points out in the end of her 1981 dissertation, "one is led to conclude that duality and polarity are, from a combinatorial perspective, essentially different phenomena".

It is one of the most prominent unsolved problems in combinatorial convexity to find an intrinsic characterization of the f-vectors of convex polytopes. Such a characterization is known for simplicial polytopes by the results of Stanley (1980a) and Billera and Lee (1981).

Oriented matroids provide a suitable framework for studying this problem for non-simplicial polytopes. Perhaps there is a matroid polytope $\mathcal{M}$ with the property that $f(\mathcal{M}) = f(\mathcal{M}')$ implies the non-realizability of $\mathcal{M}'$. An early hope that Lawrence polytopes could produce such a counter-example did not materialize. Quite to the contrary, this circle of ideas motivated the following result.

9.3.11 Proposition (Bayer and Sturmfels 1990).   *For any oriented matroid $\mathcal{M}$, the f-vector of the Lawrence polytope $\Lambda(\mathcal{M})$ is a function of the underlying matroid $\underline{\mathcal{M}}$.*

This proposition is a polytopal analogue of the Las Vergnas-Zaslavsky Theorem 4.6.1, which states that $f(\mathcal{L}(\mathcal{M}))$ is a function of $\underline{\mathcal{M}}$. The proof of Proposition 9.3.11 goes as follows. By Proposition 9.3.5, it suffices to prove that $\underline{\mathcal{M}}$ determines the f-vector of the cover poset $\mathcal{U}(\mathcal{M})$. It can be seen that each rank level set of $\mathcal{L}(\mathcal{M}(I))$ is mapped into a single rank level set of $\mathcal{U}(\mathcal{M})$ by $\phi : F \mapsto (F, I)$. Hence each $f_i(\mathcal{U}(\mathcal{M}))$ is the sum of a finite number of terms of the form $f_j(\mathcal{L}(\mathcal{M}(I)))$. The Las Vergnas-Zaslavsky theorem implies that $f_j(\mathcal{L}(\mathcal{M}(I)))$ is determined by $\underline{(\mathcal{M}(I))} = \underline{\mathcal{M}}(I)$, which depends only on $\underline{\mathcal{M}}$ (and I).

For the case when $\mathcal{M}$ is uniform, an explicit formula for $f(\Lambda(\mathcal{M}))$ in terms of the rank and number of vertices of $\mathcal{M}$ is given in Bayer and Sturmfels (1990). Also, a further generalization of Proposition 9.3.11 to flag vectors of Lawrence polytopes is proved.

9.4 Cyclic and neighborly matroid polytopes

While the Lawrence polytopes discussed in the previous section are far from being in general position, here we are concerned with an important family of uniform matroid polytopes. The class of neighborly polytopes and its most prominent representatives, the cyclic polytopes, receive much attention in combinatorial geometry because of their strong extremal properties. The upper bound theorem of McMullen (1971b), for instance, states that the number of j-dimensional faces of a d-polytope with n vertices is maximal for the cyclic d-polytope with n vertices. It is the aim of this section to study cyclic and neighborly polytopes from the point of view of oriented matroids.

We begin by rephrasing some basic facts about cyclic polytopes. Consider the rank r oriented matroid $C^{n,r}$ on E_n which is defined through its chirotope $\chi^{n,r}$ as

$$\chi^{n,r}(e_1, e_2, \ldots, e_r) \quad := \quad +1 \quad \text{whenever} \quad e_1 < e_2 < \ldots < e_r.$$

We call $C^{n,r}$ the *alternating* oriented matroid of rank r on n vertices.

9.4.1 Proposition. *The alternating oriented matroid $C^{n,r}$ is realizable. If $r \geq 3$, then $C^{n,r}$ is a matroid polytope.*

Proof. Any n distinct points on the moment curve $\mathbb{R} \to \mathbb{R}^{r-1}, t \mapsto (t, t^2, \ldots, t^{r-1})$ in affine $(r-1)$-space provide a realization of $C^{n,r}$. More precisely, pick any sequence of increasing real numbers $t_1 < t_2 < \cdots < t_n$ and consider the $n \times r$-matrix

$$\begin{pmatrix} 1 & t_1 & t_1^2 & \ldots & t_1^{r-1} \\ 1 & t_2 & t_2^2 & \ldots & t_2^{r-1} \\ \vdots & \vdots & \vdots & \ddots & \vdots \\ 1 & t_n & t_n^2 & \ldots & t_n^{r-1} \end{pmatrix}.$$

Every maximal minor of this matrix is a Vandermonde determinant and therefore easily seen to be positive.

In order to show that $C^{n,r}$ is a matroid polytope, we describe its set of signed circuits. Since the oriented matroid $C^{n,r}$ is uniform, every $(r+1)$-element subset of E_n supports a pair of circuits. Translating the chirotope into signed circuits, we see that the circuit supported by $r+1$ points $e_1 < e_2 < e_3 < \ldots < e_{r+1}$ equals the sign vector $X = e_1 \overline{e_2} e_3 \overline{e_4} e_5 \overline{e_6} e_7 \ldots$. This implies the inequality $|X^+| \geq \lfloor \frac{r+1}{2} \rfloor$ for every circuit $X = (X^+, X^-)$ of $C^{n,r}$. If $r \geq 3$, then $|X^+| \geq 2$. In this case we can use Corollary 9.1.3 to conclude that every singleton $\{e\}$ is a face of $C^{n,r}$, which means that $C^{n,r}$ is a matroid polytope. $\square$

The alternating sign pattern in the circuits of $C^{n,r}$ was first studied by Bland and Las Vergnas (1978). In the literature the term "alternating" is sometimes also used for certain weak images of $C^{n,r}$ whose circuits still have the alternating sign pattern. Here we are dealing exclusively with uniform alternating oriented matroids. For a detailed discussion and many results on non-uniform alternating oriented matroids we refer to da Silva's thesis (1987).

The following proposition characterizes the reorientation class of the alternating matroid in terms of inseparability graphs (cf. Section 7.8).

9.4.2 Proposition (Cordovil and Duchet 1987). *A uniform oriented matroid $\mathcal{M}$ of rank $r \geq 2$ on $n \geq r + 2$ elements is reorientation equivalent to the alternating matroid $C^{n,r}$ if and only if its inseparability graph is an n-cycle.*

Proof. Two consecutive vertices $i, i + 1$ of the alternating matroid polytope $C^{n,r}$ have opposite signs in all circuits containing them. So, $(i, i + 1)$ is a contravariant pair for all $i = 1, 2, \ldots, n-1$. If r is odd, then also the pair $(1, n)$ is contravariant in $C^{n,r}$; if r is even, then $(1, n)$ is covariant in $C^{n,r}$. In both cases, the inseparability graph $\mathrm{IG}(C^{n,r})$ is the n-cycle $(1, 2, 3, \ldots, n - 1, n, 1)$.

Conversely, let $\mathcal{M}$ be any uniform rank r oriented matroid on E_n such that $\mathrm{IG}(C^{n,r})$ is the n-cycle $(1, 2, \ldots, n-1, n, 1)$. After suitable reorientations we may assume that $(i, i+1)$ is contravariant for all $i = 1, 2, \ldots, n-1$. Let χ be the chirotope of $\mathcal{M}$ with $\chi(1, 2, \ldots, r) = +$. Suppose $\mathcal{M} \neq C^{n,r}$, and let $(\lambda_1, \ldots, \lambda_r)$ be the lexicographically earliest index r-tuple such that $\chi(\lambda_1, \ldots, \lambda_r) = -$. Writing $\lambda_0 = 0$, there exists some $i \in \{1, \ldots, r\}$ such that $\lambda_{i-1} < \lambda_i - 1$. Since $(\lambda_i - 1, \lambda_i)$ is a contravariant pair of $\mathcal{M}$, we have

$$\chi(\lambda_1, \ldots, \lambda_{i-1}, \lambda_i - 1, \lambda_{i+1}, \ldots, \lambda_r) = \chi(\lambda_1, \ldots, \lambda_{i-1}, \lambda_i, \lambda_{i+1}, \ldots, \lambda_r)$$
$$= -.$$

This is a contradiction to the choice of $(\lambda_1, \ldots, \lambda_r)$, and therefore $\mathcal{M} = C^{n,r}$. □

A rank r matroid polytope C on E_n will be called *cyclic* if its face lattice $\mathcal{F}_{lv}(C)$ equals the face lattice $\mathcal{F}_{lv}(C^{n,r})$ of the alternating oriented matroid $C^{n,r}$. From the definition of $\chi^{n,r}$ we obtain an explicit description of the positive cocircuits and hence the facets of cyclic matroid polytopes. This description is well known in polytope theory.

9.4.3 Proposition (Gale's evenness condition). *A hyperplane F of $C^{n,r}$ is a facet if and only if any two elements of $E_n \setminus F$ are separated by an even number of elements from F in the sequence $(1, 2, 3, \ldots, n)$.*

Next we consider the oriented matroid version of the geometric fact that all subpolytopes of even-dimensional (= odd rank) cyclic polytopes are again cyclic.

9.4.4 Proposition (Cordovil and Duchet 1987). *A matroid polytope of odd rank is cyclic if and only if it is alternating.*

Proposition 9.4.4 follows as a corollary from Theorem 9.4.13 which will be proved later in this section. In order to see that Proposition 9.4.4 fails for even rank, we consider the rank 4 oriented matroid $\widehat{C}$ of affine dependences on the six points $x_1 := (0, 0, 0)$, $x_2 := (1, 1, 1)$, $x_3 := (2, 4, 8)$, $x_4 := (3, 9, 27)$, $x_5 :=$

$(4, 16, 57)$ and $x_6 := (5, 25, 125)$. The oriented matroid $\widehat{C}$ is obtained from $C^{6,4}$ by switching the orientation of the invertible basis 2345. Since the basis 2345 does not contain any facet of $C^{6,4}$, this particular mutation has no effect on the face lattice and we have $\mathcal{F}_{lv}(\widehat{C}) = \mathcal{F}_{lv}(C^{6,4})$. Therefore $\widehat{C}$ is a uniform cyclic matroid polytope which is not alternating.

While Propositions 9.4.1 and 9.4.4 imply that all cyclic matroid polytopes of odd rank are realizable, in the even rank case there exist also non-realizable cyclic matroid polytopes.

9.4.5 Proposition (Richter and Sturmfels 1991). *There exists a non-realizable rank* 4 *cyclic matroid polytope* $\mathcal{M}$ *with* 10 *vertices.*

Proof. Let $\mathcal{M}_c$ denote the rank 4 oriented matroid on E_{10} of affine dependences among the row vectors of the following 10×3-matrix

$$\begin{pmatrix} 2.7 & 2 & 6 \\ 3 & 1 & 1 \\ 1 & 1 & 0 \\ -1 & 1 & 0 \\ -3 & 1 & 1 \\ -4 & 0 & 1 \\ -1 & -1 & -1 \\ 1 & -1 & -1 \\ 4 & 0 & 1 \\ 3.7 & 0 & -3 \end{pmatrix}.$$

The facets of $\mathcal{M}_c$ are 123, 120, 134, 145, 156, 167, 178, 189, 190, 230, 340, 450, 560, 670, 780, 890, that is, $\mathcal{M}_c$ is a cyclic matroid polytope by Proposition 9.4.3. Here the tenth row is labeled "0". The minor $\mathcal{M}_c \setminus \{1, 0\}$ by deletion of the first and the last point is a matroid polytope combinatorially equivalent to a cube: the facets of $\mathcal{M}_c \setminus \{1, 0\}$ are 2345, 2389, 2569, 3478, 4567, 6789. Furthermore, this "matroid cube" has two additional four-point planes, namely 2578 and 3469.

Now let $\widetilde{\mathcal{M}}_c$ be the oriented matroid obtained from $\mathcal{M}_c$ by redefining the orientation of a suitable degenerate tetrahedron, e.g. 2578, to be $+1$ rather than 0. Since this has no effect on the face lattice, the matroid polytope $\widetilde{\mathcal{M}}_c$ is cyclic as well.

We establish a final polynomial in order to show that $\widetilde{\mathcal{M}}_c$ is not realizable. The bracket polynomial

$$\begin{aligned} & \underline{[2345]}\,[2679][2789][3479] \\ + & \underline{[2569]}\,[2347][2789][3479] \\ - & \underline{[2578]}\,[2679][2349][3479] \\ - & \underline{[3469]}\,[2347][2789][2579] \\ + & \underline{[3478]}\,[2679][2579][2349] \\ + & \underline{[6789]}\,[2347][2579][2349] \end{aligned}$$

is an algebraic dependence among the maximal minors $[ijkl]$ of any 8×4-matrix.

Under any realization of $\widetilde{\mathcal{M}_c}$ precisely the underlined terms would be zero. This is a contradiction, hence $\widetilde{\mathcal{M}_c}$ is a non-realizable cyclic matroid polytope. $\square$

We remark that with the same final polynomial one can also produce a non-realizable *uniform* cyclic matroid polytope. To this end assign $+1$ and -1 in such a way to the underlined non-bases that all summands in the above final polynomial have the same sign.

Cordovil and Duchet (1987) have obtained the following structure theorem as an even-dimensional counterpart to Proposition 9.4.4.

9.4.6 Proposition (Cordovil and Duchet 1987). *A rank $2k$ matroid polytope $\mathcal{M}$ on E_n is cyclic if and only if $\{1,n\}$ is a contravariant pair and the contractions $\mathcal{M}/1$ and $\mathcal{M}/n$ are both alternating cyclic rank $2k-1$ polytopes.*

An interesting application of oriented matroid methods to cyclic polytopes is the following higher-dimensional generalization of the classical "Ramsey type theorem" of Erdős and Szekeres (1935); see Grünbaum (1967, Exercise 7.3.6) and Cordovil and Duchet (1987). Here we call an $(r-1)$-polytope *cyclic* if its associated rank r matroid polytope is cyclic.

9.4.7 Proposition. *Let n,d be integers with $n \geq d+1 \geq 3$. There exists an integer $N = N(n,d)$ such that every set of N points in general position in affine d-space contains the n vertices of a cyclic d-polytope. Conversely, if a combinatorial type of polytopes has this property, then it is a cyclic polytope.*

Proof. We recall the classical Ramsey theorem: given any two integers $r \leq n$, there exists an integer N such that for every 2-coloring of $\binom{E_N}{r}$ there is an n-element subset V of E_n such that $\binom{V}{r}$ is a monochromatic subset of $\binom{E_N}{r}$.

In order to prove Proposition 9.4.7, we let $r = d+1$, and we choose $N = N(n,r)$ to be the above Ramsey number. Now let $\mathcal{M}$ be any uniform rank r oriented matroid on E_N and let χ denote its chirotope. An r-subset $\{e_1, e_2, \ldots, e_r\}$ with $e_1 < e_2 < \ldots < e_r$ is colored blue if $\chi(e_1, \ldots, e_r) = +1$, and it is colored red if $\chi(e_1, \ldots, e_r) = -1$. By Ramsey's theorem, there exists an n-element subset V of E_N such that $\binom{V}{r}$ is monochromatic. This means that the restriction $\mathcal{M}(V)$ is an alternating (hence cyclic) matroid polytope.

The second assertion follows from the fact that all minors of the alternating oriented matroid $C^{N,r}$ are again alternating (hence cyclic). Suppose that some matroid polytope $\mathcal{M}$ has the property to be a minor by deletion of every "configuration" of sufficiently large size N. In particular, $\mathcal{M}$ is then a minor by deletion of $C^{N,r}$. But this means that $\mathcal{M}$ is alternating, hence cyclic. $\square$

We call a curve $\mathcal{C} \subset \mathbf{R}^d$ in affine d-space *alternating* if every hyperplane meets $\mathcal{C}$ in at most d points. This is equivalent to saying that the acyclic oriented matroid of every finite subset of $\mathcal{C}$ is alternating. The moment curve is the paradigm of an alternating curve, and it is commonly used to define the cyclic polytopes. However, the convex hull of a finite subset of any other alternating curve is a cyclic polytope as well.

This raises the question whether the converse is true: do the vertices of every cyclic polytope lie on an alternating curve? This is clearly false for odd dimensions (even rank). For instance, the vertices $x_1, \ldots, x_6$ of the rank 4 cyclic polytope $\widehat{C}$, considered in connection with Proposition 9.4.4, do not lie on an alternating curve. Perles conjectured that the vertices of every even-dimensional cyclic polytope do lie on an alternating curve. This conjecture was proved using oriented matroid methods.

9.4.8 Proposition (Sturmfels 1987d). *Let $(x_1, x_2, \ldots, x_n) \in (\mathbb{R}^d)^n$ be any affine realization of the alternating oriented matroid $C^{n,d+1}$. Then there exists an alternating curve $C \subset \mathbb{R}^d$ which contains $\{x_1, \ldots, x_n\}$.*

A real $(n - r) \times r$-matrix $\mathbf{A}$, where $n > r \geq 1$, is said to be *totally positive* if all $\binom{n}{r} - 1$ subdeterminants of $\mathbf{A}$ are positive. Total positivity plays an important role in theoretical economics and various branches of mathematics. See the survey of Ando (1987) and the book by Karlin (1968) for details and references. The next proposition shows that there is a close connection between totally positive matrices and cyclic polytopes.

9.4.9 Proposition (Sturmfels 1988b). *Let $x_1, x_2, \ldots, x_n$ be the vertices of an alternating cyclic $(r - 1)$-polytope, and let $\mathbf{A}$ be an $(n - r) \times r$-matrix of homogeneous coordinates of $x_{r+1}, \ldots, x_n$ with respect to the ordered linear basis $((-1)^{r+1}x_r, (-1)^r x_{r-1}, \ldots, -x_2, x_1)$ of $\mathbb{R}^r$. Then $\mathbf{A}$ is totally positive, and, conversely, every totally positive matrix is obtained from some alternating polytope in this way.*

We have seen in the proof of Proposition 9.4.1 that all circuits X of the alternating rank r oriented matroid $C^{n,r}$ satisfy the inequality $|X^+| \geq \lfloor \frac{r+1}{2} \rfloor$. Using Corollary 9.1.3, this implies that *any* $\lfloor \frac{r-1}{2} \rfloor$ vertices of $C^{n,r}$ form a face. In particular, already in dimension 4 (i.e. in rank 5) we have the striking counter-intuitive property that any two vertices of a cyclic polytope are connected by an edge.

More generally, we call a rank r matroid polytope $\mathcal{M}$ *neighborly* provided any $\lfloor \frac{r-1}{2} \rfloor$ vertices form a face of $\mathcal{M}$. The most interesting aspects of neighborly polytopes occur only in even dimensions (= odd ranks). In the remainder of this section we will therefore exclusively consider neighborly matroid polytopes of odd rank $r = 2k + 1$.

9.4.10 Remark. A matroid polytope $\mathcal{M}$ of rank $2k + 1$ is neighborly if and only if $|X^+| = |X^-| = k + 1$ for all circuits X of $\mathcal{M}$. In particular, all neighborly matroid polytopes of odd rank are uniform.

For a long time the cyclic polytopes, discovered by Gale (1956, 1963) but implicitly present already in work of Carathéodory (1904, 1911), were the only known neighborly $2k$-polytopes, and Gale proved that for $n \leq 2k+3$ every neighborly $2k$-polytope is indeed cyclic; see Grünbaum (1967). Grünbaum showed that for all $k \geq 2$ there exist non-cyclic neighborly $2k$-polytopes with $2k + 4$ vertices. Moreover, with his "sewing" process Shemer (1982) showed that the number $g(2k + \beta, 2k)$ of combinatorial types of neighborly $2k$-polytopes with $2k+\beta$ vertices grows superexponentially as $\beta \to \infty$ ($k \geq 2$ fixed) and as $k \to \infty$ ($\beta \geq 4$ fixed). Here we include a matroidal proof for Grünbaum's result.

9.4.11 Proposition.

(i)   Let $\mathcal{M}$ be a neighborly rank $2k+1$ matroid polytope with $n \leq 2k+3$. Then $\mathcal{M}$ is isomorphic to $C^{n,2k+1}$.

(ii) For all $k \geq 2$ there is a realizable neighborly rank $2k + 1$ matroid polytope with $2k + 4$ vertices which is not isomorphic to a cyclic matroid polytope.

Proof of (ii). Consider the rank $2k+1$ matroid polytope $\mathcal{M}$ with $2k+4$ vertices whose affine Gale diagram $\mathcal{M}^*$ is depicted in Figure 9.4.1.

It can be verified from the affine Gale diagram that all circuits X of $\mathcal{M}$ (or cocircuits X of $\mathcal{M}^*$) satisfy $|X^+| = |X^-| = k + 1$. Hence $\mathcal{M}$ is neighborly.

On the other hand, $\mathcal{M}$ is not isomorphic to the cyclic matroid polytope $C^{2k+4,2k+1}$ because their inseparability graphs are not isomorphic. The graph $\mathrm{IG}(C^{2k+4,2k+1})$ of sign-invariant pairs of $C^{2k+4,2k+1}$ is a $(2k + 4)$-cycle by Proposition 9.4.2, while the inseparability graph $\mathrm{IG}(\mathcal{M}) = \mathrm{IG}(\mathcal{M}^*)$ of the oriented matroid in Figure 9.4.1 has two connected components $\{1, 2, 3, 4\}$ and $\{5, 6, \ldots, 2k + 2, 2k + 3\}$. $\square$

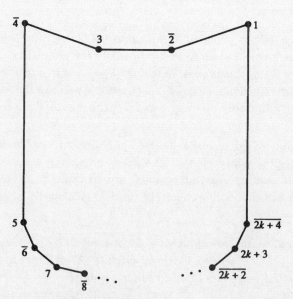

Figure 9.4.1: Affine Gale diagram of a non-cyclic neighborly
rank $2k + 1$ matroid polytope with $2k + 4$ vertices.

We will give a matroidal proof for an important theorem of Shemer (1982) which states that all neighborly matroid polytopes of odd rank are rigid (i.e., uniquely determined by their face lattice); see also Sturmfels (1988d). As a direct consequence of this we get Proposition 9.4.4. We will need the following lemma.

9.4.12 Lemma. *Let M be a neighborly rank $2k + 1$ matroid polytope with $n > 2k + 2$ vertices. A set $A = \{a_1, a_2, \ldots, a_{k+1}\} \subset E_{n-1}$ is a non-face of $M \setminus n$ if and only if there is an $i \in \{1, \ldots, k + 1\}$, such that both A and $(A \setminus \{a_i\}) \cup \{n\}$ are non-faces of M.*

Proof. (if): Suppose that A and $(A \setminus \{a_i\}) \cup \{n\}$ are non-faces of M. Since A is closed in M, Proposition 9.1.2 implies that M/A is acyclic, and hence $(\emptyset, E_n \setminus A)$ is a maximal vector of M/A. So, there exists a maximal vector X of M such that $X^+ \subseteq A$. Since X^+ is a non-face of M, Remark 9.4.10 implies $X^+ = A$ and therefore $X^- = \{1, \ldots, n\} \setminus A$. Using the same argument, also the sign vector Y defined by

$$Y^+ := (A \setminus \{a_i\}) \cup \{n\} \qquad Y^- := (\{1, \ldots, n-1\} \setminus A) \cup \{a_i\}$$

is a maximal vector of M. Applying vector elimination to X and Y, we see that either $X \setminus n$ or $Y \setminus n$ is a vector of M. But $Y \setminus n$ cannot be a vector of M because $|(Y \setminus n)^+| = k$. Here Remark 9.4.10 would contradict the neighborliness of M. Therefore the sign vector $X \setminus n$ is a vector of M and also of $M \setminus n$. Consequently $X^+ = (X \setminus n)^+ = A$ is a non-face of $M \setminus n$.

(only if): If A is a non-face of $M \setminus n$, then

$$Z = (A, \{1, \ldots, n-1\} \setminus A)$$

is a vector of $M \setminus n$ and hence of M. We can add the new point n to Z^+, thereby obtaining another vector $\tilde{Z}$ of M. Writing $\tilde{Z}$ as a conformal union of circuits of M, we obtain a representation of $(\tilde{Z})^+ = A \cup n$ as a union of $(k+1)$-element non-faces, one of which is A. Hence there exists another non-face of M which has the form $(A \setminus a_i) \cup n$. $\qquad\square$

9.4.13 Theorem (Shemer 1982). *Every neighborly rank $2k + 1$ matroid polytope is rigid.*

Proof. Let M be a neighborly rank $2k + 1$ matroid polytope with n vertices. Using induction on n, we will prove that the circuits of M are determined uniquely by the non-faces of M. This will imply Theorem 9.4.13.

If $n = 2k + 2$, then M has only one circuit X. In this case, X^+ and X^- are the two unique minimal (with respect to inclusion) non-faces of M. Now let $n > 2k + 2$ and assume that the claim is true for all neighborly rank $2k + 1$ matroid polytopes with fewer vertices. By Remark 9.4.10, all circuits of M are supported by $(2k + 2)$-sets of vertices. Let X be such a $(2k + 2)$-set (= circuit of $\underline{M}$). Since $n > 2k + 2$, there exists an element $e \notin \underline{X}$.

Consider the neighborly matroid polytope $M \setminus e$. The sign vector X is also a circuit of $M \setminus e$, and, using the induction hypothesis, the sets X^+ and X^- are determined by the non-faces of $\mathcal{M} \setminus e$. On the other hand, Lemma 9.4.12 implies that the non-faces of $\mathcal{M} \setminus e$ are determined by the non-faces of $\mathcal{M}$. Hence X^+ and X^- are determined by the non-faces of $\mathcal{M}$ and the proof is complete. $\square$

Although rigidity of simplicial polytopes is a rather strong property, it does not characterize the even-dimensional neighborly polytopes among all simplicial polytopes.

9.4.14 Proposition (Sturmfels 1988d). *For every $j \geq 4$ there exists a non-neighborly rigid uniform matroid polytope C_j of rank $3j - 3$ with $3j$ vertices.*

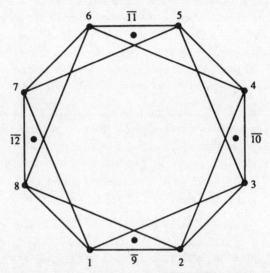

Figure 9.4.2: Affine Gale diagram of the rigid matroid polytope C_4.

Proof. Consider the alternating oriented matroid $C^{2j,3}$ and add j new vertices by the lexicographic extensions

$$2j + i \quad := \quad [(2i - 1)^-, 2i^-, (2i + 1)^-] \pmod{2j} \qquad i = 1, \ldots, j.$$

Let $\widetilde{\mathcal{M}}_j$ denote the resulting rank 3 oriented matroid on E_{3j}. Geometrically speaking, we insert a negatively signed point $2j + i$ inside the circle $C^{2j,3}$ close to the midpoint of the segment $\overline{2i - 1, 2i}$. The dual $C_j := \widetilde{\mathcal{M}}_j^*$ of $\widetilde{\mathcal{M}}_j$ is a uniform realizable rank $2j - 3$ matroid polytope. The smallest member C_4 of this family is depicted in Figure 9.4.2.

In order to prove Proposition 9.4.14, we first note that C_j is not neighborly because the j-element set $\{2j + 1, \ldots, 3n\}$ is not a face of C_j and $j < \lceil \frac{3j-1}{2} \rceil$ for all $j \geq 4$.

Now consider the following set of bases of $C_j^* = \widetilde{\mathcal{M}}_j$.

$$\mathcal{R} := \{ (1, 2, 3), (2, 3, 4), \ldots, (2j - 1, 2j, 1), (2j, 1, 2) \} \quad \cup$$
$$\{ (2j + i, k, l) : i = 1, \ldots, j, \, k \in \{1, \ldots, 2j\}, \, l \in \{2i - 1, 2i\} \}.$$

Each basis in $\mathcal{R}$ is contained in a cofacet of $\mathcal{C}_j$ of the form $\{\,2j+i,\ 2i-1,\ 2i,\ k\,\}$. This implies that the facets of $\mathcal{C}_j$ uniquely determine the orientation in $\mathcal{C}_j^*$ of each basis of $\mathcal{R}$.

On the other hand, it can be observed from Figure 9.4.2 that the chirotope of $\mathcal{C}_j^*$ is completely determined by its restriction to $\mathcal{R}$. This shows that $\mathcal{C}_j$ is rigid. $\qquad\square$

9.5 The Steinitz problem and its relatives

The classification of all combinatorial types of convex polytopes of given dimension and number of vertices has a long tradition in geometry. For polytopes of dimension 3, the first non-trivial case, this problem is essentially solved by the famous Steinitz theorem.

In this section we will use the term $(d-1)$-*sphere* to mean the face lattice of a piecewise linear regular cell complex with the intersection property, whose space is the $(d-1)$-dimensional sphere (see Section 4.7 for the definitions). A $(d-1)$-sphere is said to be *polytopal* if it is the face lattice of a convex d-polytope. Now Steinitz's classical result can be stated as follows.

9.5.1 Theorem (Steinitz and Rademacher 1934). *Every 2-sphere is polytopal.*

Mani (1972) and Kleinschmidt (1976b) have proved the same characterization for face lattice of polytopes which have only few more vertices than their dimension.

9.5.2 Theorem (Mani 1972, Kleinschmidt 1976b). *Every $(d-1)$-sphere with at most $d+3$ vertices is polytopal.*

Grünbaum and Sreedharan (1967) were the first to observe that the immediate analogue to Steinitz's Theorem 9.5.1 is not true for $(d-1)$-spheres with $d+4$ or more vertices whenever $d \geq 4$. In their enumeration of simplicial 3-spheres with 8 vertices they found two spheres which are not polytopal. Using oriented matroid methods, we can prove the following slightly stronger result.

9.5.3 Proposition (Bokowski and Sturmfels 1987). *There exists a triangulated 3-sphere $\mathcal{B}$ with 8 vertices which is not the face lattice of any matroid polytope.*

Proof. The 3-dimensional *Barnette sphere* $\mathcal{B}$ on 8 vertices is defined by the following list of facets:

1237	1238	1245	1247	1258	1346	1348	1367	1458	1467
2356	2357	2368	2457	2568	3468	3567	4567	4568.	

Suppose that there exists a rank 5 matroid polytope $\mathcal{M} = (E_8, \chi)$ such that $\mathcal{F}_{lv}(\mathcal{M}) = \mathcal{B}$. This implies that the following 12 ordered quintuples are bases of the underlying matroid $\underline{M}$ and, moreover, that all 12 signed bases have the same orientation in $\mathcal{M}$:

[12367]   [12357]   [12547]   [32547]   [32567]   [32561]
                    [12347]
[12543]   [12547]   [12647]   [13647]   [13642].

To see this, we note that [12367] and [12357] are signed bases with the same orientation because 1237 is a facet of $\mathcal{M}$. Similarly, [12357] and [12347] are equally oriented bases because 1237 is a facet, [12347] and [12547] are equally oriented bases because 1247 is a facet, and so on. Restricting our attention to the six underlined bases, and writing χ for the chirotope of $\mathcal{M}$, we get

$$\chi(12345) \cdot \chi(12367) = -\chi(12346) \cdot \chi(12357) = \chi(12347) \cdot \chi(12356) = -1.$$

This is a contradiction to the chirotope axiom (B2).   □

There have been many attempts to find a generalization of Steinitz's Theorem 9.5.1 to higher dimension, and thus to solve the *Steinitz problem*. However, so far all proposed characterizations of polytopal lattices failed because these suggested conditions were also satisfied by certain non-polytopal spheres. For a survey of results related to the Steinitz problem from the "pre-oriented-matroid era" of combinatorial convexity see Ewald, Kleinschmidt, Pachner and Schulz (1979). In this context we also mention an asymptotic result of Kalai (1988b) stating that the number of combinatorial types of simplicial spheres by far excedes the number of combinatorial types of polytopes. So, in a very precise sense, most spheres are not polytopal.

Around 1980 J. Bokowski began to approach the Steinitz problem from a new perspective. Rather than aiming for a (combinatorial) characterization of all polytopal lattices, he searched for an (algebraic) algorithm which decides for any specific lattice **L** whether **L** is polytopal. For this purpose Bokowski introduced "sign patterns of simplices consistent with the Grassmann-Plücker relations" and thus rediscovered the chirotope axioms for oriented matroids.

It follows by duality from Theorem 8.2.4 that a 3-sphere with 8 vertices is polytopal if and only if it is the face lattice of a matroid polytope. In contrast, we have seen from the Lawrence construction in Theorem 9.3.7 that there exist matroid polytopes of rank ≥ 12 whose face lattices are non-polytopal. Hence the questions remain whether there exist non-polytopal matroid polytopes of lower rank and whether there exist non-polytopal matroid polytopes in the uniform case. These two problems are answered by the following result of Bokowski. Two algebraic proofs for Theorem 9.5.4 are given in Altshuler, Bokowski and Steinberg (1980) and Bokowski and Sturmfels (1987).

9.5.4 Proposition.   *There exists a non-realizable neighborly rank 5 matroid polytope with 9 vertices.*

Proof. Consider the 3-dimensional sphere $\mathcal{S}$ on 9 vertices which is defined by the following list of facets:

1256	1259	1269	1456	1458	1469	1478	1479	1578	1579
2356	2359	2367	2378	2389	2678	2689	3478	3479	3489
3567	3579	4568	4689	5678.					

This sphere has the number 963 in Altshuler's classification of simplicial 3-spheres with 9 vertices. It is not quite neighborly: two edges 13 and 24 are missing. Thus Shemer's Theorem 9.4.13 is not quite applicable. However, a proof that there exists exactly one rank 5 matroid polytope, $AB(9)$, that has the face lattice $\mathcal{S}$, was given by Bokowski (1978); see also Altshuler, Bokowski and Steinberg (1980) and Antonin (1982). For the existence part we give the complete list of all signed cocircuits of such an oriented matroid, which we denote by $AB(9)$:

$$
\begin{array}{ccccccccc}
\overline{12}345 & \overline{123}4\overline{6} & \overline{1234}7 & \overline{1234}8 & 12349 & \overline{123}5\overline{6} & 12357 & \overline{123}5\overline{8} & \overline{123}5\overline{9} \\
\overline{123}6\overline{7} & \overline{123}6\overline{8} & \overline{123}69 & \overline{123}78 & \overline{123}7\overline{9} & \overline{123}8\overline{9} & 124\overline{56} & \overline{124}5\overline{7} & \overline{124}5\overline{8} \\
\overline{1245}9 & \overline{124}6\overline{7} & 12468 & \overline{124}6\overline{9} & \overline{124}7\overline{8} & 124\overline{79} & 12489 & 12567 & \overline{125}6\overline{8} \\
12569 & 1\overline{2}5\overline{78} & \overline{125}7\overline{9} & \overline{125}8\overline{9} & \overline{126}78 & \overline{126}7\overline{9} & \overline{126}8\overline{9} & \overline{127}89 & \overline{134}56 \\
\overline{134}57 & \overline{134}5\overline{8} & \overline{134}5\overline{9} & 134\overline{67} & 13468 & \overline{134}6\overline{9} & \overline{134}78 & 134\overline{79} & \overline{134}89 \\
\overline{135}67 & \overline{135}6\overline{8} & \overline{135}6\overline{9} & 13578 & \overline{135}7\overline{9} & \overline{135}8\overline{9} & \overline{136}78 & \overline{136}7\overline{9} & \overline{136}8\overline{9} \\
\overline{137}89 & 14567 & 1\overline{4}56\overline{8} & 14569 & 14\overline{578} & 145\overline{79} & 14589 & \overline{146}78 & 146\overline{79} \\
14\overline{689} & 14789 & 15678 & \overline{156}79 & \overline{156}89 & \overline{157}89 & \overline{167}89 & 234\overline{56} & 234\overline{57} \\
\overline{234}58 & 234\overline{59} & \overline{234}67 & \overline{234}68 & 23469 & \overline{234}78 & \overline{234}79 & \overline{234}89 & 235\overline{67} \\
23568 & \overline{235}69 & \overline{235}78 & 23579 & 235\overline{89} & \overline{236}78 & 23679 & 236\overline{89} & \overline{237}89 \\
\overline{245}67 & \overline{245}68 & \overline{245}69 & \overline{245}78 & \overline{245}79 & \overline{245}89 & 24678 & \overline{246}79 & \overline{246}89 \\
\overline{247}89 & \overline{256}78 & 256\overline{79} & 25689 & 25789 & 26789 & 34\overline{567} & 345\overline{68} & 345\overline{69} \\
34578 & 34\overline{579} & 345\overline{89} & \overline{346}78 & 34\overline{679} & 34689 & 34789 & 35678 & 356\overline{79} \\
356\overline{89} & 35789 & 367\overline{89} & 45\overline{678} & 45\overline{679} & 45689 & 45789 & 46789 & 567\overline{89}
\end{array}
$$

From this list we can read off that the pseudohyperplane arrangement associated with $AB(9)$ has exactly eight simplicial cells. The set of these 4-simplices (or mutations) of $AB(9)$ equals

$$\mathrm{Mut}(AB(9)) = \{12569, 13579, 14568, 14578, 23567, 23678, 24689, 34789\}.$$

For instance, we see that the basis 12569 is a simplicial cell because the complementary set 3478 is consistently signed in all five basic cocircuits $\overline{13478}$, $\overline{23478}$, 34578, 34678, and 34789.

If $AB(9)$ were realizable, then it would have at least nine simplicial cells by Shannon's Theorem 2.1.5. So, the above list proves that $AB(9)$ is not realizable. Since $AB(9)$ is rigid, its face lattice $\mathcal{S} = \mathcal{F}_{lv}(AB(9))$ is not polytopal. $\qquad\Box$

The following corollary is proved by duality and successive lexicographic extensions applied to the Altshuler-Bokowski oriented matroid $AB(9)$.

9.5.5 Corollary. *For all $d \geq 4$ and $n \geq d + 5$ there exists a rank $d + 1$ matroid polytope whose face lattice is a non-polytopal simplicial $(d - 1)$-sphere with n vertices.*

The construction which establishes Corollary 9.5.5 is somewhat unsatisfactory because the non-polytopality of all lattices in this infinite family is caused by one common non-realizable substructure $AB(9)$. This raises the following question.

Does there exist a *finite* collection $\{T_1, T_2, \ldots, T_K\}$ of matroid polytopes which has the following property: a given lattice is polytopal if and only if it is the face lattice of a matroid polytope none of whose minors is isomorphic to some T_i? The non-Steinitz conjecture says that there is no such finite excluded minor characterization for polytopal lattices.

9.5.6 Conjecture (Non-Steinitz Conjecture).   *For some fixed $r \geq 5$, there exists an infinite family $\mathbf{F}$ of rank r matroid polytopes such that, for each $\mathcal{M} \in \mathbf{F}$, the lattice $\mathcal{F}_{lv}(\mathcal{M})$ is not polytopal but both the deletion $\mathcal{M}\backslash v$ and the contraction $\mathcal{M}/v$ of every vertex v are realizable. It is also conjectured that one can even take $r = 5$, and that all the matroid polytopes $\mathcal{M}$ can be taken to be uniform.*

An attack on this conjecture (for $r = 7$, non-uniform) by Sturmfels (1987c) turned out to be invalid. However, the recent universality theorem for 4-polytopes by Richter-Gebert (1996b), see Section A.1, gets close to also proving Conjecture 9.5.6 in the non-uniform case. Only in the case of bounded *corank* do we have a non-Steinitz theorem for uniform matroid polytopes, as follows.

9.5.7 Theorem (Sturmfels 1988a).   *There exists an infinite family $\mathbf{G}$ of uniform matroid polytopes $\mathcal{M}$ with $(\mathrm{rank}(\mathcal{M}) + 3)$ vertices such that, for each $\mathcal{M} \in \mathbf{G}$, $\mathcal{F}_{lv}(\mathcal{M})$ is not polytopal but both $\mathcal{M} \backslash v$ and $\mathcal{M}/v$ are realizable for every vertex v of $\mathcal{M}$.*

The situation for $r = 4$ is different: The proof of Theorem 9.5.1 given in Steinitz and Rademacher (1934) implies that every combinatorial type of 3-polytope

(i)   can be realized with rational vertex coordinates, and

(ii)   satisfies the isotopy property (cf. Section 8.6).

To make this remark more precise we introduce the following definitions. Given any lattice $\mathbf{L}$, we define its *convex realization space* $\mathcal{R}(\mathbf{L})$ as the union of the realization spaces $\mathcal{R}(\mathcal{M})$ of all matroid polytopes $\mathcal{M}$ with $\mathcal{F}_{lv}(\mathcal{M}) = \mathbf{L}$.

9.5.8 Corollary (Steinitz and Rademacher 1934, Grünbaum 1967).   *The convex realization space $\mathcal{R}(\mathbf{L})$ of every 2-sphere $\mathbf{L}$ is contractible and it contains rational points.*

Using the results of Section 9.3, we can show that Corollary 9.5.8 does not generalize to higher dimensions.

9.5.9 Theorem.   *Let $\mathcal{M}$ be any oriented matroid, and let $\mathbf{L}_{\mathcal{M}} := \mathcal{F}_{lv}(\Lambda(\mathcal{M}))$ denote the face lattice of its Lawrence extension.*

(1)   *The convex realization space $\mathcal{R}(\mathbf{L}_{\mathcal{M}})$ of the sphere $\mathbf{L}_{\mathcal{M}}$ is homotopy equivalent to the realization space $\mathcal{R}(\mathcal{M})$ of the oriented matroid $\mathcal{M}$.*

(2)   *For every ordered subfield K of the real numbers, $\mathcal{R}(\mathbf{L}_{\mathcal{M}})$ contains K-rational points if and only if $\mathcal{R}(\mathcal{M})$ contains K-rational points.*

Proof. The Lawrence extension $\Lambda(\mathcal{M})$ is a rigid oriented matroid by Proposition 9.3.6; and therefore $\mathcal{R}(\Lambda(\mathcal{M})) = \mathcal{R}(\mathbf{L}_{\mathcal{M}})$. In Section 9.3 we defined the dual $\Lambda(\mathcal{M})^*$ by a sequence of lexicographic extensions from $\mathcal{M}^*$. Therefore the realization space $\mathcal{R}(\mathcal{M}^*)$ is homotopy equivalent to $\mathcal{R}(\Lambda(\mathcal{M})^*)$ by Lemma 8.2.1 and Proposition 8.2.2. On the other hand, realization spaces of oriented matroids are invariant under duality, and we have $\mathcal{R}(\mathcal{M}) = \mathcal{R}(\mathcal{M}^*)$ and $\mathcal{R}(\Lambda(\mathcal{M})) = \mathcal{R}(\Lambda(\mathcal{M})^*)$. This implies that $\mathcal{R}(\mathcal{M})$ and $\mathcal{R}(\mathbf{L}_{\mathcal{M}})$ are homotopy equivalent. The second assertion follows from the easy observation that lexicographic extensions preserve K-realizability for every ordered field K. $\quad\square$

Theorem 9.5.9 shows that the isotopy property fails for higher-dimensional polytopes. Moreover, Mnëv's universality theorem for oriented matroids (Theorem 8.6.6) carries over to convex polytopes.

9.5.10 Corollary (Mnëv 1988). *Let $V \subset \mathbb{R}^s$ be any semialgebraic variety. Then there exists a lattice $\mathbf{L}$ whose convex realization space $\mathcal{R}(\mathbf{L})$ is homotopy equivalent to V.*

In particular, the Lawrence extension of the oriented matroid S_3^{14} in Proposition 8.6.2 defines the face lattice of a 16-polytope with 28 vertices whose convex realization space has two connected components. It is known that such "disconnected polytopes" exist already in dimension 4 (Mnëv 1988, Bokowski and Guedes de Oliveira 1990).

On the computational side, Theorem 9.5.9 enables us to determine the complexity of the Steinitz problem.

Algorithmic Steinitz problem – Decision version (ASP)

- **Input:** A lattice $\mathbf{L}$
- **Output:** "YES" if $\mathbf{L}$ is polytopal over the real numbers, "NO" otherwise.

As with realizability of oriented matroids, we may require additional information in the output. In particular, if $\mathbf{L}$ is polytopal, we are usually interested in explicit vertex coordinates. The complexity of such a coordinatization can be of exponential size, as we see by combining Theorems 8.7.4 and 9.5.9. On the other hand, if $\mathbf{L}$ fails to be polytopal, there is an algebraic certificate for this in form of a final polynomial, see Section 8.5.

9.5.11 Corollary. *The algorithmic Steinitz problem (ASP) is polynomially equivalent to the existential theory of the reals (ETR). In particular, (ASP) is NP-hard.*

Proof. The polytopality of a lattice $\mathbf{L}$ can be expressed as a system of equations and inequalities over the real numbers. See Proposition 2.6 in Bokowski and Sturmfels (1989b) for details. Here the number of equations and inequalities needed is proportional to the size of the lattice $\mathbf{L}$, and they can clearly be computed in polynomial time. This defines a polynomial reduction of (ASP) to (ETR).

For the converse recall that by Theorem 8.7.2 the problem (ETR) is polynomially equivalent to realizability of oriented matroids (ROM). Hence it suffices to describe a polynomial reduction of (ROM) to (ASP). Given an oriented matroid $\mathcal{M}$, the face lattice $\mathbf{L}_{\mathcal{M}}$ of its Lawrence extension can be computed in polynomial time from $\mathcal{M}$. By Theorem 9.5.9, $\mathcal{M}$ is realizable if and only if $\mathbf{L}_{\mathcal{M}}$ is polytopal.   □

We close this section with the remark that both Corollaries 9.5.10 and 9.5.11 can be generalized to simplicial polytopes. We refer to Bokowski and Sturmfels (1989b, Theorem 6.5) or Mnëv (1988) for a construction which assigns to each uniform rank 3 oriented matroid $\mathcal{M}$ a simplicial sphere whose realization space is homotopy equivalent to $\mathcal{R}(\mathcal{M})$. Again, this assignment can be computed in linear time, and hence also the restriction of the algorithmic Steinitz problem (ASP) to simplicial complexes is polynomially equivalent to the existential theory of the reals (ETR).

9.6 Polyhedral subdivisions and triangulations

We recall that a *(polyhedral) subdivision* of a convex d-polytope P is a collection Δ of d-dimensional polytopes such that $\cup\Delta = P$ and such that $T_1, T_2 \in \Delta$ implies that $T_1 \cap T_2$ is a face of both T_1 and T_2. If all the cells T_i are d-simplices, then Δ is a *triangulation* of P. See Section 4.7 for more about the basic definitions.

In this section we define subdivisions of oriented matroids, following the work of Billera and Munson (1984c). These model the subdivisions of a convex polytope whose vertices lie in an *a priori* specified set of points.

It is our objective to illustrate the applicability of oriented matroid techniques in the study of subdivisions and triangulations of polytopes. This subject and its topological ramifications are likely to become important areas of future research, and therefore we shall emphasize basic techniques and open problems rather than proofs of known results – see also Section A.3.

9.6.1 Definition. Let $\mathcal{M}$ be an acyclic oriented matroid of rank r on a set E. A non-empty collection Δ of subsets of E (called *cells*) is a *subdivision of $\mathcal{M}$* if it satisfies:

(S1) all cells $\sigma \in \Delta$ have rank r in $\mathcal{M}$;

(S2) for every single element extension $\mathcal{M} \cup p$ of $\mathcal{M}$ and every $\sigma, \tau \in \Delta$, $p \in \text{conv}_{\mathcal{M} \cup p}(\sigma) \cap \text{conv}_{\mathcal{M} \cup p}(\tau)$ implies $p \in \text{conv}_{\mathcal{M} \cup p}(\sigma \cap \tau)$;

(S3) if $\sigma, \tau \in \Delta$, then $\sigma \cap \tau$ is a common face of the two restrictions $\mathcal{M}(\sigma)$ and $\mathcal{M}(\tau)$;

(S4) if $\sigma \in \Delta$, then each facet of $\mathcal{M}(\sigma)$ is either a facet of $\mathcal{M}$ or is contained in precisely two cells of Δ.

A subdivision Δ of $\mathcal{M}$ is said to be a *triangulation* if each $\sigma \in \Delta$ is a basis of $\mathcal{M}$.

The triangulations of $\mathcal{M}$ are the finest subdivisions in the following sense. We say that a subdivision Δ_1 is *finer* than another subdivision Δ_2, in symbols $\Delta_1 \preceq \Delta_2$, provided each cell of Δ_1 is contained in some cell of Δ_2. Clearly, each subdivision of $\mathcal{M}$ is finer than the *trivial subdivision* $\Delta = \{E\}$. We define the *secondary poset of* $\mathcal{M}$ to be the set $\Sigma(\mathcal{M})$ of all non-trivial subdivisions of $\mathcal{M}$, partially ordered by refinement. Later in this section we will discuss topological properties of this poset. Let us begin by showing that our definitions of subdivisions and triangulations of oriented matroids agree with the usual geometric definitions in the realizable case.

9.6.2 Proposition. *Let $\mathcal{M}$ be the oriented matroid of affine dependences of a spanning subset $\mathbf{X} = \{x_1, x_2, \ldots, x_n\}$ in real affine $(r-1)$-space. A collection Δ of subsets of E_n is a subdivision (resp. triangulation) of $\mathcal{M}$ if and only if $\Delta' := \{\mathrm{conv}\{x_i : i \in \tau\} : \tau \in \Delta\}$ is a subdivision (resp. triangulation) of the polytope $P = \mathrm{conv}(\mathbf{X})$.*

Proof. For each $\tau \in \Delta$ we abbreviate $\tau' := \mathrm{conv}\{x_i : i \in \tau\} \subset \mathbb{R}^{r-1}$. First suppose that Δ' is a subdivision of P. Since $\dim(\tau') = \mathrm{rank}(\tau) - 1$, we have (S1). Property (S3) is equivalent to $\sigma' \cap \tau'$ being a common face of σ' and τ', and (S4) is equivalent to the fact that Δ' is a polyhedral complex whose boundary agrees with the boundary of P. To prove (S2), let $\mathcal{M} \cup p$ be any single element extension of $\mathcal{M}$ and suppose $p \in \mathrm{conv}_{\mathcal{M} \cup p}(\sigma) \cap \mathrm{conv}_{\mathcal{M} \cup p}(\tau)$, i.e., there exist circuits C_1, C_2 of $\mathcal{M} \cup p$ such that $C_1^+ = C_2^+ = \{p\}$ and $C_1^- \subset \sigma$ and $C_2^- \subset \tau$. The cells σ' and τ' are weakly separated by an affine hyperplane which contains their common face $\sigma' \cap \tau'$, i.e., there exists a covector D of $\mathcal{M} \cup p$ with $\sigma \cup \tau \subset D^0$, $\sigma \backslash \tau \subset D^+$ and $\tau \backslash \sigma \subset D^-$. Since D is orthogonal to both C_1 and C_2, we conclude that $p \in D^0$ and $\underline{C_1}, \underline{C_2} \subset (\sigma \cap \tau) \cup \{p\}$. Hence $p \in \mathrm{conv}_{\mathcal{M} \cup p}(\sigma \cap \tau)$.

Conversely, suppose that Δ is a subdivision of the oriented matroid $\mathcal{M}$, and let $\sigma', \tau' \in \Delta$. By (S3) the polytope $\mathrm{conv}\{x_i : i \in \sigma \cap \tau\}$ is a face of both σ' and τ'. By (S2) this polytope equals $\sigma' \cap \tau'$. Hence $\sigma' \cap \tau'$ is a face of both σ' and τ', as desired. Now suppose $\cup \Delta' \neq P$. Since each τ' is closed in P, there exists $x \in \mathrm{int}(P) \backslash \cup \Delta'$. By (S1), we can also choose a sufficiently generic $y \in \cup_{\tau' \in \Delta'} \mathrm{int}(\tau')$ such that the segment $[x, y]$ intersects each facet of a cell $\tau' \in \Delta$ in at most one point. The set of such intersection points is finite and non-empty. So we can pick $\tau' \in \Delta$ and a facet F of τ' such that the intersection point $z \in F \cap [x, y]$ satisfies $[x, z] \subset \mathrm{int}(P) \backslash \cup \Delta'$. Since $F \cap \mathrm{int}(P) \neq \emptyset$, F is not a facet of P, and so (S4) implies that F must be contained in two cells σ' and τ' of Δ. By (S2), F is the intersection of σ' and τ'. But this implies $[x, z] \cap \sigma' \neq \emptyset$, which is a contradiction to our choice of τ'. This implies $\cup \Delta' = P$, and the proof of Proposition 9.6.2 is complete. $\square$

A main construction of oriented matroid triangulations is by means of single element liftings. The following result is a generalization of Theorem 4.4 in Billera and Munson (1984c). Its proof follows from the fact that the (Las Vergnas) face lattice $\mathcal{F}_{lv}(\mathcal{M})$ is a piecewise linear $(r-1)$-sphere, by Theorem 4.3.5.

9.6.3 Proposition. *Let $\widehat{M}$ be a rank $r+1$ oriented matroid on $E \cup q$ such that the contraction $M = \widehat{M}/q$ is acyclic of rank r. Then the set Δ_q of all facets of $\widehat{M}$ not containing q is a subdivision of M.*

A subdivision Δ_q of M which is defined by a single element lifting as in Proposition 9.6.3 is called a *lifting subdivision*. We now show that not all subdivisions of oriented matroids (not even of realizable ones) are lifting subdivisions.

9.6.4 Proposition. *There exists a realizable acyclic oriented matroid M of rank 4 on 8 elements and a triangulation Δ of M which is not a lifting triangulation.*

Proof. It is shown in Barnette (1970) that the 3-dimensional Barnette sphere $\mathcal{B}$ (cf. Proposition 9.5.3) with 8 vertices can be realized as a Schlegel diagram with respect to one of its facets, say $\sigma = \{\sigma_1, \sigma_2, \sigma_3, \sigma_4\}$. Let $\mathbf{X} \subset \mathbb{R}^3$ be the set of vertices of this Schlegel diagram and let M be its rank 4 oriented matroid of affine dependences of $\mathbf{X}$. By Proposition 9.6.2, the set $\Delta := \mathcal{B} \setminus \{\sigma\}$ is a triangulation of M.

Suppose $\Delta = \Delta_q$ were a lifting triangulation, defined by a rank 5 oriented matroid $\widehat{M}$ on $E \cup q$ with $M = \widehat{M}/q$. The matroid polytope $\widehat{M}$ has precisely four facets which contain q, namely, $\{q, \sigma_2, \sigma_3, \sigma_4\}$, $\{\sigma_1, q, \sigma_3, \sigma_4\}$, $\{\sigma_1, \sigma_2, q, \sigma_4\}$ and $\{\sigma_1, \sigma_2, \sigma_3, q\}$. This implies that the face lattice of the rank 5 matroid polytope $\widehat{M} \setminus q$ equals the Barnette sphere $\mathcal{B}$. This is impossible by Proposition 9.5.3. Hence Δ is not a lifting triangulation. $\quad\square$

In spite of Proposition 9.6.4, lifting subdivisions are most useful for polytope theory. To illustrate this let us take a look at an important special case. We say that a lifting $\widehat{M}$ of M is *lexicographic* if the dual $\widehat{M}^*$ is obtained from M^* by a single element lexicographic extension. Recall from Section 7.2 that such an extension is coded by a string $[e_1^{\alpha_1}, e_2^{\alpha_2}, \dots, e_k^{\alpha_k}]$, where $\alpha_1, \dots, \alpha_k \in \{+, -\}$ and $\{e_1, \dots, e_k\}$ is an independent set in M^* (hence the complement of a spanning set in M). A triangulation Δ of M is called *lexicographic* if it is defined by a lexicographic lifting $q = [e_1^{\alpha_1}, \dots, e_k^{\alpha_k}]$ as above.

The following recursive geometric description of lexicographic triangulations is due to Lee (1991). Lee's description in the case of point configurations generalizes easily to acyclic oriented matroids. Let Δ be any subdivision of M, for instance the trivial subdivision $\Delta = \{E\}$, and let $e \in E$. We will define two new subdivisions both of which are refinements of Δ, as follows.

The subdivision $p_e^-(\Delta)$ of M by *pulling the point e in Δ* consists of the following set of cells. For each $\tau \in \Delta$,

(1) if $e \notin \tau$, then $\tau \in p_e^-(\Delta)$;

(2) if $e \in \tau$, then $p_e^-(\Delta)$ contains all sets of the form $F \cup \{e\}$ where F is a facet of $M(\sigma)$ not containing e.

The subdivision $p_e^+(\Delta)$ of $\mathcal{M}$ by *pushing e in* Δ consists of the following set of cells. For each $\tau \in \Delta$,

(1) if $e \notin \tau$, then $\tau \in p_e^+(\Delta)$;

(2) if $e \in \tau$ and $\operatorname{rank}(\tau \setminus e) = r - 1$, then $\tau \in p_e^+(\Delta)$;

(3) if $e \in \tau$ and $\operatorname{rank}(\tau \setminus e) = r$, then $p_e^+(\Delta)$ contains $\tau \setminus e$ together with all sets of the form $F \cup \{e\}$ where F is a facet of $\mathcal{M}(\tau \setminus e)$ such that e lies beyond F (cf. Proposition 9.2.2).

9.6.5 Proposition (Lee 1991). *A subdivision Δ of $\mathcal{M}$ is lexicographic if and only if it can be obtained from the trivial subdivision $\{E\}$ by a sequence of pulling and pushing operations.*

The proof of Proposition 9.6.5 is left to the reader as Exercise 9.26. It is easy to see that every subdivision of an acyclic rank 2 oriented matroid is lexicographic. The same result holds for oriented matroids of low corank, but it is much harder to prove in this case.

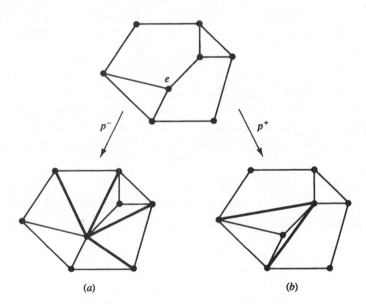

Figure 9.6.1: (a) Pulling and (b) pushing e in a
(non-lexicographic) subdivision Δ.

9.6.6 Proposition (Lee 1991). *Let $\mathcal{M}$ be an acyclic rank r oriented matroid on at most $r + 2$ points. Then each subdivision of $\mathcal{M}$ is lexicographic.*

For an application of lexicographic triangulations to computational algebraic geometry see Sturmfels (1991). In Section 5 of that article it is shown that the lexicographic triangulations of a set of lattice points $\mathcal{A}$ correspond to the (purely and reverse) lexicographic Gröbner bases of the projective toric variety defined by $\mathcal{A}$.

We remark that lifting subdivisions of oriented matroids with at least $r + 3$ points will generally not be lexicographic (see Exercise 9.25). The case $n = r + 3$ was analyzed by Azaola and Santos (1999) and Azaola (1999).

It is an important problem to understand the topological structure of the secondary poset $\Sigma(\mathcal{M})$ of all subdivisions of an oriented matroid $\mathcal{M}$. This question is made precise in Problems 9.28–9.30. For details on the topological significance of these problems see Billera, Kapranov and Sturmfels (1994), and Section A.3.

If $\mathcal{M}$ is a realizable oriented matroid, then each realization $\mathbf{X}$ of $\mathcal{M}$ gives rise to a "nice" subposet $\Sigma_{\mathbf{X}}(\mathcal{M})$ as follows. Let $\widehat{\mathcal{M}}$ be any lifting of $\mathcal{M}$ such that the realization $\mathbf{X}$ lies in the image of the contraction map of realization spaces $\kappa : \mathcal{R}(\widehat{\mathcal{M}}) \to \mathcal{R}(\mathcal{M})$. (This means that the lifting $\mathcal{M} \to \widehat{\mathcal{M}}$ can be carried out geometrically for the specific realization $\mathbf{X}$.) Any lifting subdivision of $\mathcal{M}$ with respect to such a lifting $\widehat{\mathcal{M}}$ is called $\mathbf{X}$-*regular*. It follows from Proposition 8.2.2 that each lexicographic subdivision of $\mathcal{M}$ is $\mathbf{X}$-regular for *every* realization $\mathbf{X}$ of $\mathcal{M}$. Let $\Sigma_{\mathbf{X}}(\mathcal{M})$ denote the subposet of $\Sigma(\mathcal{M})$ consisting of all $\mathbf{X}$-regular subdivisions.

9.6.7 Theorem (Gel'fand, Kapranov and Zelevinsky 1991). *The poset $\Sigma_{\mathbf{X}}(\mathcal{M})$ is the face poset of an $(n - r)$-dimensional convex polytope, called the secondary polytope of the configuration $\mathbf{X}$.*

See also Billera, Filliman and Sturmfels (1990) for a proof of Theorem 9.6.7 using oriented matroid duality. For a general study of the behavior of the polytopal lattice $\Sigma_{\mathbf{X}}(\mathcal{M})$ under the matroidal operations of duality, deletion and contraction we refer to Billera, Gel'fand and Sturmfels (1990). We summarize the relationship between the different classes of subdivisions in the following corollary.

9.6.8 Corollary. *Let $\mathbf{X}$ be an affine point configuration and $\mathcal{M}$ its oriented matroid. Then we have the following inclusions of classes of subdivisions:*

$$\text{lexicographic} \quad \subset \quad \mathbf{X}\text{-}\textit{regular} \quad \subset \quad \textit{lifting} \quad \subset \quad \textit{all subdivisions}.$$

In general, all three inclusions are strict.

Exercises

9.1   Show that an acyclic rank r oriented matroid $\mathcal{M}$ is not uniquely determined by its convex hull operator $\mathrm{conv}_{\mathcal{M}}(\cdot)$.

9.2   Find a non-realizable matroid polytope which has the face lattice of the dodecahedron.

9.3   Find a realizable matroid polytope which has the face lattice of RS(8) (cf. Section 1.5).

9.4   Let $\mathcal{M}$ be a matroid polytope whose face lattice $\mathcal{F}_{lv}(\mathcal{M})$ is a simplicial sphere. Show that there exists a uniform matroid polytope $\mathcal{M}'$ with $\mathcal{F}_{lv}(\mathcal{M}') = \mathcal{F}_{lv}(\mathcal{M})$.

9.5 Determine the number of (labeled) matroid polytopes which have the same face lattice as the cyclic d-polytope with 8 vertices, for $d = 2, 3, 4, 5, 6, 7$.

9.6 Let $\mathcal{M}$ be the rank 2 oriented matroid on E_8 defined by the matrix

$$ A = \begin{pmatrix} 1 & 1 & 1 & 1 & 1 & 1 & 1 & 1 \\ 1 & -2 & -3 & 4 & 4 & -5 & 5 & 6 \end{pmatrix}. $$

Show that $\mathcal{M}^*$ is a matroid polytope and compute its face lattice $\mathcal{F}_{lv}(\mathcal{M}^*)$.

9.7 Show that the rank $r + s + 1$ matroid polytope corresponding to the product $\Delta_r \times \Delta_s$ of an r-simplex with an s-simplex is rigid. (In Example 9.1.7 we proved this for $r = 3, s = 1$.)

9.8 Show that Theorem 9.2.1 implies the classical theorems of Carathéodory, Weyl, Krein-Milman, Radon, Helly and Hahn-Banach in the realizable case.

9.9 Let $\mathcal{M}/8$ be the rank 1 oriented matroid on E_7 obtained by contracting the element 8 of the oriented matroid $\mathcal{M}$ in Exercise 9.6, and consider the resulting extension of $(\mathcal{M}/8)^* = \mathcal{M}^* \setminus 8$ by the point 8.

(a) List all facets F of $\mathcal{M}^* \setminus 8$ and decide whether 8 lies beyond or beneath F.

(b) Compute all faces of $\mathcal{M}^*$ using Proposition 9.2.3; check with your result in Exercise 9.6.

9.10 Prove Proposition 9.3.2.

9.11 Let $\mathcal{M} = \mathcal{M}_0 \oplus p$ be an oriented matroid with a coloop.

(a) Show that $\mathcal{M}$ has an adjoint if and only if $\mathcal{M}_0$ has an adjoint.

(b) Show that $\Lambda(\mathcal{M}) = \Lambda(\mathcal{M}_0) \oplus p \oplus p'$, that is, the Lawrence polytope $\Lambda(\mathcal{M})$ is a two-fold pyramid over the Lawrence polytope $\Lambda(\mathcal{M}_0)$.

(c) Show that $\Lambda(\mathcal{M})$ has a polar if and only if $\Lambda(\mathcal{M}_0)$ has a polar.

9.12 What is the smallest rank of a matroid polytope without a polar ?

(Answer: $r = 5$, $n = 9$; see Bokowski and Schuchert 1995)

9.13* We conjecture that the converse of the Billera-Munson theorem is false: there exists an oriented matroid $\mathcal{M}$ with adjoint such that $\Lambda(\mathcal{M})$ does not have a polar. Can you prove this?

9.14 Let $\mathcal{M}$ be a uniform rank 3 oriented matroid on 6 elements. Compute the f-vector of its Lawrence polytope $\Lambda(\mathcal{M})$.

9.15* Does there exist a matroid polytope whose f-vector is not the f-vector of any polytope ?

9.16 Show that every contravariant pair of a matroid polytope is an edge. Is it true that every covariant pair fails to be an edge ?

9.17* Are all weak images of the alternating matroids $C^{n,r}$ realizable ?

In the following exercises "neighborly" means "neighborly and of odd rank."

9.18 Show that Lemma 9.4.12 is false if we do not assume that $\mathcal{M}$ is neighborly.

9.19* Does there exist a neighborly matroid polytope $\mathcal{M}$ whose realization space $\mathcal{R}(\mathcal{M})$ is disconnected ?

9.20 Compute the inseparability graph of the Altshuler-Bokowski oriented matroid $\texttt{AB}(9)$ in Section 9.5. Does there exist a (realizable) neighborly matroid polytope $\mathcal{M}$ with $\texttt{IG}(\mathcal{M}) = \emptyset$? (Hint: Consider the sphere #416 in the list of Altshuler (1977); see Bokowski and Sturmfels 1987, Section 5).

9.21* Prove or disprove the following conjecture: Every uniform matroid polytope is a contraction of a neighborly matroid polytope.

9.22 List all combinatorial types of 3-polytopes with up to 8 vertices and decide for each type whether it is rigid.

9.23* Does there exist an irrational 4-polytope ? In other words, does there exist a 3-sphere $\mathbf{L}$ such that $\mathcal{R}(\mathbf{L}) \neq \emptyset$ but $\mathcal{R}_{\mathbb{Q}}(\mathbf{L}) = \emptyset$?

9.24* Are the convex realization spaces of the simplicial polytopes in Billera and Lee (1981) contractible ? This would show that simplicial polytopes with non-trivial convex realization spaces cannot be discriminated by their f-vector.

9.25 Construct a non-lexicographic lifting triangulation of an acyclic rank 3 oriented matroid with 6 points. Also construct a non-lexicographic lifting triangulation of a rank 4 matroid polytope with 8 points.

9.26 Prove that a subdivision of an acyclic oriented matroid is lexicographic if and only if it arises from a sequence of pulling and pushing operations (Proposition 9.6.5).

9.27 Construct a minimum size example of a lifting triangulation Δ of an acyclic oriented matroid $\mathcal{M}$ which fails to be $\mathbf{X}$-regular for *all* realizations $\mathbf{X}$ of $\mathcal{M}$.

9.28* Let $\mathbf{X}$ be an affine point configuration and $\mathcal{M}$ its oriented matroid. Is the subposet $\Sigma_{\mathbf{X}}(\mathcal{M})$ a strong deformation retract of the secondary poset $\Sigma(\mathcal{M})$? In particular, does $\Sigma(\mathcal{M})$ have the homotopy type of a sphere? (This is a special case of the "generalized Baues conjecture" of Billera, Kapranov and Sturmfels (1991); see Section A.3.)

9.29* Let $\Sigma_{lift}(\mathcal{M})$ be the subposet of $\Sigma(\mathcal{M})$ consisting of all lifting subdivisions. Is $\Sigma_{lift}(\mathcal{M})$ a strong deformation retract of $\Sigma(\mathcal{M})$?

9.30 Proposition 9.6.3 defines a map from the extension lattice $\mathcal{E}(\mathcal{M}^*)$ of the dual oriented matroid to the poset $\Sigma(\mathcal{M})$ of subdivisions of M.
  (i) Show that this map is neither injective nor surjective in general.

(Santos 1997a)

  (ii)* Does this map induce a homotopy equivalence ?

9.31 Show that $\Sigma_{lift}(\mathcal{M})$ is isomorphic to $\mathcal{E}(\mathcal{M}^*)$ in the special case where $\mathcal{M}$ is a Lawrence polytope. Conclude that the secondary polytope of a Lawrence polytope is combinatorially equivalent to a zonotope.

9.32* For $d = 3, 4$, what is the minimum integer $f_d(n)$ ("grid size") such that each simplicial d-polytope with n vertices is combinatorially equivalent to a polytope whose vertices have integer coordinates between $-f_d(n)$ and $f_d(n)$. It is open whether the functions $f_3(n)$ and $f_4(n)$ grow exponentially; it can be shown that $f_3(n)$ grows at most exponentially, and $f_4(n)$ grows at most double-exponentially.

9.33* Let $\mathcal{M}$ be an oriented matroid on a set E and $e \in E$. Consider the following game between two players WHITE and BLACK playing alternately. At the beginning of the game e is painted red, all other elements are unpainted. At each move WHITE paints an unpainted element white and either reverses its sign or not. At each move BLACK paints an unpainted element black. The objective of WHITE is to capture e in the convex hull of white elements, and that of BLACK to prevent WHITE's success.

Conjecture. If $E \backslash e$ contains two disjoint bases, then WHITE playing second has a winning strategy.

(This conjecture is from Hamidoune and Las Vergnas (1986), where it is proved for graphic and cographic oriented matroids.)

9.34 Let $\mathcal{M}$ be an acyclic oriented matroid on a set E. Define the *interior* int($\mathcal{M}$) *of* $\mathcal{M}$ as the intersection of all positive cocircuits. For $A \subseteq E$ define the *relative interior* of A by relint(A) = int($\mathcal{M}(\text{conv}(A))$), and the *interior of* A by int(A) = relint(A) if $r_{\mathcal{M}}(A) = r(\mathcal{M})$ and int(A) = $\emptyset$ otherwise. For $k \geq 1$, define the *k-interior* of A by $\text{int}_k(A) = \{e \in E :$ there is $S \subseteq A$ with $r_{\mathcal{M}}(S) \geq k$ such that $e \in \text{relint}(S)\}$.

(i) Show that if $e \in \text{int}(A)$, then there is a set $B \subseteq A$ with $|B| \leq 2r(\mathcal{M})$ such that $e \in \text{int}(B)$.

(ii) "Bonnice-Klee Theorem for oriented matroids":
Generalizing Reay (1965), show that if $e \in \text{int}_k(A)$, then there is a set $B \subseteq A$ with $|B| \leq \max(r(\mathcal{M}), 2k - 2)$ such that $e \in \text{int}(B)$.

(Bienia and Las Vergnas 1990)

9.35 Let $\mathcal{M}$ be a totally cyclic oriented matroid on a set E (Definition 3.4.7). A subset $A \subseteq E$ such that $\text{conv}_{\mathcal{M}}(A) = E$ is called a *positive spanning set*.

(i) Show that A is a positive spanning set if and only if $r_{\mathcal{M}}(A) = r(\mathcal{M})$ and A is a union of positive circuits.

(ii) Show that given a positive spanning set A there is a positive spanning set B such that $|B| \leq 2r(\mathcal{M})$. Establish the relationship between this and the statement of Exercise 9.34(i).

(iii) Characterize equality in (ii).

(iv)* Given an integer $k \geq 1$, a *positive k-spanning set* is a subset $A \subseteq E$ such that $A \backslash E$ is a positive spanning set for all $S \subseteq A$ with $|S| < k$.

Problem: find a function $f(k, r(\mathcal{M}))$ such that for every positive k-spanning set A there is a positive k-spanning set $B \subseteq A$ with $|B| \leq f(k, r(\mathcal{M}))$. (See Marcus (1984) for the realizable case.) (Bienia and Las Vergnas 1990)

9.36 Show that if a point set has a non-regular subdivision, then it must have a non-regular triangulation that refines it. (Santos 1999a)

10

Linear Programming

Chapter 10 gives an introduction to linear programming on oriented matroids. It does not presuppose any experience with linear programming – but for the operations research practitioner it might offer an alternative view on linear programming from a matroid theory point of view. Our aim is to give the non-expert a smooth, *geometric* access to the fundamental ideas of oriented matroid programming, as developed in Bland (1974, 1977a). This necessitates extra care at the points where the terminologies from linear programming and from combinatorial geometry clash.

This chapter is intended to demonstrate that the oriented matroid framework can add to the understanding of the *combinatorics* and of the *geometry* of the simplex method for linear programming. In fact, the oriented matroid approach gives a geometric language for pivot algorithms, interpreting linear programs as *oriented matroid search problems*. We find that *locally consistent* information (orientation of the edges at a vertex) imply the existence of *global extrema* in pseudoarrangements. We believe that these techniques and results are applicable also to problems in other areas of mathematics.

The oriented matroid framework deals with (pseudo)linear programs, where

- positive cocircuits correspond to feasible vertices, and
- positive circuits correspond to bounding cones.

Both of these are described by oriented matroid bases, corresponding to temporary coordinate systems. Pivot algorithms are now modeled by basis exchanges, and the duality of linear programming becomes a manifestation of oriented matroid duality.

The benefits of geometric understanding are of course not one-sided: the linear programming framework offers insight into the structure of oriented matroids, and the pivot algorithms of linear programming provide important *search techniques* for oriented matroids. The failure of some of these search techniques in

some "badly behaved" oriented matroids is of considerable interest. We will detail the surprising connection between extension problems and cycling in oriented
matroid programs developed in Edmonds and Fukuda (1982) and Edmonds and
Mandel (1982).

For the purpose of geometric motivation we start in Section 10.1 by constructing combinatorial models for polyhedra in affine space, etc. This is not hard
– as the Topological Representation Theorem demonstrates, oriented matroids
provide an excellent model of the combinatorial structure given by a set of hyperplanes through the origin of a real vector space $\mathbb{R}^n$. Similarly, they can also be
used to model *affine* hyperplane arrangements, such as the arrangements given
by linear programming problems. Research in this direction was initiated by
Rockafellar (1969), who observed that many properties of the simplex method
only depend on combinatorial sign patterns in the tableaux. Rockafellar's paper
suggested that a purely combinatorial formulation of the simplex method be developed. This led Bland (1974) to his development of oriented matroids and the
simplex method for oriented matroid programming (Bland 1977a).

Bland's study had considerable impact on linear programming theory as it
gave a combinatorial abstraction of pivot algorithms for linear programming.
The "smallest subscript rule", now known as "Bland's Rule", was developed
in this context. This rule has a very simple finiteness proof in the realizable
case, which, however, fails for general oriented matroids. Bland also gave the
first constructive proof for the linear programming duality theorem on oriented
matroids, although the smallest subscript rule does not yield this.

The first finite *simplex* pivot rule for oriented matroid programming was given
much later by Todd (1985). This made possible a constructive proof of oriented
matroid programming duality that parallels the usual proof in the linear case:
by providing a finite simplex pivot algorithm, that is, a pivot rule that achieves
optimality for bounded feasible programs after finitely many steps, while never
losing feasibility once it is attained.

In Section 10.1 we will describe the general setting of affine oriented matroids
and oriented matroid programs as well as the geometric description of simplex
algorithms: the simplex algorithm for linear programming maximizes a linear
function over a polyhedron by starting at a vertex and then improving this
current solution by moving to the next vertex with a better value along an edge
of the polyhedron until optimality or unboundedness of the problem becomes
obvious. This description is translated to oriented matroids.

Section 10.2 will then supply the algebraic analogue by formalizing systems of
linear equalities in non-negative variables via tableaux and pivots. Although the
geometry behind simplex tableaux is essential, this section is quite independent
from 10.1 and therefore offers an alternative approach to the field. Also, it
refines the geometric visualization of linear programs to the practical setting of
simplex algorithms: they proceed from basis to basis of the problem by simple
one-element exchanges (pivots).

Simplex tableaux are used to represent the situation at each step of a pivot algorithm. The setting of oriented matroids leads to a nice and elegant description of this approach. Bland's "smallest subscript" rule and his ALGORITHM A pivot rule are developed here, and their combination yields a complete and constructive proof of the Duality Theorem for oriented matroid programming. Both these pivot rules have interesting "defects", which cannot appear in realizable linear programs: the smallest subscript rule can cycle even on non-degenerate programs, whereas ALGORITHM A can become infeasible from a feasible start. Thus these two methods do not define a finite *simplex* algorithm. Edmonds and Fukuda's (1982) famous example for these phenomena is constructed in Section 10.4. The construction of a genuine *simplex* algorithm is completed at the end of Section 10.3 with the description of "Todd's Rule" (without a proof).

The "bad" behavior of pivot algorithms on some oriented matroid programs is partially explained in Section 10.5 by the failure of a certain intersection property. In particular, we discuss a theorem by Edmonds and Mandel (1982) which states that non-degenerate cycling can occur on an oriented matroid if and only if it fails to satisfy the Euclidean intersection property.

In closing, we would like to mention another interesting development. In a remarkable *tour de force*, Todd (1984, 1985) has demonstrated that "linear complementarity problems" (see Murty 1988 for a survey) can be successfully modeled, characterized and solved on oriented matroids. This encompasses the case of linear programs as well as the case of (convex) quadratic programs, which have a similarly nice and complete duality theory. Both cases are covered by the case of a *bisymmetric, positive semidefinite* linear complementarity problem. We recommend Morris and Todd (1988), Klafszky and Terlaky (1989b) and Fukuda and Terlaky (1992) for explanations of the oriented matroid model, for basic facts and duality theorems, and for a discussion of the suprisingly simple "criss-cross" algorithm for linear complementarity problems.

10.1 Affine oriented matroids and linear programs

Our approach to linear programming in oriented matroids ("oriented matroid programming") in the following two sections will be guided by two goals.

 - We will give a purely affine, primal treatment (as first developed by Edmonds and Fukuda (1982)), which describes linear programming geometrically as the maximization of a linear function over a polyhedron given by a system of inequalities. For this we will try to stay close to the geometry of polyhedra in affine space.
 - At the same time, we will try to make the primal/dual structure of linear programming transparent. The duality between feasibility and boundedness is carefully developed. At the end of this line of thought, the duality theory for oriented matroid programming is supposed to be completely natural. For this, we will write linear and oriented matroid programs in a way that makes the primal and the dual programs have *exactly* the same form (in particular, they will *both* be *maximization* problems), which makes

the situation even more symmetric than it usually appears to be in linear programming texts.

We will start with an oriented matroid definition of inequality systems and polyhedra in *affine space*, due to Edmonds and Fukuda (1982) and Edmonds and Mandel (1982). The geometric idea, which in the non-realizable case is based on the Topological Representation Theorem, was already discussed in Section 4.5.

10.1.1 Definition (affine oriented matroids, polyhedra, cones).

(i)   An *affine (oriented) matroid* is a pair $(\mathcal{M}, g)$, where $\mathcal{M} = (E, \mathcal{L})$ is an oriented matroid given by its collection $\mathcal{L}$ of covectors, and $g \in E$ is not a loop. Its *affine space* is
$$\mathcal{A} = \mathcal{A}(\mathcal{M}) = \{Y \in \mathcal{L} : Y_g = +\},$$
The *plane at infinity* of the affine oriented matroid is
$$\mathcal{A}^\infty = \{Y \in \mathcal{L} : Y_g = 0\}.$$

(ii) A *polyhedron* in $\mathcal{A}$ is a set
$$P(S) = \{Y \in \mathcal{A} : Y_e \geq 0 \text{ for } e \in S\}$$
for a subset $S \subseteq E \setminus g$. Its *boundary at infinity* is
$$P^\infty(S) = \{Y \in \mathcal{A}^\infty : Y_e \geq 0 \text{ for } e \in S\}.$$

(iii) A *(simplicial) cone* in $\mathcal{A}$ is a polyhedron $P(I)$ for some $I \subseteq E$ such that $I \cup \{g\}$ is independent. Such a cone $P(I)$ is *bounded* with respect to $f \in E$ if $Z_f \leq 0$ for all $Z \in P^\infty(I)$.

Here $\mathcal{A}$ is non-empty and contains a cocircuit since g is not a loop: for this take any basis B of $\mathcal{M}$ that does contain g, and consider the cocircuit $c^*(g, B)$. The set $\mathcal{A}$ is not the set of covectors of an oriented matroid, because $\mathcal{A} \cap (-\mathcal{A}) = \emptyset$. However, $\mathcal{A}^\infty$ is the set of covectors of the contraction of $\mathcal{M}$ by g (with g adjoined as a loop.) Observe that $\mathcal{M}$ (and thus also $\mathcal{A}^\infty$) is determined by $\mathcal{A}$: in fact the topes of $\mathcal{M}$ are given by
$$\mathcal{T}(\mathcal{M}) = \mathcal{T}(\mathcal{A}) \cup \mathcal{T}(-\mathcal{A}),$$
where $\mathcal{T}(\mathcal{A})$ denotes the set of topes in $\mathcal{A}$, which is the set of maximal elements of $\mathcal{A}$ in the partial order induced from $\mathcal{L}$, that is, $\mathcal{T}(\mathcal{A}) = \max(\mathcal{A}) = \mathcal{T} \cap \mathcal{A}$. This allows us to denote an affine oriented matroid by its affine space, that is, we will write $\mathcal{A} = (\mathcal{M}, g)$ without too many pangs of conscience.

We use $(\mathcal{M}, g)$ as a model for (real) affine space, where the positive cocircuits with $Y_g = +$ correspond to feasible vertices. With the next lemma, positive circuits of $(\mathcal{M}, g)$ can very conveniently be interpreted as inconsistent inequality systems. This provides a criterion for deciding whether a polyhedron $P(S)$ is empty.

10.1.2 Lemma (Bland 1974). *Let $P(S)$ be a polyhedron in $(\mathcal{M}, g)$. Then $P(S)$ is empty if and only if there is a positive circuit $X \in \mathcal{C}(\mathcal{M})$ with $\underline{X} \subseteq S \cup g$ and $X_g = +$.*

Proof. This is Corollary 3.4.6, applied to the restriction $\mathcal{M}(S \cup g)$ and the element g.   □

Note that the proof of Lemma 10.1.2/Corollary 3.4.6 is *not constructive*. By this we mean that it does not offer any algorithm (other than exhaustive enumeration) for finding a suitable circuit X. Such an algorithmic proof will follow from Corollary 10.2.18, which can be derived without the use of Lemma 10.1.2.

10.1.3 Definition (oriented matroid program).

(1)   An *oriented matroid program* is a triple
$$(P) = (\mathcal{M}, g, f),$$
where $\mathcal{M}$ is an oriented matroid on $E = E_n \cup \{f, g\}$, g is not a loop, f is not a coloop, and $f \neq g$.

(Our convention is that the ground set $E = \{1, 2, \ldots, n, g, f\}$ has the fixed linear ordering $1 < 2 < \ldots < n < f < g$.)

Thus an oriented matroid program encodes the task to maximize an element f (assumed not to be a coloop) over the polyhedron $P(E_n)$ of an affine oriented matroid $(\mathcal{M}, g)$:
$$(P): \quad \max_{P(E_n)} f.$$

(2)   The *dual program* associated with (P) is the oriented matroid program
$$(P^*) = (\mathcal{M}^*, f, g).$$

(3)   A *minor* of an oriented matroid program $(\mathcal{M}, g, f)$ is an oriented matroid program $(\mathcal{M}', g, f)$ such that $\mathcal{M}' = \mathcal{M}'(E')$ is a minor of $\mathcal{M}$.

(This implies the requirements that $g \in E' \subseteq E$ is not a loop and $f \in E'$ is not a coloop of $\mathcal{M}'$.)

A *(single element) extension* of $(\mathcal{M}, g, f)$ is an oriented matroid program $(\widetilde{\mathcal{M}}, g, f)$, where $\widetilde{\mathcal{M}}$ is a (single element) extension of $\mathcal{M}$.

(4)   The oriented matroid program $(\mathcal{M}, g, f)$ is *non-degenerate* if $\mathcal{M}$ is uniform.

An oriented matroid program is supposed to model a linear program of the type
$$\max \ \mathbf{c}^T \mathbf{x} \ - d$$
$$A\mathbf{x} \ \leq \mathbf{b} \qquad\qquad (P)$$
$$\mathbf{x} \ \geq \mathbf{0}$$
where f denotes the linear function $f(\mathbf{x}) = \mathbf{c}^T \mathbf{x} - d$, and g is used to linearize the problem, such that $\mathcal{A} = \{g = 1\}$ is affine space. For this recall the example developed in Section 1.7, which is displayed in Figure 10.1.1. We will do an explicit translation in Section 10.2.

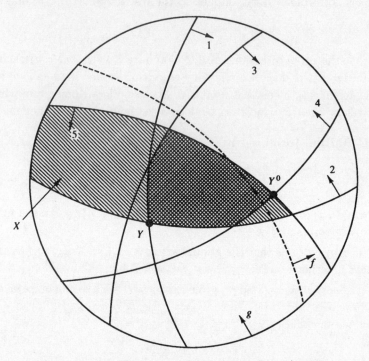

Figure 10.1.1: An oriented matroid program of rank 3.

We restrict our attention to the modeling of *maximization* problems. In fact, *up to the sign of the objective function*, the dual problem to (P) is

$$\max \ (-\mathbf{b})^T \mathbf{y} \ + d$$
$$(-A^T)\mathbf{y} \ \leq -\mathbf{c} \qquad\qquad (P^*)$$
$$\mathbf{y} \ \geq \ \mathbf{0}$$

which is a maximization problem exactly of the same form as (P). (P) is *dual* to (P^*), because the feasible vectors $\mathbf{y}$ for (P^*) describe non-negative combinations of inequalities of (P) that provide upper bounds for (P) – and dually. The following result states that *linear programming duality is a special case of oriented matroid duality*.

10.1.4 Proposition (Bland 1974). *The arrangements of halfspaces associated with (P) and (P^*) define a dual pair of oriented matroids, with the roles of f and g exchanged (as required by Definition 10.1.3(2)). Thus, if the oriented matroid program associated with (P) is $(\mathcal{M}, g, f)$, then the oriented matroid program corresponding to (P^*) is $(\mathcal{M}^*, f, g)$.*

This can be seen directly from the geometry of the dual programs, see Exercise 10.8. It will again be apparent in Section 10.2 when we study the tableaux for (P) and (P^*). Finally the validity of Corollary 10.1.11 and Theorem 10.1.12 below will prove that we have picked the "right" definition of dual oriented matroid programs.

Note that the set N_0 of variables of (P) is in natural bijection with the set of inequalities of (P^*) given by $-A^T \mathbf{y} \leq -\mathbf{c}$. Similarly, the set B_0 of variables of (P^*) corresponds to the set of inequalities $A\mathbf{x} \leq \mathbf{b}$, which identifies the ground sets of the associated oriented matroids, with $E_n = B_0 \uplus N_0$.

The only "non-standard twist" we have put into the definition of (P^*) is a missing minus sign in front of the objective function $(-\mathbf{b})^T \mathbf{y} + d$ of the dual program. This leads to a slightly altered duality theorem: if (P) and (P^*) are both feasible, then they have optimal solutions with *opposite* objective function values. However, since the "objective function" f cannot be evaluated for an oriented matroid program, we will only see in the duality theorem for oriented matroid programs that primal and dual optimal solutions have objective function values with opposite sign.

In the following, the (trivial) cases of oriented matroid programs in which f is a loop or parallel to g will be carried along, although it could be excluded for all practical purposes. It corresponds to linear programs with zero or constant objective function.

An important observation is the identification of *minors* of oriented matroid programs with (smaller) subproblems. Namely, deletion of a matroid element corresponds to deleting the corresponding inequality from the program, whereas contraction of an element corresponds to requiring equality for the corresponding inequality, so that the corresponding (slack) variable is fixed to be zero. We will discuss this again in Remark 10.3.6.

The following definitions formulate oriented matroid programming as an optimization problem over a polyhedron in affine space.

10.1.5 Definition (feasible region). Let $(\mathcal{M}, g, f)$ be an oriented matroid program.

A covector $Y \in \mathcal{A}(\mathcal{M})$ is *feasible* (or a *solution*) if $Y_e \geq 0$ for all $e \in E_n$.

The *feasible region* of the oriented matroid program (P) is the polyhedron

$$P(E_n) = \{Y \in \mathcal{A} : Y_e \geq 0 \text{ for } e \in E_n\}$$

of feasible covectors. Its *face at infinity* is

$$P^\infty(E_n) = \{Y \in \mathcal{A}^\infty : Y_e \geq 0 \text{ for } e \in E_n\}.$$

Finally,

$$\overline{P}(E_n) = P(E_n) \uplus P^\infty(E_n)$$

is the *projective closure* of the feasible polyhedron.

10.1.6 Lemma. *If Y, Y' are cells in the feasible region, then*

$$S(Y, Y') \subseteq \{f\}$$

and $Y \circ Y', Y' \circ Y \in P(E_n)$.

Also, if $Y \in P(E_n)$ and Y'' is a non-zero face of Y (that is, $Y'' \in \mathcal{L}\backslash 0$ with $Y'' \circ Y = Y$), then $Y'' \in P(E_n)$ if $Y''_g = +$ and $Y'' \in P^\infty(E_n)$ otherwise.

Thus if $P(E_n)$ is non-empty it can be thought of as a single cell of $\mathcal{A}(\mathcal{M}\backslash f)$ together with all of its faces, possibly subdivided by the pseudoplane corresponding to f, as sketched in Figure 10.1.1. Also, since every cell is the join of its vertices (every covector of $\mathcal{L}$ is the composition of the cocircuits that conform to it), we observe that if $P(E_n)$ is non-empty, then it contains a vertex (cocircuit).

For an oriented matroid program the element f does not define a *global* objective function: all that is known is a halfspace of the form $f(\mathbf{x}) \geq 0$. This does not provide the data for comparing two arbitrary points in space (cocircuits in $\mathcal{A}$) according to their value of f. It only allows us to make *local* comparisons along a pseudoline.

10.1.7 Definition (increase, optimality).   Let $(\mathcal{M}, g, f)$ be an oriented matroid program and $\mathcal{A} = (\mathcal{M}, g)$ the associated affine oriented matroid.

(i)   The *directions* in $\mathcal{A}$ are the covectors $Z \in \mathcal{A}^\infty$. We say that f *increases* (respectively *decreases* or *is constant*) in direction Z if $Z_f = +$ (respectively, $Z_f = -$ or $Z_f = 0$).

(ii)   For a feasible covector $Y \in P(E_n)$ a covector $Z \in \mathcal{A}^\infty$ is a *feasible direction* if $Y \circ Z \in P(E_n)$.

(iii)   A feasible covector Y^0 is *optimal* (or an *optimal solution*) if there is no feasible increasing direction for Y^0.

As in the case of linear programming in $\mathbb{R}^n$, we have to deal with infeasible, bounded feasible and unbounded oriented matroid programs.

10.1.8 Definition (feasibility and boundedness).   An oriented matroid program $(P) = (\mathcal{M}, g, f)$ is

(i)   *feasible* if $P(E_n)$ is non-empty, and *infeasible* otherwise.

(ii)   *bounded* if $P(E_n)$ is contained in a bounded cone, and *unbounded* otherwise.

Geometrically, we can think of the bounded cone as an "umbrella" (Roos 1984) over $P(E_n)$: the tip is the highest point of the umbrella, and thus yields an upper bound for the polyhedron contained in the umbrella.

The main concepts are illustrated in Figure 10.1.1, using the oriented matroid program from Figure 1.7.2. Here the cocircuit $Y = (00+++|-+)$ (corresponding to a feasible vertex) proves feasibility: it lies in the feasible region, which is shaded dark. The circuit $X = (0+00+|+-)$, yielding the lightly shaded bounded cone $S_{25g}^+ = S_2^+ \cap S_5^+ \cap S_g^+$, proves boundedness. (The notation for the sign vectors refers to the ordering $1 < \ldots < n < f < g$ given by Definition 10.1.3(1).)

In the same figure, the cocircuit (vertex) $Y^0 = (+++00|++)$ is optimal, and the circuit $X^0 = (000++|+-)$, corresponding to the cone S_{45g}^+, proves this.

The *philosophy* behind Definition 10.1.8 has two components. The first is to carefully prepare the stage for the duality underlying linear (and oriented matroid) programming. The second part is some general linear programming philosophy:

(i)   the feasibility of a linear program does not depend on the objective function f;

(ii)   the boundedness of a linear program does not depend on the right hand side **b**.

To achieve this consistently, we have to allow for both *infeasible, bounded* programs, like

$$\max \ x_1 + x_2$$
$$x_1 + x_2 \ \leq -1$$
$$x_1, x_2 \ \geq 0,$$

(which after a change in the right hand side becomes bounded feasible), and *unbounded infeasible* programs, like

$$\max \ x_1 + x_2$$
$$x_1 \qquad \leq -1$$
$$x_1, x_2 \ \geq 0,$$

(which after a change in the right hand side becomes unbounded feasible).

The following (less geometric, but more systematic) reformulation of Definition 10.1.8 shows that the requirements (i) and (ii) are satisfied.

10.1.9 Lemma.

(i) $(P) = (\mathcal{M}, g, f)$ *is feasible if and only if there is a cocircuit $Y \in \mathcal{C}^*(\mathcal{M})$ such that $Y_g = +$, and $Y_e \geq 0$ for all $e \in E_n$ (or, equivalently, if there is a positive cocircuit in $\mathcal{M}\backslash f$ that contains g).*

(ii) $(P) = (\mathcal{M}, g, f)$ *is bounded if and only if there is a circuit $X \in \mathcal{C}(\mathcal{M})$ such that $X_f = +$, and $X_e \geq 0$ for all $e \in E_n$ (or, equivalently, if there is a positive circuit in $\mathcal{M}/g$ that contains f).*

Proof. Part (i) follows from Lemma 10.1.6: if there is a feasible covector, then it is a conformal product of cocircuits, and at least one of these has $Y_g = +$ and is thus feasible. Restricting the feasible covector to $\mathcal{M}\backslash f$, we can guarantee that the cocircuit Y is of the form $Y = (Y(E\backslash f), Y_f)$ for a positive cocircuit $Y(E\backslash f)$ of $\mathcal{M}\backslash f$.

For part (ii), suppose $P(E_n) \subseteq P(I)$ for a bounded cone $P(I)$. Then $I \cup g$ is independent, but $I \cup \{g, f\}$ cannot be independent, for if it were independent, then we could find a cocircuit Z with $Z_e = 0$ for all $e \in I \cup g$ and $Z_f > 0$, contradicting the fact that we have a bounded cone. We may assume that $I \cup g$ is a basis, otherwise we augment it arbitrarily. Now consider the basic circuit $X := c(f, I \cup g)$. It satisfies $X_f = +$ by definition. For $e \in I$ we consider the basic cocircuit $Y := c^*(e, I \cup g) \in P^\infty(I)$. Now we have $\underline{X} \cap \underline{Y} \subseteq \{e, f\}$, with $X_f = Y_e$ by construction. From boundedness we get $Y_f \leq 0$, and thus $X_e = -Y_f \geq 0$.

The rest follows from dualizing part (i).   □

At this point, observe that from the orthogonality between circuits and cocircuits we get the following corollary.

10.1.10 Corollary.

(i) $(\mathcal{M}, g, f)$ is infeasible if and only if there is a positive circuit $X \in \mathcal{C}(\mathcal{M})$ with $X_g = +$, $X_f = 0$.

(ii) $(\mathcal{M}, g, f)$ is unbounded f and only if there is a positive cocircuit $Y \in \mathcal{C}^*(\mathcal{M})$ with $Y_f = +$, $Y_g = 0$.

Proof. The "if"-parts are clear from Lemma 10.1.9 and the orthogonality of circuits and cocircuits.

The "only if"-part for (i) is Lemma 10.1.2, applied to $(\mathcal{M}\backslash f, g)$. From this, (ii) follows from duality.   □

10.1.11 Corollary.   Let (P) be an oriented matroid program, and (P^*) its dual.

(i) (P) is feasible if and only if (P^*) is bounded.

(ii) (P) is bounded if and only if (P^*) is feasible.

We now continue with simple observations on the sets of optimal solutions of oriented matroid programs. For this, let $(\mathcal{M}, g, f)$ be an oriented matroid program.

(i)   If $Y_1, Y_2 \in P(E_n)$ are optimal solutions, then so is $Y^1 \circ Y^2$.

(ii)   If Y is an optimal solution and $Y' \in \mathcal{A}$ satisfies $Y' \leq Y$ (that is, $Y' \circ Y = Y$), then Y' is optimal as well.

(iii) For non-degenerate oriented matroid programs, the set of optimal solutions is either empty or it consists of a unique cocircuit.

This shows that the set of optimal solutions is a face of the feasible polyhedron $P(E_n)$ (together with all of its faces in $\mathcal{A}$), and that if there is an optimal solution, then there is an optimal cocircuit.

An effective criterion for optimality is given by the following result.

10.1.12 Theorem (optimal solutions, "complementary slackness"; Bland 1977a).   Let $(\mathcal{M}, g, f)$ be an oriented matroid program, and let $Y \in \mathcal{A}$ be feasible. Then

(1) Y is optimal if and only if there is a dual feasible circuit $X \in \mathcal{C}$ that satisfies $\underline{Y} \cap \underline{X} \subseteq \{f, g\}$. Such an X is then optimal for the dual program $(\mathcal{M}^*, f, g)$.

(2)   If Y is an optimal solution (covector) for $(\mathcal{M}, g, f)$ and the vector X is an optimal solution for the dual program, then $\underline{Y} \cap \underline{X} \subseteq \{f, g\}$.

Proof.

(1) Let $Y \in \mathcal{A}$ be a covector, and let $X \in \mathcal{C}$ be such that $\underline{X} \cap \underline{Y} \subseteq \{f, g\}$. Assume X is dual feasible, that is, $X_f = +$ and $X_e \geq 0$ for $e \in E_n$. Let Z be an increasing direction ($Z_f = +$, $Z_g = 0$). By orthogonality of X and Z, we get that there exists an $h \in E_n$ such that $Z_h = -$ and $X_h = +$. The complementarity condition $\underline{Y} \cap \underline{X} \subseteq \{f, g\}$ now implies $Y_h = 0$, so that $(Y \circ Z)_h = -$, that is, Z is not a feasible direction for Y.

For the converse, let $Y \in \mathcal{A}$ be an optimal solution. Let

$$S := \{e \in E_n : Y_e = 0\}.$$

From optimality of Y we get that the polyhedron $P(S)$ in the contraction $(\mathcal{M}/g, f)$ is empty: any covector Z with $Z_g = 0$, $Z_f = +$ satisfies $Z_h = -$ for some $h \in S$.

Now by Lemma 10.1.2 there is a positive circuit $X \in \mathcal{C}(\mathcal{M}/g)$ with $X_f = +$ and $\underline{X} \subseteq S \cup f$. Thus $\mathcal{C}(\mathcal{M})$ has a circuit $\tilde{X}$ with $\tilde{X}_f = +$, $\underline{\tilde{X}} \subseteq S \cup \{f, g\}$ and $\tilde{X}^- \subseteq \{g\}$. (The argument also works if f is a loop in $\mathcal{M}/g$.)

In order to see that X is dual optimal, we apply the first statement in (1) to the dual program and with reversed roles of X and Y.

(2) Let Y be an optimal covector for $(\mathcal{M}, g, f)$, let X be a vector that is optimal for the dual program, and let $h \in \underline{X} \cap \underline{Y} \cap E_n$. By (1), applied to the dual program, there exists an optimal cocircuit $\tilde{Y}$ for $(\mathcal{M}, g, f)$ such that $\underline{X} \cap \underline{\tilde{Y}} \subseteq \{f, g\}$. In particular, this means $\tilde{Y}_g = Y_g = +$ and $\tilde{Y}_h = 0$, $Y_h = +$.

By (strong) covector elimination of g between $-Y$ and $\tilde{Y}$ we now obtain a covector Z with $Z_g = 0$, $Z_h = -$, $Z^+ \subseteq \tilde{Y}^+ \cup \{f\}$, $Z^- \subseteq Y^+ \cup \{f\}$. Now orthogonality of X and Z (with $X_h = +$, $X_f = +$, $X^- \subseteq \{g\}$, $Z^+ \cap \underline{X} \subseteq \{f\}$) implies $Z_f = +$.

Thus Z is an increasing direction for Y (with $Z_f = +$, $Z_g = 0$ and $(Y \circ Z)^- \subseteq \{f\}$), so that Y cannot be optimal. $\qquad\square$

The hard part is now to show

(i)   that every feasible bounded oriented matroid program has an optimal solution,

(ii) how to find such an optimal solution, and

(iii) how to decide whether a given oriented matroid program is feasible and bounded.

The answer to problem (i) is given by the following theorem.

10.1.13 Main Theorem of Oriented Matroid Programming (Bland and Lawrence). *If $(\mathcal{M}, g, f)$ is a feasible, bounded oriented matroid program, then it has an optimal solution $Y \in \mathcal{A}$.*

This was first proved, independently and around the same time, by Lawrence (1975) and by R.E. Bland. Both proofs are non-constructive, using induction on the size of the ground set. Such proofs can be found in Bland (1977a),

Folkman and Lawrence (1978), Edmonds and Fukuda (1982) and in Fukuda and Matsui (1989). However, from a linear programming viewpoint it is more natural to devise an abstraction of Dantzig's (1963) simplex algorithm for linear programming which also solves problems (ii) and (iii). This abstraction was achieved by Bland (1977a), where the first constructive proof was given.

10.1.14 Simplex algorithm – geometric version.

Input: An oriented matroid program $(P) = (\mathcal{M}, g, f)$.

Output: An optimal cocircuit Y for (P), or a proof that (P) is infeasible or unbounded.

Phase I: (Feasibility Problem) Find a feasible cocircuit $Y^0 \in \mathcal{A}$ for (P). If (P) is infeasible, output this information and STOP.

(We will describe a way to solve this problem later – see Section 10.2.)

Phase II: (Optimization Problem)

(II.1) Test whether Y^0 is optimal. In this case, output Y^0 and STOP.

(II.2) Find a feasible edge at Y^0 in a direction in which f increases.

(A *pivot rule* here decides which direction to choose.)

(II.3) Test whether (P) is bounded in this direction; if not, output this information and STOP.

(II.4) Construct a "better" cocircuit Y^1 as the second endpoint of the chosen edge emanating from Y^0, put $Y^0 := Y^1$, and GOTO (II.1).

We will now check the correctness of this approach, that is, show that if Y^0 is not optimal, then the edge required in **(II.2)** exists and unboundedness can be tested in **(II.3)**.

In Section 10.2 we will then see how to do the test for optimality in **(II.1)** with the "simplex criterion" and to *find* the edge Z and the "better" cocircuit Y^1 in **(II.2)** and **(II.4)** by a "pivot step" (basis exchange).

10.1.15 Lemma. *If a cocircuit $Y^0 \in P(E_n)$ is feasible but not optimal, then there is a feasible direction $Z \in \mathcal{A}^\infty$ with $Z_f = +$, such that $Y^0 \circ Z$ is an edge (that is, the rank of $Y^0 \circ Z$ in $\mathcal{L}$ is 2). Furthermore the second endpoint of $Y^0 \circ Z$ is a cocircuit Y^1 such that either $Y_g^1 = +$, in which case Y^1 is feasible, or $Y_g^1 = 0$, in which case (P) is unbounded.*

Proof. If Y^0 is not optimal, then there is a $Z \in \mathcal{A}^\infty$, with $Z_f = +$, $Z_g = 0$, such that $Y^0 \circ Z \in P(E_n)$. We then have $Y^0 \neq Y^0 \circ Z$. Put $S := \{e \in E_n : Y_e^0 = 0\}$.

Now the restriction $\mathcal{M}(S \cup \{g, f\})$ has full rank $r = r(\mathcal{M})$, so the minor $\bar{\mathcal{M}} := \mathcal{M}(S \cup \{g, f\})/g$ has rank $r - 1$.

In the minor $\bar{\mathcal{M}}$, the image of Z is a positive covector $\bar{Z}$ containing f, hence there is also a positive cocircuit $\bar{Z}^0$ with this property. This $\bar{Z}^0$ extends to a cocircuit Z^0 in $\mathcal{M}$ with $Z_f^0 = +$ and $Z_g^0 = 0$.

The zero set $R := \{e \in S : \bar{Z}^0_e = 0\}$ of $\bar{Z}^0$ has rank $r - 2$ in $\bar{\mathcal{M}}$, because Z^0 is a cocircuit of $\bar{\mathcal{M}}$. R has the same rank in $\mathcal{M}$. Now we observe that $R \subseteq \{e \in E_n : (Y^0 \circ Z)_e = 0\}$. Thus we know that $Y^0 \circ Z$ is an edge.

This means that at the non-optimal vertex Y^0 there is a feasible *edge* $Y^0 \circ Z \in \overline{P}(E_n)$. The second endpoint Y^1 of this edge (with $Y^0 \circ Z = Y^0 \circ Y^1 = Y^1 \circ Y^0$) is also in $\overline{P}(E_n)$, thus either feasible or unbounded. □

Note that an alternative proof of this lemma follows from the consideration of bases, in Proposition 10.2.12.

Now we consider the graph obtained from the cocircuits in $\mathcal{A}$ and the edges between them by directing all the edges that correspond to increases in the direction given by f (that is, from Y^0 to Y^1 in the notation of Lemma 10.1.15).

For this, let $[Y^0, Y^1]$ be an edge (that is, $Y^0 \vee Y^1 = Y^0 \circ Y^1 = Y^1 \circ Y^0$, where Y^0 and Y^1 lie in $\mathcal{A}$). We get a vertex $Z := Z[Y^0, Y^1] \in \mathcal{A}^\infty$ which is determined by following the pseudoline $\overline{Y^0 Y^1}$ "to infinity", in direction from Y^0 to Y^1. Here the intersection of the pseudoline $\overline{Y^0 Y^1}$ with $S_g = \mathcal{A}^\infty$ consists of two opposite vertices, of which Z is the one which satisfies $Y^0 \circ Y^1 = Y^0 \circ Z$. Formally, this Z is the cocircuit obtained by elimination of g from Y^1 and $-Y^0$. (This elimination is unique because Y^0 and Y^1 form a modular pair.)

10.1.16 Definition (graph of a program). Let G_f be the graph whose vertices are the vertices (cocircuits) in $\mathcal{A}$ and whose edges are the edges (covectors of rank 2) in $\mathcal{A}$ between them.

Some of the edges of G_f are directed, as follows. If $Y^0 \vee Y^1$ is an edge of $\mathcal{A}$ for two cocircuits $Y^0, Y^1 \in \mathcal{A}$, then let $Z = Z[Y^0, Y^1]$ be the cocircuit obtained by elimination of g from Y^1 and $-Y^0$. Now

- if $Z_f = +$, then the *increasing* edge $[Y^0, Y^1]$ is directed from Y^0 to Y^1,
- if $Z_f = -$, then the *decreasing* edge $[Y^0, Y^1]$ is directed from Y^1 to Y^0,
- if $Z_f = 0$, then the *horizontal* edge $[Y^0, Y^1]$ is not directed.

Thus in general G_f is a graph with both directed and undirected edges (for example, the edges that lie in the pseudosphere S_f are undirected), as in Figure 10.1.2.

We will be looking for a *sink* in $G_f|_{P(E_n)}$, that is, for a vertex that is not the tail end of a directed edge. Such a vertex is either optimal, if there is no feasible increasing edge emanating from it (by the characterization of Lemma 10.1.15), or it shows unboundedness, if there is a feasible increasing edge which is not in G_f because it is unbounded, i.e. whose second endpoint $Z \in \mathcal{A}^\infty$ is "at infinity".

To search for a sink, we can start at an arbitrary feasible vertex, and always move along feasible increasing edges. Since $G_f|_{P(E_n)}$ is finite, we either find a sink, or we run into a directed cycle. The next observation is crucial.

10.1.17 Lemma. *If $\mathcal{M}$ is realizable, then G_f does not contain a cycle $\mathbf{C}$ such that*

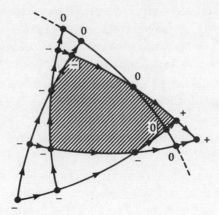

Figure 10.1.2: The graph of the program of Figure 10.1.1.

- **C** *contains a directed edge, and*
- *all directed edges of* **C** *are directed consistently with an orientation of the cycle.*

Proof. In this case, the edges are directed according to increase of the global linear objective function f. Thus the values of the objective function at the vertices of a partially directed path increase weakly, so the path cannot return to its starting point after passing its first directed edge.   □

We will see in Section 10.4 that G_f can have directed cycles in the non-realizable case, which makes the search for a sink more difficult. Thus the choice of suitable edges is crucial to obtain a finite simplex method. In the tableau version of the next section, this will correspond to the design of a suitable pivot rule. In Section 10.3 we will describe Bland's "smallest subscript rule", which does stay feasible and can be used for "Phase I" on oriented matroids. Interestingly enough, this pivot rule is finite for "Phase II" on *linear* programs (that is, for realizable oriented matroids), but exhibits infinite, non-degenerate cycling – that is, it may follow a directed cycle in G_f – in the general case (Section 10.4).

Bland (1977a) therefore constructed a different recursive pivot rule ALGORITHM A, which does lead to an algorithmic proof of Theorem 10.1.13, but does not always stay feasible in the non-realizable case (and thus is not a proper *simplex method*).

We will now give a geometric sketch of the approach taken by a simple version ("ALGORITHM DC") of ALGORITHM A. This gives an informal, recursive description of how to find an optimal or unbounded cocircuit for any feasible oriented matroid program – and thus a proof of Main Theorem 10.1.13. More details, and the duality theorem, will be available with the tableau version in the following two sections.

Proof of Main Theorem 10.1.13. Let $(\mathcal{M}, g, f)$ be an oriented matroid program, and assume that a feasible cocircuit Y^0 is given. Now choose a constraint element $h \in E_n$ such that $Y_h^0 = 0$. Geometrically, this means that Y^0 is a vertex of the feasible polyhedron P, and h corresponds to an inequality that is satisfied with equality at Y^0, or to a pseudohyperplane S_h through Y^0.

The key point is now that the minors of an oriented matroid program (see Definition 10.1.3(3)) can be considered as subproblems, which leads to a recursive approach. So, by induction on n, we know how to solve the oriented matroid program obtained by restricting to the pseudohyperplane S_h – this is $(\mathcal{M}/h, g, f)$. If this is unbounded, then so is $(\mathcal{M}, g, f)$. Otherwise we find a cocircuit Y^1 that is optimal for $(\mathcal{M}/h, g, f)$, that is, Y^1 is a maximal vertex of the polyhedron $P/h := P \cap S_h$.

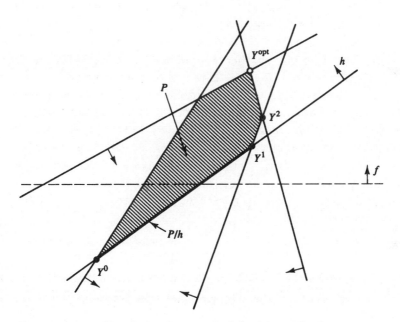

Figure 10.1.3: Sketch for the proof of the Main Theorem 10.1.13.

Now Y^1 is either optimal for $(\mathcal{M}, g, f)$ – and we are done – or one can leave S_h by an increasing edge from Y^1, to get a new feasible cocircuit Y^2 with $Y_h^2 = +$. Note that Y^2 determines a cocircuit in the program $(\mathcal{M} \backslash h, g, f)$, which is obtained by removing the constraint element h (respectively the pseudosphere S_h) from the program. Again by induction we know how to solve this program. The key lemma below states that every optimal solution to $(\mathcal{M} \backslash h, g, f)$ is also optimal for $(\mathcal{M}, g, f)$, whereas unboundedness of $(\mathcal{M} \backslash h, g, f)$ implies that $(\mathcal{M}, g, f)$ is unbounded as well. $\qquad\Box$

10.1.18 Lemma (Bland 1977a). *Let $(\mathcal{M}, g, f)$ be an oriented matroid program, $h \in E_n$, and let $Y \in \mathcal{C}^*(\mathcal{M})$ be a cocircuit with $Y_h = 0$, such that Y is optimal for $(\mathcal{M}/h, g, f)$, but not for $(\mathcal{M}, g, f)$.*

(i)   *If $Y' \in \mathcal{C}^*(\mathcal{M} \backslash h)$ is optimal for $(\mathcal{M} \backslash h, g, f)$, then the cocircuit in $(\mathcal{M}, g, f)$ corresponding to Y' is optimal for $(\mathcal{M}, g, f)$.*

(ii)   *If $Y' \in \mathcal{C}^*(\mathcal{M} \backslash h)$ proves unboundedness of $(\mathcal{M} \backslash h, g, f)$, then the corresponding cocircuit in $\mathcal{C}^*(\mathcal{M})$ proves unboundedness for $(\mathcal{M}, g, f)$.*

Proof.

(i) We only would have problems if Y' was infeasible for $(\mathcal{M}, g, f)$, with $Y_h' = -$. In this case Y and $_{-h}Y'$ would both be optimal solutions for the program $(_{-h}\mathcal{M}, g, f)$. Furthermore, the circuit X proving optimality of Y has to contain h, because Y is not optimal for $(\mathcal{M}, g, f)$. But this implies $h \in \underline{X} \cap \underline{Y}$, which contradicts Theorem 10.1.12(2).

(ii) Similarly, in this case $_{-h}Y'$ would prove unboundedness of $(_{-h}\mathcal{M}, g, f)$, while Y is optimal: this is impossible. $\square$

Of course, one can argue whether this proof is "constructive": it does give a procedure, but its description is recursive and no appropriate data structure is given. A more formal setting will be developed in the next section. There we will show how the transformation from the geometric approach to the computational framework is achieved for linear and for oriented matroid programs. Only after this transformation the simplex method becomes an "effective" algorithm.

There are several good explanations for the notorious *efficiency of the simplex method* (Shamir 1987). The key to any numerical treatment is an effective encoding of the geometric approach outlined in the previous section into a computational framework with

 – data structures that allow effective representation of the current information,
 – rules that describe reasonably simple updates after every iteration, and
 – effective criteria to do the necessary tests that determine the action taken in each step.

In the simplex method,

 – the first is achieved by encoding the vertices of the polyhedron by bases of the associated tableau,
 – the second is given by the treatment of the iteration steps as one element basis exchanges (pivots), and
 – the third is achieved by pivot rules of various complexity and effectiveness, together with the simplex criterion to check optimality.

Of course, *efficiency* is used for oriented matroids in a sense that is quite distinct from the one dictated by large scale programs in "real life programming": simplex tableaux are not a suitable way to represent the typically very sparse data of large practical (linear) problems from real world applications.

In the oriented matroid context, however, they tell you the *local information* (that is, "where you are", "what is going on" and "what to do next") in a nice and efficient way. The tableaux contain more information than is needed for the local choices made by the usual pivot rules, and the revised simplex algorithm would exploit this. On the other hand, not the whole problem (oriented matroid program) is encoded in the set of basic circuits given by a tableau: this is an important difference between linear programming and oriented matroid programming.

For a more realistic implementation, one would assume that the oriented matroid is given by a basis oracle, that is, by a subroutine that for every input of an r-tuple $N \subseteq E^r$ would output the sign $\chi(N)$. (See Schrijver (1986) for an extensive discussion of oracles, and Bachem and Reinhold (1989) for a specific discussion of oriented matroid oracles.)

Such an oracle $\mathbf{B}$ in particular tells us whether a given N (with $g \in N$ and $f \notin N$) is a basis, and in this case by repeated calls constructs the basic circuits $c(N, e)$ for $e \notin N$ and the basic cocircuits $c^*(N, e)$ for $e \in N$. Thus, by the cryptomorphism between basis and cocircuit axiomatics (Chapter 3), the basis oracle $\mathbf{B}$ is *polynomially equivalent* to an oracle $\mathbf{T}$ which for every N constructs the corresponding tableau (see Section 10.2) if N is a basis, and an error message otherwise.

One important fact is that if we have a basis oracle for a program $(P) = (\mathcal{M}, g, f)$, then we can use it to simulate a basis oracle for the dual program, but also for related programs, like the one that is obtained if we replace f by a loop, combine (P) and (P^*), etc. Operations of that type are used to achieve different "standard forms", for example to create artificial problems that solve the Phase I problem, see Section 10.2, or to construct the associated linear complementarity problem. Similarly, the simulation of basis oracles for minors of oriented matroid programs can be used for some pivot algorithms like Bland's ALGORITHM A.

10.2 Pivot steps and tableaux

A geometric description of oriented matroid programming problems and the simplex algorithm was presented in Section 10.1. In the following we will show how the tableau simplex method "implements" this approach. Here it is helpful if the reader is familiar with some basic features of the tableau simplex method for linear programming. The standard references for this are Dantzig (1963), Chvátal (1983) and Schrijver (1986).

The *simplex algorithm* implements the geometric approach of Section 10.1 by proceeding *from basis to basis* rather than from vertex to vertex (cocircuit to cocircuit). The key ideas and steps are the following.

- Consider a linear programming problem
    $$\max \ \mathbf{c}^T \mathbf{x} - d, \ A\mathbf{x} \leq \mathbf{b}, \ \mathbf{x} \geq \mathbf{0}.$$
- Introducing slack variables, we can get to a system of the form
    $$\max \ \mathbf{c}^T \mathbf{x} - d, \ A\mathbf{x} = \mathbf{b}, \ \mathbf{x} \geq \mathbf{0}.$$
- The original and the slack variables are now treated interchangeably.

- However, from the slack variables we have an identity matrix of full rank as a submatrix of A, which describes a basis of (the column space of) A.
- Pivot steps proceed from one such basis of A to the next, in single element exchanges. For simplicity we will update A by row operations so that we maintain an identity matrix of full rank in A corresponding to the current basis, possibly with interchanged columns.
- Corresponding to the current basis we maintain a current vertex (cocircuit), which stays feasible (positive) once feasibility has been reached.
- The rows of the matrix always form a set of basic cocircuits for the matroid given by the columns, corresponding to the current basis.

10.2.1 Definition (standard form for linear programs).

(i) A *linear program in standard form* is a linear program of the form

$$\max \ \mathbf{c}^T\mathbf{x} \ - d$$
$$A\mathbf{x} \ \leq \mathbf{b} \qquad\qquad (LP_0)$$
$$\mathbf{x} \ \geq \mathbf{0}.$$

Here A is a real matrix $A \in \mathbb{R}^{m \times n}$, where m (the number of inequalities) and n (the dimension of the problem) are our principal parameters for the problem size. Correspondingly we get column vectors $\mathbf{x} \in \mathbb{R}^n$, $\mathbf{b} \in \mathbb{R}^m$ and $\mathbf{c} \in \mathbb{R}^n$, and a constant $d \in \mathbb{R}$.

(ii) The inequality system of (LP_0) is transformed into an equality system

$$\max \ \mathbf{c}^T\mathbf{x} - d$$
$$A\mathbf{x} + \mathbf{u} \ = \mathbf{b} \qquad\qquad (LP_1)$$
$$\mathbf{x}, \mathbf{u} \ \geq \mathbf{0}$$

by introducing a vector $\mathbf{u} \in \mathbb{R}^m$ of non-negative *slack variables*.

Then "*special variables*" f and g are introduced, where $f := \mathbf{c}^T\mathbf{x} - d$ encodes the objective function, and g is used to linearize the problem, by encoding affine space by $g \equiv 1$:

$$\max \ f: \quad g \ \equiv 1$$
$$A\mathbf{x} + \mathbf{u} \quad - g\mathbf{b} \ = \mathbf{0} \qquad\qquad (LP_2)$$
$$-\mathbf{c}^T\mathbf{x} \quad + f + gd \ = 0$$
$$\mathbf{x}, \mathbf{u} \ \geq \mathbf{0}.$$

(iii) Renaming the variables by putting $x_{n+i} := u_i$, for $1 \leq i \leq m$, yields the *tableau form* of a linear program

$$\max \ f: \quad g \ \equiv 1$$
$$\begin{pmatrix} A & I & 0 & -\mathbf{b} \\ -\mathbf{c}^T & \mathbf{0}^T & 1 & d \end{pmatrix} \begin{pmatrix} \mathbf{x} \\ f \\ g \end{pmatrix} = \mathbf{0} \qquad\qquad (LP_3)$$
$$\mathbf{x} \geq \mathbf{0}$$

where the matrix

$$T_0 = \begin{pmatrix} A & I & 0 & -\mathbf{b} \\ -\mathbf{c}^T & \mathbf{0}^T & 1 & d \end{pmatrix}$$

is called the *initial tableau* for (LP_0).

10.2.2 Proposition. *The oriented matroid program associated with a linear program in standard form*

$$\max \ \mathbf{c}^T \mathbf{x} - d, \quad A\mathbf{x} \le \mathbf{b}, \quad \mathbf{x} \ge \mathbf{0} \tag{LP}$$

is the triple $(\mathcal{M}, g, f)$, *where* $\mathcal{M}$ *is the oriented matroid of rank* $n+1$ *on* $E := \{1, 2, \dots, m+n, g, f\}$ *given by the arrangement of halfspaces* $\mathbf{a}_i^T \mathbf{x} \le \mathbf{b}_i g$, $x_i \ge 0$, $g \ge 0$ *and* $\mathbf{c}^T \mathbf{x} \le dg$ *in* $\mathbb{R}^{n+1}$.

The linear subspace corresponding to the oriented matroid $\mathcal{M}$ as in Section 1.2(d) is the kernel of the matrix T_0, that is, the subspace $A\mathbf{x} + \mathbf{u} - g\mathbf{b} = \mathbf{0}$, $-\mathbf{c}^T\mathbf{x} + f + gd = 0$ of dimension $n+1$ in $\mathbb{R}^{m+n+2}$. Hence the oriented matroid of linear dependences between the columns of T_0 is the dual oriented matroid $\mathcal{M}^*$ of rank $m+1$. (See also the comments preceding Definition 10.2.4.)

We will now proceed as follows. First we consider a small example of a linear program in two variables (corresponding to Figure 1.7.2, see also Figure 10.1.1), and show how the simplex algorithm solves it, with Bland's extremely simple pivoting rule implemented. The objective of this is to review the simplex algorithm in a notational framework that afterwards allows us to demonstrate a completely analogous treatment of the associated oriented matroid program.

10.2.3 Example. Let (LP_0) be the "2-dimensional" linear programming problem of Section 1.7, given by

$$\begin{aligned}
\max \ x_1 + \ x_2 - 3 & \\
-x_1 + \ x_2 &\le 1 \\
x_1 - 2x_2 &\le 2 \\
x_1 + 2x_2 &\le 4 \\
x_1, x_2 &\ge 0.
\end{aligned} \tag{LP_0}$$

To get (LP_0) into the "standard form" required by the standard version of the Simplex Algorithm, we introduce non-negative slack variables, transforming (LP_0) into the equivalent program

$$\begin{aligned}
\max \ x_1 + \ x_2 - 3 & \\
-x_1 + \ x_2 + x_3 &= 1 \\
x_1 - 2x_2 \qquad + x_4 &= 2 \\
x_1 + 2x_2 \qquad\qquad + x_5 &= 4 \\
x_1, x_2, x_3, x_4, x_5 &\ge 0.
\end{aligned} \tag{LP_1}$$

To get down to sign properties, and to an even more symmetric formulation, we introduce two extra, distinguished variables: $f := x_1 + x_2 - 3$ allows us to treat the objective function as a variable, whereas a variable g linearizes the equations – where affine space, the setting of our LP problem, is considered as the set $\{g = 1\}$. This transforms (LP_1) into

$$
\begin{array}{lll}
 & \max f, & g = 1 \\
-x_1 + x_2 + x_3 & & -\ g = 0 \\
x_1 - 2x_2 \quad\quad + x_4 & & -2g = 0 \\
x_1 + 2x_2 \quad\quad\quad\quad + x_5 & & -4g = 0 \\
-x_1 - x_2 & & +f + 3g = 0 \\
& x_1, x_2, x_3, x_4, x_5 \geq 0.
\end{array}
\qquad (LP_2)
$$

Thus the optimization problem (LP_0) is equivalently given by the 3-dimensional subspace of $\mathbb{R}^7$ that is defined by the four equations of (LP_2). Here the standard coordinates in $\mathbb{R}^7$ are $x_1, \dots, x_5, f, g$, where f and g play special roles. The *polyhedron* $P(E_5)$ is the intersection of the halfspaces $\{x_i \geq 0\}$ with the 3-subspace. This identifies the hyperplanes of the arrangement with the variables of the problem.

It is customary in linear programming to record the coefficients of a linear program like (LP_2) in a tableau (matrix) T, as follows:

	(1)	(2)	(3)	(4)	(5)	(f)	(g)
(3)	-1	1	1	0	0	0	-1
(4)	1	-2	0	1	0	0	-2
(5)	1	2	0	0	1	0	-4
(f)	-1	-1	0	0	0	1	3

$T =$ (to the left of the tableau)

The sign vectors of the rows of this tableau are the *basic circuits* of the oriented matroid program, with respect to the current oriented matroid basis $[12g]$.

We have the following conventions and interpretations:

(i) T is a (4×7)-matrix – in general, we will get an $(m+1) \times (n+m+2)$-matrix from a linear program as given in (LP_0).

(ii) By construction, T has full rank. In fact, there is an identity matrix of full rank contained in T corresponding to the variables x_3, x_4, x_5 and f. This corresponds to a basis of the column space of T. We will denote this basis by $B = [345f]$, and the tableau by $T = T[B] = T[345f]$. This basis is ordered – the ordering explicitly being given by the labels of the rows of T – and it satisfies $f \in B$, $g \notin B$.

(iii) The columns of $T[B]$ are indexed by $1, 2, \dots, 5, f, g$ (which is shorthand for the variables of the problem), the rows by the basis variables $3, 4, 5, f$. Here the last column and the last row of T correspond to g and f, respectively, and are graphically set aside as somewhat special.

(iv) The entries of the tableau $T[B]$ will be denoted by t_{be} ($b \in B$, $e \in E$). Here we write $E = \{1, \dots, 5, f, g\}$ for the set of variables which indexes the columns, and $B = \{3, 4, 5, f\}$ for the set of rows, corresponding to the basis variables. Thus $t_{be} = \delta_{be}$ (Kronecker delta) for $e \in B$.

In our example, the values t_{bg} for $b \in B \backslash f$ are all negative. This corresponds to the fact that the solution $\mathbf{x}^B$ to (LP_0) given by B, namely by

$$\mathbf{x}_e^B = \begin{cases} -t_{eg} & \text{for } e \in B, \\ 1 & \text{for } e = g, \\ 0 & \text{otherwise,} \end{cases}$$

is feasible.

In our example, this is the vertex

$$(x_1, \ldots, x_5, f, g) = (0, 0, 1, 2, 4, -3, 1).$$

In order to avoid the confusion that arises quite naturally at this point, we will now clarify the precise relationship of the bases of a linear program and those of the associated oriented matroid.

Terminology. The oriented matroid $\mathcal{M}$ (of rank $n+1$) given by an inequality system $Ax \leq \mathbf{b}$ of a standard linear program 10.2.1(i) is dual to the oriented matroid $\mathcal{M}^*$ given by the columns of the corresponding simplex tableau T_0 as in Definition 10.2.1(iii).

Thus if we follow the standard linear programming terminology, talking about bases of linear programs, this will lead to a very common but *unnecessary confusion* between the bases of the two oriented matroids involved.

In our presentation we will resolve the problem by distinguishing *bases* of the oriented matroid $\mathcal{M}$ from the admissible bases for the simplex tableaux, called *T-bases*, which are bases of $\mathcal{M}^*$ and hence *cobases* of $\mathcal{M}$.

Thus in the following we will try to consistently use the prefix "*T-*" (for "tableau") whenever we refer to linear programming objects and terminology.

10.2.4 Definition (*T-bases, T-nonbases, tableaux*). Let $(P) = (\mathcal{M}, g, f)$ be an oriented matroid program.

(i)   A *T-basis* of (P) is a cobasis (basis complement) $B \subseteq E$ of $\mathcal{M}$ such that $f \in B$ and $g \notin B$. We will assume for the following that B is ordered, with f coming last, that is, $B = [b_1, \ldots, b_m, f]$. The associated *T-nonbasis* of (P) is the oriented matroid basis $N := E \setminus B$ of $\mathcal{M}$, satisfying $f \notin N$, $g \in N$.

(ii) The *oriented matroid tableau* corresponding to B is the $(m+1) \times (n+m+2)$ matrix

$$T[B] = (\tau_{be}) \in \{+, -, 0\}^{B \times E}$$

whose rows are the basic circuits

$$c(b, N) \qquad \text{for } b \in B$$

corresponding to the basis N of $\mathcal{M}$, written as row vectors. Thus the entries of the matrix $T[B]$ are

$$\tau_{be} = c(b, N)_e \qquad \text{for } e \in E, b \in B.$$

Similarly one can read off the basic cocircuits associated with N from the columns of $T[B]$, with

$$c^*(e, N)_b = \begin{cases} -\tau_{eb} & \text{for } b \in B, \\ \delta_{eb} & \text{for } b \in N. \end{cases}$$

for every $e \in N$ (where δ_{eb} denotes the Kronecker δ-function).

Note that for a linear program, the oriented matroid tableau is derived from the corresponding simplex tableau by simply taking signs. (Also, the trivial column corresponding to f is missing in your linear programming textbooks.)

Here the initial tableau for a linear program in standard form (Definition 10.2.1(iii)) is

$$T_0 = T[n+1, \ldots, n+m, f].$$

Thus the starting tableau for the oriented matroid program of Section 1.7 and Example 10.2.3 is

	(1)	(2)	(3)	(4)	(5)	(f)	(g)
(3)	−	+	+	0	0	0	−
(4)	+	−	0	+	0	0	−
(5)	+	+	0	0	+	0	−
(f)	−	−	0	0	0	+	+

$T[345f] =$

For the sake of analysis and notation it is often convenient to reorder the rows and the columns of a tableau $T[B]$. We will, however, agree that the last row of the tableau always corresponds to the objective function (labeled by $f \in B$) and the last two columns always correspond to f and g, in this order.

These conventions allow us, for an arbitrary T-basis B, to write its tableau as

	$N \backslash g$	$B \backslash f$	(f)	(g)
$B \backslash f$	A	I	0	$-\beta$
(f)	$-\gamma$	0	1	δ

$T[B] =$

Perhaps it is helpful now to give the geometric interpretation of some of the information given by a simplex tableau. For this, let $(P) = (\mathcal{M}, g, f)$ be a program, B be a T-basis, $N = E \backslash B$ the associated basis of the oriented matroid $\mathcal{M}$. Then the tableau encodes the geometric situation at the cocircuit

$$Y^g := c^*(g, N),$$

which is *numerically* given by the negative of the last column of $T[B]$

$$Y^g_e = c^*(g, N)_e = \begin{cases} -\tau_{eg} & \text{for } e \in B, \\ + & \text{for } e = g, \\ 0 & \text{otherwise.} \end{cases}$$

Geometrically, Y^g is given as the unique cocircuit such that

$$Y^g_g = + \qquad \text{(cocircuit in affine space)},$$
$$Y^g_e = 0 \qquad \text{for } e \in N \backslash g.$$

In fact, $N\backslash g$ defines a *coordinate system* in $\mathcal{A}$, whose center (all coordinates equal to 0) is Y^g. Its *coordinate hyperplanes* are the flats $H(e) := \{Y \in \mathcal{A} : Y_e = 0\}$ for $e \in N\backslash g$.

Every $e_0 \in N\backslash g$ defines a *direction*, with the *coordinate axis* $A(e)$ of all covectors such that $Y_e = 0$ for $e \in N\backslash\{g, e_0\}$, and $Y_g = Y_{e_0} = +$.

In particular, there is a unique cocircuit Y^{e_0} marking the "end" of the coordinate axis at the "coordinate plane at infinity", given by the e_0-column of $T[B]$:

$$Y_e^{e_0} = c^*(e_0, N)_e = \begin{cases} -\tau_{ee_0} & \text{for } e \in B, \\ + & \text{for } e = e_0, \\ 0 & \text{otherwise.} \end{cases}$$

This interprets the columns of $T[B]$. To understand the rows, we either dualize (see the construction of the dual tableau in Corollary 10.2.9), or interpret the rows as unsolvable strict inequality systems, given by N together with one element of B. Thus the b-row (for $b \in B$) records which orthants (of the coordinate system given by N) have empty intersection with the halfspace given by b.

As an exercise for the reader, we recommend to use the circuit and the cocircuit interpretation to get two ways to derive the tableau $T[345f]$ for our Figure 10.1.1. In Figure 10.2.1 we indicate the cocircuits Y^e ($e \in N$) and some circuits X^b ($b \in B$). The figure represents the same oriented matroid program as Figures 1.7.2 and 10.1.1. Corresponding to the T-basis $B = [345f]$, bounded triangles corresponding (with $X_g = -$) to the basic circuits X^3 and X^4, and the vertices locating the basic cocircuits Y^1, Y^2, Y^g, are indicated.)

We will now spend some time analyzing and interpreting tableaux of this type. The following Lemma and Proposition describe the way in which the (feasible) cocircuits of the geometric simplex method of Section 10.1 are encoded by "(feasible) T-bases" for the tableau simplex method.

10.2.5 Lemma. *For every cocircuit $Y \in A(\mathcal{M}\backslash f)$ there is a (usually not unique) T-basis B_Y such that for $N_Y := E\backslash B_Y$:*

$$Y = c^*(g, N_Y).$$

Conversely, every T-basis B determines a unique cocircuit $Y^B \in A(\mathcal{M}\backslash f)$ by $Y^B := c^*(g, N)$. This Y^B can be read off from the tableau $T[B]$ by completing the negative of the last column with zeroes:

$$Y_e^B = \begin{cases} \beta_e & \text{for } e \in B\backslash f, \\ -\delta & \text{for } e = f, \\ 0 & \text{for } e \in N\backslash g, \\ + & \text{for } e = g. \end{cases}$$

Proof. If $Y \in A(\mathcal{M}\backslash f)$ is a cocircuit, then $Y^0 := \{e \in E : Y_e = 0\}$ is a hyperplane of $\underline{\mathcal{M}}$ with $g \notin Y^0$. Thus $Y^0 \cup g$ contains a basis N_Y of $\mathcal{M}$, which automatically satisfies $g \in N_Y$, and thus $Y = c^*(g, N_Y)$ for $B_Y := E\backslash N_Y$. The converse is clear. $\quad\square$

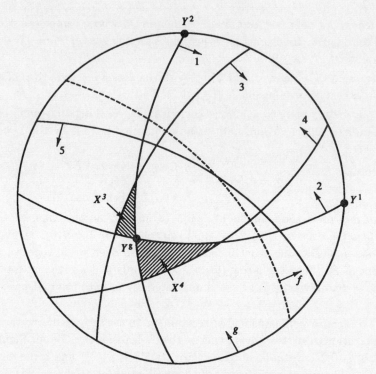

Figure 10.2.1: An oriented matroid program of rank 3.

In particular, we can read off from the last column $\begin{pmatrix} -\beta \\ \delta \end{pmatrix}$ of a tableau $T[B]$ whether the corresponding cocircuit Y^B is feasible.

10.2.6 Proposition (feasible tableau).   *Let $(P) = (\mathcal{M}, g, f)$ and let B be a T-basis for (P).*

(i)   *The cocircuit Y^B is feasible if and only if $\beta \geq 0$ (that is, $-\beta_b = c(b, N)_g \in \{0, -\}$ for all $b \in B \backslash f$). In this case, the T-basis B and the tableau $T[B]$ are called feasible.*

(ii)   *For every feasible cocircuit $Y \in \mathcal{A}(\mathcal{M} \backslash f)$, there is a feasible T-basis B_Y such that $Y = c^*(g, N_Y)$.*

(iii) *If there is a row $A_{b.}$ indexed by some $b \in B \backslash f$ such that $A_{b.} \geq 0$, $-\beta_b = +$, then (P) is infeasible.*

Dualizing Proposition 10.2.6, we get a characterization of bounded oriented matroid programs in term of their T-bases.

10.2.7 Proposition (bounded tableau).   *Let $(P) = (\mathcal{M}, g, f)$ and let B be a T-basis for (P). Then*

(i)   *If $\gamma \leq 0$ (that is, $-\gamma_e = c(f, N)_e = \tau_{fe} \in \{+, 0\}$ for all $e \in N \backslash g$), then (P) is bounded. In this case, the T-basis B and the tableau $T[B]$ are called bounded or dual feasible.*

(ii) *For every bounded program* (P) *with positive circuit* X, *there is a dual feasible* T-*basis* B_X, *such that the circuit* $X = c(f, N_X)$ *proving boundedness appears as the last row of the tableau* $T[B_X]$.

(iii) *If there is a column* $A_{.e}$ *indexed by some* $e \in N \backslash g$ *such that* $A_{.e} \leq 0$, $-\gamma_e = -$, *then* (P) *is unbounded.*

Note that in the case of non-degenerate oriented matroid programs, there is a bijection between feasible cocircuits and feasible bases (Proposition 10.2.6(ii)), respectively between bounding circuits and bounded bases (Proposition 10.2.7(ii)), because in this case every circuit/cocircuit determines the corresponding basis uniquely.

Note that Propositions 10.2.6(i) and 10.2.7(i) are translations of the characterizations of Lemma 10.1.9 from geometric statements into tableaux language; similarly Propositions 10.2.6(iii) and 10.2.7(iii) are restatements of the "if"-parts of Corollary 10.1.10.

The aim of all our efforts is to find a basis that is both feasible and dual feasible. The following corollary explains why: it serves as the optimality criterion for all simplex algorithms.

10.2.8 Corollary ("simplex criterion").   *If a* T-*basis* B *of an oriented matroid program* (P) *is both feasible and dual feasible, then*

$$Y = c^*(g, N)$$

is an optimal solution. In this case we call the T-*basis optimal.*

Conversely, for every optimal cocircuit Y *of* (P) *there is an optimal* T-*basis* B_Y *with* $Y = c^*(g, N_Y)$.

Thus a simplex tableau of the form

$$T[B] = \quad \begin{array}{c|c|c|c|c} & N\backslash g & B\backslash f & (f) & (g) \\ \hline B\backslash f & A & I & 0 & -\beta \\ \hline (f) & -\gamma & 0 & 1 & \delta \end{array}$$

is

$\longrightarrow$ feasible if $-\beta \leq 0$,

$\longrightarrow$ dual feasible if $-\gamma \geq 0$,

$\longrightarrow$ optimal if both conditions hold.

Compare this to the optimal tableau for the linear program of Example 10.2.3, given by

	(1)	(2)	(3)	(4)	(5)	(f)	(g)
(1)	+	0	0	−	+	0	−
(2)	0	+	0	+	+	0	−
(3)	0	0	+	+	+	0	−
(f)	0	0	0	+	+	+	−

$$T[123f] =$$

But much more can be read off from the tableaux. In particular, they allow us to conclude infeasibility or unboundedness from a "positive row" or a "negative column", according to Propositions 10.2.6(iii) and 10.2.7(iii).

Now let $(P) = (\mathcal{M}, g, f)$ be an oriented matroid program, B a T-basis, N the associated T-nonbasis. Then we call the T-basis $B^* := N$ of $(P^*) = (\mathcal{M}^*, f, g)$ the *dual T-basis* corresponding to B, and $N^* := B$ the corresponding *dual T-nonbasis*. With the duality between circuits and cocircuits we get an immediate construction of dual tableaux from (primal) tableaux, and the feasibility and optimality criteria translate.

10.2.9 Corollary (dual tableaux). *Let* $(P) = (\mathcal{M}, g, f)$ *be an oriented matroid program, B a T-basis. If the tableau for (P) is*

	$N\backslash g$	$B\backslash f$	(f)	(g)
$B\backslash f$	A	I	0	$-\beta$
(f)	$-\gamma$	0	1	δ

$$T[B] =$$

then with $B^* = N = E\backslash B$ the corresponding dual tableau for $(P^*) = (\mathcal{M}^*, f, g)$ is

	$B\backslash f$	$N\backslash g$	(g)	(f)
$N\backslash g$	$-A^T$	I	0	γ
(g)	β	0	1	$-\delta$

$$T^*[B^*] =$$

From the tableaux one reads off that $T[B]$ is feasible if and only if $T^*[B^*]$ is bounded, and conversely.

In particular, $T[B]$ is optimal if and only if $T^*[B^*]$ is optimal.

If we define τ_{be} also for T-nonbasic b, by

$$\tau_{be} = \begin{cases} - & \text{for } b \in N, \ e = b, \\ 0 & \text{for } b \in N, \ e \neq b, \end{cases}$$

then this allows us to write

$$T^*[B^*] = (-\tau_{eb})_{b \in B^* = N, e \in E}$$

for the dual tableau, where the primal tableau is

$$T[B] = (\tau_{be})_{b \in B, e \in E}.$$

Propositions 10.2.6 and 10.2.7 are cornerstones of the approach taken by the simplex method. *Phase I* of the algorithm finds a *feasible* basis – and this can effectively be checked in the last column of the tableau, by Proposition 10.2.6. After that in *Phase II* the algorithm successively "improves" the current basis by single element exchanges (pivot steps), maintaining feasibility, until *dual feasibility* is achieved, which is checked in the last row of the tableau, using Proposition 10.2.7. At this point Corollary 10.2.8 guarantees that the problem is solved.

The approach of the *dual simplex method* is to *first* achieve dual feasibility, and *then* to improve until feasibility is reached. Again, 10.2.6 – 10.2.8 provide the necessary checks. In fact, the dual simplex method is equivalent to the (primal) simplex method applied to the dual program, see below.

We will now describe the "pivot step" that forms a single iteration of the simplex algorithm.

10.2.10 Definition (pivot steps). Let $(\mathcal{M}, g, f)$ be an oriented matroid program.

Let B be a T-basis, and let $b_0 \in B \backslash f$ and $e_0 \in N \backslash g = E \backslash (B \cup g)$. Then

$$B' := (B \backslash b_0) \cup e_0$$

is a new T-basis if and only if $\tau_{b_0 e_0} = c(b_0, N)_{e_0} \neq 0$.

In this case the transition from B to B' as the "current basis", corresponding to the transformation

$$T[B] \longrightarrow T[B']$$

of the corresponding tableaux, is called a *pivot step*. We refer to the transition $T[B] \longrightarrow T[B']$ as a *pivot in position* (b_0, e_0).

Here b_0 is called the *leaving variable* and e_0 is called the *entering variable* of the pivot, because it leaves, respectively enters, the T-basis.

10.2.11 Definitions. Let $(\mathcal{M}, g, f)$ be an oriented matroid program and consider a pivot

$$T[B] \longrightarrow T[B'].$$

(1)   The pivot is *degenerate* if $Y^B = Y^{B'}$, that is, if and only if

$$\tau_{b_0 g} = c(b_0, N)_g = 0,$$

and *non-degenerate* otherwise.

(2)   The pivot is *horizontal* if $\gamma_{e_0} = -\tau_{f e_0} = 0$.

(3)   The pivot is *feasible* if it proceeds from a feasible basis B to another feasible basis B'.

We say that a pivot *preserves feasibility* if it is feasible, or if B is infeasible.

(4)   The pivot is *strictly increasing* if it is non-degenerate, not horizontal (that is, $\tau_{b_0 e_0}$, $\tau_{f e_0}$ and $\tau_{b_0 g}$ are all non-zero) and either exactly one or all three of $\tau_{b_0 e_0}$, $\tau_{f e_0}$ and $\tau_{b_0 g}$ are positive. So a feasible pivot is strictly increasing if and only if $\tau_{b_0 g} = -Y_{b_0}^B = -$, $\tau_{b_0 e_0} = +$, and $\tau_{f e_0} = -$.

Similarly, a feasible pivot $[-\tau_{b_0 g} \geq 0]$ is *weakly increasing* if $\tau_{b_0 e_0} = +$ and $\tau_{f e_0} = -$; so it is either strictly increasing or degenerate, but cannot be horizontal.

(5)   Finally, a *simplex pivot* is a pivot that preserves feasibility, and which in the feasible case is (at least) weakly increasing.

The geometric picture underlying these definitions is as follows. The current oriented matroid basis N defines a coordinate system in the affine space $\mathcal{A}$, with origin $Y^B = c^*(g, N)$, as described above. The first choice of a pivot step is that of a coordinate $e_0 \in N \backslash g$, which determines a positive coordinate axis

$$A^+(e_0) = \{Y \in \mathcal{A} : Y_e = 0 \text{ for } e \in N \backslash \{g, e_0\}, \text{ and } Y_{e_0} = +\}.$$

This describes the *direction* of the pivot step. We consider the cocircuit $Y^{e_0} = c^*(e_0, N)$, that is, the vertex at which the positive coordinate axis $A^+(e_0)$ intersects the hyperplane $\mathcal{A}^\infty$ at infinity. Depending on whether this vertex has positive, zero or negative objective function value $Y_f^{e_0} = c^*(e_0, N)_f = -\tau_{f e_0}$, the objective function increases, is constant or decreases in the direction of $A^+(e_0)$.

How far the pivot step goes on the axis $A(e_0)$ depends on the choice of the element $b_0 \notin N \cup g$ that replaces e_0. We distinguish several cases. The first is that the new "current vertex" $Y^{B'}$ coincides with the old one, which happens exactly if the new coordinate hyperplane contains the origin of the old coordinate system, that is, if $Y^B \in \{Y \in \mathcal{C}^* : Y_{b_0} = 0\}$, or $\tau_{b_0 g} = 0$. This describes a degenerate pivot.

If this is not the case, then Y^B and $Y^{B'}$ are joined by a sequence of edges of G_f (Definition 10.1.16) on the coordinate axis $A(e_0)$. These edges are directed from Y^B to $Y^{B'}$ if the pivot is (strictly) increasing, undirected if the pivot is horizontal, and directed from $Y^{B'}$ to Y^B if the pivot is decreasing.

To distinguish these situations, suppose first that $A^+(e_0)$ points in increasing direction, $\tau_{f e_0} = -$. Then the pivot is increasing if and only if it moves to a cocircuit on $A^+(e_0)$, that is, if Y^B and Y^{e_0} are separated by the hyperplane $H(b_0)$, so that $Y_{b_0}^B = -\tau_{b_0 g}$ and $Y_{b_0}^{e_0} = -\tau_{b_0 e_0}$ have opposite signs.

Similarly in the second case, where $A^+(e_0)$ points in negative direction ($\tau_{f e_0} = +$) we derive that $\tau_{b_0 g}$ and $\tau_{b_0 e_0}$ have to have the same sign for an increasing pivot.

In particular, a non-degenerate pivot is increasing if and only if it "moves" along directed edges of G_f (in the right direction). A pivot along undirected edges is horizontal by definition. Pivots can be both degenerate and horizontal. However, horizontal pivots are not considered by primal simplex algorithms.

For a non-degenerate *feasible* pivot, the pivot step automatically moves to a new cocircuit on $A^+(e_0)$. Thus the pivot is increasing if $A^+(e_0)$ points to positive

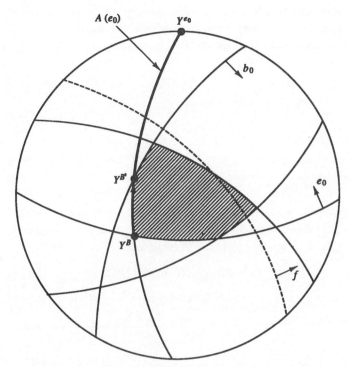

Figure 10.2.2: A strictly increasing, feasible pivot.

direction, with $Y^{e_0} = -\tau_{f e_0} = +$. Feasibility then also implies $Y^B_{b_0} = -\tau_{b_0 g} = +$ and $Y^{e_0}_{b_0} = -\tau_{b_0 e_0} = +$.

In *linear* programming, the transformation $T[B] \longrightarrow T[B']$ is achieved by doing the row operations on the matrix $T[B]$ that are necessary to create a unit vector in the e_0-column: first the b_0-row is divided by $t_{b_0 e_0}$, and then suitable multiples are subtracted from the other rows.

In *oriented matroid* programming, the update is likewise simple. As in the linear case, the goal is to do row operations to clear the e_0-column. This amounts to strong circuit exchanges:

$c(e_0, N') = \tau_{b_0 e_0} c(b_0, N)$

$c(b, N') =$ the circuit obtained from $c(b, N)$ and $c(b_0, N)$ by eliminating e_0,

   which positively contains b,

   $=$ the unique circuit X whose support is contained in

   $c(b, N) \cup c(b_0, N)$ that satisfies $X_b = +$ and $X_{e_0} = 0$ (for $b \neq e_0$).

The key to the simplex algorithm is a suitable choice of e_0 and b_0 for the pivot steps, which

(i)   guarantees that feasibility is preserved,

(ii) promises an increase in the objective function, and

(iii) achieves finiteness of the algorithm.

10.2.12 Proposition (simplex pivots; Bland 1977a). *Let B be a feasible, but not optimal T-basis, and let $e_0 \in N \backslash g$ with $\tau_{fe_0} = -\gamma_{e_0} = -$. Then exactly one of the following alternatives holds.*

(1) $\tau_{be_0} \in \{0, -\}$ *for all* $b \in B \backslash f$.

   In this case the program is unbounded (or dual inconsistent), and we say that $T[B]$ proves unboundedness.

(2)   *There is a* $b_0 \in B \backslash f$ *with* $\tau_{b_0 e_0} = +$ *and such that* $B' = (B \backslash b_0) \cup e_0$ *is again feasible.*

   Furthermore, if the program is non-degenerate, then b_0 is uniquely determined by e_0.

Proof. Once e_0 is chosen, we only consider the cocircuits Y with $Y_e = 0$ for $e \in N \backslash \{g, e_0\}$, which reduces everything to an affinely 1-dimensional problem, that is, to the rank 2 oriented matroid program

$$(\overline{P}) = \Big(\mathcal{M}/(N \backslash \{g, e_0\}), g, f\Big).$$

Now if (in situation (1)) the basic cocircuit $Y^0 := c^*(e_0, N)$ given by

$$Y_e^0 = \begin{cases} -\tau_{ee_0} & \text{for } e \in B, \\ + & \text{for } e = e_0, \\ 0 & \text{otherwise}, \end{cases}$$

is positive, then the program (P) is unbounded, and the e_0-column of $T[B]$ is a proof for this.

Otherwise (situation (2)) the linear program $(\overline{P})$ is bounded, and thus there is an optimal T-basis $\overline{B} = \{b_0, g\}$: this follows from the simple analysis of rank 2 programs, or from the Main Theorem 10.1.13 together with Corollary 10.2.8. Figure 10.2.3 depicts the three possibilities that can occur for a 1-dimensional program like $(\overline{P})$, corresponding to unboundedness, non-degenerate and degenerate pivot.

If (P) is non-degenerate, then so is $(\overline{P})$, and thus $\overline{B}$ is unique. The corresponding cocircuit

$$\overline{Y}' = c^*(b_0, N')$$

is optimal for $(\overline{P})$ and thus extends to a feasible cocircuit Y' of (P).   □

We have now, finally, collected all the parts that, when properly put together, constitute the simplex algorithm. (Compare this to the geometric version in Algorithm 10.1.14.)

10.2.13 Simplex Algorithm – tableau version.

Input: An oriented matroid program $(P) = (\mathcal{M}, g, f)$.

Output: A tableau $T[B]$ that is either optimal or proves that the program is infeasible or unbounded.

Phase I: (feasibility problem) Find a feasible T-basis B for (P), and construct $T[B]$. If (P) is infeasible, output a tableau that proves this and STOP.

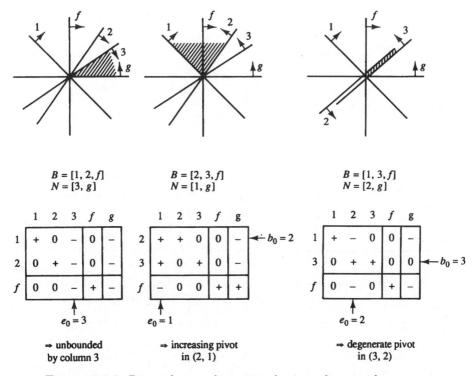

Figure 10.2.3: Pivots for a 1-dimensional oriented matroid program.

(This will, like Phase II, be solved by a suitable pivot algorithm, see Proposition 10.3.5)

Phase II: (Optimization Problem)

(II.1) (Optimality test, by 10.2.8)

If $\tau_{fe} \geq 0$ for all $e \in N \backslash g$ then $T[B]$ is optimal. Output $T[B]$ and STOP.

(II.2) (Column selection)

Choose $e_0 \in N \backslash g$ such that $\tau_{fe_0} = -$.

(II.3) (Boundedness test, by 10.2.7)

If $\tau_{be_0} \leq 0$ for all $b \in B \backslash f$, then $T[B]$ is unbounded. Output $T[B]$ with this information and STOP.

(II.4) (Row selection, by 10.2.12)

Otherwise we can choose $b_0 \in B \backslash f$ such that $B' = (B \cup e_0) \backslash b_0$ is again feasible (which implies $\tau_{b_0 e_0} = +$).

(II.5) (Update, by 10.2.10)

Put $B := B'$, construct the new tableau $T[B]$, and GOTO **(II.1)**.

Of course there is still a lot of freedom in the steps **(II.2)** and **(II.4)**. The description of the simplex algorithm will be completed once we have specified the

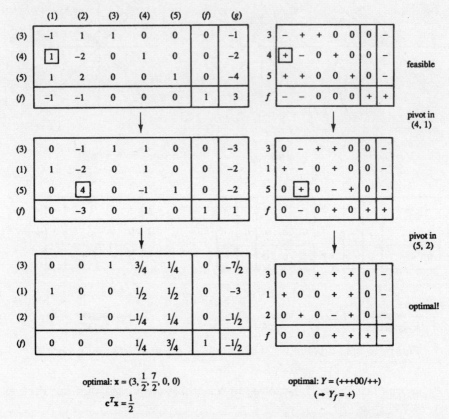

Figure 10.2.4: Simplex Algorithm (with smallest subscript rule)
on the program of Section 1.7/Example 10.2.3.

pivot rules that guide it. This leaves many different variants (leading to different *versions of the simplex algorithm*).

Figure 10.2.4 shows how the Simplex Algorithm (with Bland's pivot rule, say) solves both the *linear* and the *oriented matroid* version of the program of Example 10.2.3. It is supposed to again supply material for the construction of oriented matroid tableaux and their interpretation. Figure 10.1.1 presents the geometric picture for this – and the reader should follow the path of the simplex algorithm with Bland's rule in this figure.

In oriented matroid programming, other than in the linear case, it is difficult to give a rule that ALGORITHM 10.2.13 into a *finite* algorithm. This will be achieved in Section 10.3.

The important fact is that a finite version of the Simplex Algorithm 10.2.13 implies the linear programming duality theorems. For this recall that the Duality Theorem for a dual pair of linear programs of the form

$$\max \ \mathbf{c}^T \mathbf{x} \ -d$$
$$Ax \ \le \mathbf{b} \qquad\qquad (P)$$
$$\mathbf{x} \ \ge 0$$

and

$$\max \ (-\mathbf{b})^T \mathbf{y} \ +d$$
$$(-A^T)\mathbf{y} \ \le -\mathbf{c} \qquad\qquad (P^*)$$
$$\mathbf{y} \ \ge \ \mathbf{0}$$

states that if (P) and (P^*) are both feasible, then they have optimal solutions, and the objective function values of any optimal solutions $\bar{\mathbf{x}}$ and $\bar{\mathbf{y}}$ of (P) and (P^*) sum to zero, $\mathbf{c}^T \bar{\mathbf{x}} - \mathbf{b}^T \bar{\mathbf{y}} = 0$.

Furthermore, complementary slackness states that for every non-zero component of an optimal solution $\bar{\mathbf{x}}$, the corresponding inequality in "$(-A^T)\mathbf{y} \le -\mathbf{c}$" has to hold with equality for optimal $\bar{\mathbf{y}}$.

We will now formulate the Duality Theorem for oriented matroid programs which in particular sharpens the "Main Theorem" 10.1.13. Its validity follows from the finiteness and the stopping criteria of the simplex algorithm.

The related Complementary Slackness Theorem was already derived in Theorem 10.1.12(2); the so-called Strong Complementary Slackness Theorem will be deferred to Exercise 10.6.

10.2.14 Duality Theorem for Oriented Matroid Programming (Lawrence 1975, Bland 1977a). *Let $(P) = (\mathcal{M}, g, f)$ and $(P^*) = (\mathcal{M}^*, f, g)$ be a pair of dual oriented matroid programs. Then exactly one of the following two numbered conditions holds:*

(1) *The program (P) is infeasible, because there is a positive circuit $X \in \mathcal{C}(\mathcal{M})$ with $X_g = +$, $X_f = 0$, or the program (P^*) is infeasible, because there is a positive cocircuit $Y \in \mathcal{C}^*(\mathcal{M})$ with $Y_f = +$, $Y_g = 0$, or both.*

(2) *There exist a circuit $\bar{X} \in \mathcal{C}(\mathcal{M})$ and a cocircuit $\bar{Y} \in \mathcal{C}^*(\mathcal{M})$ such that $\bar{X}_f = +$, $\bar{X}^- \subseteq \{g\}$, $\bar{Y}_g = +$, $\bar{Y}^- \subseteq \{f\}$, and $\underline{X} \cap \underline{Y} \subseteq \{f, g\}$, implying that $\bar{Y}$ is optimal for (P) and $\bar{X}$ is optimal for (P^*).*

10.2.15 Corollary (Basis version of the Duality Theorem) (Bland 1977a). *Let $(P) = (\mathcal{M}, g, f)$ be an oriented matroid program that is both feasible and dual feasible. Then $(\mathcal{M}, g, f)$ has an optimal T-basis B, such that (with $N := E \backslash B$):*

$\bar{Y} := c^*(g, N)$ *is an optimal solution for (P), and*

$\bar{X} := c(f, N)$ *is an optimal solution for (P^*).*

Observe at this point that the conclusion of the Duality Theorem – namely, condition (1) – trivially holds if g is a loop or f is a coloop of $\mathcal{M}$. These cases were excluded in our definition of an oriented matroid program and are therefore not

covered by the Duality Theorem 10.2.14. They will however, not be excluded in
the reformulation offered below as Corollary 10.2.16. In contrast, the condition
$f \neq g$ is essential and cannot be dropped.

There is a certain tradition of phrasing duality results in form of coloring
lemmas. This goes back to Minty's (1966) 3-coloring lemma for digraphs, a
generalization of which is Corollary 10.2.17 below. We give a few examples,
starting with a 5-coloring theorem that translates the Duality Theorem 10.2.14.
These results will not be needed for the following sections.

10.2.16 Corollary (Folkman and Lawrence 1978). *Let* $\mathcal{M} = \mathcal{M}(E)$
be an oriented matroid whose ground set $E = \{f, g\} \cup W \cup R \cup B$ *is partitioned
("colored") into one faded-green element* f, *one green element* g, *and subsets of
white elements* W, *red elements* R *and blue elements* B.

Then exactly one of the following two numbered alternatives holds:

(1)   *There exists a circuit* $X \in \mathcal{C}(\mathcal{M})$ *with* $g \in X \subseteq \{g\} \cup W \cup R$, *with* $X^- \subseteq R$,
        or there exists a cocircuit $Y \in \mathcal{C}^*(\mathcal{M})$ *with* $f \in Y \subseteq \{f\} \cup W \cup B$, *with*
        $Y^- \subseteq B$.

(2)   *There exist a circuit* $X \in \mathcal{C}(\mathcal{M})$ *and a cocircuit* $Y \in \mathcal{C}^*(\mathcal{M})$ *such that*

$$f \in X \subseteq \{f\} \cup W \cup R \quad \text{with} \quad X^- \subseteq R,$$

and

$$g \in Y \subseteq \{g\} \cup W \cup B \quad \text{with} \quad Y^- \subseteq B.$$

Proof. Consider the oriented matroid program $(P) = ((\mathcal{M} \backslash B)/R, g, f)$, and apply the Duality Theorem 10.2.14.   □

As a special case, we get the generalization of Minty's (1966) 3-coloring Lemma
to oriented matroids, as follows. (See also Theorem 3.4.4.)

10.2.17 Corollary (Bland and Las Vergnas 1979). *Let* $\mathcal{M}$ *be an oriented
matroid and let* $E = W \cup R \cup B$ *be a partition of its ground set* E *into "red",
"blue" and "white" elements, where* $g \in E$ *is a red element* $(g \in R)$. *Then*

(1)   *either there exists a circuit* $X \in \mathcal{C}(\mathcal{M})$ *such that* $g \in X \subseteq W \cup R$, *with*
        $X^- \subseteq R$, *(all elements of* X, *including* g, *are white or red, the negative
        ones being all red),*

(2)   *or there exists a cocircuit* $Y \in \mathcal{C}^*(\mathcal{M})$ *such that* $g \in Y \subseteq W \cup B$ *with*
        $Y^- \subseteq B$, *(all elements of* Y, *including* g, *are white or blue, the negative
        ones being all blue), but not both.*

Proof. Consider the oriented matroid $\mathcal{M}'$ on $E' := E \cup f$ obtained by adjoining
a loop f to $\mathcal{M}$, and repaint g green.

Then $\{f\}$ is a positive circuit, f is not contained in a cocircuit, and the result
follows by applying Corollary 10.2.16 to $\mathcal{M}'$.   □

Now, further specializing to $R := E$, $W = B = \emptyset$, we recover the "Farkas
Lemma" Corollary 3.4.6:

10.2.18 Corollary (Bland and Las Vergnas 1979). *Let $\mathcal{M}$ be an oriented matroid on E and let $e_0 \in E$. Then*

either e_0 is contained in a positive circuit ($X \in \mathcal{C}(\mathcal{M})$ with $X \geq 0$, $X_{e_0} = +$),
or in a positive cocircuit ($Y \in \mathcal{C}^*(\mathcal{M})$ with $Y \geq 0$, $Y_{e_0} = +$),
but not both.

This corollary in fact only needs Phase I of the simplex algorithm for its proof: it states that whenever a program is infeasible, we can find a positive circuit that proves this.

We note that Corollary 10.2.18 is obvious for digraphs: if in a digraph D the arc e_0 is not contained in a directed cycle, then the set of vertices from which the tail of e_0 can be reached on a directed path defines a directed cut that contains e_0.

This yields a simple lemma for network flow theory on digraphs. Network flow problems can also be formulated as oriented matroid programming problems and solved by the (network) simplex method for (digraphic) oriented matroids. We only touch on this subject in Exercise 10.3.

10.3 Pivot rules

This section is devoted to pivot rules for oriented matroid programming. We start with Bland's (1977a) extremely simple "smallest subscript rule". This rule does not admit degenerate cycling, and can therefore be used to solve the Phase I problem. Interestingly enough, the "smallest subscript rule" exhibits *non-degenerate cycling*, see Section 10.4. (This is a phenomenon that cannot occur for linear programming in $\mathbb{R}^n$.) Thus it does not yield a finite algorithm for Phase II.

This is, instead, provided by Bland's (1977a) ALGORITHM A and its variants. We begin with one very simple variant, the ALGORITHM DC, which is based on deletion and contraction. The recursive nature of ALGORITHM DC requires a careful discussion of suitable data structures. From this, we see how to generalize ALGORITHM DC to ALGORITHM A. (A more general framework for recursive algorithms is developed in Jensen 1985.) We will also present a specialization of ALGORITHM DC, called the "restricted Edmonds-Fukuda rule". A special feature of ALGORITHM DC and its variants is that they do solve the Phase II problem, but can *become infeasible* for some intermediate steps. (This does not occur for realizable programs.)

The problem of designing a genuine *simplex* pivot rule for oriented matroid programs was solved by Todd (1985), drawing on his analysis of the linear complementarity problem in an oriented matroid framework. We will describe his pivot rule, but without proofs. Our survey of pivot rules will close with the extremely simple "criss-cross" rule that was independently found by Terlaky (1987) and Wang (1987). This rule is self-dual, and it does not need a feasible basis to start with. In fact, it completely mixes primal and dual feasibility, until it finally reaches optimality.

10.3.1 Definition (pivot rules, cycling).

(i)   A *pivot rule* is an algorithm that, for every tableau $T[B]$ that is not *terminal* (i.e., does not satisfy any of the termination criteria of the Simplex Algorithm 10.2.13), specifies a pivot step

$$T[B] \longrightarrow T[B'].$$

A *simplex (pivot) rule* furthermore guarantees that if B is a feasible T-basis, then so is B'. Thus a simplex pivot rule is an implementation of steps **(II.2)** and **(II.4)** of Algorithm 10.2.13.

(ii)   A pivot rule is *finite* if it does not exhibit *cycling*, that is, if there is no oriented matroid program with a T-basis B_0 for which it specifies a sequence of pivots

$$T[B_0] \longrightarrow T[B_1] \longrightarrow \ldots \longrightarrow T[B_k]$$

with $B_0 = B_k$ $(k > 0)$.

We use the term *degenerate cycling* if all the pivots in the cycle are degenerate, and *non-degenerate cycling* otherwise.

Many of the pivot rules used in *linear* programming depend on the size of the reduced costs c_e, numerical properties, minimum-ratio tests and so on. They therefore do not directly generalize to oriented matroids. We thus have to consider *combinatorial* rules that only depend on sign properties.

An extremely simple such rule was found by Bland (1977a,b). We will first consider the feasible (Phase II) case.

10.3.2 Definition (Bland 1977a,b).   The *smallest subscript rule* (*"Bland's rule"*) selects in

(II.2)   the smallest $e_0 \in N \backslash g$ with $\tau_{fe_0} = -$, and in

(II.4)   the smallest $b_0 \in B \backslash f$ with $\tau_{b_0 e_0} = +$ such that $B' := (B \backslash b_0) \cup e_0$ is feasible.

(Here "smallest" is with respect to the linear order $1 < \ldots < n < g < f$ on the set E of variables.)

10.3.3 Theorem (Bland 1977a,b).   The *smallest subscript rule does not exhibit degenerate cycling.*

Proof. Consider a degenerate cycle, let $D \subseteq E_n$ be the set of variables that leave (and enter) a T-basis during that cycle, and let $d = \max(D)$.

Let $T[B_i]$ be a tableau in the cycle from which d *enters* the T-basis. Then the last row of $T[B_i]$ is the circuit $X := c(f, N_i)$ with

(10.3.1) $X_f = +, \quad X_d = -, \quad X_e \geq 0$ for all $e < d$.

Let $T[B_j]$ be a tableau in the cycle from which d *leaves* the T-basis, via a pivot at (d, e_0), say. Then the e_0-column of $T[B_j]$ extends to the cocircuit $Y := -c^*(e_0, N_j)$ with

(10.3.2) $Y_f = -, \quad Y_d = +, \quad Y_{e_0} = -.$

Orthogonality $X \perp Y$, with $X_f Y_f = -$, implies that there is a variable $h \in E$ with $X_h Y_h = +$. By construction of X and Y we get $h \in N_i$ from $X_h \neq 0$, and $h \in B_j \cup e_0$ from $Y_h \neq 0$, thus $h \in N_i (\cap B_j \cup e_0) \subseteq D$, and hence $h \leq d$. Furthermore $X_d Y_d = -$ implies $h < d$, and thus $X_h = Y_h = +$ from (10.3.1). Also $h \neq e_0$, from (10.3.2), and therefore $h \in B_j$.

Now all the pivots in the cycle are degenerate, hence the basic cocircuits $Y^0 = c^*(g, N_l)$ remains constant during the cycle. Degeneracy of the pivots implies $Y_b^0 = 0$ for $b \in D$ and so in particular $Y_b^0 = -\tau_{bg} = 0$ for $b \in B_j \cap D$ in the tableau $T[B_j]$. Hence in $T[B_j]$ we find $\tau_{he_0} = Y_h = +$ and $\tau_{hg} = \tau_{dg} = 0$, and thus Bland's rule would pivot in (h, e_0) rather than in (d, e_0). □

We have still not given a correctness proof for the Simplex Algorithm 10.2.13. We will provide it in two parts. The Phase II part is done later in this section (Theorem 10.3.10), when suitable pivot rules are developed.

Here we will show that Bland's smallest subscript rule an be adapted to solve Phase I of the oriented matroid programming problem. For this, one uses (Phase II of) a "dual simplex method", that is, one applies the simplex method to the dual program $(\mathcal{M}^*, f, g)$ and records the corresponding T-bases and pivots for $(\mathcal{M}, g, f)$.

10.3.4 Definition (dual simplex algorithm). A simplex algorithm applied to the dual program $(\mathcal{M}^*, f, g)$ is called a *dual simplex algorithm* for $(\mathcal{M}, g, f)$.

Thus a dual simplex algorithm is a pivot algorithm that for any oriented matroid program $(\mathcal{M}, g, f)$ starts with an arbitrary T-basis, in Phase I after finitely many pivots reaches a dual feasible T-basis (or concludes that none exists) and in Phase II transforms the dual feasible T-basis into an optimal or infeasible T-basis/tableau.

10.3.5 Proposition (Bland). *Let $(\mathcal{M}, g, f)$ be an oriented matroid program. From this, construct the "auxiliary problem" $(\mathcal{M}_0, g, f_0)$, where $\mathcal{M}_0$ is obtained from $\mathcal{M}$ as $\mathcal{M}_0 := (\mathcal{M} \backslash f) \cup f_0$, by deleting f and instead adjoining a loop f_0.*

Then the dual simplex method with Bland's smallest subscript rule, applied to the program $(\mathcal{M}_0, g, f_0)$, solves the Phase I problem for $(\mathcal{M}, g, f)$.

Proof. For the dual program $(\mathcal{M}_0^*, f_0, g) = ((\mathcal{M}^*/f) \oplus f_0, f_0, g)$ the element f_0 is a coloop. So $(\{f_0\}, \emptyset)$ is a positive cocircuit, and this is the only cocircuit that contains f_0 positively. Thus every T-basis is feasible, and every pivot is degenerate.

Therefore we can apply Phase II of the dual simplex algorithm (with the smallest subscript rule) to $(\mathcal{M}_0^*, f_0, g)$, using an arbitrary starting tableau. By Theorem 10.3.3, it is finite. It terminates

- either with an optimal T-basis, yielding a circuit $X \in \mathcal{C}(\mathcal{M}_0)$ with $X_g = +$, $X_e \geq 0$ for all $e \in E_n$, which is a feasible cocircuit for $(\mathcal{M}_0, g, f_0)$ and thus determines a feasible cocircuit for $(\mathcal{M}, g, f)$,

– or with a T-basis demonstrating unboundedness of $(\mathcal{M}_0^*, f_0, g)$, which produces a cocircuit $Y \in \mathcal{C}^*(\mathcal{M}_0^*)$ such that $Y_{f_0} = 0$, $Y_g = +$ and $Y_e \geq 0$ for all $e \in E_n$, which is a circuit of $\mathcal{M}_0$, and hence a circuit of $\mathcal{M}$, which shows that $(\mathcal{M}, g, f)$ is infeasible.   □

Instead of modifying the oriented matroid program to create an auxiliary problem to which the standard algorithm can be applied, one could as well maintain the original program and modify the algorithm, by describing the action of a dual simplex algorithm on tableaux that ignores the entry of the f-column. The result is equivalent to our approach.

Now, a key result (see Section 10.4) is that *the smallest subscript rule exhibits nondegenerate cycling*, and thus does not yield a finite algorithm for Phase II. Therefore, we start with a discussion of ALGORITHM DC ("deletion and contraction") which – from a matroid theory point of view – might be *the most natural algorithm for linear programming*. First we will give a recursive, then an explicit description. After that we will generalize to Bland's ALGORITHM A, which is more flexible, and specialize to the so-called "restricted Edmonds-Fukuda rule", which on non-degenerate problems is completely determined by the initial data.

We will now only consider the Phase II problem for oriented matroid programming. So ALGORITHM DC starts with a feasible T-basis, and in finitely many steps produces a *terminal T-basis*, that is, a feasible T-basis that is either optimal or demonstrates unboundedness.

10.3.6 Remark. The minors of an oriented matroid program are canonically identified with certain *subproblems*:

– The *deletion* of $h \in E_n$ corresponds to the deletion of the corresponding inequality from the program, the corresponding variable loses its nonnegativity condition. This means that $(\mathcal{M} \backslash h, g, f)$ is a smaller problem, whose T-bases are the sets $B \backslash h$ for which B is a T-basis of $(\mathcal{M}, g, f)$ that contains h. So deleting h is equivalent to fixing h to be T-basic in $(\mathcal{M}, g, f)$, and effectively only considering the tableaux obtained by deleting the row and the column corresponding to h from $T[B]$.

– The *contraction* of h means that the corresponding inequality of the problem has to stay tight, the corresponding variable is set to 0. So $(\mathcal{M}/h, g, f)$ is a smaller problem, whose T-bases B are those T-bases of $(\mathcal{M}, g, f)$ that do not contain h. Hence considering the subproblem $(\mathcal{M}/h, g, f)$ is equivalent to a discussion of $(\mathcal{M}, g, f)$ for which h is fixed to be T-nonbasic, which amounts to the deletion of the h-column from the tableau $T[B]$.

Now the basic proceeding of ALGORITHM DC is as follows – compare this to the geometric description in Section 10.1, and to the sketch in Figure 10.1.3. The recursive nature of the algorithm leads to the unusual property that the pivot selection is 'history dependent', in that it depends not only on the current T-basis, but also on an extra data structure.

10.3.7 "ALGORITHM DC" – recursive description. Let $(\mathcal{M}, g, f)$ be an oriented matroid program and let a feasible T-basis B_0 be given. Now select a (T-nonbasic) variable $h \in E_n$ with negative reduced cost $\tau_{fh} = -$. First, fix the element h to stay in the T-nonbasis. This amounts to the same thing as to delete the h-column from the tableau, or to consider the oriented matroid program $(P/h) = (\mathcal{M}/h, g, f)$, for which B_0 is a feasible T-basis as well.

By induction on n, we know a finite pivoting method that produces a sequence of T-bases $B_0 \to B_1 \to \ldots \to B_k$ for (P/h), with B_k terminal. All these B_i are also T-bases for (P). Also, if B_k is unbounded for (P/h), then it is as well for (P). If B_k is optimal for (P/h), then for (P) it is either optimal ($\tau_{fh} \geq 0$), or it is inconsistent ($\tau_{fh} = -$ and the whole h-column is non-positive), or by Proposition 10.2.12 there is a feasible pivot that produces a new T-basis $B_{k+1} = (B_k \backslash h') \cup h$ that contains h. In this case we now fix h to stay in the T-basis, which reduces the problem to considering $(P \backslash h) := (\mathcal{M} \backslash h, g, f)$. By induction on n, we again know how to produce a sequence of pivots $B'_{k+1} = (B_{k+1} \backslash h) \longrightarrow B'_{k+2} \longrightarrow \ldots \longrightarrow B'_{k+l}$, where B'_{k+l} is a terminal T-basis for $(P \backslash h)$. This describes a sequence of pivots $B_{k+1} \longrightarrow B_{k+2} \longrightarrow \ldots \longrightarrow B_{k+l}$ for (P), between the T-bases $B_{k+i} := B'_{k+i} \cup h$.

Now Lemma 10.1.18, which we restate below in tableau version, guarantees that B_{k+l} is a terminal T-basis for (P), and we are done.

10.3.8 Lemma (Bland 1977a). *Let $(\mathcal{M}, g, f)$ be an oriented matroid program, $e \in E_n$ and B a T-basis with $e \notin B$ that is optimal for $(\mathcal{M}/e, g, f)$, but not for $(\mathcal{M}, g, f)$. Then*

- *if B' is optimal for $(\mathcal{M} \backslash e, g, f)$, then $B' \cup e$ is optimal for $(\mathcal{M}, g, f)$,*
- *if B' is unbounded for $(\mathcal{M} \backslash e, g, f)$, then $B' \cup e$ is unbounded for $(\mathcal{M}, g, f)$.*

In fact, a more general version of the algorithm is obtained if we note that the condition $\tau_{fh} = -$ for the selection of h is not necessary: one can choose an arbitrary T-nonbasic variable for h. With this, we can *always* choose an element h as above, until we get down to a subproblem of rank 1 (with $N = \{g\}$), where every tableau is automatically terminal.

Now we will go from the recursive description to an explicit version of AL-GORITHM DC. Observe for this that ALGORITHM DC works its way through a binary search tree.

We record the current status of the algorithm by the T-basis B (with $N = E \backslash B$) together with a string of distinct *currently fixed variables* $F = (h_1, \ldots, h_t)$, with $h_i \in E_n$.

- If $h_1 \in N$, then this means that we currently *assume* that we are in the case where h_1 is T-nonbasic, and $(h_2, \ldots, h_t)$ records the current situation of B with respect to the smaller subproblem $(\mathcal{M}/h_1, g, f)$.
- If $h_1 \in B$, then we already *know* that $(\mathcal{M}, g, f)$ has a terminal tableau with $h_1 \in B$, and $(h_2, \ldots, h_t)$ describes the current state of affairs of the T-basis $B \backslash h_1$ with respect to the subproblem $(\mathcal{M} \backslash h_1, g, f)$.

Similarly, $h_i \in N$ means that we currently *assume* that h_i is T-nonbasic for the subproblem $\mathcal{M}/(N \cap \{h_1, \ldots, h_{i-1}\}) \backslash (B \cap \{h_1, \ldots, h_{i-1}\})$, whereas $h_i \in B$ implies that we *know* that h_i can be assumed to be T-basic in this subproblem.

Thus the recursive pivot rule of ALGORITHM DC can explicitly be described as follows.

10.3.9 "Algorithm DC" – explicit description. (Edmonds and Fukuda 1982)   ALGORITHM DC is the following pivot algorithm for the **Phase II** problem of oriented matroid programming.

Input: An oriented matroid program $(P) = (\mathcal{M}, g, f)$, and a feasible T-basis B.

Output: A terminal tableau $T[B]$.

(II.0) **(Initialization)**

Compute $T[B]$.

Put $F := (h_1, \ldots, h_t)$, such that $\underline{F} = \{h_1, \ldots, h_t\} = N \backslash g$.

(F is the string of all T-nonbasic variables, in arbitrary order.)

(II.1) **(Optimality test)**

If $\tau_{fh} \geq 0$ for all $h \in N \cap \underline{F}$, then $T[B]$ is optimal.

Output $T[B]$ and STOP.

(II.2) **(Column selection)**

Let i_0 be the highest index ($1 \leq i_0 \leq t$) such that $\tau_{fh_{i_0}} = -$.

(So, h_{i_0} is the rightmost element in the string F that is T-nonbasic with negative reduced cost.)

Put $e_0 := h_{i_0}$.

(II.3) **(Boundedness test)**

If $\tau_{be_0} \leq 0$ for all $b \in B \backslash \{h_1, \ldots, h_{i_0}, f\}$, then $T[B]$ is unbounded.

Output $T[B]$ with this information and STOP.

(II.4) **(Row selection)**

Choose $b_0 \in B \backslash \{h_1, \ldots, h_{i_0}, f\}$ such that the tableau $T[B'] = (\tau'_{be})$ for $B' := (B \cup e_0) \backslash b_0$ satisfies $\tau'_{bg} \leq 0$ for all $b \in B \backslash \{h_1, \ldots, h_{i_0}, f\}$ (which implies $\tau_{b_0 e_0} = +$).

(II.5) **(Update)**

Put $B := B'$, $N := E \backslash B'$, recompute $T[B]$.

Put $F := (h_1, \ldots, h_{i_0}, h'_1, \ldots, h'_s)$, where

$$\{h'_1, \ldots, h'_s\} = \{h_{i_0+1}, \ldots, h_t, b_0\} \cap N,$$

and put $t := i_0 + s$.

Now GOTO **(II.1)**.

10.3.10 Theorem.   ALGORITHM DC *stops with a terminal tableau after finitely many steps.*

Proof. Correctness and finiteness of the algorithm are clear for the recursive description. For the explicit version, they are implied by the following two properties of the T-basis B and the string of fixed variables $F = (h_1, \ldots, h_t)$ that are maintained by the algorithm. For convenience, again use $B_F := B \cap \underline{F}$ and $N_F := N \cap \underline{F}$ for the sets of variables that are fixed to be T-basic respectively T-nonbasic.

(1) $B \backslash B_F$ is an optimal T-basis for $(\mathcal{M}/N_F \backslash B_F, g, f)$.

(2) Let i_0 be such that $h_{i_0} \in B$, and denote $F(i) := (h_1, \ldots, h_i)$ for $1 \leq i \leq t$. Then $(\mathcal{M}/N_{F(i_0-1)} \backslash B_{F(i_0-1)}, g, f)$ has a terminal T-basis that contains h_{i_0}.

Property (1) is clear from the update **(II.5)**: all columns with negative reduced costs are contained in F.

Property (2) is Lemma 10.3.8, applied to the subproblem

$$(\mathcal{M}/N_{F(i_0-1)} \backslash B_{F(i_0-1)}, g, f)$$

in the situation when h_{i_0} enters the T.

This also uses an induction on n that guarantees the existence of a terminal T-basis for the proper subproblems of $(\mathcal{M}/N_{F(i_0-1)} \backslash B_{F(i_0-1)}, g, f)$.

Now if ALGORITHM DC terminates in step **(II.3)**, then the current tableau is unbounded – this is guaranteed by Lemma 10.3.8. If ALGORITHM DC terminates in step **(II.1)**, then property (1) implies that the current basis is optimal.

Finiteness of ALGORITHM DC follows from the finiteness of the binary search tree, in which all pivots go "from left to right", say. Formally, we can associate to F a string $\chi(F) := (\chi_1, \ldots, \chi_t)$, with $\chi_i = 1$ if $h_i \in B$, and $\chi_i = 0$ otherwise. Then update **(II.5)** implies that the vector $\chi(F)$ increases lexicographically in every pivot. □

ALGORITHM DC does not determine the pivots to be taken uniquely – both for the leaving and for the entering variable they leave many choices (in contrast to the smallest subscript rule, which determines its pivots uniquely).

The finiteness proof of Theorem 10.3.10 shows that there are still possible modifications to consider, which lead to different data structures and to more general pivot rules. For this we note that, without losing the properties (1) and (2) of the preceding proof, we can perform the following three types of operations on the string F:

– delete the last element h_t from F if

either $h_t \in B$,

or $h_t \in N$ and $\tau_{f,h_t} \geq 0$,

– append h to the end of F, if $h \in N \backslash F$,

– permute consecutive T-nonbasic elements in $\underline{F}$.

This leads to more flexible data structures and algorithms, without sacrificing the finiteness proof of Theorem 10.3.10. In particular, the third operation allows us to store the maximal sequences of consecutive T-nonbasic elements in F as (unordered) sets. This generalizes ALGORITHM DC to Bland's ALGORITHM A:

10.3.11 "Algorithm A". ALGORITHM A is the generalization of ALGORITHM DC which instead of the string F maintains a string F^A of the form

$$F^A = (N_0, r_1, N_1, r_2, \ldots, r_q, N_q),$$

where $N = N_0 \uplus N_1 \uplus \ldots \uplus N_q$ is a partition of the set N of T-nonbasic variables, and $r_1, r_2, \ldots, r_q$ are the currently fixed T-basic elements.

With this, the **initialization step** of ALGORITHM A is canonical:

(II.0)$_A$ Compute $T[B]$, and put $F^A := (N)$, we get more flexibility in the **column selection**:

(II.2)$_A$ Choose e_0 arbitrarily in the set $N_{i_0} \cap \{h : \tau_{fh} = -\}$, for the largest possible i_0 such that this set is non-empty, and the corresponding **update** is

(II.5)$_A$ Put $B := B'$, $N := E \backslash B'$, recompute $T[B]$.

Put $F^A = (N_0, r_1, N_1, r_2, \ldots, r_{i_0}, N_{i_0} \backslash \{e_0\}, e_0, N_{i_0+1})$, where $N_{i_0+1} := (N \cup b_0) \backslash (N_0 \cup N_1 \cup \ldots \cup N_{i_0})$.

There seems to be a fair amount of confusion in the literature around these pivot rules. This starts with Bland's ALGORITHM A, for which Bland (1977a) only gives a recursive description. Edmonds and Fukuda (1982) provide an explicit description and data structure for this algorithm, which they call "ALGORITHM B". At the same time they describe a variation (with a different data structure) of this algorithm (namely ALGORITHM DC, 10.3.9), which they in turn call "ALGORITHM A". Edmonds and Fukuda did not discuss the connection between Bland's recursive description and the data structures they used.

The question whether Bland's Algorithms preserves feasibility was answered in the negative by Edmonds and Fukuda, constructing $(\mathbf{EFM}(8), g, f)$ – see Section 10.4. They did not settle the same question for their ALGORITHM DC, although the same example provides a counter-example – see Example 10.4.1/Figure 10.4.9.

Later, a still more restrictive pivot rule was described by Clausen (1987), which he, apparently misinterpreting Edmonds and Fukuda's work, called the "Edmonds-Fukuda rule". This pivot rule (see Algorithm 10.3.12, below) does *not* become infeasible on $(\mathbf{EFM}(8), g, f)$. However, an example for infeasibility of the Edmonds-Fukuda rule was later constructed by Jensen (1985) – see Example 10.4.4/Figure 10.4.12.

10.3.12 The "restricted Edmonds-Fukuda rule" (Clausen 1987). The "restricted Edmonds-Fukuda rule" is the following specification of the update **(II.5)** in ALGORITHM DC (10.3.9):

(II.5)$_{EF}$ Put $B := B'$, $N := E \backslash B'$, recompute $T[B]$.

Put $F := (h_1, \ldots, h_{i_0}, h'_1, \ldots, h'_s, b_0)$, where $(h'_1, \ldots, h'_s)$ is the sequence of elements of $\{h_{i_0+1}, \ldots, h_t\} \cap N$ *in the same order in which they appear in* $(h_{i_0+1}, \ldots, h_t)$, and put $t := i_0 + s + 1$.

Now GOTO **(II.1)**.

Observe that the restricted Edmonds-Fukuda rule is completely deterministic on non-degenerate programs: after the initialization **(II.0)** (which arbitrarily selects a linear ordering for N) there is no more choice in the execution of the algorithm.

The problem of designing a *finite* pivot rule that *preserves feasibility* was solved by Todd (1985), drawing on his analysis of the linear complementarity problem in an oriented matroid framework (Todd 1984). Todd's rule also solves the Phase I problem for oriented matroid programming. We will now treat only a special version which allows a formulation on tableaux $T[B]$ (the general version of Todd's rule as well as most of Todd's proofs operate on the larger tableaux generated by the linear complementarity approach). We will not, however, give the corresponding correctness proofs (Theorem 10.3.14) — Todd's (1985) proofs rely heavily on the linear complementarity framework and do not fit into our discussion.

The *linear* version of Todd's pivot rules relies on minimum ratio tests for the leaving and for the entering variable. Such ratio tests cannot be read off directly from an oriented matroids tableau. However, they can be simulated by "attempted pivots" in places like (b_0, g), after which the necessary information can be read off – even though they are not legal pivots.

The notational framework is as follows. We treat an oriented matroid program $(P) = (\mathcal{M}(E), g, f)$, where the ground set is

$$E = \{x_1, \dots, x_n, u_1, \dots, u_m, f, g\}.$$

This notation corresponds to that of Definition 10.2.1: the u_i are the elements ("variables") in the T-basis $B_0 = \{u_1, \dots, u_m, f\}$ of a starting tableau, whereas the x_j are the corresponding T-nonbasis variables, $N_0 = \{x_1, \dots, x_n, g\}$. Todd's rules distinguish between x_i- and u_j-variables.

For the following discussion, recall that a T-basis B is feasible if and only if $-\tau_{bg} \geq 0$ for all $b \in B \backslash f$. We define it to be *lexico-feasible* (lexicographically feasible) if

$$(-\tau_{bg}, \tau_{bu_m}, \dots, \tau_{bu_1}) \succeq (0, 0, \dots, 0) \quad \text{for all } b \in B \backslash f,$$

where $\succeq$ denotes lexicographic order.

Equivalently, the T-basis B is lexico-feasible if and only if the first non-zero entry of $(-\tau_{bg}, \tau_{bu_m}, \dots, \tau_{bu_1})$ is positive.

In particular, every lexico-feasible basis is feasible. Todd's pivot rule for Phase I constructs a lexico-feasible T-basis. (The geometry of the algorithm guarantees that the first feasible basis also is lexico-feasible.) His Phase II then achieves optimality while preserving lexico-feasibility.

Also recall the convention

$$\tau_{be} = \begin{cases} - & \text{for } b \in N, \, e = b, \\ 0 & \text{for } b \in N, \, e \neq b. \end{cases}$$

10.3.13 Definition (Todd 1985). Let $(P) = (\mathcal{M}, g, f)$ be an oriented matroid program on the ground set

$$E = \{x_1, \dots, x_n, u_1, \dots, u_m, f, g\},$$

with the starting tableau $T[B_0]$, where $B_0 = \{u_1, \ldots, u_m, f\}$.

Let $T[B]$ be the current tableau. Then *Todd's Rule* selects

in Phase I:

first the leaving variable b_0 such that $\tau_{b_0 g} = +$

and among the $b_0 \in B \backslash f$ which satisfy this, one which causes the tableau, after a pivot in position (b_0, g), to satisfy

$$(\tau_{b u_m}, \ldots, \tau_{b u_1}) \succ 0$$

for all $b \in B \backslash f$.

(If $\tau_{bg} \le 0$ for all $b \in B \backslash f$, then B is lexico-feasible, and Phase I is finished.)

then the entering variable e_0 such that $\tau_{b_0 e_0} = -$,

and among the $e_0 \in N \backslash g$ which satisfy this, one which causes the tableau, after the pivot in position (e_0, b_0), to satisfy

$$(-\tau_{x_n e}, \ldots, -\tau_{x_1 e}) \succ 0$$

for all $e \in N \backslash g$.

(If $\tau_{b_0 e} \ge 0$ for all $e \in N \backslash g$, then B is infeasible by Proposition 10.2.6(iii).)

in Phase II:

(II.2) the entering variable e_0 such that $\tau_{fe} = -$,

and among the $e_0 \in N \backslash g$ which satisfy this, one which causes the tableau, after a pivot in position (f, e_0), to satisfy $(-\tau_{x_n e}, \ldots, -\tau_{x_1 e}) \succ 0$ for all $e \in N \backslash \{f, g\}$. (If $\tau_{fe} \ge 0$ for all $e \in N \backslash g$, then B is optimal by Corollary 10.2.8.)

(II.4) the leaving variable b_0 such that $\tau_{b_0 e_0} = +$,

and among the $b_0 \in B \backslash f$ which satisfy this, one which causes the tableau, after the pivot in (b_0, e_0), to satisfy $(-\tau_{bg}, \tau_{b u_m}, \ldots, \tau_{b u_1}) \succ 0$ for all $b \in B \backslash g$.

(If $\tau_{b e_0} \le 0$ for all $b \in B \backslash f$, then B is unbounded by Proposition 10.2.7 (iii).)

10.3.14 Theorem (Todd 1985). *Algorithm 10.2.13 together with Todd's rules 10.3.13 yields a genuine simplex algorithm for oriented matroid programming.*

In particular, Todd's rules preserve feasibility, and they do not exhibit cycling.

Thus starting with an arbitrary T-basis of $(P) \overset{\bullet}{=} (\mathcal{M}, g, f)$, the Simplex Algorithm with Todd's Phase I rule after a finite number of steps either produces a lexico-feasible basis or proves infeasibility. Starting with a lexico-feasible T-basis, the Simplex Algorithm with Todd's Phase II rule after a finite number of steps either produces a lexico-feasible, optimal T-basis or proves unboundedness.

Our discussion concludes with Terlaky's (1987) self-dual criss-cross method, which was independently found by Wang (1987).

10.3.15 Criss-Cross Method. Let $(\mathcal{M}, g, f)$ be an oriented matroid program, let B_0 be an arbitrary T-basis, and let $1 < 2 < \ldots < n < f < g$ be a fixed ordering of the ground set. Then the *criss-cross method* solves $(\mathcal{M}, g, f)$ by iterating the following pivoting procedure.

For the current T-basis B, determine the smallest index i $(1 \leq i \leq n)$ that corresponds

- either to an infeasible restriction, with $i \in B$ and $\tau_{ig} = +$,

- or to a variable with negative reduced costs, such that $i \in N$ and $\tau_{fi} = -$.

    In the first case, perform a pivot in (i, j), where j $(1 \leq j \leq n)$ is the smallest index with $j \in N$ and $\tau_{ij} = -$.

    In the second case, perform a pivot in (k, i), where k $(1 \leq k \leq n)$ is the smallest index with $k \in B$ and $\tau_{ki} = +$.

It is surprising that this simple rule yields a *finite* pivoting procedure. The proof essentially generalizes that of Bland's Rule, Theorem 10.3.3. Details can be found in Terlaky (1987).

A further remarkable fact about Terlaky's criss-cross method is the ease with which it generalizes to the setting of linear complementarity problems – see Klafszky and Terlaky (1989b) and Fukuda and Terlaky (1992).

10.4 Examples

We will now construct an oriented matroid that serves as an example for various phenomena that occur for non-realizable oriented matroids of rank at least 4. This oriented matroid EFM(8) of rank 4 on 8 points first appeared in Edmonds and Fukuda (1982) and Edmonds and Mandel (1982). Our analysis of this example will show that it is *not* isomorphic to the non-realizable oriented matroid RS(8) discussed in Section 1.5, because RS(8) has fewer simplicial topes.

10.4.1 Example (Edmonds and Fukuda 1982, Edmonds and Mandel 1982). Let $\mathcal{M}$ be the uniform oriented matroid on $\{1, \ldots, 6, g\}$ of the central hyperplane arrangement defined in $\mathbb{R}^4$ by the row vectors of

$$
\begin{array}{c}
1 \\
2 \\
3 \\
4 \\
5 \\
6 \\
g
\end{array}
\left(
\begin{array}{cccc}
-1 & -\epsilon & \epsilon & 1 \\
\epsilon & -1 & -\epsilon & 1 \\
-\epsilon & \epsilon & -1 & 1 \\
1 & 0 & 0 & 0 \\
0 & 1 & 0 & 0 \\
0 & 0 & 1 & 0 \\
0 & 0 & 0 & 1
\end{array}
\right)
$$

where $0 < \epsilon < \frac{1}{3}$. This yields an affine matroid $(\mathcal{M}, g)$ with the halfspaces corresponding to the system of inequalities

$$\begin{array}{rrrl} x & +\epsilon y & -\epsilon z & \leq 1 \\ -\epsilon x & + y & +\epsilon z & \leq 1 \\ \epsilon x & -\epsilon y & + z & \leq 1 \\ x & & & \geq 0 \\ & y & & \geq 0 \\ & & z & \geq 0 \end{array}$$

This arrangement is obtained by a small (ϵ-) perturbation of the arrangement of the facet hyperplanes of the unit cube in $\mathbb{R}^3$. In the "sphere at infinity" it induces a line arrangement as in Figure 10.4.1, obtained by perturbing the arrangement given by the (doubled) coordinate hyperplanes.

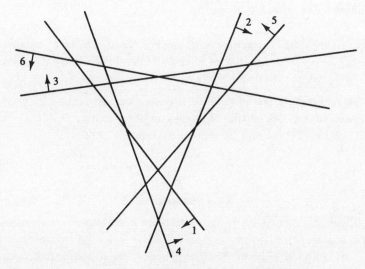

Figure 10.4.1: Affine picture of $\mathcal{A}^\infty(\mathcal{M})$.

With this one can see that the following rule defines a localization on $\mathcal{M}(\mathcal{C}^*)$, as characterized by Theorem 7.1.8.

$$\sigma(Y) = \begin{cases} Y_g & \text{if } Y_g \neq 0, \\ -h(Y) & \text{if } Y = \begin{cases} \pm(-0+0+-0), \\ \pm(+-0-0+0), \\ \pm(0+-+-00), \end{cases} \\ h(Y) & \text{otherwise,} \end{cases}$$

where $h(Y)$ denotes the sign of $h(x,y,z,u) = x+y+z$ on the ray corresponding to the cocircuit Y.

Let EFM(8) be the single element extension of $\mathcal{M}$ on $\{1,\ldots,6,g,f\}$ specified by σ.

Figure 10.4.2 shows this cocircuit signature in the plane at infinity (that is, in the plane $\{Y_g = 0\}$ resp. $\{u = 0\}$). This picture defines the single element extension $(\mathcal{M}/g) \cup f$. With this EFM(8) is completely determined by the require-

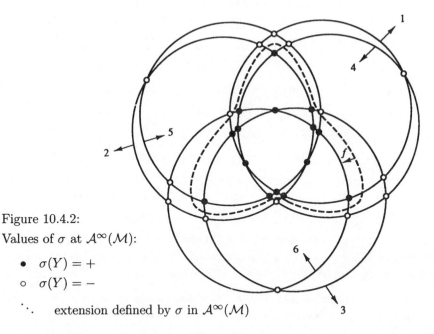

Figure 10.4.2:

Values of σ at $\mathcal{A}^\infty(\mathcal{M})$:

- $\bullet$ $\sigma(Y) = +$
- $\circ$ $\sigma(Y) = -$
- $\cdots$ extension defined by σ in $\mathcal{A}^\infty(\mathcal{M})$

ment that f and g are covariant in $\mathrm{EFM}(8)$ – this is equivalent to the assignment $\sigma(Y) = Y_g$ whenever $Y_g \neq 0$.

An alternative definition of $\mathrm{EFM}(8)$ is as follows. Let $\mathcal{M}'$ be the arrangement in $\mathbb{R}^4$ given by

$$
\begin{array}{c@{}c}
\begin{array}{c}
1 \\ 2 \\ 3 \\ 4 \\ 5 \\ 6 \\ g \\ f
\end{array} &
\left(
\begin{array}{cccc}
-1 & -\epsilon & \epsilon & 1 \\
\epsilon & -1 & -\epsilon & 1 \\
-\epsilon & \epsilon & -1 & 1 \\
1 & 0 & 0 & 0 \\
0 & 1 & 0 & 0 \\
0 & 0 & 1 & 0 \\
0 & 0 & 0 & 1 \\
1 & 1 & 1 & N
\end{array}
\right)
\end{array}
$$

where $0 < \epsilon \leq \frac{1}{3}$ and $N > \frac{3}{\epsilon}$. Then in affine space f is the halfspace $x + y + z \geq -N$.

Now $\mathrm{EFM}(8)$ is obtained from $\mathcal{M}'$ by reversing the three bases (mutations)

$$[16fg], \quad [24fg], \quad [35fg].$$

This means that the topes of $\mathrm{EFM}(8)$ are those of $\mathcal{M}'$, except that

$$\pm(\overline{1}2\overline{3}4\overline{5}6\overline{f}g) \text{ is replaced by } \pm(12\overline{3}4\overline{5}6f\overline{g}),$$
$$\pm(\overline{1}2345\overline{6}\overline{f}g) \text{ is replaced by } \pm(\overline{1}23\overline{4}5\overline{6}f\overline{g}), \text{ and}$$
$$\pm(1\overline{2}3\overline{4}56\overline{f}g) \text{ is replaced by } \pm(1\overline{2}3\overline{4}\overline{5}6f\overline{g}).$$

A complete list of cocircuits for $\mathrm{EFM}(8)$ is given by Figure 10.4.3.

$1\overline{23}45$	$1\overline{23}46$	$1\overline{23}4f$	$1\overline{23}4g$	$12\overline{35}6$	$123\overline{5}f$	$123\overline{5}g$	$123\overline{6}f$
$123\overline{6}g$	$123fg$	$1\overline{245}6$	$124\overline{5}f$	$124\overline{5}g$	$124\overline{6}f$	$124\overline{6}g$	$124\overline{fg}$
$125\overline{6}f$	$125\overline{6}g$	$125fg$	$126fg$	$1\overline{345}6$	$134\overline{5}f$	$134\overline{5}g$	$134\overline{6}f$
$134\overline{6}g$	$13\overline{4}fg$	$13\overline{56}f$	$135\overline{6}g$	$135fg$	$136fg$	$1\overline{456}f$	$1\overline{456}g$
$1\overline{45}fg$	$1\overline{4}6fg$	$156fg$	$2\overline{345}6$	$234\overline{5}f$	$234\overline{5}g$	$234\overline{6}f$	$234\overline{6}g$
$234fg$	$23\overline{56}f$	$235\overline{6}g$	$23\overline{5}fg$	$236fg$	$24\overline{56}f$	$245\overline{6}g$	$24\overline{5}fg$
$246fg$	$2\overline{56}fg$	$3\overline{456}f$	$345\overline{6}g$	$345fg$	$34\overline{6}fg$	$356fg$	$456fg$

Figure 10.4.3: Cocircuits of EFM(8).

To obtain these data we have used a revised version of the Manipu program by Bokowski and Sturmfels (1985b).

Let us continue with some more information on EFM(8): first we get some symmetries of EFM(8) that are symmetries of the cube and survive the "twisting" in the construction of EFM(8): in fact EFM(8) has

- a $\mathbb{Z}_3$ symmetry generated by $(123)(456)$, which is lifted from the corresponding permutation (xyz) of the coordinates,
- a $\mathbb{Z}_2$ symmetry generated by $(14)(26)(35)(fg)$, which also comes from a symmetry of the cube.

The next observation is not obvious, but easily verified with help of some computational tools:

EFM(8) *is self-dual:* in fact

$$(\text{EFM}(8))^* \xleftrightarrow{(23)(45)} {}_{-456g}\text{EFM}(8),$$

that is, the dual oriented matroid of EFM(8) coincides with EFM(8) after relabeling such that the elements 2 and 3 are exchanged, 4 and 5 are exchanged, and the signs of 4,5,6 and g are reversed.

This of course allows us to easily (by hand) construct a list of all circuits of EFM(8) from the list of cocircuits above. On the other hand, there are two ways to check the self-duality. The first is to construct the list of all circuits of EFM(8) from the list of cocircuits using the orthogonality definition, as in Figure 10.4.4.

$1\overline{23}45$	$1\overline{234}6$	$1\overline{23}4f$	$1\overline{23}4\overline{g}$	$12\overline{35}6$	$123\overline{5}f$	$123\overline{5}g$	$123\overline{6}f$
$123\overline{6}g$	$123f\overline{g}$	$1\overline{245}6$	$124\overline{5}f$	$124\overline{5}g$	$124\overline{6}f$	$124\overline{6}g$	$124f\overline{g}$
$125\overline{6}f$	$125\overline{6}g$	$125f\overline{g}$	$12\overline{6}f\overline{g}$	$1\overline{345}6$	$134\overline{5}f$	$134\overline{5}g$	$134\overline{6}f$
$134\overline{6}g$	$134f\overline{g}$	$13\overline{56}f$	$135\overline{6}g$	$13\overline{5}f\overline{g}$	$136f\overline{g}$	$1\overline{456}f$	$1\overline{456}g$
$145\overline{f}\overline{g}$	$1\overline{4}6f\overline{g}$	$156f\overline{g}$	$2\overline{345}6$	$234\overline{5}f$	$23\overline{45}g$	$234\overline{6}f$	$234\overline{6}g$
$23\overline{4}f\overline{g}$	$23\overline{56}f$	$235\overline{6}g$	$23\overline{5}f\overline{g}$	$236f\overline{g}$	$24\overline{56}f$	$245\overline{6}g$	$24\overline{5}f\overline{g}$
$24\overline{6}f\overline{g}$	$2\overline{56}fg$	$3\overline{456}f$	$345\overline{6}g$	$345f\overline{g}$	$34\overline{6}f\overline{g}$	$356f\overline{g}$	$456\overline{f}\overline{g}$

Figure 10.4.4: Circuits of EFM(8).

Then one can check whether this coincides with the list constructed from the list of cocircuits after relabeling and reorienting as specified by the isomorphism. The second possibility is to construct the chirotope, i.e., one lists the signs of all bases as in Figure 10.4.5, from which the self-duality can be checked directly.

$1234-$	$1235-$	$1236-$	$123f-$	$123g-$	$1245+$	$1246-$
$124f+$	$124g+$	$1256-$	$125f-$	$125g-$	$126f+$	$126g+$
$12fg+$	$1345+$	$1346+$	$134f+$	$134g+$	$1356-$	$135f-$
$135g-$	$136f-$	$136g-$	$13fg-$	$1456-$	$145f+$	$145g+$
$146f+$	$146g+$	$14fg+$	$156f-$	$156g-$	$15fg-$	$16fg-$
$2345+$	$2346+$	$234f+$	$234g+$	$2356+$	$235f+$	$235g+$
$236f-$	$236g-$	$23fg+$	$2456-$	$245f-$	$245g-$	$246f+$
$246g+$	$24fg-$	$256f+$	$256g+$	$25fg+$	$26fg-$	$3456-$
$345f-$	$345g-$	$346f-$	$346g-$	$34fg-$	$356f-$	$356g-$
$35fg-$	$36fg+$	$456f+$	$456g+$	$45fg+$	$46fg-$	$56fg+$

Figure 10.4.5: Signed bases of EFM(8).

This list is another way to define the oriented matroid EFM(8). Using Theorem 7.3.9, we now determine all mutations (tetrahedra) of EFM(8) (see Figure 10.4.6).

1246	125g	134g	1356	145f	16fg
2345	236g	24fg	256f	346f	35fg

Figure 10.4.6: Mutations of EFM(8).

Of course this list contains (underlined) the three mutations of $\mathcal{M}'$ used for the second construction of EFM(8).

Thus EFM(8) has 12 mutations, corresponding to 24 tetrahedra in the pseudosphere arrangement representing EFM(8).

This in particular implies that EFM(8) is *not* isomorphic to the rank 4 uniform oriented matroid RS(8) on 8 points by Roudneff and Sturmfels (1988): RS(8) has only seven mutations (and thus violates Shannon's theorem for the realizable case, which states that every hyperplane arrangement has at least as many simplicial regions as twice the number of hyperplanes.) Recently, Bokowski and Richter-Gebert (1990b,c) have completely classified all the uniform oriented matroids of rank 4 on 8 points, see Theorem 8.3.4. It turns out that there are 2628 orientation classes, at 24 of which are not realizable. RS(8) is the only example with only 7 mutations.

Now we consider the oriented matroid program

$$(P) := (\text{EFM}(8), g, f).$$

Its feasible region is combinatorially a cube, and the linear program still looks very much like

$$\max x + y + z$$
$$0 \le x \le 1$$
$$0 \le y \le 1$$
$$0 \le z \le 1.$$

In particular, the unique optimal solution is still given by the cocircuit $(000+++|++)$ that corresponds to $x = y = z = 1$.

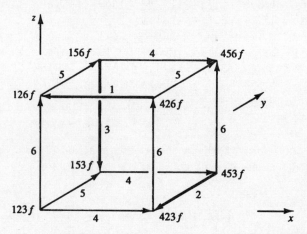

Figure 10.4.7: The edge graph of the feasible region of EFM(8).

However, by the special assignment of σ for six cocircuits at infinity (respectively the reversal of three bases) the orientation of three edges of the cube has been reversed, as indicated in Figure 10.4.7. (In this figure, the vertices are marked by the corresponding T-bases, the directed edges are labeled by the corresponding entering variables.)

This implies that there is a directed cycle $\mathbf{C}$ in the boundary of the feasible region, and thus, by Lemma 10.1.17, we have proved the following result.

10.4.2 Proposition. EFM(8) *is not realizable.*

Also Figure 10.4.7, which at every vertex of the feasible region displays the corresponding T-basis, shows that Bland's pivot rule, if started at any feasible but not optimal basis of the program (vertex of the cube), enters the non-degenerate cycle between the six T-bases

$$\ldots \longrightarrow [234f] \longrightarrow [246f] \longrightarrow [126f] \longrightarrow [156f] \longrightarrow [135f] \longrightarrow [345f] \longrightarrow \ldots$$

corresponding to a cycle between the bases (vertices)

$$\ldots \longrightarrow [156g] \longrightarrow [135g] \longrightarrow [345g] \longrightarrow [234g] \longrightarrow [246g] \longrightarrow [126g] \longrightarrow \ldots$$

and thus Proposition 10.4.3 follows.

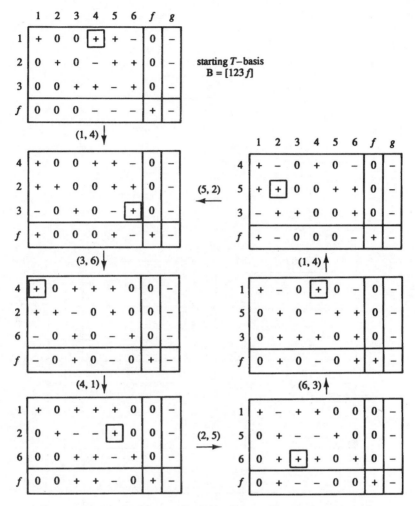

Figure 10.4.8: Tableaux for Bland's smallest subscript rule
cycling on EFM(8).

10.4.3 Proposition. *Bland's rule exhibits non-degenerate cycling on* $(\text{EFM}(8), g, f)$.

The corresponding tableaux are shown in Figure 10.4.8.

What about other pivot rules? For example, if we specify ALGORITHM DC such that the T-nonbasic variables are always inserted into the stack F in decreasing order, then it becomes infeasible on $(\text{EFM}(8), g, f)$ if initialized with $B_0 = [123f]$ and $F = 654$ – see Figure 10.4.9. Since ALGORITHM DC is a specialization of Bland's ALGORITHM A, we conclude that also Bland's ALGORITHM A does not preserve feasibility in general.

So ALGORITHMS A and DC do not stay feasible on $(\text{EFM}(8), g, f)$ for arbitrary initialization. It can be checked, however, that the restricted Edmonds-Fukuda

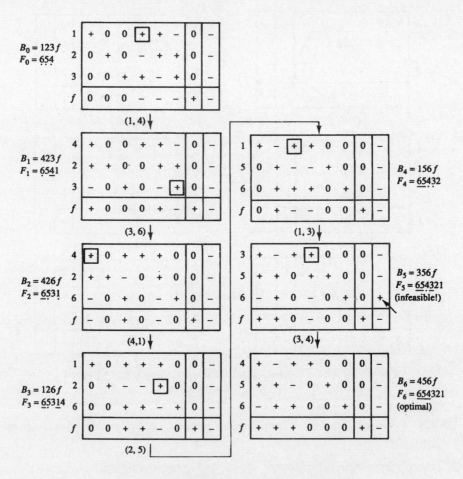

Figure 10.4.9: Tableaux for ALGORITHM DC
becoming infeasible on $(\mathcal{M}, g, f)$.

(In the string F, the T-basic elements are underlined, the T-nonbasic elements with negative cost are underdotted.)

rule preserves feasibility on $(\text{EFM}(8), g, f)$. The feasibility question for this rule was settled by Jensen (1985), as follows.

10.4.4 Example (Jensen 1985). Let $J(9)$ be the oriented matroid that arises from $\text{EFM}(8)$ by the lexicographic single element extension $7 := [1^+, 3^+, 4^-, g^+]$. Thus $J(9)$ is a uniform matroid of rank 4 on the set $E = \{1, 2, \ldots, 6, 7, f, g\}$. Note that the new element 7 forms an inseparable pair with 1, by construction.

Now consider the oriented matroid program $(J(9), g, f)$: the new "inequality" 7 cuts off the edge of the feasible region of $(\text{EFM}(8), g, f)$ that is determined by 1 and 3. Figure 10.4.10 depicts the feasible region of $(J(9), g, f)$, together with the orientation of its edges (the graph of the feasible region). Figure 10.4.11 provides the situation in the "plane at infinity" $\mathcal{A}^\infty(J(9)) = J(9)/g$, which allows us to check these orientations.

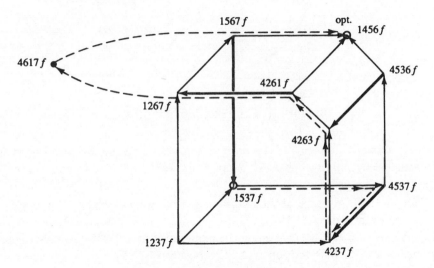

Figure 10.4.10: Graph of the feasible region of $(J(9), g, f)$.

Finally, Figure 10.4.12 shows the tableaux for the behavior of the restricted Edmonds-Fukuda rule on $(J(9), g, f)$, initialized with $B_0 = [1357f]$ and $F = 462$. Its path is indicated in Figure 10.4.9: note that the initial T-basis B_0 is already optimal for the subproblem $(J(9)/4, g, f)$, hence after the first pivot the element 4 is fixed in the T-basis, and the restricted Edmonds-Fukuda rule only treats the subproblem $(J(9)\backslash 4, g, f)$. The sixth T-basis $B_5 = [4176f]$ is feasible for this subproblem – but infeasible for $(J(9), g, f)$.

After one more pivot, the restricted Edmonds-Fukuda rule correctly reaches the optimal T-basis $B_6 = [4156f]$.

The cycling of various pivot rules on $\text{EFM}(8)$ is a crucial deviation from the situation in the linear case: non-degenerate cycling cannot occur in realizable oriented matroids. This suggests a study of "what exactly goes wrong". One

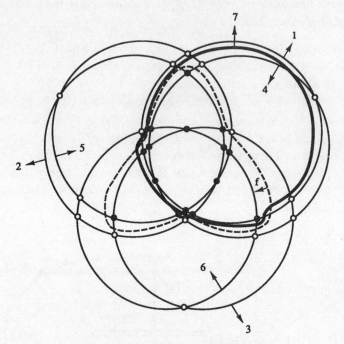

Figure 10.4.11: Sketch of $\mathcal{A}^{\infty}(\mathtt{J}(9))$.

possible answer is in terms of the *intersection properties* (see Section 7.5) which are satisfied by "nicely behaved" oriented matroids and, for example, prevent non-degenerate cycling.

Thus, in the framework of the following section, we will explain the cycling of the simplex method on $(\mathtt{EFM}(8), g, f)$ by the failure of the "Euclidean intersection property $\mathtt{IP_3}$" for $\mathtt{EFM}(8)$, that is, by the non-existence of a certain single element extension of $\mathtt{EFM}(8)$ (Definition 10.5.2/Proposition 10.5.8).

For the intersection properties of Section 7.5, $\mathtt{EFM}(8)$ can in fact serve as a "counter-example to almost everything". Here we only note how the "Levi's intersection property $\mathtt{IP_1}$" (which is stronger than $\mathtt{IP_3}$) fails on $\mathtt{EFM}(8)$.

For this, let Y', Y'' be two vertices of $f \cap g$ (that is, cocircuits of $\mathtt{EFM}(8)$ with $Y'_f = Y'_g = Y''_f = Y''_g = 0$ and $Y' \neq \pm Y''$). Then if Y is a vertex of the directed cycle, there is no extension of the pseudosphere arrangement $\mathtt{EFM}(8)$ by a pseudosphere that contains Y, Y' and Y'', by Proposition 10.5.8 below. Thus "three points do not determine a (pseudo)plane" (Goodman and Pollack 1981b). As independently observed by Edmonds and Mandel (1982), the Edmonds-Fukuda example even provides a uniform counter-example.

10.4.5 Proposition.   *There are three cocircuits Y, Y', Y'' in $\mathcal{C}^*(\mathtt{EFM}(8))$ such that there is no non-zero localization with $\sigma(Y) = \sigma(Y') = \sigma(Y'') = 0$.*

Thus the natural generalization of the Levi Enlargement Lemma 6.3.4 fails in rank 4.

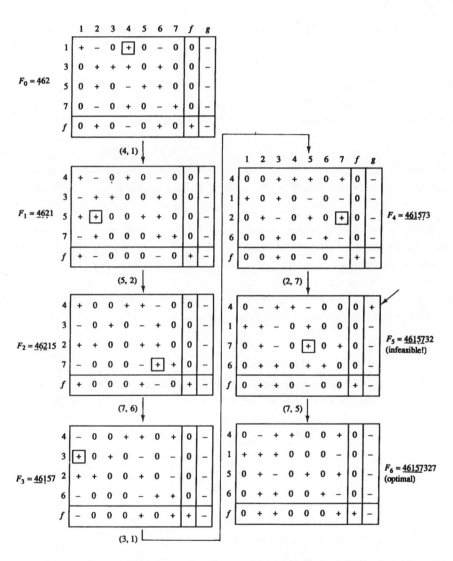

Figure 10.4.12: Tableaux for the restricted Edmonds-Fukuda rule
becoming infeasible on $(J(9), g, f)$.

From this, we get to another important consequence of the "bad properties" of the oriented matroid program $(\mathcal{M}, g, f)$. Namely, it shows that the
technique of *(topologically) sweeping arrangements* does not generalize to pseudoarrangements. This technique has been very successfully employed in various
computational geometry applications: see Edelsbrunner and Guibas (1989) for
the case of pseudoline arrangements, and Edelsbrunner, O'Rourke and Seidel
(1986) for the case of hyperplane arrangements.

10.4.6 Corollary.   *"Topological sweeping" does not generalize to pseudoarrangements of rank $r \geq 4$.*

Proof. The affine arrangement of pseudoplanes realizing $(\text{EFM}(8), g)$ cannot be
swept by a pseudoplane "parallel" to H_f (that is, coinciding with H_f at infinity),
because such a pseudoplane cannot sweep over a vertex in the directed cycle of
G_f. $\square$

However, topological sweeping can be generalized to "Euclidean pseudoarrangements", see the next section. This in particular covers the cases of pseudoline arrangements (rank 3) and of realizable hyperplane arrangements.

The reader should, however, note at this point that the "geometric" construction, description and analysis of examples like $\text{EFM}(8)$ are restricted to very small,
symmetric examples of low rank. This means that in general examples have to be
treated with better tools (that is, with the help of a computer) and in particular
the (non-)realizability proofs have to be devised algebraically. The appropriate
machinery for this is that of Chapter 8.

10.5 Euclidean matroids

The "bad" behavior of certain oriented matroids such as $\text{EFM}(8)$ suggests the
problem of characterizing "good" oriented matroids. The first step in this direction was taken by Edmonds and Mandel (1982) and Edmonds and Fukuda
(1982), with the definition of *Euclidean oriented matroids* and the proof that
those are exactly the oriented matroids on which pivot algorithms for oriented
matroid programming cannot have non-degenerate cycles. Later the work of
Bachem and coworkers related to *adjoints* of oriented matroids (Bachem and
Kern 1986a) led to introducing the hierarchy of *intersection* and *extension properties* (Bachem and Wanka 1989a). See Section 7.5, where "Euclideanness" appears as "intersection property IP_3".

We will now start our discussion with a definition of the "Euclidean property"
of oriented matroids. It goes back to Edmonds and Mandel (1982), who define
and work with "Euclidean affine oriented matroids". This class, however, has
certain drawbacks: the Euclidean property is not preserved under the single
element extensions whose existence it requires. Furthermore, there are no good
duality properties.

Therefore we follow Edmonds and Fukuda (1982) and consider "Euclidean-ness" as a property of oriented matroid programs rather than of affine matroids. Then the extensions required by the Euclidean property preserve this property. As an extra benefit, we get a duality result in Corollary 10.5.9.

The basic motivation is as follows: the problems with non-degenerate cycling come from the fact that not every two vertices of an oriented matroid program can be "compared" with respect to the objective function. They could, however, if there was a way to (consistently) define the "level set" of the objective function given by a vertex, or, better, to extend the program by hyperplanes parallel to the objective function.

The geometric basis for this is the "Euclidean postulate": for every hyperplane H and every point v in affine space there exists a unique hyperplane H' through v, parallel to H. The following two definitions translate the existence part of the "Euclidean postulate" into oriented matroid language.

10.5.1 Definition (Edmonds and Mandel 1982). Two elements $f, e \in E \backslash g$ of an affine oriented matroid $\mathcal{A} = (\mathcal{M}, g)$ are *parallel* if they are parallel ($\{f, e\}$ is a circuit) in $\mathcal{M}/g$.

Equivalently, e and f are parallel if the corresponding hyperplanes $\mathcal{H}_e$ and $\mathcal{H}_f$ "coincide at infinity", that is, if $Y_e = Y_f$ for all cocircuits $Y \in \mathcal{A}^\infty$ ($Y_g = 0$).

Here and in the following the setting of our discussion is that of affine space, modeled by affine oriented matroids (as in Section 10.1). This in particular allows us to interpret the elements of $E \backslash g$ as pseudo-hyperplanes, via the Topological Representation Theorem.

We will assume for the rest of this section that $\mathcal{M}$ is a simple matroid. Thus, for $e \in E \backslash g$, $\{e, g\}$ is independent. With this every $\mathcal{H}_e$ (and in particular $\mathcal{H}_f$) is a genuine (pseudo)hyperplane in affine space $\mathcal{A} = (\mathcal{M}, g)$. Tacitly, the element e will be identified with the corresponding hyperplane $\mathcal{H}_e = \{W \in \mathcal{A} : W_e = 0\}$ in affine space.

In this language the construction of "new" hyperplanes in affine space is achieved via single element extensions $\mathcal{M} \longrightarrow \widetilde{\mathcal{M}} = \mathcal{M} \cup p$, where the new hyperplane *goes through* Y if $(Y, 0) \in \widetilde{\mathcal{C}^*}$, corresponding to $\sigma(Y) = 0$ for the localization that defines the extension.

We will frequently abuse notation, writing Y for $(Y, \sigma(Y))$. Thus, putting $Y_p := \sigma(Y)$, we interpret $Y_p = 0$ to mean that Y *lies on (the hyperplane) p*, $Y_p = +$ says that Y *is on the positive side of p*, and so on.

10.5.2 Definition (Edmonds and Mandel 1982).

(1) An oriented matroid program $(\mathcal{M}, g, f)$ is *Euclidean* if for every cocircuit Y^0 in affine space ($Y^0 \in \mathcal{C}^*$, $Y_g^0 = +$) there exists a single element extension $\widetilde{\mathcal{M}} = \mathcal{M} \cup p$ such that the element p is parallel to f in $(\mathcal{M}, g)$, and $(Y^0, 0)$ is a cocircuit of $\widetilde{\mathcal{M}}$.

In other words, $(\mathcal{M}, g, f)$ is Euclidean if it can be extended by a hyperplane $\mathcal{H}_p$, parallel to $\mathcal{H}_f$, through any given vertex of $(\mathcal{M}, g)$.

(2)   An affine matroid $\mathcal{A} = (\mathcal{M}, g)$ is *Euclidean* if and only if $(\mathcal{M}, g, f)$ is Euclidean for all $f \in E \backslash g$.

We will now study the combinatorial structure of the extension problem given by an oriented matroid program $(\mathcal{M}, g, f)$ together with a cocircuit $Y^0 \in \mathcal{A}$. The main tool will be the graph G_f of $(\mathcal{M}, g, f)$ as introduced by Definition 10.1.16.

The first observation – directly from the definition – is that if f and f' are parallel in $(\mathcal{M}, g)$, then they induce the same graph $G_f = G_{f'}$. In fact this allows us to reduce the situation (and most of oriented matroid programming) to the case where f and g are inseparable.

Now assume that the extension problem given by $(\mathcal{M}, g, f)$ and Y^0 in fact has a solution, that is, there is a localization $\sigma : \mathcal{C}^* \longrightarrow \{+, -, 0\}$ such that

(1) $\sigma(Y^0) = 0$

(2)   for all cocircuits Y such that $Y_g = 0$, we have $\sigma(Y) = Y_f$.

10.5.3 Lemma.   *Let $(\mathcal{M}, g, f)$ be an oriented matroid program, $Y^0 \in \mathcal{A}$ a cocircuit, and let $\sigma : \mathcal{C}^* \longrightarrow \{+, -, 0\}$ specify an extension through Y^0, parallel to $\mathcal{H}_f$. Let $W = Y^1 \vee Y^2$ be an edge of G_f (given by two conformal cocircuits $Y^1, Y^2 \in \mathcal{A}$), then*

(i)   *if W is undirected, then $\sigma(Y^1) = \sigma(Y^2)$,*

(ii)   *if W is directed from Y^1 to Y^2, then $\sigma(Y^1) \preceq \sigma(Y^2)$ according to the order "$- \prec 0 \prec +$", where $\sigma(Y^1) = \sigma(Y^2) = 0$ cannot occur.*

Proof. Let $\pm Z^0 \in \mathcal{A}^\infty$ be the directions (cocircuits) determined by the affine line $\mathcal{A}(W)$ through Y^1 and Y^2. We can adapt the sign such that $W = Y^1 \vee Y^2 = Y^1 \circ Z^0 = Y^2 \circ (-Z^0)$.

(i) Suppose W is undirected. By Definition 10.1.16, this means $\sigma(Z^0) = 0$. The affine pseudoline $\mathcal{A}(W)$ has signature of Type I $(\sigma(Y^1) = \sigma(Y^2) = 0)$ or of Type II $(\sigma(Y^1) = \sigma(Y^2) \neq 0)$, see Figure 7.1.6.

(ii) Suppose W is directed, that is, $\sigma(Z^0) = +$. Then W is directed from Y^1 to Y^2, and $\mathcal{A}(W)$ has a signature of Type II or III. In both cases $\sigma(Y^1) \preceq \sigma(Y^2)$ is easy to see, where $\sigma(Y^1) = \sigma(Y^2) = 0$ cannot appear. $\square$

"Euclideanness" is a valuable concept not only because it describes an important geometric property, but also because it has an effective characterization, derived from Theorem 7.1.8. The link is provided by the notions of directed and undirected paths in G_f.

10.5.4 Definition.   A *path* in G_f from Y^0 to Y^k is an alternating sequence
$$\mathbf{P} = [Y^0, W^1, Y^1, W^2, \dots, Y^{k-1}, W^k, Y^k]$$
of vertices (cocircuits) $Y^0, Y^1, \dots, Y^{k-1}, Y^k$ and edges $W^1, W^2, \dots, W^k$, such that

– $W^i = Y^{i-1} \vee Y^i$ for all i, and

–   if W^i is directed, then it is directed from Y^{i-1} to Y^i.

A path $\mathbf{P}$ is *directed* if at least one edge W^i in $\mathbf{P}$ is directed, and *undirected* otherwise.

A path $\mathbf{P}$ is a *cycle* if $Y^0 = Y^k$ and $Y^0, \ldots, Y^{k-1}$ are pairwise distinct.

With this terminology, Lemma 10.1.17 states that for realizable oriented matroid programs, G_f does not contain directed cycles. From Lemma 10.5.3 we get that if $\mathbf{P}$ is a cycle, then $Y_f^0 = Y_f^1 = \ldots = Y_f^k$, and if the cycle $\mathbf{P}$ is directed, then $Y_f^i \neq 0$ for all i.

The following fundamental theorem by Edmonds and Mandel (1982) now characterizes Euclideanness. Even more, it gives a precise characterization of the conditions under which a specific extension of an oriented matroid program by a hyperplane parallel to f through Y exists.

In Edmonds and Fukuda's terminology, $(\mathcal{M}, g, f)$ is a "Bland oriented matroid program" (a "BOP") if G_f does not contain a directed cycle. Thus the following characterization states that $(\mathcal{M}, g, f)$ is Euclidean if and only if it is a BOP.

10.5.5 Theorem (Edmonds and Mandel 1982). *Let $(\mathcal{M}, g, f)$ be an oriented matroid program, and $Y \in \mathcal{A}$ a cocircuit (so $Y_g = +$).*

Then there is an extension of $(\mathcal{M}, g)$ by a hyperplane through Y^0 parallel to $\mathcal{H}_f$ if and only if Y is not contained in a directed cycle of G_f.

In particular, $(\mathcal{M}, g, f)$ is Euclidean if and only if G_f contains no directed cycles.

Proof. Let $\mathbf{P} = [Y^0, W^1, Y^1, W^2, \ldots, Y^{k-1}, W^k, Y^k]$ be a directed cycle through Y^0. Furthermore, let W^ℓ be the first directed edge in $\mathbf{P}$. Then if σ defines an extension of $(\mathcal{M}, g, f)$ by a hyperplane through Y^0, parallel to $\mathcal{H}_f$, then we get

$0 = \sigma(Y^0) = \ldots = \sigma(Y^{\ell-1})$ by Lemma 10.5.3(i), hence

$0 = \sigma(Y^{\ell-1}) \prec \sigma(Y^\ell) = +$ by Lemma 10.5.3(ii), and thus

$\sigma(Y^k) = +$ by monotonicity of σ along the path $\mathbf{P}$ by Lemma 10.5.3. This contradicts $Y^0 = Y^k$.

Conversely, if Y^0 is not contained in a directed cycle, then we can define a localization $\sigma : \mathcal{C}^* \longrightarrow \{+, -, 0\}$ as follows. For any cocircuit Y,

(i)   if $Y_g = +$, then

$$\sigma(Y) := \begin{cases} + & \text{if there is a directed path from } Y^0 \text{ to } Y, \\ 0 & \text{if there is an undirected path from } Y^0 \text{ to } Y, \\ - & \text{otherwise;} \end{cases}$$

(ii)   if $Y_g = 0$, then $\sigma(Y) := Y_f$, and

(iii)   if $Y_g = -$, then $\sigma(Y) := -\sigma(-Y)$.

Note that (i) is consistent if and only if Y^0 is not contained in a directed cycle. Now it is easy to check that $\sigma(Y^1) \preceq \sigma(Y^2)$ if there is a path from Y^1 to Y^2, because paths can be concatenated. Therefore, in particular σ is monotone on every affine pseudoline. This implies that σ satisfies condition (2) of the single element extension Theorem 7.1.8, and thus defines the required extension. $\square$

This theorem has various interesting consequences. We start with a simple corollary on minors of oriented matroid programs (see Definition 10.1.3(3)).

10.5.6 Corollary (Edmonds and Mandel 1982, Edmonds and Fukuda 1982).   *Every minor of a Euclidean oriented matroid program is Euclidean.*

Proof. For this we note that the graph $G_f(\mathcal{M}/e, g, f)$ is a subgraph of $G_f(\mathcal{M}, g, f)$ for $e \in E_n$, and so is $G_f(\mathcal{M}\backslash e, g, f)$ after a possible contraction of series edges. Thus we are done by Theorem 10.5.5.   □

10.5.7 Proposition.   *Every oriented matroid program of rank at most 3 is Euclidean.*

Proof. This is a reformulation of the Levi Enlargement Lemma (Proposition 6.3.4).   □

For an example, note that the oriented matroid program $(P) = (\text{EFM}(8), g, f)$ constructed in Section 10.4 is not Euclidean, because its graph G_f contains a directed cycle (Figure 10.4.7). This example is minor minimal, and in fact of minimal rank and size: every program of smaller rank is Euclidean by Proposition 10.5.7, and every program of less than eight points is realizable and hence Euclidean.

Our intuition is that every cycle in the graph G_f of a non-Euclidean oriented matroid program corresponds to the cycling of a pivot algorithm for a suitable pivot rule. (In fact, if the cycle is contained in the boundary of the feasible region, then this should correspond to cycling of a genuine simplex method.) This is true, however, only in the case of non-degenerate programs. In the general case, we cannot exclude undirected edges in the cycle, which corresponds to horizontal pivots. This forces one to include horizontal pivots, and arbitrary degenerate pivots, in the following result.

10.5.8 Proposition (Edmonds and Fukuda 1982).   *An oriented matroid program $(\mathcal{M}, g, f)$ is non-Euclidean if and only if there is a "directed" cycle*
$$T[B^0] \longrightarrow T[B^1] \longrightarrow \dots \longrightarrow T[B^k] \longrightarrow T[B^0],$$
all of whose pivots are (strictly) increasing or degenerate or horizontal, and at least one of whose pivots is strictly increasing.

Proof. The "if"-part is clear with the geometric interpretation of pivot steps (after Definition 10.2.11).

For the "only if"-part, for every (directed, or undirected) edge $Y \longrightarrow Y'$ of G_f we can easily construct T-bases B and B' such that the (strictly increasing, or horizontal) pivot $T[B] \longrightarrow T[B']$ follows the edge, with $Y^B = Y$ and $Y^{B'} = Y'$. From this the result follows, because if B and B' are T-bases such that $Y^B = Y^{B'}$, then there obviously is a sequence of degenerate pivots from $T[B]$ to $T[B']$.   □

10.5.9 Corollary (Edmonds and Fukuda 1982). *An oriented matroid program* $(P) = (\mathcal{M}, g, f)$ *is Euclidean if and only if its dual program* $(P^*) = (\mathcal{M}^*, f, g)$ *is Euclidean.*

Proof. Let $T[B^0] \longrightarrow T[B^1] \longrightarrow \ldots \longrightarrow T[B^k] \longrightarrow T[B^0]$ be a directed cycle of tableaux for (P). This immediately yields a cycle of tableaux and pivot steps

$$(10.5.1) \qquad T^*[N^0] \longrightarrow T^*[N^k] \longrightarrow \ldots \longrightarrow T^*[N^1] \longrightarrow T^*[N^0]$$

for (P^*). From the construction of the dual tableaux (Corollary 10.2.9) we can read off that

- $T[B^0] \longrightarrow T[B^1]$ is degenerate iff $T^*[N^1] \longrightarrow T^*[N^0]$ is horizontal,
- $T[B^0] \longrightarrow T[B^1]$ is horizontal iff $T^*[N^1] \longrightarrow T^*[N^0]$ is degenerate,
- $T[B^0] \longrightarrow T[B^1]$ is strictly increasing iff $T^*[N^1] \longrightarrow T^*[N^0]$ is strictly decreasing.

Thus, by reversing (10.5.1), we can construct a strictly increasing cycle for (P^*). The duality result now follows from Theorem 10.5.5 and Proposition 10.5.8. $\qquad\square$

10.5.10 Proposition. *Let* $(\mathcal{M}, g, f)$ *be a Euclidean oriented matroid program, and let* $(\mathcal{M} \cup p, g, f)$ *be an extension by a hyperplane parallel to* f. *Then* $(\mathcal{M} \cup p, g, f)$ *is again Euclidean.*

Proof. For this we note that $G_f = G_p$: the objective functions f and p induce the same graph, as was already observed before Lemma 10.5.3.

Now if G_p contains a directed cycle, then Y_p is constant and non-zero for its vertices Y, as observed after Definition 10.5.4. Hence the cycle also is a directed cycle of $G_f(\mathcal{M}, g, f)$. $\qquad\square$

Finally we can use the characterization of the Euclidean property to show that it implies a seemingly stronger extension property, see Exercise 10.1.

10.5.11 Proposition (Edmonds and Mandel 1982). *Let* $(\mathcal{M}, g, f)$ *be a Euclidean oriented matroid program. Then for every covector* $Y \in \mathcal{A}$, *there is an extension of* $(\mathcal{M}, g)$ *by a hyperplane* $\mathcal{H}_p$ *through* Y, *that is, such that* $(Y, 0) \in \mathcal{A}(\mathcal{M} \cup p)$.

Exercises

10.1 Prove Proposition 10.5.11. For this, distinguish the case where $W \leq Y$ for a directed edge $W \in \mathcal{A}$ (in which case the localization σ can be explicitly described, as in the proof of Theorem 10.5.5), and the other case, where a hyperplane through Y also goes through a vertex of Y.

10.2   Draw figures that explain the geometric meaning of
        - a degenerate pivot
        - a horizontal pivot
        - a tableau proving unboundedness.

        Pay particular attention to the situation "at infinity".

10.3   Let $D = (V, A)$ be a finite digraph, let $f, g \in A$ be distinct arcs of D
        such that g is not a loop and f is not a coloop.

        If $\mathcal{M}$ is the oriented matroid corresponding to D, interpret $(\mathcal{M}, g, f)$,
        its solutions, bases, feasibility, optimality, pivots, etc. Describe a finite
        algorithm that constructs an optimal solution or decides that none exists.

10.4   A polyhedron given by $A\mathbf{x} \leq \mathbf{b}$ is easily transformed to "standard form"
        $A\mathbf{x} = \mathbf{b}$, $\mathbf{x} \geq \mathbf{0}$ by introducing slack variables.

        Describe an oriented matroid construction which achieves this. In other
        words, for an affine oriented matroid $(\mathcal{M}(E), g)$, show how to construct
        an affine oriented matroid $(\mathcal{M}'(E'), g)$ on $E' = E \cup N$ such that E is a
        basis of $\mathcal{M}'$ and $\mathcal{M}'/N = \mathcal{M}$.

10.5   Show that the Hahn-Banach Theorem 9.2.1(6) fails for a suitable reori-
        entation of $\mathrm{EFM}(8)$.

        Namely, $A := \{1, 2, 6, f\}$ and $B := \{3, 4, 5, g\}$ are convex sets in
        $_{-123g}\mathrm{EFM}(8)$ that cannot be separated.

10.6   Let $(P) = (\mathcal{M}, g, f)$ and $(P^*) = (\mathcal{M}^*, f, g)$ be a pair of dual oriented
        matroid programs. Prove the "Strong Complementary Slackness Theo-
        rem": If (P) and (P^*) both are feasible, then they have optimal solutions
        $\bar{Y}$ and $\bar{X}$ that additionally satisfy $\underline{\bar{X}} \cup \underline{\bar{Y}} = E$.

        (Note: here $\bar{Y}$ is a solution covector, not necessarily a cocircuit, and
        similarly $\bar{X}$ is a solution vector, not necessarily a circuit.)

                                                    (Edmonds and Fukuda 1982)

10.7   Two fundamental facts in the theory of inequality systems are the

        Farkas Lemma: $A\mathbf{y} \leq \mathbf{b}$ has a solution if and only if there is no $\mathbf{x}$ with
        $\mathbf{x}^T A = \mathbf{0}$, $\mathbf{x} \geq \mathbf{0}$, $\mathbf{x}^T \mathbf{b} < 0$,

        and the

        Gordan Lemma: $A\mathbf{y} < \mathbf{0}$ has a solution if and only if there is no $\mathbf{x} \neq \mathbf{0}$
        with $\mathbf{x} \geq \mathbf{0}$, $\mathbf{x}^T A = \mathbf{0}$.

        Formulate oriented matroid generalizations and derive them from Corol-
        lary 10.2.18.

10.8   Show that the linear programs

$$\max \ \mathbf{c}^T \mathbf{x} \ - d$$
$$A\mathbf{x} \ \leq \mathbf{b} \qquad\qquad (P)$$
$$\mathbf{x} \ \geq \mathbf{0}$$

and

$$\max (-\mathbf{b})^T\mathbf{y} + d$$
$$(-A^T)\mathbf{y} \leq -\mathbf{c} \qquad (P^*)$$
$$\mathbf{y} \geq \mathbf{0}$$

considered in Section 10.1 to motivate the definition of oriented matroid programs, do define a pair of dual oriented matroid programs.

10.9 Give a constructive proof for the following "4-Coloring Theorem 3.4.4(4P)", that is, show how to use oriented matroid programming to find the required circuit or cocircuit.

For all $e \in E$ and for all partitions $E = W \cup R \cup B \cup G$ with $e \in W \cup R$, exactly one of the following alternatives holds:

– either there is a $X \in \mathcal{C}(\mathcal{M})$ with $e \in X$ and $X^+ \subseteq W \cup B$, $X^- \subseteq R \cup B$,

– or there is a $Y \in \mathcal{C}^*(\mathcal{M})$ with $e \in Y$ and $Y^+ \subseteq W \cup G$, $Y^- \subseteq R \cup G$.

(Bland and Las Vergnas 1979)

10.10 In the primal and dual oriented matroid tableaux as given in Corollary 10.2.9, interpret the sign δ geometrically.

10.11 Draw the oriented matroid program corresponding to

$$\max \; y \quad -\tfrac{3}{2}$$
$$x - y \leq 0$$
$$-x - 2y \leq -2 \qquad (LP_0)$$
$$x, y \geq 0,$$

and solve it

– with the Simplex Algorithm and Todd's rules 10.3.13 (one needs two pivots for feasibility and two further ones for optimality),

– with the Dual Simplex Algorithm and the smallest subscript rule for Phase I and ALGORITHM A for Phase II. In the latter case, how many pivots are needed in Phase I ? In Phase II ?

10.12 The effectiveness of the "criss-cross method" depends very much on a "good" variable ordering:

(a) Show that the criss-cross method needs eight pivots to solve (EFM(8), g, f). Construct the corresponding (nine) tableaux.

(b) Show that with a good ordering of the variables, namely "$6 < 4 < 5 < 1 < 2 < 3 < f < g$", the criss-cross method solves (EFM(8), g, f) with only three pivots. (Terlaky)

10.13* Give a simple (geometric) correctness and finiteness proof for Todd's rules (or simpler simplex pivot rules) which can be formulated directly on the program $(\mathcal{M}, g, f)$.

Appendix: Some Current Frontiers of Research

Currently – in 1999, six years after the first edition of this book appeared – substantial research is being done on a wide variety of aspects and questions; among them are several deep problems of oriented matroid theory that were thought to be both hard and fundamental, and are now gradually turning out to be just that. Here we provide some sketches and a few pointers to the (recent) literature, for a few selected topics.

A.1 Realization spaces

Mnëv's universality theorem (Theorem 8.6.6; Mnëv 1988) states that every primary semialgebraic set defined over $\mathbb{Z}$ is "stably equivalent" to the realization space of some oriented matroid of rank 3. In other words, the semialgebraic sets of the form

$$\mathcal{R}(X) \quad := \quad \{Y \in \mathbb{R}^{3 \times n} : \text{sign}(\det(X_{i,j,k})) = \text{sign}(\det(Y_{i,j,k}))$$
$$\text{for all } 1 \leq i < j < k \leq n\},$$

for real matrices $X \in \mathbb{R}^{3 \times n}$, can be arbitrarily complicated, both in their topological and their arithmetic properties. The even stronger universal partition theorem was announced in Mnëv (1991). It states that every partition of $\mathbb{R}^k$ generated by a set of polynomial equations (in k variables, with integral coefficients) is "stably equivalent" to a part of the stratification given by the determinants of the (3×3)-minors of a $(3 \times n)$-matrix.

These results are fundamental and far-reaching. For example, via oriented matroid (Gale) duality they imply universality and universal partition theorems for d-polytopes with $d + 4$ vertices.

For some time no complete proofs were available. This changed only recently with the complete proof of the Universal Partition Theorem by Günzel (1996) and by Richter-Gebert (1995c). Richter-Gebert (1995b, Sect. 2.5) (1996b) has also – finally! – provided a suitable notion of "stable equivalence" of semialgebraic sets (and of families of semialgebraic sets), which is *weak enough* to make the universality theorems true, and *strong enough* to imply both homotopy equivalence and arithmetic equivalence (i.e., it preserves the existence of K-rational points in the semialgebraic set for every subfield K of $\mathbb{R}$), as follows:

A.1.1 Definition (Stable equivalence). Let $V \subseteq \mathbb{R}^n$ and $W \subseteq \mathbb{R}^{n+d}$ be semialgebraic sets with $\pi(W) = V$, where $\pi : \mathbb{R}^{n+d} \to \mathbb{R}^n$ is the canonical projection that deletes the last d coordinates. V is a *stable projection* of W if W has the form
$$W = \left\{ (\mathbf{v}, \mathbf{v}') \in \mathbb{R}^{n+d} : \mathbf{v} \in V, \phi_i(\mathbf{v}) \cdot \mathbf{v}' > 0; \ \psi_j(\mathbf{v}) \cdot \mathbf{v}' = 0 \text{ for } i \in X; \ j \in Y \right\}.$$
Here X and Y denote finite (possibly empty) index sets. For $i \in X$ and $j \in Y$ the functions ϕ_i and ψ_j have to be polynomial functions
$$\phi_i = (\phi_i^1, \dots, \phi_i^d) : \mathbb{R}^n \to (\mathbb{R}^d)^* \quad \text{with } \phi_i^k \in \mathbb{Z}[x_1, \dots, x_n] \quad \text{and}$$
$$\psi_j = (\psi_j^1, \dots, \psi_j^d) : \mathbb{R}^n \to (\mathbb{R}^d)^* \quad \text{with } \psi_j^k \in \mathbb{Z}[x_1, \dots, x_n],$$
that associate to every element of $\mathbb{R}^n$ a linear functional on $\mathbb{R}^d$. Two semialgebraic sets V and W are *rationally equivalent* if there exists a homeomorphism $f : V \to W$ such that both f and f^{-1} are given by rational functions.

Two semialgebraic sets V and W are *stably equivalent* if they are in the same equivalence class with respect to the equivalence relation generated by stable projections and rational equivalence.

These concepts provided foundations for a substantial and surprising piece of progress: ending a long quest, Richter-Gebert (1995b, 1996b) established a universality theorem (and universal partition theorem) for the realization spaces of 4-dimensional polytopes. (See also Richter-Gebert and Ziegler (1995) for a sketch, and Günzel (1997) for a second proof.)

For still another, very recent, interesting universality result concerning the configuration spaces of planar linkages, see Kapovich and Millson (1998b, 1999) and Jordan and Steiner (1998). These papers trace the history of their result back to a universality theorem by Kempe (1875)!

In his work, Richter-Gebert also showed that the Steinitz problem for 3-spheres has no solution: there is no finite list of substructures ("minors") for the cellular 3-spheres such that a sphere is polytopal if and only if it contains none of the "forbidden minors."

A.1.2 Theorem (Richter-Gebert 1996b). *There is an infinite sequence of cellular 3-spheres (whose facets have the combinatorial types of non-simplicial 3-polytopes) that are not realizable, but such that any deletion of a vertex results in a cell complex that appears in the boundary complex of a convex 4-polytope (without additional vertices).*

Richter-Gebert has two beautiful, totally different proofs for this result, in Section 9.4 and in Section 15.3 of Richter-Gebert (1996b). It is not quite clear, however, whether one of them would also allow us to establish the non-Steinitz conjecture in the very strong version that we have stated as Conjecture 9.5.6.

Here are two major challenges that remain in this area:

– To construct and understand the smallest oriented matroids with non-trivial realization spaces. The smallest *known* examples are Suvorov's (1988) oriented matroid of rank 3 on 14 points with a disconnected realization space (see p. 365), and Richter-Gebert's (1996a) new example Ω_{14}^+ with the same parameters, which additionally has rational realizations and a non-realizable symmetry.

– To provide universality theorems for *simplicial* 4-dimensional polytopes — including a non-Steinitz Theorem, cf. Conjecture 9.5.6. (The polytope of Bokowski, Ewald and Kleinschmidt (1984) is still the only simplicial example known with a non-trivial realization space; see also Bokowski and Guedes de Oliveira (1990).)

A.2 Spaces of oriented matroids

The consideration of *spaces of oriented matroids* brings several very different lines of thinking into a common topological framework. Given a set S of oriented matroids, we obtain a partial order on S by weak maps, and from this we obtain a topological space by taking the order complex (see p. 200). This simplicial complex can be viewed as a combinatorial analog of a vector bundle. Just as a vector bundle represents a continuous parametrization of a set of vector spaces, this topological space can be viewed as a parametrization of elements of S. Such spaces have arisen in several contexts:

– If S is the set of non-trivial single element extensions of a fixed oriented matroid $\mathcal{M}$, the resulting space is the *extension space* $\mathcal{E}(\mathcal{M})$ of $\mathcal{M}$.

– If S is the set of all rank r oriented matroids on a fixed set of n elements, this space is the *MacPhersonian* $\mathrm{MacP}(r, n)$.

– If S is the set of all rank r strong map images of a fixed oriented matroid $\mathcal{M}$, this space is the *combinatorial Grassmannian* $\mathcal{G}(r, \mathcal{M})$.

(In fact, this example essentially encompasses the previous two: the extension space $\mathcal{E}(\mathcal{M})$ of a oriented matroid $\mathcal{M}$ is a double cover of $\mathcal{G}(r(\mathcal{M}) - 1, \mathcal{M})$, while if $\mathcal{M}$ is the unique rank n oriented matroid on a fixed set of n elements, then $\mathcal{G}(r, \mathcal{M}) = \mathrm{MacP}(r, n)$.)

Extension spaces are closely related to zonotopal tilings (via the Bohne–Dress Theorem) and to oriented matroid programs: see Sturmfels and Ziegler (1993). The MacPhersonian and combinatorial Grassmannian arise in MacPherson's theory of *combinatorial differential manifolds*, and *matroid bundles*, in which oriented matroids serve as combinatorial analogs to real vector spaces. We refer to MacPherson (1993) and Anderson (1999b) for introductions to this setting.

Among the basic conjectures in the field are:

– For any realizable rank n oriented matroid $\mathcal{M}^n$, the topology of $\mathcal{G}(k, \mathcal{M}^n)$ should be similar to that of the real Grassmannian $G(k, \mathbb{R}^n)$.

In particular, Anderson and Davis (1999) have shown that a realization of an oriented matroid $\mathcal{M}^n$ determines a canonical homotopy class of maps $c : G(k, \mathbb{R}^n) \to \mathcal{G}(k, \mathcal{M}^n)$, and it is hoped that c is a homotopy equivalence, or at least leads to an isomorphism in cohomology with various coefficients.

– The extension space $\mathcal{E}(\mathcal{M}^n)$ should have the homotopy type of an $(n-1)$-sphere if $\mathcal{M}^n$ is realizable. (This is essentially a special case of the above.)

There are substantial grounds for pessimism on both conjectures. For instance, there are examples of non-realizable $\mathcal{M}^n$ such that $\mathcal{G}(n-1, \mathcal{M}^n)$ and $\mathcal{E}(\mathcal{M}^n)$ are not even connected (Mnëv and Richter-Gebert 1993). In addition, Mnëv's universality theorem implies that for realizable $\mathcal{M}^n$ the inverse images under c of points in $\mathcal{G}(r, \mathcal{M}^n)$ can have arbitrarily complicated topology. However, substantial progress has been made on the topology of $\mathcal{E}(\mathcal{M})$ for realizable $\mathcal{M}$ (Sturmfels and Ziegler 1993) and on the topology of $\mathcal{G}(k, \mathcal{M})$ under various conditions: for small values of k (Babson 1992); for the first few homotopy groups of the MacPhersonian (Anderson 1998); and for mod 2 cohomology (Anderson and Davis 1999). Three related survey articles are Mnëv and Ziegler (1993), Anderson (1999b), and Reiner (1999).

The analogy between oriented matroids and real vector bundles leads to an intriguing and useful interplay between topology and combinatorics. On the one hand, appropriate combinatorial adaptations of classical topological methods for real vector bundles prove that for realizable $\mathcal{M}^n$ the map $c : G(k, \mathbb{R}^n) \to \mathcal{G}(k, \mathcal{M}^n)$ induces split surjections in mod 2 cohomology (Anderson and Davis 1999). On the other hand, any real vector bundle over a triangulated base space can be "combinatorialized" into a matroid bundle (MacPherson 1993, Anderson and Davis 1999), giving a combinatorial approach to the study of bundles. The most notable success in this direction has been Gel'fand and MacPherson's (1992) combinatorial formula for the rational Pontrjagin classes of a differential manifold.

The topological problems discussed in this section have close connections to classical problems of oriented matroid theory, such as Las Vergnas' conjectures that every oriented matroid has at least one mutation (Conjecture 7.3.10) and that the set of uniform oriented matroids of rank r on a given finite set is connected under performing mutations. In fact, if these conjectures are false, then the "top level" of the MacPhersonian, given by all oriented matroids without circuits of size smaller than r and at most one circuit of size r, cannot be connected. As for the Las Vergnas conjecture, Bokowski (1991a) and Richter-Gebert (1993b) have the strongest positive resp. negative partial results; more work is necessary.

Further work also remains in the understanding of weak and strong maps, cf. Section 7.7. For example, one still has to derive structural information from the failure of Las Vergnas' strong map factorization conjecture (disproved by Richter-Gebert 1993b), and to derive criteria for situations where factorization is possible.

A.3 Triangulations

Triangulations of polytopes and of point configurations form an active field of study, and the oriented matroid point of view is essential for many of the current investigations. The link is established by the concept of "triangulations of oriented matroids." However, the setting of Definition 9.6.1 is very hard to work with: conditions that have to hold "for every single element extension $\mathcal{M} \cup p$ of $\mathcal{M}$" are extremely difficult to check, since this amounts to a complete search of the extension space $\mathcal{E}(\mathcal{M})$, whose structure is not well understood. This problem was remedied by work of Anderson (1994) and Santos (1997a), who established simpler characterizations of oriented matroid triangulations.

The generalized Baues conjecture (GBC) of Billera, Kapranov and Sturmfels (1994) describes a certain "poset of cellular strings" $\Sigma(P, Q)$ for any projection $\pi \colon P \longrightarrow Q$ of polytopes, and asks whether its homotopy type is that of a sphere of dimension $\dim(P) - \dim(Q) - 1$. See Reiner (1999) for an extensive survey and status report. In particular, the general version was disproved by Rambau and Ziegler (1996), while the special case of Exercise 9.28 is still wide open; it arises when P is a simplex, $\mathcal{M}$ is the oriented matroid of the images of its vertices under π, and so $\Sigma(P, Q) = \Sigma(\mathcal{M})$. So we still don't know whether for every d-dimensional configuration of n points, the poset of polyhedral subdivisions has the homotopy type of an $(n - d - 2)$-sphere. For $n = d + 4$ this is true (Azaola 1999) and the triangulation graph is 3-connected (Azaola and Santos 1999). Triangulations of cyclic polytopes, and the GBC for the natural projections of cyclic polytopes $C_d(n) \longrightarrow C_{d'}(n)$, for $d \geq d'$, were studied intensively and successfully by Rambau (1997), Edelman, Rambau and Reiner (1997), Rambau and Santos (1997), and Athanasiadis, Rambau and Santos (1999).

If the GBC for triangulations is true, then the triangulations of any finite point set in general position are connected by bistellar flip operations. In particular, every geometric triangulation must admit *at least one* bistellar flip. This is not known in general. Theorem 9.6.7 implies that every **X**-regular triangulation has at least $n - d - 1$ bistellar flips. The same is true for general triangulations in dimension 2 and for triangulations in convex position in dimension 3 (De Loera, Santos and Urrutia 1999), but not in general. Indeed, Santos (1997b) constructed triangulations on n points in $\mathbb{R}^4$, for arbitrarily large n, for which the number of geometric bistellar flips is bounded by a constant.

In general, search and optimization problems over the set (space) of all triangulations of a given point set are very hard. Recent results show that computing the minimal or maximal number of facets in any triangulation of a given configuration is a hard problem – see De Loera (1999) and Richter-Gebert (1999). A configuration **X** found by Ohsugi and Hibi (1997) has the surprising property that no triangulation with a maximal, or with a minimal, number of facets is **X**-regular; see Firla and Ziegler (1999). Nevertheless, there is a rather successful integer programming set-up for optimization of triangulations, based on De Loera, Hoşten, Santos and Sturmfels (1996). See Tajima (1998) for further work in this direction.

A.4 Zonotopal tilings

The Bohne–Dress Theorem 2.2.13 provides a bijection between the zonotopal tilings of a fixed d-dimensional zonotope $\mathcal{Z}$ and the single element liftings of the realizable oriented matroid associated with $\mathcal{Z}$. This theorem has turned out to be, at the same time,

- fundamental; see e.g. the connection to extension spaces of oriented matroids, as in Sturmfels and Ziegler (1993),

- "intuitively obvious" (just draw pictures!), and

- surprisingly hard to prove; see Bohne (1992a) and Richter-Gebert and Ziegler (1994).

Just recently, however, Huber, Rambau and Santos (1999) have given a new and substantially different proof of the Bohne–Dress theorem. In particular, for any real matrix A there are bijections

$$\{ \text{ zonotopal tilings of the zonotope } \mathcal{Z}(A) \ \}$$
$$\updownarrow$$
$$\{ \text{ subdivisions of the Lawrence polytope } \Lambda(A) \ \}$$
$$\updownarrow$$
$$\{ \text{ single element extensions of the dual oriented matroid } \mathcal{M}^*(A) \ \}$$
$$\updownarrow$$
$$\{ \text{ single element liftings of } \mathcal{M}(A) \ \}.$$

The first bijection follows from the "Cayley trick"; see Huber, Rambau and Santos (1999). The second, more difficult, one was already established by Santos (1997a): It implies that the extension space conjecture (cf. Section A.2) is a special case of the GBC for triangulations. The third bijection is just oriented matroid duality (see p. 281).

A separate, simpler proof for the case of rank 3 – pseudoline arrangements are in bijection with zonotopal tilings of a centrally symmetric $2n$-gon – is contained in the work by Felsner and Weil (1998).

The Bohne–Dress theorem provides connections to several other areas of study. On the one hand, the classification and enumeration of rhombic tilings of a hexagon relates to the theory of plane partitions and symmetric functions; see e.g. Elnitzky (1993, 1997), Bailey (1997), and Edelman and Reiner (1996a). On the other hand, there is a definite need for a better understanding of zonotopal tilings of the entire plane (or of $\mathbb{R}^d$). Two different approaches have been started by Bohne (Kapitel 5 of Bohne 1992a) and by Crapo and Senechal (1995), but no complete picture has yet emerged. This is of interest, for example, in view of the mathematical problems posed by quasiperiodic tilings and quasicrystals; see Senechal (1993, 1995) and Lagarias (1999).

A.5 Realization algorithms

The realizability problem – given a "small" oriented matroid, find a realization or prove that none exists – is a key problem not only in oriented matroid theory, but also for various applications, such as the classification of "small" simplicial spheres into polytopal and non-polytopal ones; see Bokowski and Sturmfels (1987, 1989b), Altshuler, Bokowski and Steinberg (1980), Bokowski and Shemer (1987). The universality theorems mentioned above tell us that the problem is hard: in fact, in terms of Complexity Theory it is just as hard as the "Existential Theory of the Reals," the problem of solving general systems of algebraic equations and inequalities over the reals. While it is not known whether the problem over $\mathbb{Q}$ is algorithmically solvable at all (see Sturmfels 1987e), there are algorithms available that, at least theoretically, solve the problem over the reals. For the general problem, the currently best result is as follows.

A.5.1 (Basu, Pollack and Roy 1997).   *Let* $\mathcal{P} = \{P_1, \ldots, P_s\}$ *be a set of polynomials in* $k < s$ *variables, each of degree at most* d *and each with coefficients in a subfield* $K \subseteq \mathbb{R}$. *Then there is an algorithm which finds a solution in each connected component of the solution set, for each sign condition on* $P_1, \ldots, P_s$, *in at most* $\binom{O(s)}{k} s \, d^{O(k)} = (s/k)^k s \, d^{O(k)}$ *arithmetic operations in* K.

However, so far this is of mostly theoretical value. What can be done for specific, small examples? Given an oriented matroid of rank 3, it seems that

– the most efficient algorithm (in practice) currently available to *find a realization* (if one exists) is the iterative "rubber band" algorithm of Bokowski described in Pock (1991).

– the most efficient algorithm (in practice) currently available to *show that it is not realizable* (if it isn't) is the "biquadratic final polynomials" algorithm of Bokowski and Richter (1990a) which uses solutions of an auxiliary linear program to construct final polynomials. (An explicit example of a non-realizable oriented matroid Ω_{14}^- without a biquadratic final polynomial was constructed by Richter-Gebert (1996a).)

Neither of these two parts is guaranteed to work; but still the combination of both parts was good enough for a (still unpublished) complete classification of all 312,356 (unlabeled reorientation classes of) uniform oriented matroids of rank 3 on 10 points into realizable and non-realizable ones (Below, Bokowski, Kortenkamp, Laffaille and Richter-Gebert 1999).

A very closely related topic is that of Automatic Theorem Proving in (plane) geometry. In fact, the question of validity of a certain incidence theorem can be viewed as the realizability problem for the (oriented or unoriented) matroid of the configuration. Richter-Gebert's thesis (1991) and Wu's book (1994) here present two recent (distinct) views of the topic, both with many of its ramifications. For an implementation of these, and many other, ideas we refer to the new Interactive Geometry Software system CINDERELLA by Richter-Gebert and Kortenkamp (1999), whose development over the last few years produced the idea of "binomial proofs" of Crapo and Richter-Gebert (1995) as well as new randomized methods for automatic theorem proving.

A.6 Random walks on arrangements

Random walks on the regions of a real hyperplane arrangement have become prominent in recent research. The theory works equally well when generalized to walks on the topes of an oriented matroid. It contains as special cases several card shuffling schemes (when specialized to the braid arrangements) and such questions as how to choose a random zonotopal tiling.

This new line of research was initiated in a paper by Bidigare, Hanlon and Rockmore (1997) and then extended by Brown and Diaconis (1998). Somewhat stronger results for the special case of arrangements in $\mathbb{R}^3$ were then proved by Billera, Brown and Diaconis (1999). Their paper is written in an expository style and gives an excellent introduction to the results of the other papers and to this whole circle of ideas. See also Brown (1999) for a more general semigroup setting that also includes some random walks on general matroids, and Diaconis (1998) for the broader context.

Let us describe the basic set-up for the case of an essential and central arrangement $\mathcal{A}$ in $\mathbb{R}^3$. Let $\mathcal{C}$ denote the cell complex induced on the unit sphere by the intersections with the planes in $\mathcal{A}$. Our walk begins at some region R_0. We choose at random a vertex v_1 of $\mathcal{C}$ and move to the region $R_1 = v_1 \circ R_0$ that is closest to R_0 (in terms of distance in the tope graph) among the regions that have v_1 in their boundary. At the next step we select at random another vertex v_2 and move from R_1 to the region $R_2 = v_2 \circ R_1$ which has v_2 on its boundary and is closest to R_1 among all such regions, and so on.

This defines a finite Markov chain for which the transition probabilities are determined by the combinatorial geometry of the cell complex $\mathcal{C}$. Bidigare, Hanlon and Rockmore (1997) determined the eigenvalues of its transition matrix. These turn out to be non-negative rational numbers with simple explicit expressions in terms of the combinatorics of $\mathcal{C}$. Their result is amplified in Brown and Diaconis (1998), where the transition matrix is shown to be diagonalizable. Also the rate of convergence to the stationary distribution is discussed in these papers, and it is surprisingly rapid: the distribution is essentially stationary after only two steps! Finally, in Billera, Brown and Diaconis (1999) a remarkable explicit expression for the stationary probability distribution is given (but only for the case of $\mathbb{R}^3$). Namely, the probability of being in a region with k sides is proportional to $k - 2$.

The generalization to oriented matroids is straightforward. In this case we can allow some probability distribution on the set of covectors that is more general than the previous one (which is uniform distribution on the set of covectors of rank 1). Start at a tope T_0. Select a covector X_1 according to the given probability distribution and move to $T_1 := X_1 \circ T_0$. Then choose another covector X_2 according to the distribution and move to $T_2 := X_2 \circ T_1$, and so on.

At the time of writing it is not clear whether the stationary distribution has a simple geometric description in any other case than the rank 3 result that we quoted from Billera, Brown and Diaconis (1999). This and many other challenging questions remain open in this new branch of oriented matroid theory.

A.7 Polyhedral 2-manifolds

We close this discussion with an update on the problem of the geometric realization of triangulated 2-manifolds – and thus with a recent remarkable success of oriented matroid theory.

It was a longstanding question of Grünbaum (Exercise 3 on p. 253 of Grünbaum 1967; see also Duke 1970), whether every oriented triangulated 2-manifold (a sphere with g handles, where g is called the *genus* of the 2-manifold) has a polyhedral embedding in $\mathbb{R}^3$. For triangulated 2-spheres (2-manifolds of genus 0) such an embedding exists by Steinitz' theorem (Theorem 9.5.1). For triangulations of the torus (genus $g = 1$) the conjecture may seem quite plausible, but the problem is wide open. As for manifolds of higher genus, it may seem utterly improbable, for example in view of the existence of orientable neighborly 2-manifolds whose genus is much higher than their number n of vertices. However, the negative answer to the question was remarkably hard to come by, and oriented matroid theory played the central role in the recent solution.

The number of triangulated 2-manifolds grows explosively with the number of vertices n. Particularly interesting examples include on the one hand those of large (vertex-transitive) symmetry. We refer to Lutz (1999) for an overview, and for a classification of all 2-manifolds with vertex-transitive symmetry group and $n \leq 15$ vertices. On the other hand, one may look at neighborly 2-manifolds (all pairs of vertices adjacent). An orientable neighborly 2-manifold with n vertices exists for all $n \equiv 0, 3, 4$ or 7 (mod 12) – see Ringel (1974). Its genus is then $g = \frac{(n-3)(n-4)}{12}$. (Some non-orientable neighborly 2-manifolds arise from p_3-maximal pseudoline arrangements, cf. Theorem 6.5.2 and Exercises 6.20–6.21, via an interesting new construction of Bokowski, Roudneff and Strempel (1997).)

If one studies the embedding question for neighborly maps, then the Möbius torus on $n = 7$ vertices is the first and prime example. This torus, which appears in Möbius (1886), was rediscovered and geometrically realized by Császár (1949/50); today we know that it can be obtained as a subcomplex of any Schlegel diagram of the cyclic polytope $C_4(7)$. The space of all realizations of the Möbius torus was described by Bokowski and Eggert (1991). The next natural examples to consider include the vertex-transitive 2-manifolds ("regular maps") of genus 3 of Dyck and of Klein, with 12 resp. 24 vertices. Both of them can be realized, see Bokowski (1989) and Brehm (1987), resp. Schulte and Wills (1985).

The next parameter set for which one gets neighborly oriented 2-manifolds is $n = 12$ and $g = 6$. The corresponding maps have been classified by Altshuler (see Altshuler, Bokowski and Schuchert 1996). Not a single polyhedral embedding of such a manifold in $\mathbb{R}^3$ is known, and it seems plausible that none exists. However, it was not until 1998 that the first non-embeddability result for such a manifold could be completed: Bokowski and Guedes de Oliveira (1998) proved – using strongly improved methods for the generation of oriented matroids, and many years of CPU time in computer enumerations – that there is no acyclic uniform oriented matroid of rank 4 on 12 points that could support an intersection-free realization of the triangulation with number 54 on Altshuler's list.

Bibliography

Abello, A., Kumar, K. (1995). Visibility graphs and oriented matroids (extended abstract), in: Proc. Graph Drawing (Princeton NJ, 1994), *Lecture Notes in Computer Science* **894**, Springer, Berlin, 147–158.

Aigner, M. (1979). *Combinatorial Theory*, Grundlehren Series, **234**, Springer-Verlag, New York. [101, 337]

Alfter, M., Hochstättler, W. (1995). On pseudomodular matroids and adjoints, *Discrete Applied Math.*, **60**, 3–11.

Alfter, M., Kern, W., Wanka, A. (1990). On adjoints and dual matroids, *J. Combinatorial Theory*, Ser. B, **50**, 208–213. [309]

Allys, L., Las Vergnas, M. (1992). Minors of matroid morphisms, *J. Combinatorial Theory*, Ser. B, to appear? [320]

Alon, N. (1986). The number of polytopes, configurations and real matroids, *Mathematika*, **33**, 62–71. [372]

Alon, N., Györy, E. (1986). The number of small semispaces of a finite set of points in the plane, *J. Combinatorial Theory*, Ser. A, **41**, 154–157. [36]

Altshuler, A. (1977). Neighborly 4-polytopes and neighborly combinatorial 3-manifolds with ten vertices, *Canadian J. Math.*, **29**, 400–420. [414]

Altshuler A., Bokowski, J., Schuchert, P. (1994a). Spatial polyhedra without diagonals, *Israel J. Math.*, **86**, 373–396.

Altshuler, A., Bokowski, J., Schuchert, P. (1994b). Sphere systems and neighborly spatial polyhedra with 10 vertices, in: First International Conference on Stochastic Geometry, Convex Bodies and Empirical Measures, Palermo 1993 (M. Stoka, ed.), *Circolo Matematico di Palermo, Suppl. Rend. Circ. Mat. Palermo, II. Ser.*, **35**, 15–28.

Altshuler, A., Bokowski, J., Schuchert, P. (1996). Neighborly 2-manifolds with 12 vertices, *J. Combinatorial Theory*, Ser. A, **75**, 148–162.   [488]

Altshuler, A., Bokowski, J., Steinberg, L. (1980). The classification of simplicial 3-spheres with nine vertices into polytopes and nonpolytopes, *Discrete Math.*, **31**, 115–124.   [358, 404, 405, 486]

Anderson, L. (1994). Topology of combinatorial differential manifolds, Ph.D. Thesis, MIT, 40 pages.

Anderson, L. (1998). Homotopy groups of the combinatorial Grassmannians, *Discrete Comput. Geometry*, **20**, 549–560.   [483]

Anderson, L. (1999a). Topology of combinatorial differential manifolds, *Topology*, **38**, 197–221.

Anderson, L. (1999b). Matroid bundles, preprint 1998, 21 pages; in: "New Perspectives in Algebraic Combinatorics," MSRI Book Series, Cambridge University Press, to appear.   [483]

Anderson, L. (1999c). Representing weak maps of oriented matroids, preprint, 10 pages.   [323]

Anderson, L., Davis, J.F. (1999). Mod 2 cohomology of the combinatorial Grassmannian, preprint, 38 pages.   [482, 483]

Anderson, L., Wenger, R. (1996). Oriented matroids and hyperplane transversals, *Advances Applied Math.*, **119**, 117–125.

Ando, T. (1987). Totally positive matrices, *Linear Algebra Appl.*, **90**, 165–219. [399]

Antonin, C. (1982). Ein Algorithmusansatz für Realisierungsfragen im E^d getestet an kombinatorischen 3-Sphären, Staatsexamensarbeit, Universität Bochum 1982.   [405]

Arnol'd, V.I. (1979). Indices of singular points of 1-forms on a manifold with boundary, convolution of invariants of reflection groups, and singular projections of smooth surfaces, *Russian Math. Surveys*, **34**, 1–42.   [77]

Aronov, B., Goodman J.E., Pollack R., Wenger, R. (1998). On the Helly number for hyperplane transversals to unit balls, preprint, 13 pages.

Assmann, S.F., Kleitman, D.J. (1984). Characterization of curve map graphs. *Discrete Applied Math.*, **8**, 109–124.   [257]

Athanasiadis, C.A., De Loera, J., Reiner, V., Santos, F. (1999). Fiber polytopes for the projections between cyclic polytopes, preprint, 28 pages; *European J. Combinatorics*, to appear.

Athanasiadis, C.A., Rambau, J., Santos, F. (1999). The generalized Baues problem for cyclic polytopes II, preprint, 12 pages.   [484]

Atiyah, M.F. (1982). Convexity and commuting Hamiltonians, *Bulletin London Math. Soc*, **14**, 1–15.   [91]

Avis, D., Fukuda, K. (1991). A basis enumeration algorithm for linear systems with geometric applications, *Applied Math. Letters*, **4**, 39–42.

Avis, D., Fukuda, K. (1992). A pivoting algorithm for convex hulls and vertex enumeration of arrangements and polyhedra, *Discrete Comput. Geometry*, **8**, 295–313.

Avis, D., Fukuda, K. (1996). Reverse search for enumeration, *Discrete Applied Math.*, **65**, 21–46.

Azaola, M. (1999). The poset of polyhedral subdivisions of $d + 4$ points is a homotopy sphere, preprint in preparation.   [412, 484]

Azaola, M., Santos, F. (1999). The graph of triangulations of $d + 4$ points is 3-connected, preprint, Santander, 47 pages.   [412, 484]

Babson, E.K. (1992). A combinatorial flag space, Ph.D. Thesis, MIT 1992/93, 40 pages.   [78, 296, 483]

Bachem, A. (1983). Convexity and optimization in discrete structures, in: "Convexity and its Appl.," Birkhäuser, 9–29.

Bachem, A., Dress, A.W.M., Wenzel, W. (1992). Five variations on a theme by Gyula Farkas, *Advances Applied Math.*, **13**, 160–185.   [122]

Bachem, A., Kern, W. (1984). Polyhedral theory in oriented matroids, in: "Mathematical Programming" (R.W. Cottle, M.L. Kelmanson, B. Korte, eds.), Proc. Int. Congress, Rio de Janeiro 1981, North Holland, pp. 1–12.

Bachem, A., Kern, W. (1986a). Adjoints of oriented matroids, *Combinatorica*, **6**, 299–308.   [236, 241, 246, 291, 292, 309, 311, 392, 472]

Bachem, A., Kern, W. (1986b). Extension equivalence of oriented matroids, *European J. Combinatorics*, **7**, 193–197.   [246, 291]

Bachem, A., Kern, W. (1989). A guided tour through oriented matroid axioms, preprint 89.64, Universität Köln.

Bachem, A., Kern, W. (1992). *Linear Programming Duality. An Introduction to Oriented Matroids*, Universitext, Springer-Verlag, Berlin–Heidelberg.

Bachem, A., Kern, W. (1993). A guided tour through oriented matroid axioms, *Acta Math. Appl. Sin.*, Engl. Ser. **9**, 125–134.

Bachem, A., Reinhold A. (1989). On the complexity of the Farkas property of oriented matroids, preprint 89.65, Universität Köln.   [433]

Bachem, A., Wanka, A. (1985). On intersection properties of (oriented) matroids, *Methods of Operations Research*, **53**, 227–229.   [309]

Bachem, A., Wanka, A. (1988). Separation theorems for oriented matroids, *Discrete Math.*, **70**, 303–310.   [382]

Bachem, A., Wanka, A. (1989a). Euclidean intersection properties, *J. Combinatorial Theory*, Ser. B, **47**, 10–19.   [309, 310, 337, 472]

Bachem, A., Wanka, A. (1989b). Matroids without adjoints, *Geometriae Dedicata*, **29**, 311–315.

Bacławski, K., White, N. (1979). Higher order independence in matroids, *J. London Math. Society*, **19**, 193–202.

Bailey, G.D. (1997). Tilings of zonotopes: Discriminantal arrangements, oriented matroids, and enumeration, Ph.D. Thesis, U. Minnesota, 108 pages.   [485]

Balinski, M.L. (1961). On the graph structure of convex polyhedra in n-space, *Pacific J. Math.*, **11**, 431–434.   [185, 186]

Barnette, D.W. (1970). Diagrams and Schlegel diagrams, in: Proc. Int. Conf. "Combinatorial Structures and their Applications" (Calgary 1969), Gordon and Breach, New York, pp. 1–4.   [410]

Barnette, D.W. (1973a). A proof of the lower bound conjecture for convex polytopes, *Pacific J. Math.*, **46**, 349–354.   [384]

Barnette, D.W. (1973b). Graph theorems for manifolds, *Israel J. Math.*, **16**, 62–72.   [185]

Barnette, D.W. (1987). Two "simple" 3-spheres, *Discrete Math.*, **67**, 97–99. [25]

Barvinok, A.I. (1988). On the topological properties of spaces of polytopes, in: Viro, O.Ya. (ed.): "Topology and Geometry — Rohlin Seminar," *Lecture Notes in Mathematics*, **1346**, Springer, Heidelberg, pp. 495–500.

Basu, S., Pollack, R., Roy, M.-F. (1996). On the combinatorial and algebraic complexity of quantifier elimination, *J. ACM*, **43**, 1002–1055.

Basu, S., Pollack, R., Roy, M.-F. (1997). On computing a set of points meeting every cell defined by a family of polynomials on a variety, *J. Complexity*, **13**, 28–37.   [357, 486]

Basu, S., Pollack, R., Roy, M.-F. (1998). Complexity of computing semi-algebraic descriptions of the connected components of a semi-algebraic set, in: Proc. 1998 Int. Symp. "Symbolic and Algebraic Computation" (O. Gloor, ed.), ACM, 25–29.

Basu, S., Pollack, R., Roy, M.-F. (1999). Computing roadmaps of semi-algebraic sets on a variety, *J. Amer. Math. Soc.*, to appear.

Bayer, M.M., Lee, C.W. (1993). Combinatorial aspects of convex polytopes, in: *Handbook on Convex Geometry* (eds. P. Gruber, J. Wills), North-Holland, Amsterdam, 484–534.   [200]

Bayer, M.M., Sturmfels, B. (1990). Lawrence polytopes, *Canadian J. Math.*, **42**, 62–79.   [195, 386, 387, 390, 394]

Becker, E. (1986). On the real spectrum of a ring and its applications to semi-algebraic geometry, *Bulletin Amer. Math. Soc.*, **15**, 19–60.   [339, 358]

Below, A., Bokowski, J., Kortenkamp, U.H., Laffaille, G., Richter-Gebert, J. (1999). Classification of the arrangements of 10 pseudolines, preprint in preparation.   [486]

Benedetti, R., Risler, J.-J. (1990). *Real Algebraic and Semi-algebraic Sets*, Hermann, Paris.   [339]

Benson, C.T., Grove, L.C. (1985). *Finite Reflection Groups*, 2nd ed., Springer-Verlag, New York.   [65, 70]

Berenstein, A., Fomin, S., Zelevinsky, A. (1996). Parametrizations of canonical bases and totally positive matrices, *Advances in Math.*, **122**, 49–149.

Berge, C., Las Vergnas, M. (1984). Transversals of circuits and acyclic orientation in graphs and matroids, *Discrete Math.*, **50**, 107–108.   [156]

Bern, M., Eppstein, P., Plassmann, P., Yao, F. (1991). Horizon theorems for lines and polygons, in: "Discrete and Computational Geometry," Papers from the DIMACS Special Year (eds. J.E. Goodman, R. Pollack, W. Steiger), DIMACS Series in Discrete Mathematics and Theoretical Computer Science, **6**, Amer. Math. Soc., Providence, RI, pp. 45–66.   [270]

Bidigare, T.P., Hanlon, P., Rockmore, D.N. (1997). A combinatorial description of the spectrum for the Tsetlin library and its generalization to hyperplane arrangements, *Duke Math. J.*, to appear.   [487]

Bienia, W., Cordovil, R. (1987). An axiomatic of non-Radon partitions of oriented matroids, *European J. Combinatorics*, **8**, 1–4.   [148]

Bienia, W., da Silva, I. P. (1990). On the inversion of one base sign in an oriented matroid, *J. Combinatorial Theory*, Ser. B, **90**, 299–308.   [302]

Bienia, W., Las Vergnas, M. (1990). Positive dependence in oriented matroids, preprint.   [415, 416]

Bienstock, D. (1990). Some provably hard crossing number problems, in: Proc. 6th Annual ACM Symposium on Computational Geometry, (Berkeley, June 1990), pp. 253–260.   [372]

Bieri, H., Schmidt, P.-M. (1991). An on-line algorithm for constructing sweep planes in regular position, in: "Computational Geometry – Methods, Algorithms and Applications" (Bern, 1991), *Lecture Notes in Computer Science*, **553**, Springer, Berlin, 27–35.

Billera, L.J., Brown, K.S., Diaconis, P. (1999). Geometry and probability in three dimensions, *Amer. Math. Monthly*, **106**, in press.   [487]

Billera, L.J., Ehrenborg, R., Readdy, M. (1997). The c-2d-index of oriented matroids, *J. Combinatorial Theory, Ser. A*, **80**, 79–105.   [195]

Billera, L.J., Ehrenborg, R., Readdy, M. (1998). The cd-index of zonotopes and arrangements, in: "Mathematical essays in honor of Gian-Carlo Rota" (B.E. Sagan, R.P. Stanley, eds.), *Progress in Mathematics* Vol. 161, Birkhäuser, Boston, 23–40.

Billera, L.J., Filliman, P., Sturmfels, B. (1990). Constructions and complexity of secondary polytopes, *Advances in Math.*, **83**, 155–179.   [412]

Billera, L.J., Gel'fand, I.M., Sturmfels, B. (1990). Duality and minors of secondary polyhedra, *J. Combinatorial Theory*, Ser. B, **57** 258–268.   [412]

Billera, L.J., Kapranov, M.M., Sturmfels, B. (1994). Cellular strings on polytopes, *Proc. Amer. Math. Soc.*, **122**, 549–555.   [296, 412, 414, 484]

Billera, L.J., Lee, C.W. (1981). A proof of the sufficiency of McMullen's conditions for f-vectors of simplicial polytopes, *J. Combinatorial Theory*, Ser. A, **31**, 237–255.   [394, 414]

Billera, L.J., Munson, B.S. (1984a). Polarity and inner products in oriented matroids, *European J. Combinatorics*, **5**, 293–308.   [242, 386, 391, 392, 394]

Billera, L.J., Munson, B.S. (1984b). Oriented matroids and triangulations of convex polytopes, in: "Progress in Combinatorial Optimization" (Proc. Waterloo Silver Jubilee Conf. 1982), Academic Press, Toronto, 17–37.   [293, 386]

Billera, L.J., Munson, B.S. (1984c). Triangulations of oriented matroids and convex polytopes, *SIAM J. Algebraic Discrete Methods*, **5**, 515–525.   [293, 386, 408, 409]

Billera, L.J., Sturmfels, B. (1992). Fiber polytopes, *Annals of Math.*, **135**, 527–549.   [52, 96]

Billera, L.J., Sturmfels, B. (1994). Iterated fiber polytopes, *Mathematika*, **41**, 348-363.

Bing, R.H. (1964). Some aspects of the topology of 3-manifolds related to the Poincaré conjecture, in: "Lectures on Modern Mathematics," Vol. II (T.L. Saaty, ed.), Wiley, New York, pp. 93–128.   [211]

Bisztriczky, T., Károlyi, G. (1999). Subpolytopes of cyclic polytopes, preprint, 6 pages; *European J. Combinatorics*, to appear.

Björner, A. (1980). Shellable and Cohen-Macaulay partially ordered sets, *Transactions Amer. Math. Soc.*, **260**, 159–183. [157]

Björner, A. (1981). On complements in lattices of finite length, *Discrete Math.*, **36**, 325–326. [216]

Björner, A. (1984a). Posets, regular *CW* complexes and Bruhat order, *Europ. J. Combinatorics*, **5**, 7–16. [157, 212, 213, 223]

Björner, A. (1984b). Orderings of Coxeter groups, in: "Combinatorics and Algebra" (C. Greene, ed.), *Contemporary Math.*, **34**, Amer. Math. Soc., 175–195. [74, 181]

Björner, A. (1984c). Some combinatorial and algebraic properties of Coxeter complexes and Tits buildings, *Advances in Math.*, **52**, 173–212. [178]

Björner, A. (1995). Topological methods, in: Handbook of Combinatorics (R. Graham, M. Grötschel, L. Lovász, eds.), North Holland/Elsevier, Amsterdam, 1819–1872. [200, 216]

Björner, A., Edelman, P.H., Ziegler, G.M. (1990). Hyperplane arrangements with a lattice of regions, *Discrete Comput. Geometry*, **5**, 263–288. [174, 183, 218]

Björner, A., Kalai, G. (1991). Extended Euler-Poincaré relations for cell complexes, in: "Applied Geometry and Discrete Mathematics – The Victor Klee Festschrift" (P. Gritzmann, B. Sturmfels, eds.), DIMACS Series in Discrete Mathematics and Theoretical Computer Science, **4**, Amer. Math. Soc., Providence, RI, pp. 81–89. [199]

Björner, A., Wachs, M. (1983). On lexicographically shellable posets, *Transactions Amer. Math. Soc.*, **277**, 323–341. [157, 207, 223]

Björner, A., Ziegler, G.M. (1988). Shellability of oriented matroids, Abstract, Workshop on *Simplicial Complexes*, Institute for Mathematics and its Applications, University of Minnesota, Minneapolis, March 1988; and Abstract, Conf. on *Ordered Sets*, Oberwolfach, April 1988. [157, 178, 188]

Björner, A., Ziegler, G.M. (1992a). Combinatorial stratification of complex arrangements, *Journal Amer. Math. Soc.*, **5**, 105–149. [94, 95, 98]

Björner, A., Ziegler, G.M. (1992b). Reflections in oriented matroids and finite Coxeter groups, in preparation. [65, 185]

Bland, R.G. (1974). Complementary orthogonal subspaces of $\mathbb{R}^n$ and orientability of matroids, Ph.D. Thesis, Cornell University, 80 pages. [26, 150, 417, 418, 420, 422]

Bland, R.G. (1977a). A combinatorial abstraction of linear programming, *J. Combinatorial Theory*, Ser. B, **23**, 33–57. [26, 417ff: Chapter 10]

Bland, R.G. (1977b). New finite pivoting rules for the simplex method, *Math. Operations Research*, **2**, 103–107. [26, 452]

Bland, R.G. (1980). Linear programming duality and Minty's lemma, preprint, Cornell University, 43 pages.

Bland, R.G., Cho, D.E. (1987). Balancing configurations in $\mathbb{R}^d$ by reflection of points, preprint, Cornell University, 60 pages.

Bland, R.G., Dietrich, B.L. (1987). A unified interpretation of several combinatorial dualities, preprint, Cornell University, 41 pages.

Bland, R.G., Dietrich, B.L. (1988). An abstract duality, *Discrete Math.*, **70**, 203–208.

Bland, R.G., Jensen, D.L. (1987). Weakly oriented matroids, preprint, Cornell University, 40 pages. [156]

Bland, R.G., Ko, C.W., Sturmfels, B. (1993). A nonextremal Camion basis, *Linear Alg. Applications*, **187**, 195–199.

Bland, R.G., Las Vergnas, M. (1978). Orientability of matroids, *J. Combinatorial Theory*, Ser. B, **24**, 94–123. [29, 107ff: Chapter 3, 150, 273, 305, 324, 331, 332, 395]

Bland, R.G., Las Vergnas, M. (1979). Minty colorings and orientations of matroids, *Annals New York Acad. Sci.*, **319**, 86–92. [119, 479, 450, 451]

Blass, A., Sagan, B.E. (1986). Bijective proofs of two broken circuit theorems, *J. Graph Theory*, **10**, 15–21. [223]

Blind, G., Blind, R. (1990). Convex polytopes without triangular faces, *Israel J. Math.*, **71**, 129–134. [97]

Blind, R., Mani, P. (1987). On puzzles and polytope isomorphism, *Aequationes Math.*, **34**, 287–297. [60]

Blumenthal, L.M. (1953). *Theory and Applications of Distance Geometry*, Oxford University Press, reprinted by Chelsea, New York (1970).

Bohne, J. (1992a). Eine kombinatorische Analyse zonotopaler Raumaufteilungen, Dissertation, Bielefeld; Preprint 92-041, SFB 343, U. Bielefeld, 100 pages. [63, 485]

Bohne, J. (1992b). A characterization of oriented matroids in terms of conformal sequences, *Bayreuther Math. Schriften*, **40**, 1–5.

Bohne, J., Dress, A.W.M., Fischer, S. (1989). A simple proof for de Bruijn's dualization principle, in: "Some elementary proofs for some elementary facts in algebra and geometry," Preprint 89-027, SFB 343, Bielefeld, 8 pages; in: Proc. Raj Chandra Bose Memorial Conference "Combinatorial Mathematics and Applications," Calcutta, India, *Sankhya*, Ser A, **54**, 77–84. [52]

Bokowski, J. (1978). A non-polytopal sphere, preprint, 4 pages. [405]

Bokowski, J. (1985). Geometrische Realisierbarkeitsfragen – Chirotope und orientierte Matroide, in: Proc. 3rd Kolloq. "Discrete Geometrie" (Salzburg 1985), 53–57.

Bokowski, J. (1987). Aspects of computational synthetic geometry; II. Combinatorial complexes and their geometric realization – an algorithmic approach, in: "Computer-aided Geometric Reasoning" (H.H. Crapo, ed.), INRIA Rocquencourt, France, June 1987.

Bokowski, J. (1989). A geometric realization without self-intersections does exist for Dyck's regular map, *Discrete Comput. Geometry*, **4**, 583–589. [488]

Bokowski, J. (1990). On the generation of oriented matroids with prescribed topes, Preprint 1291, TH Darmstadt, 11 pages.

Bokowski, J. (1991a). On the Las Vergnas conjecture concerning simplicial cells in pseudo-plane arrangements, preprint, 5 pages. [483]

Bokowski, J. (1991b). On the geometric flat embedding of abstract complexes with symmetries, in: "Symmetry of discrete mathematical structures and their symmetry groups, a collection of essays" (K.-H. Hoffmann, R. Wille, eds.) *Research and Exposition in Mathematics*, Vol. 15, Heldermann Verlag, Berlin, pp. 1–48.

Bokowski, J. (1993a). Oriented matroids, in: Handbook on Convex Geometry (P. Gruber, J. Wills, eds.), North-Holland, Amsterdam, 555–602.

Bokowski, J. (1993b). Computational synthetic geometry. Computer aided decisions for finding polyhedral embeddings, *Rendiconti Semin. Mat. Messina*, Ser. II **17**, No. 2, Suppl., 1-7.

Bokowski, J. (1994). On recent progress in computational synthetic geometry, in: "Polytopes: Abstract, Convex and Computational" (T. Bisztriczky, P. McMullen, and A. Weiss, eds.), Proc. NATO Advanced Study Institute, Toronto 1993, Kluwer Academic Publishers, pp. 335–358.

Bokowski, J. (1995). On the construction of equifacetted 3-spheres, in: Proc. Conf. "Invariant Methods in Discrete and Computational Geometry," Williamstadt, Curaçao 1994 (N. White, ed.), Kluwer Academic Publishers, Dordrecht, 301–312.

Bokowski, J. (1996). Finite point sets and oriented matroids: Combinatorics in geometry, in: Proc. "Learning and Geometry: Computational Approaches," College Park MD 1991 (D.W. Kuecker, C.H. Smith, eds.), Progress in Computer Science and Applied Logic, Vol. 14, Birkhäuser Boston, pp. 67–96.

Bokowski, J., Brehm, U. (1985). A new polyhedron of genus 3 with 10 vertices, in: Papers Int. Conf. "Intuitive Geometry" (K. Böröczky, G. Fejes Tóth, eds.), Siófok/Hungary 1985, *Colloquia Math. Soc. János Bolyai*, **48**, 105–116.

Bokowski, J., Eggert, A. (1991). All realization of Möbius' torus with 7 vertices, *Structural Topology*, **17**, 59–78.   [488]

Bokowski, J., Ewald, G., Kleinschmidt, P. (1984). On combinatorial and affine automorphisms of polytopes, *Israel J. Math.*, **47**, 123–130.   [482]

Bokowski, J., Garms, K. (1987). Altshuler's sphere M_{425} is not polytopal, *European J. Combinatorics*, **8**, 227–229.   [358]

Bokowski, J., Guedes de Oliveira, A. (1990). Simplicial convex 4-polytopes do not have the isotopy property, *Portugaliae Mathematica*, **47**, 309–318.   [407, 482]

Bokowski, J., Guedes de Oliveira, A. (1994). Invariant theory-like theorems for matroids and oriented matroids, *Advances in Math.*, **109**, 34–44.

Bokowski, J., Guedes de Oliveira, A. (1998). On the generation of oriented matroids, preprint, 15 pages.   [488]

Bokowski, J., Guedes de Oliveira, A., Richter, J. (1991). Algebraic varieties characterizing matroids and oriented matroids, *Advances in Math.*, **87**, 160–185.

Bokowski, J., Guedes de Oliveira, A., Thiemann, U., Veloso da Costa, A. (1996). On the cube problem of Las Vergnas, *Geometriae Dedicata*, **63**, 25–43.

Bokowski, J., Kollewe, W. (1992). On representing contexts in line arrangements, *Order*, **8**, 393–403.

Bokowski, J., Mock, S., Streinu, I. (1999). The Folkman Lawrence Theorem of oriented matroids. An elementary proof in the rank 3 case, preprint in preparation.   [277]

Bokowski, J., Richter, J. (1990a). On the finding of final polynomials, *European J. Combinatorics*, **11**, 21–34.   [358, 360, 363, 486]

Bokowski, J., Richter-Gebert, J. (1990b). On the classification of non-realizable oriented matroids, Part I: Generation, Preprint 1283, TH Darmstadt, 17 pages. [336, 350, 465]

Bokowski, J., Richter-Gebert, J. (1990c). On the classification of non-realizable oriented matroids, Part II: Properties, preprint, TH Darmstadt, 22 pages.   [465]

Bokowski, J., Richter-Gebert, J. (1992). A new Sylvester-Gallai configuration representing the 13-point projective plane in $\mathbb{R}^4$, *J. Combinatorial Theory*, Ser. B, **54**, 161–165. [272]

Bokowski, J., Richter-Gebert, J., Schindler, W. (1992). On the distribution of order types, *Computational Geometry: Theory and Applications*, **1**, 127–142.

Bokowski, J., Richter, J., Sturmfels, B. (1990). Nonrealizability proofs in computational geometry, *Discrete Comput. Geometry*, **5**, 333–350.

Bokowski, J., Roudneff, J.-P., Strempel, T.-K. (1997). Cell decompositions of the projective plane with Petrie polygons of constant length, *Discrete Comput. Geometry*, **17**, 377–392. [488]

Bokowski, J., Schuchert, P. (1995). Altshuler's sphere M_{963}^9 revisited, *SIAM J. Discrete Math.*, **8**, 670–677. [413]

Bokowski, J., Schuchert, P. (1995). Equifacetted 3-spheres as topes of nonpolytopal matroid polytopes, in: "The László Fejes Tóth Festschrift" (I. Bárány, J Pach, eds.), *Discrete Comput. Geometry*, **13**, 347–361.

Bokowski, J., Shemer, I. (1987). Neighborly 6-polytopes with 10 vertices, *Israel J. Math.*, **58**, 103–124. [486]

Bokowski, J., Sturmfels, B. (1985a). Problems of geometrical realizability – oriented matroids and chirotopes, Preprint 901, TH Darmstadt, 20 pages.

Bokowski, J., Sturmfels, B. (1985b). Programmsystem zur Realisierung orientierter Matroide, *Preprints in Optimization*, Universität Köln, Programmdokumentation 85.22, 33 pages. [464]

Bokowski, J., Sturmfels, B. (1986a). On the coordinatization of oriented matroids, *Discrete Comput. Geometry*, **1**, 293–306. [355, 356]

Bokowski, J., Sturmfels, B. (1986b). Reell realisierbare orientierte Matroide, *Bayreuther Math. Schriften*, **21**, 1–13.

Bokowski, J., Sturmfels, B. (1987). Polytopal and nonpolytopal spheres – an algorithmic approach, *Israel J. Math.*, **57**, 257–271. [358, 391, 403, 404, 414, 486]

Bokowski, J., Sturmfels, B. (1989a). An infinite family of minor-minimal nonrealizable 3-chirotopes, *Math. Zeitschrift*, **200**, 583–589. [351]

Bokowski, J., Sturmfels, B. (1989b). *Computational Synthetic Geometry*, *Lecture Notes in Mathematics*, **1355**, Springer, Heidelberg. [80, 89, 354, 363, 407, 408, 486]

Borovik, A.V., Gelfand, I.M. (1994). WP-matroids and thin Schubert cells on Tits systems, *Advances in Math.*, **103**, 162–179.

Borovik, A.V., Gelfand, I.M., White, N.L. (1996). Boundaries of Coxeter matroids, *Advances in Math.*, **120**, 258–264.

Borovik, A.V., Gelfand, I.M., White, N.L. (1998). On exchange properties for Coxeter matroids and oriented matroids, *Discrete Math.*, **179**, 59–72.

Bourbaki, N. (1968). *Groupes et Algèbres de Lie*, Chap. 4,5, et 6, Hermann, Paris.   [65, 70, 71, 72, 74]

Brehm, U. (1987). Maximally symmetric polyhedral realizations of Dyck's regular map, *Mathematika*, **34**, 229–236.   [488]

Brown, K.S. (1989). *Buildings*, Springer, New York.   [65, 69, 70, 76]

Brown, K.S. (1999). Semigroups, rings, and Markov chains, preprint, 49 pages; *J. Theoretical Probability*, to appear.

Brown, K.S., Diaconis, P. (1998). Random walks and hyperplane arrangements, preprint, Cornell, 39 pages; *Annals Probability*, to appear.   [487]

Bruggesser, H., Mani, P. (1971). Shellable decompositions of cells and spheres, *Math. Scand.*, **29**, 197–205.   [178, 205]

Brylawski, T.H., Ziegler, G.M. (1993). Topological representation of dual pairs of oriented matroids, Special issue on "Oriented Matroids" (J. Richter-Gebert, G. M. Ziegler eds), *Discrete Comput. Geometry*, **10**, 237–240.

Buchi, J.R., Fenton, W.E. (1988). Large convex sets in oriented matroids, *J. Combinatorial Theory*, Ser. B, **45**, 293–304.   [152, 153]

Buck, R.C. (1943). Partitions of space, *Amer. Math. Monthly*, **50**, 541–544.   [221]

Camion, P. (1963). Matrices totalement unimodulaires et problèmes combinatoires, Thesis, Univ. Brussels.

Camion, P. (1968). Modules unimodulaires, *J. Combinatorial Theory*, **4**, 301–362.   [49, 305]

Canham, R.J. (1971). Ph.D. Thesis, Univ. of East Anglia, Norwich.

Canny, J. (1988). Some algebraic and geometric computations in PSPACE, in: Proc. 20th ACM Symposium on Theory of Computing (Chicago, May 1988).   [357]

Cappell, S. E., Goodman, J.E., Pach, J., Pollack, R., Sharir, M., Wenger, R. (1990). The combinatorial complexity of hyperplane transversals, in: Proc. 6th Annual ACM Symposium on Computational Geometry (Berkeley, June 1990), pp. 83–91.

Cappell, S.E., Goodman, J.E., Pach, J., Pollack, R., Sharir, M., Wenger, R. (1994). Common tangents and common transversals, *Advances Math.*, **106**, 198–215.

Carathéodory, C. (1907). Über den Variabilitätsbereich der Koeffizienten von Potenzreihen, die gegebene Werte nicht annehmen, *Math. Annalen*, **64**, 95–115; reprinted in: *Gesammelte Mathematische Schriften, Band III* (H. Tietze, ed.), C. H. Beck'sche Verlagsbuchhandlung, München 1955, pp. 54–77. [399]

Carathéodory, C. (1911). Über den Variabilitätsbereich der Fourier'schen Konstanten von positiven harmonischen Funktionen, *Rendiconto del Circolo Matematico di Palermo* **32**, 193–217; reprinted in: *Gesammelte Mathematische Schriften, Band III* (H. Tietze, ed.), C. H. Beck'sche Verlagsbuchhandlung, München 1955, pp. 78–110. [399]

Chaiken, S. (1995). Oriented matroid pairs, theory and an electric application, in: "Matroid Theory" (J. E. Bonin, J. G. Oxley, B. Servatius, eds.), *Contemporary Math.*, **197**, Amer. Math. Soc., 313–331.

Chakerian, G.D. (1970). Sylvester's problem on collinear points and a relative, *American Math. Monthly*, **77**, 164–167. [249, 250, 272]

Cheung, A.L.C. (1974). Adjoints of a geometry, *Canadian Math. Bulletin*, **17**, 363–365. [241, 291, 311]

Cheung, A.L.C., Crapo, H.H. (1976). A combinatorial perspective on algebraic geometry, *Advances in Math.*, **20**, 388–414. [29]

Chvátal, V. (1983). *Linear Programming*, Freeman, New York. [433]

Clausen, J. (1987). A note on the Edmonds-Fukuda pivoting rule for simplex algorithms, *European J. Operational Research*, **29**, 378–383. [458]

Clausen, J., Terlaky T., (1987). On the feasibility of the Edmonds-Fukuda pivoting rule for oriented matroid programming, preprint, 16 pages.

Collins, C. (1975). Quantifier elimination for real closed fields by cylindrical algebraic decomposition, in: "Automata Theory and Formal Languages" (H. Brakhage, ed.), *Lecture Notes in Computer Science*, **33**, Springer, Heidelberg, 134–163. [357]

Cooke, G.E., Finney, R.L. (1967). *Homology of Cell Complexes*, Princeton University Press. [202]

Cordovil, R. (1980a). Sur les orientations acycliques des géométries orientées de rang 3, Actes du *Colloque International sur la Theorie des Graphes et la Combinatoire* (Marseilles 1981), *Annals Discrete Math.*, **9**, 243–246.

Cordovil, R. (1980b). Sur l'évaluation $t(M; 2, 0)$ du polynôme de Tutte d'un matroïde et une conjecture de B. Grünbaum relative aux arrangements de droites du plan, *European J. Combinatorics*, **1**, 317–322.

Cordovil, R. (1981). Quelques propriétés algébriques des matroïdes, Thèse, Université Paris VI, 167 pages.

Cordovil, R. (1982a). Sur les matroïdes orientés de rang 3 et les arrangements de pseudodroites dans le plan projectif réel, *European J. Combinatorics*, **3**, 307–318. [170, 236, 263, 275]

Cordovil, R. (1982b). Sur un théorème de séparation des matroïdes orientés de rang 3, *Discrete Math.*, **40**, 163–169.   [383]

Cordovil, R. (1983a). The directions determined by n points in the plane: a matroidal generalization, *Discrete Math.*, **43**, 131–137.

Cordovil, R. (1983b). Oriented matroids of rank three and arrangements of pseudolines, *Annals Discrete Math.*, **17**, 219–223.   [181]

Cordovil, R. (1983c). Oriented matroids and geometric sorting, *Canadian Math. Bulletin*, **26**, 351–354.   [373]

Cordovil, R. (1985). A combinatorial perspective on non-Radon partitions, *J. Combinatorial Theory*, Ser. A, **38**, 38–47. (Erratum, ibid., **40**, 194.)   [148, 173]

Cordovil, R. (1986). On the number of lines determined by n points, preprint, 16 pages.

Cordovil, R. (1987). Polarity and point extensions in oriented matroids, *Linear Algebra and its Appl.*, **90**, 15–31.   [246, 310, 382]

Cordovil, R. (1994a). On the homotopy of the Salvetti complexes determined by simplicial arrangements, *European J. Combinatorics*, **15**, 207–215.   [95]

Cordovil, R. (1994b). On the center of the fundamental group of the complement of an arrangement of hyperplanes, *Portugaliae Math.*, **51**, 363–373.

Cordovil, R. (1998). The fundamental group of the complement of the complexification of a real arrangement of hyperplanes, *Advances Applied Math.*, **21**, 481–498.

Cordovil, R. (1999). A commutative algebra for oriented matroids, preprint, 9 pages.

Cordovil, R., da Silva, I.P. (1985). A problem of McMullen on the projective equivalence of polytopes, *European J. Combinatorics*, **6**, 157–161.   [385]

Cordovil, R., da Silva, I.P. (1987). Determining a matroid polytope by non-Radon partitions, *Linear Algebra and its Appl.*, **94**, 55–60.

Cordovil, R., Duchet, P. (1986). Séparation par une droite dans les matroïdes orientés de rang 3, *Discrete Math.*, **62**, 103–104.

Cordovil, R., Duchet, P. (1987). Cyclic polytopes and oriented matroids, preprint, 17 pages; revised version 1998, 24 pages, *European J. Combinatorics*, to appear. [396, 398]

Cordovil, R., Duchet, P. (1990). On the sign invariance graphs of uniform oriented matroids, *Discrete Math.*, **79**, 251–257. [324, 325, 326]

Cordovil, R., Fachada, J.L. (1995). Braid monodromy groups and wiring diagrams, *Bolletino U. M. I.*, **9-B**, 399–416. [95]

Cordovil, R., Fukuda, K. (1993). Oriented matroids and combinatorial manifolds, *European J. Combinatorics*, **14**, 9–15. [185, 186]

Cordovil, R., Fukuda, K., Guedes de Oliveira, A. (1991). On the cocircuit graph of an oriented matroid, preprint, 14 pages; revised version, ETH Zürich 1998. [186]

Cordovil, R., Guedes de Oliveira, A. (1992). A note on the fundamental group of the Salvetti complex determined by an oriented matroid, *European J. Combinatorics*, **13**, 429–437.

Cordovil, R., Guedes de Oliveira, A., Las Vergnas, M. (1996). A generalized Desargues configuration and the pure braid group, *Discrete Math.*, **160**, 105–113. [95]

Cordovil, R., Guedes de Oliveira, A., Moreira, M.L. (1988). Parallel projection of matroid spheres, *Portugaliae Mathematica*, **45**, 337–346.

Cordovil, R., Las Vergnas, M., Mandel, A. (1982). Euler's relation, Möbius functions and matroid identities, *Geometriae Dedicata*, **12**, 147–162. [199]

Cordovil, R., Moreira, M.L. (1993). A homotopy theorem on oriented matroids, *Discrete Math.*, **111**, 131–136. [182, 184, 185]

Courcelle, B., Olive, F. (1998). Une axiomatisation au premier ordre des arrangements de pseudodroites euclidiennes, preprint, 18 pages; *Annales Institut Fourier*, to appear.

Coxeter, H.S.M. (1935). The complete enumeration of finite groups of the form $R_i^2 = (R_iR_j)^{k_{ij}} = 1$, *J. London Math. Soc.*, **10**, 21–25. [70]

Coxeter, H.S.M. (1962). The classification of zonohedra by means of projective diagrams, *J. de Math. Pures Appl.*, **41**, 137–156. [52, 55, 63]

Coxeter, H.S.M. (1973). *Regular Polytopes* (third edition), Dover, New York. [70, 71, 185]

Crapo, H.H. (1965). Single-element extensions of matroids, *J. Research Nat. Bureau Standards*, **69B**, 55–65. [281, 282, 283, 285]

Crapo, H.H., Laumond, J.-P. (1989). Hamiltonian cycles in Delaunay complexes, in: "Geometry and Robotics" (J.-D. Boissonnat et J.-P. Laumond, eds.), *Lecture Notes in Computer Science*, **391**, Springer, Heidelberg, pp. 292–305.   [33]

Crapo, H.H., Penne, R. (1994). Chirality and the isotopy classification of skew lines in projective 3-space, *Advances in Math.*, **103**, 1–106.

Crapo, H., Richter-Gebert, J. (1995). Automatic proving of geometric theorems, in: Proc. Conf. "Invariant Methods in Discrete and Computational Geometry," Williamstadt, Curaçao 1994 (N. White, ed.), Kluwer Academic Publishers, Dordrecht, 167–196.   [486]

Crapo, H., Richter-Gebert, J. (1992-95). CINDERELLA Computer Interactive Drawing Environment, NeXTStep software project.

Crapo, H.H., Rota, G.-C. (1970). *Combinatorial Geometries (Preliminary Edition)*, MIT Press.   [282, 284]

Crapo, H., Senechal, M. (1995). Tilings by related zonotopes, preprint, 14 pages; *International J. Mathematical and Computer Modelling*, to appear.   [52, 485]

Crippen, G.M., Havel, T.F. (1988). *Distance Geometry and Molecular Conformation*, Research Studies Press, Taunton, England.   [33]

Császár, A. (1949/50). A polyhedron without diagonals, *Acta Scient. Math. (Szeged)*, **13**, 140–142.   [488]

Csima, J., Sawyer, E.T. (1993). There exist $6n/13$ ordinary points, *Discrete Comput. Geometry*, **9**, 187–202.   [272]

Danaraj, G., Klee, V. (1974). Shellings of spheres and polytopes, *Duke Math. J.*, **41**, 443–451.   [206, 211]

Dantzig, G.B. (1963). *Linear Programming and Extensions*, Princeton University Press.   [428, 433]

da Silva, I.P. (1987). Quelques propriétés des matroïdes orientés, Dissertation, Université Paris VI, 131 pages.   [156, 173, 396]

da Silva, I.P. (1988). An axiomatic for the set of maximal vectors of an oriented matroid based on symmetry properties of the set of vectors, preprint, 18 pages. [148]

da Silva, I.P.F. (1993a). On fillings of $2N$-gons with rhombi, *Discrete Math.*, **111**, 137–144.

da Silva, I.P.F. (1993b). On inseparability graphs of matroids having exactly one class of orientations, preprint, 7 pages.

da Silva, I.P.F. (1994). An intersection property defining series-parallel networks, preprint, 30 pages.

da Silva, I.P. (1995). Axioms for maximal vectors of an oriented matroid, a combinatorial characterization for the regions of an arrangement of pseudohyperplanes, *European J. Combinatorics*, **16**, 125–145. [156]

Daverman, R.J. (1986). *Decompositions of Manifolds*, Academic Press, Orlando, FL. [210]

de Berg, M., van Kreveld, M., Overmars, M., Schwarzkopf, O. (1997). Computational Geometry. Algorithms and Applications, Springer-Verlag, Berlin Heidelberg. [29]

Deligne, P. (1972). Les immeubles des groupes de tresses généralisés, *Inventiones Math.*, **17**, 273–302. [95, 169, 182, 184]

De Loera, J.A. (1995). Triangulations of polytopes and computational algebra, Ph.D. Thesis, Cornell University, 178 pages.

De Loera, J.A. (1996). Nonregular triangulations of products of simplices, *Discrete Comput. Geometry*, **15**, 253–264.

De Loera, J.A. (1999). Computing minimal and maximal triangulations of convex polytopes, preprint, 18 pages. [484]

De Loera, J.A., Hoşten, S., Santos, F., Sturmfels, B. (1996). The polytope of all triangulations of a point configuration, *Documenta Mathematica*, **1**, 103–119. [484]

De Loera, J.A., Morris, W.D. jr. (1997). *Q*-matrix recognition via secondary and universal polytopes, preprint, 24 pages; *Math. Programming*, to appear.

De Loera, J.A., Santos, F., Urrutia, J. (1999). The number of geometric bistellar neighbors of a triangulation, *Discrete Comput. Geometry*, **21**, 131–142. [484]

Delsarte, P., Kamp, Y. (1989). Low rank matrices with a given sign pattern, *SIAM J. Discrete Math.*, **2**, 51–63.

Deza, M., Fukuda, K. (1986). On bouquets of matroids and orientations, Publ. R.I.M.S., Kyoto University, Kokyuroku, **587**, 110–129.

Diaconis, P. (1998). From shuffling cards to walking around the building: An introduction to modern Markov chain theory, *Documenta Math. J. DMV*, Extra Vol. ICM 1998, Vol. I, 187–204. [487]

Dillencourt, M.B. (1989). An upper bound on the shortness exponent of inscribable polytopes, *J. Combinatorial Theory*, Ser. B, **46**, 66–83. [33]

Dreiding, A., Dress, A.W.M., Haegi, H. (1982). Classification of mobile molecules by category theory, *Studies in Physical and Theoretical Chemistry*, **23**, 39–58. [33]

Dreiding, A., Wirth, K. (1980). The multiplex. A classification of finite ordered point sets in oriented d-dimensional space, *Math. Chemistry*, **8**, 341–352.   [33]

Dress, A.W.M. (1986a). Duality theory for finite and infinite matroids with coefficients, *Advances in Math.*, **59**, 97–123.

Dress, A.W.M. (1986b). Chirotopes and oriented matroids. *Bayreuther Math. Schriften*, **21**, 14–68.

Dress, A.W.M., Scharlau R. (1987). Gated sets in metric spaces, *Aequationes Mathematicae*, **34**, 112–200.   [217]

Dress, A.W.M., Wenzel, W. (1988). Endliche Matroide mit Koeffizienten, *Bayreuther Math. Schriften*, **26**, 37–98.

Dress, A.W.M., Wenzel, W. (1989). Geometric algebra for combinatorial geometries, *Advances in Math.*, **77**, 1–36.   [330]

Dress, A.W.M., Wenzel, W. (1990). On combinatorial and projective geometry, *Geometriae Dedicata*, **34**, 161–197.

Dress, A.W.M., Wenzel, W. (1991). Grassmann-Plücker relations and matroids with coefficients, *Advances in Math.*, **86**, 68–110.

Dress, A.W.M., Wenzel, W. (1992a). Perfect matroids, *Advances in Math.*, **91**, 158–208.

Dress, A.W.M., Wenzel, W. (1992b). Valuated matroids, *Advances in Math.*, **93**, 214–250.

Duchet, P. (1987). Convexity in combinatorial structures, in: "Abstract analysis," Proc. 14th Winter School, Srni/Czech. 1986, *Suppl. Rend. Circ. Mat. Palermo* II. Ser., **14**, 261–293.

Duke, R.A. (1970). Geometric embeddings of complexes, *Amer. Math. Monthly*, **77**, 597–603.   [488]

Edelman, P.H. (1980). Meet-distributive lattices and the anti-exchange closure, *Algebra Universalis*, **10**, 290–299.   [376]

Edelman, P.H. (1982). The lattice of convex sets of an oriented matroid, *J. Combinatorial Theory*, Ser. B, **33**, 239–244.   [152, 376]

Edelman, P.H. (1984a). The acyclic sets of an oriented matroid, *J. Combinatorial Theory*, Ser. B, **36**, 26–31.   [180, 194, 219]

Edelman, P.H. (1984b). A partial order on the regions of $\mathbb{R}^n$ dissected by hyperplanes, *Trans. Amer. Math. Soc.*, **283**, 617–631.   [172, 181, 182, 183]

Edelman, P.H., Greene, C. (1987). Balanced tableaux, *Advances in Math.*, **63**, 42–99.   [270]

Edelman, P.H., Jamison, R.E. (1985). The theory of convex geometries, *Geometriae Dedicata*, **19**, 247–270.   [377]

Edelman, P.H., Rambau, R., Reiner, V. (1997). On subdivision posets of cyclic polytopes, MSRI Preprint 1997-030; *European J. Combinatorics*, to appear. [484]

Edelman, P.H., Reiner, V. (1996a). Free arrangements and rhombic tilings, *Discrete Comput. Geometry*, **15**, 307–340; erratum, **17** (1997), 359.   [485]

Edelman, P.H., Reiner, V. (1996b). The higher Stasheff-Tamari posets, *Mathematika*, **43**, 127–154.

Edelman, P.H., Walker, J.W. (1985). The homotopy type of hyperplane posets, *Proc. Amer. Math. Soc.*, **94**, 329–332.   [181, 216]

Edelsbrunner, H. (1987). *Algorithms in Computational Geometry*, Springer-Verlag, Berlin.   [29]

Edelsbrunner, H., Guibas, L.J. (1989). Topologically sweeping an arrangement, *J. Computer and System Sciences*, **38**, 165–194.   [472]

Edelsbrunner, H., Mücke, E.P. (1988). Simulation of simplicity: A technique to cope with degenerate cases in geometric algorithms, in: Proc. Fourth Annual ACM Symposium on Computational Geometry, pp. 118–133.   [371]

Edelsbrunner, H., O'Rourke, J., Seidel, R. (1986). Constructing arrangements of lines and hyperplanes with applications, *SIAM J. Computing*, **15**, 341–363. [270, 307, 373, 472]

Edelsbrunner, H., Seidel, R. (1986). Voronoi diagrams and arrangements, *Discrete Comput. Geometry*, **1**, 25–44.   [29]

Edelsbrunner, H., Seidel, R., Sharir, M. (1993). On the zone theorem for hyperplane arrangements, *SIAM J. Computing*, **22**, 418–429.   [308, 333]

Edelsbrunner, H., Welzl, E. (1985). On the number of line separations of a finite set in the plane, *J. Combinatorial Theory*, Ser. A, **38**, 15–29.   [36]

Edmonds, J. (1970). Submodular functions, matroids and certain polyhedra, in: "Combinatorial Structures and their Applications," (H. Hanani, N. Sauer and J. Schönheim, eds.) Gordon and Breach, New York, pp. 69–87.   [91]

Edmonds, J., Fukuda, K. (1982). Oriented matroid programming, Ph.D. Thesis of K. Fukuda, University of Waterloo, 223 pages.   [144, 147, 296, 297, 417ff: Chapter 10]

Edmonds, J., Lovász, L., Mandel, A. (1980). Solution, *Math. Intelligencer*, **2**, 107.   [249]

Edmonds, J., Mandel, A. (1978). Topology of oriented matroids, Abstract 758-05-9, *Notices Amer. Math. Soc.*, **25**, A-510.   [175, 200]

Edmonds, J., Mandel, A. (1982). Topology of oriented matroids, Ph.D. Thesis of A. Mandel, University of Waterloo, 333 pages.   [19, 41, 49, 141, 143, 144, 145, 146, 157ff: Chapters 4 and 5, 272, 277, 290, 296ff: Sect. 7.2, 311, 351, 418ff: Chapter 10]

Elnitzky, S. (1993). Rhombic tilings of polygons and classes of reduced words in Coxeter groups, Ph. D. Thesis, University of Michigan, 1993.   [485]

Elnitzky, S. (1997). Rhombic tilings of polygons and classes of reduced words in Coxeter groups, *J. Combinatorial Theory* Ser. A, **77**, (1997), 193–221.   [485]

Erdős, P., Purdy, G. (1995). Some extremal problems in combinatorial geometry, in: Handbook of Combinatorics (R. Graham, M. Grötschel, L. Lovász, eds.), North-Holland/Elsevier, Amsterdam, 809–874.   [39, 272]

Erdős, P., Szekeres, G. (1935). A combinatorial problem in geometry, *Compositio Math.*, **2**, 463–470.   [398]

Etienne, G., Las Vergnas, M. (1998). External and internal elements of a matroid basis, *Discrete Math.*, **179**, 111–119.

Ewald, G., Kleinschmidt, P., Pachner, U., Schulz, C. (1979). Neuere Entwicklungen in der kombinatorischen Konvexgeometrie, in: "Contributions to Geometry" (J. Tölke, J. Wills, eds.), Proc. Geometry Symposium Siegen 1978, Birkhäuser, Basel, pp. 131–163.   [404]

Faigle, U. (1985). Orthogonal sets, matroids, and theorems of the alternative, *Bolletino U.M.I.*, **4-B**, 139–153.

Falk, M.J. (1983). Geometry and topology of hyperplane arrangements, Ph.D. Thesis, University of Wisconsin, Madison, 103 pages.

Falk, M.J. (1990). On the algebra associated with a geometric lattice, *Advances in Math.*, **80**, 152–163.   [95]

Falk, M.J. (1993). Homotopy types of line arrangements, *Inventiones Math.*, **111**, 139–150.

Felsner, S. (1996). On the number of arrangements of pseudolines, in: Proc. 12th Annual ACM Symposium on Computational Geometry, pp. 30–37.   [270]

Felsner, S. (1997). On the number of arrangements of pseudolines, *Discrete Comput. Geometry*, **18**, 257–267.   [270]

Felsner, S., Kriegel, K. (1998). Triangles in Euclidean arrangements, preprint, FU Berlin, 10 pages; *Discrete Comput. Geometry*, to appear.

Felsner, S., Weil, H. (1998). Sweeps, arrangements and signotopes, Preprint B 98-6, FU-Berlin. [485]

Felsner, S., Weil, H. (1999). A theorem on higher Bruhat orders, *Discrete Comput. Geometry*, to appear.

Felsner, S., Ziegler, G.M. (1999). Zonotopes associated with higher Bruhat orders, preprint, TU and FU Berlin, 11 pages. [42]

Fenchel, W. (1983). Convexity through the ages, in: "Convexity and its Applications" (P. Gruber, J. Wills, eds.), Birkhäuser, Basel, pp. 120–130. [250]

Fenton, W.E. (1982). Axiomatic of convexity theory, Ph.D. Thesis, Purdue Univ., 98 pages.

Fenton, W.E. (1987). Completeness in oriented matroids, *Discrete Math.*, **66**, 79–89.

Finashin, S.M. (1988). Configurations of seven points in $\mathbb{P}^3$, in: Viro, O.Ya. (ed.): "Topology and Geometry – Rohlin Seminar," *Lecture Notes in Mathematics*, **1346**, Springer, Heidelberg, pp. 501–526.

Finschi, L., Fukuda, K., Lüthi, H.-J. (1999). Towards a unified framework for randomized pivoting algorithms in linear programming, in: "Operations Research Proceedings 1998," 113–122.

Firla, R.T., Ziegler, G.M. (1999). Hilbert bases, unimodular triangulations, and binary covers of rational polyhedral cones, *Discrete Comput. Geometry*, **21**, 205–216. [484]

Folkman, J., Lawrence, J. (1978). Oriented matroids, *J. Combinatorial Theory*, Ser. B, **25**, 199–236. [17, 19, 107, 150, 152, 157ff: Chapters 4 and 5, 253, 275, 428, 490]

Forge, D., Ramírez Alfonsín, J.L. (1998a). Straight line arrangements in the real projective plane, *Discrete Comput. Geometry*, **20**, 155–161. [279]

Forge, D., Ramírez Alfonsín, J.L. (1998b). Connected coverings and an application to oriented matroids, *Discrete Math.*, **81**, 109–122. [154]

Forge, D., Ramírez Alfonsín, J.L. (1998c). On reconstructing arrangements from their sets of simplices, preprint, 24 pages.

Forge, D., Ramírez Alfonsín, J.L. (1998d). On counting the k-face cells of cyclic arrangements, preprint, 15 pages.

Forge, D., Ramírez Alfonsín, J.L. (1999). Simplices of the union of oriented matroids, preprint, 13 pages.

Forge, D., Schuchert, P. (1996). A set of 10 points in dimension 4 not projectively equivalent to the vertices of a convex polytope, preprint, 6 pages. [385]

Fukuda, K. (1982). – see Edmonds and Fukuda (1982).

Fukuda, K. (1986). Oriented matroids and linear programming, (in Japanese), Proceedings of the 15th Symposium of the Operations Research Society of Japan, pp. 8–14.

Fukuda, K., Handa, K. (1985). Perturbations of oriented matroids and acycloids, preprint, 18 pages.

Fukuda, K., Handa, K. (1993). Antipodal graphs and oriented matroids, *Discrete Math.*, **111**, 245–256.   [174]

Fukuda, K., Matsui, T. (1989). Elementary inductive proofs for linear programming, In *Kokyuroku* **680**, Research Institute for Mathematical Sciences, Kyoto University.   [428]

Fukuda, K., Matsui, T. (1991). On the finiteness of the crisscross method, *European Journal of Operational Research*, **52**, 119–124.

Fukuda, K., Saito, S., Tamura, A. (1991). Combinatorial face enumeration in arrangements and oriented matroids, *Discrete Applied Math.*, **31**, 141–149.   [221]

Fukuda, K., Saito, S., Tamura, A., Tokuyama, T. (1991). Bounding the number of k-faces in arrangements of hyperplanes, *Discrete Applied Math.*, **31**, 151–165. [198]

Fukuda, K., Tamura, A. (1988a). Local deformation and orientation transformation in oriented matroids, *Ars Combinatoria*, **25A**, 243–258.   [296, 299, 300, 333]

Fukuda, K., Tamura, A. (1988b). Local deformation and orientation transformation in oriented matroids II, Research Reports on Information Sciences B-212, Tokyo Institute of Technology, 22 pages.   [296]

Fukuda, K., Tamura, A. (1989). Characterizations of *-families, *J. Combinatorial Theory*, Ser. B, **47**, 107–110.

Fukuda, K., Tamura, A. (1990). Dualities in signed vector spaces, *Portugaliae Mathematica*, **47**, 151–165.   [146]

Fukuda, K., Tamura, A., Tokuyama, T. (1993). A theorem on the average number of subfaces in arrangements and oriented matroids, *Geom. Dedicata*, **47**, 129–142.

Fukuda, K., Terlaky, T. (1989). A general algorithmic framework for quadratic programming and a generalization of Edmonds-Fukuda rule as a finite version of Van de Panne-Whinston method, preprint, 18 pages.

Fukuda, K., Terlaky, T. (1992). Linear complementarity and oriented matroids, *J. Oper. Research Soc. Japan*, **35**, 45–61.   [419, 461]

Fulkerson, D.R. (1968). Networks, frames, blocking systems, in: "Mathematics of the Decision Sciences," Part I, (G.B. Dantzig, A.F. Vienot eds.) Lectures in Applied Mathematics, **2**, Amer. Math. Soc., Providence, RI, pp. 303–334. [151]

Fulman, J. (1998). Descent algebras, hyperplane arrangements, and shuffling cards, preprint, 15 pages; *Proc. Amer. Math. Soc.*, to appear.

Gale, D. (1956). Neighboring vertices on a convex polyhedron, in: "Linear Inequalities and Related Systems" (H. W. Kuhn and A. W. Tucker, eds.), *Annals of Math. Studies* **38**, Princeton University Press, 255–263. [399]

Gale, D. (1963). Neighborly and cyclic polytopes, in: "Convexity" (V. Klee, ed.), *Proc. Symposia in Pure Mathematics*, Vol. VII, Amer. Math. Soc., Providence RI, 225–232. [399]

Galitzer, A.J. (1997). On the moduli spaces of polygonal linkages on the 2-sphere, Ph.D. Thesis, University of Maryland, 57 pages.

Gärtner, B. (1991). Set systems of bounded Vapnik-Chervonenkis dimension and a relation to arrangements, Diplomarbeit, FU Berlin, 74 pages.

Gärtner, B., Welzl, E. (1994). Vapnik-Chervonenkis dimension and (pseudo-) hyperplane arrangements, *Discrete Comput. Geometry*, **12**, 399–432.

Gel'fand, I.M. (1986). General theory of hypergeometric functions, *Soviet Math. Doklady*, **33**, 573–577. [78]

Gel'fand, I.M., Goresky, R.M., MacPherson R.D., Serganova, V. (1987). Combinatorial geometries, convex polyhedra and Schubert cells, *Advances in Math.*, **63**, 301–316. [83, 84, 86, 89]

Gel'fand, I.M., Kapranov, M.M., Zelevinsky, A.V. (1991). Discriminants of polynomials in several variables and triangulations of Newton polyhedra, *Leningrad Math J.*, **2**, 449–505. [412]

Gel'fand, I.M., MacPherson, R.D. (1982). Geometry in Grassmannians and a generalization of the dilogarithm, *Advances in Math.*, **44**, 279–312. [91]

Gel'fand, I.M., MacPherson, R.D. (1992). A combinatorial formula for Pontrjagin classes, *Bulletin Amer. Math. Soc.*, **26**, 304–309. [78, 296, 319, 321, 483]

Gel'fand, I.M., Rybnikov, G.L. (1990). Algebraic and topological invariants of oriented matroids, *Soviet Math. Doklady*, **40**, 148–152. [95, 99]

Gel'fand, I.M., Rybnikov, G.L., Stone, D.A. (1995). Projective orientations of matroids, *Advances in Math.*, **113**, 118–150.

Gel'fand, I.M., Serganova, V.V. (1987). On the general definition of a matroid and a greedoid, *Soviet Math. Doklady*, **33**, 6–10. [79]

Gerards, B., Hochstättler, W. (1993). Onion skins in oriented matroids, RUT-COR Research Report 14-93, Rutgers University; Preprint 93.138, Mathematisches Institut, Universität zu Köln, 3 pages.

Gioan, E., Las Vergnas, M. (1999). Bases and orientations in oriented matroids, in preparation.

Gonzalez-Sprinberg, G., Laffaille, G. (1989). Sur les arrangements simples de huit droites dans $\mathbb{R}P^2$, *C.R. Acad. Sci. Paris*, **309**, Ser. I, 341–344.

Goodman, J.E. (1980). Proof of a conjecture of Burr, Grünbaum and Sloane, *Discrete Math.*, **32**, 27–35.   [236, 260, 262, 263, 264, 274, 275, 277]

Goodman, J.E., Pollack, R. (1980a). On the combinatorial classification of non-degenerate configurations in the plane, *J. Combinatorial Theory*, Ser. A, **29**, 220–235.   [36, 351]

Goodman, J.E., Pollack, R. (1980b). Proof of Grünbaum's conjecture on the stretchability of certain arrangements of pseudolines, *J. Combinatorial Theory*, Ser. A, **29**, 385–390.   [36, 259, 344, 345]

Goodman, J.E., Pollack, R. (1981a). A combinatorial perspective on some problems in geometry, *Congressus Numerantium*, **32**, 383–394.

Goodman, J.E., Pollack, R. (1981b). Three points do not determine a (pseudo-) plane, *J. Combinatorial Theory*, Ser. A, **31**, 215–218.   [470]

Goodman, J.E., Pollack, R. (1982a). Helly-type theorems for pseudolines arrangements in P^2, *J. Combinatorial Theory*, Ser. A, **32**, 1–19.   [383]

Goodman, J.E., Pollack, R. (1982b). A theorem of ordered duality, *Geometriae Dedicata*, **12**, 63–74.   [36]

Goodman, J.E., Pollack, R. (1982c). Convexity theorems for generalized planar configurations, in: "Convexity and Related Combinatorial Geometry" (Proc. 2nd Univ. Oklahoma Conf.), Marcel Dekker, pp. 73–80.   [383]

Goodman, J.E., Pollack, R. (1983). Multidimensional sorting, *SIAM J. Computing*, **12**, 484–503.   [154, 270, 307, 372, 373]

Goodman, J.E., Pollack, R. (1984a). On the number of k-subsets of a set of n points in the plane, *J. Combinatorial Theory*, Ser. A, **36**, 101–104.

Goodman, J.E., Pollack, R. (1984b). Semispaces of configurations, cell complexes of arrangements, *J. Combinatorial Theory*, Ser. A, **37**, 257–293.   [36, 37, 236, 260, 262, 263]

Goodman, J.E., Pollack, R. (1985a). A combinatorial version of the isotopy conjecture, in: "Discrete Geometry and Convexity," (J.E. Goodman, E. Lutwak, J. Malkevitch, R. Pollack, eds.), *Annals New York Acad. Sciences*, **440**, 12–19.

Goodman, J.E., Pollack, R. (1985b). Geometric sorting theory, in: Proc. Conf. "Discrete Geometry and Convexity," (J.E. Goodman, E. Lutwak, J. Malkevitch, R. Pollack, eds.), *Annals New York Acad. Sciences*, **440**, 347–354.   [263]

Goodman, J.E., Pollack, R. (1985c). Polynomial realizations of pseudolines arrangements, *Comm. Pure Applied Math.*, **38**, 725–732.

Goodman, J.E., Pollack, R (1985d). The λ-matrix: a computer-oriented model for geometric configurations, in: Proc. 3. Kolloq. "Diskrete Geometrie," Salzburg, 119–128.

Goodman, J.E., Pollack, R. (1986a). Upper bounds for configurations and polytopes in $\mathbb{R}^d$, *Discrete Comput. Geometry*, **1**, 219–227.   [372, 375]

Goodman, J.E., Pollack, R. (1986b). There are asymptotically far fewer polytopes than we thought, *Bulletin Amer. Math. Soc.*, **14**, 127–129.

Goodman, J.E., Pollack, R. (1988). Hadwiger's transversal theorem in higher dimensions, *Journal Amer. Math. Soc.*, **1**, 301–309.

Goodman, J.E., Pollack, R. (1989). New bounds on higher dimensional configurations and polytopes, in: Proc. Third Int. Conf. "Combinatorial Mathematics" (G.S. Bloom, R.L. Graham, J. Malkevitch, eds.), *Annals New York Acad. Sciences*, **555**, 205–212.

Goodman, J.E., Pollack, R. (1991). The complexity of point configurations, *Discrete Applied Math.*, **31**, 167–180.

Goodman, J.E., Pollack, R. (1993). Allowable sequences and order types in discrete and computational geometry, in: "New Trends in Discrete and Computational Geometry" (J. Pach, ed.), Springer-Verlag, Heidelberg, 103–134.   [36]

Goodman, J.E., Pollack, R., Sturmfels, B. (1989). Coordinate representation of order types requires exponential storage, in: Proc. 21st Annual ACM Symposium on Theory of Computing (Seattle), pp. 405–410.   [370]

Goodman, J.E., Pollack, R., Sturmfels, B. (1990). The intrinsic spread of a configuration in $\mathbb{R}^d$, *Journal Amer. Math. Soc.*, **3**, 639–651.   [371]

Goodman, J.E., Pollack, R., Wenger, R. (1993). Geometric transversal theory, in: "New Trends in Discrete and Computational Geometry" (J. Pach, ed.), *Algorithms and Combinatorics*, **10**, Springer-Verlag, Berlin Heidelberg, 163–198.

Goodman, J.E., Pollack, R., Wenger, R., Zamfirescu, T. (1991). Every arrangement extends to a spread, in: Proc. Third Annual Canadian Conference on Computational Geometry, pp. 191–194.

Goodman, J.E., Pollack, R., Wenger, R., Zamfirescu, T. (1992). There is a universal topological plane, in: Proc. Eighth Annual ACM Symp. Computational Geometry, Berlin, June 1992, pp. 171–176.

Goodman, J.E., Pollack, R., Wenger, R., Zamfirescu, T. (1994a). Every arrangement extends to a spread, *Combinatorica*, **14**, 301–306.

Goodman, J.E., Pollack, R., Wenger, R., Zamfirescu, T. (1994b). Arrangements and topological planes, *American Math. Monthly*, **101**, 866–878.

Goodman, J.E., Pollack, R., Wenger, R., Zamfirescu, T. (1996). There are uncountably many universal topological planes, *Geometriae Dedicata*, **59**, 157–162.

Greene, C. (1977). Acyclic orientations (Notes), in: "Higher Combinatorics" (M. Aigner, ed.), Reidel, Dordrecht, 65–68. [195, 220]

Greene, C., Zaslavsky, T. (1983). On the interpretation of Whitney numbers through arrangements of hyperplanes, zonotopes, non-Radon partitions and orientations of graphs, *Transactions Amer. Math. Soc.*, **280**, 97–126. [195, 198, 220, 221]

Grigor'ev, D.Y., Vorobjov, N.N. (1988). Solving systems of polynomial equations in subexponential time, *J. Symbolic Computation*, **5**, 37–64. [357, 371]

Gritzmann, P., Sturmfels, B. (1993). Minkowski addition of polytopes: Computational complexity and applications to Gröbner bases, *SIAM J. Discete Math.*, **6**, 246–269. [52]

Grötschel, M., Lovász L., Schrijver, A. (1988). *Geometric Algorithms and Combinatorial Optimization*, Algorithms and Combinatorics, **2**, Springer-Verlag, Berlin.

Grünbaum, B. (1967). *Convex Polytopes*, Interscience, London. [351, 355, 377, 379, 383, 398, 400, 406]

Grünbaum, B. (1970). The importance of being straight, in: Proc. 12th Biannual Intern. Seminar of the Canadian Math. Congress (Vancouver 1969), 243–254. [15, 17]

Grünbaum, B. (1972). Arrangements and spreads, *CBMS Regional Conference Series in Math.*, **10**, Amer. Math. Soc., Providence, R.I. [39, 199, 247ff: Chapter 6, 349, 354, 488]

Grünbaum, B., Sreedharan, V. (1967). An enumeration of simplicial 4-polytopes with 8 vertices, *J. Combinatorial Theory*, **2**, 437–465. [403]

Guedes de Oliveira, A. (1988). Projecção paralela em matroides orientados, M.Sc. Thesis, University of Porto, Portugal. [154]

Guedes de Oliveira, A. (1989). Oriented matroids and projective invariant theory, Dissertation, TH Darmstadt. [330]

Guedes de Oliveira, A. (1993). Oriented matroids: An essentially topological algebraic model, in: Actas do III encontro de algebristas portugueses (Coimbra, Portugal 1993), Coimbra: Departamento de Matematica, Universidade de Coimbra, 117–129.

Guedes de Oliveira, A. (1995). On the Steinitz exchange lemma, *Discrete Math.*, **137**, 367–370.

Guibas, L., Salesin, D., Stolfi, J. (1989). Epsilon geometry: building robust algorithms from imprecise computations, in: Proc. Fifth Annual ACM Symposium on Computational Geometry, pp. 208–217.   [371]

Günzel, H. (1996). The universal partition theorem for oriented matroids, *Discrete Comput. Geometry*, **15**, (1996), 121–145.   [480]

Günzel, H. (1998). On the universal partition theorem for 4-polytopes, *Discrete Comput. Geometry*, **19**, (1997), 521–552.   [481]

Günzel, H., Hirabayashi, R., Jongen, H.T. (1994). Multiparametric optimization: on stable singularities occurring in combinatorial partition codes, *Control & Optimization*, **23**, 153–167.

Gutierrez Novoa, L. (1965). On n-ordered sets and order completeness, *Pacific J. Math.*, **15**, 1337–1345.   [151]

Halsey, E. (1971). Zonotopal complexes on the d-cube, Ph.D. Thesis, Univ. Washington, Seattle.   [52]

Hamidoune, Y.O., Las Vergnas, M. (1984). Jeux de commutation orientés sur les graphes et les matroïdes, *C.R. Acad. Sci. Paris*, Ser. A, **298**, 497–499.

Hamidoune, Y.O., Las Vergnas, M. (1986). Directed switching games on graphs and matroids, *J. Combinatorial Theory*, Ser. B, **40**, 237–269.   [153, 415]

Hamidoune, Y.O., Las Vergnas, M. (1997). Directed switching games, II. The arborescence game, *Discrete Math.*, **165/166**, 395–402.

Handa, K. (1985). The faces of an acycloid, preprint, 12 pages.   [156]

Handa, K. (1987). The faces and coboundaries of an acycloid, in: *Topology and Computer Science*, 535–545.   [156]

Handa, K. (1990). A characterization of oriented matroids in terms of topes, *European J. Combinatorics*, **11**, 41–45.   [150, 156]

Handa, K. (1993a). On conditions for an acycloid to be matroidal, preprint, 14 pages.

Handa, K. (1993b). Topes of oriented matroids and related structures, *Publ. Res. Inst. Math. Sci.*, **29**, 235–266.

Hansen, S. (1965). A generalization of a theorem of Sylvester on the lines determined by a finite point set, *Math. Scand.*, **16**, 175–180.   [272]

Hansen, V.L. (1989). *Braids and Coverings: Selected Topics*, London Math. Soc. Texts, Vol. 18, Cambridge University Press.   [263]

Harbecke, W. (1981). Zur algorithmischen Behandlung des Steinitz-Problems, Diplomarbeit, Universität Bochum.   [357]

Harborth, H. (1984). Two-colorings of simple arrangements, in: Proc. 6th Hungarian Colloq. Combinatorics (Eger 1981), North-Holland, pp. 371–378.   [271]

Harborth, H. (1985). Some simple arrangements of pseudolines with a maximum number of triangles, in: Proc. Conf. "Discrete Geometry and Convexity," New York 1982, (J.E. Goodman, E. Lutwak, J. Malkevitch, R. Pollack, eds.), *Annals New York Acad. Sciences*, **440**, 31–33.

Hartmann, M., Schneider, M.H. (1995). Max-balanced flows of oriented matroids, *Discrete Math.*, **137**, 223–240.

Heintz, J., Roy, M.-F., Solerno, P. (1990). Sur la complexité du principe Tarski-Seidenberg (in French), *Bulletin Soc. Math. France*, **118**, 101–126.   [357]

Heintz, J., Roy, M.-F., Solerno, P. (1993). On the theoretical and practical complexity of the existential theory of the reals, *Comput. J.*, **36**, 427-431.   [357]

Heinz, G. (1998). The concept of cutting vectors for vector systems in the Euclidean plane, preprint; *Beiträge zur Algebra und Geometrie*, to appear.

Heinz, G. (1999a). Lower and upper bounds on the total weight of semi-rich acyclic arrangements of oriented lines in the plane, preprint, Chemnitz, 18 pages.

Heinz, G. (1999b). The concept of cutting vectors for arrangements of semicircles, preprint, Chemnitz, 11 pages.

Heinz, G. (1999c). Order types of vector systems in the Euclidean plane, preprint, 12 pages.

Henk, M., Richter-Gebert, J., Ziegler, G.M. (1997). Basic properties of convex polytopes, in: CRC Handbook on Discrete and Computational Geometry (J.E. Goodman, J. O'Rourke, eds.), CRC Press, Boca Raton FL, 243–270.

Higgs, D.A. (1966). A lattice order for the set of all matroids on a set, *Canadian Math. Bulletin*, **9**, 684–685.

Hiller, H. (1982). *Geometry of Coxeter Groups*, Pitman, Boston.   [74, 83]

Hochstättler, W. (1990). Shellability of oriented matroids, in: Proc. Conf. "Integer Programming and Combinatorial Optimization" (R. Karman, W.R. Pulleyblank, eds.), Univ. of Waterloo Press, Waterloo, pp. 275–281.   [178]

Hochstättler, W. (1992a). Seitenflächenverbände orientierter Matroide, Dissertation, Universität zu Köln 1992, 89 pages.

Hochstättler, W. (1992b). A non-visiting path, nested cones and onion skins, Report 92-126, Mathematisches Institut, Universität zu Köln, 8 pages.

Hochstättler, W. (1993a). Nested cones and onion skins, *Applied Math. Letters*, **6**, 67–69.

Hochstättler, W. (1993b). A note on the weak zone theorem, *Congressus Numerantium*, **98**, 95–103.

Hochstättler, W. (1995). Oriented matroids from wild spheres, Report 95.200, Mathematisches Institut, Universität zu Köln, 14 pages.

Hochstättler, W. (1997). A lattice theoretic characterization of oriented matroids, *European J. Combinatorics*, **18**, 563–574.

Hochstättler, W., Kromberg, S. (1996a). Adjoints and duals of matroids linearly representable over a skewfield, *Math. Scand.*, **78**, 5–12.

Hochstättler, W., Kromberg, S. (1996b). A pseudoconfiguration of points without adjoint, *J. Combinatorial Theory*, Ser. B, **68**, 277–294.

Hochstättler, W., Nešetřil, J. (1995). Linear programming duality and morphisms, Report 95.209, Universität zu Köln, 16 pages.

Hodge, W.V.D., Pedoe, D. (1947). *Methods of Algebraic Geometry*, Cambridge University Press. [80]

Hoffmann, C., Hopcroft, J., Karasick, M. (1988). Towards implementing robust geometric algorithms, in: Proc. Fourth Annual ACM Symposium on Computational Geometry, (Urbana, Illinois). [370]

Huber, B., Rambau, J., Santos, F. (1999). The Cayley trick, lifting subdivisions and the Bohne-Dress theorem on zonotopal tilings, preprint, 21 pages. [485]

Hudson, J.F.P. (1969). *Piecewise Linear Topology*, Benjamin, New York. [209, 210]

Huijari, G.P. (1986). An optimal algorithm for the coordinatization of oriented matroids, *Discrete Comput. Geometry*, in press? [371]

Humphreys, J.E. (1972). *Introduction to Lie Algebras and Representation Theory*, Graduate Texts in Mathematics, **9**, Springer, New York. [71, 72]

Humphreys, J.E. (1975). *Linear Algebraic Groups*, Graduate Texts in Mathematics, **21**, Springer, New York. [79]

Humphreys, J.E. (1990). *Reflection Groups and Coxeter Groups*, Cambridge University Press. [65, 70]

Hund, U. (1985). Pseudosphärenarrangements zu orientierten Matroiden, Diplomarbeit, Universität Münster, 79 pages.

Ingleton, A.W. (1971). Representations of matroids, in: "Combinatorial Mathematics and its Applications" (D.J.A. Welsh, ed.), Academic Press, 149–167. [305]

Jacobson, N. (1964). *Lectures in Abstract Algebra*, Volume III, van Nostrand, Princeton. [353]

Jaggi, B. (1992). Punktmengen mit vorgeschriebenen Distanzen und ihre Konfigurationsräume, Dissertation, Universität Bern.

Jaggi, B., Mani-Levitska, P. (1988). A simple arrangement of lines without the isotopy property, preprint, Universität Bern, 28 pages. [363]

Jaggi, B., Mani-Levitska, P., Sturmfels, B., White, N. (1989). Costructing uniform oriented matroids without the isotopy property, *Discrete Comput. Geometry*, **4**, 97–100. [366]

Jambu, M., Terao, H. (1984). Free arrangements of hyperplanes and supersolvable lattices, *Advances in Math.*, **52**, 248–258. [77]

Jamison, R.E. (1982). A perspective on abstract convexity: Classifying alignments by varieties, in: "Convexity and Related Combinatorial Geometry" (D.C. Kay and M. Breem, eds.), Proc. Second Univ. Oklahoma Conf. 1980, Dekker, New York, pp. 113–150. [377]

Jamison, R.E. (1985). A survey of the slope problem, in: "Discrete Geometry and Convexity" (J.E. Goodman, E. Lutwak, J. Malkevitch, R. Pollack, eds.), *Annals New York Acad. Sciences*, **440**, 34–51. [41]

Jaritz, R. (1996). Orientierung und Ordnungsfunktionen in kombinatorischen Geometrien, Preprint Math/Inf/96/2, Universität Jena, 15 pages.

Jaritz, R. (1997). Oriented matroids in terms of order functions, *Beiträge zur Algebra und Geometrie*, **38**, 249–260.

Jensen, D.L. (1985). Coloring and duality: combinatorial augmentation methods, Ph.D. Thesis, Cornell University, 371 pages. [451, 458, 469]

Jordan, D., Steiner, M. (1998). Configuration spaces of mechanical linkages, preprint, 16 pages; revised version 1999, 18 pages. [481]

Joswig, M., Ziegler, G.M. (1999). Neighborly cubical polytopes, Preprint, 20 pages; Lanl Preprint math.CO/9812033. [97]

Kahn, J. (1987). On lattices with Möbius function $\pm 1, 0$, *Discrete Comput. Geometry*, **2**, 1–8.

Kalai, G. (1988a). A simple way to tell a simple polytope from its graph, *J. Combinatorial Theory*, Ser. A, **49**, 381–383.   [60]

Kalai, G. (1988b). Many triangulated spheres, *Discrete Comput. Geometry*, **3**, 1–14.   [404]

Kalhoff, F.B. (1996). Oriented rank three matroids and projective planes, Preprint 96-06, Universität Dortmund, 23 pages.

Kapovich, M., Millson, J.J. (1995). On the moduli space of polygons in the Euclidean plane, *J. Differential Geometry*, **42**, 430–464.

Kapovich, M., Millson, J.J. (1996). The symplectic geometry of polygons in Euclidean space, *J. Differential Geometry*, **44**, 479–513. plane, Journal of Diff. Geometry, Vol. 42 (1995) N 1, 133–164.

Kapovich, M., Millson, J.J. (1997). On the moduli space of a spherical polygonal linkage, preprint, 16 pages; *Bull. Canadian Math. Soc.*, to appear.

Kapovich, M., Millson, J.J. (1998a). On representation varieties of Artin groups, projective arrangements and the fundamental groups of smooth complex algebraic varieties, preprint, 75 pages; *Publ. Math. IHES*, to appear.

Kapovich, M., Millson, J.J. (1998b). Universality theorems for configuration spaces of planar linkages, Lanl preprint math.AG/9803150, University of Utah, 45 pages.   [481]

Kapovich, M., Millson, J.J. (1999). Moduli spaces of linkages and arrangements, in: "Advances in Geometry" (J.-L. Brylinski et al., eds.), *Progress in Mathematics* Vol. 172, Birkhäuser, 237–270.   [481]

Kapranov, M.M., Voevodsky, V.A. (1991a). Combinatorial-geometric aspects of polycategory theory: Pasting schemes and higher Bruhat order, *Cahiers de Topologie et de la Géometrie Differentielle*, **32**, 11–28.

Kapranov, M.M., Voevodsky, V.A. (1991b). Free n-category generated by a cube, oriented matroids, and higher Bruhat orders, *Funct. Anal. Appl.*, **25**, No.1, 50–52; translation from *Funkts. Anal. Prilozh.*, **25**, No.1, 62–65.

Karlander, J. (1992). A characterization of affine sign vector systems, preprint, KTH Stockholm, 24 pages.   [154, 187]

Karlin, S. (1968). *Total Positivity*, Vol. I, Stanford University Press.   [399]

Karzel, H. (1969). Konvexität in halbgeordneten projektiven und affinen Räumen, *Abh. Math. Sem. Univ. Hamburg*, **33**, 231–242.   [151]

Keijsper, R., Pendavingh, R., Schrijver, A. (1998). Adjacency, inseparability, and base orderability in matroids, preprint, CWI Amsterdam.

Kelly, L.M., Moser, W.O.J. (1958). On the number of ordinary lines determined by n points, Canadian J. Math., 10, 210–219.   [39]

Kelly, L.M., Rottenberg, R. (1972). Simple points in pseudoline arrangements, Pacific J. Math., 40, 617–622.   [39, 272]

Kempe, A.B. (1875). On a general method of describing plane curves of the n-th degree by linkwork, Proc. London Math. Soc., 7, 213–216.   [481]

Kern, W. (1982). Adjoints und Polare von orientierten Matroiden, Diplomarbeit, Universität Erlangen.

Kern, W. (1985). Verbandstheoretische Dualität in kombinatorischen Geometrien und orientierten Matroiden, Dissertation, Universität Köln, 57 pages.

Kingan, S. (1999). Bibliography of matroids, http://members.aol.com/matroids/biblio.htm   [101]

Klafszky, E., Terlaky, T. (1986). A new approach to the feasibility problem for oriented matroids (in Hungarian), Alkamazott Mat. Lapok, 12, 279–282.

Klafszky, E., Terlaky, T. (1987). Remarks on the feasibility problem of oriented matroids, Annales Universitatis Scientiarum Budapestiensis de Rolando Eötvös nominatae, Sectio Computatorica, 7, 155–157.

Klafszky, E., Terlaky, T. (1989a). Oriented matroids, quadratic programming and the criss-cross method (in Hungarian), Alkalmazott Mat. Lapok, 14, 365–375.

Klafszky E., Terlaky T. (1989b). Some generalizations of the criss-cross method for the linear complementarity problem of oriented matroids, Combinatorica, 9, 189–198.   [419, 461]

Klafszky, E., Terlaky, T. (1992). Some generalizations of the criss-cross method for quadratic programming, Optimization, 24, 127–139.

Klee, V., Kleinschmidt, P. (1995). Convex polytopes and related complexes, in: Handbook of Combinatorics (R. Graham, M. Grötschel, and L. Lovász, eds.), North-Holland/Elsevier, Amsterdam, pp. 875–917.   [200, 377]

Kleiman, S., Laksov, D. (1972). Schubert calculus, Amer. Math. Monthly, 79, 1061–1082.   [83]

Kleinschmidt, P. (1976a). On facets with non-arbitrary shapes, Pacific J. Math., 65, 511–515.   [25, 403]

Kleinschmidt, P. (1976b). Sphären mit wenigen Ecken, Geometriae Dedicata, 5, 307–320.

Kleinschmidt, P., Onn, S. (1996). Signable posets and partitionable simplicial complexes, Discrete Comput. Geometry, 15, 443–466.

Klin, M.H., Tratch, S.S., Treskov, V.E. (1989). A graph-theoretic interpretation of 2D-configurations, in: Proc. 4th All-Union Conf. "Methods and Programmes of Solv. Optimiz. Problems on Graphs and Networks," Vol. 1, Novosibirsk (1989), 87–89 (in Russian).

Klin, M.H., Tratch, S.S., Zefirov, N.S. (1990). 2D-configurations and clique-cyclic orientations of the graphs $L(K_p)$, *Reports in Molecul. Th.*, **1**, No. 2, 149–163. [33]

Knuth, D.E. (1992). *Axioms and Hulls, Lecture Notes in Computer Science*, **606**, Springer, Berlin Heidelberg. [30, 270]

Kollewe, W. (1992). Representation of data by pseudoline arrangements, Preprint Nr. 1505, Technische Hochschule Darmstadt, 8 pages.

Kortenkamp, U.H. (1997). Every simplicial polytope with at most $d+4$ vertices is a quotient of a neighborly polytope, *Discrete Comput. Geometry*, **18**, 455–462.

Kromberg, S. (1995). Adjoints, Schiefkörper und algebraische Matroide, Dissertation, Universität Köln, 149 pages.

Lagarias, J.C (1999). Geometric models for quasicrystals. I. Delone sets of finite type, *Discrete Comput. Geometry*, **21**, 161–191. [485]

Larman, D. (1972). On sets projectively equivalent to the vertices of a convex polytope, *Bulletin London Math. Soc.*, **4**, 6–12. [385]

Las Vergnas, M. (1974). Matroïdes orientables, preprint, 80 pages (unpublished, announced in Las Vergnas (1975a).) [150]

Las Vergnas, M. (1975a). Matroïdes orientables, *C.R. Acad. Sci. Paris*, Ser.A, **280**, 61–64. [124, 125, 136, 150, 152, 161, 193, 296, 376, 378, 382]

Las Vergnas, M. (1975b). Coordinatizable strong maps of matroids, preprint, 123 pages. [319]

Las Vergnas, M. (1975c). Sur les extensions principales d'un matroïde, *C.R. Acad. Sci. Paris*, Ser.A, **280**, 187–190. [313]

Las Vergnas, M. (1977). Acyclic and totally cyclic orientations of combinatorial geometries, *Discrete Math.*, **20**, 51–61. [195, 198, 222]

Las Vergnas, M. (1978a). Bases in oriented matroids, *J. Combinatorial Theory*, Ser. B, **25**, 283–289. [124, 125, 153]

Las Vergnas, M. (1978b). Extensions ponctuelles d'une géométrie combinatoire orienté, in: "Problèmes Combinatoires et Théorie des Graphes" (Actes Coll. Orsay 1976), Colloques internationaux, C.N.R.S., No. 260, pp. 265–270. [113, 137, 281ff: Sects. 7.1 and 7.2, 313]

Las Vergnas, M. (1978c). Sur les activités des orientations d'une géométrie combinatoire, in: "Codes et Hypergraphes" (Actes Coll. Math. Discrète Bruxelles, Cahiers Centre Et. Rech. Opér.), **20**, pp. 293–300.   [198]

Las Vergnas, M. (1980a). Convexity in oriented matroids, *J. Combinatorial Theory*, Ser. B, **29**, 231–243.   [132, 152, 161, 166, 167, 170, 194, 198, 305, 376, 378, 382]

Las Vergnas, M. (1980b). On the Tutte polynomial of a morphism of matroids, *Annals Discrete Math.*, **8**, 7–20.

Las Vergnas, M. (1984a). Oriented matroids as signed geometries real in corank 2, in: "Finite and Infinite Sets" (Proc. 6th Hungarian Combinatorial Conf. Eger 1981), North-Holland, 555–565.   [113, 137, 141]

Las Vergnas, M. (1984b). The Tutte polynomial of a morphism of matroids. II Activities of orientations, in: "Progress in Graph Theory" (Proc. Waterloo Silver Jubilee Conf. 1982), Academic Press, Toronto, pp. 367–380.   [198, 222]

Las Vergnas, M. (1984c). A correspondence between spanning trees and orientations in graphs, in: "Graph Theory and Combinatorics" (Proc. Cambridge Comb. Conf. 1983), Academic Press, London, pp. 233–238.   [223]

Las Vergnas, M. (1986a). Order properties of lines in the plane and a conjecture of G. Ringel, *J. Combinatorial Theory*, Ser. B, **41**, 246–249.   [41, 366]

Las Vergnas, M. (1986b). Hamilton paths in tournaments and a problem of McMullen on projective transformations in $\mathbb{R}^d$, *Bulletin London Math. Soc.*, **18**, 571–572.   [381, 385]

Las Vergnas, M. (1990). Acyclic reorientations of weakly oriented matroids, *J. Combinatorial Theory*, Ser. B, **49**, 195–199.   [222]

Las Vergnas, M., Roudneff, J.-P., Salaün, I. (1991). Regular polytopes and oriented matroids, preprint, 18 pages.   [332, 333]

Lawrence, J. (1975). Oriented matroids, Ph.D. Thesis, University of Washington, Seattle, 67 pages.   [26, 150, 427, 449]

Lawrence, J. (1982). Oriented matroids and multiply ordered sets, *Linear Algebra Appl.*, **48**, 1–12.   [125, 124, 128]

Lawrence, J. (1983). Lopsided sets and orthant-intersection by convex sets, *Pacific J. Math.*, **104**, 155–173.   [148, 155]

Lawrence, J. (1984a). Shellability of oriented matroid complexes, preprint, 8 pages.   [176, 178]

Lawrence, J. (1984b). Some dual pairs of simple oriented matroids by concatenation, preprint, 7 pages.   [316, 317, 348]

Lawrence, J. (1999). Total polynomials of oriented matroids, preprint, 13 pages; *European J. Combinatorics*, to appear.

Lawrence, J., Weinberg, B. (1980). Unions of oriented matroids, *Annals Discrete Math.*, **8**, 29–34.   [315, 346]

Lawrence, J., Weinberg, B. (1981). Unions of oriented matroids, *Linear Alg. Appl.*, **41**, 183–200.

Lee, C.W. (1991). Regular triangulations of convex polytopes, in: "Applied Geometry and Discrete Mathematics – The Victor Klee Festschrift" (P. Gritzmann, B. Sturmfels, eds.), DIMACS Series in Discrete Math. and Theoretical Computer Science, **4**, Amer. Math. Soc., Providence, RI, pp. 443–456.   [410, 411]

Lee, C.W. (1997). Triangulations of polytopes, in: CRC Handbook on Discrete and Computational Geometry (J. E. Goodman, J. O'Rourke, eds.), CRC Press, Boca Raton FL, 271–290.

Lee, J. (1986). Subspaces with well-scaled frames, Thesis, Cornell University.

Lee, J. (1990). The incidence structure of subspaces with well-scaled frames, *J. Combinatorial Theory*, Ser. B, **50**, 265–287.

Lee, J. (1996). Orienting matroids representable over both $GF[3]$ and $GF[5]$, Preprint, 7 pages; *European J. Combinatorics*, to appear.

Lee, J., Scobee, M. (1997/98). A characterization of the orientations of ternary matroids, preprint, U. Kentucky, 19 pages.

Levi, F. (1926). Die Teilung der projektiven Ebene durch Gerade oder Pseudogerade, *Ber. Math.-Phys. Kl. Sächs. Akad. Wiss.*, **78**, 256–267.   [252, 259, 260, 271]

Lickorish, W.B.R. (1991). Unshellable triangulations of spheres, *European J. Combinatorics*, **12**, 527–530.   [211]

Linhart, J. (1993a). On the total weight of arrangements of halfspaces, *Geometriae Dedicata*, **43**, 165–172.

Linhart, J. (1993b). Arrangements of oriented hyperplanes, *Discrete Comput. Geometry*, **10**, 435–446.

Linhart, J., Yang, Y. (1996). Arrangements of arcs and pseudocircles, *Beiträge zur Algebra und Geometrie*, **37**, 391–398.

Liskovets, V. (1998). Reductive enumeration under mutually orthogonal group actions, *Acta Applic. Math.*, **52**, 91–120.

Ljubić, D., Roudneff, J.-P., Sturmfels, B. (1989). Arrangements of lines and pseudolines without adjacent triangles, *J. Combinatorial Theory*, Ser. A, **50**, 24–32.   [17]

Lombardi, H. (1990). Nullstellensatz réel effectif et variantes, *C. R. Acad. Sci. Paris*, Série I, **310**, 635–640.   [360]

Lovász, L., Schrijver, A. (1981). Remarks on a theorem of Rédei, *Stud. Sci. Math. Hung.*, **16**, 449–454.   [41]

Lundell, A.T., Weingram, S. (1969). *The Topology of CW Complexes*, Van Nostrand, New York.   [202]

Lutz, F.H. (1999). Triangulated manifolds with few vertices and vertex-transitive group actions, Dissertation, TU Berlin, 131 pages.   [488]

Maclagan, D. (1998). Boolean term orders and the root system B_n, Lanl preprint math.CO/9809134; *Order*, to appear.

MacLane, S. (1936). Some interpretations of abstract linear independence in terms of projective geometry, *Amer. J. Math.*, **58**, 236–241.   [274, 354]

MacPherson, R.D. (1993). Combinatorial differential manifolds, in: "Topological Methods in Modern Mathematics," Proc. of a Symp. in Honor of John Milnor's Sixtieth Birthday, SUNY Stony Brook, June 1991 (L. R. Goldberg, A. V. Phillips, eds.), Publish or Perish, Houston TX, 203–221.   [78, 296, 319, 483]

Mandel, A. (1978). Decision process for representability of matroids and oriented matroids, Research Report CORR 78-40, University of Waterloo, November 1978.   [354]

Mandel, A. (1982). – see Edmonds and Mandel (1982).

Mani, P. (1972). Spheres with few vertices, *J. Combinatorial Theory*, Ser. A, **13**, 346–352.   [403]

Marcus, D.A. (1981). Minimal positive 2-spanning sets of vectors, *Proceedings Amer. Math. Soc.*, **82**, 165–172.

Marcus, D.A. (1984). Gale diagrams of convex polytopes and positive sets of vectors, Discrete Applied Math., **9**, 47–67.   [416]

Martini, H. (1985). Some results and problems around zonotopes, in: *Colloquia Mathematica Societatis János Bolyai*, **48** (Proc. Sciófok 1985), 383–418.   [52]

Mason, J.H. (1971). Geometrical realization of combinatorial geometries, *Proceedings Amer. Math. Soc.*, **30**, 15–21.   [236]

Maurer, S. (1973). Matroid basis graphs, I, *J. Combinatorial Theory*, **14**, 216–240.   [132, 133]

Mazur, B. (1986). Arithmetic on curves, *Bulletin Amer. Math. Soc.*, **14**, 207–259.   [354]

McDiarmid, C. (1987). General percolation and oriented matroids, in: "Random Graphs '85," Lect. 2nd International Seminar (Poznan, Poland 1985), *Annals Discrete Math.*, **33**, 187–197.

McKee, T.A. (1980). Logical and matroidal duality in combinatorial linear programming, in: "Combinatorics, Graph Theory and Computing," Proc. 11th Southeastern Conf., Boca Raton, Florida 1980, Vol. II, *Congr. Numerantium*, **29**, 667–672.

McMullen, P. (1971a). On zonotopes, *Transactions Amer. Math. Soc.*, **159**, 91–109. [52, 63, 97]

McMullen, P. (1971b). The numbers of faces of simplicial polytopes, *Israel J. Math.*, **9**, 559–570. [384, 395]

McMullen, P. (1975). Space-tiling zonotopes, *Mathematika*, **22**, 202–211. [60]

McMullen, P. (1979). Transforms, diagrams and representations, in: "Contributions to Geometry" (J. Tölke, J. Wills, eds.), Proc. Geometry Symposium Siegen 1978, Birkhäuser, 92–130. [379]

McMullen, P. (1984). Volumes of projections of unit cubes, *Bulletin London Math. Soc.*, **15**, 278–280. [63]

McNulty, J. (1993). Ports and oriented matroids, *Congressus Numerantium*, **96**, 11–20.

McNulty, J. (1994). Two new classes of non-orientable matroids, preprint, 41 pages. [337]

Mei, P.S. (1971). Axiomatic theory of linear and convex closure, Ph.D. Thesis, Purdue University, 152 pages. [151]

Melchior, E. (1940). Über Vielseite der projektiven Ebene, *Deutsche Math.*, **5**, 461–475. [272]

Milenkovic, V.J. (1988). Verifiable implementations of geometric algorithms using finite precision arithmetic, Ph.D. Thesis, Carnegie Mellon University. [371]

Milenkovic, V.J., Nackman, L.R. (1990). Finding compact coordinate representations for polygons and polyhedra, in: Proc. Sixth ACM Annual Symposium on Computational Geometry (Berkeley, June 1990), pp. 244–252. [371]

Miller, D.A. (1983). A class of topological oriented matroids with some applications to non-linear programming, Ph.D. Thesis, University of California, Berkeley, 105 pages.

Miller, D.A. (1987). Oriented matroids from smooth manifolds, *J. Combinatorial Theory*, Ser. B, **43**, 173–186. [246, 260]

Minty, G.J. (1966). On the axiomatic foundations of the theories of directed linear graphs, electrical networks and network-programming, *J. Math. and Mechanics*, **15**, 485–520. [151, 450]

Mnëv, N.E. (1985a). Realizability of combinatorial types of convex polyhedra over fields. *J. Soviet Mathematics*, **28**, 606–609. [363]

Mnëv, M.N. (1985b). On manifolds of combinatorial types of projective configurations and convex polyhedra, *Soviet Math. Doklady*, **32**, 335–337.

Mnëv, M.N. (1988). The universality theorems on the classification problem of configuration varieties and convex polytopes varieties, in: Viro, O.Ya. (ed.): "Topology and Geometry – Rohlin Seminar," *Lecture Notes in Mathematics*, **1346**, Springer, Heidelberg, pp. 527–544. [89, 259, 363, 368, 369, 407, 408, 480]

Mnëv, N.E. (1991). The universality theorem on the oriented matroid stratification of the space of real matrices, in: "Discrete and Computational Geometry: Papers from the DIMACS Special Year" (J.E. Goodman, R. Pollack, W. Steiger, eds.), DIMACS Series in Discrete Mathematics and Theoretical Computer Science, **6**, Amer. Math. Soc., Providence RI, pp. 237–243. [480]

Mnëv, N.E., Richter-Gebert, J. (1993). Two constructions of oriented matroids with disconnected extension space, *Discrete Comput. Geometry*, **10**, 271–285. [296, 483]

Mnëv, N.E., Ziegler, G.M. (1993). Combinatorial models for the finite-dimensional Grassmannians, *Discrete Comput. Geometry*, **10**, 241–250. [78, 296, 321, 483]

Möbius, A.F. (1886). *Gesammelte Werke, Band II* (F. Klein, ed.), Verlag M. Sändig, Wiesbaden, 1886; reprint: Verlag S. Hirzel, Stuttgart, 1967. [488]

Morris, W.D. jr. (1986). Oriented matroids and the linear complementary problem, Ph.D. Thesis, Cornell University, 111 pages.

Morris, W.D. jr. (1993). A non-realizable lopsided set of the 7-cube, *Note di Matematica*, **13**, 21–32.

Morris, W.D. jr. (1994). LCP degree theory and oriented matroids, *SIAM J. Matrix Anal. Appl.*, **15**, 995–1006.

Morris, W.D. jr., Todd, M.J. (1988). Symmetry and positive definiteness in oriented matroids, *European J. Combinatorics*, **9**, 121–129. [419]

Morrison, R.T., Boyd, R.N. (1987). *Organic Chemistry*, 5th ed., Allyn & Bacon, Boston. [34]

Mumford, D. (1988). *The Red Book of Varieties and Schemes, Lecture Notes in Mathematics*, **1358**, Springer, Heidelberg. [88]

Munkres, J.R. (1984). *Elements of Algebraic Topology*, Addison-Wesley, Menlo Park. [200]

Munson, B.S. (1981). Face lattices of oriented matroids, Ph.D. Thesis, Cornell University, 135 pages. [383, 384, 386, 391]

Murty, K.G. (1988). *Linear Complementarity, Linear and Nonlinear Programming*, Heldermann Verlag, Berlin. [419]

Nef, W., Schmidt, P.-M. (1990). Computing a sweeping-plane in regular ("general") position: A numerical and a symbolic solution, *J. Symbolic Computation* **10**, 633–646.

Oda, T. (1988). *Convex Bodies and Algebraic Geometry: an Introduction to the Theory of Toric Varieties*, Springer-Verlag, New York. [90]

Ohsugi, H., Hibi, T. (1997). A normal $(0, 1)$-polytope none of whose regular triangulations is unimodular, *Discrete Comput. Geometry*, **21**, 201–204. [484]

Onn, S. (1997). Signable and partitionable posets, *European J. Combinatorics*, **18**, 921–938.

Orlik, P. (1989). *Introduction to Arrangements*, CBMS Regional Conf. Series in Math., **72**, Amer. Math. Soc., Providence, R.I. [77, 93]

Orlik, P., Solomon, L. (1980). Combinatorics and topology of complements of hyperplanes, *Inventiones Math.*, **56**, 167–189. [93, 94]

Orlik, P., Terao, H. (1992). *Arrangements of Hyperplanes*, Grundlehren Series, **300**, Springer-Verlag, Berlin–Heidelberg. [93]

O'Rourke, J., Streinu, I. (1997). Vertex-edge pseudo-visibility graphs: Characterization and recognition, in: Proc. 13th Annual ACM Symposium on Computational Geometry (Nice, June 1997), pp. 119-128.

Ossona de Mendez, P. (1994). Orientations bipolaires, Doctoral Thesis, École de Hautes Études en Science Sociales, Paris, 113 pages.

Oxley, J. (1992). *Matroid Theory*, Oxford University Press, Oxford 1992. [101]

Pachner, U. (1986). Konstruktionsmethoden und das kombinatorische Homöomorphieproblem für Triangulationen semilinearer Mannigfaltigkeiten, *Abh. Math. Sem. Univ. Hamburg*, **57**, 69–86. [211]

Paris, L. (1993). Universal cover of Salvetti's complex and topology of simplicial arrangements of hyperplanes, *Trans. Amer. Math. Soc.*, **340**, 149–178. [95]

Penne, R. (1992). Lines in 3-space. Isotopy, chirality and weavings, Dissertation, University of Antwerp, 238 pages.

Penne, R. (1993). Configurations of few lines in 3-space. Isotopy, chirality and planar layouts, *Geometriae Dedicata*, **45**, 49–82.

Penne, R. (1994). Some nonrealizable line diagrams, *J. Intell. Robot. Syst.*, **11**, 193–207.

Penne, R. (1996). Moves on pseudoline diagrams, *European J. Combinatorics*, **17**, 569–593.

Perrin, R. (1881/82). Sur le problème des aspects, *Bull. Soc. Math. France*, **10**, 103–127. [150]

Philipp, M. (1994). Combinatorial Properties of Arrangements of Halfspaces, Dissertation, Universität Salzburg, Austria, 109 pages.

Pocchiola, M., Vegter, G. (1994). Order types and visibility types of configurations of disjoint convex plane sets, Extended abstract, Preprint LIENS-94-4, Dept. Mathématiques Informatique, École Normale Supérieure, Paris, 13 pages.

Pocchiola, M., Vegter, G. (1996a). Pseudo-triangulations: Theory and applications, in: Proc. 12th Annual ACM Symposium on Computational Geometry, pp. 291–300.

Pocchiola, M., Vegter, G. (1996b). The visibility complex, *Internat. J. Comput. Geometry Applications*, **6**, 279–308.

Pocchiola, M., Vegter, G. (1996c). Topologically sweeping visibility complexes via pseudo-triangulations, *Discrete Comput. Geometry*, **16**, 419–453.

Pocchiola, M., Vegter, G. (1996d). Minimal tangent visibility graphs, *Comput. Geom. Theory Applications*, **6**, 303–314.

Pocchiola, M., Vegter, G. (1999). On polygonal covers, in: "Advances in Discrete and Computational Geometry" B. Chazelle, J. Goodman, R. Pollack, eds.), *Contemporary Math.*, **223**, Amer. Math. Soc., 257–268.

Pock, K.P. (1991). Entscheidungsmethoden zur Realisierbarkeit orientierter Matroide, Diplomarbeit, TH Darmstadt, 56 pages. [486]

Pollack, R., Roy, M.-F. (1993). On the number of cells defined by a set of polynomials, *C. R. Acad. Sci. Paris*, Ser. I, **316**, 573–577.

Polymeris, A. (1984). Sign vectors. Polar consistency, preprint, 31 pages.

Pretzel, O. (1995). Orientations of chain groups, *Order*, **12**, 135–142.

Prieß-Crampe, S. (1967). Archimedisch angeordnete projektive Ebenen, *Math. Zeitschrift*, **99**, 305–348.

Quillen, D. (1978). Homotopy properties of the poset of non-trivial p-subgroups of a group, *Advances in Math.*, **28**, 101–128. [216]

Rado, R. (1952). An inequality, *J. London Math. Soc.*, **27**, 1–6.   [97]

Rambau, J. (1996). Polyhedral subdivisions and projections of polytopes, Dissertation, TU Berlin, 122 pages.

Rambau, J. (1997). Triangulations of cyclic polytopes and higher Bruhat orders, *Mathematika*, **44**, 162–194.   [484]

Rambau, J., Santos, F. (1997). The generalized Baues problem for cyclic polytopes, I, preprint, 22 pages; *European J. Combinatorics*, to appear.   [484]

Rambau, J., Ziegler, G.M. (1996). Projections of polytopes and the Generalized Baues Conjecture, *Discrete Comput. Geometry* **16**, 215–237.   [414, 484]

Ramírez Alfonsín, J.L. (1999). Spatial graphs and oriented matroids: the trefoil, *Discrete Comput. Geometry*, **22** (1999), 149–158.

Ramírez Alfonsín, J.L. (1998). A special arrangement with minimal number of triangles, *Information Processing Letters*, **67**, 273–276.   [279]

Ramírez Alfonsín, J.L. (1999a). Lawrence oriented matroids and a problem of McMullen on the projective equivalences of polytopes, *European J. Combinatorics*, to appear.   [385]

Ramírez Alfonsín, J.L. (1999b). On linked spatial representations, preprint, 11 pages.

Randell, R. (1989). Lattice-isotopic arrangements are topologically isomorphic, *Proceedings Amer. Math. Soc.*, **107**, 555–559.   [94]

Reay, J.R. (1965). A new proof of the Bonnice-Klee theorem, *Proceedings Amer. Math. Soc.*, **16**, 585–587.   [415]

Rédei, L. (1934). Ein kombinatorischer Satz, *Acta Scient. Math. (Szeged)*, **7**, 39–43.   [386]

Reiner, V. (1999). The generalized Baues problem, preprint, 35 pages; in: "New Perspectives in Algebraic Combinatorics," MSRI Book Series, Cambridge University Press, to appear.   [483, 484]

Renegar, J. (1992). On the computational complexity and geometry of the first-order theory of the reals, Parts I-III, *J. Symbolic Comput.*, **13**, 255–299, 301–327, 329–352.   [357]

Richter, J. (1988). Kombinatorische Realisierbarkeitskriterien für orientierte Matroide, (Diplomarbeit, TH Darmstadt, 112 pages); *Mitteilungen Math. Seminar Gießen*, **194** (1989).   [244, 350, 268]

Richter-Gebert, J. (1991). On the realizability problem for combinatorial geometries – decision methods, Dissertation, TH Darmstadt.   [486]

Richter-Gebert, J. (1992). New construction methods for oriented matroids, Dissertation, KTH Stockholm, 102 pages. [304]

Richter-Gebert, J. (1993a), Non-euclidean uniform oriented matroids have biquadratic final polynomials, *Combinatorica*, **13**, 259–268.

Richter-Gebert, J. (1993b). Oriented matroids with few mutations, *Discrete Comput. Geometry*, **10**, 251–269. [261, 304, 321, 483]

Richter-Gebert, J. (1993c). Combinatorial obstructions to the lifting of weaving diagrams, *Discrete Comput. Geometry*, **10**, 287–312.

Richter-Gebert, J. (1993d). Line arrangements and zonotopal tilings: a little printer exercise, *HyperSpace*, **2**, 8–17. [52]

Richter-Gebert, J. (1995a). Mechanical theorem proving in projective geometry, *Annals of Mathematics and Artificial Intelligence*, **13**, 139-172.

Richter-Gebert, J. (1995b). Realization spaces of 4-polytopes are universal, Habilitationsschrift, TU Berlin, 111 pages;
http://www.math.tu-berlin.de/~richter [481]

Richter-Gebert, J. (1995c). Mnëv's universality theorem revisited, *Séminaire Lotaringien de Combinatoire*, 15 pages. [480]

Richter-Gebert, J. (1996a). Two interesting oriented matroids, *Documenta Mathematica*, **1**, 137–148. [482, 486]

Richter-Gebert, J. (1996b). *Realization Spaces of Polytopes, Lecture Notes in Mathematics*, **1643**, Springer, Berlin Heidelberg. [406, 481]

Richter-Gebert, J. (1998). Testing orientability for matroids is NP-complete, preprint, 12 pages; special issue in honor of Henry Crapo (J. Kung, ed.), *Advances Applied Math.*, to appear.

Richter-Gebert, J. (1999). Finding small triangulations of polytope boundaries is hard, preprint. [484]

Richter-Gebert, J., Kortenkamp, U.H. (1999). *Cinderella – The Interactive Geometry Software*, Software and 143 pp. manual, Springer-Verlag Heidelberg;
http://www.cinderella.de [486]

Richter, J., Sturmfels, B. (1991). On the topology and geometric construction of oriented matroids and convex polytopes, *Transactions Amer. Math. Soc.*, **325**, 389–412. [42, 342, 346, 397]

Richter-Gebert, J., Ziegler, G.M. (1994). Zonotopal tilings and the Bohne-Dress theorem, in: Proc. "Jerusalem Combinatorics '93" (H. Barcelo, G. Kalai, eds.), *Contemporary Math.*, **178**, Amer. Math. Soc., 211–232. [485]

Richter-Gebert, J., Ziegler, G.M. (1995). Realization spaces of 4-polytopes are universal, *Bulletin Amer. Math. Soc.*, **32**, 403–412. [481]

Richter-Gebert, J., Ziegler, G.M. (1997). Oriented matroids, in: CRC Handbook on Discrete and Computational Geometry (J. E. Goodman, J. O'Rourke, eds.), CRC Press, Boca Raton FL, 111–132.

Ringel, G. (1956). Teilungen der Ebene durch Geraden oder topologische Geraden, *Math. Zeitschrift*, **64**, 79–102.   [41, 259, 260, 266, 267, 269, 349, 363]

Ringel, G. (1957). Über Geraden in allgemeiner Lage, *Elemente der Math.*, **12**, 75–82.   [266, 267, 277]

Ringel, G. (1974). *Map Color Theorem*, Grundlehren Series, **234**, Springer-Verlag, New York.   [488]

Rockafellar, R.T. (1969). The elementary vectors of a subspace of $\mathbb{R}^n$, in: "Combinatorial Mathematics and its Applications" (Proc. Chapel Hill Conf.), Univ. North Carolina Press, pp. 104–127.   [26, 151, 418]

Ronan, M. (1989). *Lectures on Buildings*, Academic Press, Boston.   [76]

Roos, C. (1984). Network programming with umbrellas, preprint 84-32, Delft University of Technology, Dept. Mathematics and Informatics, 68 pages.   [424]

Roudneff, J.-P. (1986). On the number of triangles in simple arrangements of pseudolines in the real projective plane, *Discrete Math.*, **60**, 245–251.   [271, 278, 279]

Roudneff, J.-P. (1987a). Matroïdes orientés et arrangements de pseudo-droites, Thèse, Paris, 159 pages.   [345]

Roudneff, J.-P. (1987b). Quadrilaterals and pentagons in arrangements of lines, *Geometriae Dedicata*, **23**, 221–227.

Roudneff, J.-P. (1988a). Arrangements of lines with a minimal number of triangles are simple, *Discrete Comput. Geometry*, **3**, 97–102.   [271]

Roudneff, J.-P. (1988b). Tverberg-type theorems for pseudoconfigurations of points in the plane, *European J. Combinatorics*, **9**, 189–198.   [267, 383]

Roudneff, J.-P. (1988c). Reconstruction of the orientation class of an oriented matroid, *European J. Combinatorics*, **9**, 423–429.   [278, 329]

Roudneff, J.-P. (1989). Inseparability graphs of oriented matroids, *Combinatorica*, **9**, 75–84.   [324, 327, 328]

Roudneff, J.-P. (1991). Cells with many facets in arrangements of hyperplanes, *Discrete Math.*, **98**, 185–191.

Roudneff, J.-P. (1996). The maximum number of triangles in arrangements of pseudolines, *J. Combinatorial Theory* Ser. B, **66**, 44–74.   [271]

Roudneff, J.-P. (1998). Partitions of points into simplices with k-dimensional intersection, preprint, 22 pages; *European J. Combinatorics*, to appear.

Roudneff, J.-P., Sturmfels, B. (1988). Simplicial cells in arrangements and mutations of oriented matroids, *Geometriae Dedicata*, **27**, 153–170.   [24, 50, 269, 302, 316, 334, 465]

Roudneff, J.-P., Wagowski, M. (1989). Characterizations of ternary matroids in terms of circuit signatures, *J. Combinatorial Theory, Ser. B*, **47**, 93–106.

Rudin, M.E. (1958). An unshellable triangulation of a tetrahedron, *Bulletin Amer. Math. Soc.*, **64**, 90–91.   [211]

Rushing, T.B. (1973). *Topological Embeddings*, Academic Press, New York. [210, 226, 248]

Sahni, S. (1974). Computationally related problems, *SIAM J. Computing*, **3**, 262–279.   [357, 371]

Saito, K. (1975). On the uniformization of complements of discriminant loci, in: Conference Notes, Amer. Math. Soc. Summer Institute, Williamstown.   [77]

Salvetti, M. (1987). Topology of the complement of real hyperplanes in $\mathbb{C}^N$, *Inventiones Math.*, **88**, 603–618.   [94, 95, 181, 184]

Salvetti, M. (1993). On the homotopy theory of the complexes associated to metrical-hemisphere complexes, *Discrete Math.*, **113**, 155–177.   [95]

Santos, F. (1994). On Delaunay oriented matroids, in: "Proc. 6th Canadian Conf. Computational Geometry" (M. Keil, ed.), University of Saskatchewan, Saskatoon, 375–380.

Santos, F. (1996). On Delaunay oriented matroids for convex distance functions, *Discrete Comput. Geometry*, **16**, 197–210.

Santos, F. (1997a). Triangulations of oriented matroids, preprint, Oxford, 77 pages.   [414, 484, 485]

Santos, F. (1997b). Triangulations with very few geometric bistellar neighbors, preprint, Santander, 20 pages; *Discrete Comput. Geometry*, to appear.   [484]

Santos, F. (1999a). On the refinements of a polyhedral subdivision, preprint, Santander, 26 pages.   [416]

Santos, F. (1999b). Applications of the polyhedral Cayley trick to triangulations of polytopes, preprint in preparation.

Schmidt, P.-M. (1991). About the precision of the coefficients of a sweeping-plane, in: "Geometrical Problems of Image Processing" (Georgenthal, 1991), *Res. Inform.* **4**, Akademie-Verlag, Berlin, 107–113.

Schmidt, P.-M. (1992). Counting and generating sequences of finite planar point sets, *Algorithms Review*, **2**, No 3, 139–151.

Schneider, R. (1987). Tesselations generated by hyperplanes, *Discrete Comput. Geometry*, **2**, 223–232.   [198]

Schoute, P.H. (1911). Analytic treatment of the polytopes regularly derived from the regular polytopes, *Verh. Konink. Acad. Wetensch. Amsterdam*, **11**, No. 3. [571]

Schrijver, A. (1986). *Theory of Linear and Integer Programming*, Wiley-Interscience, Chichester.   [122, 433]

Schuchert, P. (1995). Matroid-Polytope und Einbettungen kombinatorischer Mannigfaltigkeiten, Dissertation, TH Darmstadt 1995; Shaker Verlag, 127 Seiten.

Schulte, E., Wills, J.M. (1985). A polyhedral realization of Felix Klein's map $\{3,7\}_8$ on a Riemann surface of genus 3, *J. London Math. Soc.*, Ser. 2, **32**, 539–262.   [488]

Scobee, M. (1999). On a conjecture concerning dyadic oriented matroids, *Electronic J. Combinatorics*, **6**, R23.

Scott, P. (1970). On the sets of directions determined by n points, *Amer. Math. Monthly*, **77**, 502–505.   [39]

Seidel, R. (1987). On the number of faces of higher order Voronoi diagrams, in: Proc. Third Annual ACM Symposium on Computational Geometry, pp. 181–187. [33]

Senechal, M. (1993). *Crystalline Symmetries. An Informal Mathematical Introduction*, Adam Hilger, Bristol, Philadelphia and New York.   [485]

Senechal, M. (1995). *Quasicrystals and Geometry*, Cambridge University Press. [485]

Shamir, R. (1987). The efficiency of the simplex method: a survey, *Management Sci.*, **33**, 301–334.   [43]

Shannon, R.W. (1974). Certain extremal problems in arrangements of hyperplanes, Ph.D. Thesis, University of Washington, Seattle, 97 pages.   [249, 271]

Shannon, R.W. (1976). A lower bound on the number of cells in arrangements of hyperplanes, *J. Combinatorial Theory*, Ser. A, **20**, 327–335.   [198, 272]

Shannon, R.W. (1979). Simplicial cells in arrangements of hyperplanes, *Geometriae Dedicata*, **8**, 179–187.   [49]

Shemer, I. (1982). Neighborly polytopes, *Israel J. Math.*, **43**, 291–314.   [400, 401]

Shephard, G.C. (1974). Combinatorial properties of the associated zonotope, *Canadian J. Math.*, **26**, 302–321.   [52, 55, 60, 62]

Shor, P. (1991). Stretchability of pseudolines is NP-hard, in: "Applied Geometry and Discrete Mathematics – The Victor Klee Festschrift" (P. Gritzmann, B. Sturmfels, eds.), DIMACS Series in Discrete Math. and Theoretical Computer Science, **4**, Amer. Math. Soc., Providence, RI, pp. 531–554.   [362, 370]

Skiena, S. (1990). Counting the k-projections of a point set, *J. Combinatorial Theory* Ser. A, **55**, 153–160.

Solomon, L. (1966). The orders of the finite Chevalley groups, *J. Algebra*, **3**, 376–393.   [73]

Solomon, L., Terao, H. (1987). A formula for the characteristic polynomial of an arrangement, *Advances in Math.*, **64**, 305–325.   [77]

Spanier, E.H. (1966). *Algebraic Topology*, McGraw-Hill, New York.   [200, 216, 229]

Sperner, E. (1949). Die Ordnungsfunktion einer Geometrie, *Math. Annalen*, **121**, 107–130.   [151]

Stanley, R.P. (1972). Supersolvable lattices, *Algebra Universalis*, **2**, 197–217.   [77]

Stanley, R.P. (1973). Acyclic orientations of graphs, *Discrete Math.*, **5**, 171–178.   [220]

Stanley, R.P. (1980a). The number of faces of simplicial convex polytopes, *Advances in Math.*, **35**, 236–238.   [394]

Stanley, R.P. (1980b). Decompositions of rational convex polytopes, *Annals of Discrete Math.*, **6**, 333–342.   [62]

Stanley, R.P. (1981). Two combinatorial applications of the Aleksandrov-Fenchel inequalities, *J. Combinatorial Theory*, Series A, **31**, 56–65.   [62]

Stanley, R.P. (1984a). On the number of reduced decompositions of elements of Coxeter groups, *European J. Combinatorics*, **5**, 359–372.

Stanley, R.P. (1984b). T-free arrangements of hyperplanes, in: "Progress in Graph Theory" (J.A. Bondy, U.S.R. Murty, eds.), Academic Press, New York, p. 539.   [77]

Stanley, R.P. (1991). A zonotope associated with graphical degree sequences, in: "Applied Geometry and Discrete Mathematics – The Victor Klee Festschrift"

(P. Gritzmann, B. Sturmfels, eds.), DIMACS Series in Discrete Mathematics and Theoretical Computer Science, **4**, Amer. Math. Soc., Providence, RI, pp. 555–570. [52, 62]

Steiger, W., Streinu, I. (1995). A pseudo-algorithmic separation of lines from pseudo-lines, *Information Processing Letters* **53**, 295-299.

Steiner, J. (1826). Einige Sätze über die Teilung der Ebene und des Raumes, *J. Reine Angew. Math.*, **1**, 349–364. [252]

Steinitz, E., Rademacher, H. (1934). *Vorlesungen über die Theorie der Polyeder*, Springer-Verlag, Berlin. [403, 406]

Stevens, B. (1993). Miro-Matroide – Unabhängigkeitsstrukturen in komplexen Matroiden, Diplomarbeit, Universität zu Köln, 141 pages.

Streinu, I. (1997). Clusters of stars, in: Proc. 13th Annual ACM Symposium on Computational Geometry (Nice, June 1997), pp. 439-441.

Strommer, T. (1977). Triangles in arrangements of lines, *J. Combinatorial Theory*, Ser.A, **23**, 314–320. [280]

Sturmfels, B. (1985). Zur linearen Realisierbarkeit orientierter Matroide, Diplomarbeit, TH Darmstadt, 167 pages. [391]

Sturmfels, B. (1986). Central and parallel projections of polytopes, *Discrete Math.*, **62**, 315–318.

Sturmfels, B. (1987a). Oriented matroids and combinatorial convex geometry, Dissertation, TH Darmstadt, 95 pages.

Sturmfels, B. (1987b). Computational synthetic geometry, Ph.D. Thesis, University of Washington, Seattle, 144 pages. [359]

Sturmfels, B. (1987c). Boundary complexes of convex polytopes cannot be characterized locally, *J. London Math. Soc.*, **35**, 314–326. [406]

Sturmfels, B. (1987d). Cyclic polytopes and d-order curves, *Geometriae Dedicata*, **24**, 103–107. [399]

Sturmfels, B. (1987e). On the decidability of Diophantine problems in combinatorial geometry, *Bulletin Amer. Math. Soc.*, **17**, 121–124. [354, 486]

Sturmfels, B. (1987f). Aspects of computational synthetic geometry; I. Algorithmic coordinatization of matroids, in H.H. Crapo (ed.): *Computeraided geometric reasoning*, INRIA Rocquencourt, France, June 1987, pp. 57–86.

Sturmfels, B. (1988a). Some applications of affine Gale diagrams to polytopes with few vertices, *SIAM J. Discrete Math.*, **1**, 121–133. [25, 380, 386, 406]

Sturmfels, B. (1988b). Totally positive matrices and cyclic polytopes, *Linear Algebra Appl.*, **107**, 275–281.   [399]

Sturmfels, B. (1988c). Simplicial polytopes without the isotopy property, I.M.A. preprint 410, University of Minnesota, 4 pages.

Sturmfels, B. (1988d). Neighborly polytopes and oriented matroids, *European J. Combinatorics*, **9**, 537–546.   [401, 402]

Sturmfels, B. (1989). On the matroid stratification of Grassmann varieties, specialization of coordinates, and a problem of N. White, *Advances in Math.*, **75**, 202–211.   [86]

Sturmfels, B. (1991). Gröbner bases of toric varieties, *Tôhoku Math. J.*, **43**, 249–261.   [411]

Sturmfels, B., White, N. (1989). Gröbner bases and invariant theory, *Advances in Math.*, **76**, 245–259.   [80, 349, 366]

Sturmfels, B., Ziegler, G.M. (1993). Extensions spaces of oriented matroids, *Discrete Comput. Geometry*, **10**, 23–45.   [295, 482, 483, 485]

Suvorov, P.Y. (1988). Isotopic but not rigidly isotopic plane systems of straight lines, in: Viro, O.Ya. (ed.): "Topology and Geometry – Rohlin Seminar," *Lecture Notes in Mathematics*, **1346**, Springer, Heidelberg, pp. 545–556.   [363, 366, 482]

Tajima, A. (1998). Optimality and integer programming formulations of triangulations in general dimensions, in: Proc. 9th Annual "International Symp. Algorithms and Computation" (ISSAC 98), *Lecture Notes in Computer Science*, Springer Heidelberg, to appear.

Tarski, A. (1951). *Decision Method for Elementary Algebra and Geometry*, 2nd revised ed., University of California Press.   [354]

Terao, H. (1980). Arrangements of hyperplanes and their freeness I, II, *J. Fac. Science Univ. Tokyo, Sci. IA*, **27**, 293–320.   [77]

Terao, H. (1981). Generalized exponents of a free arrangement of hyperplanes and Shephard-Todd-Brieskorn formula, *Inventiones Math.*, **63**, 159–179.   [77]

Terlaky, T. (1985a). The criss-cross method and its applications (in Hungarian), Ph.D. Thesis, Hungarian Academy of Sciences, Budapest.

Terlaky, T. (1985b). A finite criss-cross method for oriented matroids, (in Hungarian) *Alkalmazott Mat. Lapok*, **11**, 385–398.

Terlaky, T. (1987). A finite criss-cross method for oriented matroids, *J. Combinatorial Theory*, Ser. B, **42**, 319–327.   [451, 460, 461]

Terlaky, T., Wang, Z. (1993). A general scheme for solving linear complementarity problems in the setting of oriented matroids, in: "Combinatorics and Graph Theory" (H.P. Yap et al., eds.), World Scientific, Singapore, pp. 244–255.

Tits, J. (1968). Le problème des mots dans les groupes de Coxeter, *Ist Naz. Alta Math., Symposia Math.*, **1**, 175–185.   [185, 268]

Tits, J. (1974). *Buildings of Spherical Type and Finite BN-pairs, Lecture Notes in Mathematics*, **386**, Springer, Berlin.   [76, 169, 171]

Tits, J. (1982). A local approach to buildings, in: "The Geometric Vein. (The Coxeter Festschrift)," (C. Davis, B. Grünbaum, F.A. Sherk, eds.), Springer-Verlag, New York, 519–547.   [76]

Todd, M.J. (1984). Complementarity in oriented matroids, *SIAM J. Algebraic Discrete Methods*, **5**, 467–487.   [294, 419, 459]

Todd, M.J. (1985). Linear and quadratic programming in oriented matroids, *J. Combinatorial Theory*, Ser. B, **39**, 105–133.   [294, 418, 419, 451, 459, 460]

Tomizawa, N. (1984a). Theory of holometry (I) – on acycloids, (in Japanese), *Papers of the Technical Group on Circuits and Systems*, Institute of Electronics and Communication Engineers of Japan, CAS 84–14.

Tomizawa, N. (1984b). Theory of holometry (II) – acycloids and greedy systems, (in Japanese), *Papers of the Technical Group on Circuits and Systems*, Institute of Electronics and Communication Engineers of Japan, CAS 84–65.

Tomizawa, N. (1984c). Theory of acycloids and holometry, (in Japanese), *RIMS kokyuroku* (Memoir of the Research Institute of Mathematical Sciences of Kyoto University) "Graph Theory and Applications," **534**, 91–138.

Tratch, S.S., Zefirov, N.S. (1987). Combinatorial models and algorithms in chemistry. The ladder of combinatorial objects and its application to the formalization of structural problems of organic chemistry (in Russian), in: "Principles of Symmetry and Systemology in Chemistry" (N.F. Stepanov, ed.), Moscow State University Publ., Moscow, pp. 54–86.   [33]

Ungar, P. (1982). $2N$ noncollinear points determine at least $2N$ directions, *J. Combinatorial Theory*, Ser. A, **33**, 343–347.   [36, 39, 40]

Vámos, P. (1978). The missing axiom of matroid theory is lost forever, *J. London Math. Soc*, **18**, 403–408.   [351]

Varchenko, A.N. (1987). Combinatorics and topology of the disposition of affine hyperplanes in real space, *Functional Anal. Appl.*, **21**, 9–19.

Varchenko, A.N. (1989). On the numbers of faces of a configuration of hyperplanes, *Soviet Math. Doklady*, **38**, 291–295.   [198]

Varchenko, A.N., Gel'fand, I.M. (1988). Heaviside functions of a configuration of hyperplanes, *Functional Anal. Appl.*, **21**, 255–270.   [99]

Vershik, A.M. (1982).   Critical points of fields of convex polytopes and the Pareto-Smale optimum with respect to a convex cone, *Soviet Math. Doklady*, **26**, 353–356.

Vershik, A.M. (1988). Topology of the convex polytopes' manifolds, the manifold of the projective configurations of a given combinatorial type and representations of lattices, in: Viro (1988), pp. 557–581.   [363]

Viro, O.Ya., ed. (1988). *Topology and Geometry – Rohlin Seminar, Lecture Notes in Mathematics*, **1346**, Springer, Heidelberg.   [363]

Volodin, I.A., Kuznetsov, V.E., Fomenko, A.T. (1974). The problem of discriminating algorithmically the standard three-dimensional sphere, *Russian Math. Surveys*, **29**, 71–171.   [212]

Wachs, M.L., Walker, J.W. (1986). On geometric semilattices, *Order*, **2**, 367–385.   [48]

Wagowski, M. (1989a). The Tutte group of a weakly orientable matroid, *Linear Algebra Appl.*, **17**, 21–24.

Wagowski, M. (1989b). Matroid signatures coordinatizable over a semiring, *European J. Combinatorics*, **10**, 393–398.

Walker, J.W. (1981). Homotopy type and Euler characteristic of partially ordered sets, *European J. Combinatorics*, **2**, 373–384.   [216]

Wang, Z. (1986). A finite algorithm for feasible circuits over oriented matroids, in: "Graph theory and its applications: East and West" (M.F. Capobianco et al., eds.), Proc. First China-USA International Conference (Jinan, China, June 1986), *Annals New York Acad. Sciences*, **576**, 602–605.

Wang, Z. (1987). A finite conformal-elimination free algorithm over oriented matroid programming, *Chinese Annals of Math.*, **8B**, 120–125.   [451, 460]

Wang, Z. (1992). A general deterministic method for oriented matroid programming, *Chinese Annals of Math.* Ser. B, **13**, 222–229.

Wang, Z. (1995). A general scheme for solving linear complementarity problems in the setting of oriented matroids, (in Chinese), *Chin. Annals Math. Ser. A*, **16**, 94–99.

Wanka, A. (1986). Matroiderweiterungen zur Existenz endlicher LP-Algorithmen, von Hahn-Banach Sätzen und Polarität von orientierten Matroiden, Dissertation, Universität Köln, 86 pages.

Wegner, G. (1984). Kruskal-Katona's theorem in generalized complexes, in: "Finite and Infinite Sets," vol. 2, *Coll. Math. Soc. János Bolyai*, **37**, North-Holland, Amsterdam, pp. 821–827.   [199]

Weil, H. (1999). Signotope. Zur Kombinatorik von Pseudohyperebenenarrangements, Diplomarbeit, FU Berlin, 89 pages.

Weinberg, L. (1992). On the generation of *d*-ordered sets: A proof based on determinant theory, *IEEE Trans. Circuits Syst., I, Fundam. Theory Appl.* **39**, 415–418.

Welsh, D.J.A. (1976). *Matroid Theory*, Academic Press, London.   [101, 305]

Welzl, E. (1986). More on *k*-sets in the plane, *Discrete Computat. Geometry*, **1**, 95–100.   [36]

Wenzel, W. (1991). Projective equivalence of matroids with coefficients, *J. Combinatorial Theory*, Ser. A, **57**, 15–45.

Wenzel, W. (1992). Combinatorial Algebra of Δ-Matroids and Related Combinatorial Geometries, Habilitationsschrift, Bielefield 1991, 148 Seiten; preprint 92/45, "Combinatorics and its Applications," Zentrum für interdisziplinäre Forschung, Universität Bielefeld 1992.

Wenzel, W. (1996). A unified treatment of the theories of matroids with coefficients and of Δ-matroids with coefficients, *Beiträge zur Algebra und Geometrie*, **37**, 161–198.

Wenzel, W. (1998). Projective equivalence of Δ-matroids with coefficients and symplecive geometries, *Geometriae Dedicata*, **69**, 141–178.

White, N. (1975). The bracket ring of a combinatorial geometry. I, *Transactions Amer. Math. Soc.*, **202**, 79–103.

White, N. (1977). The basis monomial ring of a matroid, *Advances in Math.*, **24**, 292–297.   [91]

White, N., ed. (1986). *Theory of Matroids*, Cambridge University Press.   [101, 110, 115, 282, 284, 313, 319, 320, 330, 331, 335, 337]

White, N., ed. (1987). *Combinatorial Geometries*, Cambridge University Press. [39, 60, 76, 89, 91, 101, 116, 193-200, 218, 221, 282, 305, 331, 332, 338, 354]

White, N. (1989). A non-uniform oriented matroid which violates the isotopy property, *Discrete Comput. Geometry*, **4**, 1–2.   [363]

White, N. (1991). Multilinear Cayley factorization, *J. Symbolic Computation*, **11**, 421–438.   [89]

White, N., ed. (1992). *Matroid Applications*, Cambridge University Press.   [93, 99, 101, 153, 207, 222, 223, 377]

White, N. (1995). The Coxeter matroids of Gelfand et al., in: "Matroid Theory" (J. E. Bonin, J. G. Oxley, B. Servatius, eds.), *Contemporary Math.*, **197**, Amer. Math. Soc., 401–409.

Winder, R.O. (1966). Partitions of N-space by hyperplanes, *SIAM J. Applied Math.*, **14**, 811–818.   [193]

Wu, W. (1994). *Mechanical Theorem Proving in Geometry*, Texts and Monographs in Symbolic Computation, Vol. 2, Springer-Verlag, Vienna.   [486]

Yuzvinsky, S. (1993). The first two obstructions to the freeness of arrangements, *Transactions Amer. Math. Soc.*, **335**, 231–244.   [77]

Zaslavsky, T. (1975a). Facing up to arrangements: face count formulas for partitions of space by hyperplanes, *Memoirs Amer. Math. Soc.*, **1**, no. 154.   [187, 193, 194, 195, 197, 198, 220, 221]

Zaslavsky, T. (1975b). Counting the faces of cut-up spaces, *Bulletin Amer. Math. Soc.*, **81**, 916–918.

Zaslavsky, T. (1975c). Combinatorial ordered geometry, Part I: Bilateral geometry, or, Generalized affine and vector space ordering, typewritten manuscript, 64 pages.

Zaslavsky, T. (1977). A combinatorial analysis of topological dissections, *Advances in Math.*, **25**, 267–285.   [187, 193, 194]

Zaslavsky, T. (1979). Arrangements of hyperplanes: Matroids and graphs, in: Proc. Tenth Southeastern Conf. Combinatorics (F. Hoffmann et al, eds.), Vol. II, Utilitas Math. Publ., Winnipeg, Manitoba, pp. 895–911.

Zaslavsky, T. (1981). The slimmest arrangements of hyperplanes: II. Base-pointed geometric lattices and Euclidean arrangements, *Mathematika*, **28**, 169–190.   [48, 198]

Zaslavsky, T. (1983). The slimmest arrangements of hyperplanes: I: Geometric lattices and projective arrangements, *Geometriae Dedicata*, **14**, 243–259.   [198]

Zaslavsky, T. (1985). Extremal arrangements of hyperplanes, in: Proc. Conf. "Discrete Geometry and Convexity" (J.E. Goodman, E. Lutwak, J. Malkevitch, R. Pollack, eds.), *Annals New York Acad. Sciences*, **440**, pp. 69–80.   [198]

Zaslavsky, T. (1991). Orientation of signed graphs, *European J. Combinatorics*, **12**, 283–291.

Zaslavsky, T. (1998). Bibliography of signed and gain graphs, *Electronic J. Combinatorics*, DS#8; published July 20, 1998, http://www.combinatorics.org/Surveys/index.html

Zeeman, E.-C. (1963). *Seminar on Combinatorial Topology*, mimeographed notes, Inst. des Hautes Études Sci., Paris.   [209, 214]

Ziegler, G.M. (1987). Algebraic combinatorics of hyperplane arrangements, Ph.D. Thesis, MIT, Cambridge MA, 168 pages. [94]

Ziegler, G.M. (1988). The face lattice of hyperplane arrangements, *Discrete Math.*, **74**, 233–238. [187]

Ziegler, G.M. (1989a). Combinatorial construction of logarithmic differential forms, *Advances in Math.*, **76**, 116–154. [77]

Ziegler, G.M. (1989b). Multiarrangements of hyperplanes and their freeness, *Contemporary Math.*, **90**, Amer. Math. Soc., 345–358. [77]

Ziegler, G.M. (1991). Some minimal non-orientable matroids of rank 3, *Geometriae Dedicata*, **38**, 365–371. [274, 279, 331, 337]

Ziegler, G.M. (1992). Combinatorial models for subspace arrangements, Habilitationsschrift, TU Berlin, April 1992.

Ziegler, G.M. (1993a). Higher Bruhat orders and cyclic hyperplane arrangements, *Topology*, **32**, 259–279.

Ziegler, G.M. (1993b). "What is a complex matroid?", *Discrete Comput. Geometry*, **10**, 313–348.

Ziegler, G.M. (1996). Oriented matroids today, Dynamic survey and updated bibliography, *Electronic J. Combinatorics*, **3**, DS#4; published April 1, 1996; revised versions June 22, 1996 and September 15, 1998. [x]

Ziegler, G.M. (1998). *Lectures on Polytopes*, Graduate Texts in Mathematics **152**, Springer-Verlag, New York 1995; Revised printing; *Updates, corrections, and more*, http://www.math.tu-berlin.de/~ziegler

Zmyslony, M. (1982). Orientierte Matroide und ihre Anwendung in der Netzwerktheorie, in: "Graphen und Netzwerke – Theorie und Anwendung" (27. Intern. Wiss. Koll. TH Ilmenau), 65–68.

Index